Surface and Underground Excavations
2nd Edition

Surface and Underground Excavations 2nd Edition

Methods, Techniques and Equipment

Ratan Raj Tatiya, PhD, BEng

CRC Press
Taylor & Francis Group
Boca Raton London New York

CRC Press is an imprint of the
Taylor & Francis Group, an **informa** business

A BALKEMA BOOK

Cover illustrations: Courtesy of Shutterstock (www.shutterstock.com)

Front cover:
Huge excavator in a coal mine, Copyright: Kodda

Back cover:
Left photo: *Construction of a tunnel boring machine, Copyright: Chindegrao*
Right photo: *Cutter head of a tunnel boring machine, Copyright: VILevi*

CRC Press
Taylor & Francis Group
6000 Broken Sound Parkway NW, Suite 300
Boca Raton, FL 33487-2742

First issued in paperback 2018

CRC Press/Balkema is an imprint of the Taylor & Francis Group, an informa business

ISBN-13: 978-0-415-62119-9 (hbk)
ISBN-13: 978-1-138-49616-3 (pbk)

Typeset by MPS Limited, Chennai, India

Library of Congress Cataloging-in-Publication Data

Tatiya, Ratan.
   Surface and underground excavations : methods, techniques and equipment / Ratan Raj Tatiya. — 2nd ed.
      p. cm.
   Summary:"This expanded second edition is a comprehensive text on the latest technologies and developments in excavation for any type of surface or underground excavation. A great number of topics is covered, as well as excavation techniques for various operations. The book has been wholly revised, and includes the latest trends and best practices as well as questions at the end of each chapter. It is now even more appealing to students in earth sciences, geology and in civil, mining and construction engineering and those with a general or professional interest in surface and underground excavations"— Provided by publisher.
   Includes bibliographical references and index.
   ISBN 978-0-415-62119-9 (hardback : alk. paper)   1. Excavation.   2. Rock excavation.
3. Underground areas.   I. Title.

   TA730.T387   2013
   624.1'52—dc23
                                                                        2012047173

Published by: CRC Press/Balkema
                P.O. Box 11320, 2301 EH, Leiden, The Netherlands
                e-mail: Pub.NL@taylorandfrancis.com
                www.crcpress.com – www.taylorandfrancis.com

Visit the Taylor & Francis Web site at
http://www.taylorandfrancis.com

and the CRC Press Web site at
http://www.crcpress.com

Dedication

To my wife Shashi, and children Anand, Gaurav and Sapna.

Contents

19   Sustainable Development   807

Acknowledgements

The author wishes to express his sincere gratitude to the educational institutes, professional societies, manufacturing companies, publishers, mining companies and other institutes, as listed below, for their permission, facilities and release of technical data, release of literature, information and material.

Alimak Company, Sweden
AngloGold Ashanti, South Africa
Atlas Copco, Sweden
Canadian Standard Association, Canada
Caterpillar, USA
CRC Mining, Australia
CSIRO, Australia
Essel Mining and Industries – Aditya Birla Group, India
Fantini, Italy
Federation of Indian Mining Institute (FIMI)
Herrenknecht GmbH, Schwanau, Germany
Hitachi, Japan
Institute of Material, Mineral and Mining (IMMM), UK
International Council of Mining and Metals (ICMM), UK
International Labor Organization (ILO) Encyclopaedia
Istituto Internazionale del Marmo, Milano, Italy
Khetri Cooper Complex, India
Krupp, Germany
Lomin, South Africa
Marini, Italy
Mineral Information Institute (MII), Denver, Colorado
Mining Engineers' Association of India (MEAI)
Mining, Geological and Metallurgical Institute (MGMI), India
Mir publishers, Russia
Modular Mining System, USA
Newmont Mining Corporation, Australia
Nitro Nobel, Sweden
Paurat, Germany
Rapid Excavations and Tunneling Conference (RETC), American Institute of Mining, Metallurgical, and Petroleum Engineering, USA (AIME)
Rio Tinto, Australia/South Africa

Robbins, USA
Sandvik Tamrock, Finland
Siemens, Germany
SME/SMME – Society for Mining, Metallurgy, and Exploration, Inc., USA
Sultan Qaboos University, Oman
United Nations Economic Commission for Europe (UNECE)
Wirth Company, Erkelenz, Germany
Wikipedia, the free encyclopaedia
World Coal Association (WCA)
Zhengzhou Coal Mining Machinery (Group) Co., Ltd., China

Preface

The excavation industry is booming and this trend will continue due to the growing world population which is likely to be doubled from 6 billion in the year 2000 to 12 billion at its present growth rate by the year 2040, and also due to changes in the lifestyle, living standards and progressive outlook within countries, with each country aiming to grow from its current level of undeveloped to developing; from developing to a developed country and from developed to superpower. IMM Australia predicts that *"Over the next 50 years the world will use 5 times the mineral resources that have been mined to the year 2000. To meet this predicted demand, the industry must grow as internationally competitive sector, underpinned by innovations and technology."*

The use of underground space in urban areas is becoming increasingly important due to scarcity of land in the densely populated/inhabited areas and due to environmental concerns. Methods, techniques and equipment are available to excavate a large volume of rocks beneath the surface efficiently. This is the reason that thousands of kilometers of tunnels for transportation (rail, road and water conveyance) and large excavations, which are known as 'caverns', are being created for civil works, storage facilities, defense installations, hydro-electric power plants, and recreation facilities. This is generating billions of cubic meters of earth-material daily, besides that generated by mining activities the world over. Due to this scenario many multi-national companies (manufacturers) are involved in producing explosives and their accessories; earth moving, rock drilling and cutting, and tunneling machines.

Thus, the excavation activities for civil and construction industries and to produce minerals from mines are vital. But the output from mines, tunnels and caverns is not restricted to rocks and ground, metallic and non-metallic ores and fuels but also includes gaseous emissions, liquid effluents, solid waste, radiation, particulate-matters, heat and noise. Equally associated with excavations are hazards such as fires, explosions, inundation, and other accidents or disasters. All these are detrimental to the health not only of direct and indirect stakeholders particularly when exceeding the permissible limits but also to the well-being of biotic and abiotic components of nature. Growing health-problems of the world's citizens and global issues (problems) such as acid rain, ozone depletion, photochemical smog, acid drainage and global warming are testimony to this fact.

We understand 'production at the desired rate' is the bread and butter of those concerned (right from shop floor worker to the highest executive); productivity brings excellence to the production. However both cannot be achieved if safety is jeopardized, pollution is at its peak workers' health is not looked after and societal welfare is neglected. Thus, Health, Safety and Environment (HSE) must be considered a critical business activity on a par with production and productivity. And a thorough balance is required between these three critical business activities together with proper care for society to achieve sustainable development, which is beneficial socially, economically and ecologically to the present as well as future generations. There is no simple and straightforward solution; but minimizing losses of various kinds could be one way of achieving it. As such loss prevention strategy should be an integral part of the procedure of running mines and tunnelling projects effectively. And to accommodate these features, Chapter 18 entitled: 'Hazards, Occupational Health and Safety (OHS), Environment and Loss Prevention' and Chapter 19 entitled 'Sustainable Development' have been added to the 17 chapters of the first edition.

Apart from this two sub-chapters: 'Mineral Inventory Evaluation together with Resources Classification by UNECE' and 'Planning for Mine Closure' have been added to Chapters 2 and 16 respectively.

Each of the 17 chapters of the 1st edition has been reviewed and text has been modified wherever required with the inclusion of latest trends and information. Thus, the 19 chapters of the 2nd edition of 'Surface and Underground Excavations' treat the latest developments and technologies in excavation operations, and cover:

- Excavation in surface and underground locales, and in any direction; horizontal (tunneling, drifting), vertical (raising, sinking, stoping) or inclined;
- Minerals' prospecting, exploration, evaluation and classification & site investigations for civil constructions; Rocks and ground characterization;
- Unit operations like drilling, explosives and blasting, mucking, support and reinforcements, haulage, hoisting and services;
- Planning, design, construction, stoping, liquidation and mine closure, including case studies on these subjects;
- The creation of large underground space for caverns (hydro-power stations, oil, gas and nuclear waste repositories), and driving tunnels using conventional methods as well as tunnel borers and latest techniques;
- Surface mining methods (dimension stone quarrying, open pit and open casts) and underground stoping methods (open, supported and caving);
- Methodologies to select a stoping method, based on economic analysis;
- Executing 'Mass Blasting', Mining at 'Ultra Depths' and Mining Difficult Deposits Using Non-Conventional Technologies;
- Fair treatment of Excavations' potential hazards, accidents, industrial hygiene, working conditions, ergonomics, environment degradation and mitigation measures, loss prevention and abnormalities diagnosis and remedial measures;
- Reasonable coverage of Automation, Application of IT including 'Entrepreneur Resources Planning (ERP)', Global competitiveness, Precision in operations, Standardization & Benchmarking.

In summary, the book offers a comprehensive text on excavations of any type be it surface or underground with or without aid of explosives using latest methods, equipment and techniques. Its introductory chapters describe rocks, minerals, mineral inventory evaluation, resources classification, prospecting, exploration & site investigations (Chapters 2 and 3). It covers operations pertaining to tunneling (Chapters 9 to 11), raising (Chapter 13), sinking (Chapter 14), drifting (Chapter 9), stoping (Chapter 16), quarrying and surface-mining (Chapter 17), underground mining, pillar blasting, liquidation and mine closure (Chapter 16). Unit operations (Chapters 4–8): drilling, explosives and blasting, mucking, haulage, hoisting, supports and reinforcement have been covered. It deals with design, planning, development (Chapter 12), and construction of surface and subsurface excavations including caverns (Chapter 15). Its new additions include: Hazards (Risks) analysis and management, safety, occupational health and surveillance, environment degradation and mitigation measures and loss prevention strategies (Chapter 18).

The concluding chapter, 19, entitled: '*Sustainable Development (SD)*' attempts to cover Principles/Guidelines for SD by ICMM and Status of SD in Mining, based on Stakeholders' Views through Survey by 'GlobalScan'. It proposes a strategy to run mines in an economically viable (beneficial) way by implementing cost-effective systems, technologies and best practices.

Thus, this 2nd edition represents a comprehensive text on mineral inventory evaluation (ore – reserves estimation) through to 'mine closure' together with the concern for occupational health and safety (OHS), environment and loss prevention, and sustainable development in mining.

The foregoing discussion reveals that this book discusses very many vital aspects of various subject areas. The author's industrial background ensures that the material is industrially relevant and his academic background ensures that the fundamental and basics required to help readers are included.

The author has had the opportunity to amass 42 years of professional experience. He has worked with multinationals from more than 40 countries; and in multi-cultural environments initially for a decade in the industry and then as senior university professor and industrial consultant for more than 25 years; he returned to the mineral industry as a senior executive in 2005 and worked as director of an engineering institute; he is presently involved in the mining and civil construction industries as a consultant.

The book contains material that is useful for engineering disciplines such as: mining, civil construction, petroleum, Environment, occupational health and safety (OHS) and geology. As such it is intended to serve as a textbook for students of undergraduate level and the first year of graduate level at schools or institutes offering courses in any of the above-mentioned disciplines. More material is presented than can be accommodated in a 2-credit course, and it should be sufficient for a 3-credit course. In addition, officials, supervisors, engineers and professionals of these disciplines should find this book beneficial.

The text likely covers syllabi at various institutes for Civil, Construction or Earth Science on subjects such as ground excavation engineering, tunnel engineering, sinking & sub-surface engineering, open-cut excavators and earth movers among others. For mining/mineral engineering students it intends to cover syllabi designed for subjects such as: unit operations, mine development, rock excavation and tunneling, underground mining, surface mining and quarrying; mine safety and loss prevention; etc.

This book features the inclusion of best practices, case studies, latest trends, global surveys and toolkits (guidelines) wherever feasible. Finally, it advocates prevention of losses of all kinds, which is a noble way of working as it results in producing goods and services with maximum productivity and least costs safely, and provides suggestions as to how an industry can survive in a world of Global Competition and

Recession? Each chapter ends with either 'The way forward' or 'Concluding Remarks' sections to conclude it. At the end of each chapter 'Questions' have been included to provide readers a better understanding of the subject matter. However, instructors/professors could create their own quizzes and questions using this material and by providing/suggesting supplementary material on the subject matter.

In the end this book is a result of appreciation from students and colleagues, support from my family members and cooperation from industries, professional societies (SME, IMMM & ICMM), an academic institute (Sultan Qaboos University), companies and organizations as mentioned in the acknowledgement listing who encouraged me by providing valuable information, at my request, and all those who helped me directly or indirectly in this endeavor. I wish to express my sincere gratitude to them.

Ratan R. Tatiya
July 2012

Conversion table

Multiply metric unit	By	To obtain English unit
kilometer (km)	0.6214	mile (mi)
meter (m)	1.0936	yard (yd)
meter (m)	3.28	foot (ft)
centimeter (cm)	0.0328	foot (ft)
millimeter (mm)	0.03937	inch (in)
sq kilometer (km^2)	0.3861	sq mile (mile^2)
hectare (ha)	2.471	acre
sq meter (m^2)	10.764	sq. foot (ft^2)
sq meter (m^2)	1550	sq inch (in^2)
sq centimeter (cm^2)	0.1550	sq inch (in^2)
cu centimeter (cm^3)	0.061	cubic inch (in^3)
cubic meter (m^3)	1.308	cubic yard (yd^3)
liter (l)	61.02	cubic inch (in^3)
liter (l)	0.001308	cubic yard (yd^3)
km/h	0.621	mph
liter (l)	0.2642	US gallon
liter (l)	0.22	imperial gallon
metric ton (t)	0.984	long ton (lg ton)
metric ton (t)	1.102	short ton (sh ton)
kilogram (kg)	2.205	pound advp. (lb)
gram (gm)	0.0353	ounce advp. (oz)
kilonewton (kn)	225	pound (force)
newton (n)	0.225	pound (force)
cu centimeter (cm^3)	0.0338	fluid ounce
kg/m^3	1.686	pounds/yd^3
kg/m^3	0.062	pounds/ft^3
kg/cm^2	14.225	pounds/in^2
kilocalorie (kcal)	3.968	Btu
kilogram-meter (kg.m)	7.233	foot-pound
meter-kilogram (m.kg)	7.233	pound-foot
metric horsepower (cv)	0.9863	Hp
kilowatt (kw)	1.341	Hp
kilopascal (kpa)	0.145	psi
bar	14.5	psi
tons/m^3	1692	pounds/yd^3
decaliter	0.283	bushel

Multiply English unit	By	To obtain metric unit
mile (mi)	1.609	kilometer (km)
yard (yd)	0.9144	meter (m)
foot (ft)	0.3048	meter (m)
inch (in)	25.4	millimeter (mm)
sq mile (mile^2)	2.590	sq kilometer (km^2)
acre	0.4047	hectare (ha)
sq foot (ft^2)	0.0929	sq meter (m^2)
sq inch (in^2)	0.000645	sq meter (m^2)
cubic yard (yd^3)	0.7645	cubic meter (m^3)
cubic inch (in^3)	16.387	cu centimeter (cm^3)
cubic foot (ft^3)	0.0823	cubic meter (m^3)
cubic inch (in^3)	0.0164	liter (l)
cubic yard (yd^3)	764.55	liter (l)
mph	1.61	km/h
Ton-mph	1.459	tkm/h
U.S. gallon	3.785	liter (l)
U.S. gallon	0.833	Imperial gallon
long ton (lg ton)	1.016	metric ton (t)
short ton (sh ton)	0.907	metric ton (t)
pound advp. (lb)	0.4536	kilogram (kg)
ounce advp. (oz)	28.35	gram (gm)
pound (force)	0.0045	kilonewton (kn)
pound (force)	4.45	newton (n)
fluid ounce (fl oz)	29.57	cu centimeter (cm^3)
pounds/yd^3	0.5933	kg/m^3
pounds/ft^3	16.018	kg/m^3
pounds/in^2	0.0703	kg/cm^2
Btu	0.2520	kilogram calorie
foot-pound	0.1383	kilogram-meter (kg · m)
horsepower (hp)	1.014	metric horsepower
horsepower (hp)	0.7457	kilowatt (kw)
psi	6.89	kilopascal (kpa)
psi	0.0689	bar
pounds/yd^3	0.0005928	tons/m^3
pounds (no. 2 diesel fuel)	0.1413	U.S. gallon
bushel	3.524	decaliter

Note: Some of the above factors have been rounded for convenience. For exact conversion factors please refer to the International System of units (SI) table. (Courtesy: CaterPillar)

Metric unit equivalents

1 km	=	1000 m
1 m	=	100 cm
1 cm	=	10 mm
1 km^2	=	100 ha
1 ha	=	10,000 m^2
1 m^2	=	10,000 cm^2
1 cm^2	=	100 mm^2
1 m^3	=	1000 liters
1 liter	=	100 cm
1 metric ton	=	1000 kg
1 quintal	=	100 kg
1 N	=	0.10197 kg · m/s^2
1 kg	=	1000 g
1 g	=	100 mg
1 bar	=	14.504 psi
1 bar	=	427 kg · m
1 cal	=	427 kg · m
		0.0016 cv · h
		0.00116 kw · h

Torque unit

1 CV	=	75 kg · m/s
1 kg/cm^2	=	0.97 atmosph.

English unit equivalents

1 mile	=	1760 yd
1 yd	=	3 ft
1 ft	=	12 in
1 sq. mile	=	640 acres
1 acre	=	43,560 ft^2
1 ft^2	=	144 in^2
1 ft^3	=	7.48 gal liq.
1 quart	=	32 fl oz
1 fl oz	=	1.8 in^3
1 sh ton	=	2000 lb
1 lg ton	=	2240 lb
1 lb	=	16 oz, avdp
1 Btu	=	778 ft lb
		0.000393 hph
		0.000293 kw · h

Torque unit

1 mechnical hp	=	550 ft-lb/s
1 atmosph.	=	14.7 lb/in^2

Power unit equivalents

kW	=	kilowatt
hp	=	Mechanical horse power
CV	=	Cheval Vapeur (steam horsepower)
		French designation for horsepower
PS	=	Pferdestärke, German designation for horsepower
I hp	=	$1.014\,CV = 1.014\,PS = 0.7457\,kW$
I PS	=	$I\,CV = 0.986\,hp$
		$0.7355\,kW$
I kW	=	$1.341\,hp$
		$1.36\,CV$
		$1.36\,PS$

Note: Some of the above factors have been rounded for convenience. For exact conversion factors please refer to the International System of units (SI) table. (Courtesy: CaterPillar)

Chapter 1

Introduction

Development and Clean Environment are two sides of the same coin.

1.1 EXCAVATIONS AND THEIR CLASSIFICATION

The meaning of the word 'excavate' is to dislodge the rock massif from its original place (in-situ). This involves two operations: digging the ground and its disposal. This can be carried out on any formation that exists within the earth's crust. This operation can create openings or excavations of different sizes, shapes and configurations at the desired location. The location could be hilly terrain, plain ground, desert, cropland, forests or any other terrain. It could be within an urban context or in the countryside and even sometimes within bodies of water or in ground saturated with water. It could commence at, above or below the ground level or datum and extend in any direction: horizontal, inclined, vertically up or down.

The purpose of creating openings is manifold and therefore, in this modern era different kinds of excavations are necessary. Broadly, based on locale, the excavations can be grouped into two main classes:

1. surface excavations (fig. 1.1).
2. subsurface or underground excavations (fig. 1.2).

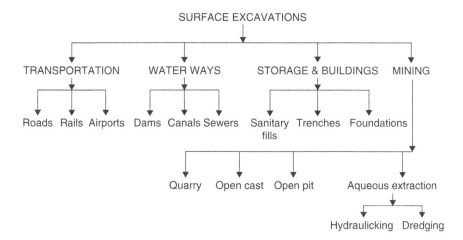

Figure 1.1   Classification of type of surface excavations (based on purpose).

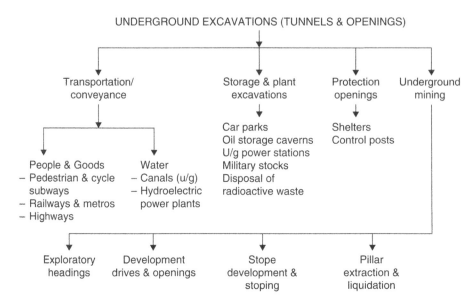

Figure 1.2   Classification of underground excavations (tunnels and large openings) based on their function and utilities.

Thus, surface and underground are two locales where the excavations can be carried out. At both these locales excavations are required principally for the following two purposes:

a. Excavations necessary to exploit minerals. From here on these will be termed mining excavations.
b. Excavations necessary to build structures including tunnels. From here on these will be termed civil works excavations, or civil excavations.

1.2   SURFACE EXCAVATIONS

The magnitude of surface excavations is many-fold than the underground excavations described in the succeeding paragraphs. It amounts to billions of cubic meters or tonnage every year. This excavation is necessary; firstly, to remove the enormous amount of rock material lying above mineral deposits as overburden and also to produce useful minerals themselves at the mines, which could be open pits, opencasts or quarries.

Secondly, to remove the enormous volume of earth-material while constructing rail routes and roadways, canals, dams and many vital civil constructions including buildings as shown in Fig. 1.1. In fact these are the development or infrastructures' building activities that are essential for the growth and prosperity of any country.

1.3   UNDERGROUND EXCAVATIONS

There are two locales where subsurface openings or tunnels are driven. The first category includes those tunnels or openings, which serve people by way of providing passage to rails, roads, navigation, pedestrian use etc. and also for conveyance of water and that serve as sewerage. These tunnels, constructed for civil works and built to have a very long life, need to be very safe with regard to their stability, ventilation, illumination and risks of getting flooded etc. Globally, an additional few thousand kilometers are driven every year.

The second category of tunnels or openings is for the purpose of exploring and exploiting mineral deposits, which are deep-seated, and cannot be mined by surface mining methods. These tunnels are small in size (cross section) but during the life of a mine, their lengths total hundreds of kilometers, and thus globally, several thousand kilometers of tunnels of this kind are required to be driven every year. For example, Mount Isa Mines, which is largest copper, silver and zinc producing company in Australia owns 975 km of underground openings (tunnels, raises and shafts). Mine workings extend to 5 km long, and 1.2 km wide, with their deepest point at 1800 m (depth) below the surface.

In addition to tunnels, large underground excavations are also mandatory mainly for two applications: first for exploitation of minerals, for which they are driven together with mine tunnels, and secondly for the purpose of storage of oil (fig. 15.4), power generation (fig. 15.3), defense utilities, storing nuclear and hazardous wastes (fig. 15.5) and many other applications (figs 15.6, 15.7). Figure 1.2 classifies the subsurface openings, which are essential and driven globally to the magnitude of millions of linear meters or billions of cubic meters every year.

Figures 1.1 and 1.2 show breakup of excavations' networks both for civil works and mining. The present trend is to go for more and more of such excavations and section 1.4 reveals the reason behind this.

1.4   IMPORTANCE OF MINERALS AND BRIEF HISTORY OF THEIR RECOVERY

Minerals are naturally occurring inorganic or organic substances and mining is the process of digging, excavating or extracting them commercially. Minerals are some of the basic natural resources and mining is as old as civilization, starting some 300,000 (B.C.) years[9] ago with the search for useful stones. According to the needs of man different minerals have been investigated during different periods of history, and hence, the cultural ages of man are associated with minerals or their derivatives and have been termed *Stone Age, Bronze Age, Iron Age and Atomic Age*. By the start of Christian era all '*Seven Metals of Antiquity*' namely copper, tin, gold, silver, lead, iron and mercury were known and mined.

After air, water and food, minerals are our basic need. They are used in manufacturing utensils, tools, appliances, machines and equipment. They provide shelter, power, energy, means of transport and communication. Their use in assuring peace and prosperity of any country is vital, as it is, during wartime in the manufacturing of weapons and warfare. Their use as jewelry, cosmetics, dye and coinage is very well established. In this materialistic world, without minerals one cannot survive. As per the analysis carried out by the Mineral Information Institute (MII, Colorado, USA),[13] a newly born American baby starts consuming about 60 kg of minerals every day till he or she becomes an old aged person and dies, on average at the age of 80 (fig. 1.3). Can you imagine how many minerals a citizen in your own country needs every day? Table 1.1 outlines the classification of minerals based on their use in our day-to-day life.

Similarly going through table 1.2[13] and figure 1.4[13] one could realize how useful minerals are to manufacture things around us, and the items of our daily

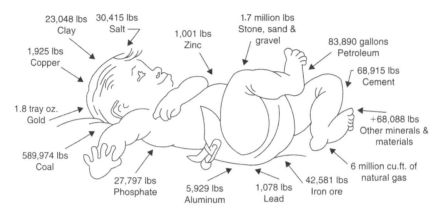

23,048 lbs   30,415 lbs
Clay   Salt

1,001 lbs
Zinc

1.7 million lbs
Stone, sand &
gravel

83,890 gallons
Petroleum

1,925 lbs
Copper

68,915 lbs
Cement

1.8 tray oz.
Gold

+68,088 lbs
Other minerals &
materials

589,974 lbs
Coal

6 million cu.ft. of
natural gas

27,797 lbs
Phosphate

5,929 lbs
Aluminum

1,078 lbs
Lead

42,581 lbs
Iron ore

Figure 1.3   A newly born baby would require 3.75 million pounds (1.7 million kg = 1700 tonnes.) of minerals, metal and fuels throughout his/her life in a developed country, such as USA. What would this requirement be in your own country?

Table 1.1   Minerals' classification based on their use in our day-to-day life.

Metal ores & metals proper	Fuels	Nonmetallic minerals
Precious metals: ores of silver, gold and platinum-group metals	*Fossil*: petroleum, natural gas, oil, shale, tar sand, coal, lignite	*Salts*: table salt, salt-peter, sodium, potassium & magnesium salts, sulfates etc.
Base metals: ores of copper, lead, zinc, tin, molybdenum etc.	*Nuclear*: ores of uranium, thorium and others	*Abrasives*: emery, columbium & tantalum, corundum, pumice, honing & polishing stones, flint etc.
Ferrous metals: ores of iron, manganese, chromium, cobalt, tungsten, vanadium, silicon etc.		*Ceramic glass & Industrial minerals*: asbestos, refectory clays, quartz & quartzite, mica, feldspar, talc & many others
Miscellaneous minor metals: ores of indium, dolomite, acid resistant & gallium, cadmium, germanium, mercury, bismuth, antimony, rare earth element etc.		*Building material*: sand, gypsum, limestone, clays, gravel, anhydrite, sandstone etc., including dimensional stones such as slate, granite, marble etc.
		Precious, colored, decorative, or ornamental stones: diamond, garnet, opal, turquoise, aquamarine, tourmaline, various types of quartz, amber, malachite, jasper etc.
		Radio active and rare minerals: radium, lithium, rubidium etc.
		Natural gases: oxygen, nitrogen, argon, and other rare gases such as methane, helium etc.
		Miscellaneous industrial minerals: barite, pyrite, graphite, mineral paints, lithographic stone, mineral wax, chalk, magnesite, mineral sulphur, trinoli, or the minerals not covered in the above list.

Table 1.2 Minerals around us that makes our life going.[13]

Items – Minerals contributing in its manufacturing	Items – Minerals contributing in its manufacturing
Automobile – 15 different minerals & metals	Linoleum – Calcium carbonate, clay,
Baby powder – Talc	wollastonitt
Cake/Bread – Gypsum, phosphates	Lipstick – Calcium carbonate, talc
Carbon paper – Bentonite, zeolite	Kitty litter – Attapulgite, montmorillonite,
Carpet – Calcium carbonate, limestone	zeolites, diatomite, pumice, volcanic ash
Caulking – Limestone, gypsum	Ink – Calcium carbonate
Concrete – Limestone, gypsum, iron oxide, clay	Pencil – Graphite, clay
Counter tops – Titanium dioxide, calcium	Plant fertilizers – Potash, phosphate,
carbonate, aluminum hydrate	nitrogen, sulfur
Computer – 33 minerals (fig. 1.4)	Porcelain figurines – Silica, limestone,
Drinking water – Limestone, lime, salt, fluorite	borates, soda ash, gypsum
Fiberglass roofing – Silica, borates, limestone,	Pots and pans – Aluminum, iron
soda ash, feldspar	Potting soil – Vermiculite, perlite, gypsum,
Fruit juice – Perlite, diatomite	zeolites, peat
Glass/Ceramics – Silica sand, limestone,	Spackling – Gypsum, mica, clay, calcium
talc, lithium, borates, soda ash, feldspar	carbonate
Glossy paper – Kaolin clay, limestone, sodium	Sports equipment – Graphite, fiberglass
sulfate, lime, soda ash, titanium dioxide	Sugar – Limestone, lime
Hair cream – Calcium carbonate	Television – 35 different minerals & metals
Household cleaners – Silica, pumice,	Toothpaste – Calcium carbonate, limestone,
diatomite, feldspat, limestone	sodium carbonate, fluorite
Jewelry – Precious and semi-precious stones,	Vegetable oil – Clay, perlite, diatomite
gold, silver	Wallboard – Gypsum, clay, perlite,
Medicines – Calcium carbonate, magnesium,	vermiculite, aluminum hydrate, borates
dolomite, kaolin, barium, iodine, sulfur, lithium	

Figure 1.4 Did you know? More than 33 minerals are required to make a computer. These vital computer ingredients are Al, Co, Cu, Au, Fe, Hg, Mo, Mn, Ni, Ag, Sn, Zn, barite, beryllium, columbium, gallium, germanium, indium, lanthanides, lithium, mica, platinum, quartz crystals, rhenium, selenium, silicon, strontium, tantalum, tellurium, tungsten, vanadium, yttrium and zirconium. All these components are hosted in a plastic enclosure, which is produced by the petroleum industry.

consumption. Minerals are not only present within the earth crust (at the subsurface and above the surface) and in the sea but also on other planets and the moon. Minerals are available in three states – solid, liquid and gas. The scope of this book is to deal with the solid state of the minerals and not their liquid and gaseous forms i.e. petroleum and its products.

No country is self sufficient in mineral reserves, but the mineral resources constitute an essential part of national economy in terms of export income, government revenues, and GDP. The living standard of any country is judged by its per capita mineral consumption; for example in early 1980's America's per capita annual mineral consumption was some 20 tons. The per capita mineral production and consumption of some of the advanced countries like US, Russia, and Australia is almost in the same ratio, or the consumption is even higher than production. The reason is that from minerals, which are the basic raw materials, value added products are prepared in these countries. This strategy has multiple positive effects on boosting the economy. Any country, which is an extremely poor mineral producer and consumer, exhibits very weak industrial growth and infrastructures.

It should be borne in mind that the ground or rock formations vary from place to place and even within the same place, hence, the same technique, method or equipment cannot be applied universally.

1.5 CURRENT STATUS OF MINERAL INDUSTRY

Today products of the mineral industry pervade the lives of all mankind. Looking at the global scenario, the progress that has been made in the process of mineral exploration and exploitation in the last five decades was not matched even during last five centuries. This was made possible by the application of advanced technology to fulfill the needs of rapid industrial growth and an increasing world population.

There is a tough competition of the minerals and metal prices in the world market, which again is a setback for the minerals any country exports or in a surplus position. Most of the developed countries of the world are far ahead with regard to mineral production, consumption, import, export, technological development, productivity and many other vital spheres than those countries which are undeveloped and developing. Even at the domestic front mineral industry's share in the gross domestic product (GDP) comparing with other industries such as petroleum, agriculture, manufacturing, trading etc. should be considerable. It is obvious, therefore, if any country has not to sink into downward economic trend, its mineral resources should be tapped, channeled and kept motivated to bridge the gap in the spheres outlined above.

1.6 EXCAVATION TECHNOLOGIES/SYSTEMS – DEVELOPMENT & GROWTH

As stated above that a mineral has got three physical states – solid, liquid and gaseous form and the solid minerals can be further divided as metals, non-metals and fuels. This aspect makes the scope of mining and hence, excavation technology very wide.

Ancient miners used basic versions of many modern tools, appliances and equipment (fig. 1.6). Rock was mostly dislodged in-situ from hammers and wedges and the

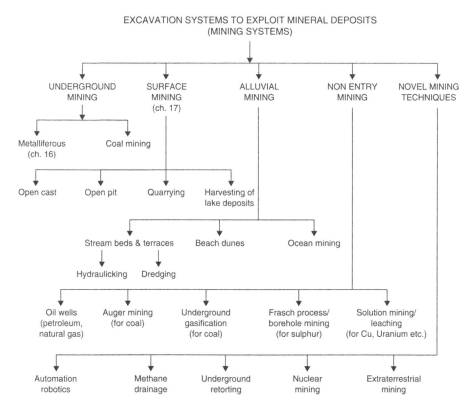

Figure 1.5   Excavation (Mining) Systems to exploit mineral (solid, liquid or gas) deposits within earth's great spheres.

resultant lumps were eased by pick or crowbar. Use of iron hammers with chisel or wedges were also made to break the rock. Iron shovels with wooden handles were used to muck out the broken spoil. In some cases the method of fire setting (i.e. first heating the hard rock and then cooling it by the cold water causing it to break), introduced in prehistoric times, was used to break down very hard rocks. Tools[17] were generally made of iron, copper or bronze, although Roman miners sometimes used stone hammers. All tools appear to have had short handles to facilitate their use in narrow cramped working places.

But during the period in between ancient times and particularly after 19th century there have been many important events, inventions and developments that have resulted the new techniques, methods and equipment. This scenario has brought the mineral industry in the forefront to feed the requirements of masses. This can be verified by the facts that as how the equipment, techniques and methods that have been described in the following chapters, could achieve this?

In the line diagram figure 1.5, the prevalent excavation technologies that are covering earth's great spheres have been shown. It includes excavating the minerals in their all the three states on commercial basis and also projects the future technologies,

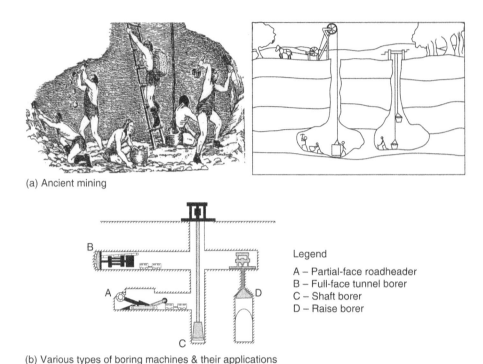

(a) Ancient mining

Legend
A – Partial-face roadheader
B – Full-face tunnel borer
C – Shaft borer
D – Raise borer

(b) Various types of boring machines & their applications

Figure 1.6   (a): Excavation of mineral during ancient times using primitive tools but they basic versions of modern tools. (b): A phenomenal growth and development in mining and excavation systems with regard to methods, techniques and equipment during this modern era. Modern equipment – boring and heading machines capable of creating excavations in any direction without aid of explosives shown.

which are in the process of development. This classification is based on the type of minerals and their occurrence based on their spatial position.

Flat and low dipping deposits outcropping to the surface or at a shallow depth are taken care by the surface mining method known as open cast mining (figs 17.7, 17.8(a)[7]). To win inclined to steeply dipping deposits open pit mining is used (fig. 1.7(a) – upper portion, fig. 17.2). Quarrying is applied to mine out the dimensional stones such as granite, marble, slate and few others (figs 17.22, 17.23).[12] The lake deposits for mining the salts is also come under surface mining and these are mined by harvesting.

Figure 1.6(a)[13] is a classic example of mining and tunneling during ancient times. With available techniques, as illustrated in figure 1.5, which are the results of consistent efforts of men from centuries; today ground and rocks could be excavated in any direction with application of modern equipment as shown in figure[5] 1.6(b). These sets of equipment are safe and productive.

Beyond break-even depths (where cost of mining is equal to the price fetched), the deposits cannot be mined by surface mining methods and it calls for the underground mining systems. The coal deposits all over the world are very widely spread and also extends beneath the surface, and that is why, they have been separated from the rest

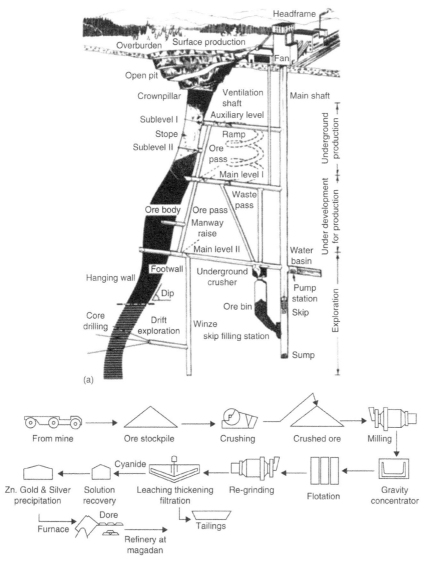

(b) Processing ore at a Gold mining and processing complex in Russia

Figure 1.7   (a): Mining the outcropping or shallow seated deposits by the application of a surface mining method (open pit mining in this case) followed by underground mining beyond the break-even depth. Illustration is typical example of iron mining in Sweden. (b): Mining and processing to obtain final product.

of the minerals and are won by underground coal mining methods (figs 16.3, 16.5 to 16.7, 16.23). This type of mining calls for a special care and attention towards ground control, fire, explosion and inundation. It needs a special type of safety culture, than mining rest of minerals, which comes under underground metal mining (chapter 16).

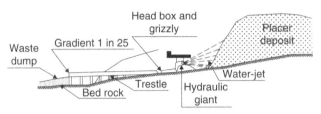

(a) Hydraulicking to mine out placer deposits

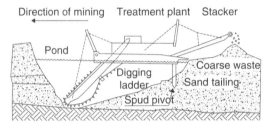

(b) Dredging to mine out deposits from water bodies

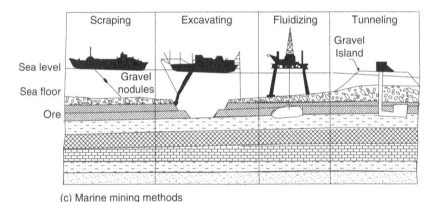

(c) Marine mining methods

Figure 1.8   Aqueous mineral excavation/extraction techniques.

In underground metal mining situation *ground fragmentation* is one of the main worries of a mining engineer. The strength of ore and its enclosing rocks, dip and thickness of the deposit govern the mining methods under this system. These methods (fig. 1.7(a)[2]) fall under three main categories – unsupported, artificially supported and caving methods, as described in chapter 16. In this mining system with the increase in depth the problem of heat, humidity, haulage, hoisting and rock burst increases.

Deposits saturated with water, and under the water bodies including ocean are recovered by alluvial mining.

Dredging – a floating vessel equipped for dislodging or excavating, loading and processing mechanisms is essentially a prevalent method for mining deposits such as clay, silt, sand, gravel and any associated minerals under a water-body (fig. 1.8(b)[3]), using a bucket-line, dragline and /or submerged water jets. The mined-out material is transported hydraulically or mechanically to a washing plant/station which may be part of the dredging rig, or a separate unit to recover the useful contents. This technique is also popular to clear and deepen water channels and floodplains.[27]

Hydraulicking[27] – a method to mine out placer (alluvial) deposits (fig. 1.8(a)), having small concentrations of metals such as: tin, gold, titanium, silver and tungsten. In this method water with high pressure is sprayed to convert loosely consolidated or unconsolidated deposit into slurry for onward processing to recover the useful contents. These hydraulic methods are primarily applicable to metal and aggregate stone deposits. In addition, coal, sandstone and mill tailings are also amenable to this method. Adequate water supply with high pressure and availability of waste disposal are the main considerations in the selection of this method. The slurry is conveyed through 'sluice boxes' and the process is known as 'sluicing'.

Scraping, Excavating, Fluidizing and Tunneling (fig. 1.8(c)[3]) could be applied to excavate deep-seated marine deposits. Application of these methods on commercial scale is yet to be established.

No entry mining is very widely undertaken; and mining of petroleum and gas is so wide that this branch of mining system has been separately dealt as petroleum and natural gas engineering.

Deep-seated coal deposits after break-even striping depth are recovered by using large *auger drills* at some of the coalmines of USA (fig. 17.8(b)[21]).

Frasch processing[11] (fig. 1.9(c)) is the main method to recover sulfur by non-conventional methods.

Application of solution mining and leaching is getting increased to recover already mined low-grade deposits of copper, gold, uranium and few others (fig. 1.9(a)[1]). In this system certain solutions are allowed to react with the ores of interest for some duration. Recovery of useful ores and metals from the old dumps is undertaken with application of *heap leaching* (fig. 1.9(b)), which was established by Rio Tinto in Spain more than 300 years ago when water percolating slowly through heaps was colored blue and copper was recovered through its precipitation onto scrap iron.[27]

The solution used to extract the soluble metal is referred to as a lixiviant.[27] The most common lixiviants used in this mining sector are dilute solutions of alkaline sodium cyanide for gold, acidic sulfuric acid for copper, aqueous sulfur dioxide for manganese and sulfuric acid-ferric sulfate for uranium ores; however, most leached uranium and soluble salts are collected by in situ mining in which the lixiviant is injected directly into the ore body without prior mechanical extraction. This latter technique enables low-grade ores to be processed without extracting the ore from the mineral deposit.

Use of bacteria is also made in some cases to boost the rate of reaction. *Thiobacillus* ferro-oxidants as iron-oxidizing species and *Thiobacillus* thio-oxidants as sulfur-oxidizing species are used. They convert ferrous to ferric and sulfide to sulfate respectively.

The health hazards unique to solution mining are the potential exposure to the chemical lixiviants[27] during transportation, leach field activities and chemical and electrolytic processing. Acid mist exposures may occur in metal electro-winning tank-houses. Ionizing radiation hazards, which increase proportionally from extraction to concentration, must be addressed in uranium mining.

Coal deposits which are of low grade, thin, deep seated, previously worked and with adverse geological conditions have been successfully recovered by converting the

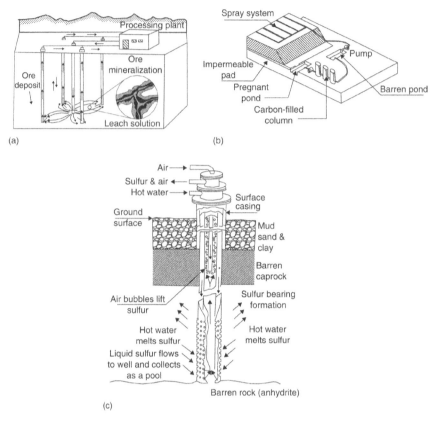

(a)   (b)

(c)

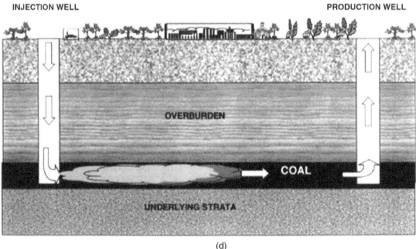

(d)

Figure 1.9   (a): In-situ leaching monitoring from surface. (b): Heap leaching to recover useful contents (metals) from low grade ore heaps. (c): Borehole mining to recover sulfur. (d) Underground coal gasification (schematic presentation).[23]

coal into gas by the technique known as *coal gasification*.[8] The process of converting in situ coal to synthetic gas (Syngas) and then bringing out the gas to the surface through pipe lines for its use to produce heat, generate power or to synthesize as a variety of chemical products such as hydrogen, methanol or synthetic natural gas, is known as underground coal gasification (UCG).[23] A coal gasification process achieved from surface gasifiers is replicated underground by drilling wells into the coal seams, injecting air or oxygen or steam or a combination of these through the injection well to gasify the in-seam coal, as shown in figure 1.9(d). When an oxidant is injected into the coal seam, several physico-chemical reactions such as oxidation, reduction and pyrolysis take place. The synthetic gas (Syngas) product is transported to the surface through the production well for its processing and utilization.

A rolling programme of UCG along a coal seam leaves highly porous cavities and stressed strata in its wake. The cooled abandoned cavities may be accessed through the existing production well or by directional drilling for long term underground storage of CO_2.

Figure 1.9(d)[23] is a schematic presentation of UCG's injection and production processes. For the success of this technique the parameters that are to be looked into include: reaction kinetics, heat transfer, gas flow, hydrology, thermally affected geophysics and several other geological controls for a consistent yield of the syngas. Many diverse disciplines are combined in this technology for transmission of gases to and from the combustion zone, gasification over long distances, controlled cavity growth, optimum power output and environmental benefits.

Some of the novel techniques that are in their initial phase; to name them they are: *methane drainage, automation and robotics, underground retorting, nuclear mining and extraterrestrial mining*. The last one refers to possibilities of exploitation of mineral resources from moon.

Methane drainage is being successfully applied in the U.K. whereas in other countries it has to prove its commercial viability. Positive indications are there for its growth on safety grounds, and also its competitiveness with natural gas.

Automation and robotics:[27] In hazardous conditions minimum human exposure is a desirable feature and these concepts/techniques have great acceptability and a bright future. However, its widespread adoption depends upon more technological ruggedness, especially for the underground regime, which in turn should enhance its economic viability.

Use of nuclear energy for peaceful purposes including mining has a great potential but its technical viability is yet to be established. A promising method of fluidizing certain hydrocarbon deposits like oil shale and tar sand is by *underground retorting*[6] (in which pyrolysis of kerogen occurs in situ); but its practicality on a commercial scale is yet to be established.

Extraterrestrial mining[27] refers to possibilities of exploitation of mineral resources from the Moon. This is amongst the 'novel methods' but colonization of outer space (most likely site: the Moon) is a must to justify risky, untried extraterrestrial mining. Launching of a US space station would revive interest in the concept.

Nuclear mining: this has radiation hazards and as such its application would be subject to devising those technologies which can deal with this hazard safely. In addition, at present the Nuclear Test Ban Treaty is in effect to apply this technique in practice.

1.7   UNIQUE FEATURES OF MINERAL INDUSTRY

Mining industry, or mineral industry in general, differs from others in several ways, as outlined below (Also ref. 18.1 '9G Evaluation'):

Long gestation or construction period: Before any mineral is mined it has to undergo several stages; starting from exploration, feasibility studies, mine development and construction. These stages take several years unlike other industries, which can be set up in a relatively shorter duration. This, longer gestation period results into higher establishment costs and requires a proper planning to take care of the escalation in costs of commodities and fluctuation in the mineral prices.

Mining – a risky business: The amount of risk is the highest comparing other industries in terms of the geological uncertainties, which means encountering low grade and less amount of reserves with abnormal disturbances than predicted during planning. This warrants for high rate of return comparing other industries.

Mine hazards: Working in a confined space underground that too in dark under heat, humid, gassy and watery conditions make the miner's job as most difficult and risky. So is the case while working in the surface mines under adverse climatic conditions The miners are also liable to occupational diseases such as asbestosis, silicosis and few others. In addition, the risks of fire, explosion, inundation and ground failure are part and parcel of this industry. This makes mining operations less attractive so far the job selection is concerned and warrants higher wages and more benefits to the workers.

Mineral reserve – a diminishing asset: The ore reserves get depleted every day once mining is begun and they cannot be replenished. Therefore, any deposit or mine has a definite life, and after which, the resources used in terms of man, machine, equipment and infrastructures have to be diverted. This results in financial losses and makes a mining venture less attractive to an entrepreneur. On the contrary any other industry can be run continuously so far the basic raw materials are fed to it.

In fact mineral resources belong to all mankind, present and future, and should not be squandered. Thus, mining has strong ethical side to it. The miner must be educated to be aware of it. This means that at the time of geological prospecting proper evaluation of ore reserves and during mining and processing, maximum recovery must be ensured.

Access to the deposit, location of mines and working faces/spots: Usually access to any deposit is difficult due to its location in a remote area. This, in turn, warrants establishment of proper infrastructures including the welfare facilities for the workers in that locality, and not only this, even within a mine the working spots/faces get change and a miner encounters a new environment of work like a soldier gets on a war front. A miner has to fight always against the nature, and that is why, the term 'winning a deposit' which is used to mine-out a deposit, is very common in mining. All these aspects make miner's job challenging. On the contrary, most of other industries can be setup anywhere as per the convenience and a worker is not required to change his working spot. On the same lines, Lineberry and Paolini[10] describes the worst case in underground mining is the *'encumbered space'*. This is due to the fact that in an underground situation the working space is inherently tight, distorted, congested, isolated and inaccessible, of poor

Table 1.3 Mining – in a 'Total System Context'.

System	Description
Ground control (esp. underground)	Support, caving, closure, bursts, subsidence
Excavating and handling	Rock penetration, fragmentation, bulk solid movement, storage and transfer
Life support (esp. underground)	Ventilation, drainage, illumination, personal safety and health
Normal support (logistics and environment)	Power, supplies, repair, maintenance, surveying, construction, management, personnel, environment

quality, and deteriorating. These adverse conditions endanger personnel, damage mobile equipment, and affect all activities. The same is true even for the surface mines.

Mining a transitory activity: Different from civil engineers who are conditioned to design and build works/structures that are robust enough to last centuries; whereas, mining engineers must design their 'permanent' works to last mine lifetime and no more. Any extension in safety or life length is an increase in cost and less profit, and this ultimately amounts raising the cutoff grade, and thereby recoverable reserves of a deposit.

Environmental impacts: Mining to processing to a get the final product is a complex operation. First ore is mined from surface or/and underground mines. Thus to produce any end-product from a mineral deposit is a long process. There are several stages after mining and they are: Concentration (crushing, grinding, separation, classification, leaching, thickening, drying, etc.), Smelting, Refining and Casting etc.[4] as shown in figure 1.7(b). Mining as well as extractive metallurgical operations are detrimental to the environment. Land degradation, water and air pollution, change in the land use, disturbance to flora and fauna in and around the area occupied by the mining lease; are some of the inherent features of mining which can not be avoided but their adverse impact can only be minimized. This may be noted that any mining venture that is not able to pay to control the environmental impact and for land reclamation at the end of operation, is not feasible.

Mining must be performed as an economic activity: Mining must be performed as an economic activity that is to say at profit. Management must be careful and efficient, particularly in this era, where consumer demand and international recession make minerals a buyer market. Mining operations must adjust to this difficult reality to survive.

Total system context: Describing mining as a 'Total System Context'[10] consists of the systems such as: Ground control, excavating and handling, life support and normal support (logistics and environment). Table 1.3 describes[7] these systems.

In order to undertake any mining activity, the process starts with creating an excavation, which results the dislodged or broken rock to be transferred for its onward processing. Ground control measures are taken simultaneously. Mine could be compared[7] to a busy city with a concern for water, light, power, communication, transportation, supplies, sewerage and construction.

Table 1.4 Different phases during mine life.[10]

Phases	Operations
Prospecting, exploration and Engineering studies	Conceive, study, and investigate; preparation of feasibility and detailed reports. Mine planning and design
Mine development and construction	Implementation, construction, development and start of mine
Mining and production	Exploiting deposit to sustain production
Liquidation	Recovery of the pillars left and final closure
Post mining	Sealing and aftercare of the mining establishments as per the prevalent regulations (laws) of the concerned country

1.7.1 Different phases of mine life

Unlike other engineering disciplines, a mining engineer has to look after the different phases of mine life, as described in table 1.4 and figure 12.5(c).

1.8 BRIEF HISTORY OF CIVIL WORK EXCAVATIONS INCLUDING TUNNELING

Quarrying is older than underground mining and certainly older than agriculture[12] this means the excavation activities started with the ancient civilization; naturally men must have made excavation for their shelter using the ancient and primitive tools described in the preceding sections.

Tracing the history[16,17,22] of the art of tunneling; it reveals that during the period 3000 B.C. to 500 A.D. in Egypt, Malta, Austria and few other places; the tunnels were driven for the purpose of mining. During period A.D. 50 to 1500 apart from mining tunnels for water supply (Greek water tunnel 1.5 km during 600 A.D.), road, military and burial purposes were driven. During 19th and 20th centuries there have been remarkable growth in numbers and total mileage of tunneling work, which were undertaken for different purposes. Some of the prominent tunnels driven during this period have been shown in table 1.5. In addition, there are many more which have been not included in this list.

In figure 1.10 some prominent underground civil structures have been illustrated. Figure 1.10(a) represents tunnels and declines/ramps to facilitate use of tyred vehicles (automobiles). Tunnels could be tracked to facilitate locomotives haulage system (figs 7.3(f), 9.12). Hydroelectric power generation requires a network of openings of varying sizes and shapes, as illustrated in figure 1.10(c)[15] and figure 15.3. It is a network of tunnels, large chambers, shafts, winzes and raises. Likewise a repository, to burry out the hazardous nuclear-waste also requires a network of excavations of different configurations, as shown in figures 1.10(b) and 15.7.

As described in preceding sections, the relationship between Underground-Mining and Tunneling is very close and very old, and this is due to the fact that mostly the Methods, Techniques and Equipment used for them are common. Approaches to drive through varying ground conditions such as soft and unstable ground, watery strata and hard rocks are the similar to a great extent. However, important distinction

Table 1.5 Some important tunnels of 20th century.[18,24]

Name	Country	Purpose	Length, km	Year of construction
Gotthard	Switzerland	Railway	16.3	1881
Simlon	Italy/Switzerland	Railway	19.8	1906
Moffat	USA	Railway	9.9	1927
Alva B. Adams	USA	Water conveyance	19.5	1946
Mont Blanc	France/Italy	Highway	12.6	1965
Seikan	Japan	Railway	53.9	1965
Mersey	UK	Highway	4.2	1986
Channel Tunnel	UK	Railway	19	1991
Channel Tunnel	France	Railway	20	1991
Eurotunnel	UK–France	Subsea Railway	50.4	1994
Lotschberg	Switzerland	Railway	34.7	2007
Koralm	Austria	2 Tubes, Railway	32.8	
Guadarrama	Spain	2 Tubes, Railway	28.4	2007
Hakkoda	Japan	Railway	26.4	2010
Pajares	Spain	2 Tubes, Railway	24.6	2010
Daisan-Shibisan	Japan	Railway	10.1	2004
Severomuyskiy	Russia	Railway	15.3	2001
Firenzuola	Italy	Railway	15.2	2009
Some tunneling projects for future	China	Mainland–Taiwan	125	Work
	Russia–Ukraine	Krasnodar–Krym	45	already
	Finland–Sweden	Umea–Vaasa	25–48	begun
	Estonia–Finland	Tallin–Helsinki	54–80	
	Ireland–Wales	Dublin–Holyhead	95	
	Japan–Korea	Tsushima–Kaikyo (Strait) and Korean Strait	120	

between them should be understood due to the fact that mine's life is limited to the extent till the mineral deposit is depleted, whereas, tunnel's life is till its purpose is served. Tunnels are driven to serve the people by way of providing passage to rails, roads, navigation, pedestrian etc. and also for conveyance of water and serve as sewerage. Thus, the tunnels constructed for civil works and having very long life, need to be very safe with regard to their stability, ventilation, illumination and risks of getting flooded etc. Tunnels are located at shallow depth and have large cross sectional areas. In general, function of tunnels and underground openings both in civil and mining are multiple, as outlined in figure 1.2. This figure illustrates use of tunnels and drives in different environments. However, following are the are some of the unique features of civil tunnels:[14]

- Perfect horizontal alignment
- Unlimited operational life
- Perfect ventilation and illumination during operational phase
- Tightly controlling the infiltration of water, gases, or ground contaminants
- Permanent lining with aesthetic look

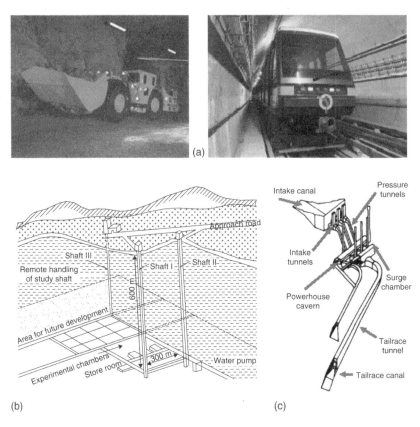

Figure 1.10 Some important civil constructions involving excavations in all directions at any locale – surface as well as sub-surface. (a): Typical tunnel driven using trackless equipment (left), and a tracked tunnel (right). (b): A typical repository – schematic presentation involving tunnels, shafts and other excavations. (c): Main components of a Hydro-power plant (ITA Australia).

• Provision for cross passages and emergency exists
• Provision for noise and vibration impacts abetment in case of railway and highway tunnels
• Special attention in earthquake prone areas.

1.9   THE CURRENT SCENARIO

1.9.1   Population growth[25,28]

With the present growth rate of about 1.8%, the world's population is going to be doubled in next 39 years, as shown in Figure 19.6 (sec. 19.3). The population problem is not just a matter of numbers but also a matter of fulfillment of our basic needs such as: clean air, water, nourishing food and shelter; managing them effectively is becoming more and more difficult.

1.9.2 Lifestyle[13,25,26]

The world's culture, customs and lifestyles must be considered while assessing minerals required per capita. In some of the developed countries people's habits and lifestyle require large consumption of goods, services and energy. The wastage of these commodities is very great, thereby resulting in mineral consumption at an abnormal rate besides resulting in environment-related problems.

In a study to find out how the consumption pattern of minerals has drastically changed within 200 years in the USA, it was found that consumption of commodities such as aluminum, cement, coal, glass, iron, phosphate, slate and few others has risen from 50 to 300 times. This is mainly due to changes in the lifestyle and living standard of average middle-class American citizens. Is it not detrimental to the environment, as compared with citizens from Asian and African continents whose lifestyle is very simple but their populations are higher?

1.9.3 Globalization[28]

Business has crossed barriers and boundaries. Most of the countries are welcoming foreign investment which was closed by many countries in the past. Multinational companies (MNCs) are looking for acquisition of natural resources such as oil, gas, minerals, etc., and are establishing industries of various kinds the world over. The major driving force of globalization is the concern to create a single global market place. Its major characteristics include:

- a single market place where free trading would be possible;
- encouraging use of more and more computerization and automation in industries and trade;
- establishing new information and communication channels; and
- resulting massive urbanization and population migration between countries.

MNCs invest huge sums of money and expect its recovery at the earliest, and hence, a higher rate of returns. In the prevalent scenario globalization is considered as a system that is having both positive and negative implications for sustainable development. Also ref. sec. 19.3.

1.9.4 Buyer's market

Except for minerals, goods and services of strategic importance, most minerals, goods and services are available at a competitive price. No monopoly in any sector exists any more and thus a quality product at a cheaper rate (competitive price) is readily available.

To understand the impact of competition, figure 1.10(d) (left and right) illustrates the delivery price of iron ore to western as well as eastern markets by the leading MNCs. Companies such as BHP Billiton, Rio-Tinto, Kumba and CVRD are having delivery (production and freight costs) of iron ore just half of some of the companies in developing countries or in Europe, who are trying hard to overcome shortcomings in areas such as basic infrastructures, systems, technical know-how etc. and also for bringing further technological excellence to perform even better in adverse natural conditions (such as depth and type of deposits etc.). But to compete in the international market one will have to get rid of all these hurdles.

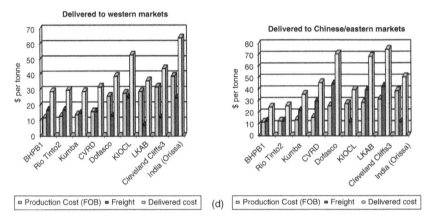

Figure 1.10   (d): The prevalent competition in the delivery price of iron ore to eastern as well as western markets by the leading MNCs.

1.9.5   Technological developments and renovations[26,28]

Looking at the global scenario, the progress that has been made in manufacturing and production of goods of mass consumption in the last five decades is far ahead of what had been achieved during the last five centuries. After the 19th century there were many important events, inventions and developments that have resulted in new techniques, methods and equipment. These include use of giant-sized equipment, application of automation, modular systems, and remote control techniques.

1.9.6   Information technology (IT) and its impacts[26,28]

Use of IT and computers in all the industrial sectors is growing and it has compelled producers to compete in the international market. Many manufacturing companies/units are merging. MNCs are looking for out-sourcing and wish to take advantage of cheaper energy/power, human resources and know-how. Ref. secs 5.16 and 18.6.6 also.

1.10   TOMORROW'S MINE & CIVIL EXCAVATIONS

Figure 1.11(a) to (c)[20] illustrate a vision of tomorrow's mine and civil structures, which will be remote controlled with the application of modern technology. Application of the information technology would be at forefront. Use of modular system to monitor various unit operations at surface as well as underground mines has already begun. In tomorrow's mines use of robotics to carryout repetitive tasks that too in hazardous and risky locales would be part of the process. Application of laser for precise survey, measurements and monitoring would play an important role. Sensors installed at strategic locations would help to monitor mine atmosphere. There will be thorough technology transfer from one field of engineering to another, for example, hydraulic fracturing is the concept used in oil fields for better oil recovery is now getting application to induce block caving in mines. Remote control would lead to more and more automation, and that in turn, would require more and more use of robotics.

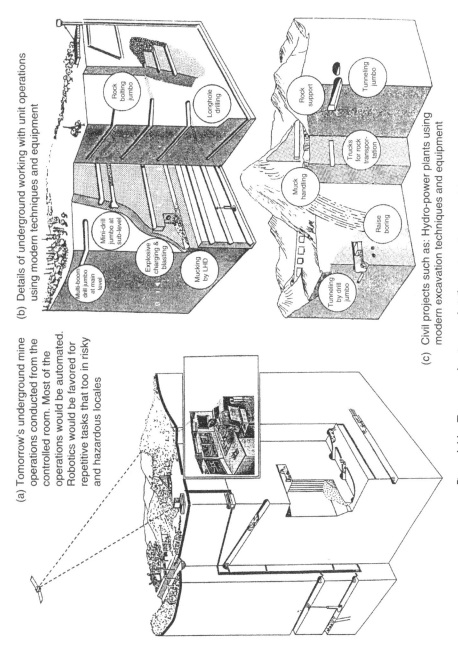

(a) Tomorrow's underground mine operations conducted from the controlled room. Most of the operations would be automated. Robotics would be favored for repetitive tasks that too in risky and hazardous locales

(b) Details of underground working with unit operations using modern techniques and equipment

Rock bolting jumbo

Longhole drilling

Multi-boom drill jumbo at main level

Mini-drill jumbo at sub-level

Explosive charging & blasting

Mucking by LHD

(c) Civil projects such as: Hydro-power plants using modern excavation techniques and equipment

Rock support

Tunneling jumbo

Trucks for rock transportation

Muck handling

Raise boring

Tunneling by drill jumbo

Figure 1.11   Tomorrow's mines and civil constructions – a vision.

1.11   THE WAY FORWARD[27]

In the prevalent scenario thorough competition in the global mineral market, declining mineral grades, higher treatment costs, stringent mining laws, privatization and restructuring are some of parameters that are compelling mine owners to reduce costs and increase productivity. As a result employment is falling in many mining areas due to increased productivity, radical restructuring and privatization. These changes not only affect mine-workers who must find alternative employment, but those remaining in the industry are required to have more skills and more flexibility. Finding the balance between the desire of mining companies to cut costs and those of workers to safeguard their jobs has been a key issue throughout the world in the mining sector. It suggests mining communities must adapt to new mining operations, as well as to downsizing or closure.

QUESTIONS

1. A newly born American baby would require 3.75 million pounds of minerals, metal and fuels throughout his/her life; what is this requirement in your own country?
2. Classify minerals based on their use in our day-to-day life.
3. Classify surface excavations based on their function and utilities.
4. Classify underground excavations (tunnels and large openings) based on their function and utilities.
5. Draw a line diagram to illustrate excavation (mining) systems to exploit mineral (solid, liquid or gas) deposits within earth's great spheres.
6. Give merits and limitations of leaching/solution mining.
7. How do the mining industry or the mineral industry in general, differ from other industries?
8. What will tomorrow's mine and civil excavations be like? Describe them.
9. Mention the processes involved to recover gold from ore produced at mines.
10. Phenomenal growth and development in mining and excavation systems with regard to methods, techniques and equipment has taken place during this modern era. Describe it.
11. List the different phases during the life of a mine.
12. List leaching types. During copper mining in the past, waste rock was stacked but it retained some percentage of copper. Name a leaching method and the solvent for the purpose of recovery.
13. List locales where excavations are created. What is the principal purpose in creating them?
14. List minerals found in your own country and which of them are being mined. Name minerals which your country is exporting and importing.
15. List some important tunnels of the 20th century.
16. List unique features of civil tunnels and how they differ from mining excavations.
17. Name important design specifications for UCG. Give the merits and limitations of UCG.
18. Name the different kinds of leaching. What are the minerals that are being recovered by the application of leaching or solution mining?

19. Name the important factors for successful in-situ leaching. Illustrate the process of in-situ leaching and heap leaching. Write down the steps for these operations. Mention the solvents that are used to recover gold, uranium and copper in the leaching process.
20. Describe the methods of mineral fluidization which are used for extracting mineral deposits, giving an example in each case of its application at present.
21. Describe the equipment used during the process of hydraulicking. Name the types of dredges available for underwater excavation of the placer deposits.
22. Describe the type of coal and other parameters suitable for underground coal gasification (UCG). Give a conceptual diagram to represent the process of UCG.
23. Name the by-products from a smelter. Name the product which needs to be thrown out from a mill.
24. Name some nuclear ores.
25. Name some of the minerals used as: (a) building materials, (b) precious stones, (c) radioactive minerals, (d) ceramic glass and industrial minerals.
26. Name the non-conventional mining method to recover sulfur. What will you need to pump to melt sulfur?
27. What environment setups can mining disturb?
28. What is mining and what should be aimed at for a successful mining operation?
29. Why are hydraulic mining methods the cheapest methods of mining?
30. Write an essay as to how minerals around us make our lives function.
31. Write and illustrate the various mechanisms that could be applied to win sea deposits.
32. Write a brief history of civil work excavations including tunneling.
33. Describe the basic stages of UCG. Mention the practices that are followed for the preparation of a reaction channel between a pair of boreholes. How is ignition initiated to establish a combustion front?
34. Describe the prevalent global scenario and how it is influencing the mineral industry? What are the challenges?
35. Describe the problems of mining deep sea deposits, at present.
36. Suggest a mining method or system for each of the following situations:
    a. Alluvial deposits.
    b. Deposits lying under water bodies at shallow depths.
    c. Deposits lying at the bottom of the sea.
    d. A low quality coal deposit not otherwise suitable for either surface or underground mining.
    e. A low grade copper ore lying at the surface as heaps.
    f. A sulfur deposit using hot water.
    g. Mining on the Moon.

REFERENCES

1. Ahaness, J.K. et al.: In-situ mining of hard rocks. In: Hartman (edt.): *SME Mining Engineering Handbook*. SMME, Colorado, pp. 1515–16.
2. Atlas Copco manual, 4th edition, 1982.

3. Cruickshank, M.J.: Marine Mining. In: Hartman (edt.): *SME Mining Engineering Handbook*. SMME, Colorado, pp. 2000, 2007.
4. Dobrymin, A.: Gold mining in Russia. *Mining Magazine*, July, 2002, pp. 26.
5. Franklin, J.A. and Dusseault, M.B.: *Rock Engineering*. McGraw-Hill, New York, 1989, pp. 484.
6. Grant, B.F.: Retorting oil shale underground: problems and possibilities. *Quart. Colorado School of Mines*, 1964, vol. 59, pp. 40.
7. Grim, E.C. and Hill, R.D.: Environmental protection of surface mining coal. *US Environment Protection Agency*, Cincinnati, Ohio, 1974, EPA-67012-74-093.
8. Hartman, H.L.: *Introductory Mining Engineering*. Johan Wiley & Sons, New York, 1987, pp. 527.
9. Lacy, W.C. and Lacy, J.C.: History of mining. In: Hartman (edt.): *SME Mining Engineering Handbook*. SMME, Colorado, pp. 5–10.
10. Lineberry, G.T. and Paolini, A.P.: equipment selection and sizing. In: Hartman (edt.): *SME Mining Engineering Handbook*. SMME, Colorado, pp. 1551–55.
11. Marsden, R.W. and Lucas, J.R.: Specialized underground extraction systems. In: Cummins & Given (eds.): *SME Mining Engineering Handbook*. AIME, New York, 1973, pp. 21: 21–103.
12. Meade, L.P.: Dimension stone: Friendsville quarry. In: Hartman (edt.): *SME Mining Engineering Handbook*. SMME, Colorado, pp. 1401–02.
13. Mineral Information Institute (MII), *Web site*, Denver, Colorado, 2001.
14. Monesees, J.E. and Hansmire, W.H.: Civil works tunnels for vehicles, water and waste-water. In: Hartman (edt.): *SME Mining Engineering Handbook*. SMME, Colorado, pp. 2124.
15. Olofsson, S.: *Applied Explosives Technology for Construction and Mining*. Applex, Sweden, 1997, pp. 160–170.
16. Sandstrom, G.E.: *The history of tunneling*. Barrier and Rockliff, London, 1963, pp. 427.
17. Shepherd, R.: *Ancient Mining*. Elsevier Applied Science for IMM, London, 1993, pp. 20–22.
18. Sinha, R.S.: Introduction. In: R.S. Sinha (edt.): *Underground Structures*. Elsevier, New York, 1989, pp. 20–21.
19. Stefanko, R. and Bise, C.J.: *Coal mining technology: Theory and practice*. SME-AIME, New York, 1983.
20. Tamrock: *Underground drilling and loading handbook*, Tuula Puhakka (edtr.), 1997, pp. 44.
21. Treuhaft, M.B.: examples of current surface mining research: operational and potential. Mining Engineering, 1984, vol. 36, no. 1, pp. 53–63.
22. Whittaker, B.N. and Frith, R.C.: *Tunneling – Design, Stability and Construction*. IMM Publication; 1990, pp. 1–17.
23. Singh, A.K. et al.: An approach to underground coal gasification in India. Pres. In: *The Mining Geological and Metallurgical Institute of India (MGMI), Centenary & 1st Asian Mining Congress*, 16–18 January, 2006, Kolkata.
24. http://www.lotsberg.net/data/rail.html
25. Kupchella, C.E. and Hyland, M.C.: *Environmental Science*. Prentice-Hall International Inc., 1993, pp. 61–68, 112–14, 191–93.
26. Tatiya, R.R.: Health, Safety and Environment (HSE) Management – Where do you stand? *Souvenir: 17th Environment Week Celebration, Indian Bureau of Mines*, 2006.
27. Thomas A. Hethmon and Kyle B. Dotson: Surface mining methods. Chap. 74 In: *International Labor Organization (ILO) Encyclopedia and CISILO Database*.
28. White Paper on: *Environmental Management Policy*, Department of Environment Affairs and Tourism, South Africa, July 1997.

Chapter 2

Rocks, minerals and mineral inventory evaluation

The mineral resources of today are the results of natural processes over billions of years but with the prevalent rate of consumption, they will be depleted within the next few centuries. Can we afford their wastage, and poor recoveries during mining and processing?

2.1   FORMATION PROCESS AND CLASSIFICATION

Rocks are aggregates of any combination of minerals (e.g. Quartz, Calcite, Galena), elements (e.g. Sulfur, Gold), solid organic material (e.g. Coal), and/or other rocks.

$$Rocks = Minerals + Elements +/- Solid Organics +/- Other Rocks$$
$$(+ sign indicates presence; - sign indicates absence) \tag{2.1}$$

The natural occurrence of a mineral or rock in sufficient amounts (quality and quantity) worth exploitation is known as a deposit. Based on their origin rocks have been classified as (figs 2.1, 2.2):

- Primary – Igneous rocks[2,4,8,9,10]
- Secondary – Sedimentary and Metamorphic rocks.[2,4,8,9,10]

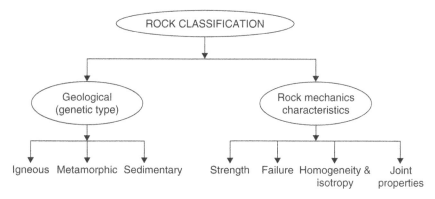

Figure 2.1   Origins and cycles of rocks.

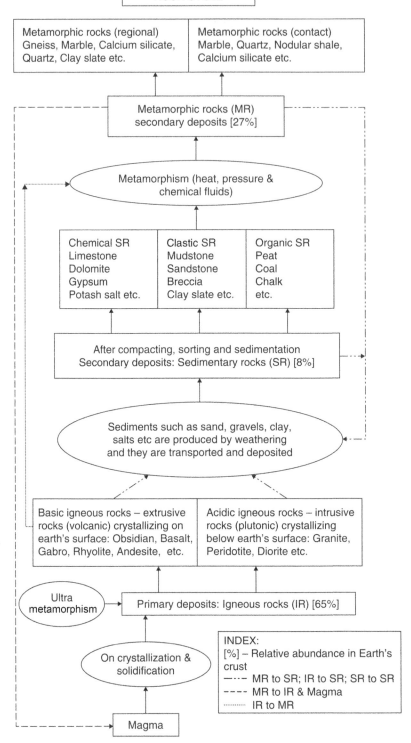

Figure 2.2 Origins and cycles of rocks.

2.1.1   Igneous rocks[8]

These are the parents of all other rocks. *Magma* is hot molten rock material generated within the Earth. When magma reaches the surface it is called *lava*. Igneous rocks are the result of cooling and crystallization of magma and lava. On crystallization the acidic magma gives rise to a mother liquor that is the source of numerous accessory minerals forming *epigenetic magmatic ore* deposits (*formed subsequently to that of the host rock*) such as contact metamorphic, pegmatite, vein deposits (shallow, intermediate and deep seated) and magmatic spring deposits on the surface. They also give rise to acidic igneous rocks such as granites, synites, quartz monazites, monazites and diorites. These are also known as *intrusive* rocks that crystallize below the earth's surface. Large, irregular *intrusive rock masses* are called *batholiths*.

Basic magma on crystallization gives rise to basic igneous rocks such as gabbros, peridotites, dunites and basalt, and also to the syngenetic magmatic ore deposits (*formed at the same time to that of the host rock*) *accessory* minerals such as oxides, native elements and sulfides. These are also known as *extrusive* rocks as they crystallize on the Earth's surface.

Intrusive igneous rocks cool slowly, producing a *coarse texture* with mineral grains visible to the naked eye. The minerals that form are determined by the chemistry of the magma and the way that it cools (relatively slowly or quickly, steadily or variably). The grains are typically interlocking, and of more-or-less the same size.

Extrusive igneous rocks (sometimes called volcanic) cool quickly, which causes very small crystals to form, if any at all. This produces *fine-grained rocks*, which without a microscope, can be identified only by color. The *minerals that form during cooling determine the color*. Like the intrusive rocks, the minerals formed reflect the chemistry of the magma. Colors vary from white to black, with pink, tan, and gray being common intermediate colors. The *texture* of these rocks can also be influenced by the amount of gas trapped in the lava when it cools.

If presence of SiO_2 content exceeds 62%, the rock is considered as Acidic; from $<62\%$–52% Intermediate; from $<52\%$–45% Basic; $<45\%$ Ultra-basic. In general igneous rocks are poorer in silica contents and richer in ferromagnesian silicates[7]. The acid rocks are more abrasive and harder than the basic ones but they are more resistant to impact.

Dikes are tabular igneous bodies formed vertically or across sedimentary bedding. Those formed horizontally or parallel to bedding are called *sills*.

2.1.2   Sedimentary rocks

The acidic igneous rocks and the basic magma deposits give rise to secondary mineral deposits formed through weathering (fig. 2.2). The weathering could be either chemical or mechanical. Chemical weathering produces soluble products that are carried downward to form secondary enriched deposits such as copper. Also these soluble products are removed by streams and deposited as chemical sedimentary rocks such as: iron, manganese, limestone, chalk, dolomite, phosphates, iron, limestone, diatomaceous earth, coal, petroleum and natural gas. Chemical weathering also produces insoluble products that are either residual conglomerates, quartzites, breccias etc. and also detrital (placer) deposits such as: gold, platinum, gems, rare minerals etc. The residuals formed chemically give rise to deposits such as: Fe, Al, Ni, Cr, Pb, kaolin,

Table 2.1 Conversion of some rocks into metamorphic rocks.

Original rock	Metamorphic rock
Mudstone/Shale	Slate
Shale	Chlorite Schist
Basalt/Gabbro	Biotite Schist
Granite/Diorite	Gneiss
Limestone/Dolomite	Marble
Quartz-rich Sandstone	Quartzite

laterites etc. and also the residuals released by the chemical disintegration give rise to minerals such as: gold, platinum, gems, rare minerals etc.

Thus, in the sedimentary rock deposition process the weathered fragments are transported via water, air or ice before they are deposited and transformed. Sediments are transformed into rocks by *cementing* them, usually by calcite, silica or iron oxides that glue the fragments together. *Compaction* is achieved when fragments are squashed together. Sedimentary rocks generally occur in layers or beds that range in thickness from few centimeters to hundreds of meters. Their texture ranges from very fine grained, to very coarse. Colors include red, brown, gray, yellow, pink, black, green and purple.

2.1.3 Metamorphic rocks[1,4,8,10]

In addition to the above any primary or secondary mineral deposit may be subjected to great pressure, heat and chemical alteration producing regional metamorphism that gives rise to metamorphic rocks such as: gneisses, schist, marble, slates, quartzites, serpentines etc. (table 2.1), and metamorphic ore deposits such as: garnets, andalusite, graphite, emery, talc, soapstone, schist, zinc, manganese, etc.

Metamorphism can also occur in areas of stress such as faulting and folding of rock or in areas of plate tectonics such as the oceanic crust colliding into the continental crust. The principal characteristic of metamorphic changes is that they occur while the rock is solid. Following are some of the important characteristics that are associated with these rocks.

Texture characteristics are very important in classifying metamorphic rocks. They range from very fine-grained to coarse-grained minerals. Metamorphic rocks can be divided into two textural groups, foliated (layered) and unfoliated (not layered).

Foliation: Parallel layers of minerals, sometimes of different composition, giving the rock a distinctive planar to platy feature (Schist, Gneiss).

Unfoliated: No preferred orientation of minerals. The rock has no preferred orientation of breakage (Quartzite and Marble).

Rock cleavage: A property of a rock that allows for easy breaking along parallel planes or surfaces. Metamorphic rocks tend to break or cleave most easily along planes parallel with foliation.

The relative abundance of the three rock groups in the earth's crust are: 65% Igneous, 27% Metamorphic and 8% Sedimentary.[10]

2.2   ROCK CYCLE & TYPE OF DEPOSITS[8,9,12]

Ore deposits, as all rocks, have an intimate relationship with their setting in the tectonic regime (the processes responsible for their formation; Greek *tekton*, a builder). Their origin can be understood in terms of earth processes. A line diagram (fig. 2.1) represents these processes. As evident from the figure, magma is the principal material of the lithosphere responsible for the deposits of all kinds. The dominant geological processes[9] involved in the formation of rocks of different types have been shown within the ellipses in the diagram.

Apart from the rock formation, the formation of *valuable ore* deposits which are the potential source of metal and minerals of our daily use, have been deposited by the geologic processes, as described in the following paragraphs. In table 2.2, the ores for which the geologic processes were responsible for their formation have been tabulated.

Placer deposits (residual and detrital deposits): These deposits composed of minerals that have been released by weathering and later on have been transported, sorted and collected by natural agencies into valuable deposits. Such minerals are usually of high specific gravity and are resistant to abrasion and weathering. Examples are gold, diamonds, platinum, tin, monazite, magnetite and ilmenite.

Metamorphic deposits: These deposits have undergone changes when subjected to high temperature, great pressure and chemical alterations by solutions. They have become warped, twisted and folded; the original minerals are rearranged and recrystallized.

Contact metamorphic deposits: Magmatic gases and solutions invade and change the rocks which they intrude forming new minerals and depositing valuable metals. The metals are mainly iron, copper, though gold, silver, lead, zinc, tungsten and tin are found in minor quantities.

Pegmatite deposits: These are found in or near igneous rocks and at the outer margin of intrusive masses. They have the composition of igneous rocks but contain a smaller range of minerals. They are derived from very thin fluids and usually coarsely crystalline. They frequently contain valuable gem minerals such as garnet, topaz, beryl, emerald, tourmaline, and sapphire.

Magma crystallization: Syngenetic igneous deposits or magmatic segregation are formed by the solidification of basic magmatic material and occurs as dikes and irregular masses. Examples are diamond, chromite, corundum etc.

Hydrothermal veins and replacements: These are formed by the precipitation from very hot vapors and solutions, probably originating from the molten rocks. Examples are tin, gold etc.

2.3   TEXTURE, GRAIN SIZE AND SHAPE[4]

The texture or fabric of a rock is the size, the shape and the arrangement of its constituents. All igneous and most metamorphic rocks are crystalline, whereas sedimentary rocks are made up of grains of fragments (known as fragmental). Crystalline rocks consist of an interlocking mosaic of crystals, whereas fragmental (detrital or clastic) rocks are made up of grains that are usually not in such close contact. Due to these

Table 2.2   Details of some important ores.[8,10,12]

Ores	Brief description
Native ores	This includes ores of certain noble metals such as gold.
Sulfide ores	Chalcopyrite ($CuFeS_2$), galena (PbS), sphalerite (ZnS), stibnite (Sb_2S_3), molybdenite (MoS_2), pyrite (FeS_2), acanthite (Ag_2S) etc.
Oxidized ores	These are comprising of oxides, carbonates and sulfates of various ferrous, non-ferrous and rare metals, for example: Fe_2O_3, $2Fe_2O_3\ 3H_2O$, MnO_2, $PbCO_3$, CuO_2, SnO_2.
Silicate ores	These are usually ores of rare and scattered elements wherein mineral is silicate or aluminosilicate, for example: beryl $Be_3Al_2(Si_6O_{18})$, zircon $ZrSiO_4$ etc.
Specific ores	
Aluminium	This is Bauxite containing 20–55% Al_2O_3 (alumina); It is a light to dark brown colored product of decomposed igneous and metamorphic rocks. It is soft to medium hard and occurs as lenses or pockets.
Apatites	These Phosphate ores contain P_2O_5 (35%). They are used to make super-phosphates.
Asbestos ore	The ores are chrysotile and serpentine. They are light to dark coloured ores occur as veins. They are soft to hard.
Chrome	Chromite occurs as chromic oxide containing 28–62% Cr and known as chrome. High grade (Cr exceeds 45%) is used in metallurgy and the low grade (Cr ranges 30–40%) in the chemical industry. This dark colored ore occurs in veins and disseminations in igneous and metamorphic rocks. It is medium-hard to hard.
Copper	The common ores are: Chalcosine (Cu_2S), bornite (Cu_2FeS_2), chalcopyrite ($CuFeS_2$) etc. containing 0.5–1% Cu. These ores also contains small amount of Au, Cd, sulfides of Fe, Zn, Ni, Pb and other elements.
Gold ore	This yellow element occurs in veins of igneous rocks, chiefly in quartzite, which is hard. It also occurs as placer deposits which are soft.
Iron	Common ores are: Magnetite (Fe_3O_4) 72% iron; Hematite (Fe_2O_3) 70% iron; Hydro-hematite ($Fe_2O_3 \cdot nH_2O$); Goethite ($Fe_2O_3 + nH_2O$); Limonite ($2Fe_2O_3 \cdot 3H_2O$) 60% iron; Siderite ($FeCO_3$). Average Fe content in these ores range 15–40%, As, S, P, Zn, are the harmful admixtures. Ore density 3–4.5 t/m^3. Most iron ores need concentration.
Manganese	Pyrolusite (MnO_2) ore contains Mn upto 63%. Mn can be 15–20% if iron and limestone are present in ore. These dark coloured ores are associated with sedimentary and metamorphic rocks as thick to thin bedding.
Mica	The common ore are: Muscovite, phlogophite, lepidolite. Mica crystals less than 4 cm^2 area are not considered. Content is measured in kg/m^3.
Molybdenum	The usual ores are: molybdenite (MoS_2), quartz-Mo, quartz-tungsten-Mo, Cu–Mo ores. Mo ranges: 0.05–1%.
Nickel	Ni occurs as Pentlandite and revdinslite ores. These ores depend upon its accompanying elements such as Cu, Co, platinum. About 90% of Ni is produced from copper-nickel sulfides.
Phosphorites	They are classed as: concretionary (12–35% P_2O_5), granular (5–16% P_2O_5), microgranular (26–28% P_2O_5). P_2O_5 could be 4–5% but usually 18–20%.

(Continued)

Table 2.2   Continued.

Ores	Brief description
Silver ore	Ore argentite is a silver sulfide of 87% silver content. This dark colored ore occurs in veins of igneous and metamorphic rocks, along with gold and nickel. Ore is medium-hard to hard.
Tin	Cassiterite (SnO_2) having 0.3–0.8% meta content. Up to 1% if favorable conditions prevail.
Tungsten	The ores are: Wolframite [(Fe, Mn)WO_4], scheelite ($CaWO_4$). Minimum content is 0.1–0.5% WO_3. Quartz-wolframite, skarn scheelite are the main ores.
Uranium ore	Ore uraninite is a mixture of uranium oxides, averaging about 86% uranium content, and is generally associated with other minerals. It occurs in veins in igneous and metamorphic rocks, and as disseminations in sedimentary rocks with medium to heavy beddings. Ores are medium-hard to hard.
Complex ores	Usually comprise of Pb and Zn, as well as Cu, Au, Ag, Tungsten, molybdenum, cadmium and sometimes antimony, bismuth and tin. Pb and Zn content is around 2% Galena (PbS) and sphalerite (zinc blend) are usual ores. Occurs as replacement, sheet and fissure ores.

features crystalline rocks are stronger, less porous, and less deformable than fragmental varieties with similar mineral composition. A quartz-feldspar granite is much stronger than quartz-feldspar sandstone.

A texture may be termed *homogeneous or heterogeneous*, depending on whether or not all parts of sample have a near-identical texture and mineral composition. It may be termed as *isotropic or anisotropic* depending on whether or not preferred orientations are visible.

2.3.1   Grain sizes and shapes[4]

The commonly accepted size designation follows in this manner: clay – finer than 0.002 mm; silt: 0.002–0.06 mm; sand: 0.06–2 mm; gravel: 2–60 mm; cobbles: 60–200 mm; and boulders coarser than 200 mm. The shapes of crystalline grains are described by terms such as equi-dimensional (1:1:1); platy or discoid (two long axes and one short); and fibrous or prolate (two short axis and one long). Fragmental grains are described as: angular, sub-angular, rounded, sub-rounded or well rounded. Rock with round shaped grains is easier to drill as in sandstone.

2.3.2   Durability, plasticity and swelling potential of rocks[4]

Slake Durability: defined as the resistance of rock to wetting and drying cycles. It can be determined by immersing samples in water and noting their rates of disintegration. All rocks are more or less affected by wetting and drying. Rocks such as granites and well-cemented quartzitic sandstones are durable because they can survive many cycles of drying and wetting without disintegration. Rocks containing clay and other minerals such as anhydrite disintegrate when exposed atmospheric wetting and drying.

2.4   THE CONCEPTS OF MINERAL RESOURCES AND RESERVES; MINERAL INVENTORY, CUTOFF GRADE AND ORES

The Sun, air, water, flora, fauna, minerals and soils are the natural resources that are available on, above and under the Earth's surface. Like other natural resources minerals are our basic need. Metals, nonmetals, and fuels are the three classes of minerals (table 1.1) and more than 3500 minerals have been identified so far.[10] Minerals are everywhere around us. For example, it is estimated that more than 70 million tons of gold is in the ocean waters. It would be too expensive to recover because it is so scattered. Minerals need to be concentrated into deposits by Earth's natural processes to be useful to us. The Earth is a huge storehouse.

The difference between the terms, *mineral resources* and *reserves*; *Geological and mineable reserves*; *minerable and commercial reserves*; have been dealt in chapter 12. Quantity of reserves, grade-wise, for a deposit is known as its mineral inventory (fig. 12.3(e)). Thus, there is a difference between mineral inventory and ore reserves. It is a well known fact that almost all minerals occur in nature as compounds. The processes and forces have concentrated these minerals in widely different amounts. When this concentration is sufficient so that it can be economically exploited and recovered, the deposit is referred as an *ore*.[49] Thus, in ore reserves estimation, the technique, method, equipment to mine-out the deposit together with its cost of mining, and selling price come into the picture; whereas in mineral inventory estimation considerations of these parameters is not required. These concepts have been dealt in chapter 12.

2.4.1   Some important ores – chemical & mineralogical composition

Based on their chemical and mineralogical composition, a brief description of common ores is given in table 2.2.

2.5   GEOLOGICAL STRUCTURES

2.5.1   Geometry of a deposit

Figure 2.3[2] describes the terms used in conjunction with any mineral deposit. *Dip* is the angle at which the strata or mineral deposits are inclined to the horizontal plane, and *strike direction*, is perpendicular to it (fig. 2.3(i)). A deposit can be flat (dipping horizontally to below 20°), inclined (dipping 20° to below 50°) and steeply inclined (dipping from 50° to vertical, figure 2.3(ii)).

Thickness of a deposit is the distance at right angles between the hanging and footwall of an inclined deposit, or between roof and floor, of a flat deposit. Hanging-wall is the upper side of an inclined vein (fig. 2.3(iii)). It is called roof in a bedded deposit; whereas footwall is the wall or rock under a vein, and for the bedded deposits, it is the floor. A deposit can be very thin, thin, thick, or very thick.

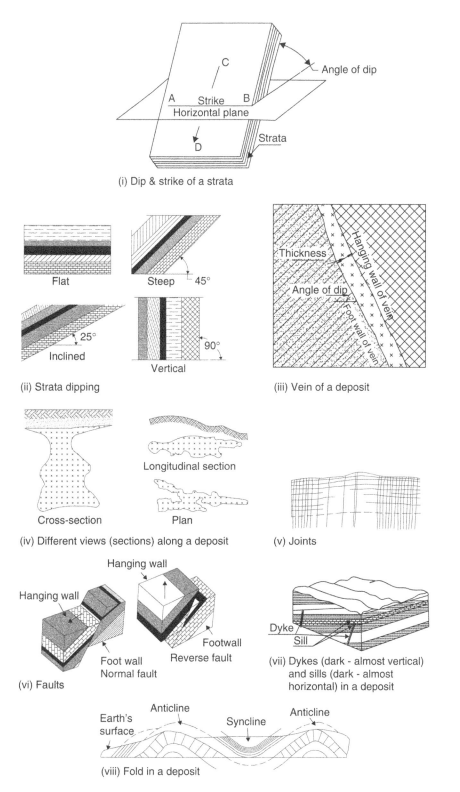

(i) Dip & strike of a strata

Flat

Steep 45°

Inclined 25°

Vertical 90°

(ii) Strata dipping

Thickness

Hanging wall of vein

Angle of dip

Foot wall of vein

(iii) Vein of a deposit

Longitudinal section

Cross-section

Plan

(iv) Different views (sections) along a deposit

(v) Joints

Hanging wall

Hanging wall

Foot wall
Normal fault

Footwall
Reverse fault

(vi) Faults

Dyke

Sill

(vii) Dykes (dark - almost vertical) and sills (dark - almost horizontal) in a deposit

Earth's surface

Anticline

Syncline

Anticline

(viii) Fold in a deposit

Figure 2.3 Schematic presentation of geological structures, and terms used to designate a deposit.

It can be *outcropping* to the surface and continue up to a shallow, moderate or great *depth*. It may also be blanketed by a hill or mountainous region, and extending below ground to a certain depth.

2.5.2   Forms of deposits

Bedded, massive and flaggy: rocks are called massive when bedding planes in them are more than 1.2 m apart; bedded when between 75 mm and 1.2 m, and flaggy when less than 75 mm apart.[2]

Vein: a zone or belt of mineralized rock lying within boundaries clearly separating it from the neighboring rocks (fig. 2.3(iii)).

Lode: in general, a lode (fig. 2.3(iv)), vein or ledge is a tabular deposit of valuable mineral between definite boundaries. Whether it is a fissure formation or not.

Seam: it is a deposit limited by two more or less parallel planes, a shape which is typical of sedimentary rocks. This is also a tabular deposit.

Any change in a mineral deposit than its normal format is known as disturbance. Section 2.5.3 describes some of the disturbances that can be encountered by the mineral deposits.

2.5.3   Structural features of rock mass

Rocks during the process of their formation and afterwards are subjected to a number of forces within the Earth's crust. There may be a single force, or combination of forces resulting from ground stresses, tectonic forces, hydrostatic forces, pore pressures, and temperatures stresses. As a result of these forces and their magnitude, rocks are continuously undergoing varying degrees of deformation, resulting in the formation of different kinds of structural features (figs 2.3, 2.4, 2.5). For example, fractures or joints may initially develop within a rock mass, followed by dislocation of the fractured rock blocks. In some circumstances, these dislocated rock blocks move faster than the adjacent blocks, resulting in larger deformation between each block. Such structural features are referred as faults, described later. Both faults and joints are the result of brittle behavior of a rock mass. Joints and faults can be easily identified from the component of displacement parallel to the structure. Joints usually have very small normal displacement, referred to as joint aperture.

Fold, anticline and syncline: the layered rocks, when subjected to stress, it commonly bends or wraps them, forming folds [figs 2.5(a), (b)]. A fold convex upward is anticline; and the one convex downward is syncline (fig. 2.3(viii)). The extent of folding and its ultimate shape depends on the intensity and duration of internal forces, as well as the properties of the rock material. In any type of rock folds may develop, however, in sedimentary and igneous rocks these structures are common. Classification of folds is given in figure 2.4.

Dike/dyke, sills, stocks and batholiths: the magma once formed tends to work its way upward through the crust, shouldering aside the overlying rocks. Some magma comes to rest and solidifies within the crust as dykes/dikes (planar bodies cutting across the

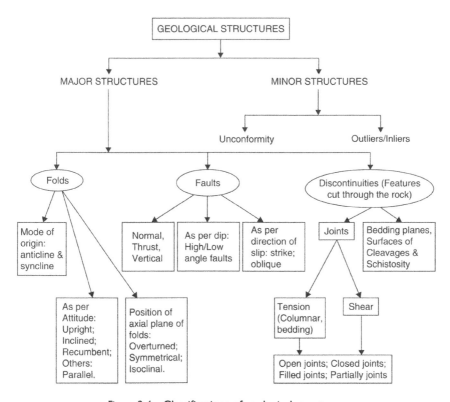

Figure 2.4   Classifications of geological structures.

beds of adjacent rocks, figure 2.3(vii)), *sills* (planar bodies parallel to beds), *stocks* (bodies essentially cylindrical or lenticular, with largest dimension essentially vertical) and *batholiths* (large rock masses with many square-miles – say, 40 or more – of surface exposure essentially rootless stocks).

Unconformity: a plane that separates rocks dissimilar in terms of origin, orientation or age is called an unconformity.

Discontinuities: these are the features such as joints, bedding planes, and surfaces of cleavage or schistosity, which cut through the rock, are known as discontinuities. The common terms and important features of discontinuities are as listed below:[5]

- Orientation – Dip & Direction
- Spacing and Frequency
- Persistence, Size and Shape
- Surface properties: Roughness and Coating
- Strength
- Aperture
- Discontinuity sets
- Block size
- Filling; Seepage

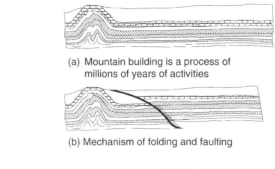

(a) Mountain building is a process of millions of years of activities

(b) Mechanism of folding and faulting

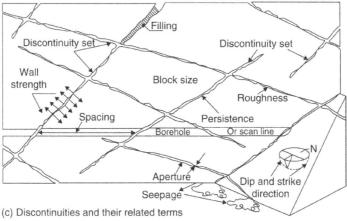

(c) Discontinuities and their related terms

Figure 2.5   Schematic presentation of folds and discontinuities with their related terms.

Genesis: *Bedding; Joints; Foliation; Schistosity; Faults; Shears*

Fault:[1,2] When the shearing stress exceeds the shearing resistance of the rocks, fractured rock blocks undergo considerable displacement along a favorable shear plane resulting in the formation of a new discontinuity, which is referred as a 'Fault'. Depending upon the internal stresses, and rock properties, these relative displacements may vary from few centimeters to several kilometers (Priest, 1993). The surface of the fracture is known as a fault plane (fig. 2.3(vi)). In literature several systems have been used to classify faults; and thus, fault could be of several types (fig. 2.4). If the fault plane is not vertical, the block of rock overlying the plane is called the hanging wall block; and the one underlying block is known as footwall block. If h/w block has moved downward relative to the f/w block, the fault is known as normal and if upward the fault is known as reverse.

Fracture: it can be a crack, joint or fault in a rock due to mechanical failure by stress. This is also known as rapture (fig. 2.3(v & vi)).

Joint: a divisional plane or surface that divides a rock and along which there has been no visible movement parallel to the plane or surface. Also a joint is fracture with no appreciable movement in the fracture plane, with or without measurable separation (fig. 2.3(v)). Joints limit the strength of rock mass and they also control bulk deformation and flow of ground water. Most flow of ground water occurs along open joints rather than through the pores in the rock material, unless the joints are widely spaced and tight, and the rock material is very porous.

Joints in a rock mass are usually developed as families of cracks with probably regular spacing, and these joint families are referred as joint sets. Formation of joints is associated with the effect of differential stresses; some joints are more prominent and well developed extending for several kilometers while others are minor joints having length only few centimeters. In order to characterize joint sets it is necessary to consider their properties such as spacing, orientation, length, gap length and the apertures (fig. 2.5(c))[5]. Joints in rocks may be open, close or filled with some material such as clay, silt etc. While studying joints their strike and dip directions are measured. Joints are classified as dip, oblique and strike joints, however, the classification based on their origin i.e. tension or shear forces is more common. According to this logic joints are classified as: Shear and Tension joints. Tension joints may develop during rock formation and even afterwards. Columnar, sheeting and mutually perpendicular joints are some common types of tension joints. Tension joints are common in igneous rocks whereas shear joints in sedimentary rocks.

Effects of fold and faults:[1,14]
If an excavation is driven or located within a folded structure:

1. Valuable deposits of economic importance may have their original proportions modified after formations, by faulting and folding.
2. Faults may truncate valuable seams (veins or lodes), or possibly conceal them or duplicate them.
3. Rocks are highly stressed locally.
4. Rocks posses decreased competence due to high state of fracturing.
5. Folding and faulting are both associated with jointing which divides the rock into blocks. Heavy support may be necessary to prevent an excavation from collapse in ground where jointing is severe.
6. Many joints and faults also provide pathways for the movement of water to excavations. Consequently it is quite common to encounter localized and significantly deep-seated weathering particularly in near surface situations.
7. Fault zone width is difficult to predict and it can vary along the length of fault. Fault gauge is of low competence and exhibits poor stand-up time.
8. Hudson and Harison[5] expect a relation of the following form:

$$Stability = \frac{1}{number\ of\ discontinuities} = \frac{1}{Engineering\ dimension} - - - \qquad (2.2)$$

2.6   PHYSICAL AND MECHANICAL CHARACTERISTICS OF ORES AND ROCKS

2.6.1   Rocks as rock mechanics

Rock mechanics deals with the behavior of rocks and nature of stresses that act, or redistributed while underground and surface excavations such as tunnels, chambers, stopes, benches or pits are created. It helps in designing of tunnels and mine openings, and their supports. Rocks mass at depth is under stress due to weight of overlying rock (super-incumbent load) and to possible stresses of tectonic origin. Presence of mine openings induces or redistributes stresses in the rock surrounding the openings, and this rock (and openings) will fail if the rock stress exceeds the rock strength. Based on this logic, in order to design the stable openings, it is important to study the pattern of stresses that are likely to act on them and strength of in-situ rock mass.

2.6.2 Rock composition

Based on rock mechanics a rock is composed of three phases: solid minerals, water and air.[4] The later two, together, fill the pore space. The following parameters describe the relative percentage of these phases:

- *Dry density or unit weight* is defined as the weight of solids divided by the weight of total specimen.
- *Porosity* is the pore volume as a percentage of total volume.
- *Degree of saturation* is the ratio of water to pore space by volume.
- *Water content* is the ratio of water to solid by weight.

Following are the six common rock forming mineral assemblages that control the mechanical properties of most of rocks encountered in engineering projects:[4]

Quartzofeldspathic: Acid igneous rocks, quartz and arkose sandstones, gneisses and granulites; usually strong and brittle.

Lithic/basic: Basic igneous rocks (basalts and gabbros), lithic and greywacke sandstone, amphibolites; usually strong and brittle.

Micaceous: Schist is the one which contains more than 50% platy minerals; and gneisses is the one that contains more than 20% mica; often fissile and weak.

Carbonate: Limestone, marble, and dolomites; weaker than category 2 and 3 and soluble over geological time spans; normally brittle, and viscous and plastic only at high temperatures and pressures.

Saline: Rock salt, potash, and gypsum; usually weak and plastic; sometimes viscous when deep seated; water soluble.

Pelitic (clay bearing): Mudstone, shales, and phyllites; often viscous, plastic, and weak.

Common rock names and their geological definitions as described by Dearman[3], and ISRM[6], are given in table 2.3.

To understand the rocks in terms of rock mechanics[11] in its simplest way, the following assumptions[11] are made:

1. Rock is perfectly elastic (stress is proportionsal to strain)
2. Rock is homogeneous (there are no significant imperfections)
3. Rock is isotropic (its elastic properties are the same in all directions).

In practice these assumptions are never true and due to this fact the experts in this subject area have developed a number of systems. The prevalent systems have attempted to take into considerations number of parameters and prominent amongst them are: presence water and hydrostatic pressure, geological features including discontinuities, ambient temperature, and few others.

Table 2.3 Common rock names and their geological definitions. (Based on Dearman, 1974; ISRM, 1981a.)

Genetic Group		Sedimentary				Metamorphic		Igneous		
Structure		Bedded				Foliated	Massive-Jointed			
		Fragmental (Detrital grains)				Crystalline or glassy (cryptocrystalline)				
		Grains are of rock, quartz, feldspar, and clay minerals	50% grains are of carbonate	50% grains are of fine-grained igneous rock	Chemical organic rocks	Quartz, feldspars, micas, acicular dark minerals	Depends on parent rock	Light-Colored minerals are quartz, feldspar, mica and feldspar like minerals		
Grain size mm	Texture		(Limestone)	(Volcanic ash)				Acid	Intermediate	Basic
60	Very Coarse grained (Rudaceous)	Grains are of rock fragments	Calcirudite	Rounded grains: agglomerate; Angular grains: Volcanic breccia	Saline rocks: halite, anhydrite, Gypsum, Limestone Dolomite Peat Lignite Coal	Gneiss: alternate bands of granular and flaky minerals	Pegmatite			
2	Coarse grained (Rudaceous)	Rounded grains: Conglomerate; Angular: grains: breccia					Quartzite Marble Granulite Hornfels Amphibolite	Granite	Diorite	Gabbro
0.06	Medium grained (Arenaceous)	Sandstone: grains are mainly mineral fragments. Quartz sandstone: 95% quartz, voids empty or cemented. Graywacke: 73% quartz, 15% fine detrital matrix, rock and feldspar fragments	Calciarenite	Tuff				Microgranite	Microdiorite	Dolerite
0.002	Fine grained (Argillaceous or lutaceous)	Mudstone Shale: fissile: 50% fine grained particles Claystone: 50% very fine-grained particles	Calcilutite (chalk)					Rhyolite	Andesite	Basalt
	Very fine grained (Argillaceous or lutaceous)									
	Glassy				Chert flint			Volcanic glasses: obsidian, pitchstone, tachylite		

Table 2.4   Common relations used to calculate rock strength.[11]

Parameters	Formula/Relation	Range
Young's modulus of elasticity	Stress/strain; $E = (s/e)$	$5-10 \times 10^6$ psi $(34-69 \times 10^3$ MPa$)$
Poisson's ratio	Lateral strain/Longitudinal strain; $\mu = (e_{lat}/e_{long})$	0.1–0.3
Unit strength, based on unconfined uniaxial tests	Force per unit area $[f = (F/A)]$ Compression f_c Tension f_t Shear f_s	5000–50,000 psi (34–345 MPa) 400–2500 psi (2.8–17 MPa) 500–4000 psi (3.4–28 Mpa)
Specific weight	$w = 62.4\,SG$ lb/ft^3 $w = 1000\,SG$ kg/m^3	SG – specific gravity
Vertical stress acting on a horizontal plane	$S_y = wL = 0.433\,SG \times L$ psi	L is depth $= 1000\,SG \times L$ Pa
Horizontal stress acting on a vertical plane	$S_x = k\,S_y$; k is constant; values varying from 0 to >1.	
Hydrostatic pressure	$S_x = S_y$; $(k = 1)$	Occurs at great depth or in wet, squeezing, and running ground

2.6.3   Rock strength[11]

A set of mechanical and physical properties such as hardness, toughness, jointing, laminations, presence of foreign inclusions and intercalation determine the rock strength. The mechanical strength is measured as compressive, tensile, bending and shear strength. The behaviour of rock can be presented by a stress–strain curve. It can be seen that initially, deformation increases approximately proportional with increasing load. Eventually a stress level is reached at which fracture is initiated, that is, minute cracks which are present in almost any material, start to propagate. With the increasing deformation, the crack propagation is stable, that is, if the stress increase is stopped, the crack propagation also stops. Further increasing the stress, however, leads to another stress level called critical energy release at which the crack propagation is unstable, that is, it continues even if the stress increase is stopped. Next the maximum load bearing capacity is reached, called strength failure and this is in fact the strength of the rock material. Strength and deformation properties of intact rock material are affected by many factors and prominent amongst them are:

- Anisotropy
- Moisture content/pore water pressure
- Confining pressure

Strength of rocks has influence on selecting a mining method as this property has got a direct bearing on selection of mining and excavation equipment and tools, also assessing the consumption pattern of material, labour productivity and cost of extraction. The common relations used to calculate rock strength is given in table 2.4.

Table 2.5   Hardness of representative minerals in Moh's, Knoop's, Vickers, Rosiwal and CERCHAR scales.[9,12]

Minerals	Moh's scale hardness	Knoop	Vickers	Rosiwal	CERCHAR
Talc $Mg_3(OH)_2Si_4O_{10}$	1 Easily crumbled with fingernail	12	20	0.82	0
Gypsum $Ca(SO_4)\cdot 2H_2O$	2 Easily scratched with fingernail	32	50	0.85	0.3
Calcite $CaCO_3$	3 Difficult to scratch with fingernail	85	125	4.08	0.8
Fluorite CaF_2	4 Easily scratched with knife	163	265	4.3	1.9
Apatite $Ca_5(Cl, F, OH)\cdot (PO_4)_3$	5 Scratched with knife	395	550	7.3	3.1
Feldspar (Orthoclase) $KAlSi_3O_8$	6 Very difficult to scratch with knife	560			
Quartz SiO_2	7 Scratches glass, can be scratched with a file of sp. Steel	710–790	1060	141	5.7
Topaz $Al_{12}Si_6O_{25}F_{10}$	8 Scratches glass, can be scratched with emery	1250			
Corundum Al_2O_3	9 Scratches glass, can be scratched with diamond	1700–2200	2300	1000	
Tungsten carbide (WC)	9+	2800			
Titanium carbide TiC	9+	3200			
Boron carbide B_4C	9+	3500			
Diamond C	10 Scratches glass	8000–8500			

2.7   SOME OTHER PROPERTIES/CHARACTERISTICS

2.7.1   Hardness of minerals[9]

Minerals have different hardness and are usually classified according to Knoop scale/Moh's scale. The Knoop hardness is determined by ability to withstand indentation by a special wedge shaped tool known as, Knoop indenter, which is used during the test. Given below in table 2.5 is the hardness for some of the minerals in Moh's and

Knoop's scales. In addition to above, other test that are conducted in laboratories are:

- Rosiwal Mineral Abrasivity Rating
- CERCHAR Abrasivity Index
- Vicker Hardness Number Rock (VHNR)

Rosiwal mineral abrasivity rating: In this test rating is based on the loss of volume while grinding the mineral specimen relative to corundum.

$$Rosiwal = [1000 \times Volume\ loss\ corundum] / [Volume\ loss\ mineral\ specimen] \qquad (2.3)$$

Cerchar abrasivity index: CERCHAR is a research center in France who developed this scratch test to rate the rock wear capacity. Its values vary from 0 to 7.3; For example: Igneous rocks 1.7–6.2; Sedimentary rocks 0.1 to 6.2; Metamorphic rocks 1.3–7.3.

Vicker hardness number rock (VHNR): This hardness number is defined as the ratio of the applied load (kilogram force) to the total (inclined) area (mm^2) of the impression. There is a linear relationship between log(Moh's) and log(VHN):

$$log(VHN) = 2.5\ log(Moh's) + 1 \qquad (2.4)$$

Hardness describes ability scratch one mineral to another. Moh's scale hardness is based on this criterion. Toughness describes the resistance to fracture that comes essentially from the tensile strength of rock. Many drilling bits are designed to induce local tensile failure within rock, as rock is much weaker in tension than it is in compression. Interlocking of mineral grains and strong mineral cement affect toughness and it is common to find that mineral cleavage is also significant, with coarse grained rocks such as gabbros (having large grains and surfaces of cleavages), drilling faster than their finer grained equivalents, e.g. dolerites. Drilling bits for tough rock have strong shoulders and small closely spaced points; those for weaker materials are lighter and carry wide-spaced and pointed teeth.

1 Abrasiveness describes the ability of rock fragments to wear away the drill bits. Rock composed essentially of quartz is very abrasive.
- Hardness – the resistance to penetrate by a pointed tool.
- Toughness – the resistance of the mass to the separation of the pieces from it, in other words, the capacity to suffer considerable plastic deformation up to the moment of breakage.
- Elasticity or resilience – the resistance to impact seen when a tool rebounds.

2.7.2 Rock breakability

According to its breakability – every rock falls into one of the following groups:[9,14]

1. Friable and flowing ground
2. Soft
3. Brittle
4. Strong and
5. Very strong.

1. Friable and flowing ground (sand, soil or peat) – consists of separate particles not bonded or weakly bonded together. Some soils of this group (fine sand and silts), when saturated with water can flow and called quicksand or running sands. Mining is extremely difficult in such soils.
2. Soft soils (clay) – although they consist of cohesive particles, easily penetrated with tool and do not greatly resist the separation of a part from the mass.
3. Brittle rock (shale, limestone, sandstone, coal etc.) is fairly hard but comparatively easily crushed, and pieces separate from the mass along numerous cracks.
4. Strong rock (strong sandstone, granite, magnetite etc.) has high resistance to penetration by a tool and to separation of piece from the mass.
5. Very strong rock (quartzite, diabase, and porphyry) has the highest resistance to penetration by a tool and to separation of piece from the mass.

Prof. Protodykonov proposed[2] a classification of rocks according to their strength and Protodykonov strength number, f, can be calculated by dividing the compressive strength, c, by 100

$$\text{i.e. } f = c/100 \tag{2.5}$$

Cleavage: Minerals with good cleavage do not concentrate into placers and give poor recovery in gravity concentration plants as these minerals are slim.

Density: Minerals with high specific gravity such as gold (SG 15–19) and tin (SG 7) separate easily from silicates (SG 2–3) in placers and gravity plants.

Magnetism: Only two common minerals, magnetite and pyrrhotite, are sufficiently magnetic to be detected by magnetometers during prospecting. They can also be separated from other minerals by hand magnets. A few other minerals are also magnetic but can be separated by high intensity electromagnets.

Conductivity: Some minerals are electrically conductive and this property is used during prospecting operations to identify the minerals. Oxidation of mineral produces earth currents, which can be detected by relatively inexpensive instruments.

Wettability: Flotation, one of the main methods of mineral beneficiation depends upon the surface chemistry of minerals, namely whether the mineral can be wetted or not. Minerals which cannot be wetted can be brought to surface by attaching air bubbles to them, of a finally grounded ore, whereas the minerals which can be wetted will sink and can be separated from there.

2.8   RELATED TERMS – ROCK AND MINERAL DEPOSITS[1,2,4,9,10]

There are some related terms to the rock and mineral deposits, which are defined below:

Magma: molten rock is magma from which igneous rocks are formed.

Primary ore deposits: these are the deposits, which were deposited during the original period/s of mineralization.

Rocks: the crust of the Earth consists of different types of rock which are composed of one, or more frequently more than one, mineral element or chemical compound. The common rock forming minerals are quartz, calcite, feldspar, hornblende, mica and chlorite.

Eruptive rocks: Eruptive rocks, which are also known as igneous, or magmatic rocks have oozed out in the molten form (magma) from the interior of Earth and crystallized. If the magma has solidified slowly under high pressure at a great depth a rock with relatively large crystals will have been formed e.g. granite.

Host rock: the wall rock of an epigenetic deposit.

Barren rock: The rock surrounding the deposit (enclosing or country rock) or included in it but containing no useful substance or insufficient amount of useful component, is termed as barren rock.

Gossan: The ferruginous deposit filling the upper part of some mineral vein, or forming superficial cover over masses of pyrite. It consists mainly of hydrated iron oxide and has resulted from the removal of sulfur as well as the copper or the sulfides originally present.

Country rock: the rock in which the ore deposit is enclosed. It is the general mass of adjacent rock as distinguished from that of a vein, or lode.

Mineral: it is a naturally occurring inorganic (sometimes organic also such as coal) substance and a mineral deposit is a natural body in the Earth's crust. Minerals have three physical states – solid, liquid and gaseous form. The solid minerals can be further divided as metals, non-metals and fuels. The physical characteristics include properties such as color, luster, form, fracture, cleavage, hardness, tenacity and specific gravity. Other characteristics include fusibility, fluorescence, magnetism and electrical conductivity.

Ore and orebody: the portion of any mineral deposit that can be mined at profit is known as ore, and body of earth containing ore is known as orebody.

Float: it consists of loose pieces of ore or particles of metal and is produced by the weathering of an outcrop.

Outcrop: a part of a rock formation that appears at the surface of the ground is known as an outcrop.

Strata: sedimentary rock layers.

Seam: a deposit limited by two more or less parallel planes, a shape which is typical of sedimentary rocks.

Lithology: the character of a rock described in terms of its structure, color, mineral composition, grain size, and arrangement of its component parts; all those visible features that in the aggregate impart individuality of the rock.

Weathering and erosion: the mechanical and chemical breakdown of rock surface of the earth crust is called weathering, and mechanical and chemical transportation of rock from one point of another is called erosion.

Reserves: the quantitative assessment of a mineral deposit within a defined boundary is known as reserves.

Intact rock: This rock contains neither joints nor hair cracks, and hence if it breaks, it breaks across sound rock. On account of damage to the rock due to blasting, spalls may drop off several hours or days after blasting. This is also known as spalling condition. Hard, intact rock may also be encountered in the popping condition involving the spontaneous and sudden detachment of rock slabs from the sides or roof.

Stratified rock: it consists of strata with little or no resistance against separation along the boundaries between strata. In such rocks spalling is very common.

Moderately jointed rock: it contains joints and hair cracks, but the blocks between the joints are locally grown together or so intimately interlocked that vertical walls do not require lateral support. In rocks of this type both spalling and popping conditions may be encountered.

Blocky or seamy rock: it consists of chemically intact or almost intact fragments, which are entirely separated from each other and imperfectly interlocked. In such rocks vertical walls may require lateral support.

Squeezing rock: it is rock with high percentage of microscopic and submicroscopic particles of micaceous minerals or clay minerals with low swelling capacity.

Swelling rock: it causes swelling of the tunnel walls on account of expansion. It is common in rocks containing the clay minerals.

2.9   MINERAL INVENTORY EVALUATION

2.9.1   Introduction

It is the purpose of this section to briefly review the mineral inventory evaluation techniques available and demonstrate a procedure to select them in order to evaluate a mineral deposit. The considerations to be made while modelling a deposit and the logical steps that should be followed to evaluate it are outlined. A distinction is made between mineral inventory and ore reserves.

The basic geological data derived from boreholes, exploratory test pits, drivages, and mine development includes information on parameters such as mineral thickness, grade, depth below surface, structural attitude of beds, rock types and lithology of the beds overlying and underlying the deposit. These data are usually supplemented by analytical information obtained from chemical, physical and structural investigations. The relatively small number of sample points in the mineralized zone, and sparseness of the sample volume makes it necessary for the available information to be interpolated between sample points and extrapolations are usually undertaken by making suitable assumptions about the parameter's continuity in the mineralized zone and building a suitable mathematical model. The mathematical model provides the values of the chosen parameters at the nodes of the superimposed grids of any desired dimensions. This may be a grid cell equal to the size of a mining block or any other criterion. The network of values for any parameter can then be used for mine planning, designing, and controlling functions. Additionally, these data can also be used to calculate the mineral inventory and grade in total and/or for any desired block.

The in situ mineral inventory or reserve which can be mined at a profit is termed an ore. Converting the mineral inventory into ore reserves thus introduces the process

Table 2.6 Calculation of grade of ore from the borehole data record at given cutoff grade (0.30% Cu).

Core recovery Interval	Thickness, m (L)	Grade % Cu (G)	L × G	Remark
0–30 m	10	1.51	15.1	
30–33 m	3	0.23		Below cutoff
33–37 m	14	0.93	13.02	
37–41 m	4	0.49	1.96	
41–45 m	4	0.13		Below cutoff
45–50 m	5	0.60	3.00	
Total	33 m		33.08	

Average grade = **33.08/33** = 1.002% (applying equation 2.6)

of mine design. Mineral computations are made at all stages of the mine's life. Mineral reserves computed during the exploration phase of the mine's life may not require such precision as needed during the production stage.

2.9.2 Grade computation from borehole data

In table 2.6 the process of core recovery details in terms of its length and grade and subsequent computation of the grade from the same hole has been illustrated, which is the first step to assess grade from a bore hole at and above the pre-specified cutoff grade. Equation 2.6 details the relation/formula used to compute the grade.

Average grade = sum of length of mineralization multiplied by grade/sum of length of mineralization

$$G = [\Sigma(L \times G)]/\Sigma L \tag{2.6}$$

where: L is length of mineralization at and above cutoff grade
G grade of mineralization at and above the given cutoff grade

2.9.3 Mineral inventory modelling/estimation techniques[15,33,35,43]

Popoff (1966) described following three principles used to calculate mineral inventories by the conventional methods.

- The rule of gradual change;
- The rule of nearest point or equal influence; and
- The rule of generalization.

Under the first two principles fall the conventional geometrical methods such as:

- Method of polygons;
- Triangle or triangular prism method;
- Cross sectional method;
- Angular bisection;

Table 2.7 Calculation of average grade and thickness of mineralization of a polygon by considering the hole that lies within it; at cutoff grade = 0.4% Cu

Interval	Thickness, m	Grade % Cu	Grade × thickness	Remark
0–30	30	0.31	0.00	Below 0.4% cutoff
30–33	3	0.47	1.41	
33–37	4	0.73	2.92	
Likewise	–	–	–	
–	–	–	–	
Total	50		35.5	

Average grade of polygon D_1 = 35.5/50 = 0.71% Cu. Thickness = 50 m.

Table 2.8 Calculation of reserves, and average grade when combining all the polygons.

Polygon	Area, m²	Thickness	A × T, m³	T. Factor	A × T/TF, tons	Grade % Cu	Ton × G
D_1	500	50	25000	0.5	50000	0.71	35500
Likewise	–	–	–	–	–	–	–
Likewise	–	–	–	–	–	–	–
Total					800500		736460

Reserve Tons = 736460; Average grade = 736460/800500 = 0.92%.

- Rectangular blocks; and
- Inverse square distance weighting (IDW) method.

The rule of generalization is really not a rule at all, and is usually arbitrarily applied as a matter of judgement reflecting past experience. The conventional methods mentioned above are the functions of geometry and distance between samples, which simplify the calculations of volume and grade. They are not functions of the mineralization characteristics, which they propose to measure (Barnes 1980, pp. 54).

2.9.3.1   Method of polygons[33]

In this method, the average grade of mineralization encountered by the sample point within the polygon is considered to accurately represent the grade of the entire material within a polygon. The method assumes that the area of influence of any sample point extends halfway to the adjacent sample points. The procedure for construction of a polygon is illustrated in figure 2.6(i), (a) which shows the holes to be considered (Drill plan), (b) their connecting lines, (c) construction of the polygon by the perpendicular bisectors of the lines between the adjacent holes and finally (d) the polygon. In this manner the polygons are drawn and joined to outline a network of polygons. These polygons can be constructed on plans, cross-section or longitudinal sections. The area of each one of them can be calculated using a planimeter and the volume calculated by multiplying the area by the thickness of the mineralization within the hole (above the cutoff grade). To obtain the tonnage the volume is multiplied by the tonnage factor. The average grade of the hole within this polygon is considered its grade. Grade – tonnage data of each of these polygons is considered to obtain the total reserves and the average grade. Tables 2.7 and 2.8 illustrate the process.

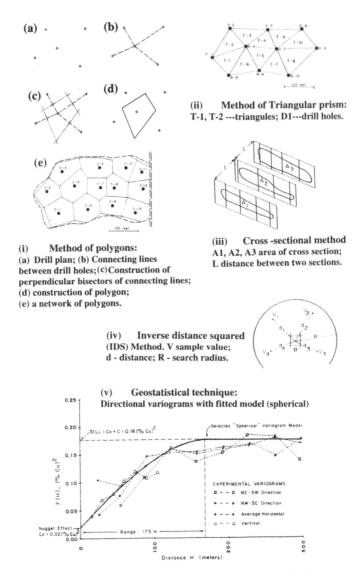

(ii)   Method of Triangular prism:
T-1, T-2 ---triangules; D1---drill holes.

(i)   Method of polygons:
(a) Drill plan; (b) Connecting lines
between drill holes; (c) Construction of
perpendicular bisectors of connecting lines;
(d) construction of polygon;
(e) a network of polygons.

(iii)   Cross -sectional method
A1, A2, A3 area of cross section;
L distance between two sections.

(iv)   Inverse distance squared
(IDS) Method. V sample value;
d - distance; R - search radius.

(v)   Geostatistical technique:
Directional variograms with fitted model (spherical)

Figure 2.6   Mineral inventory estimation techniques.[15,18,19,33,43]

2.9.3.2   *Triangle or triangular prism method*

This is a modification of the polygon method. In this method a series of triangles are
constructed with the drill holes at the apices. The advantage of this method is that all
the three points are considered in the calculation of the thickness and grade parameters
for each triangular reserve block. Figure 2.6(ii), illustrates construction of a network
of triangles and tables 2.9 to 2.11 descibe the process of reserve and grade computation from the drill hole/bore hole data.

Table 2.9   Details of holes with regard to their thicknesses and average grades.

Drill hole no.	Thickness, m	Average grade %
D-1	15	0.93
D-2	25	0.77
Likewise	–	–
D-4	33	1.05
Likewise	–	–

Calculating area of triangle T-1, Area = 400 m².

Table 2.10   Calculation of average grade and thickness of mineralization of a triangle by considering all the holes at the apices.

Drill hole no.	Thickness, m	Average grade %	Grade × Thickness
D-1	15	0.93	13.95
D-4	33	1.05	34.65
D-5	25	0.72	18.00
	73		66.60

Average grade = 66.60/73 = 0.91%.
Tonnage = 400 × (73/3) × (1/2) = 4866.6.

Table 2.11   Calculation of reserves and average grade when combining triangles.

Triangle	Drill holes	Reserve tons	Av. grade	Tons × grade
T-1	D1, D-4, D-5	4866.6	0.91	4428.6
T-2	D-1, D-2, D-4	–	–	–
Likewise				
Total		250000		262500

Tonnage = 250000. Average grade = 262500/250000 = 1.05%Cu.

2.9.3.3   *Cross-sectional method*[33,43]

This is a geometric method which is based on a gradual change principle. In order to estimate reserves by this method, first of all the cross-sections at a specified interval along the orebody, as shown in figure 2.6(iii) are prepared. Using a planimeter or otherwise the area of each section is calculated. The average grade of each of these sections is then calculated using length-weighted sample value, using the area-weighted sample value average, or arithmetic average of the sample values. The volume of each of these blocks can be calculated by using the following relation.

$$\text{Vol.} = (A_1 + A_2)L/2 \quad \text{and} \quad \text{Tonnage} = \text{vol.} \times \text{tonnage factor.}$$

Summing the results and computing the weighted average the reserves and grade can be assessed.

2.9.3.4   Inverse Square Distance Weighting (IDW) method[19,20,33,36]

This uses the principle of gradual change for making value estimates. This interpolation technique uses straightforward mathematics for weighing the influence of all surrounding samples upon the block being evaluated (fig. 2.6(iv)). An estimate value B of a block is given by:

$$B = \{(v_1/d_1) + (v_2/d_2) + \text{-----} + v_n/d_n)\} / \{(1/d_1) + (1/d_2) + \text{--------} (1/d_n)\} \qquad (2.7)$$

It can be also used in anisotropic situations using the following relation:

$$B = \{(v_1/d_1^2) + (v_2/d_2^2) + \text{-----} + v_n/d_n^2)\} / \{(1/d_1^2) + (1/d_2^2) + \text{--------} (1/d_n^2)\} \qquad (2.8)$$

2.9.3.5   Classical statistics[15,22,28,29,42]

Classical statistics and tests have been used in mineral reserve evaluation for many years. However, these methods assume that samples taken from an unknown population are randomly selected and are independent of each other. In addition, they assume that data have either a normal (Gaussian) or a log-normal distribution (David, 1977). In the context of a mineralized zone, this implies that the position from which any sample was taken is not important. Sample assays taken from holes drilled in close proximity to one another, within a mineralized zone, obviously should not be random or independent. Closely spaced samples should demonstrate some correlation, in other words reflect some degree of continuity in the mineralization. If this is not the case either there is no continuous mineralization or the samples have been taken over an excessively large spacing.

Despite the limitations of classical statistics in mineral reserve estimation, much can be gained from studies of sample distribution in terms of providing estimates of ore tonnage, grade and metal content (Barnes 1980, p xx). Traditional statistical decision theory has played an important role in orebody modelling and ore evaluation methods (Hazen, 1967). Excellent descriptions of statistical methods, approaches, and calculations with respect to mineral reserve estimation have been given by Sichel (1952, 1973), Krige (1962, 1978), Hazen (1967) and Koch and Link (1970, 1971).

2.9.3.6   Geostatistics[15,18,24,29,30,36,39,40]

The geostatistical grade estimation method is known as "kriging". It was named after the pioneer of the technique in South African gold mines, D.G. Krige, in 1950 by Matheron, to designate a best linear unbiased estimator for assigning values to mining blocks using geostatistical techniques. Like the IDW method, kriging also involves assigning grades to a uniformly-spaced grid of points as a weighted average of nearby sample grades. However, in kriging, weighting factors are not assigned as an arbitrary function of distance. Instead a mathematical model of sample grade variation within the orebody is first determined by a variogram study of the samples themselves. Next, the variogram model and the sample locations, including distance separating the samples themselves, are used to set up a series of linear equations which are solved to determine the optimum weighting factors. In *simple kriging* there is no constraint imposed that the sum of the kriging weights is unity, but in *ordinary kriging* the constraint is imposed that the sum of the kriging weights given directly to the samples is unity. Raymond (1982, pp. 20) described kriging as an improvement over the IDW method in several respects:

1. The most important advantage is that, with kriging, weights are not chosen arbitrarily (as explained above).
2. With kriging, overweighting to clusters of samples is overcome by considering not only the distances from samples to the point to be estimated, but also the distance separating the samples themselves.
3. By selecting only the nearest samples, optimum estimates are provided by kriging even in the areas of wide sample spacing.
4. With kriging, expected estimation error may also be calculated for each point estimate or for the estimate of larger blocks of material. This calculation provides a method of checking variograms and an appreciation of the reliability of the estimates. Although determination of estimation error is often quoted as the major advantage of kriging, it is in fact of secondary interest in production estimates. The major requirement is in the best grade estimate from the available information.

The merits of kriging over other methods with regard to improved estimation have been discussed by Royle[39] (1979, pp. 101) and Rendu (1979, pp. 207). Full description of theory has been dealt with by O' Leary (1980), David (1977, 1988), Barnes (1980), Journel et al. (1978), Royle (1975), Iskas et al. (1989) and Clark (1979, 1993).

However, Royle (1990) has outlined kriging as an estimation system that allocates weights to sample data within and in the vicinity of an area or volume with which something is to be estimated – call it the 'estimation area'; the area within which data are used for estimation is called the search area; in its usual applications this is centered on the estimation area.

The weights given to the samples are called kriging coefficients or kriging weights. Their values alter as the sampling pattern changes, so that kriging is a weighted moving average process in which the weights vary according to sampling pattern and to the spatial variation exhibited by the variable being estimated. The weights are allocated to the samples so that the mean squared estimation error is minimized. This minimized estimation error is called the kriging variance and it can be calculated for any sampling pattern and estimation area configuration.

If a given sampling pattern, fixed relative to what is being estimated, is moved to an infinite number of locations over the field within which estimates are being made, the kriging variance is defined as the mean squared difference between the estimates and the actual values.

Because the variogram (fig. 2.6(v)) depends only on the difference between the sample values and not on their absolute values, the kriging variance of linear kriging systems are independent of sample values, as are kriging weights. For globally unbiased estimates over the whole field the requirement is that the kriging weights sum to unity.

2.9.3.7 *Non-linear estimation techniques in geostatistics*[31,32,34,36,40,44,48]

There are several non-linear estimation techniques available in geostatistics. Most common amongst them is *log-normal kriging*. When the grade distribution of the samples is log-normal, the use of log-normal kriging is often recommended (e.g. Rendu, 1978). Log-normal kriging involves variogram calculations and kriging with logarithms of grade (or grade modified by a constant). Raymond (1982) states that sometimes the logarithmic variogram is quite variable from one area to another. By assuming one variogram for the total area, an additional error is introduced in converting from logarithmic to arithmetic grades. A naive application of the log-normal approach will inevitably lead to erroneous results as the method is highly distribution-dependent (Kwa, 1984).

Instead of log-normal kriging, cross-validation techniques can be used to circumvent any bias in the estimation (O' Leary, 1986, personal communication). A common practice currently used for the grade and tonnage estimation of a mineral deposit is to estimate the grade of a panel (or block) and sum up the tonnage of the panels whose estimated averaged grade is higher than a cutoff grade. If the recoverable tonnage and grade is made on these panels, a bias is made, over-estimation of tonnage for cutoffs below the mean, which leads to a result of the so-called "disappearing tonnage problem". To solve this problem Matheron (1975) provided a more promising non-linear technique known as *disjunctive kriging* (DK) (Young, 1982). The objective of this technique is to estimate the density function of the grade distribution of selective mining units, within a panel, from the grades of the nearby samples. Using this density function a complete grade-tonnage curve can be calculated for the panel. This is an area where future research could be undertaken to evaluate a mineral deposit taking into account the density function at varying grades within it.

The limitations of the method have been expressed by Young (1982) – "the positiveness of the estimated conditional density function may not be guaranteed for any location and DK estimates may face the problems of yielding odd recoveries and figures such as negative recoveries and decreasing ore grade with higher cutoff grades".

Verly (1983) states that a more promising non-linear technique is the multivariate *Gaussian kriging* approach developed at Stanford University. This method is very attractive in the sense that it is more mathematically robust and is fairly easy to implement. However, there are some problems: namely the "despising" of zero values and of the considerable computer run time needed (Kwa, 1984).

In addition to the non-linear techniques, there exist some non-parametric methods. These are *indicator kriging* (Journal, 1983) and its variant, *probability kriging* (Sullivan, 1983). Both methods are easy to implement but suffer from order relation problems (Kwa, 1984). Sullivan (1983) and Davis (1983) showed how this problem can be solved using quadratic programming and an approximation technique has been suggested by Journal (1983). According to Sullivan, the relative merits of probability kriging over indicator kriging are firstly, that it provides a more accurate estimate (lower estimation variance) and that of the smoothing problems that affect indicator kriging are eliminated. However, disadvantages are that more variograms must be calculated and modelled for each value of cutoff target.

Other types of kriging that are not widely used (Nobel, 1992) include *universal kriging, co-kriging and soft kriging*. Universal kriging is a method to incorporate trends (drifts) into the kriging equations. If the trends are defined according to a secondary variable, it is known as universal kriging with exogenic drift. Theoretically, the problem cannot be solved (Royle, 1990) as the nature of drift needs to be known in order to determine the underlying variogram and vice versa.

Where two variables with enough correlation have been measured some advantage may be gained by using *co-kriging;* both variables are used together to estimate each variable in turn, providing a richer set of data. In addition to the variogram of each of the variables, a cross-variogram is modelled which shows how the two variables co-vary in space. However, where both variables have been used equally well there is little to be gained by co-kriging (Royle, 1990). Co-kriging is best used where one variable, which can be measured only with great expense or difficulty, is correlated with another variable that is easily measured at low cost. A common example is uranium deposits where low cost radiometer data from boreholes are correlated with expensive uranium assays taken from high cost drill cores.

When co-kriging is used with qualitative secondary variables (known as soft data) such as alternation, rock type or other geologic features, it is known as *soft kriging.* It is a recent development. All kriging systems described so far use 'hard' data i.e. samples within which the magnitudes of the variables have been measured. Tawo and Dowd (1990), based on a case study undertaken, conclude that soft kriging has the advantage of distinguishing between information types but the main drawback of the approach is in the adequate assessment of guesses made by the experts; which vary from one individual to another. In addition, it is computationally costly, although when data are sparse its benefit outweighs this cost.

Other recent developments to evaluate a mineral deposit include the application of *Expert System* and *Artificial Neural Network (ANN)* techniques. It is hoped that many useful case studies on their applications will appear in due course of time.

2.9.4 IMPORTANT CONSIDERATIONS FOR EVALUATION OF THE MINERAL INVENTORY

2.9.4.1 *Homogeneity and mode of origin*[27]

King et al. (1982) presented a useful diagram (fig. 2.7(a)) and table 16.1 to present the proportion of ore in a mineral inventory as a function of homogeneity. It can be seen how the magnitude of the confidence limits decreases with an increase in the coefficient of variation. In general it can be stated that the more homogeneous a deposit, the less difficult it is to evaluate. A gold deposit can be considered to be more difficult to evaluate than a coal deposit. A copper deposit cannot be considered easy to evaluate.

The degree of difficulty associated with evaluating a mineral inventory can be represented by a diagram (fig. 2.7(b)) given by Caras et al. (1983). From figure 2.7(a) it is evident that moving from concordant (sedimentary and biological, indigenous) deposits to more discordant (migratory, exotic) deposits, the difficulty associated with evaluating orebodies increases.

Referring to figure 2.7 and table 16.1 it is possible to assess without mathematics whether the mineral to be evaluated involves low or high risk in estimating the grade.

2.9.4.2 *Geological and mineralogical boundaries*[15]

Mineral deposits rarely are completely homogeneous and often consist of subsets separated by geological and mineralogical boundaries. Sometimes different mineral

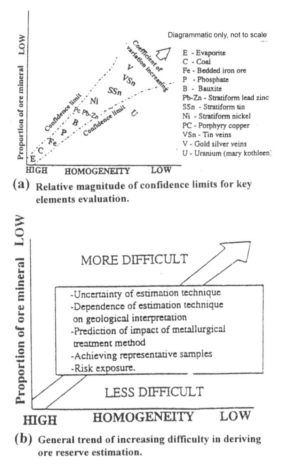

(a) Relative magnitude of confidence limits for key elements evaluation.

(b) General trend of increasing difficulty in deriving ore reserve estimation.

Figure 2.7   Important considerations while evaluating a mineral deposit (Not to scale).[17]

populations are difficult to distinguish and may be individually lumped together – often without serious consequences, but more often with detrimental effects upon analytical results (Barnes 1980, pp. 62). Before applying any mineral estimation technique it is, therefore, important to identify different populations.

Mineralogical boundaries can be classified as: gradational boundaries; mineral suite zonation boundaries; mineralogical changes on account of host rock environment; oxidation, leached and enrichment boundaries. Changes in mineralogical continuity on account of these boundaries really constitute different population subsets which must be identified before applying any extension function. In practice a variogram for each of these subsets should be calculated. If their variograms are very similar, they can be considered as a single population but if zones are intrinsically different, it is important to establish a boundary between them.

In some deposits there may be a difference in mineralogy and mineralogical continuity with change in the host rock. In a situation like this a rock identifier should be used to separate data belonging to a particular rock type.

The geological boundaries are usually defined by a change in lithology and/or stratigraphy, or the structure, for example wall rock, fault and fold boundaries. These boundaries represent sharp environmental changes in terms of mineral zones, usually ore-waste or mineral–non-mineral. For example, in porphyry coppers there is commonly an association between lithology and higher and lower metal content, making interpretation of lithological outlines an essential feature in the process of mineral estimation. In any situation, assistance from the geologist is essential to establish different populations, boundaries or subsets before evaluating a deposit.

2.9.5   Computation of the mineral inventory

It is apparent that without use of computers geostatistics would be just a theory to be taught in the universities, and without its significant contribution to the mineral industry. It involves mathematical functions which are difficult and time consuming if solved manually. At present many kriging programs have been written and are available throughout the mineral industry. This analysis is usually carried out following the steps outlined below:

- Statistical Analysis
- Structural Analysis
- Estimation of Grade-Kriging and
- Miscellaneous.

2.9.5.1   Logical steps followed

Having fed the basic data set into the computer, the following procedure is adhered to in order to calculate the mineral inventory. This includes reading borehole data on a from-to basis, converting assay values measured on a from-to basis into a series of point values with known coordinates. The assay values are further composited over a specified interval. The choice of composite interval is usually dictated by the type of mineral deposit and the mining method selected to mine it. This compositional change is necessary to consider the data with their equal "support". The sampling data can be composited at a regular interval using either rectangular integration or linear interpolation techniques (Davis, 1973). The data sets, sometimes, need to be split into the rock types they belong to. This becomes important if the deposit has different populations.

2.9.5.2   Graphical presentation of data[43]

Using the common plotting library and some of the graphics software available at a computer centre, the geographical locations of the boreholes and the sampling points can be presented graphically. The same routines can be used to represent the results of an analysis afterwards.

2.9.5.3   Statistical analysis and cumulative probability distribution[22,28,42,43]

Following the graphical presentation of input data, a statistical analysis is carried out to assess:

- General statistics of the deposit
- Grade distribution and
- Presence of one or more populations within the data set.

The cumulative probability distribution is a fairly sensitive measure of the continuity of local mineralization. A smooth curve with a similar shape throughout suggests homogeneous mineralization. An abrupt change in the curve suggests a geological discontinuity limiting mineralization. Based on the probability distribution and geological reasoning the data set, if needed, is further split to represent different geological regimes or populations.

2.9.5.4   Structural analysis – the semi-variogram[15,18,19,20,47]

Variogram: The variation that exists among samples some distance apart within a continuous mineral deposit is a measure their spatial correlation. It is obvious that the difference between the values of samples taken very close to each other will be a minimum but as the distance between samples increases their difference in values, on average, becomes greater. The variogram, as defined by Barnes (1980) is an arithmetically simple graph plotting the average difference between sample values at specified distances or lags apart. Since the difference between two paired values can be positive or negative, therefore, a classical statistical technique is applied in which the difference is squared and summed and then divided by twice the number of pairs found. The results is called geostatistical variance $\gamma(h)$ and represented by the formula:

$$\gamma(h) = (1/2n) \sum_{i}^{n} \{(z(x_i + h) - z(x_i)\}^2$$

(2.9)

where $z(x_i)$ is the grade at sample point x, $z(x_i+h)$ is the grade at x+h meters/feet away, h is the distance vector (directional), and n is the number of data pairs counted along the directional lag (h). This is called the experimental variogram.

Experimental semi-variograms in different directions are computed to identify the continuity of the mineralization in these directions. Any anisotropy, if it exists, is noted. By the process of computing variograms in different directions as well as vertically, as shown in figure 2.6(v), one can readily determine not only the mineralogical trend (anisotropy) but also the magnitude of the directional changes in the zone of influence. Knowing quantitatively the mineralogical range in three dimensions, it is relatively simple (Barnes, 1980) to assign directional anisotropic factors that will give proper weighting to samples relative to their location from the point or block being evaluated. For example, if the range of influence along the trend is twice as great as the range normal to trend, we can multiply the distance in the normal direction by a factor of two to restore 'geometric isotropy' in terms of major trend directions.

When not enough sample pairs are available in the directions other than those of drilling, experimental semi-variograms in these directions show 'noise'. An average experimental semi-variogram using an angle of regularization of 180 degrees, in most of the cases, is therefore evolved.

Several intrinsic schemes or mathematical models have been proposed over the years to represent a probability function that describes the behavior of various experimental semi-variograms. These are: the Linear model, the DeWijsian model and Spherical model (Clark, 1982).

David (1973, pp. 103) stated that the spherical model occurs naturally in all the deposits where grades become independent of each other once a given distance, or range, is reached. Spherical models are usually applicable in most of the sedimentary deposits and porphyry copper deposits. In figure 2.6(v), the spherical model has been used to fit the experimental semi-variogram.

2.9.5.5   Trend surface analysis[43]

Sometimes a semi-variogram exhibits a trend or drift in the data set. In order to deal with this problem, any trend in the data set should be removed. This can be achieved by fitting a simple surface to the sample data by trend surface analysis. The grade at any given point can be considered to be a function of drift component and residual component. Deducting the drift component from the data set the residual component can be obtained. The residual values can be used to compute the semi-variogram.

2.9.5.6   Checking the variogram model

The technique referred as *point kriging cross validation* can be used to test the validity of semi-variogram parameters. In this process a point value is estimated, not the block value (explained later). The kriging method remains the same except that each sample value is taken in turn, removed from the data set and its value kriged from the nearest surrounding sample values. In order for the model parameters to be valid, two criteria should be met:

- a close correlation between the mean of the kriged estimates and mean of the actual value.
- a close correlation between the average kriging variance and the average variance of estimation.

2.9.5.7   Block Kriging

Volume-variance effect: The volume variance effect refers to the inverse relationship between the distribution variance and the volume of the blocks. The volume-variance effect is characterized by Krige's relationship that takes into account the variance of blocks in the deposit, the variance of samples in the deposit and the variance of samples in the block. The volume variance relation is unimportant where the entire deposit is above the cutoff grade or where the ore is mined non-selectively. But, in practice, the cutoff grade is higher and only a portion of the deposit is mined; hence, volume variance relation is an important parameter that should be looked into while deciding the size of a block.

2.9.5.8   Block dimensions[15,21,37]

In underground mines, particularly where there is large variation in grade, blocks of small size are essential. In general, it can be stated that the more the precision of ore boundaries warranted the smaller should be the size of the block. For example, short term planning requires a good knowledge of grade in relatively small blocks (Diehl and David, 1982). Since selectivity is an important point of the study (designing the stopes), a small block of 1 m size is required (Deraisme, et al. 1984).

If a block or panel is relatively large, i.e. encompassing several samples, one can have more confidence in predicting its overall grade than for a small block size in which no samples have been taken (Barnes, 1980, pp. 49). This is an important aspect which should be taken into account when planning the underground diamond drilling fans. Close sampling is essential for stope design purposes. Availability of data decides the smallness of the block (Royle, 1977). In order to model a deposit, based on the source sampling information and the type of estimation such as global or local, the block size can be decided, for example in a global estimation the block size of – 30 × 30 × 30 m was used. And for the same deposit during local estimation a block size of – 10 × 10 × 10 m was used. However, for stope designing purposes small size panels (across the deposit) or blocks are warranted.

2.9.5.9 Kriging procedure

- The kriging parameters as outlined below are used.
- Number of samples required to krige a block: this range can vary, usually minimum 4 and maximum 15 samples (based on the usual practice) are used to krige a block.
- Radius of sample search: this parameter should be specified, preferably within the range obtained for model variogram.

Model variogram: for each of the mutually perpendicular directions, the nugget variance (Co), the transition amplitude (C) and range (A) should be specified. If anisotropy exists, the ratio of anisotropy should also be specified. Once the input parameters are specified the following procedure is adopted to each block in order:

1. Calculation of the average variability of samples contained within the dimensions of the block – the extension variance.
2. Location of the nearest 4–15 sample values to the block center within the range prescribed.
3. If sufficient samples are not located (within the search range), the procedure is repeated for the next block.
4. Establishment of the kriging matrices:
   a. set up the S-matrix which considers the expected variability each of the nearest surrounding values and themselves.
   b. set up the T-matrix which considers the expected variability between each of the nearest surrounding values and the block center.
5. Calculate the kriged estimate and its kriging variance.
6. Locate the next block and repeat the same procedure till the last block is kriged.

2.9.6 Graphical presentation of the kriged results[43]

The kriging results could be presented in the form of plans and sections showing the mineralized zones and grade tonnage curves to indicate tonnage and grade contained within a volume of ground for a series of given cutoff grades. Figure 16.36 details mineral inventory along a section; which has been assessed by the application of

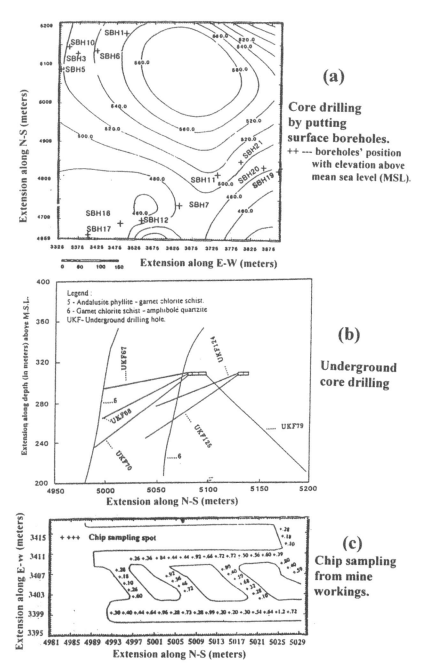

(a)

**Core drilling
by putting
surface boreholes.**
++ --- boreholes' position
      with elevation above
      mean sea level (MSL).

(b)

**Underground
core drilling**

(c)

**Chip sampling
from mine
workings.**

Figure 2.8   Sources of sampling information while evaluating a mineral deposit which is likely to be
mined by underground mining.[43]

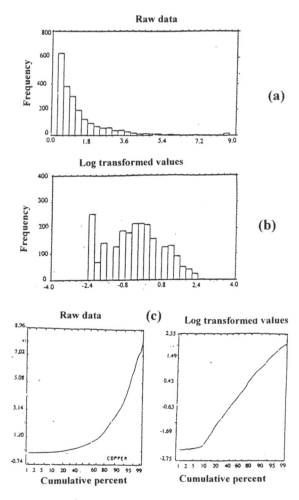

Figure 2.9   Statistical analysis during evaluation of a mineral deposit. Histograms & cumulative probability plots for the raw as well as log-transformed values for a copper deposit are shown.[43]

geostatistics. Figure 2.8 is a graphical presentation of sampling data which could be gathered through various operations such as: diamond core drilling at surface, core drilling from u/g and chip sampling. Figure 2.9 presents statistical analysis of the sample data on grade and their cumulative probability plots. It could be done for raw as well as log transformed values, as shown in figure 2.10 illustrates as how orebody profiles are ultimately drawn from the evaluation that could be undertaken using the various techniques that have been discussed above.

2.9.7   Grade-tonnage calculation and plotting the curves[43,46]

In sec. 12.2.1, 12.2.2 and 12.2.3 the difference between mineral reserves and resources, details of mineral inventory computation and establishment/computation

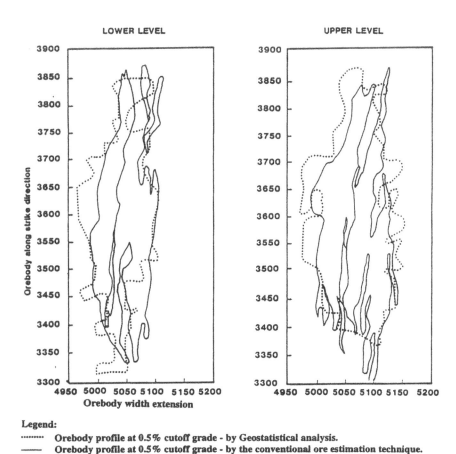

Figure 2.10   Plotting the results of ore evaluation graphically. Comparison of orebody profiles as obtained with those obtained through the cross section method and geostatisitical method for copper deposit have been shown.[43]

of Grade-Tonnage curves have been discussed. Please refer to figures 12.3 to 12.5 and table 12.1 which detail these aspects.

2.9.8   Selection of a suitable mining/stoping method

The methodology presented in section 16.5.3 undergoes the following features:

- *Estimating mineral inventory from the basic geological information, which has been discussed in the preceding sections (2.9.1 to 2.9.7).*
- *Formulating a basis for cutoff grade decisions to predict mineable ore reserves (sec. 16.5.3).*
- *Evaluating stope boundaries for different stoping methods (sec. 16.5.3).*
- *Designing stopes and their economic evaluation for the different mining methods through the algorithms developed (sec. 16.5.3).*
- *Selection of stoping method (sec. 16.5.3).*

2.10   RESOURCES CLASSIFICATION BY UNECE[49]

Apart from what has been discussed in sections 12.2.1 to 12.2.3; the classification which has been proposed by the UNECE under the subject area: Classification of Mineral reserves and resources – Solid fuel and Mineral Commodities in 1997, and discussions that were held from time to time till 2–4 May 2012, are presented below. It details various aspects of this classification.

Economic Commission for Europe

Committee on Sustainable Energy

Expert Group on Resource Classification

Third session
Geneva, 2-4 May 2012
Item 9 of the provisional agenda

High-level mapping of the United Nations International Framework Classification for Reserves/Resources – Solid Fuels and Mineral Commodities of 1997 to the United Nations Framework Classification for Fossil Energy and Mineral Reserves and Resources 2009

Note by the secretariat

The following tables and discussions are provided as a guide to the relationship between the United Nations International Framework Classification for Reserves/Resources – Solid Fuels and Mineral Commodities of 1997 (UNFC–1997) (ENERGY/WP.1/R.77[1]) and the United Nations Framework Classification for Fossil Energy and Mineral Reserves and Resources 2009 (UNFC–2009) (ECE Energy Series No.39). The comparison focusses on the changes to the definitions of the individual categories for each of the three axes of the UNFC system. The UNFC–2009 category definitions reflect general principles rather than more specific and detailed requirements of UNFC–1997, such as the existence of a specific type of report. In most cases, but not all, the intention is that the two definitions are aligned in terms of level of knowledge and/or confidence.

Since UNFC–1997 addressed solid fuels and mineral commodities only, this document makes reference to the definitions found in the Committee for Mineral Reserves International Reporting Standards (CRIRSCO) Template of 2006[2] to which UNFC-2009 is aligned for such commodities. Particular care is required to avoid assuming that reserve and resource terminology can be used as a correlation tool between the two versions of the UNFC.[3] Reference must be made to the individual category definitions, as reproduced in this document, in every case.

---

[1] United Nations Economic Commission for Europe (ECE) multilingual publication in Chinese, English, French, German, Portuguese, Russian and Spanish.

[2] CRIRSCO Template (2006) available at: http://www.crirsco.com/crirsco_template_v2.pdf.

[3] For example, the terms Measured, Indicated and Inferred Mineral Resource were applied to 331, 332 and 333 respectively under UNFC-1997, whereas under the UNFC-2009 definitions they are now aligned with 221, 222 and 223.

High-Level Mapping of UNFC-1997 to UNFC-2009: E axis

Category	1997 Definition	2009 Definition	2009 Supporting Explanation	Discussion
E1	Quantities, reported in tonnes/volume with grade/quantity, demonstrated by means of a Prefeasibility Study, Feasibility Study or Mining Report, in order of increasing accuracy, that justify extraction under the technological, economic, environmental and other relevant conditions, realistically assumed at the time of the determination	Extraction and sale has been confirmed to be economically viable.[4]	Extraction and sale is economic on the basis of current market conditions and realistic assumptions of future market conditions. All necessary approvals/contracts have been confirmed or there are reasonable expectations that all such approvals/contracts will be obtained within a reasonable timeframe. Economic viability is not affected by short-term adverse market conditions provided that longer-term forecasts remain positive.	No material change, other than being based on principles rather than a specific (defined) type of report.
E2	Quantities, reported in tonnes/volume with grade/quantity, demonstrated by means of a Prefeasibility Study, Feasibility Study or Mining Report, in order of increasing accuracy, not justifying extraction under the technological, economic, environmental and other relevant conditions, realistically assumed at the time of the determination, but possibly so in the future.	Extraction and sale is expected to become economically viable in the foreseeable future.[3]	Extraction and sale has not yet been confirmed to be economic but, on the basis of realistic assumptions of future market conditions, there are reasonable prospects for economic extraction and sale in the foreseeable future.	No material change, unless E2 (1997) included quantities for which there were no reasonable prospects for economic extraction and sale in the foreseeable future (i.e. they would fail the CRIRSCO definition for Mineral Resources). These would now be classified as E3 (2009).
E3	Quantities, reported in tonnes/volume with grade/quantity, estimated by means of a Geological Study to be of intrinsic economic interest Since the Geological Study includes only a preliminary evaluation of Economic Viability, no distinction can be made between economic and potentially economic.	Extraction and sale is not expected to become economically viable in the foreseeable future or evaluation is at too early a stage to determine economic viability.[4]	On the basis of realistic assumptions of future market conditions, it is currently considered that there are not reasonable prospects for economic extraction and sale in the foreseeable future; or, economic viability of extraction cannot yet be determined due to insufficient information (e.g. during the exploration phase).Also included are quantities that are forecast to be extracted, but which will not be available for sale.	No material change, since E3 (1997) would be consistent with: "economic viability of extraction cannot yet be determined due to insufficient information". E3 (2009) also includes uneconomic quantities and those that will be extracted but not sold.

---

[4] The phrase "economically viable" encompasses economic (in the narrow sense) plus other relevant "market conditions", and includes consideration of prices, costs, legal/fiscal framework, environmental, social and all other non‑technical factors that could directly impact the viability of a development project.

Category	1997 Definition	2009 Definition	2009 Supporting Explanation	Discussion
F1	A Mining Report is understood as the current documentation of the state of development and exploitation of a deposit during its economic life including current mining plans. It is generally made by the operator of the mine. The study takes into consideration the quantity and quality of the minerals extracted during the reporting time, changes in Economic Viability categories due to changes in prices and costs, development of relevant technology, newly imposed environmental or other regulations, and data on exploration conducted concurrently with mining. It presents the current status of the deposit, providing a detailed and accurate up-to date statement on the reserves and the remaining resources. A Feasibility Study assesses in detail the technical soundness and Economic Viability of a mining project, and serves as the basis for the investment decision and as a bankable document for project financing. The study constitutes an audit of all geological, engineering, environmental, legal and economic information accumulated on the project. Generally, a separate environmental impact study is required. Cost data must be reasonable accurate (usually within ± 10%), and no further investigations should be necessary to make the investment decision. The information basis associated with this level of accuracy comprises the reserve figures based on the results of Detailed Exploration, technological pilot tests and capital and operating cost calculations such as quotations of equipment suppliers.	Feasibility of extraction by a defined development project or mining operation has been confirmed.	Extraction is currently taking place; or, implementation of the development project or mining operation is underway; or, sufficiently detailed studies have been completed to demonstrate the feasibility of extraction by implementing a defined development project or mining operation.	F1 (2009) is based on the principle of having undertaken sufficient studies to demonstrate that the project can proceed, rather than linking it to a specific (defined) type of study. In some cases, projects may be implemented on the basis of a Prefeasibility study and these would change from F2 (1997) to F1 (2009). Any mining project that satisfies the CRIRSCO definition for Mineral Reserves would be F1 (2009).
F2	A Prefeasibility Study provides a preliminary assessment of the Economic Viability of a deposit and forms the basis for justifying further investigations (Detailed Exploration and Feasibility Study). It usually follows a successful exploration campaign, and summarizes all geological, engineering, environmental, legal and economic information accumulated to date on the project. In projects that have reached a relatively advanced stage, the Prefeasibility Study should have error limits of ± 25%. In less advanced projects higher errors are to be expected. Various terms are in use internationally for Prefeasibility Studies reflecting the actual accuracy level. The data required to achieve this level of accuracy are reserves/resources figures based on Detailed and General Exploration, technological tests at laboratory scale and cost estimates e.g. from catalogues or based on comparable mining operations. The Prefeasibility Study addresses the items listed under the Feasibility Study, although not in as much detail.	Feasibility of extraction by a defined development project or mining operation is subject to further evaluation	Preliminary studies demonstrate the existence of a deposit in such form, quality and quantity that the feasibility of extraction by a defined (at least in broad terms) development project or mining operation can be evaluated. Further data acquisition and/or studies may be required to confirm the feasibility of extraction.	No material change, other than being based on principles rather than a specific (defined) type of report. However, note that any mining project that satisfies the CRIRSCO definition for Mineral Resources would be F2 (2009) rather than F3 (1997).

Category	1997 Definition	2009 Definition	2009 Supporting Explanation	Discussion
F3	A Geological Study is an initial evaluation of Economic Viability. This is obtained by applying meaningful cut-off values for grade, thickness, depth, and costs estimated from comparable mining operations. Economic Viability categories, however, cannot in general be defined from the Geological Study because of the lack of detail necessary for an Economic Viability evaluation. The resource quantities estimated may indicate that the deposit is of intrinsic economic interest, i.e. in the range of economic to potentially economic. A Geological Study is generally carried out in the following main stages: Reconnaissance, Prospecting, General Exploration and Detailed Exploration (for definition of each stage see below). The purpose of the Geological Study is to identify mineralization, to establish continuity, quantity, and quality of a mineral deposit, and thereby define an investment opportunity.	Feasibility of extraction by a defined development project or mining operation cannot be evaluated due to limited technical data.	Very preliminary studies (e.g. during the exploration phase), which may be based on a defined (at least in conceptual terms) development project or mining operation, indicate the need for further data acquisition in order to confirm the existence of a deposit in such form, quality and quantity that the feasibility of extraction can be evaluated.	No material change, other than being based on principles rather than a specific (defined) type of report. See comments for F2.
F4	N/a.	No development project or mining operation has been identified.	In situ (in-place) quantities that will not be extracted by any currently defined development project or mining operation.	A new category for 2009.
G1	Detailed Exploration involves the detailed three-dimensional delineation of a known deposit achieved through sampling, such as outcrops, trenches, boreholes, shafts and tunnels. Sampling grids are closely spaced such that size, shape, structure, grade, and other relevant characteristics of the deposit are established with a high degree of accuracy. Processing tests involving bulk sampling may be required. A decision whether to conduct a Feasibility Study can be made from the information provided by Detailed Exploration.	Quantities associated with a known deposit that can be estimated with a high level of confidence.	For in situ (in place) quantities, and for recoverable estimates of fossil energy and mineral resources that are extracted as solids, quantities are typically categorised discretely, where each discrete estimate reflects the level of geological knowledge and confidence associated with a specific part of the deposit. The estimates are categorised as G1, G2 and/or G3 as appropriate	No material change provided that the level of geological knowledge and confidence is equivalent to that defined by CRIRSCO for a Measured Mineral Resource
G2	General Exploration involves the initial delineation of an identified deposit. Methods used include the surface mapping, widely spaced sampling, trenching and drilling for preliminary evaluation of mineral quantity and quality (including mineralogical tests on laboratory scale if required), and limited interpolation based on indirect methods of investigation. The objective is to establish the main geological features of a deposit, giving a reasonable indication of continuity and providing an initial estimate of size, shape, structure and grade. The degree of accuracy should be sufficient for deciding whether a Prefeasibility Study and Detailed Exploration are warranted.	Quantities associated with a known deposit that can be estimated with a moderate level of confidence.	For recoverable estimates of fossil energy and mineral resources that are extracted as fluids, their mobile nature generally precludes assigning recoverable quantities to discrete parts of an accumulation. Recoverable quantities should be evaluated on the basis of the impact of the development scheme on the accumulation as a whole and are usually categorised on the basis of three scenarios or outcomes that are equivalent to G1, G1+G2 and G1+G2+G3.	No material change provided that the level of geological knowledge and confidence is equivalent to that defined by CRIRSCO for an Indicated Mineral Resource. Otherwise, the quantities should be assigned as G3 (2009). Also, note that G2 can apply to a part of a deposit that also has a part that satisfies G1, since the level of "exploration" may not be the same over the whole deposit.

Category	1997 Definition	2009 Definition	2009 Supporting Explanation	Discussion
G3	Prospecting is the systematic process of searching for a mineral deposit by narrowing down areas of promising enhanced mineral potential. The methods utilized are outcrop identification, geological mapping, and indirect methods such as geophysical and geochemical studies. Limited trenching, drilling, and sampling may be carried out. The objective is to identify a deposit which will be the target for further exploration. Estimates of quantities are inferred, based on interpretation of geological, geophysical and geochemical results.	Quantities associated with a known deposit that can be estimated with a low level of confidence.		No material change provided that the level of geological knowledge and confidence is equivalent to that defined by CRIRSCO for an Inferred Mineral Resource. Otherwise, the quantities should be assigned as G4 (2009). Also, note that G3 can apply to a part of a deposit that also has parts that satisfy G1 and/or G2, since the level of "exploration" may not be the same over the whole deposit.
G4	A Reconnaissance study identifies areas of enhanced mineral potential on a regional scale based primarily on results of regional geological studies, regional geological mapping, airborne and indirect methods, preliminary field inspection, as well as geological inference and extrapolation. The objective is to identify mineralized areas worthy of further investigation towards deposit identification. Estimates of quantities should only be made if sufficient data are available and when an analogy with known deposits of similar geological character is possible, and then only within an order of magnitude.	Estimated quantities associated with a potential deposit, based primarily on indirect evidence.	Quantities that are estimated during the exploration phase are subject to a substantial range of uncertainty as well as a major risk that no development project or mining operation may subsequently be implemented to extract the estimated quantities. Where a single estimate is provided, it should be the expected outcome but, where possible, a full range of uncertainty in the size of the potential deposit should be documented (e.g. in the form of a probability distribution). In addition, it is recommended that the chance (probability) that the potential deposit will become a deposit of any commercial significance is also documented	No material change. Note that although the UNFC 2009 definitions are written so that they can be applied at the level of an individual deposit (even at the exploration stage, as is commonly done in the petroleum sector), they may also be applied at a regional scale to document resource potential for a geological province, for example. Such applications are discussed in the Specifications to UNFC 2009 (in preparation).

2.11 THE WAY FORWARD

Kriging is an improvement over the conventional methods in several respects. By selecting only the nearest samples, optimum estimates are provided by kriging even in the areas of widely sampled spacing. With kriging, expected estimation error may also be calculated for each point estimate or for the estimate of larger blocks of material. This calculation provides a method of checking the variogram and an appreciation of the reliability of the estimates. Although determination of estimation error is often quoted as the major advantage of kriging, it is in fact of secondary interest in production estimates. The major advantage is the best grade estimate from the available information.

QUESTIONS

1. Based on rock mechanics what is the rock composed of? List those ingredients.
2. Classify the geological structures.

3. Comment – 'Mineral resources of today are the results of natural processes of billions of years but with the prevalent rate of consumption, they will be depleted within the next few centuries'. Is it true? How we can preserve these resources?

4. Define term: hardness. List different ways (scales) to express minerals hardness. How does it influence the rock drilling process?

5. Define these terms: Deposit, Magma, Eruptive rocks, Host Rock, Gossan, Country Rock, Ore and Orebody, Outcrop, Strata, Seam, Lithology, Weathering and Erosion, Intact Rock, Stratified Rock, Moderately Jointed Rock, Blocky or Seamy Rock, Squeezing Rock, Swelling Rock, Lode, Vein, Bedded Deposit, Massive and Flaggy Deposit.

6. Define: Fold, anticline and syncline; Dike/dyke, sills, stocks and batholiths; Density, Cleavage, Unconformity, Discontinuities, Fault, Joint, Fracture.

7. Describe physical & mechanical characteristics of ores and rocks.

8. Describe Rosiwal mineral abrasivity rating.

9. Draw a rock cycle. Which are the main (dominant) geologic processes responsible for this cycle?

10. Draw schematic presentation of geological structures, and terms used to designate a deposit.

11. Give classifications of rocks based on their origin. Describe each one of them briefly.

12. Give Prof. Protodykonov's rock classification.

13. How a mineral deposit can occur in nature.

14. How does the presence of fold and faults influence the underground excavations driving process and stability. What precautions should be observed while driving through such structures?

15. List important characteristics that are associated with the metamorphic rocks.

16. List principal assumptions made to understand the rocks in terms of rock mechanics in its simplest way.

17. List prominent factors which influence the strength and deformation properties of intact rock material

18. List some important ores known to you.

19. List the natural resources known to you.

20. List the processes responsible for the formation of mineral deposits.

21. List the six common rock forming mineral assemblages that control the mechanical properties of most of rocks encountered in engineering projects.

22. Rock is composed of what?

23. What constitutes the texture or fabric of a rock?

24. What describes the geometry of a deposit?

25. What should be your aim when mining a mineral deposit?

26. Where are the mineral deposits of your own country mainly concentrated? Also write the nearby areas/towns where these deposits are located using a mineral map of your country or any other source of information. According to its breakability – every rock falls into one of the following groups – list them.

27. Write Hudson and Harison's relation to describe relation between number of discontinuities and stability of excavations.

28. There are four cross sections having the areas 6500 m², 5800 m², 1400 m² and 7000 m² respectively. The distance between the two adjacent cross sections is

20 m. Using the cross sectional method, calculate ore reserves using the tonnage factor to be 2.5 ton/m^3.

29. There are a number of methods available to evaluate a mineral inventory from the basic geological data. List them. Also describe each one's merits and limitations and also where each one could be a best choice?

30. Selection of mineral estimation technique/method depends upon number of parameters/considerations – list them.

31. The geostatistical technique (kriging) appears to be an improvement over conventional methods in several respects. List them.

32. Calculate the grade of ore from the following borehole data record if the cutoff grade is 0.30% Cu.

Core recovery Interval	Thickness, m	Grade % Cu
0–30 m	10	1.51
30–33 m	3	0.23
33–37 m	14	0.93
37–41 m	4	0.49
41–45 m	4	0.13
45–50 m	5	0.60

33. Calculate average grade and tonnage of this triangular prism, if its area is 550 m^2. Assume ore density to be 2.5 ton/m^3. Details of hole with regard to its thickness and average grade are given in the table below:

Drill hole no.	Thickness, m	Average grade %
D-1	25	0.93
D-2	43	1.05
D-3	35	0.72

34. Give a relation to calculate reserves by Inverse Distance Weighing (IDW) method. Given below in the table are the details of area and grade of the polygons having mineralization, calculate the reserves and average grade when combining all the polygons. Use 0.4 m^3 per ton as the specific gravity or tonnage factor.

Polygon	Area m^2	Thickness	Grade % Cu
P$_1$	655	44 m	0.70
P$_2$	788	55 m	0.90
P$_3$	877	77 m	0.80

35. What are the highlights/main features of the classification which has been recommended by UNECC. Is it followed in your country?

REFERENCES

1. Blyth, F.G.H. and Freitas, M.H.: *Geology for Engineers*. ELBS, 1988, pp. 254.
2. Boky, B.: *Mining*. Mir Publishers, Moscow, 1988, pp. 11–17.
3. Dearman, W.R.: The charactrization of rocks in civil engineering practice in Britain. *Proc. Colloq. Geologie de l'Ingnieur*, Belgium, 1974, pp. 1–75.
4. Franklin, J.A. and Dusseault, M.B.: *Rock Engineering*. McGraw-Hill, New York, 1989, pp. 14–46.
5. Hudson, J.A. and Harrison, J.P.: *Engineering Rock Mechanics*. Pergamon, 1997, pp. 116.
6. ISRM: ISRM Commission on classification of rocks and rock masses, M. Rocha, Coordinator, *Int. J. Rock Mech. Mi. Sc.* Geomech. Abstr. 18(1), 1981(a), pp. 85–110.
7. Jimeno, C.L; Jimeno, E.L. and Carcedo, F.J.A.: *Drilling and Blasting of Rocks*. Balkema, A.A., Netherlands, 1995, pp. 30–35.
8. Lewis, R.S. and Clark, G.B.: *Elements of Mining*. 3rd ed; Wiley, New York, 1964, pp. 7–10.
9. Matti, H.: *Rock Excavation Handbook*. Sandvik – Tamrock, 1999, pp. 12–52.
10. Mineral Information Institute (MII), 2001. Denver, Colorado, *Web site*.
11. Panek, l.A.: *Stresses About Mine Openings in a Homogeneous Rock Body*. Edward Bros. Ann Arbor, 1951, pp. 50.
12. Peele, R.: Mining Engineering Handbook, 3rd ed; Wiley, New York, 1941, pp. 1:15–49, 2:18–28.
13. Taylor, H.K. General background theory of cutoff grade. *Trans. Inst. Min. Metal.* London (Section A mineral Industry), 1972, pp. A160–79.
14. Whittaker, B.N. and Frith, R.C.: *Tunnelling – Design, Stability and Construction*. IMM Publication; 1990, pp. 19–51.
15. Barnes, M. (1980) Computer-assisted mineral appraisal and feasibility. New York, AIME.
16. Biswas, S.K. et al.: Geostatistical approach to ore reserve calculation – A case study, Symposium of Innovations and New Technologies for copper production in India, 1985, pp. 9–16.
17. Carras, S.N.: Comparative ore reserve methodologies for gold mine evaluation, the Aust. I.M.M. Perth and Kalgoorlie Branches, Regional Conference on: "Gold mining, Metallurgy and Geology", 1984, pp. 1–12.
18. Clark, I.: Practical geostatistics, London, Applied Science Publication, 1986.
19. David, M.: Geostatistical ore reserve estimation, Amsterdam, Elsevier, Scient. Publ. 1977, 1993.
20. David, M.: Grade tonnage curve: use and misuse in ore reserves estimation, Trans. Inst. Min. Metall. (Section A: Mineral Industry), 1972, Vol. 81, pp. 129–132.
21. Diehl, P. and David, M.: Geostatistical concepts and algorithms for ore reserve classification, 17th APCOM symposium, 1982, pp. 413–423.
22. Hazen, S.W.: Some statistical Techniques for analysing mine and mineral deposit sample and assay data, Bulletin 621, US Bureau of Mines, 1967.
23. Hinde, C. et al., Application of interactive graphics mine planning system at Malanjkhand open pit copper mine in India, 18th APCOM symposium, IMM, London, 1984, pp. 1077–1082.
24. Issaks, E.H. and Srivastva, R.M.: An introduction to applied geostatistics, Oxford Uni. Press; New York; 1989, pp. 279.
25. Journal, A.G.: Recoverable reserves estimation – the geostatistical approach, Mining Engg. 1985, pp. 563–568.
26. Journal, A.G. and Huijbregts, C.J.: Mining Geostatistics, London, Academic Press, 1978.
27. King, H. F.; McMhon, D.W and Butjtor, G.J.: A guide to the understanding of ore reserves estimation, Supplement to proceedings no. 281, A.I.M.M. Parkville Vic; 1982, pp. 21.

28. Koch, G.S. et al. Statistical Analysis of Geological data, Vol. 1 and 2, Wiley and Sons, New York, 1970–71.
29. Krige, D.G.: Statistical application of mine valuation, *Jour. Inst. Mine Survey of Afr.* 1962, 12(2), 45–84, 12(3), 95–136.
30. Krige, D.G.: log-normal de-Wijsian geostatistics in ore evaluation, *Jour. S.Afr. Inst. Min. Metall*, Monograph series, 1978, 1–50.
31. Kwa, B.L. et al. Indicator approach to the mineral reserve estimation of gold deposits in Navada, *18th APCOM Symposium*, 1984, pp. 343–366.
32. Matheron, G.: A simple substitute for conditional expectation: disjunctive kriging, in: "Geostat 75", 1975, pp. 237–251.
33. Noble, Alan C.: Ore resources estimation, SME mining engineering hand book, 1992, pp. 345–358.
34. O'Leary, J.: Ore resources estimation methods and grade control at Scully mine, Canada – An integrated geological/geostatistical approach, Mining Magazine, April, 1980, pp. 300–314.
35. Popoff, C.C.: Computing reserves of the mineral deposits: Principles and conventional methods, Information circular 8283, US Bur. of mines, 1966.
36. Raymond, Gray F. (1982) Geostatistical grade estimation of Mount Isa's copper orebodies, *proceedings – Australasian Inst. of Mining and Met.* Vol. 284, pp. 17–39. Rendu, J. (1979) Kriging, logarithmic kriging and conditional expectations: comparison of theory with actual results, *16th APCOM Symposium*.
37. Royle, A.G.: A practical guide to Geostatistics, Mining Sciences Dept. University of Leeds, U.K. 1975.
38. Royle, A.G.: Global estimates of ore reserves, Trans. Inst. Min. Metall. (Section A: Mineral Industry), 1977, vol. 86, pp. A9–a17.
39. Royle, A.G.: Why geostatistics? *Engg. and Mining Jou.* 1979, pp. 101.
40. Royle, A.G.: Kriging, Luma, University of Leeds, 1990, pp. 39–47.
41. Sichel, H.S.: New methods in statistical evaluation of mine sampling data, *Trans. Inst. Min. Metal.* London, 1952, 61, 261–288.
42. Sichel, H.S.: Statistical valuation of diamondiferous deposits, *Jour. S. Afr. Inst. Min. Metal.* London, 1973, 61, 261–288.
43. Tatiya, R.R.: PhD thesis, Imperial College of Science and Technology, London University, London, 1987.
44. Tawo, E.E. and Dowd, P.A. Using different sources of information to improve the quality of an estimate – the soft constrained kriging approach, Luma, University of Leeds, 1990, pp. 147–154.
45. Verly, G. The Multigaussian Approach and its application to the estimation of local reserves, *Jour. of International Association for Mathematical Geology*, 1983, vol. 15, pp. 263–290.
46. Vickers, E.L.: Marginal analysis – its application in determining the cutoff grade, Min. Engg. 1961, 13(6):578–582.
47. Watson, M.I. A study in variogram estimation in no. 2 seam at Delma Colliery, Chamber of Mines South Africa, Johannesburg, 1977, pp. 117.
48. Young, D.S. Development and approach of Disjunctive Kriging model: Discrete Gaussian model, 17th APCOM symposium, 1982, pp. 544–562.
49. http://www.unece.org/fileadmin/DAM/energy/se/
50. Sydney Allison. Processing Ore. In: Chap. 74 – International Labor Organization (ILO) Encyclopedia and CISILO database *(website accessed in 2012)*.

Chapter 3

Prospecting, exploration & site investigations

Mineral wealth belongs to every one of us; no matter where it is located. And who owns it? Its proper exploration, systematic development and exploitation and judicious utilization are our moral responsibilities.

3.1 INTRODUCTION

This chapter aims to describe two aspects: First, the procedures and techniques required to search a mineral deposit, and second, to investigate about the ground, as much as possible within the given constraints; before undertaking any tunneling, or excavation project. The first part aims to establish a mineral inventory for the purpose of carrying out the feasibility studies for declaring a deposit worth mining or not. The second part aims to examine the suitability of ground or site for a particular civil construction, or an excavation project including tunnels. There are many aspects common to both; for example, the terminology used, methods and techniques employed, tests conducted, and equipment deployed.

3.2 PROSPECTING AND EXPLORATION[7]

Prospecting means searching of minerals, and therefore, it is carried out first of all. Even if a mineral is found, the prospecting is continued till it gives enough information for the preliminary appraisal of any mineral deposit; so that decision can be taken whether to carry out further exploration work or not? Figure 3.1[2] outlines a flowchart to undertake the exploration tasks. Prospecting includes three stages:[7]

1. Finding signs of the mineral in the locality, or general indications that it may be there
2. Finding the deposit
3. Exploring the deposit.

3.2.1 Finding signs of the mineral in the locality or general indications

Finding signs of the mineral in the locality or general indications that it may be there, can be established by finding some of the signs, listed below:

- Exposure of the mineral at the ground surface
- The topography relief

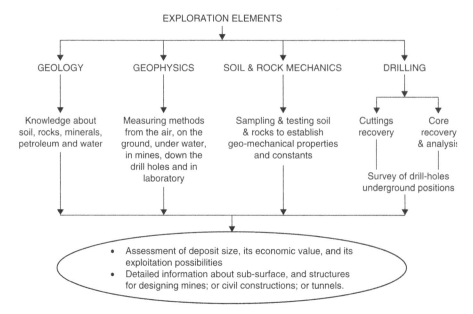

EXPLORATION ELEMENTS

GEOLOGY   GEOPHYSICS   SOIL & ROCK MECHANICS   DRILLING

Knowledge about   Measuring methods   Sampling & testing soil   Cuttings   Core
soil, rocks, minerals,   from the air, on the   & rocks to establish   recovery   recovery
petroleum and water   ground, under water,   geo-mechanical properties   & analysis
                       in mines, down the   and constants
                       drill holes and in
                       laboratory

                                             Survey of drill-holes
                                             underground positions

- Assessment of deposit size, its economic value, and its
  exploitation possibilities
- Detailed information about sub-surface, and structures
  for designing mines; or civil constructions; or tunnels.

Figure 3.1   Line diagram of exploration elements.

- Fragments of the mineral at the ground surface
- Traces of the ancient mine workings
- Vegetation
- Egress of underground water.

3.2.1.1   *Geological studies*

Any mineral deposit is a geologic body, which when hidden, can be successfully found
and exploited only through the utilization of the full geologic principles. Geological studies help in selection of the area to be explored i.e. target area. Compilation and interpretation of data at any stage of exploration to direct the exploration program further
are the key tasks that are included in the geological studies. Aerial photography, topography map, information collected through the direct methods of prospecting, and the
one through the indirect methods (geophysical and geo-chemical) are used for the geological interpretation, and to identify the target area.

3.2.1.2   *Geo-chemical studies*

This means determination of the relative abundance of the elements, which may occur
in rocks, soils, water, air, gossans, plants or stream sediments. Through systematic collection and analysis of appropriate samples, geo-chemical anomalies (either actual element or the one which is usually associated with element being sought) are determined.
In Canada, geologists have trained their dogs to sniff out their exploration clues.
German Shepherds have been taught to use their excellent sense of smell to find sulfides of Pb, Zn, Ni, Mo, Cu and Ag (fig. 3.4(g)).[12] This together with other information
simplifies the process of selecting the target area.

3.2.2   Finding the deposit or preliminary proving

During this stage an attempt is made to establish some of these parameters:

- Area/location of the deposits and its shape
- Depth of deposits, dip and strike directions
- Thickness details
- Type of surrounding rocks i.e. as over burden, h/w, f/w
- Grade, mineralogical and chemical composition
- Quantity and variation with respect to depth etc.

This can be achieved by the application of:

 (i) Geophysical techniques (briefly outlined below) and
(ii) By putting exploratory headings such as; trenches, pits, adits and drives.

3.2.2.1   *Geophysical methods/studies/surveys*[3,8,11,14]

Geophysics is the study of physics of the earth with regard to its physical properties, composition and structure. In mineral exploration various physical properties of earth, such as: electrical, gravitational, magnetic, compositional, mechanical and thermal are measured by a variety of methods to detect directly or indirectly areas which are *anomalous* as related to their surroundings (term *anomaly* is defined as *a statistically significant departure from the normal values*). Geophysical surveys include the subsurface surveys, surface surveys, fixed wing or helicopter surveys in the atmosphere, and the surveys from the satellites orbiting the earth above the atmosphere. A survey may be carried out prior to, during and after prospecting drilling. This type of survey generally has two objectives: to cut the total exploration costs, and to ensure that the prospecting drilling has the highest chance of success. More often than not, it is necessary to use a combination of two or more methods to acquire sufficient data for a reliable interpretation. An interpretation of the geophysical survey results, together with geological and drilling data, can provide a firm basis for deciding whether to continue or abandon an exploration project.

Gravity surveys: Normal earth gravity is 981 cm/sec^2; any variation in this parameter is noted and after applying the correction due to latitude, elevation, topography and tidal change (i.e. change in gravity w.r.t. time). These surveys are of limited use in geotechnical evaluation of the orebody and the surrounding rock. The gravity meters measure the density at a particular point that is influenced by the density of materials all around the measured points. In figure 3.3(a) the presence of a dense body, which increases the force of gravity diverts the lines away from the vertical, as shown by the solid arrows. As distance from the dense body increases, the magnitude of gravitational field including its deviation from the vertical diminishes, and eventually disappears.

Electromagnetic surveys: These surveys are based on the concept that when an electric current is subjected to a primary alternating field, the induced current creates a secondary field. The resultant field therefore differs in amplitude and phase from the primary field, and these differences can be detected and measured. These surveys are used mainly to find the mineral deposits and to map the geological structure. Among

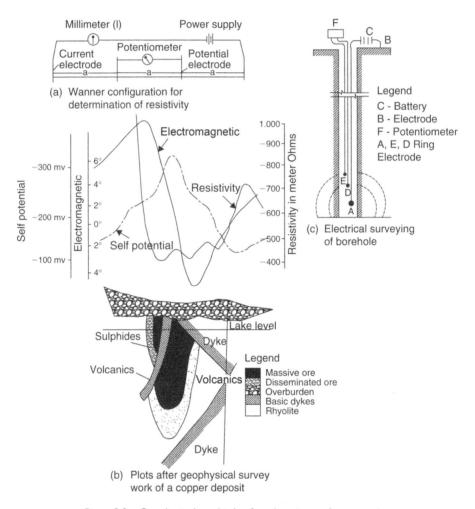

Figure 3.2   Geophysical methods of exploration and prospecting.

the minerals successfully located by this method are the sulfide ores of copper and lead, magnetite, pyrite, some manganese ores, and graphite. The graphs obtained for a copper deposit as a result of self-potential, electro-magnetic and resistivity surveys are shown in figure 3.2(b). All these plots show anomaly over the copper deposit when compared with its surroundings.

Electrical resistivity: The method is useful to measure the resistivity of overburden and rock material. As shown in figure 3.2(a), in a commonly used method four equal spaced electrodes (known as Wanner configuration) are used. A current flow is established between the outer electrodes and the voltage drop is measured between the inner electrodes. Some machines read directly the ohms. The electrical resistivity is calculated by the equation: $\sigma = 2\pi a(E/I)$ where: σ is the resistivity in ohmmeters, a – the electrode's spacing in meters, E – the voltage drop and I – the current in amperes. Electrical well logging as shown in figure 3.2(c) is also based on this principle.

Electrical well logging is the measurement of the resistance of the rock in that part of the borehole, which is unsupported by the casing. In to a well three effectively insulated conductors of different lengths are lowered. The bared ends of the conductors are fitted with heavy lead rings. The ring A on the longest cable acts as an electrode supplying current to the ground. The other electrode, B, is at the surface. Current source is the battery C in the circuit AB. Rings D and E are electrodes in the measuring circuit and are joined to the potentiometer F at the surface, measuring potential difference between points D and E. Thus, measuring the current in the circuit AB and the potential difference between D and E, the resistance of the ground can be calculated.

In this method significant developments have taken place, but more research and improvement still need to be made. This method undoubtedly carries a great potential for identifying water-bearing zones.[13]

Magnetic surveys: These surveys make use of the variations in the earth's magnetic field caused by the magnetic properties of subsurface bodies. In ore prospecting, these variations are specially useful in locating magnetite, pyrhotite, and ilmenite. In oil exploration magnetic surveys may be of value when structural features obscure the sedimentary formations overlying the oil. Magnetic logging in drill holes is also used to obtain information for directing further drilling. Magnetometers are used to undertake this survey. A magnetometer can read directly the total magnetic field or its horizontal or vertical components (fig. 3.3(d)). This instrument can be used to locate the intrusive dikes, faults and lithologic boundaries. Contour maps showing the lines of equal magnetic intensity are prepared to know the trend. Proton magnetometers calibrated in gammas (gamma = $(10)^{-5}$ orested) and with digital reading are nowadays very common. Effective magnetic surveys may be conducted on the ground, from the shipboard or from an aircraft.

Seismic method – reflection: In this system shock waves induced in the ground, penetrate to a surface or discontinuity that reflects the shock wave back to the surface. The method has applications in oil exploration to locate salt domes, anticlines etc. and also it is useful to find out the continuity of the sedimentary rock units. The arrangement is shown in figure 3.3(b).

Seismic method – refraction: This method is capable of gathering information to a depth of 150 m or more. Generally the higher the seismic velocity, the more competent is the rock. The method also assumes that seismic velocity of the individual layers increases as the depth increases. A shock wave is induced in the ground and the travel time to pick up the points (geophones) spaced along a line is measured with a very accurate timing device (± 0.001 sec.). The instruments used for this purpose are multi-channel seismographs with 24 or more geophones to a single geophone timer (Hansen and Lachel, 1982). The energy source to generate a shock wave could be: a hammer striking a steel plate, the dropping of a heavy weight, or the detonation of an explosive. The arrangement is shown in figure 3.3(c).

Concerning ground quality characterization, the most important recent development probably has been in the field of seismic tomography.[13] Since its introduction for this purpose about 18 years ago, the method has been continuously improved regarding signal source and receiver, as well as interpretation, and today, it can provide invaluable information on subsurface ground conditions. Most commonly, the tomographic method is applied between drill holes (often from core drilling) as shown in the

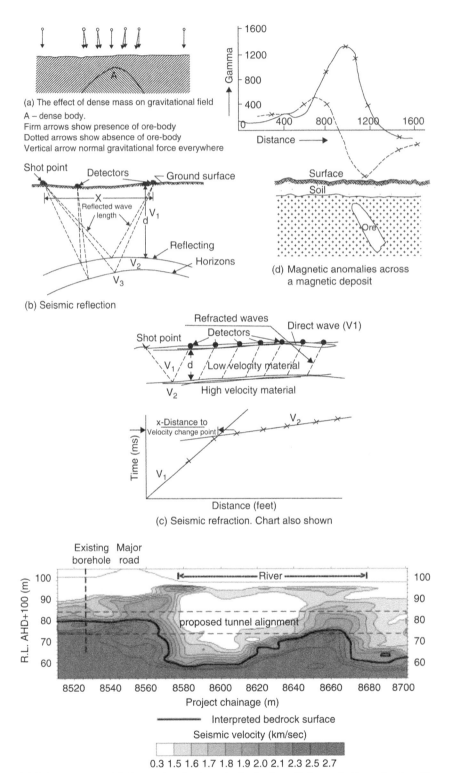

(a) The effect of dense mass on gravitational field

A – dense body.
Firm arrows show presence of ore-body
Dotted arrows show absence of ore-body
Vertical arrow normal gravitational force everywhere

(b) Seismic reflection

(d) Magnetic anomalies across a magnetic deposit

(c) Seismic refraction. Chart also shown

(e) Seismic tomographic image showing irregular bedrock profile beneath Alexandra Canal to the west of Sydney Airport and along the initially proposed NSR tunnel alignment.

Figure 3.3 Geophysical methods of exploration and prospecting.

figure 3.3(e) (referred to as crosshole tomography).[13] Alternatively, it may be applied between underground openings, or between one underground opening and the surface (in subsea tunnels it is sometimes applied between a drillhole ahead of the face and the sea bottom).

Nuclear surveys: In these surveys areas having radiation intensity considerably higher than the normal background for the area are located. These surveys are used for the search of uranium and thorium, and indirectly to locate minerals, which are associated with radioactive substances. Nuclear methods are also used for minerals analysis of specimens, and in drill holes.

Geothermal surveys: Heat difference between ore and their host rocks, or between thermal water and their surroundings, are detectable by geothermal methods. They may be used to locate ore deposits boundaries. Most geothermal measurements are made in drill holes.

Remote sensing: This includes aerial photography, side looking airborne radar (SLAR), false color infrared (IR) photography, thermal IR photography, and multi special scanning from satellites or high altitude aircraft. The use of such imagery is mainly to locate lineaments (distinct features) and their lengths and orientations. The lineaments located by means of such techniques should be always checked on ground, as there are many reasons for lineaments other than those associated with rock structure.

Table 3.1 summarizes the geophysical techniques, which are directly or indirectly helpful in exploring the mineral deposits.

Continuous progress has been made in geophysical data processing and interpretation, making the results more accurate and reliable.[13] However, seismic refraction or other routine geophysical methods do not automatically give high quality results in all geological environments. Particularly, there are limitations across deep clefts due to side reflection. When a high degree of accuracy is needed in such cases the topography surveys and prospecting drilling (boring) should be carried out. Generally, the highest value of geophysical pre-construction investigation undoubtedly is obtained when combine its results with the other investigations.

3.2.2.2   Putting exploratory headings[2,7]

For this purpose prospecting or exploratory trenches, pits, adits and drives are driven. The type of entry will depend upon the geometry and location of the deposit w.r.t. surface datum. As shown in figure 3.4(a),[7] if the over burden is thin and the dip is steep, trenches can be dug for exploration. Vertical pits, as shown in figures 3.4(b)[7] and (d), can prospect more gently dipping deposits with small over burden cover. Boreholes, as shown in figure 3.4(c)[7] can prospect the flat and shallow to deep-seated deposits. For the steeply dipping bedded deposits, sometimes pitting, cross-cutting, driving and borehole drilling, as shown in figures 3.4(e)[7] and 3.4(f)[2] are essential.

Due to the uncertainty of projecting geological information obtained from surface mapping towards the depth, excavation of adits or shafts may be required as part of the site investigation program.[13] This is most relevant in very complex geology and/or when very detailed information on the rock mass conditions is required. Sometimes, the main purpose may also be in-situ measurement (for instance of rock stresses) or testing (e.g. the shear strength of discontinuities).

Table 3.1 Geophysical exploration techniques and their applications.[2,14]

Geophysical exploration techniques	Direct	Indirect
Resistivity	Massive sulphides (e.g. sulphides of Fe, Cu, Pb, Ni, Co, Mo); quartz, calcite, sand & gravel, special clays rock salt	Water exploration; bulk materials; base metals; phosphates; potash; uranium; coal; natural stream; detailed tectonics; determination of deposits and profile of bedrock; location of buried channels; geological mapping.
Induced polarization	Disseminated sulfide deposits; oxides of manganese; Zn; Cu; As; coal seams	Associated minerals; e.g. Pb, Zn; Mo; Ag; Au
Self potential	Sulfide and graphite mineralization zones; e.g. Sulfides of pyrite, pyrhotite; Cu; Mn ore	Associated minerals; e.g. Pb, Zn; Ni; Ag; Au
Seismic	Depth of bedrock; geological mapping; buried channels; faults; sand and gravel deposits	Oil and gas; water; tin; diamonds; heavy minerals; coal; rock quality and splitting characteristics.
Magnetic	Magnetic pyrhotite; hematite; totno-magnetite; geological mapping; tracing the course of dikes and intrusions	Iron ore; chromite; copper ore; gold (associated with intrusive rocks); kimberlites; oil and gas (from a study of depth to 'magnetic basement'); e.g. thickness of sedimentary sequence
Electromagnetic	Conductive sulfides and oxides; conductive orebodies; manganese oxides; magnetite and graphite	Kimberlites; associated minerals; shear zones; conductivity mapping
Gravity	Dense sulfides and oxides; conductive orebodies; manganese oxide; magnetite and graphite	Location of intrusions and faults; oil and gas (from a study of thick sedimentary basins)
Nuclear	Uranium; thorium; coal; lignite; monazite; phosphates; other radio active minerals	Geological mapping; mineral analysis; oil and gas; water content; rock density (from back-scattering of artificially produced γ rays)
Geothermal	Thermal springs	Natural stream; boron; sulphur; cavities; thermal borehole logging used to identify coal seams
Magneto-telluric	Geological study of rock strata; oil reservoirs; geo-thermal reservoirs; deep-seated orebodies	Oil; natural stream; boron; sulphur; and brines; complex sulfide ores

3.2.3 Exploring the deposits or detailed proving – prospecting drilling[2]

The objective of prospecting drilling is to get samples from depths below the surface. The two basic methods for this purpose are core recovery and cuttings recovery. In the former a core (hollow) bit is used and in the latter a full-hole (solid) bit is used and

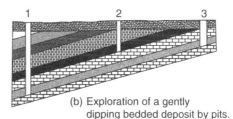

(a) Exploration of a steeply dipping bedded deposit by trenching

(b) Exploration of a gently dipping bedded deposit by pits. 1, 2, 3 sequence of pitting

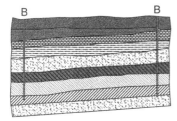

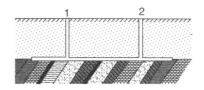

(c) Exploration of gently dipping bedded deposit by boreholes (B)

(d) Exploration of a steeply dipping bedded deposit under thick overburden. ab, bc, 1, 2 pits & crosscuts

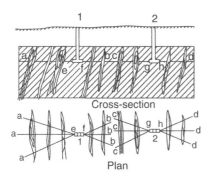

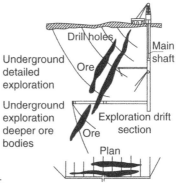

(e) Exploration of vein deposits by pits 1, 2; with crosscuts ef, gh; and by boreholes bf, ea, gc and hd drilled from the faces of cross cuts

(f) Investigating an orebody by diamond core drilling from surface and underground

(g) In Canada geologists have trained dogs to sniff out exploration clues. German shepherd have been taught to use their excellent sense of smell to find sulfide of Pb, Zn, Ni, Mo and Ag

Figure 3.4   Exploration by trenching, pitting and diamond core drilling.

cuttings are collected. Both methods can be used on the surface or underground. Line diagram 3.5 illustrates this aspect.

Modern prospecting rigs make extensive use of pneumatics, hydraulics and electronics. In a conventional rig, the drill rod string made up of three- or six-meter lengths of steel or aluminum rods; is handled by derrick, hoist, tackle, rod brake and rod tools. This arrangement demands lots of power and reduces the time of productive drilling. In modern rigs hydraulics are frequently used for the rod handling. This eliminates the need for all separate accessories. It cuts non-productive drilling time, and improves the working conditions for the driller.

A hydraulic or mechanical chuck holds the drill rod string firmly so that it may be rotated at the desired speed. Drilling rotation is always clock-wise. The feed frame/mechanism applies the necessary force to give the right pressure on the bit for effective cutting. The flush pump passes the water, or any other flushing fluid down through the rod string and past the core barrel and core bit. This cools the bit, carries cuttings up to the surface outside the drill rods, reduces the friction between drill string and hole wall, and by building up hydrostatic pressure helps in stabilizing hole's wall. Aluminum rods having half the weight that of steel rods vibrate less than them, and core bit life is longer when these rods are used.

The bit cuts out a core of rock and this moves up into the core barrel until the barrel is filled. A standard core barrel takes three meters of core. When the core barrel is full, the rod string must be hoisted and unthreaded until it reaches the surface, where it is emptied. This operation also provides opportunity to replace the bit if necessary. It is advantageous to use wire-line drill rods, and a wire-line core barrel. When such a core barrel is filled it is hoisted up by cable inside the drill rod string; the string remains in place behind the drill bit. Drilling restarts. Since rod pulling is no longer necessary each time the core barrel is filled, productive drilling time goes up. Wire line drilling, however, requires large diameter steel rods, gives somewhat smaller core, and often may have a lower rate of penetration than that achieved with conventional core barrel and small diameter diamond bit. The choice of core bit[2] will depend mainly on formation hardness, flushing medium, and type of drill. Diamond bits may be either surface set or impregnated. Hard metal bits, for soft formations, may have either cemented carbide inserts or a matrix impregnated with cemented carbide particles.

A reaming shell, with diamonds or cemented carbide inserts, is usually placed behind the drill bit to keep the hole diameter correct and to help to reduce the rod string vibrations. There are two diamond core-drilling standards in general usage, one based on inch dimensions and other based on the metric standard.

Core drilling is among the routine methods for subsurface exploration.[13] Most commonly, NX-size core drill is used, representing a hole diameter of 76 mm (3″) and a core diameter of 54 mm (2 1/8″).[11] The drilling often has multiple purposes, of which the following are in most cases the most important:

• Verification of the geological interpretation.
• To obtain more information on rock type boundaries and degree of weathering. This includes samples for ore deposits.
• To supplement information on orientation and character of weakness zones.
• To provide samples for laboratory analyses.
• Hydrogeological and/or geophysical testing.

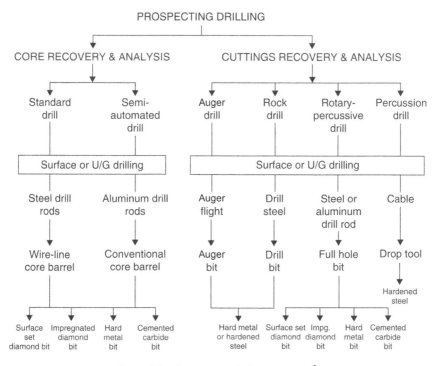

Figure 3.5 Prospecting drilling methods.[2]

In the case of tunneling projects, the drilling often is carried out with the prime purpose to investigate major faults or weakness zones assumed to be crucial for the stability and ground water conditions of the opening. The drill-hole will also give valuable information (fig. 3.6) about the adjacent rock mass. A parameter closely linked to core drilling is the RQD-value, representing the total length of recovered core pieces greater than or equal to 10 cm (4″) divided by the length of the attempted core run, expressed as a percentage (sec. 3.5).

Apart from drillhole testing, the recent development in this field involves directional drilling; making it possible to have core drilling, practically, in any direction; for example, along the alignment of a planned tunnel.

3.3 PHASES OF PROSPECTING AND EXPLORATION PROGRAM[3]

The concept applied is whole to part; i.e. the search starts by undertaking prospecting activities of a region, which could be sometimes several thousand square kilometers; by taking decisions based on the choosing the areas of interest and rejecting those that are not suitable for that point of time. There are two stages,[2] prospecting and exploration, as shown in figure 3.7.[3] During first stage, the work is carried out in two phases, the regional appraisal and detailed reconnaissance of the favorable areas. During the second stages also the work is carried out in two phases (3 and 4). The third phase is devoted to the surface appraisal of the target area; and during last

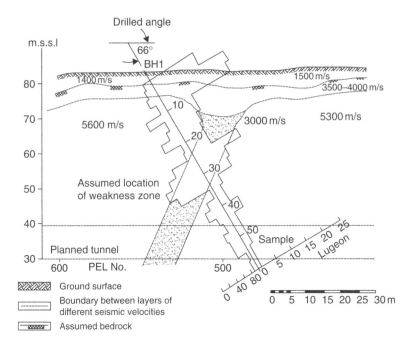

Figure 3.6   Results of core drilling along an exploratory borehole. The type of information (data) such as seismic refraction, RQD and Lugeon* values along the hole have been shown.

*Lugeon (a unit devised to quantify the water permeability of bedrock and the hydraulic conductivity resulting from fractures; it is named after Maurice Lugeon, a Swiss geologist who first formulated the method in 1933.) Lugeon value is defined as the loss of water in litters per minute and per meter borehole at an average pressure of I MPa.

phase, three-dimensional sampling and evaluation process follows. During this endeavor the areas decrease[2] from 2,500–25,000 km^2 in phase-1 to 2.5–125 km^2 in phases 2 and 3; and finally to 0.25–50 km^2 in phase 4. The compilation of prospecting and exploration activities is similar to shown in table 3.2, for civil work sites. The tasks involve fieldwork, laboratory testing, and office works as shown figure 3.7.

3.4   SITE INVESTIGATIONS FOR CIVIL CONSTRUCTIONS, OR ANY EXCAVATION PROJECT INCLUDING TUNNELS AND CAVERNS[15]

Success in any construction or civil work including driving tunnels lies if a proper forecast about the regimes of soil, water, rocks, gases (if any) which are likely to encountered, is made before such activities are undertaken. Technical know-how, and experts are available to design and construct even if the sites available have adverse geological setups provided they are made aware of them in advance. In want of proper specifications about the site delays, cost overruns, disputes and other unwanted scenarios are very common. If latent adverse geological features remain undetected during both the design and construction phases, the potential of failure during operation in the following years remains.

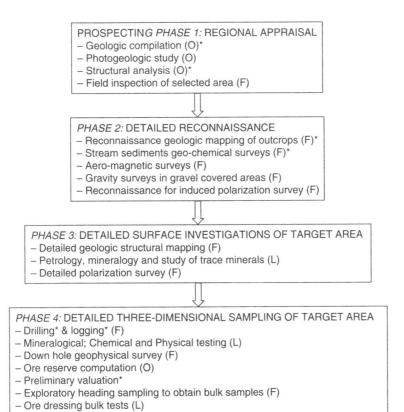

Figure 3.7   Details of activities and methods employed during prospecting and exploration based on a case study of a copper deposit. O – Office; L – Lab.; F – Field; * – Activity or method, which is indispensable.

Table 3.2   Applicability of main investigating methods to assess ground conditions.[13]

Investigation method/factors to investigate	Desk study	Field mapping	Core drilling	Geophysics	Exploratory headings	Field testing	Lab. testing
Rock types	x	x	x	(x)	x	–	x
Mechanical properties	(x)	(x)	(x)	(x)	x	x	x
Weathering	(x)	(x)	x	x	x	–	–
Soil cover	x	x	x	x	x	–	–
Jointing	(x)	x	(x)	–	x	(x)	(x)
Fault/weakness zones	x	x	x	x	x	–	(x)
Rock stresses	(x)	–	–	–	(x)	x	–
Ground water conditions	(x)	(x)	(x)	x	x	x	–

x: the method well suited; (x): the method is partly (sometimes) suited; –: the method is not suited.

PHASES OF SITE INVESTIGATIONS

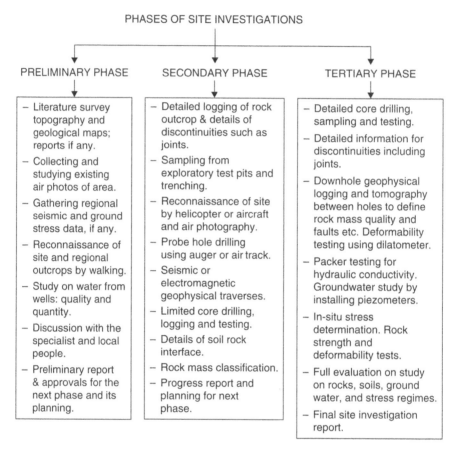

PRELIMINARY PHASE	SECONDARY PHASE	TERTIARY PHASE
– Literature survey topography and geological maps; reports if any. – Collecting and studying existing air photos of area. – Gathering regional seismic and ground stress data, if any. – Reconnaissance of site and regional outcrops by walking. – Study on water from wells: quality and quantity. – Discussion with the specialist and local people. – Preliminary report & approvals for the next phase and its planning.	– Detailed logging of rock outcrop & details of discontinuities such as joints. – Sampling from exploratory test pits and trenching. – Reconnaissance of site by helicopter or aircraft and air photography. – Probe hole drilling using auger or air track. – Seismic or electromagnetic geophysical traverses. – Limited core drilling, logging and testing. – Details of soil rock interface. – Rock mass classification. – Progress report and planning for next phase.	– Detailed core drilling, sampling and testing. – Detailed information for discontinuities including joints. – Downhole geophysical logging and tomography between holes to define rock mass quality and faults etc. Deformability testing using dilatometer. – Packer testing for hydraulic conductivity. Groundwater study by installing piezometers. – In-situ stress determination. Rock strength and deformability tests. – Full evaluation on study on rocks, soils, ground water, and stress regimes. – Final site investigation report.

Figure 3.8   Phases of site investigations.[5]

Like mineral prospecting and exploration; for civil construction the investigation can be divided into three phases: *Preliminary, Secondary or Intermediate, and Tertiary or Final* (fig. 3.8). During the initial phase it is reconnaissance and compilation of available information and data. Fieldwork, sampling, testing, large spaced drilling, and enhancement in the datasets collected during initial phase are planned for the second phase. During the last phase detailed exploration, testing, analysis, compilation for the data sets required for different regimes (soil, water, rocks and gases, etc.) are ready. Decisions and further planning are based on the report prepared during this phase.

3.5   ROCKS AND GROUND CHARACTERIZATION[4,5,6,9,10,12]

3.5.1   Rock strength classification[5]

It is important to identify the type of ground or rocks likely to be encountered based on their strength. Useful guidelines have been proposed by various authors and a comparison of uniaxial compressive strength of rocks to distinguish soils and rocks based on their strength has been carried out by Bieniawski in 1984, as shown in figure 3.9.[5] By considering the classifications proposed by Coates (1964), Deere and Millar (1966), Geological Society (1970), Jennings (1973), Bieniawski (1973) and ISRM (1979) this comparison was made. Thus, based on this logic one can assess ground/rock conditions as per its strength. But this criteria will not be sufficient to characterize the ground, as other parameters one needs to include are geological discontinuities such as joints, cracks, fissures etc., and also the presence of water. To take care of all these parameters and assess the ground quality, work has been done by a number of authors and they have designated them as: '*Rock mass classification, rock mass rating, or rock structure rating*'. Salient points of some of these studies are described in the following sections.

3.5.2   Rock mass classifications

Of many rock mass classifications in existence today, six require special attention (table 3.3[6]) because they are the most commonly known, namely, those proposed by: Terzaghi (1946), Luuffer (1958), Deere (1964), Wickham et al. (1972), Bieniawski (1973) and Barton et al. (1974).

Terzaghi (1946) in his rock mass classification for tunnels, divided rocks in the following 7 categories:[5]

1. Stratified rocks
2. Intact rocks
3. Moderately jointed rocks (e.g. vertical walls requiring no support)
4. Blocky rocks (e.g. vertical walls requiring support)
5. Crushed rocks
6. Squeezing rocks (low swelling capacity)
7. Swelling rocks (high swelling capacity)

As described by *Coates*[9] (1964), one of the main deficiencies of this classification is that it does not give information on the strength or permeability of the rock mass. The classification given by Coates (1964) includes the following five main characteristics of rocks including those related with rock strength:

1. Uniaxial compressive strength [weak (<35 MPa); Strong (35–175 MPa, homogeneous and isotropic rocks); Very strong (>175 MPa, homogeneous and isotropic rocks)]
2. Pre-failure deformation behavior of rocks: Elastic; Viscous.
3. Failure characteristics of the rocks: Brittle; Plastic
4. Gross homogeneity: Massive; Layered (e.g. sedimentary rocks); and
5. Continuity of rocks mass: solid (joint spacing >1.8 m); blocky (joint spacing <1.8 m); broken (pass through a 75 mm sieve).

3.6   ROCK QUALITY DESIGNATION (RQD)[10,12]

Deere (1964), based on core recovery during drilling from a drill hole, proposed the following relation to calculate RQD; only core pieces that are 100 mm or greater

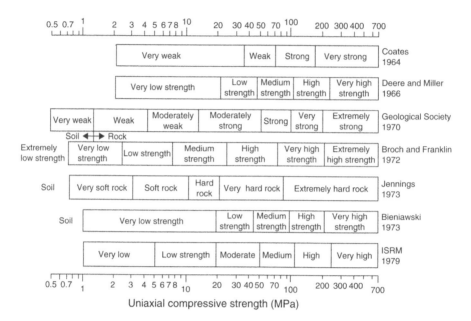

Uniaxial compressive strength (MPa)

Figure 3.9   Ground and rocks classification based on their strength.

Table 3.3   Major engineering rock mass classification systems.[6]

Rock mass classification system	Originator	Application
Rock loads	Terzaghi (1946)	Tunneling with steel support
Standup time	Luuffer (1958)	Tunneling
NATM	Rabcewicz, Pacher miller (1964)	Tunneling
RQD Index	Deere (1964)	Core logging, tunneling
RSR concept	Wickham et al. (1972)	Tunneling
RMR concept	Bieniawski (1973) and modified by many others	Mining and tunneling
Q system	Barton et al. (1974)	Tunnels, chambers
Strength-size	Franklin (1975)	Tunneling
Basic geotechnical description	International Society of Rock Mechanics	General applications

in length are included. This system is useful for rough estimation of the rock mass behavior.

$$RQD = \frac{\sum Length\, of\, core\, pieces > 10cm\, length}{Total\, Length\, of\, core\, run} \; - - - -$$ (3.1)

RQD (%)	Classification
90–100	Excellent
75–90	Good
50–75	Fair
25–50	Poor
>25	Very poor

3.6.1   Q (rock mass quality) system[4,12]

Barton, Liena and Lunde developed this system of rock mass classification at the Norwegian Geotechnical Institute in 1974.[4] It is based on study of some 1000 tunnel case histories. Six parameters, listed below, were used to calculate Q, using following relation:

$$Q = (RQD / J_n) \times (J_r / J_a) \times (J_w / SRF) \tag{3.2}$$

RQD – Rock quality designation
  J_n – Number of joint sets indicating the 'freedom' of rock mass
  J_r – Roughness of most unfavorable joint set
  J_a – Degree of alteration or filling of the most unfavorable joint set
  J_w – Degree of joint seepage, or joint water reduction factor
 SRF – Stress reduction factor, which calculates load reduction due to excavation, apparent stress, squeezing and swelling.

3.6.2   Geomechanics classification (RMR system)

The system was developed by *Bieniawski* in 1973.[5,6] Following parameters are used to classify this geomechanics classification or a rock mass rating (RMR) system, as given by

1. Uniaxial compressive strength (range of values 0 to 15)
2. Rock quality designation (range of values 3 to 20)
3. Spacing of discontinuities (range of values 5 to 20)
4. Condition of discontinuities (range of values 0 to 30)
5. Ground water condition (range of values 0 to 15)
6. Orientation of discontinuities (range of values 0 to 260)

The geomechanic classification is presented in table 3.4. In this table five parameters have been grouped into five ranges of values (classification parameters and their ratings). A higher rate indicates better rock mass condition. These ratings are adjusted based on the discontinuities' orientation with respect to direction of tunneling and mine openings (Item B, table 3.4). Charts[6] A through E (fig. 3.10) are used to evaluate average condition of each of the discontinuities. Chart D is used if either RQD or discontinuity spacing data is lacking. The rock mass classification is determined from the total ratings, as shown by item C, in table 3.4. Meaning of rock mass classes has been illustrated as item D in table 3.4. In this table rock mass have been

Table 3.4   Geomechanic classification. Tables A through D allocating ratings of different aspects. Meaning of mass classes shown in table (D).

A. CLASSIFICATION PARAMETERS AND THEIR RATINGS

	Parameter	Ranges of values						
1	Strength of intact rock material — Point-load strength index (MPa)	>10	4–10	2–4	1–2	For this low range, uniaxial compressive test is preferred		
	Uniaxial compressive strength (MPa)	>250	100–250	50–100	25–50	5–25	1–5	<1
	Rating	15	12	7	4	2	1	0
2	Drill core quality RQD (%)	90–100	75–90	50–75	25–50	<25		
	Rating	20	17	13	8	3		
3	Spacing of discontinuities	>2 m	0.6–2 m	200–600 mm	60–200 mm	<60 mm		
	Rating	20	15	10	8	5		
4	Condition of discontinuities	Very rough surfaces Not continuous No separation Unweathered wall rock	Slightly rough surfaces Separation <1 mm Highly weathered wall	Slightly rough surfaces Separation <1 mm Highly weathered wall	Slickensided surfaces or Gouge <5 mm thick or Separation 1–5 mm Continuous	Solt gouge >5 mm thick or Separation >5 mm Continuous		
	Rating	30	25	20	10	0		
5	Groundwater — Inflow per 10 m tunnel length (L/min)	None	<10	10–25	25–125	>125		
	Ratio joint water pressure/Major principal stress	0	<0.1	0.1–0.2	0.2–0.5	>0.5		
	General conditions	Completely dry	Damp	Wet	Dripping	Flowing		
	Rating	15	10	7	4	0		

B. RATING ADJUSTMENT FOR DISCONTINUITY ORIENTATIONS

Strike and Dip Orientations of Discontinuities		Very Favorable	Favorable	Fair	Unfavorable	Very Unfavorable
Ratings	Tunnels and mines	0	−2	−5	−10	−12
	Foundations	0	−2	−7	−10	−25
	Slopes	0	−5	−25	−50	−60

C. ROCK MASS CLASSES DETERMINED FROM TOTAL RATINGS

Rating	100 ← 81	80 ← 61	60 ← 41	40 ← 21	<20
Class No.	I	II	III	IV	V
Description	Very good rock	Good rock	Fair rock	Poor rock	Very poor rock

D. MEANING OF ROCK MASS CLASSES

Class No.	I	II	III	IV	V
Average stand-up time	20 yr for 15-m span	1 yr for 10-m span	1 wk for 5-m span	10 h for 2.5-m span	30 min for 1-m span
Cohesion of rock mass (kPa)	>400	300–400	200–300	100–200	<100
Friction angle of the rock mass (deg)	>45	35–45	25–35	15–25	<15

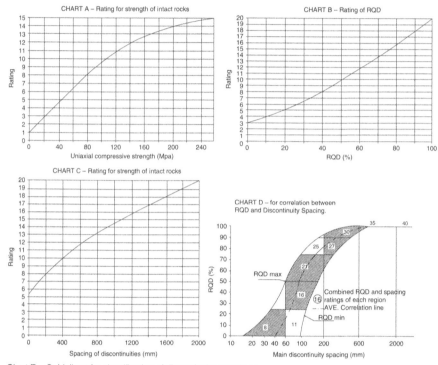

Chart E – Guidelines for classification of discontinuity conditions.

Parameters	RATINGS				
Discontinuity length (persistence/continuity)	<1 m 6	1–3 m 4	3–10 m 2	10–20 m 1	>20 m 0
Separation (aperture)	None 6	<0.1 mm 5	0.1–1.0 mm 4	1–5 mm 1	>5 mm 0
Roughness	Very rough 6	Rough 5	Slightly rough 3	Smooth 1	Slickensided 0
	Hard filling			Soft filling	
Infilling (gouge)	None 6	<5 mm 4	>5 mm 2	<5 mm 2	>5 mm 0
Weathering	Unweathered 6	Slightly weathered 5	Moderately weathered 3	Highly weathered 1	Decomposed 0
Note: Some conditions are mutually exclusive. For example, if infilling is present, it is irrelevant what the roughness may be, since its effect will be overshadowed by the influence of gouge.					

Figure 3.10 Geomechanic classification to evaluate average condition of discontinuities using charts A to E.

classified in five groups I to V. For each group average stand-up time, cohesion of the rock mass and friction angle has been specified.

The RMR (table 3.4) can be used to calculate the probable support load (P), by using the equation given by *Unal* (1983)

$$P = (100 - RMR)\ \gamma B/100 \tag{3.3}$$

γ is rock density (kg/m^3) and B is tunnel width (m).

Table 3.5 Behavior of rock mass in tunneling based on Q system.

Rock mass quality Q	Behavior of rock mass in tunneling
1000 ← 400	Exceptionally good
400 ← 100	Extremely good
100 ← 40	Very good
40 ← 10	Good
10 ← 4	Fair
4 ← 1	Poor
1 ← 0.1	Very poor
0.1 ← 0.01	Extremely poor
0.01 ← 0.001	Exceptional poor

Instability modes[16]

The rock mass behavior or possibility of instability modes for an underground excavation depends the magnitude of stress relative to rock-mass strength and structure (i.e. degree of jointing and joint persistence). Nine distinctly different behavior modes are illustrated in fig. 3.11; ranging from elastic behavior to essentially intact rock in a low-stress environment to plastic behavior of highly jointed rock mass under high stresses.

3.6.3 Rock structure rating (RSR)

This concept was established by *Wickham* et al. (1972)[5,6] and the parameters considered have been grouped in two sections: Geologic and construction. The geologic parameters include:

1. Rock mass
2. Joint pattern
3. Dip and strike
4. Discontinuities
5. Faults, shears and folds
6. Ground water
7. Rock material properties
8. Weathering or alteration.

The construction parameters include:

1. Direction of drive
2. Size of tunnel
3. Method of excavation.

These authors have presented two rating systems as RSR no. 1 and RSR no. 2. RSR no. 1 establishes rating based the nine geologic parameters mentioned above and is more specific to the historic and geologic information, which is available prior to construction. RSR no. 2 is a general approach, which considers the following, three parameters: A, B and C.

- Parameter A: general rock structure appraisal relating to rock type competency or degree of folding of the rock mass. The range of values is 8 to 30.

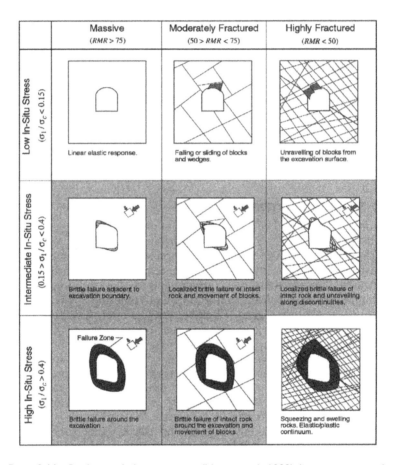

Figure 3.11 Rock mass behavior matrix (Martin et al., 1999) (www.smenet.org).

- Parameter B: related joint pattern to the drivage direction, range of values 12 to 50.
- Parameter C: general evaluation of ground water as well as sum of A and B. The range of values is 5 to 20.

The tables given by these authors enable to calculate values of A, B and C.

While applying these ratings to the ground to be evaluated, no matter which theory is chosen but the very concept of rock mass classification enables the designer to gain a better understanding of the influence of geologic and other parameters. This leads to better engineering judgment and better communication in the matter and ultimately leads to cost saving and better execution of the project. Application of these theories is made to select tunnel supports (Chapters 8, 9).

Rock mass characteristics are of significance importance in designing many structures in the following manner:[5]

- Modulus of elasticity: essential for the design of tunnels, chambers mines and dam foundations.

- Compressive strength: important in design of mine pillars
- Shear strength: important in rock slopes, foundation and dam abutments
- Tensile strength: important in mine roofs
- Frictional properties (cohesion and friction angle): important in fractured masses, yield zones, residual strength and rock bolt design
- Post failure modulus: important in longwall mining and pillar design
- Bearing capacity: important for mine floor and foundations
- Thermo-mechanical response: important in nuclear waste disposal.

Rock stresses are measured using any of the following three techniques:[13]

- The overcoring technique
- Flatjack testing
- Hydraulic fracturing.

The first two are normally carried out in underground openings (although triaxial overcoring in a few cases have been carried out from the surface in 10–20 m deep drill-holes), and thus at the pre-construction investigation stage, are restricted mainly to exploratory audits. As a result of the considerable developments in methodology during the last decade, the hydraulic fracturing technique is being applied currently in drill-holes of 100–200 meters depths and more.

3.7 GEOLOGICAL AND GEOTECHNICAL FACTORS[16]

Table 3.6 displays ranking of geological and geotechnical factors influencing key issues. Factors having critical or major influence should be given due importance.

3.8 THE WAY FORWARD

Do you know?[18]
- Gold is used in the **electronics industry** to make more than 10 billion tiny electrical contacts every year.
- Gold has been used as a **currency** for over 5000 years.
- The **largest gold nugget ever** discovered weighed 70.92 kilograms (approx. 160 lbs) and was found in Victoria, Australia.
- Gold has **medicinal** and **healing** properties – it can be used in treating rheumatoid arthritis, chronic ulcers and tuberculosis.
- Gold can be beaten wafer thin – technically that's 0.00001 mm thick.
- One ounce of gold can be beaten into 16 square meters of **gold leaf**.
- One ounce can be **drawn** into eight kilometers (five miles) of gold **wire**.
- Related to South Africa:
  - The **deepest mine in the world** is Western Deep Levels gold mine on the Far West Rand, now approaching a depth of 4 kilometers (approx. 13,000 feet). Mining does not as yet take place at that depth.
  - On average, gold in South African mines, is found in only **5.1 parts per million** from rock extracted at **depths** of up to 3.5 kilometers (approx. 2 miles) below the surface.

Table 3.6   Ranking Geological and Geotechnical factors[17] influencing key issues. 1 – Critical Influence; 2 – Major Influence; 3 – Minor Influence (www.smenet.org).

GEOLOGICAL/ GEOTECHNICAL FACTORS	Planning			Design				Construction						
	Routing	Alignment	Portal Site	Tunnel Size	Excavation Stability	Tunnel Shape	Lining	Tunneling Method	Rate of Advance	Material Handling	Water Inflow	Water Pressure	Lining	Blasting
Stratigraphic/Structural														
Stratigraphic Sequence	2	2	2	3	1	1	1	1	1	2	2	3	3	1
Lithology	2	1	2	3	1	1	1	1	1	1	1	1	1	1
Folding	2	2	2	3	2	3	3	1	2	3	2	3	3	2
Faulting	1	1	1	2	1	2	1	1	1	2	1	1	1	1
Tectonic														
Seismicity	3	3	3	3	2	3	3	3	3	3	3	3	3	3
Crustal Instability	3	3	3	3	2	3	3	3	3	3	3	3	3	3
Capable Faulting	1	1	1	3	1	3	1	1	2	3	3	3	3	3
In situ Stresses	3	3	3	1	1	1	1	1	1	3	3	1	1	2
Volcanic Activity	2	2	2	3	2	3	3	3	3	3	3	3	3	3
Mechanical														
Rock Mass Strength	3	3	3	1	1	1	1	1	1	1	3	3	1	1
Deformation Moult	3	3	3	2	1	1	1	3	3	3	3	3	1	2
Discontinuities	3	3	3	2	1	1	1	2	2	2	1	1	1	1
Hydrological														
Groundwater Regime	2	2	2	3	2	3	2	1	1	2	1	1	1	2
Miscellaneous														
Mass wasting	1	1	1	3	2	3	3	3	3	3	3	3	3	3
Avalanches	1	1	1	3	2	3	3	3	3	3	3	3	3	3
Karatification	1	1	2	3	1	3	3	3	3	3	1	1	2	3
Gas	1	1	2	3	3	3	3	3	3	3	3	3	3	3

○ The total **volume of rock cut away** each year in South Africa's gold mines would make a railway tunnel 3500 kilometers (approx. 2200 miles) long reaching between London and Leningrad.

○ For every ton of rock mined, nearly 15 tons of **ventilation air** is pumped underground.

○ The **volume of water** pumped daily from the mines would fill 3 million domestic bath tubs to the brim.

○ **Cooling plants** on South African gold mines have a capacity equal to nearly 3.5 million domestic refrigerators.

○ Virgin **rock temperatures** higher than 52°C (126°F) have been recorded in South African gold mines.

○ The South African gold mining industry consumes enough **electricity** – over 23 million megawatt hours – to power a city with 3 million inhabitants.

QUESTIONS

1. Core drilling is among the routine methods for underground exploration. List the common sizes (diameters) of cores recovered.
2. Define the following terms: mineral, mineral deposit, ore, exploration, prospecting, anomaly, geophysics.
3. Detail the activities and methods employed during prospecting and exploration of a base metal deposit such as copper.
4. Draw a flow-chart of exploration techniques and their applications.
5. Give a line diagram showing the steps and processes that are involved in a mineral exploration. How can different deposits be explored by putting the exploratory heading (i.e. drives/crosscuts)? Illustrate them.
6. Give a line diagram to describe the 'prospecting drilling methods'.
7. In tunneling projects, what is the prime purpose of carrying out the prospecting drilling?
8. List some geophysical prospecting methods. What benefits can we get from their use? Mention how cores can be recovered during the prospecting drilling using a diamond drill. What instruments are used for gravity surveys and also for magnetic surveys?
9. List major engineering rock mass classification systems, mentioning the name of originator in each case; also specify their field of application.
10. List site investigation phases for civil construction (or any excavation) projects including tunnels and caverns. Tabulate activities that are usually covered/ undertaken during these phases.
11. List the methods that can be applied for prospecting drilling. Give a suitable sketch of a diamond drill rig.
12. List the tasks involved in compilation of prospecting and exploration activities during a tunneling project to assess the ground conditions.
13. Does mineral wealth belong to every one of us, no matter where it is located? And who owns it? How could you ensure its proper exploration, systematic development and exploitation and judicious utilization?
14. Name three stages of prospecting and briefly describe each one of them.
15. Tabulate geological and geotechnical factors influencing key issues which you would consider during planning, design and construction phases of a tunneling project.
16. Tabulate geophysical exploration techniques and their applications.
17. Tabulate ground and rocks classification based on their strength.
18. What are geo-chemical anomalies, how they are helpful in mineral prospecting?
19. What are the various stages of prospecting, briefly list them. List geophysical prospecting methods. Briefly describe the utility of each one giving some examples.

20. Which methods of geophysical prospecting can be applied from the air? Where will nuclear seismic methods be suitable? With respect to prospecting when can geophysical surveys be carried out? What is the utility of these geophysical methods? While undertaking core drilling, is it possible to compile data on: seismic refraction, RQD and Lugeon values?

REFERENCES

1. Agoshkov, M.; Borisov, S. and Boyarsky, V.: *Mining of Ores and Non-metallic Minerals.* Mir Publishers, Moscow, 1988, pp. 14
2. *Atlas Copco Manual*, 4th edition, 1982, pp. 229–35, 292–302.
3. Bailly, P.A. and Still, A.R.: Exploration for mineral deposits. In: Cummins & Given (eds.): *SME Mining Engineering Handbook.* AIME, New York, 1973, pp. 5: 2–9.
4. Barton, N.; Lien, R. and Lunde, J.: Engineering classification of rock masses for design of tunnel support. *Rock mechanics*, vol. 6, no. 4, 1974, pp. 183–236.
5. Bieniawski, Z.T.: Ground control. In: *SME Mining Engineering Handbook*, Hartman (edt.). SMME, Colorado, 1992, pp. 897–911.
6. Bieniawski, Z.T.: *Rock mechanics design in mining and tunneling.* A.A. Balkema, 1984, pp. 97–132.
7. Boky, B.: *Mining.* Mir Publishers, Moscow, 1967, pp. 11–27.
8. Bruce, Charles.: Orebody evaluation. In: *Underground Mining Methods Handbook*, W.A. Hustrulid (edt.), SME-AIME, New York, 1982, pp. 3–4, 40–43.
9. Coats, D.F.: Classification of rocks for rock mechanics. *International journal of rock mechanics & mining science*, vol. 1, 1964, pp. 421–29.
10. Deere, D.U.: Technical description of rock cores for engineering purposes. *Rock mechanics and engineering geology*, vol. 1, no. 1, 1964, pp. 17–22.
11. Hartman, H.L.: *Introductory Mining Engineering.* John Wiley & Sons, New York, 1987, pp. 40–61.
12. Matti, H.: *Rock Excavation Handbook.* Sandvik – Tamrock, 1999, pp. 20–52.
13. Nilsen, Bjorn and Ozdemir, Levent.: Recent development in site investigations and testing for hard TBM projects. *In: Proc. RETC*, AIME, 1999, pp. 715–31.
14. Rogers, G.R.: Geophysical surveys. In: Cummins & Given (eds.): *SME Mining Engineering Handbook.* AIME, New York, 1973, pp. 5: 2–5, 24–34.
15. Whittaker, B.N. and Frith, R.C.: *Tunnelling – Design, Stability and Construction.* IMM Publication; 1990, pp. 19–35.
16. Kaiser and Tannant, D.D.: The role of shotcrete in hard rock mines. In: Hustrulid and Bullock (eds.): *SME Underground Mining Methods.* Colorado, 2001, pp. 583 (www.smenet.org).
17. Railing, G.E.: Methods of objective ground assessment. *Proc. RETC*, AIME, 1983, pp. 76–79 (www.smenet.org).
18. http://www.bullion.org.za

Chapter 4

Drilling

Drilling or boring is a prime operation in excavation technology without which exploration, development, exploitation and liquidation of mineral deposits could not succeed.

4.1 INTRODUCTION – UNIT OPERATIONS

The unit operations are referred to as the basic operations that need to be carried out to dig out or excavate ground and dispose off the spoil so generated to a particular destination during mining, and tunneling operations. These operations are mandatory during any phase of mine-life, i.e. development, exploitation (or stoping) and liquidation, in order to mine out a deposit by the application of any of the mining methods in practice. These basic operations can be grouped into two classes: dislodging rock from the rock massif or deposit which is known as *primary breaking*, and *handling of material* so generated. In any production cycle these unit operations need to be carried out but apart from these operations some ancillary activities are also carried out, and these are referred as the auxiliary operations.

Production cycle during Tunneling (or mine development) = *shot hole drilling + blasting + mucking + hauling + hoisting (optional)*

Similarly, the Production cycle during stoping operations in mines, or large excavations in civil and construction projects = *blasthole drilling + blasting + mucking + hauling + hoisting (optional)*

The term cycle implies that the mining and tunneling operations are cyclic or repetitive in nature but efforts are being made to make the process continuous where the mineral/ rock/ground after breaking moves without interruption. Truly speaking this should be the ultimate aim but we are far away from such a system. A number of techniques can fragment the rock but the prominent amongst them is drilling and blasting. In the following paragraphs details of one of the unit operations, DRILLING' are discussed and the rest of the unit operations have been dealt in the next few chapters.

4.2 PRIMARY ROCK BREAKING

Detaching the large rock mass from its parent deposit is known as rock breakage. Since prehistoric time man devised several ways to achieve this task and he made the greatest technological advance in mining history when eventually he discovered

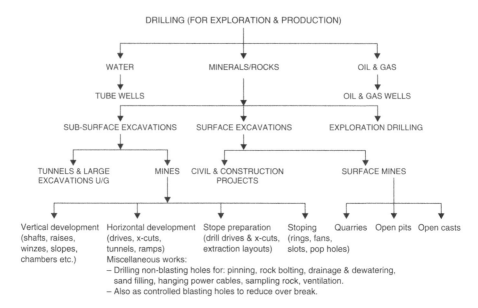

Figure 4.1   Use of drilling for various purposes.

explosive and used it for rock breaking purposes. Application of explosive in the rock is carried out by means of drilling holes, which are known as *shot holes, blastholes or big blastholes* depending upon their length and diameter. Holes of small diameter (32–45 mm) and short length (upto 3 m) are termed as *shot holes*, and they are drilled during tunneling and drivage work in mines. The *blastholes* are longer (exceeding 3 m to 40 m or so) and larger in diameter (exceeding 45 mm to 75 mm or so), and that are drilled as cut-holes in tunnels and drives, and in the stopes. Besides their use in surface mines, recently use of very large diameter (exceeding 75 mm) long holes, known as *big blastholes*, have begun for the raising and stoping operations in underground mines too. This terminology to designate production drilling will be used throughout this book. Thus, to dislodge the rock from its rock-massif use of suitable drills, explosives and blasting techniques, is made. This task can also be accomplished without using explosives, which will be dealt in the following chapters. This chapter deals with the drills and their utilities in detail.

4.3   DRILLING

Drilling (with a few exceptions such as: exploration, to provide drainage, in fixing rock bolts, in stabilizing slopes and to test foundations), is employed in mining and tunneling for placement of explosives. Figure 4.1, illustrates the application of this operation.

4.4   OPERATING COMPONENTS OF THE DRILLING SYSTEM

There are four main functional components of a drilling system, working in the following manner to attack the rock as illustrated in figure 4.2(d).[2]

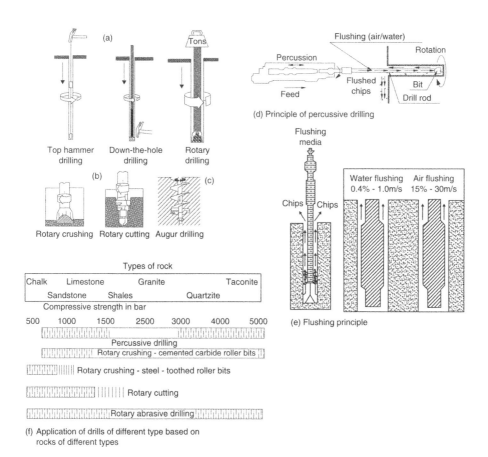

Figure 4.2   Drilling principles, mechanism and flushing systems. Functional components of a drilling system also shown. (Courtesy: Atlas Copco)[1]

1. *The drill*: it acts as prime mover converting the original form of energy that could be fluid, pneumatic or electric into the mechanical energy to actuate the system.
2. *The rod (or drill steel, stem or pipe)*: it transmits the energy from prime mover to the bit or applicator.
3. *The bit*: it is the applicator of energy attacking the rock mechanically to achieve penetration.
4. *The circulation fluid*: it cleans the hole, cools the bit, and at times stabilizes the hole. It supports the penetration through removal of cuttings. Air, water or sometimes mud can be used for this purpose. It flushes the cuttings as per the principle illustrated in figure 4.2(e).[1] Figure 4.2(e) also shows the flushing velocities.

4.5   MECHANICS OF ROCK PENETRATION

Using the drills the rock is attacked mechanically, as shown in figure 4.2(a),[1] either by percussive or rotary actions. Combinations (roller bit, rotary-percussion) of these two

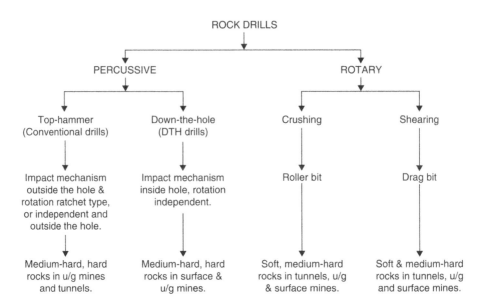

Figure 4.3   Rock drills classification based on application of mechanical energy to rock.

methods are also used. The resulting action of the bit in each case is almost similar, i.e. crushing and chipping; what differs mainly is that the crushing action predominates in percussion drilling and chipping action in the rotary drilling, and a hybrid action in the combination of the two systems. Based on this logic the drills are manufactured *as percussive*, *rotary-percussive* and *rotary*. A classification based on this logic is presented in figure 4.3.

4.5.1   Top-hammer drilling

In this system the top-hammer's piston hits the shank adapter and creates a shock wave, which is transmitted through the drill string to the bit (fig. 4.2(a)). The energy is discharged against the bottom of the hole and the surface of the rock is crushed into drill cuttings. These cuttings are in turn transported up the hole by means of flushing air that is supplied through the flushing hole in the drill string. As the drill is rotated the whole bottom area is worked upon. The rock drill and drill string are arranged on a feeding device. The feed force keeps the drill constantly in contact with the rock surface in order to utilize the impact power to the maximum.

In good drilling conditions use of these drills, is an obvious choice due to low energy consumption and investments on drill-strings.[1] In surface mines and civil construction sites 76–127 mm (3″–5″) hole diameters is the usual range.[1]

4.5.2   Down-the-hole (DTH) drilling

In this system the down-the-hole hammer and its impact mechanism operate down the hole. The piston strikes directly on the bit, and no energy is lost through joints in the drill string (fig. 4.2(a)). The drill tubes (rods, steels) convey compressed air to the impact mechanism and transmit rotation torque and feed force. The exhaust air blows the holes and cleans it and carries the cuttings up the hole. The drills, which are known by the various trade names such as 'down-the-hole drill', 'in-the-hole-drill' have been, referred here as DTH drills.

DTH drills differ from the conventional drills by virtue of placement of the drill in the drill string. The DTH drill follows immediately behind the bit into the hole, rather than remaining on the feed as with the ordinary drifters and jackhammers. Thus, no energy is dissipated through the steel or couplings, and the penetration rate is nearly constant, regardless of the depth of the hole. Since the drill must operate on compressed air and tolerates only small amounts of water, cuttings are flushed either by air with water-mist injection, or by standard mine air with a dust collector.

This is a very simple method for the operators for deep and straight hole drilling.[1] In surface mines 85–165 mm (3.4″–6.5″) hole diameters is the usual range.[1]

4.5.3   Rotary drilling

Rotary crushing is a drilling method, which was originally used for drilling oil wells, but it is nowadays also employed for the blast hole drilling in large open pits and hard species of rocks. It is used for a rock having the compressive strength up to 5000 bar (72,500 psi). In rotary drilling energy is transmitted via the drill rod, which rotates at the same time as the drill bit is forced down by high feed force (fig. 4.2(a)). All rotary drilling requires high feed pressure and slow rotation. The relationship between these two parameters varies with the type of rock. In soft formations low pressure and higher rotation rate and vice versa, are the logics usually followed. In general, if the rock hardness is less than 4.0 on Moh's scale, the rotary drilling has established its advantages, except when the rock is abrasive. The rotary drills can be operated using either compressed air or electrical power.

When drilling is done by *rotary crushing* method, the energy is transmitted to the drill via a pipe which is rotated, and presses the bit against the rock (fig. 4.2(b)). The cemented carbide buttons press the rock and break off the chips, in principle in the same manner, as percussive drilling.

When drilling is done *by rotary cutting* method the energy is transmitted to the insert via a drill tube, which is rotated and presses the inserts against the rock. The edge of insert then generates a pressure on the rock and cracks off the chips (fig. 4.2(b)).

Thus, rotary drilling is unbeatable in difficult drilling conditions, as it gives high productivity and good penetration rates in such conditions.[1] In surface mines and civil construction sites 90–165 mm (3.5″–6.5″) hole diameters is the usual range.[1]

4.5.4   Augur drill

The augur drill (fig. 4.2(c)) is the simplest type of rotary drill in which a hollow-stem auger is rotated into the ground without mud or flushing. The continuous-flight augurs convey the cuttings continuously to the surface. This also works on the rotary cutting principle.

4.5.5 Rotary abrasive drilling

(This has been dealt with in chapter 3) Figure 4.2(f) provides a guideline for the application of various types of rock drills, working on the different principles, for rocks of different compressive strength.

4.6 ROCK DRILL CLASSIFICATION[1,3,4,5,7]

In order to meet the variety of conditions encountered in rock drilling several distinct types of drills have been developed with the passage of time as illustrated in figures 4.5 and 4.6. In general, rock drills may be classified as either hand held or mounted, as illustrated by a line diagram shown in figure 4.4. The hand held drills include an electric drill (fig. 4.5(a)), jackhammer (fig. 4.5(b)), jackdrills or jacklegs (fig. 4.5(c)) and stoper (fig. 4.5(d)). The mounted drills are known as 'drifters' (fig. 4.6(a)). Table 4.1,[4] shows that each type of pneumatic drill is available in several sizes from different manufacturers and the range of parameters within which they can operate.

The jackhammers or sinkers are used for general mine utility (services such as: pinholes, anchor holes, pop holes), and shaft or winze sinking purposes. They can be classified according to their weight as light, medium or heavy duty. The weight ranges from 7 to 30 kg.

The rock drill jackleg is made by clamping a pusher leg to jack hammer to support the weight of the machine and to feed the tool forward in horizontal or upward direction (fig. 4.5(c)). These are generally classified as per cylinder bore size as medium or heavy duty. The bore size ranges from 60 to 83 mm. Due to their lightweight and versatility these are very effective in small sized drifting, tunneling or heading, and stoping operations.

Hand held and pusher leg mounted drills, even today, remains a widely accepted choice in most of the drivage work. But so far as rate of drilling and precision is concerned this technique has got limitations as in this case the feed thrust has to be balanced by the man's weight and strength. Pusher leg drilling is an arduous as well as a skilled task, and rock drill performance has to be matched with physique and skill of the operator.

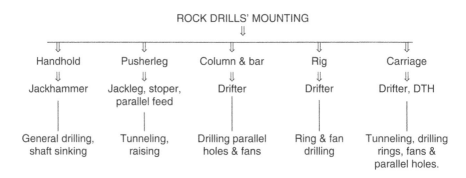

Figure 4.4 Rock drills classification based on their mountings.

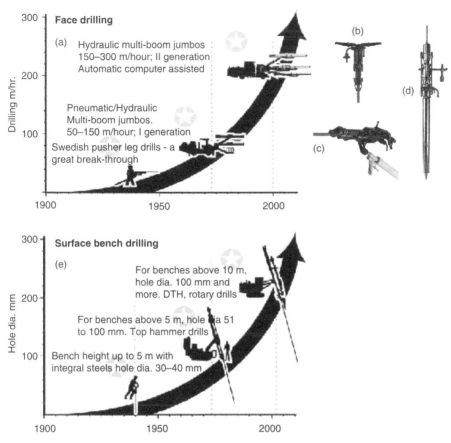

Figure 4.5 Historical review of rock drilling technology. Top: Development of rock drills jackleg to fully automatic multi-boom hydraulic jumbos. Bottom: Sinkers to DTH, Rotary and Top-hammer drills. Right top: Sinker/jack hammer; bottom; jackleg drill; right-most: Stoper for drilling in upward direction. (Courtesy: Atlas Copco)[1]

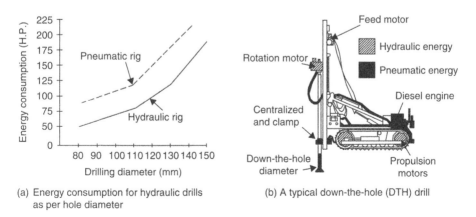

(a) Energy consumption for hydraulic drills as per hole diameter

(b) A typical down-the-hole (DTH) drill

Figure 4.6 Comparison of hydraulic and pneumatic power.

Table 4.1   Rock drills' specifications, in general.[4]

Drill	Class	Weight or cylinder bore diameter	Air consump. Cfm#	Impact H.P	Hole diameter range, mm	Hole depth range, m	Use
J/H or Sinker	V. light	>18 kg	50–70	2	19–32	0.3–0.6	Maintenance
	Light	18–25 kg	70–100	2–2.5	32–38	0.6–1.2	General utility
	Medium	25–30 kg	100–120	2.5–3	35–41	1.2–2.4	General utility
	Heavy	over 30 kg	120–140	Over 3	38–44	1.2–3.7	Shaft sinking
Jack drill or jackleg	Medium	60–66 mm	150–160	3–4	32–41	1.2–3.7	Drifting, stoping,
	Heavy	68–83 mm	180–210	4–5	38–44	1.2–3.7	Drifting, stoping
Stoper	Light	>34 kg	150–170	3–3.5	32–38	1.2–3.7	Raising
	Medium	34–45 kg	170–190	3.5–4	35–41	1.2–3.7	Raising, stoping
	Heavy	over 45 kg	190–210	4–5	38–44	1.2–3.7	Raising, stoping
Drifter	Light	83–102 mm	200–300*	>6	38–44	1.2–2.4	Small tunnels
	Medium	102–114 mm	300–400*	6–10	41–51	2.4–3.7	Medium tunnels
	Heavy	Over 114 mm	400–500*	10	44–57	3.7–30.5	Large tunnels and longholing

\# – Air consumption for riffle-bar drills with water flushing. The water pressure should be equal to or slightly less than air pressure. The air pressure for all these drills is 620–690 kPa (90–100 psi).

* – add 0.071 cum./sec (150 cft) for independent rotation.

The stoper is a jackhammer, which is rigidly attached to a pneumatic cylinder for drilling holes in the upward direction. The cylinder or leg may be in line with the drill, or it may be offset from the drill line, to provide a short overall length for drilling in low height workings. The stoper is often designed with a jackdrill as its base, and therefore, it is available in the same bore size range as the jackdrills. Its weight ranges from 34 kg to more than 45 kg. It is used for raising and stoping operations for drilling either vertical or steeply inclined up holes.

In case of conventional drills apart from the limitations listed above, in a specific situation where rapid drilling cycles are required, particularly in small headings the problem of over crowding these machines arises, and practically a limit comes when both the men and the machines are unable to function effectively. Looking into these problems further mechanization in drilling operations has been brought about by the introduction of drilling jumbos as shown in figure 4.5.

These jumbos usually consist of high performance rock drills called drifters (fig. 4.5), mounted on a feed system, which is supported by a boom. The feed alignment to ensure drilling of parallel holes is based upon a moving parallelogram mechanism with the links either mechanically or hydraulically operated or positioned.

All the controls are lever or valve operated. In order to ensure several drills to be operated by one operator, mechanisms are available to stop the drill when the hole is completed and to return to its original position, completely automatic. All that an operator has to do is to reposition the boom for its next hole and start drilling.

Standard self propelled rigs fitted with single or multi boom are available for their use in drivage work, tunnels and slopes with their mountings on rail bogies, rubber tired or tracked vehicles, as shown in figure 4.5. The drifter rock-drill is too heavy and powerful to be supported by a man, so it is mounted on a hydraulic boom, a column mounting or a portable mounting having crawler or wheel chassis. Drifters are used for the drivage work particularly the horizontal development work such as drifting, cross-cutting and tunneling. They are classified as light, medium and heavy duty, as per the bore size, which ranges from 83 to more than 115 mm.

These jumbos are designed for the specific tasks related to drilling at the development headings, tunnels and stopes. Today these jumbos are available for various purposes as described in the following sections.

4.6.1   Tunneling/development drill jumbos[1,3,4,5,7]

The development mine entries such as levels, drifts, crosscuts and sublevels; and civil tunnels may have varying cross sectional areas and gradient, and to suit these varying conditions different types of drill jumbos have been developed. For example, a jumbo of about 2.1 m width is used in the mines with narrow drifts, tight turns and frequent crosscuts. This type of jumbo is commonly known as Mini-bore jumbo.

A jumbo[5] with 2.0 m overall width is used for sublevel driving but it is capable to drill the faces as large as 9 m wide $\times$ 4.5 m high.[5]

A jumbo having 2.4 m overall width is most common for its use in hard rock mining. It can also be used for rock bolting, room and pillar stoping operations. This jumbo can be used for drifts and tunnel configuration in the range of 3.7 m $\times$ 3 m to 9.8 m $\times$ 6.7 m.

A typical jumbo, includes an energy system, operator's station, chassis and boom equipped with a drifter or drill. Conceptual diagrams of hydraulic drifters have been shown in figure 4.5.

4.6.2   Shaft jumbos[5]

It is a compact unit designed to suit the narrow space available during shaft sinking operations. This jumbo usually consists of a column with a horizontal top platform to which the drill booms are attached. The vertical column acts as the air header and the top platform serve as storage space for the hydraulic power unit.

4.6.3   Ring drilling jumbos

To achieve longhole drilling with maximum speed, accuracy and safety at low cost there has been consistent improvement in the design of ring drills. Earlier column and bar mounted ring drills have been replaced by single ring drill or the double (twin) drill ring jumbos with either skid or pneumatic-tyred undercarriages. The feed mechanism is either screw, chain or cable type. The mechanism for rotation is usually independent from the one governing the percussion, so that feed can be regulated depending upon the rock conditions. The controls are separated from the machine and can be placed remotely. The ring drilling work consists of drilling the blastholes radially from a drill drive keeping the drill at a fixed position in a plane that may be vertical, horizontal or inclined.

4.6.4   Fan drilling jumbos[1,4,5,7]

In sublevel caving the drilling work is considerably high and this calls for deployment of highly productive drills. One to three boom jumbos are available for this purpose. A fan shaped drill hole pattern is drilled using these jumbos.

4.6.5   Wagon drill jumbos

The main consideration in cut and fill stoping operations while selecting drilling equipment is the firmness of the fill, width and height of the working area of the stope. The pneumatic tyred three wheels, or four-wheel chassis carriers are used to mount the drills for this purpose. Air motors propel these carriers. Such drills are known as wagon drills.

4.6.6   DTH drill jumbos[1,4,5,7]

Besides their use as non-blasting holes (to provide free face), the principal application of these drills is in the primary blasthole drilling. Prior to the advent of these drills, extensive development work was required in the stopes before the production drilling could be started. Sublevels were required to have the access to the fan or ring drills to the stopes; which in turn amounts to more development work. To utilize a DTH drill only top heading and draw points are necessary.

Various configurations of DTH drills are available. The basic energy source for this drill is compressed air but other functions are powered either by the compressed air or by air powered hydraulic power pack. The pneumatic rigs utilize several basic air motors and conventional feed systems and they are more familiar. The hydraulic systems are better with respect to the speed, force and accuracy, and becoming common for underground applications. The DTH drills are mounted either on crawler track figure 4.6 or rubber tyre vehicle and the tramming power is provided either by the pneumatic or hydraulic motors. Mostly, these drills are towed to the working spots by other vehicles. A spindle in the rotary head that is mounted on the feed rotates the drill rods and the drill. The torque is supplied either by pneumatic or hydraulic motors. The rotation speed is variable and it ranges from 0 to 50 rpm. A DTH drill consists of a replaceable shell or jacket, containing a piston that oscillates back and forth to strike directly on the shank end of the bit. Most DTH drills are without valve, using ports to control the movement of the piston. The exhaust air is ported through the bit, providing the flushing air that cleans the face and conveys the cuttings to the collar of the hole. Based on the required hole size, these drills are available in various sizes. The usual compressed air pressure to operate these drills is upto 250 psi (1725 kPa). The common sizes of the bits used with these drills are 102 to 165 mm diameters. The flat faced button bit is a very common bit that is mostly used but drop center, X and cross bits could also be used.

The type of drills (described above) used on jumbos include percussive drills, rotary drills, rotary percussive drill and auger drills. Applications of these drill jumbos during the stoping operations in underground metal mining operations have been illustrated in figure 16.33(b) and their selection is usually governed by the rock strength.

4.6.7 Roof bolting jumbos

These jumbos are meant for roof bolting. In some designs of multi boom jumbos one or two booms are exclusively meant for rock bolting so that along with the face drilling rock bolting of the immediate roof can be undertaken.

4.7 MOTIVE POWER OF ROCK DRILLS

In addition to the above-mentioned basis, there are several other ways to classify the rock drills. Depending upon the motive power they can be classified as pneumatic, electrical and hydraulic rock drills, as illustrated in figure 4.7.

4.7.1 Electric drills

These are used for the rotary drilling of the shot holes (fig. 4.5(a)). These can be hand or column mounted. Handhold drills are suitable for drilling in soft rocks having weight: 15–25 kg; motor rating 1–1.5 kw; and rotation speed 300–900 rpm. Column mounted drills can be used for the blast-hole drilling and for the core drilling, e.g. diamond drills for exploration and prospecting purposes.

4.7.2 Pneumatic drills

These are the most commonly used drills in metal and non-metal mines, and tunnels. These are low cost, simple in design and suitable for rough handling and use. These drills suffer from the disadvantage of low efficiency in terms of input compressed air power. Also these drills are noisy and their exhausts generate mist and fog. They are suitable for any degree of toughness of the rock.

4.7.3 Hydraulic drills

Introduction of hydraulic drills in underground mines and tunnels is recent. Presently more than dozen reputed manufacturers are in the arena giving considerably different designs. Initially rotary hydraulic drills came up and later on the rotary percussive drills.

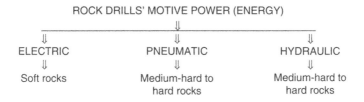

Figure 4.7 Rock drill classification based on their motive power.

The hydraulic drills operate by the intermittent application of high-pressure hydraulic oil to a double acting piston; the frequency of oil application is controlled by the movement of piston or by the action of a sliding or rotating valve or by a combination of both. Both may operate by the oil pressure, or by the piston or the valve. These drills have separate rotation motors giving adjustable rotation speeds maximum of 30 rpm. Some of the models are also fitted with reversible rotation mechanisms.

One to four boom jumbos are available for use in civil tunnels, and development and stoping operations in underground mines. The advantages claimed by their use are:

- 45% of the energy input in this case is converted into useful work as compared to 15% in the case of conventional drills.
- High penetration rates, longer bit and steel life.
- High drilling productivity compared with the pneumatic drills.
- Reduced noise level from 3–17 dB than the silenced pneumatic drills.
- Some of the models give adjustable piston stroke; thereby larger strokes with less frequency can be used in tough rocks and whereas a short stroke with high frequency can be chosen for the brittle rocks.

These drills require high capital investment, skilled operators and high degree of engineering and maintenance skills. Their use in small mines or tunnels with capital scarcity situations cannot be justified; but new mining ventures and tunnels aiming at high output could find them beneficial.

As per the mechanical action upon the hole bottom they can be classified as percussion, rotary and rotary-percussion types of hydraulic drills.

4.8   DRILLING ACCESSORIES[1,4,7]

As shown in figure 4.2(d) to drill a hole, apart from the rock drill some drilling accessories are required. These are integral drill steels, extension rods, shank adapters, sleeve couplings and bits. An integral drill steel as shown in figure 4.8 consists of a rod with a forged shank at one end and a forged bit with cemented carbide inserts at the other end. Thus, each steel is of a specific length and cannot be extended. Once the first drill steel has drilled all the way into the rock, it is withdrawn and replaced by the longer one to drill further into the rock. The integral drill steels are available in increasing lengths with reduced diameters, as shown in table 4.2. Thus the smallest drill section

Table 4.2   Integral drill steel series (size specifications).[7]

22 mm integral chisel bit hollow hexagonal drill steel			
Length, mm	Bit diameter, mm	Length, mm	Bit diameter, mm
800	34	800	36
1600	33	1600	35
2400	32	2400	34
3200	31	3200	33
4800	29		

has the largest diameter and its diameter is selected as per the size (diameter) of the explosive cartridge. The most common integral drill steel is chisel-type and other types include multiple-insert steels, button steels, double chisel steels and cross edge bit steels, as shown in figure 4.8.

4.8.1   Extension drill steels

Threaded rods can be joined together to form a string, which can be used to drill long holes (longer than the length of one rod). These rods have male threads and they are coupled to others with the use of couplings having internal female threads. There are two types of rods: (i) shank rods – these are the rods with an integral shank and a bit end. (ii) Full section extension rods – they can be round or hexagonal having threads at both ends. In long hole drilling first a shank adapter is inserted into the drill. This transmits the impact energy and rotation from the drill to drilling string. These accessories have been shown in figure 4.8.[1,4,7]

4.8.2   Bits[1,4,7]

It is a part of the drilling equipment that performs the crushing work. The part of the bit in contact with the rock is made of cemented carbide in the form of buttons or inserts. The threaded rod is normally screwed into the bit until it bottoms. The impact energy is then transmitted between the end of the rod and the thread bottom of the bit. The flushing medium is supplied through the flushing hole in the rod and is distributed through flushing holes in the center and/or at the sides of the bit front.

Button bits (fig. 4.8) have more wear resistant cemented carbide than insert bits. These bits are available from diameters of 35 mm and upwards. Insert bits are available in a wide variety of designs having diameters from 35 mm and upwards. Cross bits and X-bits are the most common insert bits (fig. 4.8). Cross-bits have an angle of 90° between the inserts, where as X bits have 75° and 105°. X bits are used for large dia. hole ⩾75 mm. The button bits allow regrinding interval 4–5 times longer than the insert bits. Insert bits are more resistant to heavy gauge wear. Retract bits are used for

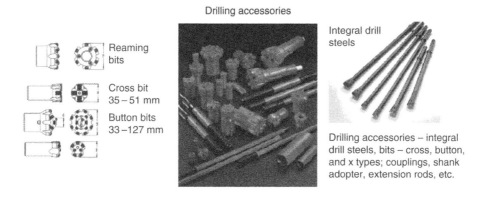

Drilling accessories

Reaming bits

Cross bit 35 – 51 mm

Button bits 33 –127 mm

Integral drill steels

Drilling accessories – integral drill steels, bits – cross, button, and x types; couplings, shank adopter, extension rods, etc.

Figure 4.8   Common drilling accessories used during development drifting, raising and tunneling operations (Courtesy: Atlas Copco; Sandvik – Tamrock).[1,7]

Table 4.3   Service life of drilling accessories during surface and underground operations (Courtesy: Atlas Copco)[1].

	Rock types			
	Civil works – bench blasting		Tunneling, drifting & underground operations	
Accessories	Abrasive	Slightly abrasive	Abrasive	Slightly abrasive
Integral drill steels				
Regrinding interval	20–25 m	150 m	20–25 m	150 m
Service life	150–200 m	600–800 m	200–300 m	700–800 m
Threaded insert bits				
Regrinding interval	20–25 m	150 m	20–25 m	150 m
Service life	200–400 m	800–1200 m	250–350 m	900–1200 m
Threaded button bits			250–550 m	1000–3000 m
dia. ⩾64 mm			(Service Life)	(Service Life)
Regrinding interval	60–100 m	300 m		
Service life	400–1000 m	1200–2500 m		
dia. <57 mm				
Regrinding interval	100–150 m	300 m		
Service life	300–600 m	900–1300 m		
DTH button bits				
Regrinding interval	40–60 m	300 m		
Service life	400–1000 m	1200–2500 m		
Extension rods				
Service life	600–1800 m			
– Pneumatic drills			1000–1500 m	
– Hydraulic drills			1600–2400 m	
Threaded integral steels				
Service life			600–800 m	
Coupling sleeves				
Service life	100% of rod service life		100% of rod service life	
Shank adaptors				
Service life				
Pneumatic rock drills	1500–2000 m		1200–1600 m	
Hydraulic rock drills	3000–4000 m		2500–3500 m	

drilling in a rock where the holes tend to cave behind the bit, making it difficult to withdraw the drilling equipment.

DTH bits are made with shanks to fit different drills. The normal size ranges 85–215 mm. Button, core crusher and full face, are the three designs available. The first one is the most common and the last one is suitable for drilling in loose rock for filling material.

When drilling is made by the rotary cutting method use is made of drag bits. Roller bits have been designed for their use during the rotary drilling. This type of bit consists of a bit-body with three movable conical rollers, known as tricone. Buttons are distributed over the three rollers in such a manner that the entire bottom of the hole works when the bit rotates. Different designs are available to suit different rocks.

In figure 4.8, the common drilling accessories used during development drifting, raising and tunneling works have been shown. For longhole drilling the common type of drilling accessory has been depicted in figure 4.8. With electric drills different accessories are needed.

4.8.3 Impact of rock-type on drilling performance[1,8]

It is important to know and understand the rock properties, listed below, to effectively utilize the rock-drills and their accessories. These parameters are related to rock types.[10]

- Abrasiveness describes the ability of rock fragments to wear away the drill bits. Rock composed essentially of quartz is very abrasive. Some of the rocks with high quartz contents are: diorite (10–20%), gneiss (15–50%), granite (20–35%), greywacke (10–25%), mica schist (15–35%), pegmatite (15–30%), phyllite (10–50%), quartzite (60–100%), sandstone (25–90%), slate (10–35%), shale (0–20%).
- Hardness – the resistance to penetration by a pointed tool.
- Rock texture – angular grains are more abrasive than round ones. Lenticular shape of grains, as in schist, poses more difficulty during drilling than round ones as in sandstone.
- Toughness – the resistance of the mass to the separation of pieces from it, in other words, the capacity to suffer considerable plastic deformation up to the moment of breakage.
- Elasticity or resilience – the resistance to impact that could be seen when a tool rebounds.
- Brittle rock (shale, limestone, sandstone, coal etc.) is fairly hard but comparatively easily crushed, and pieces separate from the mass along numerous cracks.
- Strong rock (strong sandstone, granite, magnetite etc.) has high resistance to penetration by a tool and to separation of pieces from the mass.
- Very strong rock (quartzite, diabase, and porphyry) has the highest resistance to penetration by a tool and to separation of pieces from the mass.
- Rocks that are more porous, have low crushing strength and are easier to drill.
- Polymineral rocks, such as granite, due to their heterogeneity are more abrasive.

4.9 SELECTION OF DRILL[4,6]

Drill selection: Drill selection for a particular application should be based on the technological and cost factors. It is considered that the lower cost is obtainable in soft rock with rotary drag-bit drilling, in medium and hard rock with rotary roller-bit and rotary-percussion drilling, and in very hard rock with percussion drilling. Use of percussive drills is very common in underground metalliferous mines and tunnels. The rotary drills are common in underground coalmines. In surface mines both types of drills have applications depending upon the rock types.

Drilling efficiency: Drilling efficiency can be measured by taking into consideration the following parameters:

- By the manner in which the drilling tool i.e. the drill acts upon the hole bottom (percussive, rotary or percussive rotary)
- The forces and the rate with which the drilling tools act upon the hole bottom
- Hole diameter and its depth
- The method and speed with which the drilling cuttings are removed from the hole.

These factors determine a type of drill required to suit a particular type of rock, as drillabilty of rocks differs widely. This factor can be determined in the following manner.

When using a percussive drill, the compressive forces prevail and shearing forces when rotary drilling. The magnitude of these forces w.r.t. drillability in a given rock is considered to be almost equal. Thus, the compressive strain σ_c and shearing strain σ_{sh} are of decisive importance. Since breaking of rock is possible only when the cuttings of rock are removed from bottom of the hole, the bulk density of rock, γ, must also be therefore accounted. The drillabity index can be assessed by the formula (4.1).[6]

$$I_d = 0.007 \, (\sigma_c + \sigma_{sh}) + 0.7 \, \gamma \tag{4.1}$$

Where: Compressive strain σ_c in kg/cm^2; Shear strain σ_{sh} in kg/cm^2; Bulk density γ in kg/dm^3.

If value of I_d is 1–5 it is easily drilled, if from 5.1 to 10 then medium drilled, difficultly drilled when I_d ranges from 10.1–15, extremely difficultly drilled if I_d ranges between 15.1 to 24.

4.10   SUMMARY – ROCK DRILL APPLICATIONS

Table 4.4   Brief description of equipment useful for sinking, raising, drifting, tunneling and excavations required for civil and construction industries.[1,8,10]

Equipment	Technical details	Field of application/utility
(1) Jack hammer, or sinker.	Available for very light to heavy duties. Weighing from less than 18 kg to over 30 kg. Capable of drilling holes of 19–44 mm dia. and lengths up to 3.7 m	For general utility and shaft sinking. General utility during civil and construction projects.
(2) Jack leg (jack hammer with pusher-leg).	More powerful than category (1). Available with cylinder bore of 60–83 mm. Capable of drilling holes of 32–44 mm dia. and lengths up to 3.7 m	Small sized drifting and tunneling projects.
(3) Stoper.	Available for very light to heavy duties. Weighing from 34 kg to over 45 kg. Capable of drilling holes of 32–44 mm dia. and lengths up to 3.7 m	Upward holes – vertical/inclined for raising operations.

(Continued)

Table 4.4   Continued.

Equipment	Technical details	Field of application/utility
(4) Drifter.	More powerful than category (1). Available with cylinder bore of 83 mm to 114 mm and more. Capable of drilling holes of 38–57 mm dia. and lengths up to 30 m, or even more.	Small to large sized drifting and tunneling projects. Drifter can drill holes in any direction (0–360°).
(5) Tunneling and drifting jumbos. Jumbos for sinking operations fitted with 'sinkers' are also available.	Drill jumbos with single- to multi-booms are available. The jumbos could be with rubber tires, tracked, or crawler mounted. Pneumatic as well as hydraulic jumbos are available. A multi boom tunneling jumbo could also be fitted to rock-bolting booms with stopers.	Tunneling and drifting jumbos for narrow to wide dimensions are available to suit a specific need. Sinking jumbos are used for mechanized drilling in shafts.
(6) Rotary percussive drills.	Light to heavy duty top hammer or DTH drills mounted on a rig. Capable of drilling holes in the range of 76–216 mm (3–8.5″) dia. and depth up to 25 m, and more.	Civil and construction projects: bench drilling, tieback anchoring, grouting, well drilling.
(7) Chipping hammers/pneumatic breakers.	They are handy as weight hardly exceeds 11 kg. Tools include Moil point, narrow chisel, wide chisel, wedge, digging blade, asphalt cutter, axe, spade, tampering stem, shank rod, roughing head, etc.	These pneumatic hammers and breakers are suitable for tough material such as frozen ground, asphalt, concrete, coal, clay, non-consolidated rocks and a few others. They could be also used for light digging jobs. They are the most suited demolition tools.
(8) Rig mounted hydraulic breakers. These breakers are attached to the different excavators and carriers.	They are suitable for the tasks that could not be done by pneumatic breakers and chipping hammers, as described under item (7).	The rig mounted hydraulic breakers are the efficient tools for many operations at surface as well as sub-surface (underground) locales. The tasks include: demolition, scaling, breaking soft rocks at roads and bridges, in process industry, splitting boulders.
(9) Motor drills and breakers.	These units are fitted with driving motor and mechanisms of drilling and allied tasks.	Civil and construction projects during repair and maintenance at dams, bridges, roads, railway tracks. For digging hard and frozen ground, tamping of backfills and splitting exposed rocks. Holes for putting signposts, pipes, earth rods, earth anchors of different lengths and diameters.

4.11 Drilling postures

In order to obtain tunnels or mine openings of different size, shapes, orientation and slopes/gradients, the holes are drilled in different directions. In the horizontal drivage work, holes drilled are almost horizontal or slightly inclined slightly upward or downward to it. This is known as breasting. During stoping operations drilling is carried out in this posture for slashing the ground and driving rooms and galleries. Pusher-leg mounted jackhammers or drifting jumbos fitted with drifters are used for drilling in this posture.

To drive an opening upward, such as raises, or during vertical development in the upward direction; vertical or steeply inclined holes are drilled upward, this is known as over-hand drilling. This posture is also carried out during some of the stoping operations such as cut & fill, shrinkage etc. Drills such as stopers or drifters are used for this purpose.

In order to drive an opening in the downward direction such as shafts, winzes etc. vertical or steeply inclined holes are drilled in the downward direction. This is termed as under-hand drilling. Drilling in this manner is also undertaken some of the stoping operations such as DTH, VCR etc. Drills such as sinkers, DTH machines and drifters are used.

Apart from drilling in horizontal, vertically up or down directions, during stoping operations drilling in all directions may be required. This is achieved by the use of fan and ring drilling jumbos, specially designed for this purpose. Suitable drifters are mounted on these jumbos.

4.12 THE WAY FORWARD

Improved fragmentation is a better way to extract minerals (ore, waste rocks, and over-burden) to save energy. The concept has already been detailed in sec. 19.5.7 and illustrated in figure 19.14. Several mining companies are actively developing fully-autonomous surface drills having advantages such as: accuracy, consistency, speed, improved reliability, less maintenance and also ability to measure rock properties in real-time and thereby achieving improved blasting and fragmentation results. It has potential for many applications in the future.

QUESTIONS

1. A drill jumbo is to be used for driving a tunnel. It has got three booms on which drifters have been mounted. This jumbo drilling is to be used at the face. If the rate of drilling is 0.33 meters/minute, the number of holes to be drilled are 53 and the size of the round is 3 m. Calculate how much time it will take to complete the drilling operation at the face.
2. Are mining and tunneling operations cyclic? What are the advantages of making mining and tunneling operations continuous, rather than cyclic? Should we strive for such a system?

3. Classify rock drills based on the application of mechanical energy to rock. Mention a field of application for each one of them.

4. Classify rock drills based on their working principles and illustrate by simple sketches. Illustrate a drilling mechanism and show its different components. Give an application of different types of rock drills known to you for the type of drivage work. Name different types of mountings and motive power to run these drills.

5. Classify rock drills giving their applications (i.e. type of mines where they could be deployed).

6. Define 'unit operations'. List those for carrying out development and also to obtain production from the mines.

7. Describe hydraulic drills. List their merits and limitations.

8. Give a line diagram to describe the use of drilling for various purposes.

9. Give or illustrate a historical review of rock drilling technology.

10. How, in practice, are rotary crushing and rotary cutting achieved?

11. Illustrate how a flushing during drilling operation could be achieved?

12. List rock properties which influence rock drill performance. Is it sensible to take care of these properties while selecting the rock-drills and their accessories?

13. List common drilling accessories used during development drifting, raising and tunneling operations.

14. List components of an integral drill steel.

15. List drill jumbos which are available for use in mines.

16. List drilling postures known to you and specifying a field of application for each one of them.

17. List rock drilling methods. Under what conditions will each be most suitable?

18. List rock drill motive power types.

19. Mention the role of operations: drilling or boring in excavation technology. Could we succeed in carrying the excavation operations without them?

20. Mention the types of jumbo drills that are available, specifying the mining tasks for which each one of them could be used.

21. Name the type of bits used with extension drill steels.

22. There are four main functional components of a drilling system, list them and mention the function of each one of them.

23. What considerations you would make when selecting a drill? How would you measure their drilling efficiency?

24. What is a drill jumbo? What are its components? Give their utility and describe them briefly: roof bolting jumbos; DTH drill jumbos; wagon drill jumbos; fan drilling jumbos; ring drilling jumbos; shaft jumbos; drifting and tunneling jumbos.

25. What is the role of explosives in drilling? Define and differentiate amongst: shot hole, blasthole or big blastholes based on the nomenclature which is usually followed.

26. Where will you find hydraulic drills used and how are they superior to conventional drills?

27. Where would you deploy: stopper, jackhammer with air-leg, handheld electric drill, multi-boom jumbo drilling parallel holes, shaft jumbo, motor drills and breakers, rig mounted hydraulic breakers, chipping hammers/pneumatic breakers.

28. Differentiate between the following:
   a. A jack hammer drill and a stoper drill.

b. A jack hammer drill and a drifter drill.

c. A drifter drill and a DTH drill.

29. Propose the most suitable drills for the following operations:

a. Drilling a round hole in a vertical raise.

b. For exploration of a mineral deposit when the core is required to be recovered.

c. For mining limestone deposits by open pit mining.

REFERENCES

1. *Atlas Copco manual*, 4th edition, 1982 and *Atlas Copco Leaflets*.
2. Hartman, H.L.: *Introductory Mining Engineering*. John Wiley & Sons, New York, 1987, pp. 117–18.
3. Henry, H.R.: Percussive Drill Jumbos. In: W.A. Hustrulid (ed.): *Underground Mining Method Handbook*. SME (AIMMPE), New York, 1982, pp. 1034–40.
4. Jimeno, C.L.; Jimenó, E.L. and Carcedo, F.J.A.: *Drilling and Blasting of Rocks*. A.A. Balkema, Netherlands, 1997, pp. 8–47.
5. Kurt, E.H.: Primary Breaking. In: W.A. Hustrulid (ed.): *Underground Mining Method Handbook*. SME (AIMMPE), New York, 1982, pp. 999–1005.
6. Rzhevsky, V.V.: *Opencast Mining Unit Operations*. Mir Publishers, Moscow, 1985, pp. 117.
7. Sandvik – Tamock: Drills and drilling accessories – *Leaflets and literature*.
8. Henry, H.R.: Percussive Drill Jumbos. In: *Underground Mining Method Handbook*, W.A. Hustrulid (ed.), SME (AIMMPE), New York, 1982, pp. 1034–40.
9. Jimeno, C.L.; Jimenó, E.L. and Carcedo, F.J.A.: *Drilling and Blasting of Rocks*. A.A. Balkema, Netherlands, 1997, pp. 24–26, 130.
10. Matti, H.: *Rock Excavation Handbook*. Sandvik – Tamrock, 1999, pp. 101, 118–21.

Chapter 5

Explosives and blasting

Unsafe acts and unsafe conditions cause accidents. Inculcating safe habits through training and education could bring accident rate near 0%.

5.1 INTRODUCTION – EXPLOSIVES

An explosive is a substance or mixture of substances, which with the application of a suitable stimulus, such as shock, impact, heat, friction, ignition, spark etc., undergoes an instantaneous chemical transformation into an enormous volume of gases having high temperature, heat energy and pressure. This, in turn, causes disturbance in the surroundings that may be solid, liquid, gas or their combination. The disturbance in the air causes air blast and this is heard as a loud bang. The disturbance in a solid structure results in its shattering and demolition. During wartime this property is utilized for destruction purposes but the same is used for dislodging, breaking or fragmentation of the rocks for quarrying, mining, tunneling, or excavation works in our day-to-day life. The energy released by an explosive does the following operations:

- Rock fragmentation
- Rock displacement
- Seismic vibrations
- Air blast (heard as loud bang).

5.2 DETONATION AND DEFLAGRATION[6]

As stated above, when an explosive is initiated, it undergoes chemical decomposition. This decomposition is a self-propagating exothermic reaction, which is known as an explosion. The gases of this explosion with an elevated temperature are compressed at a high pressure. This sudden rise in temperature and pressure from ambient conditions results into a shock or detonation waves traveling through the unreacted explosive charge. Thus, *detonation* (fig. 5.1(a))[6] is the process of propagation of the shock waves through an explosive charge. The velocity of detonation is in the range of 1500 to 9000 m/sec. well above the speed of sound. *Deflagration* (fig. 5.1(a)) is the process of burning with extremely rapid rate the explosive's ingredients, but this rate or speed of burning, is well below the speed of sound.

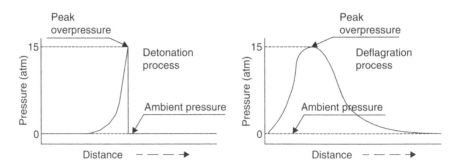

Figure 5.1(a)   Conceptual diagrams – Detonation and deflagration phenomenon.

Table 5.1   Common ingredients of explosives.[7,11]

Items	Ingredients
Explosive	Fuels + Oxidizers + Sensitizers + Energizers + Miscellaneous Agents
Common fuels	Fuel oil, carbon, aluminum, TNT
Common oxidizers	AN, Sodium nitrate, Calcium carbonate etc.
Common sensitizers	NG, TNT, Nitro-starch, aluminum etc.
Common energizers	Metallic powders
Common miscellaneous agents	Water, thickeners, gelatinizers, emulsifiers, stabilizers, flame retarders etc.
Main *elements* of these ingredients	Oxygen, Nitrogen, Hydrogen and Carbon, plus certain metallic elements such as: aluminum, magnesium, sodium, calcium etc.

5.3   COMMON INGREDIENTS OF EXPLOSIVES

Explosives are manufactured using fuels, oxidizers, sensitizers, energizers and few other substances in varying percentages. Given in table 5.1 is an account of the type of ingredients usually used.

5.4   CLASSIFICATION OF EXPLOSIVES[7,8]

Explosives have wide applications in mining and tunneling operations to carry out rock fragmentation for the differing conditions; hence, a wide range of these products is available. Given below is the general classification of explosives. The line diagram shown in figure 5.1(b) depicts this aspect.

5.4.1   Primary or initiating explosives[8,14]

Primary explosives may be defined as those explosive substances, which respond to stimuli like shock, impact, friction, flame etc. and pass from the state of deflagration (a high rate of burning) to detonation. Example: mercury fulminates, lead styphanate, Di-Azo-Nitrophenol (DDNP), Tetrazene etc. It is used in the manufacturing of the

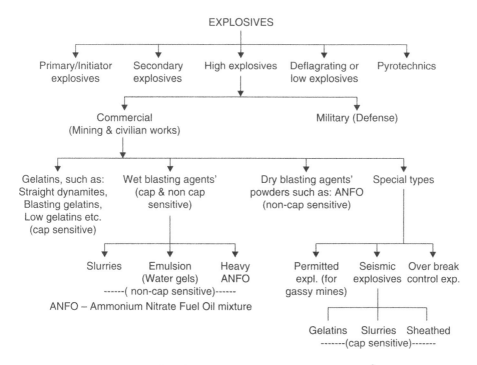

Figure 5.1(b)   General classifications of the explosives.[8]

detonators, detonating fuses and boosters. The mixture of lead styphanate, lead oxide and aluminum powder, known as A.S.A mixture, is also used as a primary explosive.

5.4.2   Secondary explosives

These are the explosive substances, which are capable of detonation, created by a primary explosive and not by the deflagration. Thus these explosives have a high rate of detonation and initiated by the primary explosives. Example: Penta Erythritol Tetra Nitrate (PETN), RDX, Tetryl etc. These explosives are used in the manufacturing of the detonators and form their base charge.

5.4.3   Pyrotechnic explosives[8]

Pyrotechnic compositions are used as a delay element in the manufacturing of the detonators and also as electric explosive devices (E.E.D), known as fuse-head or 'match-head' or Squibb. Pyrophoric metals like zirconium or cerium, oxidizing agents like lead peroxide, red lead, potassium chlorate, peroxides of barium and lead, and fuels like silicon, charcoal are used in delay elements and EEDs.

5.4.4   Low explosives[8,14]

The earliest known explosives belong to this class. These are commercially known as gunpowder or black powder. It is a mechanical mixture of ingredients such as

Table 5.2   Basic properties of nitroglycerin based explosives products.[7] (1 fps i.e. ft/sec = 0.3049 m/sec)

NG based Explosives	Sp. Gravity	Detonation velocity, fps	Water resistance	Fume quality
Straight dynamite	1.3–1.4	9000 to 19000	Poor to good	Poor
Extra dynamite	0.8–1.3	6500 to 12500	Poor to fair	Fair to good
Blasting gelatin	1.3	25000	Excellent	Poor
Straight gelatin	1.3–1.7	11000 to 25000	Excellent	Poor to good
Extra gelatin	1.3–1.5	16000 to 20000	Very good	Good to v. good
Semi gelatin	0.9–1.3	10500 to 12000	Fair to very good	Very good

charcoal (15%), sulfur (10%) and potassium nitrate, KNO_3, (75%). It is initiated by ignition (deflagration) and decomposition is slow. Its flame propagates slowly, a few m/sec. and burning particles are liable to remain in contact with the surrounding atmosphere for a considerable duration. It produces considerable amount of noxious gases rendering its use unsuitable for underground mines. It has a heaving effect on the rocks and gets spoiled by water.

5.4.5   Commercial explosives – high explosives

These are the explosive substances, which cannot be initiated easily by the stimulus such as impact, friction or flame but with the application of a shock pressure or a detonation wave. Example: Tri-Nitro-Toluene (TNT), Nitroglycerin (NG) and slurry explosives. The various NG based explosives and their properties have been presented in table 5.2.[6] These explosives can be classified as commercial and military explosives.

5.4.5.1   *Gelatin explosives*

Nitroglycerin: It is produced by the reaction of glycerin and nitric acid. It is an oily fluid. It is so sensitive that by shock of any nature it can explode. To make it suitable for its industrial use either it must be absorbed in an inert material or it must be gelatinized. Explosives containing NG, are available three consistencies: Gelatinous, semi-gelatinous and powdery. Higher NG contents renders explosives gelatinous; lower NG content up to 10% powdery. NG based explosives (fig. 5.3(b))[7,8,11] can be divided into three classes:

- Dynamites (straight dynamite, ammoniac dynamite)
- Blasting gelatin
- Semi gelatin.

5.4.5.1.1   **Dynamites (straight dynamite, ammonia dynamite)**

The NG based explosives were called dynamites. The first commercial explosive in NG was absorbed in natural mineral kieselghur, was termed as '*Straight Dynamite*'. Later on ammonium nitrate was introduced and a mixture of AN, NG, $NaNO_3$, and fuel element was marketed as Ammonia Dynamite[7,8] (fig. 5.3(b)).

Table 5.3(a)   Composition of some slurries and emulsion (wet blasting agents).[4,10,12,15,20]

Ingredients	Aluminum sensitized slurry	Water gel slurry	Explosives sensitized slurry	Emulsion
Fuel/Sensitizer	Al – 10%	Amine nitrate – 13%	TNT or Nitrostrach – 25%	Wax or Oil – 6%
Water	Water – 15%	Water – 15%	Water – 25%	Water – 14%
Oxidizer	NH_4NO_3 – 44%	NH_4NO_3 – 63%	NH_4NO_3 – 44%	NH_4NO_3 – 76%
Thicker	Guar gum – 1%	Guar gum – 1%	Guar gum – 1%	Guar gum – 1%
Others oxidizers	$Ca(NO_3)_2$ – 25% Ethylene glycol – 5%	$NaNO_3$ – 5% Ammonium perchlorate – 3%	$NaNO_3$ – 15%	
Micro balloons				Hollow micro-balloons – 2%

Table 5.3(b)   Composition and Velocity of detonation (VOD) of commercial explosives and blasting agents.

	INGREDIENTS					
Explosive/ Blasting Agent	Oxidizer	Oxidizer Size	Fuel	Form	Sensitizer	VOD km/sec
NG based (Dynamites)	Solid nitrate salts	0.2 mm	Solid metal absorbents	Solid	Liquid NG, voids/bubbles, friction	4.0
ANFO	Solid nitrate salts	2 mm	Liquid diesel oil	Solid	Voids/friction	3.2
Slurry	Solid/liquid (salt solutions, nitrate salts)	0.2 mm	Solid/liquid, Al, carbonaceous	Solid/ Liquid	Fine Al, bubbles	3.3
				Liquid	Bubbles	5 to 6
Emulsion	Liquid salt solutions	0.001 mm	Liquid oils, waxes			

5.4.5.1.2   Blasting gelatin

This is the most powerful explosive containing 92% NG and 8% Nitrocellulose (NC) which contains 12.2% nitrogen. Chalk, zinc oxide, air bubbles, acetone etc. are added to make the composition suitable for blasting purposes.

5.4.5.1.3   Semi gelatin

These are also termed as low NG, or high AN explosives due to the fact that in these explosives NG is mixed with NC, to form gel matrix which is mixed with AN in various proportions. Starch and wood meal are used as fuels. Straight gelatin and per-mitted explosive that are used in coalmines also fall in this category.

5.4.5.2   *Wet blasting agents*

Blasting agent is a mixture of fuel and oxidizer. It is not classified as an explosive, and cannot be detonated by a detonator (no. 8). A *Dry Blasting Agent* is a granular, free running mix of solid oxidizer (usually AN); prilled into porous pellets, into which a liquid fuel or propellant is absorbed. The typical example is ANFO. Main ingredients required to produce wet blasting agents have been shown in table 5.3. The main ingredients to produce slurries and emulsion explosives have been also shown in this diagram.

The blasting agents that contain more than 5% water by weight are referred to as *wet blasting agents*; within this category falls slurry explosives, water gels, emulsions, and heavy ANFO.

5.4.5.2.1 Slurry explosives

Slurry explosive is defined as a semi-solid or pasty suspension of oxidizers, fuel, sensitizers etc. in a thickener like guar gum. Inorganic cross-linking agents are added to prevent the segregation of solid and liquid on storage. The final product is the cross-linked water gel. The oxidizers commonly used are nitrates and perchlorates of ammonium, sodium and calcium. The fuels are glycol, starch, sugar, coal powder, sulfur etc. TNT, Nitro starch, finally divided aluminum powder; air bubbles are used as sensitizers. Micro-balloons of 3–4 microns size are also used as sensitizers. Slurry explosives are replacing the gelatin explosives due to the following characteristics they possess:

• Built in safety against fire, friction and impact
• Water compatibility
• Reductions in the production of toxic gases like CO and NO (fumes).

5.4.5.2.2 Emulsions

Emulsions are a two-liquid phase containing microscopic droplets of aqueous nitrates of salts (chiefly AN) dispersed in fuel oil, wax or paraffin using emulsifying agent. Micro-spheres, microscopic glass or plastic air filled bubbles, and AN droplet form the oxidizer. It is mostly mixed at site prior to charging into holes. It is also available in cartridge packs.

5.4.5.2.3 Heavy ANFO

Heavy ANFO is 45 to 50% AN emulsion mixed with prilled ANFO. This is done to increase density of ANFO. It is mostly mixed at site prior to charging into holes; but it is also available in cartridge packs.

5.4.5.3 *Dry blasting agents*

Powder explosives: One of the major applications of the prilled [in the form of small spherical prills (like balls)] ammonium nitrate coated with an anti-caking agent is in the manufacture of powder explosives. These are used as powder or in the form of cartridges. Caking, bad fumes, poor water compatibility and low density are some its drawbacks. Its low cost, ease in manufacturing, handling and use has made it widely acceptable in surface as well as non-coal u/g mines. A detailed description is given in the following paragraphs.

5.4.5.3.1 Explosive ANFO[16]

Ammonium nitrate: Ammonium nitrate (AN) is well known for its military and civilian use. It has been used extensively in both world wars to manufacture Amatol which is 80% + 20%, or 50% + 50% mixture of AN and TNT. It is an excellent fertilizer. Explosive properties of AN was known accidentally when a shipload of fertilizer grade AN blew up suddenly due to a fire accident. It was considered a potential blasting agent since then.

Ammonium nitrate, which was earlier known as an oxidizer in the manufacturing of explosives, has become the principal ingredient of the commercial explosives in use, in the mining industry. Today due to some of its inherent properties AN based explosives are in use all over the world. In the commercial explosives AN percentage varies in the range of 10–95%. In all the principal classes of explosives i.e. NG based, dry and wet blasting agents AN is used. When AN is mixed with 5–6% fuel oil, the mixture is known as ANFO. ANFO has become an indispensable explosive for most of the surface mines and underground non-coal mines.

5.4.5.3.2   ANFO mixing[16]

Most porous prilled AN absorbs oil up to 6–7%. Excess oil collects at the bottom of the container. Since 6% is approximately the stoichiometric value of oil, preparation of this blasting agent is comparatively simple. Accurate compounding and thorough homogeneous mixing is all that is required to give a superior product.

Earlier practice which is still followed is to pour AN in the dry holes and then pour fuel oil in it. In large dia. holes with large primer this gives moderately satisfactory results. This practice is now replaced by surface mixing. Care should be taken to put in the correct amount and give sufficient soaking time. The most satisfactory method, however, is to provide some type of mechanical mixer with careful control of quantities. A uniform product can be obtained by mixing in any of the conventional mixers. A mixer should minimize frictional heating and crushing of prilled material. From safety consideration, a mixer should provide minimum confinement.

Addition of an oil soluble dye "waxoline red" of 1 gm. per liter gives a slight pink coloration to the mixed product and also helps to achieve a uniform mixing denoted by color index. Too much of the dye mars the judgment of uniform mixing. It has been found that with oil of 0.82 S.G. 100 kg of AN needs 7.6 liters; to give a mixture of 6% by weight. It is again stressed that smaller the diameter of hole, the care while mixing should be more to get better results. It has been recommended that under extreme cold climate a little preheating of oil at least 20°C below the flash point gives better mixing because of increased inter molecular activity.

5.4.5.3.3   ANFO loading

ANFO loading in large diameter down holes is hardly any problem because the mixed ANFO can be directly poured inside the holes. For large diameters, mixed ANFO, is also available in a cartridge form, which can be loaded like any other explosive. Only for small diameter holes, loading has to be done by pneumatic means for quick, compact and thorough loading. The loading equipment is known as Anoloaders. The loaders are broadly of two types:

1. Pressure type (fig. 5.2(b))[4]
2. Ejector type, whose principle of working is important to note at this stage (fig. 5.2(a))[4]
3. Combined type (combining pressure and ejecting features).

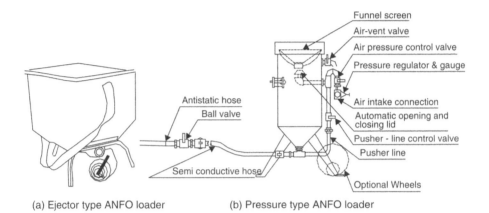

(a) Ejector type ANFO loader   (b) Pressure type ANFO loader

(c) Charging robot including ANFO vessel. The charging robot can be operated from drivers cabin or external via cable. (Courtesy: GIA Industries.)

Figure 5.2   Explosive charging loaders of different types.

5.4.5.4 *Pneumatic loaders and principles of loading*[4,16]

5.4.5.4.1   Pressure type loaders

These are heavy duty and fast loading transportable machines and can load effectively up to 25 meters of vertically up holes. In this particular type unlike the other ejector type a plug of ANFO mixture is forced to reach the hole against a positive pressure as a continuous column, whereas in the ejector type the ANFO particles are blown along with a continuous current of air. In the ejector type, the ANFO particles bombard against the loaded front and get broken, resulting in compaction, but in the pressure type the prills crack against the positive pressure from one side and as soon as ANFO leaves the loading hose, it is broken into smaller fragments. The pressure type loader (fig. 5.2(b)) essentially consists of: a tank with cap, discharge valve, air cylinder and remote control unit, and gauges.

In pressure type loaders, the loading rate is directly dependent upon the pressure suitable for loading. An increased pressure develops in blowing of ANFO out very fast, whereas, a decrease in pressure results in locking of ANFO inside the loading hose. It is a matter of experience as it is to be decided based on the parameters such as: length of hole and its direction, condition of the mixed ANFO whether it is moist or dry, prill sizes' uniformity, and whether ANFO is powdery or not.

5.4.5.4.2   Ejector type loader[4]

This is a portable air operated loader (fig. 5.2(a)) for ANFO loading in holes of small length and diameters. It loads with a high density and with high speed when operated properly. This model has no 'blocking problem' as in case of pressure type where because of pressure from one side and resistance from other side ANFO gets blocked sometimes in the loading hose. This type has a specially designed ejector, hence the name.

Assembly: It has an anti-static loading hose 20 to 25 mm diameter and 3 to 4 m long with a protective spring to prevent excessive flexing of the hose. A clamp tightens the hose with the diffuser ejector. It has a compressed air inlet of 25 cm. It has also a permanent grounding lug on the ejector body for earth connection. A self-adhesive color tape is fixed on the discharge end of loading hose to note the length of hose inserted in the hole.

Operation: Safety glasses and gloves must be worn while working with the loader. The operator should be on one side of the hole and never to the front directly. It operates on direct air pressure and on the squeeze of hand operated valve. No air lubricator should be used with anoloader air supply.

5.4.5.4.3   Combined type (combining pressure and ejecting features)

By combining the application of both the mechanisms, the loader [(refer figs. 5.1(a) & (b)] can achieve double the rate of loading than that in pressure type and can cope with hole length up to 30 m effectively.

5.4.5.5   *Safety aspects*[16]

- Safety record of ANFO explosives is excellent as compared to conventional explosives. But it cannot be said that accidents cannot and will not occur. ANFO mixture is non-cap-sensitive. This mixture is also insensitive to friction and impact tests as generally applied high explosives. However, a strong booster is used so that satisfactory initiation is resulted. This in itself does not create any hazard if the transport, loading and firing is conducted in the same manner as recommended for other explosives. Their propagation through an air gap is poor. Air pockets up to 25 to 50 mm may result in failure of blast. Cap-sensitive ANFO has also been developed by repeated temperature cycling through the 90°-phase change region. Evidently these mixtures can be treated as cap-sensitive high explosives. But unfortunately these cap-sensitive explosives are very costly.
- Danger of fire is perhaps the greatest hazard with ANFO, although small quantities of ANFO are difficult to burn. For safety aspects, ANFO is recommended to be used in the same manner as other high explosives.
- ANFO has a tendency of spontaneous heating in pyrite bearing ores. ANFO has been found to react with pyretic ores at as low as 85°C resulting in an exothermic reaction. Addition of 0.5 to 1% of calcium carbonate, urea, zinc oxide or magnesium oxide decreases the reactivity. However, the percentage of pyrites in the ore that affects this phenomenon is 5 to 30% by weight. Use of ANFO in pyretic ore bodies should be done with care specially in hot ground conditions. Exothermic reaction can produce temperature exceeding thermal initiation of charge.

- During handling AN has a corrosive action on human skin resulting in black patches and scaling of skin. So a rubber glove is recommended for ordinary use. AN dissolves on the skin moisture and goes into the subcutaneous area causing irritation.
- Electrostatic hazards of pneumatic loading are very important and are dealt with in the following paragraph.

5.4.5.6 Static hazards associated with ANFO loading[16]

Static charge is built up on the loading hoses and equipment and the problem has been the greatest so far in the usage of ANFO underground.

A certain degree of safety can definitely be achieved by using conductive, or semi-conductive hose, and the established standards so far. The salient points may be summarized as follows:

a. The loading hose should be semi-conductive with a resistance high enough to insulate any stray current yet conductive enough to bleed off any static charge build up. Such loading hoses are called 'LO STAT' and a yellow stripe runs throughout their length to identify them.
b. The electrical characteristic of the loading hose should be uniform throughout its length. Resistance lying between 17,000 and 67,000 ohms/m; electrical capacity of typical PVC hose is 4 p.f. and available discharge energy is 24 MJ. Basically it should have sufficient resistance to corrosion from oil and be stiff enough to avoid too much kinking.
c. Except in case of non-electric detonators like Anodets etc. bottom priming should not be done and priming should be done at the collar at the end of loading allowing sufficient time for the hole and operator to get discharged of any electric static charge.
d. The entire system and the operator should be effectively grounded with the earth. Only approved semi-conductive loading hose should be used. In case of hoses lined with a non-conductive material pneumatic loading should not be adopted.
e. Maximum resistance of the drill hole to the ground must be less than 10 Mega ohms and the maximum resistance of the loader operator to the ground must be less than 100 Mega ohms, while electric detonators are to be used.
f. After every loading operation some time should be allowed for the charge to leak away and the detonator should be placed from the collar side, and the operator should also ground himself before handling the detonators.
g. A detonator continuity test must be made religiously. This will prevent discharge between the shell and the lead wires, which may be of corona type, or a simple spark.
h. Leg wires should be shunted and not connected to ground during pneumatic loading.
i. Effective earth line should be connected with the entire assembly of ANFO loader.
j. Synthetic fibers like nylon, teryline etc. should not be on the body of the persons doing ANFO loading since they have tendency to accumulate and retain charge. For similar reasons rubber-soled boots should also not be used.

The quantity of electric charge has been seen to depend largely upon humidity conditions and general conductivity of the rocks. When the rock is fairly conductive, the charge is dissipated as soon as developed.

As discussed earlier 'Anodet' or antistatic detonators have obviated the use of electric detonators in the ANFO blasting system, as such use of the ordinary electric detonators should be avoided.

5.4.5.7 *Special types of explosives*

5.4.5.7.1 Permitted explosives[14]

These explosives have been designed to use in u/g coalmines to avoid methane-coal dust explosion. These are available in granular, gelatinous and slurry forms. For wet coal mines the gelatinous type is more suited. The VOD of these explosives is in the range of 6000 to 16,000 ft/sec. (1830 to 4880 m/sec.) A cooling agent is incorporated in all permitted explosives. Common amongst them are sodium chloride, potassium chloride and ammonium chloride.

A 3 mm thick cover (sheath) of sodium bicarbonate, when wrapped all along the length of the cartridge, this is known as *Sheathed explosive*. This is also a permitted type of the explosive, which can be used in coalmines.

5.4.5.7.2 Seismic explosives[10]

In seismic exploration work for mineral discovery, shots are fired in the earth crust, often under high heads of water, so that the resulting ground vibration reading can be recorded by the geophones in some selected areas. The trade name given varies from company to company e.g. Seismograph high explosive, Petrogel, Geogel etc.

5.4.5.7.3 Overbreak control explosives[10]

For smooth, perimeter blasting (sec. 9.3.3) special types of explosives often give a poor coupling ratio, having the diameter of cartridge in the range of 16–22 mm. The trade name given varies from company to company e.g. 'Smoothite', 'Kleen-cut' etc.

5.4.6 Military explosives[8]

These explosives are less sensitive to impact and shock compared to dynamites. They show high brisance or shattering effect. Brisance is described as the ability of explosive to shatter and fragment steel, concrete and other very hard structures. Their velocity of detonation is in the range of 7000 to 9000 m/sec. comparing the same for the commercial explosives, which is up to 5000 m/sec. These are known by the names such as: TNT, PETN, RDX, TETRYL etc. They have high detonation pressures of the order of 17 million p.s.i (117.1 million kps (1 p.s.i. = 6.89 kps (kilopascal). The components are either melted or poured or casted into shells or suspended. These explosives feature following characteristics:

- Maximum power/unit volume
- Minimum weight/unit power
- High velocity of detonation
- Long term stability under adverse storage conditions
- Insensitivity to shock on firing and impact.

Common properties of military explosives[8] have been shown in table 5.4.

Other military explosives include the mixtures such as ammonium salt of picric acid (Picramate), dinitro toluene (DNT), ethylene diamine dinitrate (EDDN), Ammonium nitrate (AN), cyclotol (RDX + TNT), composition 'B' (RDX +

Table 5.4   Common properties of military explosives.[8]

Explosive	Density gms.c.c	VOD m/sec	Gas vol. lit/kg.	Detonation kcal/kg	Temp. °C	Sp. pressure kg/cm²	Lead block expansion mm
Trinitro toluene – TNT	1.63	6950	685	1085	3630	3749	310
Penta erythritol tetremotrate – PETN	1.77	8300	780	1408	3560	–	500
Cyclo trimethylene trinitremine – RDX	1.73	8500	910	1390	3380	4150	485
Trinitro phenyl methyl nitromine – TETRYL	1.06	7500	710	1320	3370	4684	450
Nitroglycerin – NG			715	1470	3153	4060	390
HBX (RDX + TNT + Wax)			782	1435	3500	–	480

TNT + Wax), Torpex (RDX + TNT + Aluminium), Ametol (TNT + AN), Pentolite (PETN + TNT), Tetrytol (Tetryl + TNT) and many more compositions. NG is normally used in making 'double based' propellants.

5.5   BLASTING PROPERTIES OF EXPLOSIVES[7,8,11,13]

Each explosive has certain specific properties or the characteristics. Its ingredients such as nitroglycerin and ammonium nitrate contents have direct influence on some of its properties such as resistance to water, detonation velocity, costs etc. These aspects were studied,[7] as shown in figure 5.3. Given below are some of the important properties, which influence the ultimate choice of an explosive.

5.5.1   Strength

It is the energy released/unit weight (known as weight strength); or per unit volume (known as bulk strength) of an explosive. It is nowadays expressed relative to ANFO at 100% i.e. taking ANFO as standard. High strength is needed to shatter the hard rocks but use of high strength explosive in the soft, weak and fractured rocks will be wastage of the excessive energy imparted by these explosives. Strength of an explosive is measured by:

- Shock generated (VOD and speed of chemical reaction)
- Gas volume
- Energy
- Detonation pressure
- Explosion temperature
- Velocity of detonation (VOD) is the measure of the shattering effect of an explosive, an important parameter for hard rock blasting. It changes with change in

diameter and density of explosives. 'Dautriche', electronic or Hess method or tests can measure VOD.

- Gas volume

The larger the gas volume of an explosive the larger will be the throw obtained. If throw is to be minimized its ingredients should be adjusted to get minimum volume of gas and maximum heat output. 'Ballistic Mortar' test and 'Trauzl block' test generally measure it.

- Energy

The oxygen balance and reactive ingredients determines the energy output of an explosive. This energy represents the temperature of explosion and hence the maximum work that can be done by an explosive is indicated by this value.

- Detonation Pressure

Based on detonation velocity and density of explosives a shock wave pressure that is built ahead of the reaction zone is known as detonation pressure. The higher the detonation pressure, the higher would be the brisance capability (i.e. the ability to break or shatter rock by shock or impact). Its value varies from 5 to 150 KB. Due to this property a primer having higher detonation pressure should be selected. Given below is the mathematical relation to express this parameter:

$$p = 2.5 \, \rho v^2 x \, 10^{-6} \tag{5.1}$$

Whereas: p = detonation pressure in kilobars (KB)
ρ = explosive density in gms/c.c
v = velocity of detonation in m/sec.

Above the critical density, detonation pressure is zero, as the cartridge does not explode.

- Explosion temperature

This parameter is calculated based on the thermodynamic data of the ingredients. In coal mines a balancing of explosion temperature and the gas volume play an important role. If explosion temperature exceeds 1000°C it can make the methane atmosphere incendive i.e. mixture of air & methane can catch fire and explode.

5.5.2 Detonation velocity

It is the velocity with which the detonation waves move through a column of explosives. Following are the factors that affect the detonation velocity:

- Explosive type,
- Diameter, confinement,
- Temperature &
- Priming.

In general, higher the velocity of detonation the better will be the shattering effect. The explosive's detonation velocity ranges from 1500–6700 m/sec.

In general, the larger the diameter the higher is the velocity of detonation until a steady state velocity is reached. For every explosive there is a minimum critical diameter at which the detonation process once initiated, will support itself in the column. The influence of hole dia. on the detonation velocity for various types of explosives has been studied, as shown in figure 5.3(a).[7]

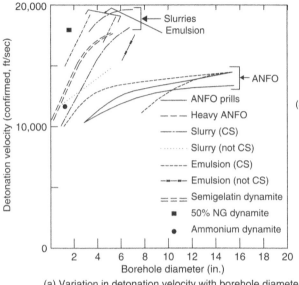

(a) Variation of detonation velocity for the few selected explosives[5] as per the blast-hole diameters. (Conv. 1″ = 25.4 mm; 1 fps = 0.3048 m/s)

ANFO prills
Heavy ANFO
Slurry (CS)
Slurry (not CS)
Emulsion (CS)
Emulsion (not CS)
Semigelatin dynamite
50% NG dynamite
Ammonium dynamite

(a) Variation in detonation velocity with borehole diameter for few selected explosives (Conv. 1 in. = 25.4 mm; 1 fps = 0.3148 m/sec)

INGREDIENTS	NONGELATINOUS	GELATINOUS	PROPERTIES
Decreasing COST / Increasing Ammonium Nitrate / Decreasing Nitroglycerin	Nitroglycerin	Blasting Gelatin	Decreasing Density / Decreasing Detonation Velocity / Decreasing Water Resistance
	Straight dynamite	Straight Gelatin	
	High density Ammo. Dynamite	Ammonia Gelatin	
	Low density Ammo. Dynamite	Semi. Gelatin	
	Dry blasting agents	Slurries	
	Increasing Water Resistance		

(b) Relation of ingredients and properties of explosives

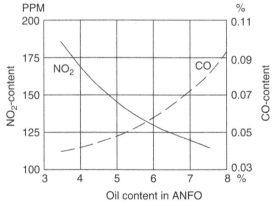

ANFO & oil content; CO & NO₂ gas quantities

(c) Influence of oil content in ANFO contributing to toxic gas production

Figure 5.3 Explosive characteristics.

5.5.3 Density

The density of explosives is in the range of 0.5 to 1.7. A dense explosive releases more energy/unit volume, hence it is useful for the hard and denser strata. For any explosive there is a *critical density*, above which, it cannot reliably detonate. For example for TNT – 1.78 gms/c.c; ANFO – above 1 gms/c.c.

5.5.4 Water resistance

A practical way to judge the ability of any explosive to resist water is its capability to withstand exposure to water without losing sensitivity or efficiency. ANFO has poor water resistance. Slurries have good water resistance, and whereas, the NG based explosives have the best water resistance, as shown in figure 5.3(b).[11]

5.5.5 Fume characteristics, or class, or medical aspects

An explosive after blasting should generate the minimum the amount of toxic gases such as carbon monooxide, oxides of nitrogen etc. It varies from $0.023\,m^3/kg$ (fume volume/ unit weight) to as high as $0.094\,m^3/kg$. In some of the NG based explosives, the fumes emitting out from it, enters into the blood circulation, causing headache.

5.5.6 Oxygen balance[7,8]

As stated above, any explosive contains oxidizing and combustible (fuels) ingredients. A proper balance of these ingredients is essential to minimize production of the toxic (poisonous) gases, e.g. an excess of oxygen produces for example nitric oxides, nitrogen peroxide and deficiency of oxygen results in the production of carbon monoxide. Also such an imbalance effects the energy generation. This can be illustrated by taking the example of ANFO explosive, which is a mixture of ammonium nitrate and fuel oil. The former acts as an oxidizer and the latter a combustible agent. While mixing them in varying percentages, the resultant reactions can be represented by the chemical reactions as under:

$$3NH_4NO_3 + CH_2 \Rightarrow 7H_2O + CO_2 + 3N_2 \quad + 0.93\ Kcal/gm.$$
$$94.5\% \qquad 5.5\% \qquad\qquad\qquad oxygen\ balanced$$

$$\text{(5.2a)}$$

$$2NH_4NO_3 + CH_2 \Rightarrow 5H_2O + CO + 2N_2 \quad + 0.81\ Kcal/gm.$$
$$92.0\% \qquad 8.0\% \qquad\qquad oxygen\ balance\ -\ negative,\ fuel\ in\ excess.$$

$$\text{(5.2b)}$$

$$5NH_4NO_3 + CH_2 \Rightarrow 11H_2O + CO + 4N_2 + 2NO \quad + 0.60\ Kcal/gm.$$
$$96.6\% \qquad 3.4\% \qquad\qquad oxygen\ balance\ -positive,\ oxidizer\ in\ excess.$$

$$\text{(5.2c)}$$

The above equations and figure 5.3(c),[13] illustrate that an oxygen balanced mixture generates minimum harmful gases and maximum energy.

Calculation of oxygen balance: Oxygen balance can be determined by following the steps outlined below:

- Write the molecular formula and molecular weight.
- Find the number of C, O, H and nitrogen atoms.
- Remove two oxygen atoms/carbon atom (CO_2); and half oxygen per hydrogen atom (H_2O formation).
- Leave nitrogen atom as nitrogen molecule (N_2).
- Note, how much oxygen is left behind (+). If not then calculate how much oxygen is required (−).

$$C_aH_bN_cO_d = aCO_2 + 0.5bH_2O + 0.5cN_2 + (d - 0.5b - 2a)O_2 \qquad (5.3)$$

Where a, b, c, and d are the number of carbon, hydrogen, nitrogen and oxygen atoms in the explosive substance.
Example: To calculate *oxygen balance of the fuel oil.*
Formula: CH_2 Molecular weight = 12 + 2 = 14
a = 1, b = 2, c = 0, d = 0
$(0 − 0.5 \times 2 − 2 \times 1) = −3$ atoms of oxygen.
14 gms of diesel oil require 48 gms of oxygen, so 1 gm of diesel oil will require = −48/14 = −3.43
So, oxygen balance of fuel oil is −3.43.

Similarly calculation of *oxygen balance of ammonium nitrate* (NH_4NO_3).
Molecular weight = (14 + 4 + 14 + 48) = 80
a = 0, b = 4, c = 2, d = 3
Oxygen balance = $(d − 0.5b − 2a)O_2 = (3 − 0.5 \times 4 − 2 \times 0) = 1$ atom of oxygen
80 gms of ammonium nitrate gives 16 gms of oxygen,
So 1 gm of ammonium nitrate will give = 16/80 = 0.2 gms of oxygen.

For ANFO to be oxygen balanced:
AN × 0.2 + fuel oil × (−3.4) = 0
Let, AN be y%
0.2 y + (100 − y)(−3.4) = 0
Or 3.6 y = 340; or y = 94.5
Thus, an oxygen balancing ANFO should contain 5.5% fuel oil and 94.5% ammonium nitrate.

Calculation of oxygen balance of Nitroglycerin:
Molecular formula: $C_3H_5N_3O_9$
Molecular weight = (12 × 3 + 1 × 5 + 14 × 3 + 16 × 9) = 227
Oxygen required or available = $(d − 0.5b − 2a) = (9 − 0.5 \times 5 − 2 \times 3)$
$= 0.5$ atoms of oxygen
So oxygen balance = Molecular weight of available oxygen/molecular weight of substance = (0.5 × 16)/227 = 0.035

Calculation of Oxygen balance of PETN
Formula: $C_5H_8N_4O_{12}$ Molecular weight $= 316$
Oxygen required $= (d - 0.5b - 2a) = (12 - 0.5 \times 8 - 2 \times 5)$
$\qquad\qquad\quad = -2$ atoms of oxygen $= -32$ gms.
Oxygen balance $= -32/316 = -0.1$

5.5.7   Completion of reaction

Achieving a complete reaction at the required speed during blasting is the next important factor, for example if a carbon atom is not oxidized to carbon dioxide but carbon monoxide, the production of energy comes down by 75% of the expected energy, as shown below. Similarly, formation of oxides of nitrogen involves the absorption of energy.

$$C + O_2 \ \Rightarrow \ CO_2 + 94\ Kcal/gm. \tag{5.4a}$$

$$0.5N_2 + 0.5O_2 \ \Rightarrow \ NO \text{ - } 22\ Kcal/gm. \tag{5.4b}$$

$$C + 0.5O_2 \ \Rightarrow \ CO + 26\ Kcal/gm. \tag{5.4c}$$

Reaction (eq. 5.4(b)) and (eq. 5.4(c)) not only produces lower energy but also yields toxic gases. In ANFO explosive if moisture content exceeds 1%, it not only causes caking of ANFO but also makes the reaction incomplete.

5.5.8   Detonation pressure

Based on detonation velocity and density of explosives a shock wave pressure, which is built ahead of reaction zone, is known as detonation pressure. Higher the detonation pressure higher would be the brisance capability. Its value varies from 5 to 150 KB. Due to this property a primer having higher detonation pressure should be selected. Using equation (5.1) detonation pressure can be assessed.

5.5.9   Borehole pressure and critical diameter

It is an important parameter, which measures the breaking and displacement property of an explosive. Its value varies from 10–60 KB (1000 to 6000 kpa).
   Critical diameter: Sensitivity of an explosive is an important property, which is measured by its ability to propagate the detonation wave. The detonation wave tends to fall or fade when diameter of explosive charge decreases. The minimum diameter of a charge, below which the detonation does not proceed, resulting in misfire, is called 'Critical Diameter'. At lower diameter even if the explosive is sensitive, the reaction in the cartridge may be incomplete.

5.5.10   Sensitivity

It is measured as the explosive's propagation property to bridge a gap between two consecutive cartridges or a column of an explosive charge e.g. if a cartridge is cut into two halves, and the resultant pieces are kept apart. By initiating one of them, with how much gap the other will be able to accept the propagation wave, if blasted unconfined in a paper tube.

Table 5.5 Some important explosives together with their density, bulk strength, weight strength and costs.[11]

Explosives	Density (g/c.c.)	Relative weight strength (ANFO = 100)	Relative bulk strength (ANFO = 100)	Relative cost/ unit volume (ANFO = 100)
AN				67
ANFO	0.85	100	100	100
ANFO (dense)	1.10	100	135	130
15% Al/ANFO	0.85	135	135	183
15% Al/ANFO, dense	1.10	135	175	237
Peletized TNT	1.0	90	106	392
1% Al/NCN slurry	1.35	86	136	397
20% TNT slurry	1.48	87	151	421
40% Dynamite	1.44	82	139	551
25% TNT slurry/ 15% Al slurry	1.60	140	264	722
95% Dynamite	1.40	138	193	824

5.5.11 Safety in handling & storage qualities

ANFO has poor storage quality being hygroscopic in nature. ANFO if handled without gloves can cause skin irritation. Also salt of some explosives under extreme temperature conditions evaporates, making its cartridges hard and deformed. By proper waxing of the explosive cartridges the effect of moisture on them can be minimized.

One of the important requirements of an explosive is that it can be stored, transported and used under the normal conditions without any risk to the persons handling it and carrying out the blasting operations. In order to have a safe manufacturing, transport, handling and the end use of an explosive, various tests are made on the ingredients and final product. The tests include impact test (fall hammer test), friction pendulum test, torpedo friction test, projectile impact test and bullet sensitivity test.[8] For example, NG powder will explode if a weight of 0.5 kg fall on it from a height of 20–30 cm. Whereas if a weight of 0.5 kg falls from about 8 m on it, a cap-sensitive slurry may explode.

5.5.12 Explosive cost

While selecting an explosive its cost plays an important role. Comparing to AN (Ammonium Nitrate), the relative cost of some of the common explosives on unit weight basis has been given in table 5.5.

5.6 EXPLOSIVE INITIATING DEVICES/SYSTEMS

Any explosive needs stimuli like shocking, friction or flaming for it to blast, or the reaction to initiate in it. The devices used to carryout these operations are known as

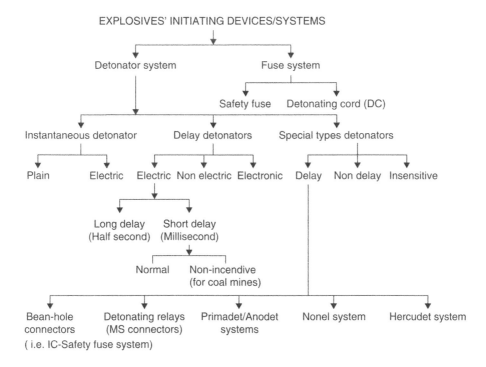

Figure 5.4 Classification of explosive initiating devices/systems.

initiating devices. The description below outlines the development and application of each of such devices/techniques to initiate an explosive. In the line diagram (fig. 5.4), a classification of explosive initiating devices/systems has been shown.

5.6.1 Detonator system[7,8,10]

5.6.1.1 *Detonators*

In order to initiate high explosives and the blasting agents, a strong shock or detonation is required. A capsule of sensitive explosive material termed a detonator can accomplish this. A detonator consists of a metal tube or shell (Cu, bronze or Al), generally 5.5 to 7.5 mm in outer diameter and a varying length depending upon whether it is instantaneous or delay type (fig. 5.5).

In a detonator a base charge PETN (Secondary explosive) is placed at its bottom. To initiate this base charge a column of primary explosive, which is a mixture of lead styphanate, lead oxide and aluminum powder, known as A.S.A mixture is placed over it. The charges are compacted under adequate pressure to give the desired strength.

Strength of a detonator: Based on the quantity of base charge and A.S.A charge quantity; the detonators are designated as detonator no. 1 to no. 8, or more in the order of increasing quantities of these charges. Thus, No. 8 cap produces much stronger pressure pulse than no. 6 cap. No. 6 detonator contains 0.35 gm of A.S.A mixture and 0.25 gm

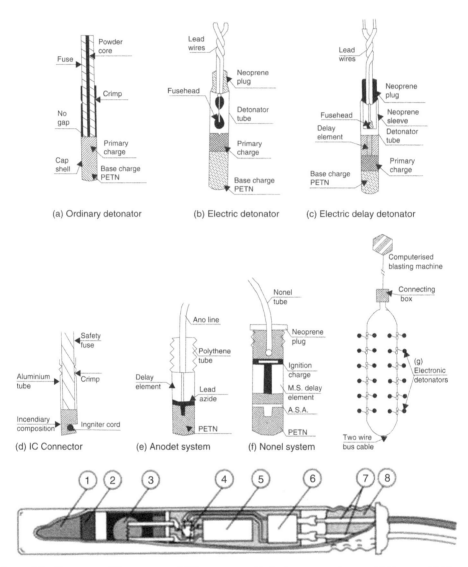

Figure 5.5   Detonators/Blasting-caps of different types. (a) Ordinary non-electric. (b) Electric. (c) Electric Delay. (d) IC connector. (e) Anodet system. (f) Nonel System (Antistatic and non-electric). (g) Electronic: 1-Base charge (PETN), 2-Primary explosive (lead-azide), 3-Mtachhead with bridge wire, 4-Integrated circuit chip, Capacitor, 6-Over voltage protection circuitry, 7-lad wires, 8-Sealing plug (Courtesy: Sandvik–Tamrock).

of PETN or tetryl. No. 8 carries large charge, 25% more than No. 6, and used to blast hard rocks.

The method of initiating the charge may be a safety fuse, as in case of plain detonator or by a fuse head as in case of electric detonator.

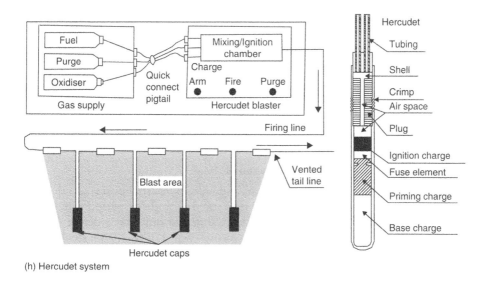

(h) Hercudet system

Figure 5.5(h) Hercudet system with detonator.

5.6.1.2 *Instantaneous detonators*[8]

5.6.1.2.1 **Plain detonator**

This is simpler in construction and made of an aluminum shell closed at the bottom and open at the other end (fig. 5.5(a)). It is used under dry and non-gassy conditions and initiated by the safety fuse that is inserted in its open end and crimped.

5.6.1.2.2 **Instantaneous electric detonators**

These detonators (fig. 5.5(b)) have the same construction as the plain detonators except that an electric explosive device, often called a fuse-head, is used to initiate primary explosive charge incorporated within it. In this detonator a bridge wire is provided, and the mouth of the tube is sealed with a plastic plug through which the insulated leg wires pass. Proper electric current, when passed through the bridge wire of the fuse head, fuses it; thereby it becomes incandescent and ignites the priming charge. The detonator is fired instantaneously, i.e. at the same time, as the current is passed.

5.6.1.3 *Delay detonators*

5.6.1.3.1 **Electric delay detonators**

These are manufactured as two varieties – Long/half second delay detonators and short/millisecond delay detonators. These detonators (fig. 5.5(c)) are longer in length than the instantaneous electric detonators as a delay element is incorporated in between the primary charge and the fuse head. Long delay detonators are available in 0–15 numbers, with a nominal half-second time interval between each delay.

In short delays the delay interval is much shorter. These types of detonators are available in a wide range of intervals using no. 6 and no. 8 strength caps. These short delays can be further classified as normal and non-incendive delays. The normal detonators are available in the range of 18–38 delays[10] each interval varying from 8 to 100 or more milliseconds. The leg wires' length is in the range of 4 to 60 ft (1.2 to 18 m). These delays are widely used in mines other than u/g coalmines. They are also used in tunnels.

The non-incendive types of detonators are used in coalmines and are made of copper tubes with copper leg wires of 4–16 ft (1.2 to 5 m) lengths. These are available in 10 delays with interval of 25–75 milliseconds. Both types can be used in wet conditions.

5.6.1.3.2 Electronic delay detonators[13]

An electronic delay detonator is very recent and in its exterior appearance it looks like a conventional detonator. The detonator is marked with delay period number from 1 to 250. This number does not indicate delay time but only the order in which the detonators will be fired. Each detonator has its own time reference, but the final delay time is determined through the interaction between the detonators and the computerized blasting machine before their firing.

Dyno Nobel is one of the companies who are manufacturing this. In this system detonators react to the dedicated blasting machine eliminating risk of unintentional initiation by any other energy source. In the manufacturing of this detonator several elements are used as the chain reaction of igniting the detonator, and detonating the charge in the drill hole. Each element involves a time delay, which is not the same for all normally equal detonators. The reason is that each element has a certain amount of scatter time. In figure 5.5(g) the internal structure of an electronic detonator has been shown. Apart from base charge (PETN) and primary explosives, lead wires and sealing plug as in the conventional detonators, the other important components include: match-head with bridge wire, an integrated chip, capacitor and an over voltage protection circuit. Not detonating without a unique activation code and protection against excessive voltage are some its unique features that allows it to be safer than the conventional detonators. The blasting machine is the central unit, which supplies the detonator with the initiation energy and allocates it the delay time. These are some of the specific features of these detonators:

- Shortest delay time is 1 ms and longest is 5.25 seconds.
- In this system a maximum up to 500 detonators can be connected to the blasting machine.
- The filter combination with toroid, gives protection against parasite currents, static electricity produced during pneumatic charging of explosives or radio frequency signals.
- They are extremely precise to the extent of 0.2 ms.
- The limitation at present is their costs, which is 10–15 times the conventional caps.

5.6.1.3.3 Non-electric delay detonators: detonating relays (ms connectors)

This system is used in conjunction with detonating cords (DC) for blasting a large number of holes and is capable of introducing millisecond intervals (delays) between holes or rows of holes. A detonating relay consists of a long aluminum tube with two mini-delay detonators on both side and having an attenuate in the center. The opening at either end can be crimped to detonating cord. These are manufactured with delay interval of 15, 17, 25, 35, 45, 50, 60 and 100 ms.[14] Use of such relays can provide advantages such as easy and safer to handle, better fragmentation, reduced ground vibration, better muck pile and reduction in overall costs. Only one detonator is required to fire a blast. Their placement in wet conditions should be avoided. The system finds its applications in surface mines and u/g metalliferous mines.

5.6.1.3.4 Primadet and anodet non-electric delay blasting systems[10,14]

To safeguard against the static charge and current hazards from the electric detonators Ensign Backford developed the primadet system. It consists of three components:

1. A blasting cap (no. 6) with delay elements (short or long delay). Short delay system with 30 delay periods ranging from 25 to 250 milliseconds.
2. A detonating cord having PETN of 4 grains/ft called primaline. One end of which is crimped into the blasting cap at the time of its manufacturing. It is available in different lengths 2 m to 15 m (6–50 ft).
3. A plastic 'J' connector for readily attaching the free end of the primaline to the trunk line.

Anodet (figs 5.5(e) and 5.9(e)) is similar to primadet and it has been developed by CIL for its use during charging ANFO pneumatically in the blastholes of 25 to 70 mm. The primaline is known as Anoline in this system. In this system to locate the primer centrally in the hole the manufacturer also supplies a plastic cap holder. The Anoline is available in the length of 3, 4, or 5 m. Anodet short delays are available from 0–30 numbers. Long delay are available from 0–15. In figure 5.9(e) procedure to use an anodet detonator together with its accessories has been illustrated.

5.6.1.3.5 The nonel system[10,13,14]

It is an invention by Nitro Nobel AB Sweden used as a nonlectric system without use of detonating cords (fig. 5.5(f)). The manufacturer supplies it as a standard pack of the following four components:

1. a Nonel tube – it is a transparent tube having 3 mm external dia. with 1.5 mm bore. Inside wall of this tube is coated with low concentration of explosive powder that posses the ability to conduct a shock wave at constant velocity.
2. a plain detonator with a delay element.
3. a connecting block, provided with a mini-detonator – which supplies the shock wave to the Nonel tube.
4. a starting gun and Nonel trunk line.

A special gun initiates the complete circuit that energies a Nonel trunk line, which in turn initiates each connecting block connected to it. The mini-detonator in the connecting block supplies the shock wave. It travels through the tube and emerges in the detonator as an intensive tongue of flame. The Nonel detonators are supplied in the range of 20 delay intervals each of 25 milliseconds, and six more each of 100–150 milliseconds.

Nonel is a closed system. Each hole is supplied with a separate Nonel unit and a simple manual operation connects each unit to the preceding one. The ignition impulse, once ignited, is transmitted from unit to unit via the connecting blocks. Several rounds may be fired in parallel. Since the system is non-electric, no balancing or instrument checking is required.

The system virtually eliminates accidents common with electrical blasting systems while at the same time radically simplifies the blasting operations.

5.6.1.3.6   Combined primadet-nonel system

Nowadays a combined system for a variety of precise non-electric hook ups (LP & MS) for underground applications is available. The primadets are connected with a Nonel shock tube instead of primaline.

5.6.1.3.7   The hercudet blasting cap system[14]

This system also eliminated all the hazards associated with the use of electric detonators. It is practically noiseless. It consists of the following three major elements (fig. 5.5(h)):

1. A special aluminium shell Hercudet detonator, having a delay element and two plastic tubes in place of two leg wires (as in an electric detonator) (fig. 5.5(h)).
2. Hercudet connectors for connecting lengths of tubing between adjacent holes in the circuit.
3. Hercudet blasting machine (with bottle and tester).

In the Hercudet system the detonators are connected in the circuit as shown in figure 5.5(h). To fire the round, the valves on the bottle box are opened to charge the blasting machine with the firing mixture of fuel and oxidizer, the 'arm' button is pressed for a short time, and then the 'fire' button. This initiates the gas mixture in the ignition chamber of the blasting machine, resulting in a detonation that proceeds at about 300 m/sec (1000 ft/sec) and initiates all the detonators.

The Hercudet system has been developed in the USA. This non-electric system does not require any detonating cord, as the other non-electric systems such as primadet, anodet etc. need. This factor obviates excessive noise resulting from most other non-electric systems.

The system consists of a delay blasting caps appearing like the conventional ones, but with hollow plastic tubes replacing wires or detonating cord. Tubes have no explosives or other filling or coatings. In use the caps are connected together via a tube circuit and when all connections are made, the hook up is checked for continuity. After thus proven, a mixture of fuel oxidizer gases is introduced to fill the tubing. A spark produced in the ignition chamber within the blasting console then causes reaction to travel at a speed of 200 m/sec. throughout the circuit activating all the caps.

5.6.1.3.8   Advantages of short delay blasting

The advantages of the blasting with the use of short delay millisecond detonators comparing the same with half second or instantaneous delays are as under:

• Reduction in ground vibrations
• Reduction in air concussion

- Reductions in over-break
- Improved fragmentation
- Better control on fly rock.

5.6.2 Fuse/cord system[10]

5.6.2.1 Safety fuse

William Blackford in 1883 introduced safety fuse to initiate gunpowder black powders.[10] Safety fuse consists of a core of fine-grained gunpowder black powder, wrapped with layers of tapes or textile yarns and waterproof coatings, to guard against moisture and shock. Its rate of burning is 600 mm/min. It is available in a coil of 915 m (3000 ft).

5.6.2.2 Detonating fuse/cord (DC)

Detonating fuse is a cord having a primary explosive, such as PETN, as its core and warping of textile fibers, wire and plastic coverings around this core. Its VOD is around 6500 m/sec. Its external diameter is in the range of 4 to 10 mm (0.15 to 0.4 inches) with a core load in the range of 8–60 grains of PETN/ft or 10–15 gms/m. Special types of DCs are available with varying core loads such as: Seismic cord with 100 gr./ft for seismic work; RDX 70 Primacord with 70 gr./ft for oil well perforating; PETN 60 plastic with 60 gr./ft for oil well servicing; Plastic Reinforced Primacord with 54 gr./ft for under-water blasting; a Detacord with 18 gr./ft and B line with 25 gr./ft for secondary blasting.

DC is safe to handle, extremely water resistant and capable of transmitting energy of a detonator to all points along its length. DC detonators are not required to be put inside the holes. Some of blasting powders like ANFO requires a greater initiating effect through out its charge column, and DC can fulfill this requirement very well. It can be initiated by using a plain or electric detonator. To blast a number of holes, the DC is inserted into the holes by lacing it to a primer cartridge or threading through a cast booster. The DC coming out from each of the holes (as a branch line) is connected to a common trunk line by strapping (taping), clove hitch or by a plastic connector. The detonator, plain or electric (of no. 6 strength) is lashed with tape, with its base pointing in the intended direction of travel of the detonation wave. The prevalent cord connections such as L joint, Double joint, Clove hitch joint and Lap joint have been shown in figure 5.6(e). Choice is governed by the blasting circuit or the design.

5.6.2.3 Igniter cords (IC)[10]

It is cord-like in appearance and when ignited the flame passes along its length at a uniform rate. These are available with three rates of ignition; Fast – @ 11.5 sec/m (3.5 sec/ft, black in color); Medium speed @ 16–31 sec/m (5–10 sec/ft, green in color) and; Slow @ 50–65 sec/m (15–20 sec/ft, red in color). They can be used in surface mines, non-gassy and metalliferous underground mines, and tunnels for lighting any number of safety fuses in a desired sequence. IC connectors are required to use this cord, as shown in figure 5.5(d).

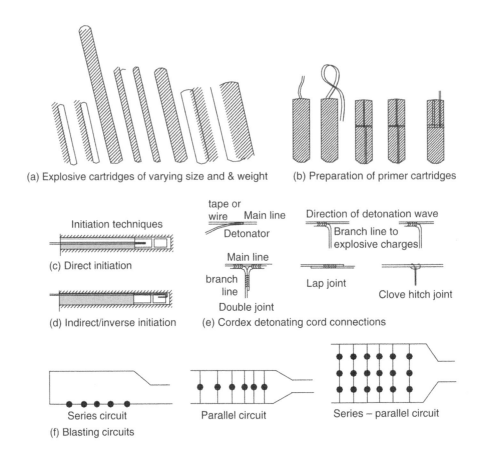

(a) Explosive cartridges of varying size and & weight

(b) Preparation of primer cartridges

Initiation techniques

(c) Direct initiation

(d) Indirect/inverse initiation

tape or wire   Main line

Detonator

Main line

branch line

Double joint

Direction of detonation wave

Branch line to explosive charges

Lap joint

Clove hitch joint

(e) Cordex detonating cord connections

Series circuit

Parallel circuit

Series – parallel circuit

(f) Blasting circuits

Figure 5.6 Explosive cartridges, initiation procedure, blasting circuits and detonation cord connections.

5.7 EXPLOSIVE CHARGING TECHNIQUES

Apart from the manual charging, use of ANFO loaders to charge the holes have been described in the preceding sections. Given below is the brief description of some other charging devices that are used.

Russian Drum type charge loader with mixer for dry granulated explosive and slurry charging.
The important features include: Dry explosive from the hopper is fed to the mixing chamber where it gets mixed with water, and conveyed though the hose to the hole to be charged. The charging tube/hose is withdrawn gradually. A typical loader of this type has the following specifications:

Hole dia.   – 60–160 mm
Loading depth, m   – up to 50 m
Inclination of hole – Any
Av. productivity   – up to 6 tons/hr
Air flow rate   – 10 m³/min
Reach in m   – up to 250 m

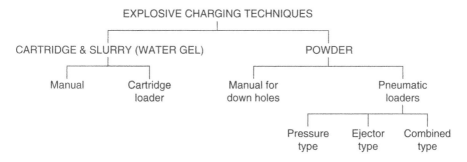

Figure 5.7   Various explosive charging techniques.

5.7.1   Water gel (slurry loader)

It is available for loading cartridges of less than 1-inch diameter. This product is liquid when manufactured but 'gels' after few hours. Use of pneumatic loading allows cartridges to the hole through the hose safely and quickly. It can be used for charging fans and rings. Given below are some its important features:

Largest cartridge size = 38 mm dia.
Loading hole size = 100 mm (max.)
Charging up holes up to = 60 m length.
Loading rate = 10 times faster than the conventional tamping stick method of loading.[4]

These loaders are useful for the pneumatic loading of the watergel cartridges into vertically up ring and fan holes. Loading is uniform and consistent. Their application in tunneling and drivage work for holes up to 3 m lengths is limited as there is no saving in time but the charging of the holes is uniform, which in turn, gives better results. The manufacturers of watergel explosives manufacture the watergel cartridge loaders. DuPont is one of them.

In order to charge the explosives of different types in the shot hole, blastholes and big blastholes, the techniques used have been summarized in the line diagram shown in figure 5.7.

5.8   BLASTING ACCESSORIES

5.8.1   Exploders[10]

These are the machines designed to fire the electric detonators. As shown in figures 5.9(b) and (c), these machines can be classified as *Generator (magneto)* (fig. 5.9(b)) *type* and *Condenser discharge type. The generator type* exploder works on the principle of an electric generator through which the current can be generated either by the rack bar mechanism or a twist handle mechanism. The current generated is used to fire the blasting caps connected in a circuit. In these types of exploders until a certain minimum pre-fixed voltage is generated, it is not transmitted to the external blasting circuit, to avoid any misfire due to insufficient current (electric power). These exploders are manually operated so that power can be generated any time, but require skill in handling and use. Their repair is simple and these are useful to fire multi-shots.

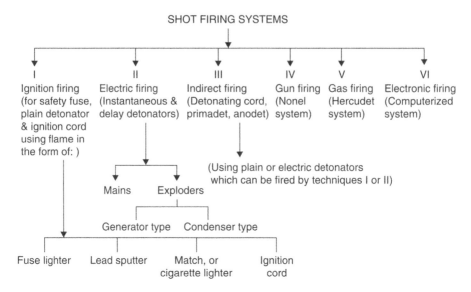

Figure 5.8   Classifications of firing systems for initiating explosives of different kinds.

Condenser discharge types (fig. 5.9(c)) of exploders are designed for multi-shot firing. Their basic source is either a low voltage dry cell battery or an electromagnetic generator. When a low voltage battery is used, first of all, the low voltage is converted to high voltage through DC to DC converter. The high voltage so generated charges the capacitor. When capacitor is fully charged a neon lamp indicates it. The voltage is discharged to the external blasting circuit connected to the exploder. It is light in weight and compact in size comparing with the magneto type of exploders of the same capacity. It is easy to operate but discharge of dry battery may affect its performance. One of its drawbacks is that the voltage from the capacitor is not fully discharged to the external circuit and some residual voltage remains in the capacitor, which in turn, may fire another circuit accidentally. The peak current can become high if few shots are fired, thereby causing the fuse head explosion and side burst of the detonators.

Electric energy from *power mains* is also used nowadays when heavy blasting for underground metal mines or in surface mines is undertaken. For this purpose the blasting cable is laid away from the service lines such as compressed air, water, ventilation ducts etc. Safety features such as a fuse box with main switch; a firing box and a short circuiting box are used when firing by mains. The circuits can be connected in series, parallel or series-parallel, as the case may be (fig. 5.6(f)).

5.8.2   Circuit testers[10]

In electric firing it is essential to check for the resistance of the circuit, its continuity and presence of any short-circuiting, if any. This is achieved by the use of *galvanometer* and Blaster's *multimeter*. A *galvanometer* (fig. 5.9(a)) is used to check the resistance of the individual detonators and the resistance of the complete circuit. A *multimeter* can be used to detect any current leakage in leg wires or blasting cables.

Blasting machines

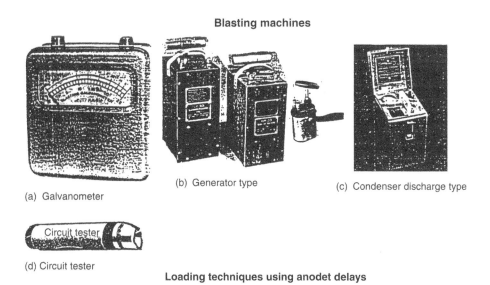

(a) Galvanometer

(b) Generator type

(c) Condenser discharge type

(d) Circuit tester

Loading techniques using anodet delays

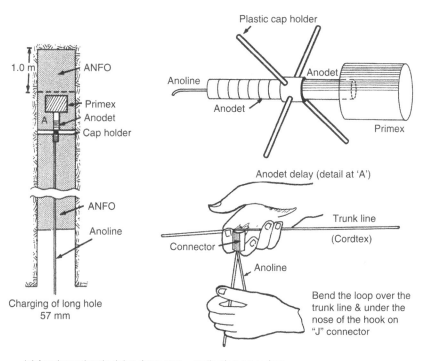

Charging of long hole
57 mm

(e) Anodet antistatic delay detonator – application procedure

Figure 5.9   Blasting accessories.

Sometimes stray current is available due to leakage of current from the external sources other than the exploder. This may prove to be dangerous. The multimeter can detect this. A dry cell battery operates the multimeter but its current is kept within the safe limits so that testing of the circuit and detonators can be carried out safely. This is designed to test voltage, resistance (ohms) and current in milliamperes. It can be used to measure voltage output from an exploder.

A *CIL Circuit tester*, manufactured by CIL, is available in a handy cylindrical housing (fig. 5.9(d)). It can test circuit resistance up to 75 ohms. However, before use of any of these testing appliances in the mines, approval from the competent safety authorities should be obtained.

5.8.3 Other blasting tools

The other blasting tools include: Crimper to crimp safety fuse into plain detonators; Pricker made of wood or a non ferrous material to prick into an explosive cartridge to prepare the primers; Knife to cut safety fuse; Stemming rod; Scraper; Flame safety lamp (in coal mines); Shot firing cable; Stop watch (when safety fuse is used) and suitable warning sign-boards or signaling arrangement.

5.9 FIRING SYSTEMS – CLASSIFICATION

Sequential firing: In many applications it is desirable to fire shots not instantly but in a sequential order. For different initiating devices/systems this is achieved in the manner described below. The line diagram presented in figure 5.8 can summarize various shot firing systems described.

5.9.1 While firing with a safety fuse

A safety fuse can be ignited by match-head, cigarette lighter or other lighters such as hot wire fuse type, pull wire fuse type etc. meant specially for this purpose. The other way is with the use of IC (Ignition cord) – which is first of all ignited by any of these lighters. To achieve a sequential firing while using a safety fuse any of these practices can be adopted:

- Cutting fuse of different length and/or lighting them in a desired order.
- By connecting standard lengths of safety fuses (exactly of same length) with IC in a desired order.

5.9.2 Firing with electric detonators

This is achieved by the use of long delay (half second) and short delay (millisecond) detonators. In coalmines non-incendive type detonators having copper tubes are used. The electric detonators charged in a face could be connected in series, parallel or series-parallel (fig. 5.6(f)). Series circuit should be preferred while firing up to 40 shots. If number exceed this, then series-parallel connections should be made.

5.9.3 Non-electric systems

Using detonating cord: To achieve delay with DC a millisecond connector is used. Its construction details are shown in figure 5.5(d). The other non-electric system includes use of Anodes, Primadets, Nonel and Hercudets. Description of this system has been dealt with in the preceding sections.

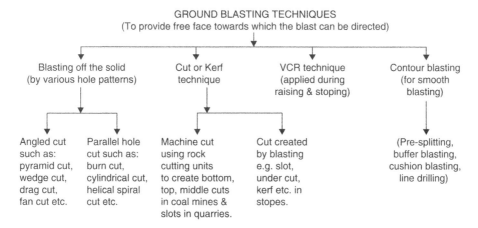

Figure 5.10   Ground blasting techniques.

5.10   GROUND BLASTING TECHNIQUES

In order to blast the in-situ ground from its original place, apart from the use of different types of explosives and their initiating devices, the techniques outlined in figure 5.10, need to be applied. Selection of these techniques is based on the type of the drivage work to be undertaken. Details of these techniques have been described in the following chapters, wherever appropriate.

5.10.1   Control/contour blasting[14,25,27,28]

In order to minimize the damage from blasts the following techniques are used:

• pre-splitting;
• cushion or trim blasting;
• smooth blasting & buffer blasting;
• line drilling.

5.10.1.1   *Pre-splitting*

The idea of pre-splitting is to isolate the blasting area from the rest of the rock formation by creating an artificial crack along the theoretical excavation plane. This technique involves drilling of a last row of holes of smaller diameter than those for the main blast at closer spacing. De-coupled explosive charge is used during blasting. This row of holes is blasted prior to the main rows, thereby creating a fracture, or joint plane in the rock mass before firing the production blasting. In this technique holes of 50–100 mm diameter are usually spaced at 10–20 times the hole diameter (fig. 5.10(a)). The guideline given in figure 17.14 could be adapted. The holes are charged with about one tenth of the normal charge. This could be blasted together with the main blast but given an initial delay, the pre-split row must be blasted first. Guidelines given by DuPont, as shown in table 9.14, could be used.

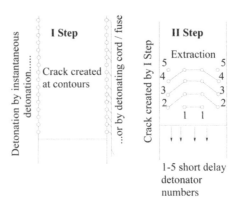

Figure 5.10(a)   Pre-splitting as applicable during bench blasting. The two stages/steps illustrated.

This technique is used mostly in surface blasting like road cuts, foundations where concrete lining or structural concrete is poured directly against the rock wall, etc. For best results, detonating cords or instantaneous detonators should be used for initiation. Noise may be a problem, as pre-split holes should not be stemmed. Another problem can be excessive ground vibrations.

5.10.1.2 *Cushion blasting*

Single row blasting with decoupled and/or deck charging; the spacing between the holes is kept greater than in the previous technique of pre-splitting. In surface mines when cushion blast holes have the same diameter as that of the production or main blast, it is known as 'trim blasting' or 'slashing'. The hole diameter varies from 50 mm (2″) to 164 mm (6.5″). The charge is usually fired with no delay, or minimum delay, between the holes. Detonating cord is the best way of initiation where noise or air blast is not a problem.

Holes are charged by tapping (tying) small diameter cartridges of 25–32 mm diameter to DC. They are spaced at a distance of 30–50 cm depending upon hole diameter The space between charges and hole walls is filled with inert material such as sand or crushed rocks. Executing the main blast including mucking the broken rock before the cushion blast is mandatory. This technique functions well in incompetent formations[28] for surface excavations and requires less drilling than pre-splitting.

5.10.1.3 *Smooth blasting & buffer blasting*

This technique is mainly used for underground tunnels, large caverns, chambers and openings. Smooth blasting was developed in Sweden during the 1950s and 1960s. In this technique extra care is taken while drilling and blasting the line holes (peripheral holes in the case of tunnels, and last row of holes during bench or surface blasting), as described in sections 9.3.3 and 17.6.4. Please refer to table 9.4 also. This technique is also known as 'buffer blasting' when used in conjunction with surface excavations.

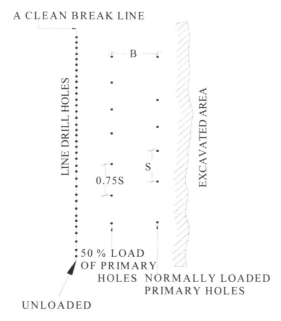

A CLEAN BREAK LINE

B

LINE DRILL HOLES

EXCAVATED AREA

0.75S

S

50 % LOAD
OF PRIMARY
HOLES   NORMALLY LOADED
PRIMARY HOLES

UNLOADED

Figure 5.10(b)   Line drilling applied during bench blasting. B – Burden. S – Spacing.

5.10.1.4   *Line drilling*

When drilling is done skin to skin i.e. without any spacing, or with very little spacing; the technique is known as 'line drilling'. This technique is also popular for 'block separation' during quarrying of dimension stones, such as marble as shown in figure 17.23(b) and described in table 17.5.

When used for the large underground tunnels and caverns, holes of 40–75 mm diameter and 100–150 mm apart are drilled along the perimeter of the required excavation, as shown in figure 5.10(b). Holes should be drilled prior to blasting the main blast. These holes are not loaded but their presence protects the surrounding rock from the damage that may be caused by the main blast. Large amounts of drilling with precision make the process costlier than other techniques. However, it is preferred where explosive and other methods do not work efficiently, or there are technical restrictions. Its use becomes mandatory where even lightly charged explosives could cause damage beyond the excavation line.[28]

5.11   SECONDARY BREAKING[17]

Generation of unwanted chunks of ore and waste rock while mining any deposit with the application of different blasting techniques is unavoidable. Dealing with these large chunks, known as boulders, either at their place of generation or on grizzlies is essential in order to facilitate the process of loading, hauling and crushing. This ultimately makes the process of muck handling safe, productive and economical (fig. 5.16(b)). Jamming of muck in the working stopes underground is another problem, which requires certain techniques to deal with.

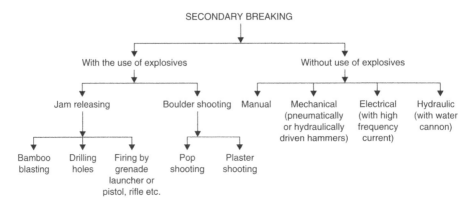

Figure 5.11 Secondary rock breaking techniques.

Secondary breaking is the process of breaking the over-sized boulders (lumps) which result during the primary blasting operations. Careful planning can minimize generation of these over-sized boulders but it cannot be completely eliminated. The over-sized boulder not only prevents the smooth flow of muck from the stopes to the draw points and ore-passes but many times chock/block their mouth. Handling of over-sized boulders gives undue strain to the loading and hauling equipment reducing their overall working life and efficiency. Optimum size of the muck eases the process of muck handling and ensures its smooth flow right from the stoping areas (or the place of its generation) to crushing units; improving overall productivity of a mine.

5.11.1 Secondary rock breaking methods

For over-sized boulders when treated with the aid of explosives, the process is known as secondary blasting. These boulders when brought to the grizzlies are reduced to the required size either by manual hammering, or by any of the modern rock breakers – mechanical or electrical. The line diagram shown in figure 5.11 has presented this classification.

5.11.1.1 *With the aid of explosives*[17]

5.11.1.1.1 Plaster shooting

In this process the boulder is shot by putting explosive over it and plastering it with a mud cap. Although the process gives higher powder factor comparing the pop shooting, its application can be justified in the stoping areas where less number of draw points are available, and time for pop shooting cannot be spared due to production pressure, or where pop shooting facilities do not prevail.

5.11.1.1.2 Pop shooting

The pop shooting ensures effective breakage of the boulder due to explosive concentration in the small diameter shot holes drilled in the boulder to be dealt with. In this technique in comparison to plaster shooting, better shattering effect with low powder factor can be achieved but a large number of draw points should be available to perform continuous drilling and blasting operations.

5.11.1.1.3   Releasing jammed muck from the draw points[1,17]

Jamming of the muck near the brows of the draw points in the troughs or funnels of the working stopes is a day-to-day problem in the mines. In order to release the jammed muck, in most of the cases, neither mucking equipment nor personnel are allowed to approach it; hence, this is tackled from a remote and safe point. Bamboo blasting is a popular method applied to release or blast the jammed muck. As shown in figure 5.12(a), this technique involves tying the explosive cartridges to one end of a bamboo of the required size, and then putting it in contact with jam to be released, and blasting the charge which ultimately releases the jam.

Many times in spite of repeated bamboo blasting, the jammed boulders do not roll down. In such circumstances at many of the mines, with prior approval of the safety authorities, the jam is released by the use of a small machine gun, rifles, throwing hand bomb or by firing a grenade launcher (fig. 5.12(c)).

5.11.2   Without aid of explosives

5.11.2.1   Mechanical rock breaking[17]

5.11.2.1.1   Manual breaking

In low output mines with surplus manpower this method of breaking boulders at grizzlies, with application of a sledgehammer manually is in practice even today. The method is slow and hazardous to the personnel carrying out this operation.

Another method of boulder breaking is the use of pneumatic hammer, which can be operated manually by one or two persons, is used on grizzlies. The pneumatic hammer in its mechanism is like a jackhammer except that it does not have the rotation mechanism and imparts only the hammering action. The operating compressed air pressures ranges from 4.5 to 6 kg/cm^2.

5.11.2.1.2   Mechanical rock breakers[17]

In mechanized mines for smooth flow of the muck to keep the grizzlies clean, specially those which are feeding the ore passes or the primary crushers, is an important task to be planned. Installation of mechanical breakers on such grizzlies has become a routine feature throughout the world. According to the output and strength of rock, any of the rock breakers (fig. 5.12(e) & (f)) described below are used.

5.11.2.1.3   Hydraulic rock breakers[17]

In this type of breakers hammering as well as boom movements are carried out with application of hydraulic power. The machine is either mounted on a concrete base or can be installed on a mobile van. A typical breaker of this type, as shown in figure 5.12(e), has these details: The breaker consists of a set of booms each articulated by hydraulic cylinders with a maximum horizontal reach of 8 m and maximum vertical

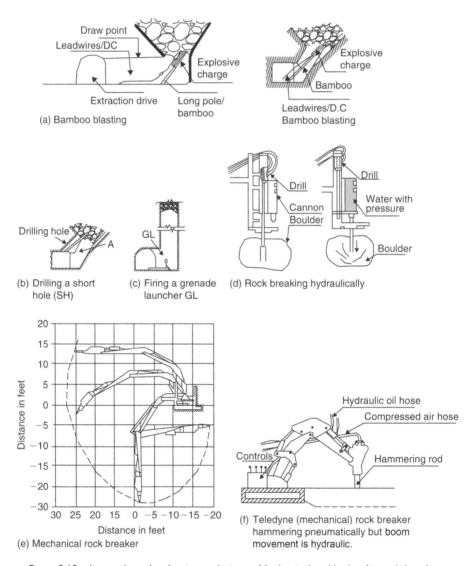

Figure 5.12 Jammed muck releasing techniques. Mechanical and hydraulic rock breakers.

reach up to 6 m. The boom can swing up to 270°. The front boom has a handle capable of 180° rotation. The hammer attached to the boom is designed for 6 blows/sec with an impact energy of 179 kg-m. per blow. The demolition tool attached is 100 mm in dia. and 600 mm long. The oil pump is run by 40 H.P. (30 kw) electric motor. The breaker is well suited to medium-hard rocks.

5.11.2.1.4   Teledyne rock breaker

This is a widely used breaker in mines for its application on grizzlies in underground as well surface mines (fig. 5.12(f)). It is suitable for hard rock and differs from the hydraulic breaker by having the hammering action pneumatically. Mechanical rock breaking is safer, efficient and economical and in use very widely.

5.11.2.2 Electrical rock breaking[15]

The electrical energy can be converted into a thermal, magnetic and mechanical power, which in this case is utilized to fracture the rock. Every rock mass depicts certain electrical properties such as resistance, inductance, and electrolytic conductivity in varying degrees. Some rocks can be classified as semi-conductive with respect to their ability to break electrically. This means they become conductive at some critical voltage level. There are several methods to induce conductivity.

The metallic ores such as magnetite, hematite, pyrite, galena, copper ores, titanium and many others increase their conductivity at some critical voltage level and carry the current through a network of conductive zones. Heating of the rock a few hundred degrees takes place only in isolated areas within the rock mass. The average temperature during fragmentation would increase by only a few degrees.

5.11.2.2.1 Rock breaking by the use of high frequency current[15]

This method of rock breaking is based on the following principle (fig. 5.13).[15] Current of a certain frequency is passed by the direct contact with the rock subjected to disintegration. Because of the action of the high frequency electric field and of the conducting current, the rock situated between contacts is heated rapidly and undergoes thermal disruption. Dielectric rocks or rocks of poor conduction thus become the conductors through the breakage channels. Continued heating of the current conducting channels generates thermoelectric tension in the rock that is sufficient to break it. The conditions of thermal breakage in various rocks depend upon their electrical and magnetic properties. Thus, each type of rock responds to a certain current frequency, usually to the order of $8000\,Hz$. The scheme corresponding to this method is shown in figure 5.13(a). The oscillatory circuit consists of inductance L_1 and capacitance C_1. The winding of the high frequency transformer consists of inductance L_1 and L_2. The contact terminals are connected to the egress of the secondary winding, made up of two to four turns, with the aid of coaxial cable. The electric circuit is formed by the secondary winding L_2, the contact terminals and the block of rock.

Figure 5.13(b) shows the principle of installation for rock breaking by simultaneous use of high and industrial frequency currents. In this case for electrical rock breaking and formation of current conducting channels, high frequency current is employed, whereas, to break the rock, alternating current of industrial frequency together with high frequency current are used. The inductance L_2 and capacitance C_2, and the contact terminals constitute the charged circuit whereas the inductance L_1 and capacitance C_1, the oscillatory circuit of high frequency.

Figure 5.13(c) shows the electrical set up of the rock breaking installation using high frequency current with impulses. The method consists of use of high frequency currents for the creation of channels whereas the disintegration of the rock is achieved by impulses received from the capacitor. Voltage of high frequency is obtained from the transformer L_1–L_2 through capacitance C_2. At the moment of the formation of breakage channels in the rocks, the relay RT interrupts the charge network whereas the capacitor C_1 is discharged through the gap D in the rock causing it to break.

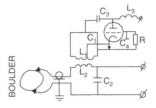

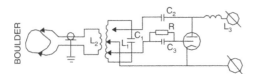

(a) Schematic presentation for rock breaking by the use of high frequency current

(b) Schematic presentation for rock breaking by current of high & industrial frequencies simultaneously

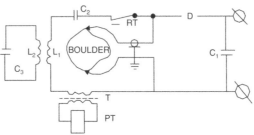

(c) Schematic presentation for rock breaking by high frequency current with impulses

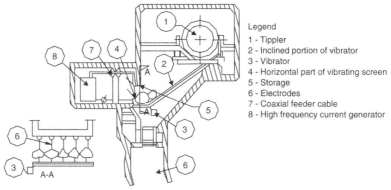

Legend
1 - Tippler
2 - Inclined portion of vibrator
3 - Vibrator
4 - Horizontal part of vibrating screen
5 - Storage
6 - Electrodes
7 - Coaxial feeder cable
8 - High frequency current generator

(d) Underground installation for secondary breaking by electrothermal method

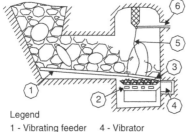

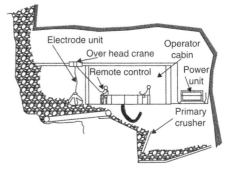

Legend
1 - Vibrating feeder   4 - Vibrator
2 - Belt conveyor   5 - Contact electrodes
3 - Vibrating screen   6 - Feeder cable

(e) Underground installation for secondary breaking by electrothermal method. Use of vibrating feeder shown

(f) Underground installation of an electric rock breaker before feeding underground crusher

Figure 5.13 Secondary breaking electrically at underground installations.

Figure 5.13(d) shows a scheme of automation of secondary breakage in a crusher with tippler. As shown in the section A-A five or six electrodes are suspended from the roof of the compartment by an insulating element for breakage of boulders.

Figure 5.13(e) shows the arrangement of crusher for secondary breaking by crushing with a vibrating feeder. It is always advantageous to carry out secondary breakage by combination of two methods: mechanical (jaw crushers) and electro-thermal (by high frequency current). This combined method of secondary breakage is termed as thermo-mechanical. Figure 5.13(f) shows an underground installation of rock breaker, the technical data for such a breaker are as under:

Electrical load – 100 KVA; Line voltage – 6000 V; Max. voltage – 2460 V; Frequency – 50 Hz.

Electrode dia. 75 mm; Electrodes – graphite; Max. size of boulder – 3 × 3 × 3 m Energy – 4.38 KWH/m^3.

The electric rock breaking[15] studies have shown the power consumption of various sizes of rock fragments as under:

- 1 KWH or less/ton. for obtaining fragment size in the range of 200–500 kg.
- 3–5 KWH/ton. for obtaining fragment size in the range of 200–50 kg.
- 0–15 KWH/ton. for obtaining fragment size less than 25 cm.

The electrical rock breakers are in use in some of the mines in Russia, US and many other countries. This has added advantages of economy, generation of no noise, dust and flying fragments.

5.11.2.3 *Hydraulic boulder splitter*

Atlas Copco[2] has developed a unit, CRAC-200, with a hydraulic cannon that shoots a water projectile into drilled holes, as shown in figure 5.12(d). The high water pressure created in the hole splits the rock. The unit consists of a rock drill, a water cannon and a feed mechanism. This set can be fixed on the floor or put on a mobile van. The splitting operation consists of drilling a hole of 34–36 mm dia. of 0.8 m depth. The cannon is swung into position over the drilled hole and then it is charged with 1.8 lit. water. The cannon forces the water projectile into the hole causing the boulder to split.

5.12 USE, HANDLING, TRANSPORTATION AND STORAGE OF EXPLOSIVES

Explosive is a commodity that cannot be allowed to be handled by any one else than an person authorized by the government, as it requires a special skill for its handling, use, transfer and storage, apart from the security reasons. Proper accountability is kept at every stage, right from receiving from the manufacturer up to its end use, to avoid any pilferage. Explosive is *very sensitive to shock, impact, jolt, friction, ignition, spark or tampering*, hence the important guide-line is that, all *precautions must be taken against all these factors during its storage, transportation, handling and use*. To safeguard against all these dangers, every country has its own rules and regulations. One will find

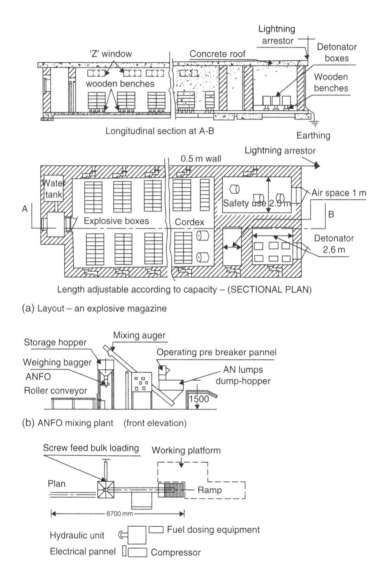

Figure 5.14   (a): Layout of an explosive magazine. (b): Layout of an ANFO mixing plant (Courtesy: Oman Mining Company, Oman).

that these regulations have been formulated by taking into consideration of these guidelines.

5.12.1   Magazine

It is a place where an explosive is stored (fig. 5.14(a)). It is constructed using specified specifications by the safety authority of any country and need to comply with certain basic design considerations. It should be located in an isolated and remote area, such as an area surrounded by hills etc. or by artificially created earth mounds. The electric over-head lines should be at least 91 m (300 ft or as specified by the safety authority) away. In general, the following guidelines are followed while constructing a magazine:

- The roof should be leak proof and the floor damp proof. Dimensions should be chosen as per the capacity.
- The doors and window should be of sufficient strength and constructed by fitting inner lining of wood. No iron nail, hinge etc. should be used. All hinges, locks etc. should be made of brass or any non-ferrous material such as copper, bronze etc. The idea is that any material that can produce a spark should not be used as a tool or construction material in the direct contact with the explosive. All doors should open outwards.
- The magazine must be fitted with an effective lightning conductor system and all iron and steel used in the construction of doors etc. should be properly bonded and earthend. Earthing should be checked periodically.
- Provision for water and fire extinguishers should be made.
- 'Z' type of ventilators should be provided near the floor and roof in the walls.
- Detonators must be stored in a separate annex, which can be accessed separately. Walls between explosive and detonator compartments should not be less than 0.9 m (3 ft) thick (or as specified by the safety authority).
- All detonators, explosive containers and fuse box etc. should be stored on wooden benches.
- Magazines should be fenced properly from all sides.
- Provision for its guarding by watchmen, round the clock, in rotation must be made. Only authorized persons should access the magazine.

In figure 5.14(a) the layout of a magazine having almost all the features as described above has been shown. In figure 5.14(b) an ANFO mixing plant at a copper mining complex in Oman has been shown.

Special vans are used to transport explosives. Containers of special design are used to transport explosives from the magazine to underground up to its place of use. Usually these explosive containers are kept in a special underground station, known as 'Reserve Station' before carrying them to the face. The blaster transports detonators separately.

5.13   EXPLOSIVE SELECTION[1]

Selection of an explosive requires a review of the type of explosives available, size and type of blastholes usually drilled, blasting theories and techniques available. Experience of the planner and past performance also plays an important role.

While blasting of any kind, the rate of release of energy, which is measured as velocity of detonation is of prime importance The relation between the borehole dia. and velocity of detonation varies as per blasthole dia. for any explosive, as shown in figure 5.3(a). Also there is a definite relation between VOD and explosive density, as shown in figure 5.3(b). Thus, to a particular dia. of blasthole a matching VOD can be selected and for a desired VOD range, a commercial explosive of a particular density can be chosen.

But during this selection: the locale, fume characteristics, degree of fragmentation, type of profile and, above all, the cost of explosives will be the main considerations.

Here the locale signifies use of explosive for surface or underground mines or tunnels, and in underground also whether for development or stoping operations. In an underground situation fume characteristics will play a very crucial role whereas in surface mines this may not be the major consideration. In underground coal mines protection against fire and explosion of the methane gas due to blasting will be the main criteria. The strength of rock, degree of fragmentation and type of free face available during a particular blast mainly govern explosive strength required. In u/g non-coal and metal mines, and tunnels where heavy blasting can be undertaken, reduction in blast vibrations plays a significant role.

A proper explosive selection, its judicious utilization and quality work in the blasting operations makes the process safer, economical and productive.

5.14   BLASTING THEORY[3,7,12,13,14]

As shown in figure 5.15(e), when a cylindrical charge is fired in a blast-hole, the detonation moves up the explosive column from the primer, a high pressure stress wave travels into rock mass. The positions of detonation waves and stress waves are as shown in this figure at different times. A horizontal section through this charged blasthole, figure 5.15(b), shows how the area surrounding the hole is divided into radial fractures at different point of time (zones 1 to 5) by the compression shockwaves. These waves from the free face are reflected back as tensile stress waves [fig. 5.15(b) I – A, B, C; II, III]. Since rocks are weaker in tension than compression, these tension waves cause more and more fracture to rock mass (fig. 5.15(c)). Desired fracture or fragmentation will occur when there is proper burden and the rock mass subjected to this phenomenon is free from the natural discontinuities such as fractures, joints etc. In any blasting operation only 3% of the explosive energy is used by the compression wave and boulders will be generated if this energy is not sufficient to return back after traveling up to the free face. The compression waves only enlarge the radial cracks but tension waves cause the rock to fragment.

The rapid expansion of the gases in the blasthole causes '*flexture or bending*' (fig. 5.15(c) and (d)). The gas pressure also causes radial cracks through the rock mass up to the burden and then its displacement. Figure 5.15 illustrates all these mechanisms.[3,7,8,12,13,14]

For a spherical charge the crater theory as described in section 13.10 should be used.

5.14.1   Adverse impacts of explosives

As stated above in section 5.1, an explosive results in fragmentation, throw, vibrations and noise. Explosives when not judiciously utilized may result in over-break (fig. 5.15(f)) or under-break. Over-break involves cost in terms of:

- Excessive drilling that might have been done.
- Excessive explosives (or more charge) used than required.

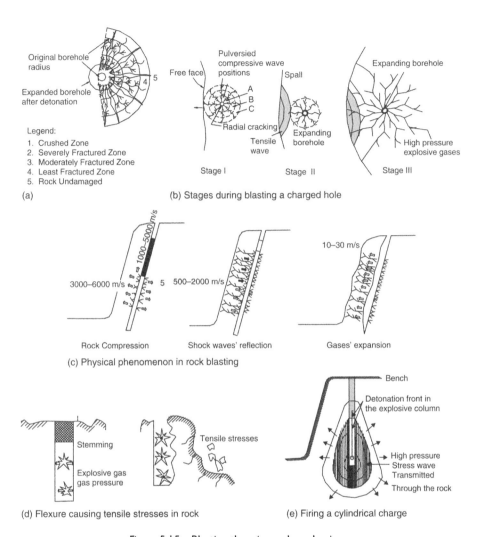

Figure 5.15   Blasting theories and mechanisms.

- Over-break causing ore dilution (or mixed with undesirable rocks) and generation of an additional amount of muck, thereby causing additional handling cost of unwanted rocks.
- Formation of loose rock and its scaling.
- More ground broken than the designed profile, requiring additional costs for its support and reinforcement.

Similarly under-break would result in reduced advance blasting round in tunnels, and irregular and under-sized excavation that will need repeat drilling, blasting, scaling, mucking and other operations. This is how productivity and costs of the operations are jeopardized when explosives, drilling and blasting operations do not properly match.

5.14.1.1   *Ground/land vibrations*

As described in the preceding sections, explosives when detonated produce shock waves. The unutilised part of shockwaves for the purpose of rock-fragmentation causes vibrations in the neighbouring/surrounding rock mass and structures that are present within a certain radius from the blast site. The impact of shock waves to the surroundings would depend upon the geology and geo-mechanical properties of the surrounding rocks. Geology includes the mean type of rocks and the presence of discontinuities if any. Fissured and/or jointed rocks could dampen vibrations.

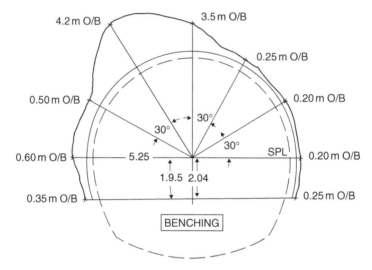

Figure 5.15(f)   An over-break at a tunnel of a Hydro-project. Boldened profile represents the over-break.

Table 5.6   Peak particle velocity and its influence on nearby structures.

Peak particle velocity, mm/sec.	Effects
600	New cracks form in rock mass
300	Fall of rock in unlined tunnels
190	Fall of plaster and serious cracking in buildings
140	Minor new cracks, opening of old cracks
100	Safe limit for lined tunnels, reinforced concrete
50	Safe limit for residential building
30	Feels severe
10	Disturbing to people
5	Some complaints likely
1	Vibrations are noticeable
0.1	Barely perceptible vibrations

Geo-mechanics refers to seismic velocity, density and elastic constants of rocks. The factors that influence this phenomenon most are:

(a) Explosive and related parameters[20,22,25]

☐ Charge weight/delay. This is measured as particle velocity i.e. the velocity at which a particle of ground vibrates when hit by a seismic wave. It is a measure of ground vibrations. It is expressed as:

$v \propto Q^a$. The value of constant 'a' as given by USBM (United States Bureau of Mines) is 0.8. v – particle velocity in mm/sec. Q is the explosive charge in kg. Also ref. sec. 16.6.4.1 and equation 16.5.

☐ The effect should also be considered in the light of soundness of structure in terms of its age, condition, and material and quality of construction.

☐ Table 5.6, illustrates peak particle velocity as a measure of vibration levels and effects on the surroundings.[22]

☐ Powder factor (explosive in kg/m³ or kg/ton): lowering the powder factor helps in reducing particle velocity to some extent but when the reduction is more it works contrary. In a trial when the powder factor was reduced by 20% from the optimum, the vibration levels when measured were two to three times higher as a consequence of poor confinement and spatial distribution of the explosive charge, causing lack of displacement and swelling energy.[25]

☐ Type of explosive: explosives with lower blast-hole pressure (refer equation 5.1) generate a lower level of vibrations. ANFO is one of those explosives. It generates fewer vibrations than slurry explosives and water-gels.

☐ Delay period; some of the studies undertaken suggest that a successive delay interval of 17 ms could eliminate the summing effect of vibrations.

☐ Blasthole diameter: larger diameter holes have adverse effects, as they could accommodate more explosives/unit length (refer eq. 5.1), and thereby sometimes it results in excessive weight/delay.

☐ Bench height: guideline[15] or as a rule of thumb; H/B >2, i.e. bench height should be greater than twice the burden to obtain good fragmentation, eliminate the toe problem and reduce ground vibrations.

(b) Burden and spacing: if the burden is excessive then the explosion gases find resistance to fragmentation and rock displacement, therefore easy escape for these gases is to be transformed into seismic vibrations. Too small a burden allows explosion gases to result in fine rock fragments and their uncontrolled throw. Part of the energy is also converted to produce more noise and air blast. Similar effects are noticed with improper 'hole spacing'. This phenomenon is also equally applicable, and follows almost the same logic during tunnel blasting.

(c) Sub-grade drilling: excessive sub-grade drilling often becomes a reason for uneven floor, large vibrations and unwanted expanses on drilling and explosive consumption. This should be, therefore, set (decided) after conducting trials. Inclined holes during bench blasting, rather than vertical ones, allows better use of energy with reduced vibrations.

(d) Stemming: too high stemming could of course give better confinement but often becomes the cause of excessive ground vibrations.

(e) Deck charging: when properly planned based on rock types, this yields better fragmentation, reasonable throw and reduced vibration level.

(f) Distance: as distance from point of blast increases the vibrations diminish, as per the following relation: $v \propto 1/D^b$. Value of 'b' as given by USBM is 1.6. This magnitude can reduce rapidly if soil as an over burden is present; because a large part of energy is used in overcoming friction between particles and in displacing them.[25]

5.14.1.2   *Air blast and noise*

Air blast is the pressure wave that is associated with the detonation of an explosive charge, whereas noise is the audible and infrasonic part of the spectrum: from 20 Hz to 20 kHz. Low frequency air vibrations, fewer than 20 Hz, are considered as air blasts. They can cause damage to the structures surrounding the blast site. The reasons for this phenomenon, as described by Linehan and Wiss[26] are:

- Ground vibrations that are caused by an explosion (the rock pressure pulse).
- Escaping gases from the blast-hole when the stemming is ejected (stemming release pulse). The under-stemmed holes or holes that are not stemmed could also increase its magnitude.
- Escaping gases through the fractures that are created during the blast (gas release pulse).
- Detonation of initiating blasting cord in the air.
- Rock displacement (air pressure pulse).
- Collision between the projected fragments.

High frequency vibrations are more prominent during a blast and they could be felt in windows, doors etc. Air blast over pressure greater than 0.7 kPa (0.1 psi) can break windows, and over pressure exceeding 7 kPa would certainly break windows,[6] and even in absence of damage, complaints and legal actions resulting from annoying levels of noise and vibrations can close the operation.

5.14.1.3   *Rock throw*

The function of explosives is not only rock fragmentation but also displacement from its original location. But if this throw is excessive, it becomes a nuisance and could damage the structures surrounding the blasting site. In urban areas, covering the blasting site using suitable means is almost mandatory. Explosives with high bubble energy, such as ANFO, produce greater rock throw than those having elevated strain energy, such as dynamites (NG based explosives). Insufficient stemming also leads to excessive throw. Bottom/indirect initiation should be preferred. The fissure and fractured rocks also facilitates throw as massive and homogeneous rocks. Proper blast design is the key to minimising the problem of excessive throw. As stated earlier throw is a desired and inherent property of explosives. In urban areas coverings of different types are used in practice to cover the blast site. Jimeno, Jimeno and Carcedo[25] suggested a covering with following features:

- economical and reusable;
- allowing gas escape;
- lightweight with high resistance, and ease in placing and shifting.

In practice, the following means are used:

1. Covering with sand: the material from a neighbouring site such as sand could be dug and spread over the blast. The recommended thickness of backfilled material should be equal to the stemming length and not below 0.8 to 1 m.[25]
2. Use of the used and discarded lengths of conveyor belts. The belts are overlapped and pinned to the ground using sandbags.
3. Metal screens or mesh, nylon nets, or rubber tires (with overlapping) are some of the other means that are used.

While covering the blast site using any of the practices mentioned above, it is necessary to make sure to protect the connections of the blasting round.

5.15 DRILLING AND BLASTING PERFORMANCE[1]

Performance of rock breakage or ground excavation with the aid of explosives can be assessed taking into consideration the indicators listed below.

5.15.1 Percentage pull

It is the ratio of length of round drilled to the effective linear advance obtained after blasting. Pull below 100% reflects inefficient drilling and blasting. This adversely affects the powder factor, which is the amount of explosive required per unit of rock blasted (i.e. explosive in kg/t or kg/m^3), and drills factor, which is the rock yield/m (i.e. t/m) of drilling.

5.15.2 Over-break factor

After blasting the face an additional breakage at the face is usually obtained as well as the designed one. The over-break factor is the ratio of the area of the face after blasting, including the over break, to the designed one. Over-break has adverse effects with regard to the face stability, cost of support, face configuration and amount of muck generation due to dilution caused. Contrary to this is under-break that can result in formation of loose, irregular face configuration, and poor drill and powder factors. Exact confirmation of the blasted face with the designed one reflects the skill of the operators. It gives optimum results.

5.15.3 Degree of fragmentation

Generation of over-sized or under-sized chunks/rock pieces has an overall impact on cost. This parameter has a direct relation with unit operations such as drilling, blasting, mucking, transportation and primary crushing. Hence, an optimum size of fragment is always warranted. The relationship between costs of these operations w.r.t. degree of fragmentation is illustrated in figure 5.16.[9]

5.15.4 Overall cost

The overall efficiency of drilling and blasting should be looked at in totality both during development and stoping operations. To choose an alternative means of rock

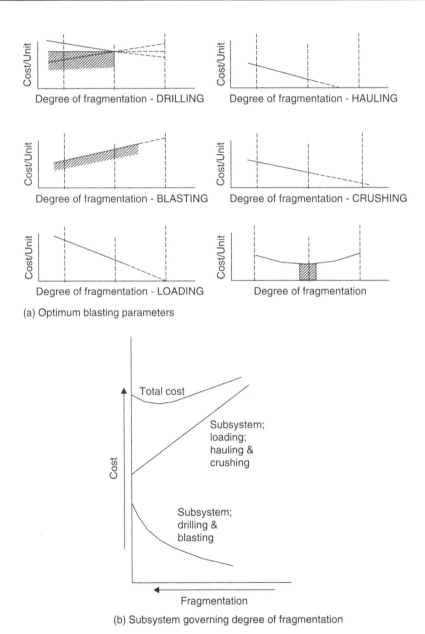

Figure 5.16   Parameters influencing degree of fragmentation and ultimately the costs.

breakage by any means other than drilling and blasting, overall cost of the operations should be calculated and compared. The mathematical relation in equation (5.5)[1] can be used for this purpose.

$$C_{tot} = C_d + C_{b1} + C_{b2} + C_m + C_{tr} + C_h + C_{cru} + C_{mis} \qquad (5.5)$$

Where: C_{tot} is total cost of mining/t of ore; and C_d, C_{b1}, C_{b2}, C_m, C_{tr}, C_h, C_{cru}, C_{mis}, are the cost/t of drilling, primary blasting, secondary blasting, mucking, haulage, hoisting, primary crushing and miscellaneous respectively.

Use of $+/-$ signs should be made when comparing the costs of two systems w.r.t. the unit operations used in this relation. $(+)$ Sign, if the cost is excessive than the one with the aid of explosives and $(-)$ sign, if it is less than that. In this manner an overall cost difference between various systems can be assessed and efficiency of the system can be judged.

Natural conditions vary from mine to mine (or one tunnel to another) and even within the same mine and, therefore, it should be bear in the mind that establishment of proper drilling and blasting practices and selection of a suitable method, design and equipment to perform these operations is a matter of experience of the planner and their proper execution through field trials and test results.

5.16 RECENT TRENDS IN EXPLOSIVES AND BLASTING TECHNOLOGY[18,24]

Explosives and Accessories

- Product placement mechanisms: in place of conventional cartridges use of bulk products for both surface and underground operations is gaining popularity. Use of delivery vehicles that boost blast accuracy and safety; high precision pumps; blending and measurement devices; robotic arms that place the product in the hole; and remote controls are some of the innovations in the explosives accessories that have taken place in the recent past.
- Yield-based outsourcing: another trend is outsourcing of blasting-related services, ranging from consulting on safety to providing comprehensive packages priced according to the volume of "shot rock" on the ground or ore processed. Operating companies tend to be highly cost-conscious. Objective-based optimization, as given in figure 5.18, would be the ultimate solution.
- Use of ANFO: blasting agent ANFO or ANFO-based emulsions is likely to continue in the days to come, while use of blasting agents (e.g., water gels) would gradually decrease due to their higher cost.
- Use of IT: blasting generally is conducted according to rules-of-thumb. With the use of IT, databases would be created to optimize blasting and reduce haulage and processing costs. All details like hole position, hole depth, nature and condition of holes, type and quantity of explosives; initiation system, sequence and delay timings would be recorded. A database of blasting results would also include information on: fly-rock, throw, muck profile, blast photographs and video recording. In fact, blast information needs to be made available online for easy retrieval and analysis. This helps in monitoring and reducing the blasting-related environmental hazards.
- Introduction of programmable blast detonators: solid-state programmable blast detonators are now commercially available (Cunningham, 2004). More precise delay timing and greater compatibility with remote-controlled loading of explosives and wireless detonation are some of the advantages claimed by this technology. However, these initiating systems have higher costs warranting their commercial viability to be established.

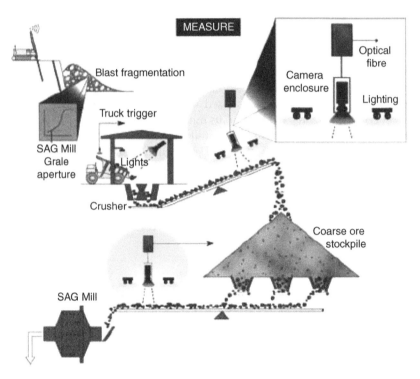

Figure 5.17   Mine-to-mill optimization concept being used by Julius Kruttschnitt Mineral Research Centre (JKMRC), Australia.[24]

- Electronic detonators: the successful use of these detonators to control ground vibrations from blasting is a classic example of research leading on to commercial deployment for the benefit of all the stakeholders (operators, regulators, consultants and local residents). Electronic detonators are set to have a major impact on the mining industry and tunneling projects.
- Mass blasting: the trend to carry out large blasts would increase, thus reducing the frequency of blasts to boost productivity and safety.
- Introduction of new tool and tackles: there would be increased use of new tools such as laser face profiling, borehole tracking for hole deviation measurements to calculate actual burdens at all points on rock face, muck pile profiling to determine throw, cast blasting, muck-pile models to determine swell & blasting effectiveness.
- Autonomous drilling: innovative practices in the area of drilling (as described in sec. 19.5.7), new bulk loading of explosives, performance measurement and the evaluation of blast outcome and productivity would be adopted. Ways to reduce generation of fines and dust from blasting operations would become available.
- Control/contour blasting: To get stable walls/slopes, contour blasting techniques (as described in sections 5.10.1, 9.3.3, tables 9.11, 9.12 & sec. 17.6.4) to reduce damage to remaining rock would be a usual practice.

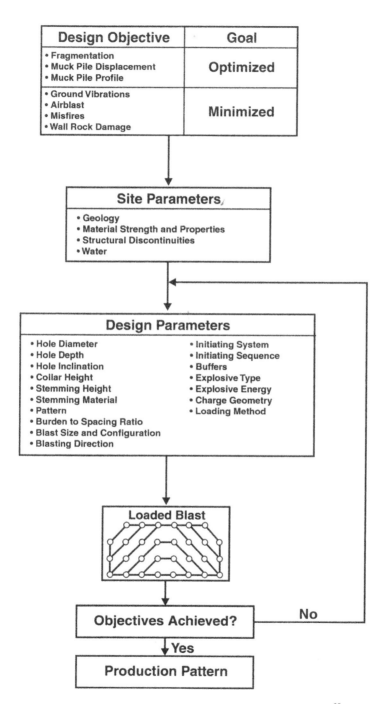

Figure 5.18   Blast design process, Atlas Powder Company (1987); Hustrulid (1999)[23] (www.smenet. org).

Blasting Results Assessment[18,24]

- Fragmentation analysis: Fragmentation analysis using image analysis methods is usual practice nowadays. However the next step which is in vogue is that fragmentation size distribution data is collected directly on a shovel, as described in sec. 19.5.7. This is possible by using extremely rugged hardware designed for the rough and tough working conditions of surface mines. The embedded computer is designed for a high-vibration environment, does not need any moving parts and is sealed for dust and moisture resistance.
- JKMRC Research on Mine-to-mill Optimization: The entire operation is taken into account, from the blasting process to the comminuting circuit performance and the interactions between them, in order to optimize the entire size reduction process, as shown in figure 5.17, Mine-to-mill optimization has been successfully applied in gold, copper, and lead/zinc operations throughout the world. There would be increased application of this concept in coal and aggregate production.

Security

- It likely that in the days to come safety and security during manufacture, storage, transport, supply, export, use or disposal of AN; the security would impose additional requirements to comply with.
- Any user whoever he or she may be would be required to undergo background checking.
- Records would be required to be maintained for all transactions, including the authorization given and also for use and disposal of explosives.
- Hence, there is a need to review explosives regulations to implement security checking for persons having access to explosives and ensure severe penalties for breaches of regulations.

5.17   CONCLUDING REMARKS

The energy that is not used in the rock fragmentation and its throw is sometimes more than 85% of that developed during the blast. This energy badly damages the structural strength of rock even sometimes outside the theoretical radius of influence. This is how rocks become weakened. Joints, fissures and cracks present earlier get wider, and even new fissures and cracks are developed. This ultimately results in the reduction of overall rock mass cohesion. This could cause over-break and even leaving fractured ground open, which could be a potential source of collapse. The recent trends in explosives and blasting technology, as detailed above in sec. 5.16, if implemented judiciously, using the blast optimum strategy which has been shown in figure 5.18[23]; could help in achieving improved results from this cheapest source of energy – explosives.

QUESTIONS

1. Calculate the detonation pressure of explosive ANFO if its density is 0.85 g/cc and the velocity of detonation is 3000 m/sec.
2. Calculate the explosive density if an explosive is capable of producing a detonation pressure of 30 kilobars and its velocity of detonation is 2900 m/sec.

3. Classify contour blasting techniques. Write down an application for each one of them. Describe them and illustrate with the aid of suitable sketches.

4. Classify detonators and write down the utility of each one of them.

5. Classify initiating devices/systems known to you. Give figures of detonators known to you. How can explosives be charged into holes? Name shot firing systems known to you.

6. Define explosives. Give a classification by a line diagram. What tasks does an explosive do?

7. Define peak particle velocity. Give a formula to calculate it. Mention applications of this concept. Is there any influence on the structures that are near the blasting site/face? Describe it.

8. Define secondary breaking. List techniques used for secondary breaking.

9. Define these terms: shothole, blasthole, big-blasthole, cap sensitive explosive, NG-based explosive, slurry explosive, blasting cap, dry powder, magazine.

10. Determine voltage necessary, if 40 electric delay detonators that have been charged in a face are to be blasted by connecting them: (a) in series, (b) in parallel. Consider the resistance of each detonator including its lead wires to be 1.5Ω, connecting wire resistance to be 5Ω and that of the blasting cable as 6Ω. Also consider the minimum firing current to be 1 A (in any series).

11. Differentiate between the following: (a) low explosive and high explosive. (b) dry blasting agent and wet blasting agent. (c) Safety fuse and detonating fuse. (d) Detonation and deflagration. (e) Primary and secondary blasting. (f) Plaster shooting and pop shooting.

12. Are the NG-based explosives called dynamites?

13. Draw the following circuits: (a) Connection of 50 'Anodets' up to exploder. (b) Connection for 'NONEL' system up to starting gun. (c) Firing one shot if 'low explosive' has been charged at the face.

14. Explosives when not judiciously used may result in over-break or under-break. Over-break involves cost in terms of what?

15. Give classification of ground-blasting techniques, specifying field of application of each one of them.

16. Give classifications of firing systems for initiating explosives of different kinds.

17. Give the correct composition (proportion) of AN & FO, while preparing ANFO, so that an oxygen-balanced mixture is obtained.

18. Give figures of any two detonators know to you. Suggest type of detonator to blast black-powder or gunpowder. To have good fragmentation and reduced vibrations which detonator will be better?

19. Give oxygen-balanced composition of ANFO. Is it suitable to use in a wet hole? Is it cap-sensitive? What type of explosive will be needed to initiate ANFO charged in a hole? How can ANFO be charged in up holes?

20. How do military explosives differ from the commercial explosives? Name some military explosives.

21. How do you manufacture ANFO? What will happen if fuel is added less or more than required?

22. If you are required to store 10,000 detonators for your mine, what provision shall you make?

23. Illustrate the influence of borehole diameter on the velocity of detonation (VOD).

24. Illustrate the techniques of explosive initiation using a primer in a hole. Mention a situation for which each one of them will be a suitable choice.

25. List blasting theories known to you and describe each one of them.

26. List common methods of rock fragmentation that are used in mines. If rock fragmentation at a mine is poor, what does this mean? How it could be improved? List the governing parameters/factors for 'rock-fragmentation' and how do you ensure the desired degree of fragmentation? On what 'Unit Operations' at surface as well underground mines, will the degree of fragmentation have influence?

27. List different methods available to fragment rocks. What methods of rock fragmentation will be suitable for a marble quarry?

28. Make a list of blaster's tools. What type of exploder will you use to blast 50 shots?

29. Make a list of the main blasting properties (characteristics) of the explosives which you will consider while selecting an explosive for a mine. Describe them briefly.

30. Make a list of the type of exploders available for use in mines. Suggest a type of exploder if 150 electric detonators connected in series are to be fired in a large underground mine producing lead and zinc ores.

31. Make a list of parameters influencing the degree of fragmentation, and ultimately overall cost, of mining.

32. Mention the name of a blasting powder which is least resistant to water, and also the explosive which is best suited for watery conditions.

33. Mention where you will use the following: (a) Ignition cord. Write down its burning or ignition rate range. (b) Safety fuse. Give its usual rate of burning and composition. (c) Detonating cord, cordex. Give its velocity of detonation. (d) Illustrate with help of a sketch, how a detonator is connected with cordex.

34. What is the minimum number of free faces required for an efficient blasting?

35. Name a low explosive know to you and what ingredients you will use to manufacture it?

36. Name a secondary explosive. Give its use. How do military explosives differ from the commercial explosives?

37. Selection of an explosive requires a review of a number of parameters. Could you list them?

38. Should we store explosive cartridges and detonators together in a magazine?

39. Suggest methods of releasing jammed or bridged muck from the draw points and ore passes.

40. Tabulate the composition of some slurries and emulsions known to you by taking into consideration the important features/ingredients.

41. The manufacturer supplies 'Nonel' as a standard pack having four components – list them. How does this system differ from the conventional electric blasting system?

42. What advantages can slurry explosive provide?

43. What are 'blasting agents?' List them. Write down an application for each one of them.

44. What are the desirable properties you will look into to select an explosive. Who is the right person to carry out blasting operations? What type of exploder will you use to blast 100 shots?

45. Name the cheapest explosive and also the costliest one. What type of explosive will be suitable to sink a shaft or well?

46. What are the main ingredients of an explosive? What is the meaning of a cap-sensitive explosive? Name some cap-sensitive explosives known to you. Name some oxidizers.

47. What are the static hazards associated with ANFO loading pneumatically. What precautions should be taken?

48. Under explosives rules, list the type of explosives that can be put in the same magazine. Can you store detonators and the same room where NG based explosive have been stored? Give a line diagram in plan and section, showing the layout of a large sized explosive magazine. What safety provisions are required to be made in such magazines?

49. What is a primer? What is known as blasting with the use of a kerf. Where is it suitable? Where can kerf be put with respect to a drive or tunnel face?

50. What is misfire? What is the best way to deal with it? What precautions should be taken during blasting? Calculate the rock yield after blasting a tunnel face of $7\,m \times 3.5\,m$ size, if the size of round is $2.8\,m$ and effective pull is 85%. Use rock density of $2.9\,t/m^3$. If the amount of explosive charged is $60\,kg$, calculate the powder factor.

51. What precautions should be taken during blasting to avoid misfire. Calculate the rock yield after blasting a tunnel face of $5\,m \times 3\,m$ size, if the size of round is $3\,m$ and effective pull is 90%. Use rock density of $3\,t/m^3$.

52. What type of circuits (connections) can be made to blast the electric delay detonators used in a development heading of an underground copper mine. Illustrate with the help of suitable sketches.

53. Where would you use the following types of explosives: (i) pyrotechnic (ii) primary (iii) secondary (iv) high explosives.

54. Where you would recommend use of: seismic explosives, permitted explosives.

55. Why is Ammonium Nitrate (AN) Fuel Oil (FO) mixture (ANFO) so popular? Write down the equation illustrating the resultant products of ANFO blasting. How ANFO can be charged in inclined upward holes in an underground mine having holes of $55\,mm$ dia, and 10–$15\,m$ long; suggest a procedure, equipment and type of detonators giving reasons for your selection.

56. Write down the methods of secondary breaking.

57. Propose explosives for the following situations:
    (i) During shaft sinking in hard rock below the water table.
    (ii) Sublevel stoping while charging 'rings' in dry conditions in an underground lead-zinc mine.
    (iii) Sublevel stoping while charging blast-holes in a 'slot' in wet conditions in an underground copper mine.

58. Write notes covering following aspects:
    (i) Electronic detonation – details as to how it differs from an electric detonator, its merits, limitations and scope.
    (ii) Safety precautions during handling, storage and transportation of explosives.
    (iii) Planning a heavy underground blast in a Pb-Zn mine using 20 tons of explosives at a time. How would you determine damage to the surroundings; write a formula for this, if any.

59. Suggest explosives for the following working conditions:
    a. An open cast limestone mine. The stratum is dry.
    b. An underground copper mine to produce medium-hard to hard ores. The deposit, in general, is dry. The mine is required to produce a bulk quantity of ore per day.
    c. An underground gassy coal mine.
    d. Holes drilled in a river bed under water.
    e. Development faces in a deep underground gold mine having hard rocks.
60. Give the field of application of the following techniques:
    a. Pre-splitting.
    b. Cushion or trim blasting.
    c. Buffer blasting.
    d. Line drilling.
    e. Smooth wall blasting.

REFERENCES

1. Agoshkov, M.; Borisov, S. and Boyarsky, V.: *Mining of Ores and Non-metallic Minerals*. Mir Publishers, Moscow, 1988, pp. 80, 96.
2. Atlas Copco – Secondary breaking; *Leaflets*.
3. Bauer, A. and Crosby W.A.: Blasting. In: B.A. Kennedy (edt.): *Surface Mining*. SMME, Colorado, 1990, pp. 548.
4. Champion, M.M.: Explosive loading equipment. In: *Underground Mining Methods Handbook*, W.A. Hustrulid (edt.). SME-AIME, New York, 1982, pp. 1084–1091.
5. Clark, G.B.: Explosives. In: *Surface Mining*, E.P. Pfleider (edt.), AIME, New York, 1964, pp. 341–354.
6. Crowl, D.A. and Louvar, J.F.: *Chemical process safety*. Prentice Hall PTR, New Jersey, 2002, pp. 253–254.
7. Dowding, C.H. and Aimone, C.T.: Rock breakage and explosives. In: Hartman (edt.): *SME Mining Engineering Handbook*. SMME, Colorado, pp. 722–735.
8. Ganpathy, B.: *Advance course on rock blasting*, IDL, Hydrabad, India, Nov. 1978.
9. Goddoy S.G. and Viera M.D.: Computerized model for design optimization blasting pattern in tunnels. *Tunneling '82*, IMM Publication; Editor Michael J.J. Jones.
10. Gregory, C.E.: *Explosives for North American Engineers*. Trans. Tech. Publ. Rockport, M.A., 1979, pp. 303.
11. Hartman, H.L.: *Introductory Mining Engineering*. John Wiley & Sons, New York, 1987, pp. 125–133.
12. Jimeno, C.L.; Jimeno, E.L. and Carcedo, F.J.A.: *Drilling and Blasting of Rocks*. A.A. Balkema, Netherlands, 1997, pp. 98–130.
13. Matti, H.: *Rock Excavation Handbook*. Sandvik-Tamrock, 1999, pp. 100–121.
14. Pradhan, G.K.: *Explosives and Blasting Techniques*. Mintech Publications, Buuvenswar, 1996, pp. 5–10, 53–80, 240–242.
15. Sarapuu, Erich.: Electro energetic rock breaking systems. *Mining Congress Jou*. Vol. 59, 1973.
16. Sundram, S.S. and Singh, B.: Blasting and its underground applications. *Technical renovations – Khetri Copper Complex*, 1978, pp. 1–15.
17. Tatiya, R.R.: Current practices of secondary breaking in underground metalliferous mines in India and abroad. *National Symp. on mechanics of mining ground*, BHU, 1982, pp. 129–136.

18. Bhandari, S.: Blasting technology for the next decade. In: International Seminar: Mine Advan. Tech. Jodhpur, India, 14–16 February 2009, pp. 229–36.

19. Dick, R.A.; Fletcher, L.R. and Andrea, D.V.: *Explosives and Blasting Procedures Manual*, USBM, IC 8925, Govt. Print. Off., Washington DC, 1983, pp. 105.

20. Franklin, J.A. and Dusseault, M.B.: *Rock Engineering*. McGraw-Hill, New York, 1989, pp. 14–34.

21. Gregory, C.E.: *Explosives for North American Engineers*. Trans. Tech. Publ. Rockport, M.A., 1979, pp. 303.

22. Hendron, A.J. and Oriard, L. L.: Specification for controlled blasting in Civil Engineering. *Proc. RETC, AIME*, 1971, pp. 1585–1609 (www.smenet.org).

23. Holmberg, R.; Hustrulid, H. and Cunningham, C.: Blast design for underground mining applications. In: Hustrulid and Bullock (eds.): *SME Underground Mining Methods*. Colorado, 2001, pp. 635–61.

24. http://www.jkmrc.uq.edu.au/ (accessed in 2009).

25. Jimeno, C.L.; Jimeno, E.L. and Carcedo, F.J.A.: *Drilling and Blasting of Rocks*. A.A. Balkema, Netherlands, 1979, pp. 24–26, 130.

26. Linehan, P. and Wiss, J.F.: Vibrations and air blast noise from surface coal mine blasting. SME, AIME Fall Meeting, 1980.

27. Matti, H.: *Rock Excavation Handbook*. Sandvik – Tamrock, 1999, pp. 101, 118–21. Sandvik-Tamrock – leaflets and literature (hard and soft copies).

28. Olofsson, S.: *Applied Explosives Technology for Construction and Mining*. Applex, Sweden, 1997, pp. 140–143, 180.

Chapter 6

Mucking, casting and excavation

Excavation of any kind generates dust, and it is injurious to health. Controlling its quality (size and shape) and quantity (concentration) could minimize this hazard.

6.1 INTRODUCTION

Material (rock or ore) fragmented with or without aid of explosives from a working face (which may be a tunnel, large underground chamber, open cut excavation at any working site or mines) is known as muck. The process of loading this muck into a conveyance for transportation away from the face is known as mucking. When this muck is discharged to an adjacent area, practically without moving the mucking equipment, and simply by swinging it, the process is known as casting. Excavation is the process of digging ground from its bank face (in-situ) and elevating by an excavator to discharge it either to a haulage unit, or an adjacent area.

Referring to a working cycle, figure 6.1, while driving tunnels of small to large size, with conventional (drilling and blasting) or tunnel borers, about one third of the cycle time is occupied by the mucking operation. This share increases to about 70%

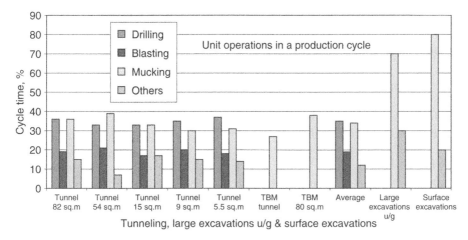

Figure 6.1 An account of mucking time in a production cycle. In a tunnel-bored (TBM) tunnel, ground support is a sequential operation, whereas, in other tunnels it goes side by side. (sq.m. – cross-sectional area of tunnel in m^2.)

when excavating large excavations underground (stopes) where the process of rock fragmentation can go side by side (simultaneously to mucking). While the scenario at the surface, where mucking, loading or casting occupies about 80 to 90% share; and thereby it becomes a most vital component of a working cycle. The importance of this basic aspect has long been realized and human endeavor has been to invent the largest possible earth moving equipment to handle the enormous amount of rock material that is lying as overburden to the useful minerals (ores) and also the enormous volumes that need to be excavated while constructing rail routes and roadways, canals, dams and many vital civil constructions including buildings. The largest earth moving equipment on earth, the Bucket Wheel Excavator, which has attained a daily production 240,000 m^3/day (12,000 m^3/hour) in Germany (fig. 6.10(b));[13] the dragline of as high as 17 0 m^3 bucket capacity[21] with an hourly output of 5700 m^3; and shovels of as large as 140 m^3 bucket capacity[21] to produce earth-material of 5200 m^3/hour, are the results of efforts to handle material at the lowest costs. All this equipment is believed to be the largest man made, dry land, and mobile structures on the earth. Creating a narrow and precise excavation at the subsurface is equally important and a variety of special sets of equipment to cater for this need have been devised.

The advantage of using bulk earth moving machines makes them attractive to take underground and therefore, use of dipper shovels, hydraulic excavators, load, haul and dumping units (LHDs) (Load-Haul Dumpers) with a bucket capacity of 10 m^3, for their use in large sized tunnels and stopes, has become a usual practice.[19] Narrow workings and tunnels of small size, which sometimes run kilometers together, require compact, efficient and small sized loaders. Sinking or digging ground downward in a very rough, tough, watery and narrow space is an operation of utmost skill that requires special types of mucking units.

Thus for mucking, ground excavation and casting, locales are many and so are the varied requirements in terms of output, capital available, size of excavation and some others. Due to this reason, today, many companies are in the arena to manufacture equipment of different range and specifications to suit industries' requirements. In this chapter efforts have been made to cover the majority of them.

6.2 MUCK CHARACTERISTICS[18]

Muck characteristics such as shape, size, volume, hardness, moisture content, angle of repose, abrasiveness, dryness and stickiness are dependent on a number of factors. These characteristics influence the material handling system. For the same size of tunnel excavation, soft ground yields less volume of excavation than a rock tunnel. The bulking factor[18] of rock excavation is 150–225%. Soft ground has a bulking factor of 105 to 160%. Bulking factor is the increase in volume which the excavated material undergoes by virtue of excavation. The material in its existing state remains in triaxial state of compression, but the excavation changes this triaxial to uniaxial state of stress that results in volume expansion. Bulking is measured by this percentage of volume expansion.

Stickiness is a very undesirable feature for muck removal. Loading, dumping and cleaning of sticky muck is a very time consuming and expensive process. Sometimes, inert material like sand has to be added to a sticky muck to make it workable.

The lump size of muck is an important characteristic for transportation by conveyor, pipelines and bucket wheel elevators. Wetness and stickiness are important considerations for all mode of transportation except those by hydraulic transportation. Abrasiveness is an important consideration for pneumatic transportation.

A good mucking and transportation system must respond and conform to the changes in muck size, shape, weight, and flow rate that might be expected during excavations for underground structures.

6.3   CLASSIFICATION

Based on the locale of their applications, mucking equipment can be classified into following two classes:

1. Underground Mucking Units
2. Surface – Excavation, Loading and Casting Units

6.4   UNDERGROUND MUCKING UNITS

For mucking from underground openings, including tunnels, several types of equipment are available. Some of the prominent sets of equipment are described below:

- Overshot loaders – Rocker shovel
- Autoloaders – Hopper loaders and LHDs
- Arm loaders
- Scrapers
- Dipper shovels and hydraulic excavators (shovels) [sec. 6.11, 6.12]

The classification of these loaders is given by way of a line diagram in figure 6.2. A suitable match of loading and transpiration units has been also shown in figure 9.11.

6.4.1   Overshot loaders[1]

These loaders pick up the muck from the face and discharge it to the rear without turning as illustrated in figure 6.3(a). In the rear either Granby cars or sinking buckets are deployed which can be replaced when filled. These loaders are track, crawler or wheel mounted and can be run using electric, compressed air or diesel power. Loaders of this type are composed of a bucket with a handle secured to rocker arm, bogie, and turntable with a winch for lifting bucket, 3 motors (two for traction and 1 motor for bucket movement) and the control mechanism. These loaders are manufactured by Eimco Company with trade names such as Eimco-21, 21B; Eimco-824; Eimco-630 etc. and in Russia[6] these are designated as ΠΠΗ type loaders.

These machines are simple in design and require minimum maintenance. Track or tyre mounted loaders find their application in faces having cross sectional area below 8 m^2 or so; whereas crawler mounted loaders such Eimco-630 (fig. 6.3(b)) are suitable for mucking from sinking shafts and drives with undulating and rough floors. However, this loader does not work well in a circular shaft having diameter less than 5.5 m.

In general, at the larger sized openings and tunnels these loaders are not efficient. Low productivity is resulted due to the fact that the operator gets fatigue very quickly

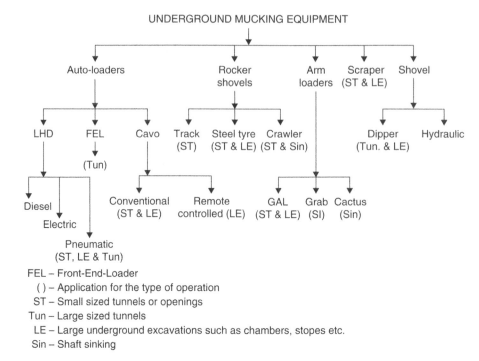

Figure 6.2 Classification of mucking equipment together with their applications for the underground excavation and tunneling operations.

by its continuous jogs and jars. The performance of these machines also depends upon the bucket capacity, which ranges from 0.2 to 0.6 m^3. The performance curves[1] are shown in figure 6.4(a).

In addition to the loaders described in the preceding paragraphs, side discharged loaders are also available with lateral unloading buckets. These types of loaders are used for horizontal workings of low height and are particularly suitable for the workings equipped with conveyors.

6.4.2 Autoloaders – Hopper loaders and LHDs

These are mucking and transporting machines under which the following two types of loaders are available:

1. Mucking and delivering.
2. Mucking and transporting.

6.4.2.1 Autoloaders – mucking and delivering

A loader of this category does all the three operations i.e. Loading, Hauling and Dumping. One of these loaders is a hopper loader having the overshot bucket loading into the hopper mounted on the same machine. When this hopper is full, the loader travels up to the discharge end, which could be a waste-pass, ore-pass or a mill-hole to discharge the muck.

Underground excavation & tunneling equipment

Rocker shovels

(a) An overhead shovel loader: mucking from development & stoping faces

(b) Eimco-630 loader mucking from sinking shafts & development faces

Auto loaders

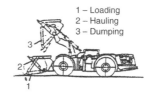

1 – Loading
2 – Hauling
3 – Dumping

(c) A cavo loader: for development & stoping faces. Unloading into ore/waste passes

(d) LHD units (Diesel, electric or pneumatic) for direct unloading into ore/waste passes. Also loading into trucks, dumpers & shuttle cars

Arm loaders

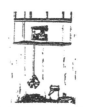

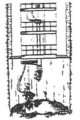

(e) A cactus-grab mucker with a riddle-style suspension carriage

(f) A riddle mucker with clamshell

(g) A cactus-grab mucker with a central column & cantilevered boom

Grab & clamshell loaders for shaft mucking

(h) A hydraulic cryderman mucker

(i) Alimak mucker with backhoe bucket

(j) A pneumatic cryderman mucker

Arm loaders for shaft mucking

(k) Gathering arm loader, muck is discharged into shuttle car

Scrapers

(l) A scraper for mucking into mill holes during stoping operations

(m) Shuttle car carrying muck received from a LHD or gathering arm loader

Figure 6.3 Mucking equipment for tunnels, shafts and underground mines.

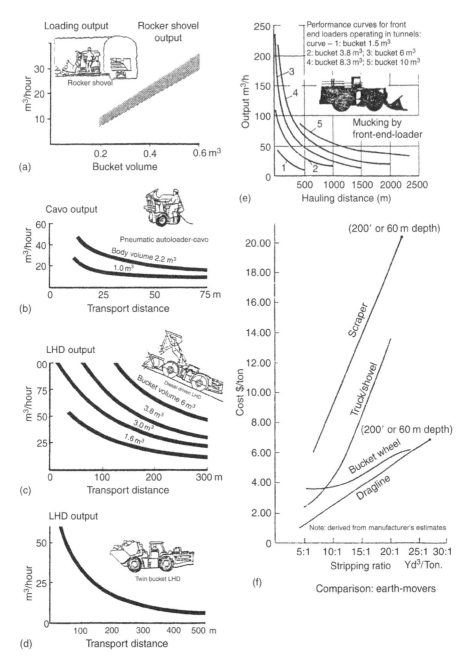

Figure 6.4 Performance curves of various loading/mucking units.

Cavo loaders[1] (figs 6.3(c) and 6.4(b)) which are pneumatically operated wheel mounted loaders with body or hoppers available in two sizes, 1 or 2.2 m³, are the examples of this kind of loaders. Performance of these loaders is a matter of body capacity and the travel distance from the mucking face to the discharge end, as illustrated in   figure 6.4(b). Remote controlled Cavo are the latest version of these loaders.

6.4.2.2   *Mucking and transporting – load haul and dump units (LHDs)*[9,11,19]

It is similar in appearance to a conventional front-end loader (described later). Although the LHD (fig. 6.3(d)) does not offer top travel speeds, it has 50% greater bucket capacity, a slightly smaller engine and generally better emission exhaust characteristics than a front end loader. This unit is so popular that more than 75% of world's underground metal mines use them to drive small and large sized tunnels, chambers and wide excavations (stopes).

6.4.2.2.1   Constructional details[11,19]

LHD being a productivity-oriented machine, great care is taken right from its manufacturing stage. Its longer, lower and narrower profile makes it particularly adaptable to development drifts where width is important and in flat-bedded deposits where the height is vital. Its greater machine length reduces maneuverability but this improves axle weight distribution and allows an increase in bucket size. A central articulation provides perfect tracking and greater maneuverability. It has heavy planetary axles and a four-wheel drive.

Although some of the smaller LHDs are available with electric motors, mostly they have diesel engines of power varying from 78 hp; for small models, to 145 hp. or more. The engines are either air or water-cooled. Service, emergency and parking brakes with fire resistant hydraulic fluids are common in LHD units. Besides a mileage indicator, a headlight, an audible warning signal, a portable fire extinguisher (within easy reach of the operator), and a canopy are some of the common fittings with these units.

6.4.2.2.2   Special provisions[12]

A substantial portion of the space envelope of an LHD is fitted with diesel exhaust treatment devices which may be, according to individual preference, water, catalytic fume diluter, or a combination of these types. Spray or bath, that in turn may be batches type or constant level type, may do exhaust treatment with water. Catalytic purifiers used are either monolithic or palletized. A safety device is also fitted to automatically shut off the fuels supply to the engine if the temperature of the exhaust gases from the conditioner exceeds 85°C, or a preset value.

6.4.2.2.3   Buckets of LHD and other dimensions

LHDs are available in buckets of various sizes (i.e. pay load) ranging from $0.8\,\mathrm{m}^3$ to $10\,\mathrm{m}^3$ with a payload of 1.5 tons to about 17 tons, but the general trend is for $1.53\,\mathrm{m}^3$ and $3.83\,\mathrm{m}^3$ LHDs. The buckets with split lips are usually fitted to these units, but in draw-point loading one piece buckets with a 20 cm lead of T_1 steel construction has proved better. When the bucket teeth and lips wear out, generally after loading 50,000 t of rock, it is sent for build up and lips replacement. Its height ranges from 1.8 m to 2.5 m and width from 1 m to 3.05 m. Range for turning radius is in between 2.4 to 5.8 m. Given in table 6.1 is the range of bucket capacity by one of the manufacturers.

Table 6.1 Details of LHD buckets[7] with pay load and power rating. Applicable for both diesel as well as power versions. (Conversion factor: $1\,m^3 = 1.308\,yd^3$.)

Payload, t	Capacity, m^3	Power, kW
3.5	1.5	63
3.5	1.6	63
4	2	63
6	3	102
8	4.5	170
9.5	5	170
14	6	204
17	8.5	240

6.4.2.2.4 LHD tyres

Treaded or smooth tyres, with or without chains are fitted to LHD units. In a majority of mines traded tyres without chains are used because the chains have proved harmful due to their cutting actions under some circumstances. Average life of a tyre is 750–1000 hours. Retreading can be done more than eight times. Tyre cost is generally 10–20% of the total operating cost. Tyre wear is mainly due to poor road surfaces, wet conditions, excessive wheel spin, incorrect operating pressure and its general misuse which may be, sometimes, due to insufficient clearance from side walls.

6.4.2.2.5 Distance, gradient and speed

The operating gradient is defined as maximum gradient against which loaded LHD units operate. Most of the mines are operating LHDs between 10–20% gradient, but by operating LHDs on a flat gradient will improve the machine's life and reduce operating costs. Though LHDs can operate to a hauling distance of more than a kilometer, for $0.8\,m^3$ bucket capacity, the economically feasible distance could be up to 75 m if operated in stoping areas and 150 m for development headings (as recommended by some of the manufacturers) (fig. 6.4(c)). This distance increases with the increase in bucket size, for example, for $10\,m^3$ bucket size LHDs; the economically feasible distance is up to 1.2 km in stoping areas and 2 km for development headings. The speed of LHDs with bucket of $3\,m^3$ and higher ranges between 8–16 km/hr with average of 13 km/hr on the level surfaces.

6.4.2.2.6 Ventilation

The higher capacity and longer tramming distance units are fitted with diesel engines. These need excessive ventilation arrangements and efficient exhaust treatment devices. Every country has its safety laws (i.e. regulations) to operate LHDs particularly with regard to ventilation standards; for example, in some countries, need to follow the following rules where LHDs are operating in underground mine roadways:

- Velocity of air current should be more than 30 m/min.
- Presence of inflammable gases in the general body of air should not exceed 0.2% or 0.5% that of toxic gases at any point.

- In general body of air the concentration of carbon monoxide (CO) should not exceed 0.01% (100 ppm) or the oxides of nitrogen should not exceed 0.001% (10 ppm); where CO is found to be 50 ppm, or oxides of nitrogen 5 ppm, steps should be taken to improve the ventilation.

For efficient application of diesel driven equipment, the following aspects also should be looked into:

- Choice of clean engines to minimize the production of the toxic gases.
- Use of oxy-catalytic exhaust scrubbers to eliminate most of the toxic gases in the engine exhaust and the odor normally associated with diesel engines.
- Daily checks on atmospheric conditions, at each diesel exhaust unit, in addition to the complete weekly tests and examinations required under the safety rules.
- Proper maintenance of engine, air intake filters and exhaust scrubbers.

6.4.2.2.7   Latest developments[1,9,11,19]
Remote controlled LHDs are successfully operating in underground stopes in unprotected areas. Availability of LHD versions with double buckets (fig. 6.4(d)) allows better productivity. The Eject-O-Dump (EOD) type buckets offered by some manufacturers facilitate the reach of the bucket ahead of tyres, by reducing the height of bucket while dumping into the dumper or truck. EOD buckets can also load high-wide trucks with a minimum reach over the side and with minimum back height requirement.

A Toro LHD at LKAB (underground iron ore mine producing about 70,000 tons/day in Sweden), which has been recently introduced is capable of tele-remote loading and automatic tramming[20]. And, while tele-remote operations are not new but automatic tramming is certainly the new. This unit is capable of loading and dumping 8000 t of muck/day.

6.4.2.3   Desirable features

6.4.2.3.1   Perfect layout
- In order to facilitate the convenient running of the trackless units, the underground roadways of adequate cross-sectional area should be laid as straight as possible with minimum curves and turnings. For one-way traffic, the vehicle's width +2 m, should be the minimum width, and for two-way traffic there should be a clear-cut space of minimum 1.5 m when the vehicles cross each other. There should be a clear space of not less than 0.3 m between the cab of a trackless unit and roof and in any other case not less than 1.8 m from the board where the driver stands.
- While using LHDs in draw-points, the draw-points should be put at an angle exceeding 90° with the extraction drive for easy manoeuvrability. The brow of the draw points should be reinforced concrete or rock bolted.

6.4.2.3.2   Suitable drainage and road maintenance
- Wet and soft roadways are not suitable for trackless mining. Water and sharp rock should be removed from the roads to achieve better tyre life. Bad roads lead to severe tyre damage as well as considerable strain on machine suspension and

chassis. Sudden shock loads transmitted through the frame can play havoc with sensitive instrumentation, shake of nut-bolts and substantially shorten a unit's working life.

• Manufacturers are supposed to design machines that could withstand the most arduous conditions. Now it is generally appreciated that regular road grading and compacting and sometimes even laying of cement or tarmac surface, could lead to saving in overall cost. Good roads means better road performance. Haul roads should be laid in a similar manner as the tracks in the track mining.

6.4.2.3.3 Well-fragmented muck

An important consideration for lowering overall maintenance cost is good fragmentation. Apparently small savings in drilling and/or explosive costs can lead to a sharp increase in maintenance and other costs that may not be immediately noticeable. Improved fragmentation means easier loading, less strain on both loading and hauling units, less down time for dealing with large rocks, less spillage etc.

6.4.2.3.4 Maintenance

Besides many other factors, production is based on the performance of maintenance crew at the face and in workshops. A planned maintenance program is most essential. Provision of well-equipped strategically placed workshops adjacent to working areas should not be forgotten.

6.4.2.3.5 Trained personnel

Using the trackless equipment, which is bulky, in a limited space underground requires skilled personnel to operate and trained personnel for its maintenance.

6.4.2.4 *Advantages*

Higher productivity: Due to better availability, better utilization at production points and requirement of less manpower to handle large tonnage, the output per man-shift has been increased wherever this system has been adopted.

Flexibility: This equipment can be used for any kind of operation such as in the development of small tunnels, large excavations including chambers, stopes and caverns.

High Production: Due to the availability of high capacity loading and hauling units and their flexibility in operation, a large output from concentrated areas can be achieved.

Mobility and Versatility: The trackless equipment scores over rail haulage in its flexibility in terms of mobility and being able to negotiate reasonable gradients. Many mines employ gradients of the order of 20% or more. These gradients are not viable with conventional rail haulage. In figure 9.11 its use in different environments i.e. combination with haulage units of different kinds, has been illustrated.

Development speed and cost: The trackless equipment being productive and capable of producing large tonnage enables the investors to reach their targets within a very

A wheel scraper

A wheel loader   A backhoe   A hydraulic excavator   A front shovel

Figure 6.5   Excavators (Courtesy: Caterpillar).

short period. This gives quick return to the investment made and reduce overall cost of production.

6.4.2.5   *Limitations*

Use of diesel equipment requires elaborate ventilation arrangements to deal with the problem of exhaust's fumes, heat and dust. High maintenance time and cost is another disadvantage and besides these highly skilled men are needed for maintenance as well as for their operation. Non-availability of original spares, sometimes, poses serious problems. The trackless system does not work effectively in the following situations, and there the use of conventional track system becomes an obvious choice.

- In a geologically disturbed and weak strata.
- For low output and low capital investment.
- In highly gassy strata.
- In deep workings and tunnels due to their restricted size.

6.4.2.6   *Manufacturers*

LHDs are manufactured by more than a dozen companies and some of the leading manufactures are given with the range of bucket capacity (in Yd^3) units they manufacture: Wagner (1–11), Eimco (1–15), Caterpillar (3.5–5), Joy (1–7), Atlas Copco (1–6), Equipment Miner (1–12), Schopf (2.1–6.3), Tam-rock (1.3–6.3), GHH and few more.

6.5   ARM LOADERS

6.5.1   Gathering-arm-loader (GAL)

Gathering-arm-loader (fig. 6.3(k)) was introduced in the beginning of 20th century. This crawler-mounted equipment finds its application primarily in coal mines but gradually non-coal mines also started using this. Dimensions of this equipment varies with individual manufactures (such as Joy; Goodman etc.). Its height (less canopy) ranges from 0.61 m to 1.22 m, width from 2.4 m to 2.7 m, and the discharge end's height ranges between 1.1 m to 2 m. The conveyor fitted with it has a width range of 0.7 m to 1 m.

This is a continuous-action mucking machine having a pair of gathering arms. These arms feed the muck upon a flight conveyor, which transfers it to a shuttle car or mine cars. It can operate with an inclination up to 10°. Excellent maneuverability and versatility in whatever the width of workings, high capacity, small size particularly the height, and ease of operation makes this equipment acceptable in coal mines. Amongst its shortcomings is its unsuitability for rock of high abrasiveness. It requires uniformly fragmented muck of small size without boulders. It has poor grabbing of fine muck.

6.5.2   Arm loaders for sinking operations[1,5]

Some of the arm loaders described below find their applications mainly during shaft sinking or driving in the downward directions. Their use is, thus, confined to these operations only.

6.5.3   Riddle mucker[5]

This loader (figs 6.3(e), (f)) was developed in the early 1950s. It consists of a hoisting and traveling mechanism that operates a clamshell suspended on cables. A pneumatic tugger hoist operates the clamshell. The carriage is suspended on rails, located near the permanent support work of the shaft. It is mostly used in rectangular shafts. The clamshell capacity ranges from 0.3 to 0.76 m^3. Output up to 30 t/hour can be achieved.

6.5.4   Cryderman mucker[5]

This loader (fig. 6.3(j)) was also developed in the early 1950s and it is operated by means of pneumatic cylinders and telescopic boom. This is suspended from an independent hoisting system usually located at the surface and used mostly for the rectangular or circular shafts. It can be used in inclined shafts also. A hydraulic version (fig. 6.3(h)) has been also developed in Canada. This can yield an output up to 80 t/hr. This is powered by a self-contained hydraulic power pack. It is suspended in the same way as its pneumatic version. It has a longer boom than the pneumatic one and the bucket capacity is of 0.57 m^3.

6.5.5   Cactus-grab muckers[5]

This is a pneumatic cactus grab (fig. 6.3(g)), which is suspended by cables. This can be used by mounting it on a carriage like a riddle mucker. This type of equipment was initially used in some of the South African mines but presently it finds its application globally. Grabs of 0.4 to 0.85 m^3, with hourly output up to 200 t are available for their use in rectangular or circular shafts. This unit is usually mounted on the multi-deck-sinking platform.

6.5.6 Backhoe mucker[5]

The Alimak Co. Sweden, has developed a machine which has a backhoe action rather than clamshell or cactus grab (fig. 6.3(i)). This is hydraulically operated equipment with a self-contained power pack. The unit can be attached to the shaft wall or can be suspended from the bottom of a sinking stage.

6.6 SCRAPERS

A scraper unit (fig. 6.3(l)) consists of a scraper, a wire rope for filling, a wire rope for pulling, a return sheave, a driving winch, a loading slide and power unit. The power unit has motors, coupling, gear systems etc. There are mainly two type of scrapers: Box and Hoe. The box type of scraper is employed for small size rocks of low specific gravity. Its capacity is in the range of $0.3–0.35 \, m^3$. The hoe-type scraper finds application in loading large size muck of higher specific gravity. It has capacity in the range of $0.3–0.5 \, m^3$. Considering the scrapers as a loading unit, it can be said that they are sufficiently versatile i.e. usable in workings of various cross-sections both horizontal and inclined (up to 35°) in rocks of different physico-chemical properties, provided that the floor is firm. Scrapers are simple in design, and require little investment. The drawbacks are frequent breakdowns of the flexible ropes used for this purpose and difficult cleaning at the walls of the workings. Its efficiency depends upon the scraping distance, type of rock – its size and strength. Use of these units for workings with cross-section up to $20 \, m^2$ is not uncommon. Other utilities include their use at sand gathering plants, coal or mineral handling plants such coal washeries and loading yards to transfer muck into the transportation units such as railway-wagons, trucks etc.

6.7 MUCKING IN TUNNELS[3,18]

The spoil (muck) handling in tunnels could be achieved with the use of any of the loading equipment described in the previous section, if the size of tunnel is up to $30 \, m^2$. But for the tunnels having larger cross-sectional areas, particularly for civil works, other sets of equipment for the purpose of mucking are deployed. Prominent amongst them are the front-end-loaders (FEL) and hydraulic shovels (described in the following sections). The later one is used for very large size of tunnels. FEL's bucket digs into the muck pile and loads itself and discharges the muck into a truck, which carries it out of the tunnel for its disposal. This system is flexible and suitable for a distance range of 300–1500 m and gradient up to 27%[18]; but it requires efficient ventilation and pollution control measures to combat heat, dust, noise and harmful gases that are generated by FEL and truck fleets. A dozer is also required to push the muck towards the face.

Range of specifications varies from model to model for their use in tunnels e.g. Caterpillar Company's loaders have length range: 6.9–13.6 m; height range: 5–9.3 m; bucket capacity: $2.1–9.2 \, m^3$ and engine H.P. rating: 65–690. Mucking performance depends upon the bucket size and hauling distance, as shown in figure 6.4(e).

6.7.1 Dipper and hydraulic shovels[3,4,6,17]

Shovels are basically used during surface mining of deposits at the open pits and open cast mines. Their proven productivity in these surface mines has tempted tunnel engineers and miners to take them below ground for mucking in large sized tunnels and openings underground. In figure 9.14(b), use of shovels for mucking in a Russian tunnel has been shown. The latest trend is use of hydraulic shovels, which have been proved successful for mucking from very large sized tunnels (fig. 9.14(a)) and even at the world's largest underground salt mine in West Germany. Machine body and configuration of boom permits it to work and travel in tight places. The 'wrist action' of the bucket allows the operator to loosen tight muck piles for loading with faster rates and ease into the transportation units.

6.7.2 Mucking in TBM driven tunnels

The description of use of tunnel boring machines, both partial heading and full boring machines, has been given in chapters 10 and 11, and it reveals that partial heading machines work on the principle of undercutting. During up-stroke, cutting is achieved, while during down stroke, the rotating cutter head draws the muck on the panzer type conveyor for its transfer to the rear.

Similarly, all modern road headers utilize the gathering arm loading system and chain conveyor in the center of the machine, which can discharge muck to track or trackless transportation units as illustrated in figures 10.3 and 10.4. A full-face tunneling machine consists of a rotating head fitted with the rock cutting tools. This head is forced into the tunnel face. A single pass is sufficient to create a round or elliptical hole (i.e. full face). The cutter-head buckets, or scoops that transfers cuttings to a conveyor belt, remove the cuttings.

6.8 SURFACE – EXCAVATION, LOADING AND CASTING UNITS[2,3,21]

Surface excavators are the digging equipment that dig consolidated, semi-consolidated ground or broken muck. The excavators used at the surface can be classified in different manners and prominent amongst them are:

1. *Based on number of buckets*: Single or multi buckets[12] (fig. 17.15). Except bucket ladder and bucket wheel excavators, all other excavators as shown in figure 6.6, have a single bucket.
2. *Based on mounting* which could be wheel tyre, crawler track (also known as Caterpillar track) or walking mechanism (in case of heavy duty draglines).
3. *Based on motive power*, which could be diesel or electric.
4. *Based on swinging mechanism which could be non-revolving* (such as front-end loaders and hydraulic excavators) *or revolving* (such as – dipper shovel, backhoe, dragline and grab).
5. *Based on continuity of operation, which could be continuous* (such as BWE and BCE) or *cyclic* (the rest all are cyclic).

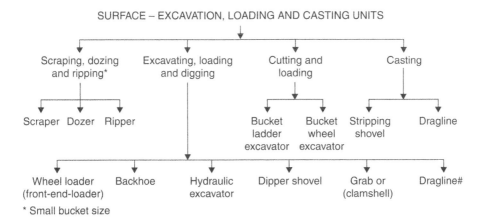

SURFACE – EXCAVATION, LOADING AND CASTING UNITS

Figure 6.6   Classification of earth excavating, digging, cutting & loading, mucking and casting units in operation at the Surface. Details of them dealt in Chapter 17.

These excavators find their application in many public and private works such as trench and canal digging, dam sites, rail routes and road construction sites, and at all the surface mines – open-pit, open-cast or quarrying to remove overburden and minerals of all kinds. A line diagram (fig. 6.6) illustrates this classification.

6.9   WHEEL LOADERS – FRONT END LOADERS[2,3]

These wheel loaders are similar to LHDs, described in the preceding sections. The difference between them lies in the fact that FELs are longer in length, taller in height and narrower in width than LHDs. This feature makes them suitable for their use at the surface where working space is unconfined and ample. They differ from dipper shovels, which have a dipper in place of a bucket to carry the load. A shovel can revolve to discharge contents of its dipper, whereas a FEL has to travel to discharge its bucket to a muck pile or haulage unit.

Wheel loaders (fig. 17.15(e)) are available with small bucket size of $0.6\,\mathrm{m^3}$ to $18\,\mathrm{m^3}$ or more (a Caterpillar's range, similarly it could be for the products of other companies). Important specifications and working range of such loaders are given with the manufacturer's handbook. These units are wheel mounted (0.6 to $18\,\mathrm{m^3}$) but some models (0.8 to $4\,\mathrm{m^3}$) are also available with crawler mounting for their use in rough and undulating terrain. Both of these versions can be primarily used for loading (i.e. mucking) and secondarily for hauling and casting.

Such loaders are used for mucking and ground excavation purposes in soft to hard formations. They are portable and move from one site to another and within the same site. Availability with a wide range of buckets and capacities are also favorable features of such units.

Output from these units/hour is matter of bucket capacity. Apart from surface quarrying, pitting, ground excavation (in general) at the construction sites for road, rail, dam or any other civil work, they find their use for mucking at large sized tunnels.

6.10   BACKHOE[2,3]

These loaders (fig. 17.15(d)) find their use where it is expedient to keep the loader at the original ground surface (level) and excavate the ground from the subsurface below this level. The situations of this kind are the trenching, canal excavations and building foundation works. These are available with wheel as well as crawler mountings. The later is for the large sized units of more than $1.5\,m^3$ bucket capacity from the stability point of view. The muck can be loaded into a haulage unit or it can be cast at the sides of the excavation. Thus, it can be used as a loading as well as casting machine.

The motions of its different organs may be mechanical, hydraulic, or a combination of hydraulic and mechanical. This equipment can be converted to a shovel, dragline or crane and vice-versa. But the hydraulic excavators are usually not convertible into any other kind of excavator. Production from a backhoe is controlled by several factors and prominent amongst them are:

• Type of ground – weathered, ripped or fragmented by blasting
• Digging depth and dumping height, angle of swing etc.
• Bucket or dipper fill factor
• Interruptions and interference at the working site by the existing utilities (power, cable, or pipelines etc.) site for dumping or casting the muck, and other local factors.

6.11   HYDRAULIC EXCAVATORS[2,3,4,6]

Use of this machine gained popularity during the 1970s. This is also known as a hydraulic shovel. It differs[6] from Front End Loader with respect to mounting (crawler in place of tyres), greater digging force and less fuel cost/unit loading and more rugged structure of the main components. As compared to a dipper shovel it has greater mobility, higher travel speed, higher cutting force and improved steerability. Hydraulic pumps and motors play an important role in the functioning of this unit. This unit can be diesel or electrically (a.c) driven. Thus, its application lies at all those locales where FEL or dipper shovel can be deployed, both at the surface as well as underground. The enormous development that has taken place in this unit is evident by the world's largest hydraulic excavator[4] – O & K's RU400. This excavator working at Syncrude, Canada is capable of filling a 290 tons capacity truck in four passes. Figures 6.7 and 17.15(d) illustrate this excavator.

Figure 6.7   A hydraulic excavator working at Civil site.

6.12 SHOVEL[2,3,17,21]

This is one of the most important excavators that is available today. It is available with very small to very large capacity buckets which are capable of excavating any type of muck or ground. This unit is known as face, crowd or dipper shovel. The one which is of large bucket capacity (usually more than 5 yd^3) is known as a strip shovel. It is used for casting the ground in the adjacent area within its reach. This feature finds its application in opencast mines, where it strips the ore deposit (coal or any other mineral) by removing the overburden and casting or back filling it in the worked out space i.e. the place where from the ore (useful mineral) has been removed. The cycle time of the strip shovel is shorter than a face shovel.

In this unit, the dipper is rigidly fixed to the boom (fig. 6.8). Its front-end portion consists of dipper, dipper handle and the boom. The dipper handle is hinged to the

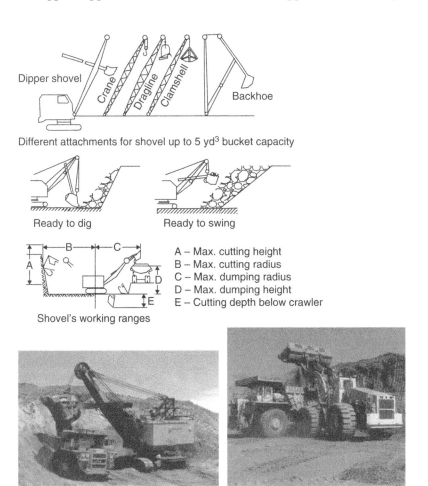

Different attachments for shovel up to 5 yd^3 bucket capacity

Ready to dig

Ready to swing

A – Max. cutting height
B – Max. cutting radius
C – Max. dumping radius
D – Max. dumping height
E – Cutting depth below crawler

Shovel's working ranges

P&H 1900 electric shovels of 8.25 m^3 bucket loading
Wabco off highway 120 tons truck.

Figure 6.8 Shovels – some details. Photo: A shovel working at an iron ore mine.

boom. The dipper is also hung by hoist cables, which pass through the sheaves mounted at the end of the boom. The dipper handle is carried on a saddle bearing (block) over the crowd shaft mounted on the boom. This connection between dipper handle and boom allows the dipper handle to rotate on the crowd shaft by the action of the hoisting cables, moving from digging to fully loaded and swinging position, as shown in figure 6.8.

The lower part of the boom is hinged to a platform and its upper part is held in position by the boom suspension cables attached to the gentry members (tension, compression and spreader). The usual inclination of the boom is 45° to the horizontal.

The front-end portion and the cabin, which houses driving and gear units and controls, are on a swinging platform (turntable) that can be moved in the horizontal plane. This mechanism enables the shovel to directly discharge the dipper into the truck (or casting directly to a spoil bank). Thus, the working cycle of a shovel is composed of digging, swinging to discharge, discharging, swinging back to the face, and lowering of the dipper to the toe of the face.

The main working ranges of this equipment are shown in figure 6.8 and they are: Maximum cutting height (A), Maximum digging radius (B), Maximum digging or discharge radius (C), Maximum dumping height (D), and cutting depth below crawler (E). These dimensions are determined by the dipper (bucket) capacity, length of boom and handle.

The shovel is a self-propelling unit moving on the Caterpillar track. The motions to the dipper and the unit as whole are imparted by the electric motors. The electric power can be supplied through a diesel generator or directly by the electric power supply mains.

The shovel up to 5 yd^3 bucket capacity is a multipurpose unit, which with the aid of different attachments and accessories, can be used as face shovel, crane, dragline, backhoe or clamshell, as shown figure 6.8. This unit is so popular and useful that numbers of manufacturers are in the arena with different models, capacities and features. In general, it can be divided into four categories:

- Small (0.5–2 m^3 bucket size), for small scale earth moving jobs in soft ground
- Medium (>2–5 m^3 bucket size)
- Large sized units (>5–25 m^3 bucket size)
- Very large sized units (Greater than 25 m^3 bucket)

To understand the different specifications and features, the main features of these four categories have been shown in table 6.2.

6.13   DRAGLINE[2,17,21]

A dragline is a single bucket excavator in which the bucket is pulled by a drag rope (hence the name – dragline) over the face towards the equipment itself, as shown in figure 6.9(a). It differs from the face shovel, described in the preceding section, in that the bucket (1) is not fixed rigidly to the boom but hangs from flexible ropes. The bucket is hanged to the rope by the lifting chains and their separating bar (fig. 6.9(a)). These chains are joined together and are attached to a load-line (4) (drag rope). They are also attached to a dumping line (8) whose other end is fixed to the front end of the

Table 6.2   Specifications[21] and working range, in general, of shovels of different capacities. (Conversion factors: 1 m = 3.281 ft; 1 m³ = 1.308 yd³.) Units of more than 20 yd³ bucket capacity are usually custom built based on the specifications of the user.

Items	Bucket size			
	0.5–2 m³	*2–5 m³*	*5–25 m³*	*40–140 m³*
Dipper capacity, m³	1	3–5	10, 15, 25	39, 59, 82, 141
Boom length, m	6.7	10.5	45, 34, 34	36, 52, 61, 67
Dipper handle length, m	4.9	6.2	24, 18, 18	
Digging radius, m	6.4	8.2	29, 20, –	35, 47, 60, 65
Max. dumping height, m	5.5	6.7	36, 24, 24	24, 34, 44, 45
Power: electric motors, Kw	80	250	1700	2000, 5500, 12000, 21000
Weight, tons	140	165	900–1000	1390, 3740, 7060, 14000
Output, m³/hour	120–150	250–300	800–1000	1560, 2250, 3100, 5300

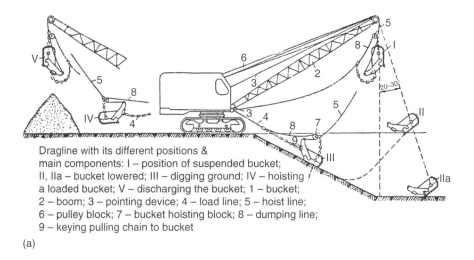

Dragline with its different positions &
main components: I – position of suspended bucket;
II, IIa – bucket lowered; III – digging ground; IV – hoisting
a loaded bucket; V – discharging the bucket; 1 – bucket;
2 – boom; 3 – pointing device; 4 – load line; 5 – hoist line;
6 – pulley block; 7 – bucket hoisting block; 8 – dumping line;
9 – keying pulling chain to bucket

(a)

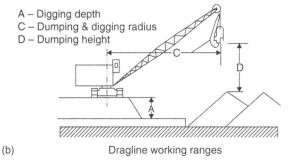

A – Digging depth
C – Dumping & digging radius
D – Dumping height

(b)   Dragline working ranges

Figure 6.9   (a) Dragline with its working details. (b) Working range of dragline.

bucket after passing over the bucket hoisting block (7). This block is at the junction of the hoist line (5) and bucket chains.

Dragline stands on the bench, which is to be dug. First, the load line (drag rope) is slackened, the bucket is lowered to the floor of the face (bench) and then pulled by the drag rope towards the equipment (dragline itself). During pulling it is filled by digging a strip of about 80–500 mm deep. This process is repeated 2–4 times, depending upon the type of material (ground). By keeping the load-line taut, the bucket is pulled by the hoist line (5), at the same time stabilizing by the dumping line or rope (8). The hoisting rope gradually raises the bucket and the drag rope is slowly slackened. As the bucket raise towards the boom, it is at the same time swung towards, the dumping site to discharge its content. Merely releasing the load line (the drag rope) discharges the bucket. After getting bucket discharged, the excavator is swung back to face and the cycle is repeated. Working range of a dragline has been illustrated in figure 6.9(b).

The dragline travels on a Caterpillar track or a walking mechanism. The usual inclination of the boom is in the range of 20–25°. It can swing to 360°. Draglines of as small as 0.6 m³ to as high as 175 m³ bucket capacity are available for use in soil and loose ground (table 6.3). Specifications[16] of two models have been tabulated below:

Details	'Page' 11.5 m³	'Marion' 29.8 m³
Boom length	64 m	67 m
Bucket capacity	11.5 m	29.8 m³
Max. digging depth	42 m	28 m
Max. dumping height	24 m	30 m
Dumping radius	66 m	61 m
Driving motors	1000 H.P.	2500 H.P.
Boom inclination	25°	30°

They are also suitable for blasted rock. Draglines are deployed directly for casting the ground without any intermediate haulage, since positioning of the bucket exactly above a transport unit is not very much practicable; spillage is more and thereby

Table 6.3 Specifications[21] and working range, in general, of draglines of different capacities. (Conversion factors: 1 m = 3.281 ft; 1 m³ = 1.308 yd³.) Units of more than 20 yd³ bucket capacity are usually custom built based on the specifications of the user.

Items	0.5–4 m³ bucket	5–20 m³ bucket	50–175 m³ bucket
Bucket capacity, m³	0.58, 0.98, 1.56, 3.9	9.4, 19.6, 47	58, 82, 94, 173
Boom length, m	12, 15, 24, 41	56, 73, 84	84, 87, 91, 94
Dumping height, m	8, 7, 9, 18	22, 32, 43	32, 34, 39, 39
Dumping radius, m	11, 13, 20, 37	53, 68, 75	81, 84, 86, 92
Max. digging depth, m		26, 34, 72	50, 49, 47, 56
Power: electric motors, kw	80D, 112D, 185D, 300	700, 2075, 5000	7500, 10500, 14000, 50000
Weight, tons	22, 32, 58, 168	640, 1450, 3050	4410, 5130, 6830, 14000
Output, m³/hour	35, 47, 63, 133	331, 682, 1616	2040, 2825, 3300, 5900

cycle time increases. In some circumstances a hopper can be used to receive the muck while transport units are positioned below them to receive muck. A unit with bucket capacity up to 5 m³ can be converted into shovel, backhoe or crane.

6.13.1 Multi bucket excavators[2,12,17]

A multi bucket excavator, as the name suggests is equipped with a mechanism that can engage a number of buckets in series for the purpose of ground excavation. This earth moving equipment has two versions:

Bucket Chain Excavator (BCE)
Bucket Wheel Excavator (BWE)

6.14 BUCKET CHAIN EXCAVATOR (BCE)[2,12,17,21]

This is also known as a bucket ladder excavator. This equipment, as shown in figure 6.10(a), has an endless chain, supported by a frame, and with a number of buckets fitted to it. When the buckets move in a forward direction, they cut the ground and unload it as they overturn at the sheave or tumbler. Thus, the process of cutting and loading is continuous and non-cyclic. The driving gear and the motors to run the system are housed in a separate platform, which can move on wheel or Caterpillars, along the face. This platform may be placed either at the upper berm as shown in figures 17.15(j) and 6.11, or it can also be placed in the lower berm; but digging in this case is rather slow and not very efficient. When the equipment is in operation, the chains move slowly at a speed in the range of 0.6–1.2 m/sec, and also the equipment itself travels slowly along the face at the rate of 4–12 m/sec. Thus, the face is continuously cut or scraped. The whole assembly of buckets and the frame enclosing it can be raised or lowered by the suspended cables. With this mechanism, the slope of the bank can be changed. To counterbalance the weight of a heavy boom or jib, a massive counter weight is provided on the other side of the machine.

These units are suitable for soft, friable and loose formations, including clay or ground that is free of large boulders or slab, stumps etc. BCEs find their application particularly in areas where the material has to be dug below the excavator operation level and where the material has to be well blended during the digging process. As far as operation mode and excavator design are concerned differentiation is made between two types: bucket chain excavators moving on crawlers and those moving on rails. The first mainly work by the block mining method, the latter mainly work in low-cut mode. In figure 6.11(b) BCE working in parallel digging mode in a German open pit has been shown. It is equipped with a rail-shifting device and its capacity is 810 m³/h.

6.15 BUCKET WHEEL EXCAVATOR (BWE)[2,12,17,21]

This unit is also a heavy-duty continuous excavator, which is capable of producing a large output. The body of an excavator, as shown in figure 6.10(a), rests on an under frame having Caterpillar tracks. This unit can turn round its vertical axis by a swinging mechanism. The bucket wheel has a number of buckets mounted on it. This wheel is

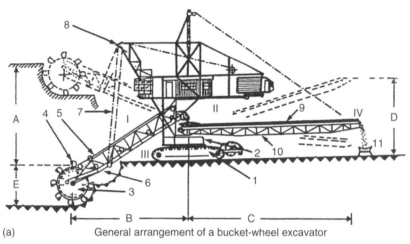

(a)   General arrangement of a bucket-wheel excavator

I – Boom (could be lowered or raised); II – structure (housing machinery, ballast and winches); III – under carriage with crawler, control equipment and cabin for switch gear, workshop and crew; IV – discharge end. 1 – underframe; 2 – Swing mechanism; 3 – Operating wheel; 4 – buckets; 5 – frame (jib); 6 – conveyor; 7 – suspension cable; 8 – Upper frame; 9 – belt conveyor; 10 – Swinging arm; 11 – conveyance to receive muck. Working ranges; A – cutting height; B – cutting radius; C – dumping radius; D – dumping height; E – cutting depth.

(b) Bucket wheel excavator – the largest mining equipment on earth. This technical giant is custom designed and built for the specific operating conditions and production rates. Currently attained and production rates are up to 240,000 m³ (bank) per day, digging heights of as much as 100 m and service weights of 13,500 tons. (Courtesy: Krupp, 2000)

Figure 6.10   (a) Bucket wheel excavater details. (b): The largest bucket wheel excavator in the world with a daily capacity of 240,000 m³ (bank), operating in the German Hambach open pit mine of Rheinbraun AG.

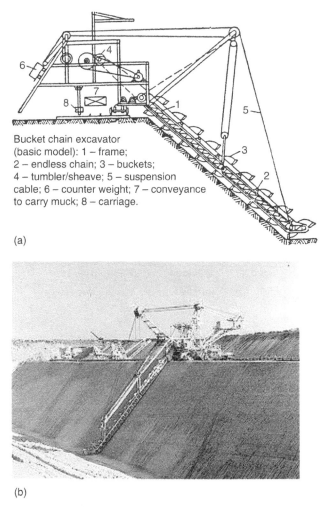

Bucket chain excavator
(basic model): 1 – frame;
2 – endless chain; 3 – buckets;
4 – tumbler/sheave; 5 – suspension
cable; 6 – counter weight; 7 – conveyance
to carry muck; 8 – carriage.

(a)

(b)

Figure 6.11   (a) Bucket chain excavator (basic model). (b) This bucket chain excavator works in parallel digging mode in a German open pit mine, and it is equipped with a rail shifting device. Capacity: 810 m³/h (Courtesy: Krupp, 2000).

mounted at the end of a frame or jib, which looks like a girder or heavy beam. A conveyor installed within the jib receives the ground cut by the wheel buckets and delivers it to another conveyor, which is fitted at the tail end of this equipment. The ground/rock from this conveyor is directly loaded into a haulage unit, which could be train, conveyor, or a fleet of trucks. The jib can be lowered or raised with the help of cables that are suspended from a boom.

Use of the world's largest BWE at a German Hambach open pit mine, Rheinbraun,[13] having a daily output of 240,000 m³ (bank) is shown in figure 6.10(b). Similar is the mine lignite deposit at Neyveli in India using bucket wheel excavators, where the stripping ratio is 11 (or 5.5 m³ per ton of lignite), i.e. 11 tons of of overburden

Table 6.4 A comparison of various excavators.

Items/parameters	Front-end-loader	Shovel	Dragline	Bucket wheel excavator
Ideal use: loading, casting	Loading haulage units	Loading haulage units, casting	Side casting of o/b	Loading as well as casting o/b
Dump yard location	Anywhere	Anywhere	Within the reach of boom	Anywhere, by using a number of units in series
Bucket movement and digging horizon	Under control and digs from the same level	Under control and digs from the same level	Control not efficient. Digs below its level	Under control, can dig from any level
Formations segregation	High	High	Low	High
Flexibility: varied ground conditions	Good to poor	Good to Poor	Good	Fair to poor
Mobility	Good	Good	Low	Low
Cost/ton: varying stripping ratios (fig. 6.4(f))	High	High	Lowest	Lower than shovel but higher than dragline
Capital cost	Low/m^3 of bucket capacity	Low/m^3 of bucket capacity	High/m^3 of bucket capacity	High/m^3 of bucket capacity
Continuity of operation	Cyclic	Cyclic	Cyclic	Continuous
For poorly fragmented muck or hard formation	Good	Good	Poor	Poor
Facilitate land reclamation	No	Strip shovel	Yes	Yes
Handling parting	Well	Well	Difficult	Difficult

need to be removed to mine 1 ton of lignite, and 13 tons of water need to be pumped to mine 1 ton of lignite. The excavators of this type can work at different benches of the same pit. A comparison of different excavators has been made in table 6.4.

In figure 6.4(f), a comparison of this kind with regard to cost of production at different stripping ratios has been made. It indicates that for high output a dragline is a better choice particularly to cast the overburden.

6.16 CALCULATIONS FOR SELECTION OF SHOVEL/EXCAVATOR[3,8]

These calculations have been described in section 17.5.1.

Table 6.5 Format/Form to estimate equipment cost/ton. and per unit (tons. or m^3) cost of production. *** – this could be 12% of tyre cost for favorable, 15% for average, or 17% for unfavorable working conditions (Anon 1981).

A. Ownership costs:

1	Calculation of Depreciation		Remarks
a.	Purchase value		As charged by the supplier ($+$)
b.	Salvage value ($-$ deduct)	15% of purchase value (a)	To be deducted ($-$)
c.	Freight	Based on equip. weight	To be obtained from the transporter ($+$)
d.	Unloading and moving costs	Usually 10% of (c)	($+$)
e.	Tyre cost (if the unit is mounted with tyres)		*–This calculation is also optional ($-$)
f.	Delivery value	$f = a + c + d - b - e$*	This is the total value of equipment
g.	Life of equipment in years		By experience & guidelines of manufacturer
h.	Total scheduled hours		Same as above
i.	Scheduled hours/year	$i = h/g$	
j.	Depreciation cost/hr	$j = f/i$	
k.	Average annual investment	$k = f (g + 1)/2 g$	Rate $= (g + 1)/2 g$, can be calculated as factor or %

2	Calculation of fixed costs other than purchase values		
l.	Interest		
m.	Taxes		
n.	Insurance		
o.	Sub total of l, m and n (usually $l + m + n =$ 18% of k)	$o = l + m + n$ (in percentage)	It varies from place to place and ranges 8 to 20% of k.
p.	Total interest, insurance, tax etc. cost/year	o k	These are annual Fixed costs
q.	Interest, insurance, taxes etc. Cost/hour	$q = p/i$	Fixed cost/hr
r.	*Total hourly ownership costs*	$r = q + j$	Sum of depreciation and fixed costs

B. Operating costs:

1	Tyre replacement cost (if applicable)		For track or Caterpillar mounted equipment not applicable
s.	Purchase price of tyres		
t.	Tyre life in hours		By experience & guidelines of manufacturer
u.	Tyre cost/hour	$u. = s/t$	

(Continued)

Table 6.5   Continued.

2	v.	Tyre repair cost/hour	-----% × Tyre cost/hr***	
3	w.	Repair and maintenance cost/hr	% × depreciation cost/hr	This % can vary, up to 60% favorable; 70% average; more than 70% unfavorable (Anon, 1976)
4	x.	Fuel or power consumption/hr	Cost per unit × number of units consumed/hr	This can also vary. For electrically driven units it will depend upon the load factor
5	y.	Miscellaneous cost (not covered above)/hr		Such as lubrication costs, which could be 1% of fuel costs
6	z.	Labor cost/hr	z = wages of operator + helpers, + 35% FB	Based on crew strength. FB – fringe benefits
7	Zz. Total operating cost/hr		$Zz = u + v + w + x + y + z$	This is total variable cost/hr
8	*Total ownership & Operating cost/hr*		Tcost = zz + r	Fixed + variable costs, gives the total cost/hr (A + B)
9	*Production cost/ton*		T cost/production per hour in tons	This is how unit costs can be calculated

6.17   TOTAL COST CALCULATIONS[3,8]

The following format or form can be used to calculate total cost per hour and also unit volume or weight (m^3, or tons). Items not pertaining to particular equipment should be excluded from the calculation.

6.18   GOVERNING FACTORS FOR THE SELECTION OF MUCKING EQUIPMENT

The important factors that need consideration while selecting mucking equipment are mainly governed by the parameters listed below.

- Transportation system: track or trackless
- Capital available
- Required output or progress/shift
- Size of opening where it is to be operated
- Size of rock fragments
- Mucking lead (fig. 6.4)
- Unit operation system: cyclic or non-cyclic.

Before selecting any equipment consideration and a thorough analysis to the following factors should be given.

- Environment factors (*Noise and vibrations; Exhaust gases; Dust; Fog and fumes*)
- Accident factors (*Vehicle's overall design; Falling rock; Danger to third person*)
- Ergonomic factors (*Ergonomic design; Possibility of social contact; Comforts it provides to the operator*)

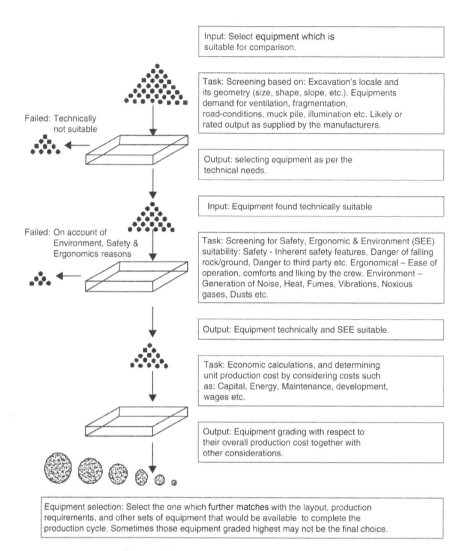

Input: Select equipment which is suitable for comparison.

Task: Screening based on: Excavation's locale and its geometry (size, shape, slope, etc.). Equipments demand for ventilation, fragmentation, road-conditions, muck pile, illumination etc. Likely or rated output as supplied by the manufacturers.

Failed: Technically not suitable

Output: selecting equipment as per the technical needs.

Input: Equipment found technically suitable

Failed: On account of Environment, Safety & Ergonomics reasons

Task: Screening for Safety, Ergonomic & Environment (SEE) suitability: Safety - Inherent safety features, Danger of falling rock/ground, Danger to third party etc. Ergonomical – Ease of operation, comforts and liking by the crew. Environment – Generation of Noise, Heat, Fumes, Vibrations, Noxious gases, Dusts etc.

Output: Equipment technically and SEE suitable.

Task: Economic calculations, and determining unit production cost by considering costs such as: Capital, Energy, Maintenance, development, wages etc.

Output: Equipment grading with respect to their overall production cost together with other considerations.

Equipment selection: Select the one which further matches with the layout, production requirements, and other sets of equipment that would be available to complete the production cycle. Sometimes those equipment graded highest may not be the final choice.

Figure 6.12   A flow-chart to select equipment.

- Technical factors (*Fragmentation; Tunnel's dimensions; Ventilation; Road conditions*)
- Economic factors (*Costs: Capital; Energy; Maintenance; Wages; Capacity*)

A flowchart given in figure 16.12 could be used a guideline to select any equipment including the mucking units.

6.19   THE WAY FORWARD

As a matter of fact the advent of equipment together with technology has brought about a drastic change in method design and its selection. This has allowed designing bulk-mining methods. But a high degree mechanization and automation as detailed in sec. 19.5.8; if not effectively utilized, results in high capital costs and creates problems of unemployment, whereas use of the primitive tools and machines lowers productivity. Experience indicates that *percentage availability and utilization of any unit or resource* has a remarkable influence on overall performance. Least delays are the key for success of any operation.

QUESTIONS

1. Based on bucket size how can you categorize shovels?
2. Based on locale, classify mucking equipment.
3. Classify earth-moving equipment that is used at the surface by giving a line-diagram. Name types of surface mines known to you. List the type of equipment available for mucking in these mines and suggest mucking equipment for each of these mines.
4. Define dragline and briefly describe its important features in terms of its field of application, merits and limitations during for handling the muck and ground at the surface mines. What is the largest size of this unit?
5. Define these terms: muck, mucking, casting, bulking factor, earth moving machines, loading, scraping, and dozing.
6. Describe and give the fields of application for these units: auto-loaders; rocker shovels; arm loaders; scraper and shovel.
7. Differentiate between LHDs and FELs. How do FELs differ from dipper shovels? What is the function of the dipper in a revolving shovel?
8. "Excavation of any kind generates dust, and it is injurious to health". Do you agree with this statement? How can dust generation be minimized?
9. Give a line diagram showing mineral/material handling systems at the surface. You can also refer to Chapter 17.
10. Give a classification of loading equipment which is used for: (i) Underground mines (coal as well as metal mines). (ii) Surface mines (open cast as well as open pit mines).
11. Give a line diagram to classify earth excavating, digging, cutting & loading, mucking and casting units in operation for excavation at the surface locale.
12. List sets of equipment that are deployed for mucking in civil works such as large tunnels, caverns and powerhouses.
13. List versions of multi-bucket excavators. For what type of formations are these units suitable?
14. Mention the items to be covered under each of these aspects: accident factors, ergonomic factors, technical factors and economic factors while selection of any equipment.
15. Name earth moving equipment that could be used in civil construction that is going on nowadays at the university campus. Write down the purpose of each of the items of equipment you listed.
16. Name the highest capacity mucking equipment working with earth, based on your knowledge also specify production from it per hour.

17. Name mucking equipment for underground tunnels and mines. Which of these loaders is most popular and used?
18. Name types of underground mines known to you. List the types of equipment available for mucking in these mines and suggest mucking equipment for their use in making tunnels or drives and in stopes.
19. Prepare a flow diagram to select a mucking unit. Mention technical considerations you made in your selection.
20. What are the latest developments in LHDs and list the desirable conditions to achieve the optimum output from these units.
21. What attachments can be made to a dipper shovel enabling it perform different operations?
22. What factors will you consider when selecting mucking and transportation equipment?
23. What is a gathering arm loader (GAL)? Where you will recommend its use?
24. What is LHD and where it is used? Also mention the range of bucket capacity it is available with.
25. What operations is the LHD equipment is capable of doing? Name the operations/places in underground mines; where you will recommend its use? What are the limiting factors for the use of this equipment? What factors affect the output from LHD? Illustrate it by means of graph.
26. What parameters you would to include when describing 'muck characteristics'?
27. Where are backhoes deployed? Production from a backhoe is controlled by several factors; list those prominent amongst them.
28. Where will you recommend the use of a shuttle car? How many motors are fitted in it and what is the function of each one of them? Give the specification of any two models of shuttle cars known to you.
29. Why are hydraulic excavators gaining popularity? How do they differ from front end loaders and dipper shovels?
30. Work out the following for LHD equipment: constructional details, special provisions, buckets of LHD and other dimensions, LHD tyres, distance, gradient and speed, ventilation requirements, advantages, limitations and list of manufacturers.
31. Write down the important features (including its production capabilities) of the largest man-made equipment on Earth, the BWE.
32. Give the range of bucket capacity in which the following equipment is available:
    a. A shovel.
    b. A dragline.
    c. A bucket wheel excavator.
33. Prepare a guide or selection chart to use the following, considering the parameters such as production rate, pit life, pit depth, type of deposit and investment:
    a. Dozer & front end loader.
    b. Dozer scraper.
    c. Shovel – trucks.
    d. Dragline.
    e. Bucket wheel excavator (BWE).

REFERENCES

1. *Atlas Copco manual*, 4th edition, 1982.
2. Brealey, S.C.; Belley, J. and Rickus, J.E.: Mineral quality determination and control in stratified deposits. *IMM's International Symposium on Surface Mining and Quarrying*, Bristol, England, 4–6 Oct. 1983, pp. 153–157.
3. *Caterpillar Performance Handbook*, 2000; Excavators; Wheel loaders; Backhoes.
4. Chadwick, J.: The largest hydraulic excavator O & K. *Mining Magazine*, 1999.
5. Dangler: Mucking equipment in sinking. In: W.A. Hustrulid (edt.): *Underground Mining Methods Handbook*, SME-AIME, New York, 1982, pp. 1263–1265.
6. Files, T.I.: Hydraulic excavators. In: B.A. Kennedy (edt.): *Surface Mining*. SMME, Colorado, 1990, pp. 1635.
7. GHH *leaflets and websites*.
8. Hartman, H.L.: *Introductory Mining Engineering*. John Wiley & Sons, New York, 1987, pp. 134–140.
9. Johnstone, H.A.: Trends in trackless mining. *Mining Magazine*, Jan. 1975.
10. Kahle, M.B. and Mosele, C.A.: Development of mining methods in Gulf Coast Lignite. *Mining Engg.* Vol. 35, 1983, pp. 1163–66.
11. Knopp, R.A.: Concept in LHD design. *Mining Magazine*, Jan. 1975.
12. *Krupp – Websites*, 2000.
13. Modular mining systems, 1998, *leaflets*.
14. Nautilus International Control & Engineering, 1998. *Leaflets*.
15. Page and Marion, *leaflets*.
16. Pokrovsky, N.M.: *Driving horizontal workings and tunnels*. Mir Publishers, Moscow, 1988, pp. 69, 295.
17. Shevyakov, L.: *Mining of mineral deposits*. Foreign language publishing house, Moscow, 1988, pp. 616–627.
18. Sinha, R.S.: Introduction. In: R.S. Sinha (edt.): *Underground Structures*. Elsevier, New York, 1989, pp. 20–21.
19. Tatiya, R.R.: Use of LHD in underground metalliferous mines. *Jour. of Institution of Engineers*, Vol. 62, 1981, pp. 70–73.
20. Toro tele remote controlled LHD system. Leaflets.
21. Weimer, W.H. and Weimer, W.A.: Surface coal mines. In: Cummins and Given (eds.): *SME Mining Engineering Handbook*, AIME, New York, 1973, pp. 17–129, 130.

Chapter 7

Transportation – haulage and hoisting

Saving (reduction) in the consumption of materials and energies are most important in achieving cost reduction and in boosting productivity which are the keys of success.

7.1 INTRODUCTION

At any mine, tunnel, or civil construction site once the rock has been dislodged from its original place, it needs to be removed immediately to its final destination at the surface which could be a waste rock muck pile or ore's stock pile to feed to the processing plant, or final dispatch destination point which is the users. From the working face of the excavation site it is loaded with the use of any of the mucking units as described in chapter 6 into a conveyance which carries it through horizontal, inclined, vertical, or combination of both horizontal and vertical routes, to the discharge point. Movement of the muck through a horizontal or inclined path is known as haulage, and through the steeply inclined to vertical path (up or down) as hoisting. While carrying out surface excavation activities, the vertical transportation (hoisting) except during quarrying operations is almost negligible whereas the haulage operation is almost mandatory. During underground mining, in most of the cases, both haulage and hoisting operations are essential and they form an important link to the routine production cycle. Any weakness or bottleneck due to these links may result in loss of production, reduction in productivity and increase in overall costs. Apart from this, safety is the most important consideration when operating any of the haulage or hoisting units in any mine or tunnels, as these units need to be operated in a confined space underground under various hazardous conditions. There has been a gradual development in the means or equipment that are being used for these operations. Until the early seventies use of haulage units running on track was very much in vogue but since then trackless transportation units are very rapidly replacing them. In this chapter, in brief, first the prevalent haulage system has been dealt with; and hoisting afterwards. Figure 7.1 illustrates the broad classification of transport system that is applicable for the rocks and minerals. In figure 7.2 this system has been classified based on continuous and batch system logic.

7.2 HAULAGE SYSTEM

This can be described under two headings – track and trackless. Track haulage includes rope and locomotive haulage, which runs on rail or track. Trackless systems includes automobile, conveyors and transportation through pipes.

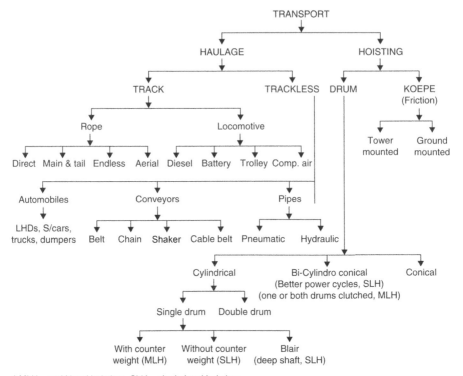

Figure 7.1   General classification of mine transportation system.

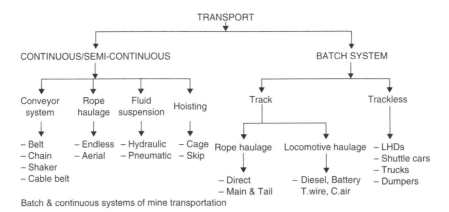

Figure 7.2   Batch and continuous system of transportation.

7.2.1   Rail or track mounted – rope haulage[12]

Rope haulage is the oldest means of transport used in underground mines and even today it is indispensable as it is used in one form of or other, (either haulage, hoisting or both) in most of the underground mines.

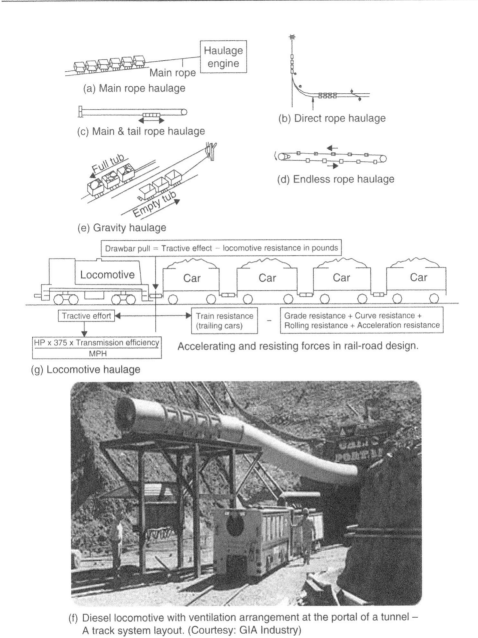

(a) Main rope haulage

(b) Direct rope haulage

(c) Main & tail rope haulage

(d) Endless rope haulage

(e) Gravity haulage

Drawbar pull = Tractive effect – locomotive resistance in pounds

Tractive effort — Train resistance (trailing cars) — Grade resistance + Curve resistance + Rolling resistance + Acceleration resistance

$$\frac{HP \times 375 \times Transmission\ efficiency}{MPH}$$

Accelerating and resisting forces in rail-road design.

(g) Locomotive haulage

(f) Diesel locomotive with ventilation arrangement at the portal of a tunnel – A track system layout. (Courtesy: GIA Industry)

Figure 7.3   Track system – underground rope haulage and locomotive haulage.

Direct (fig. 7.3(a), 7.3(b)), main and tail (fig. 7.3(c)) and endless (fig. 7.3(d)) are the principal designs of the rope haulage that are used. A comparative statement shown in table 7.1 briefly outlines the important features of each of these systems.

Table 7.1 Comparison of Rope haulage systems.

Parameters	Direct	Rope haulage Endless	Main & tail
System	Loaded tubs are hauled up the gradient using power, whereas empty tubs are hauled down the gradient using gravity. Bad track may cause derailment.	System consists of an endless rope running on two tracks. On one side of rope loaded mine cars and on the other empty cars run. Rope moves with a speed of 0.4 to 1 m/sec in one direction only. Rope tensioning devices required.	System consists of double drums for two ropes (which may be of different dia.); one is attached at the leading end of a train of cars and the other at its tail end. The train runs on a single track and loaded cars are hauled up the gradient.
Haulage roads required	Uniformly graded (not below 5–6°), straight single tracked road.	Straight roads laid with double track. Gradient below 1 in 10.	Roads may be undulating single tracked.
Applications, output & mode	For low output in u/g mines usually at the development districts. Batch system.	Low but continuous output, installed at the main haulage levels in u/g mines.	Low output from the main haulage roads those are undulating and rough in u/g mines.

7.2.1.1   *Rope haulage calculations*[12]

Drawbar pull: The pull or force required to overcome the resistance to motion is known as Drawbar pull. The resistance to motion is equivalent to the frictional resistance (+/−) the gradient resistance.

Frictional Resistance $F = \mu W$, on a horizontal plane
$= \mu W \cos\theta$, on an inclined plane having inclination θ
with horizontal.
But for all practical purposes its highest value, $F = \mu W$, is considered. (7.1)

Gradient Resistance $= W \sin\theta$, But for all practical purposes
$\sin\theta = \tan\theta$, as such
Gradient Resistance $= \tan\theta$, which means. (7.2)
$= Wi$, where $i=1/n$, n is gradient value

Tractive Effort = Drawbar Pull = $\mu W + Wi$ (7.3a)

Tractive Effort or Drawbar pull required for:

7.2.1.1.1   Direct rope haulage system
Tractive Effort = Drawbar pull
= Pull required to draw loaded mine cars + Pull required to draw rope
$= (F + G) + (f_1 + g_1) = (\mu W + Wi) + (\mu_1 w_1 + w_1 i)$ (7.3b)

7.2.1.1.2   Endless rope haulage system

Tractive Effort = Drawbar pull

= Pull required to draw loaded mine cars + Pull required
to draw empty cars + pull required to draw rope

$$= (F + G) + (f \text{-} g) + [(f_1 + g_1) + (f_1 \text{-} g_1)]$$
$$= F + f + 2f_1 + G \text{-} g \ = (\mu W + \mu w + 2\mu_1 w_1 + Wi \text{-} wi) \qquad (7.3c)$$

Whereas: F – Frictional resistance of loaded cars;

    G – Gradient resistance of loaded cars;

    f – frictional resistance of empty cars;

    g – gradient resistance of empty cars;

    W – weight of loaded cars;

    w – weight of empty cars;

    w_1– weight of rope;

    i – gradient of road;

    μ – coefficient of friction between minecar's wheel and track;

    μ_1 – coefficient of friction between rope and rollers;

    g_1 – gradient resistance of rope;

    f_1 – frictional resistance of rope.

$$I.H.P = (Tractive\ effort\ in\ kg.\ x\ Haulage\ speed\ in\ m/sec.)\ /\ 75 \qquad (7.4)$$

$$B.H.P = I.H.P/\eta \qquad (7.5)$$

Whereas: I.H.P = Indicated horse power;

    B.H.P = Brake horse power;

    η = overall efficiency

7.2.1.2   Scope and applications of rope haulage

This system is simple in construction and maintenance. It can suit any gradient and curvature of roadways. Even gravity can assist it to the extent that in hilly terrain it could be installed as shown in figure 7.3(e). This type of arrangement, when installed in an incline, it is known as self-acting-incline. The main drawbacks of rope haulage are the large amount of manual work required at the terminals; absence of complete mechanization and automation of all operations, less reliability and complicated work involved in negotiating branches and junctions. This makes the system inadequate for modern mines and tunnels. This system could be used for low capacity mines having a low degree of mechanization. It finds its application as an auxiliary haulage for material transport in some mines. However, on steep gradients, where belt conveyors cannot be deployed, its use is almost mandatory; and it is used in conjunction with cage and skip hoisting operations in shafts and inclines.

7.2.2   Locomotive haulage[3,5,8,12]

Rail transport finds its application as gathering and main haulage in the underground mines and tunnels. The rope haulage system works on track and so does the locomotive haulage. The locomotive haulage (fig. 7.3(g)) is best suited as a long distance haulage with a gradient in the range of 1 in 200 to 300. However, gradients up to 1 in 30 for a short distance can also be negotiated by this system. The system is flexible comparing to rope and belt conveyor systems. Good roads, efficient maintenance, large output and adequate ventilation are the basic requirements for the success of this system. Well-drained, properly graded and minimum turning with smooth curves constitutes a good road. Laying the rails of suitable size (i.e. weight per meter or yard) with proper fittings and alignment is the key to the success of this system. The weight of rails varies depending upon the weight of locomotive and the number of wheels it has (which could be either four or six). The range is 15–50 kg/m (30–100 lb./yard) for locomotive weight that varies from 5 to 100 tons.

Basically locomotives for underground use are either diesel or electric power driven. Under the electric system – battery, trolley wire, combine trolley-battery and compressed air driven locomotives can be listed. The last one is almost outdated nowadays. Elaborate ventilation requirements in underground gassy coalmines restrict the use of diesel locomotives. Similarly use of trolley wires locomotives is also restricted in such mines due to the risk of fire and explosion that can be caused by the bare trolley wire and electric spark. The battery locomotive is, therefore, a better choice for all types of mines and tunnels, particularly when it is used as a gathering haulage. The trolley wire and diesel locomotives find their applications in the main roads and civil tunnels.

7.2.2.1   Electric locomotives[5,8,12]

Trolley wire locomotives: Direct current (DC) Trolley wire locomotives are used in mines and tunnels. These locomotives are simple in design but capable of bearing heavy loads even under rough and adverse conditions. Supply of power is external and can be unlimited. However, its movement is confined within the trolley wire's network. The trolley wire must be hung at a uniform elevation above the rail and aligned with the track. This is achieved by fixing hangers at an interval of 7–9 m all along the track and about 4–5 m interval on the curves and turnings. The amount of sag should be less than 1%. The power for electric traction is obtained from this over-head conductor (made from hand drawn cadmium copper or hand drawn copper, having cross section in either of figure eight or some variation of it) or from single or duel conductor cable reels and automatic trolley pole receivers. The power return circuit for these locomotives is through the rails (except for duel-conductor cable reel locomotives), and therefore, the rails must be properly bonded for the efficient and safe operation of the locomotives. The polarity of trolley wire may be either positive or negative based on its design. The driving unit of these locomotives consists of 2 or 4 electric motors driving axles through suitable gearing arrangement. The H.P. ranges from 60 to 400; and speed from 10–40 km/hr. The voltage range is 220 v to 500 v. GIA industries supply locomotives weighing 2–20 tons.[8]

The requirement for perfect bonding of the rails (to act as return conductor), chances of fire and explosion due to naked/bare trolley wire particularly in gassy coal mines, and accidents due to electric shocks are some of its limitations. In some of the American and German mines use is made of duel conductor cable reel. This system has advantages, such as: elimination of rail bonding and the risk of accidental firing of the blasting circuit due to stray current. Risk due to electric shock is also at a minimum. Thus, using rail as a return conductor results the operation to be simple and economical but requires much attention on its safety aspects.

7.2.2.2   Battery locomotives[5,8,12]

These electric locomotives receive power from the storage batteries, such as: lead-acid or Ni/Cd, carried on board the locomotive. As stated above this type of locomotive has universal applications as main line, gathering or marshaling locomotive. These locomotives weigh 2 to 25 tons and can run at speeds in the range of 8 to 30 km/hr. In battery type locomotives lead-acid type of batteries having life up to 4 years or 1250 discharge cycles are used. When fully charged voltage/cells are 2 v. A specific gravity of 1.28 g/c.c. for the electrolyte (sulfuric acid) should be maintained for the best results. Alkaline batteries that are bulky, costly and with a life up to 10 years can be used for this purpose. A voltage 1.2 v/cell is obtained from these batteries. Hydrogen – an explosive gas – is given off when charging the batteries. Over heating of a cell during charging may cause fire. However, if proper care and precautions are taken at the charging station, these hazards can be minimized.

7.2.2.3   Combination locomotives[5,8,12]

The trolley locomotives fitted with an auxiliary storage battery are in use in many countries including USA and Germany. This system enables the locomotive to run on the battery at places where there is no trolley wire.

In some trolley wire locomotive designs an electric cable reel is also included to allow its use beyond the trolley wires' layout. A small motor drives this reel. Combined trolley and diesel locomotives are also available.

Battery and trolley wire locomotives are fitted with traction type series DC motors due to their excellent starting torque. Motors are totally enclosed and capable to withstand an over load up to 300% without damage for a short duration. The motors are started in series and run up to half speed with minimum external resistance in the circuit and then in parallel up to full speed.

On small locomotives drum type cam operated controllers are used with five speeds in each direction and also with a separate interlocking reversing drum. The segments and contacts are renewable. Removal of the reversing handle renders the controller 'dead', since it can be removed only on the 'off' position.

7.2.2.4   Diesel locomotives[5,8,12]

This type of locomotive (fig. 7.3(f)) for use in underground mines and tunnels is built of all metallic construction, and all other parts, which are liable to cause fire are substantially shrouded by steel covers. In addition, the following features are incorporated to minimize the risk of fire.

These locomotives have some of these provisions: air filter to prevent carbon particles suspended in air to enter into the engine; flame-trap to trap any flame due to back fire of the engine; exhaust conditioner to cool the exhaust gases; water cooling jacket together with a temperature gauge to stop the engine in the event of over heating; high compression mechanism to fire the engine in place of electric spark plugs; and flame proof fittings. GIA industries, Sweden manufacture 2–40 tons locomotives.[8]

Table 7.2   Comparison of various types of locomotives used in mines and tunnels[5,8,12].

Parameters	Diesel locomotive (DL)	Battery locomotive (BL)	Trolley wire loco. (TW)	Compressed air loco. (CA)
Readiness to service	Needs transfer of diesel from surface to u/g	Needs change of battery	Needs switches and rectifiers	Needs transfer of comp. air from surface to u/g tank
Limitation of travel	By the power it carries, self contained	By the power it carries, self contained	Restricted within the layout of trolley wires	Within radius of 5 km
Running conditions	Least over load capacity. High tractive effort	Has over load capacity, good tractive effort	High overload capacity with better speed. Steep gradient negotiable	Can take over load
Reliability	Max. breakdowns	More than DL	Least break downs	Reliable
Maintenance	Max. maint; needs exhaust conditioner, flame trap and their replacement	Least mnt.	Most skilled job is done at the power station, hence least maint. required	Care is needed regarding air leakage
Safety	Health and fire hazards due to exhaust emission	Safer than DL. but batteries are not flame proof, emits H_2 during charging	Danger of shock, leakage and igniting fire-damp due to spark	Safer
Fixed cost	Least	Higher	Highest	More than DL & BL
Running cost	Maximum	Least	Less	Costliest
Working conditions & suitability	Non gassy mines and tunnels u/g	Good as level as well as gathering haulage	Non gassy mines for higher output	Limited to u/g metal mines
Utilization factor	Lower than BL and T.W.	Good	Good, it can wok continuously	Requires recharging after short travel
Power/wt. ratio	7 h.p/ton	2.6 H.P./ton	8–15 H.P./ton	–
Power consumption	n.a.	About 0.2 kw./t-km	About 0.2 kw./t-km	0.6 kwh/t.-km

7.2.2.5   *Compressed air locomotives*[12]

Earlier use of these locomotives was mainly in metal mines where use of compressed air used to be very extensive. These are of two types: the high-pressure type, which is charged with special air cylinders (150–200 atg) and low-pressure type, which are charged by ordinary compressed air network of the mines. The main advantage of compressed air locomotives is that they are absolutely flame proof particularly in roadways of highly gassy mines where use of other types of locomotives is prohibited by law. Low efficiency (10–12%) and high power consumption are the main drawbacks and which is why nowadays they are almost obsolete, as other types, as shown by way of a comparative statement in table 7.2, can perform better.

Table 7.3   Value of C for differing conditions.

Rail conditions	C – for un-sanded rails	C – for sanded rails
Clear dry rails, starting and accelerating	0.3	0.4
Clear dry rails, continuous running	0.25	0.35
Clear dry rails, locomotive braking	0.20	0.30
Wet rails	0.15	0.25

7.2.2.6   *Other fittings*

In general, the fittings that should be included with a mine-locomotive are: emergency brake, sanding device, speed gauge, km, recorder, headlights, red light at the rear, audible warning signal, fire extinguisher within easy reach of the operator, operator's seat and a portable lamp for emergency.

Spring applied, fully self-adjusting caliper brakes provide full operational and emergency braking features.

In a modern mine a control room via radio contact coordinates the movement of trains throughout the mine. Some mines have a computerized monitoring, operating and signaling system. In Germany, the construction of 'satellite mines' provide the opportunity to apply similar techniques underground, as developed for high speed surface railways with regard to locomotives, trains and man-riding cars.

7.2.2.7   *Locomotive calculations*[3,5,8,12]

Tractive effort: It is the total force delivered by the motive power of locomotive, through the gearing, at wheel treads. When this force is greater than the product of locomotive weight and the coefficient of adhesion between the wheels and rails, the wheels will slip i.e. it will roll. This can be numerically expressed as:

Total or maximum tractive effort $T_E = W_L \, C$ $\qquad$ (7.6)

Where: C is the coefficient of adhesion whose value depends upon the condition of track, and whether it is sanded or not. Given in table 7.3, are the values of C (Bise, 1986).

Drawbar pull: This is the force exerted on the coupled load by a locomotive through its drawbar, or coupling, and is the sum of the tractive resistance of the coupled load. The drawbar pull that a locomotive is capable of developing is determined by subtracting the tractive effort, from the sum of the tractive resistance of the locomotive. This resistance is offered by several sources: rolling resistance, which the entire train offers is equal to weight of the train in tons. (i.e. weight of locomotive + weight of mine cars with pay load) multiplied by a frictional coefficient μ, which could be 10–15 kg/ton (20–30 lb/ton); Curve resistance which can be ignored, gradient resistance and the force required to provide acceleration to the motion (as given in the formulae specified below).

$\qquad$ *Drawbar pull* $= R_0 \, W_T$ $\qquad$ (7.7)

$\qquad$ *Running resistance/t* $R_0 = \mu +/- \{(1/n) \times 1000)\} +/- (a/g)$ $\qquad$ (7.8)

Total Tractive Effort $T_E = R_0 (W_L + W_T)$ (7.9)

Also, Total Tractive Effort $T_E = T_0 W_L$
Eq. (7.9) = Eq. (7.10); $T_0 W_L = R_0 (W_L + W_T)$ (7.10)

Thus, *Drawbar Pull* $R_0 W_T = T_0 W_L - R_0 W_L$
$$= W_L (T_0 - R_0)$$ (7.11)

Or Trailing Load (or weight of train which can be hauled)

$W_T = \{W_L (T_0 - R_0) / (R_0)\}$ (7.12)

B.H.P = (Tractive effort in kg. x speed of locomotive in m/sec.) / (75 x η)
$= (T_E \ x \ v) / (75 \ x \ η)$ (7.13)

Where: T_0 – Tractive effort/t in kg.
T_E – Total tractive effort in kg.
R_0 – Running resistance/t in kg.
W_L – Weight of locomotive in tons.
W_T – Weight of trailing load i.e. weight of train in tons.
η – Efficiency of the system
μ – Frictional resistance/ton in kg.
n – Denominator value of gradient i.e. 1/n, use sign + for up; – for down gradient.
a – acceleration of train in m/sec²; Use (+) for acceleration and (−) for retardation
g – acceleration due to gravity in m/sec² = 9.81
v – Locomotive speed in m/sec.

The above calculation indicates that weight of locomotive is important in order to pull the load. Locomotive weight of 10 tons/h.p. is reasonable for its trouble free operation (Roger et al. 1982).

7.3 TRACKLESS OR TYRED HAULAGE SYSTEM

7.3.1 Automobiles[2,6,7,11]

These trackless units operate on roads and are tyre wheel mounted. Mainly the following units are available for use in surface and underground mines:

1. LHD – when used as transporting unit, the hauling distance should not exceed 150 m
2. Shuttle cars – the limiting distance for these units are in the range of 1–2 km
3. Low Profile Trucks/Dumpers for their use in u/g mines.

7.3.2 LHD

Full description of LHDs has been given in chapter 6.

7.3.3 Shuttle car[12]

This vehicle was brought into mines in the early 1950s and still finds its application in coal and non-coal mines. A chain and flight conveyor fitted in the center of its body transfers the muck from its rear end towards the front one at the time of its loading and it discharges its muck on to a grizzly, conveyor or minecar, when unloading it. This vehicle shuttles between the loading face and its discharge end, hence, the name shuttle car, and it is not required to turn around. Battery, cable reeled electric power or diesel could operate it. The battery version could not find an application, hence, in non-coal mines diesel operated and in coal mines electrically run shuttle cars are used. In later types movement of the car is restricted to a distance below 250 m (max. length of cable). However, for better results its travel distance should be within a 100 m lead to avoid the waiting time.

The car can be loaded by a continuous miner, gathering arm loader and by a LHD in exceptional circumstances. These cars are designed to operate on A.C. (440, 550 or 950 v) or D.C. (250 v). For mines having steeper gradient than 15% A.C. powered shuttle cars are preferred. These cars are equipped with an elevating front conveyor that allows it to discharge the muck into a mine car or belt feeder without the use of ramps. Two electric motors (A.C. or D.C.) one for traction and other for conveyor are fitted to this unit. In some designs two traction motors are provided. Given below are the specifications' (Breithaupt, 1982) range for some of its important parameters: Height 0.7–1.2 m (28–52 in.), Length 7.5 to 8 m (25–27 ft.), Conveyor width 2.2 to 1,6 m (48–64 in.); Overall width 2.4 to 3.2 m (96–126 in.), Capacity 4 to 11.5 m³ (105–410 ft³). Max. rating 7.5–12 tons. Traction motor's h.p. 20–35 (D.C.); 40–50 (A.C.); Conveyor motor's h.p. 60–110 (D.C.); 65–140 (A.C.); Total h.p. range 60–110.

7.3.4 Underground trucks[2,6,7,8,11]

Use of these units (fig. 7.4(a)) began in the early 1970s in underground mines. Both two-wheel drives and four-wheel drive trucks are in use. A two-wheel drive truck finds its application in level to 12% up gradient mine roadways. The road should be with hard surface. It should not be very slippery and soft. Four-wheel drive trucks can be used for rough, slippery and even at the steeper gradients than 12%. Trucks can be classified into three types: Tip dumpers, Telescopic dumpers and Push-plate dumpers.

Tip Dumper: This truck is designed to lift its rear body. While lifting the body at the time of unloading the muck is discharged by gravity to grizzlies, ore passes, waste passes or any other dumping point. The capacity range of these trucks is 5–40 tons.

Telescopic Dumper: This truck is designed to accommodate maximum payload within its space and are compact to negotiate low back heights. It is fitted with a telescopic bed. Loading starts with the telescopic bed in the rear position, and as the load accumulates, the telescopic bed is drawn forward, moving the muck towards the front, then rest of the box (truck body) is filled. While discharging the muck, the telescopic bed is moved towards rear, thereby unloading half the muck and the other half is ejected, out of the truck, by a push plate. The usual capacity ranges between 10–25 tons.

Push-Plate Dumper: This unit is similar to telescopic type dumper but in this case in place of two stages pushing, to discharge the muck, it is accomplished by a single stroke. These trucks are usually available with a capacity range of 10–25 tons.

(a)

A rigid off highway truck   Articulated dumper   Articulated truck

(b)

Figure/photos 7.4   (a) Low profile trucks entering into a tunnel – a track-less system layout (Courtesy: GIA Industry). (b) Some transportation units (Courtesy: CaterPillar).

Kiruna, Wagnor, Eimco, Mormet, Kelbl, Caterpiller and few other companies man-ufacture the low profile dumpers having capacity up to 50 tons. In figure 7.4 Kiruna's low profile diesel trucks have been shown. Electric trucks are successfully working in the mines of Australia, Canada, Sweden, Spain, China, and few others; and so are the diesel ones. This company also supplies 35-ton trucks. The electric trucks are envi-ronment-friendly as they are less noisy and do not produce exhaust fumes.

The truck traffic control calls for an uninterrupted operation of the trucks. In this regard the patterns of approach and spotting of the trucks by an excavator (loader) plays an important part for effective utilization of the trucks. If the drive or tunnel width ⩾1.8 truck length the truck can take a turn near the face. But in narrow work-ings it is necessary to provide recesses measuring 5 m × 3.5 m every 80–100 m of the roadway length to ensure turning of the truck.

Specially shaped chains of high strength steel are wrapped over the tyres to guard against their rapid wear over the rough ground and to eliminate their skidding.

When rail and truck haulage are compared, it can be noted that trucks are more versatile, more productive when large capacity dump trucks are used and their traffic is simple to organize. In underground situations restrictions on haul distance due to ventilation problems and complicated maneuvering particularly in narrow workings are some of the limitations. Rail haulage is advantageous when large tonnage for longer distance needs to be handled.

7.3.4.1   *Trackless or tyred haulage system*[2,6,7,8,11]

These trackless units operate on roads and are tyre wheel mounted. Mainly following units are available for their use in surface and underground mines:

1. LHD – when used as transporting unit, the hauling distance should not exceed 150 m.
2. Shuttle cars – the limiting distance for these units are in the range of 1–2 km.
3. Off Highway Trucks/Dumpers, & Low Profile Trucks/Dumpers for use in u/g mines – these units are available as:
4. Normal size: having capacity of 10, 20, 25, 35, 70 and 120 tons.
5. Giant size: having capacity of 135, 160, 180, 225, 270 and 315 tons (fig. 7.16).

These trucks/dumpers could be of rear discharge type, bottom discharge type or side discharge type. In mines off-highway trucks for carrying heavy loads on abnormally uneven surfaces, with slow speed and for a shorter hauling distances, are used. Their speed is limited to 80 km/hr.

7.3.4.1.1   *Load Distribution:*

- *Load on rear axle = (A/C) x Payload* (7.14a)

- *Load on front axle = (B/C) x Payload* (7.14b)

- *Load/tyre = load on a particular axle/ number of tyres on that axle* (7.14c)

Wherea: A – distance from front axle to center of payload
       B – distance from rear axle to center of payload
       C – wheel base distance.

Rimpull is the weight on the driving wheels. It is a function of truck's gross weight and the total resistance to the motion which is equal to the gradient resistance + rolling resistance.

- *Power required = GVW x effective gradient* (7.15)

Where: GVW is the gross vehicle weight in kg or lbs.
Trucks are rated based on the weight of empty truck + weight of material to be loaded.

- *Effective gradient = road gradient + rolling resistance (%)* (7.16)

- *Usable power = (weight on the driving wheels) x coefficient of traction* (7.17)
                (could be obtained from standard tables)

To find available power use the characteristic curves as shown in figure 7.5(a); and to determine the performance of a truck haulage system, the following four items must be considered:

1. Haulage capacity in tons. to be carried.
2. Cycle time, which is the sum of:
   - Loading
   - Hauling

- Dumping and
- Return time of truck.

3. Hourly production rate:
- *Cycles/hr. = (Effective time available/hr. in minutes)/(Time taken for one cycle in minutes)* (7.18)

- *Hourly output = (Number of cycles/hr) x (capacity of truck in tons.)* (7.19)

4. Apply correction factor depending upon the job conditions such as fill factor, load factor etc.

Use performance curves supplied by the manufacturers. These are

I. Rim pull speed gradability curves (fig. 7.5(a)). On the same plot the rim pull (i.e. power available) that will be available at varying gradients is also plotted. Thus, from plot I, at the varying gradients the rim pull can be obtained.
II. Curve is the plot of Time v/s Distance of travel (one way) on varying gradient (i.e. gradient + rolling resistance). This plot is made available for loaded (fig. 7.5(b)) as well as empty trucks (fig. 7.5(c)).
III. Rolling resistance factor chart (fig. 7.4(d)), showing the industry accepted standards of rolling resistance factor (20 lbs per ton = 1%).

How to find available power at any gradient?

1. First locate point of intersection of GVW and gradient line.
2. Moving this point on left hand side on the Y-axis will give the rim pull (This is the available power).

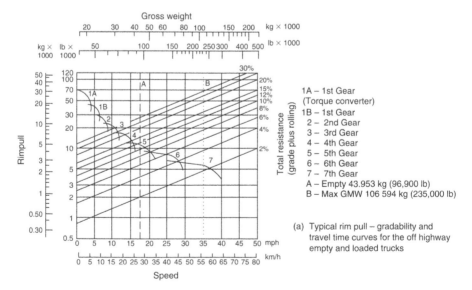

(a) Typical rim pull – gradability and travel time curves for the off highway empty and loaded trucks

Figure 7.5 Typical rim pull – gradability and travel time curves for the off highway empty and loaded trucks.

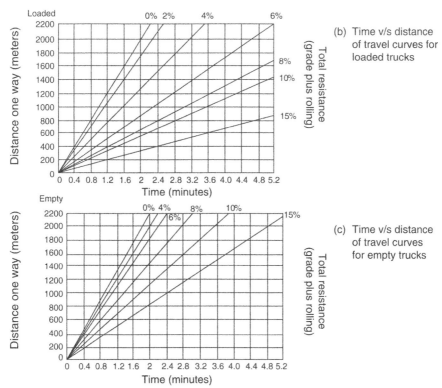

(b) Time v/s distance
of travel curves for
loaded trucks

(c) Time v/s distance
of travel curves
for empty trucks

Road conditions	lb per ton	kg/t
A hard, smooth, stabilized surfaced roadway without penetration under load; watered; maintained	40	(20)
A firm, smooth, rolling roadway with dirt or light surfacing; flexing slightly under load or undulating; maintained fairly regularly watered	65	(35)
Snow, packed	50	(25)
Snow, loose	90	(45)
A dirt roadway; rutted; flexing under load; little if any maintenance; no water 1 in (25 mm) or 2 in (50 mm) tyre penetration	100	(50)
Rutted dirt roadway; soft under travel; no maintenance; no stabilization; 4 in (100 mm) to 6 in (150 mm) tyre penetration	150	(75)
Lose sand or gravel	200	(100)
Soft, muddy, rutted roadway; no maintenance	200 to 4000	(100 to 200)

(d) Rolling resistance factor chart.

Figure 7.5   Continued.

3. When moving towards the Y-axis it intersects to a particular gear, drop this point of intersection vertically down to the X-axis; this will give the speed of the vehicle.

1. Estimate the cycle time and production from a 1,038,000 lbs. (62,600 kg) GVW off-highway truck with 80,000 lbs. (36,300 kg) on its rear wheels when loaded to its

rated capacity. It is to be operated on a 5000 ft. (1524 m) level haul road. The road flexes under load, is rutted and has little maintenance. Assume the following:

Loading time = 1.30 min.
Maneuvering and dumping time = 0.75 min.
Truck capacity = 35 tons.
Job efficiency = 50 min/60 min.
Coefficient of traction = 0.4

Solution:

Grade   = 0% since it is operated on a leveled road.
Rolling resistance   = 100/20 = 5% (from chart(d) fig. 7.5.)
Total effective grade   = 5% + 0% = 5%
Cycle time: loading   = 1.30 min; dumping = 0.75 min; haul time = 3 min. (from
                                 fig. 7.5(b));
return time   = 1.82 min. (from fig. 7.5(c)); total cycle time = 6.87 min.
cycles/hr.   = 50/6.87 = 7.25
Tons. hauled/hr.   = 7.25 × 35 = 254 tons.
Power required   = GVW × effective gradient = 1,038,000 × 5% = 6900 lbs.
The available power can be determined from figure 7.5(a), which is equal to 7000 lbs.
Usable power   = Wt. on driving wheels × coefficient of traction
                                 = 80,000 × 0.4 = 32,000 lbs.

Thus, the vehicle can operate under the given conditions.

2. Instead of the route given in problem 1, an alternative route of 3000 ft. at a gradient of 1 in 5 (adverse) is used, keeping all other data as given in problem 1. Find out which route should be selected based on the lowest cycle time and hourly output.

Grade   = 5% since it is operated on this gradient
Rolling resistance   = 100/20 = 5% ((from chart (d) fig. 7.5.)
Total effective grade while hauling up   = 5% + 5% = 10%
Total effective grade while returning   = 5% − 5% = 0%

Gross vehicle weight on haul   = 1,038,000 lbs.
Gross vehicle weight on return   = 1,038,000 − 35 × 2000 = 68,000 lbs.

Power required while hauling up = 1,038,000 × 0.1 = 13,800 lbs.
Power required on return   = 68,000 × 0 = 0

Power available 14,000 lbs (from figure 7.5(a)); Thus vehicle can be operated under the given conditions.

Cycle time: loading = 1.30 min; dumping = 0.75 min; haul time = 3.7 min. (from
                      figure 4);
return time   = 1.0 min. (from figure 5)
total cycle time   = 6.75 min.
cycles/hr.   = 50/6.75 = 7.38
Tons. hauled/hr.   = 7.38 × 35 = 258 tons.

Route 2 works out to be better as cycle time is less and production obtained is more than alternative 1. Thus, vehicle should be operated on the alternative route.

7.4 CONVEYOR SYSTEM[12]

7.4.1 Belt conveyors

Widely used amongst the conveyor system of haulage are the belt conveyors (fig. 7.6(a)) having their applications both for surface as well as underground mines. A belt conveyor is basically an endless strap stretched between two drums. The belt carries the material and transmits the pull. A belt conveyor system essentially consists of a steel structure all along its length. To this structure are mounted carrying idlers (fig. 7.6(b)), which can be from 2–5, and return idlers. The spacing between carrying idler varies from 1.2 to 2.1 m and from 2.4 to 6.1 m for return idlers. In order to achieve a trough shape to accommodate more and more fragmented or loose material, the carrying idlers mounted on the sides are fixed at 20–35° to the horizontal. A driving unit installed at its one of ends having a motor, gears, driving drum (fig. 7.6(c)) and other fittings to start, stop, or run the belt. At the other end a take-up pulley is

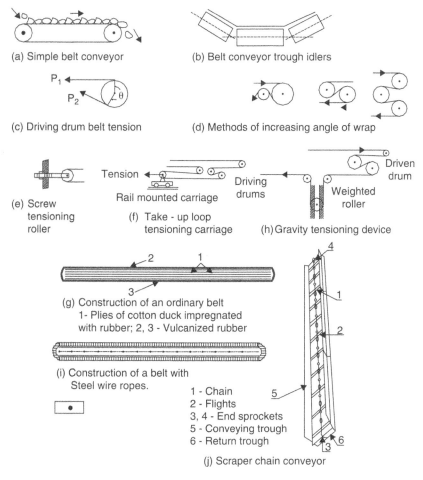

(a) Simple belt conveyor

(b) Belt conveyor trough idlers

(c) Driving drum belt tension

(d) Methods of increasing angle of wrap

(e) Screw tensioning roller

(f) Take - up loop tensioning carriage

(h) Gravity tensioning device

(g) Construction of an ordinary belt
1- Plies of cotton duck impregnated with rubber; 2, 3 - Vulcanized rubber

(i) Construction of a belt with Steel wire ropes.

1 - Chain
2 - Flights
3, 4 - End sprockets
5 - Conveying trough
6 - Return trough

(j) Scraper chain conveyor

Figure 7.6 Conveyor haulage system.

fitted to provide necessary tension to the system. At the discharge end a belt cleaner cleans any adhered material.

The belt conveyers can be classified as stationary, mobile or portable based on their mobility from one place to another. As per their path they can be classified as horizontal, inclined or their combination. Belt conveyors' inclination to horizontal, β, depends upon friction between belt surface and the material to be conveyed, the manner in which material is loaded on to it, and the static angle of the repose of the material to be conveyed. Although β can be up to 40°, if belts of special design are used, up to 18° is very common.

The belt consists of layers of plies that are made of rough woven cotton fabrics (fig. 7.6(h)). The process of vulcanization using natural and synthetic rubber bonds them. Sometimes plies are made of extra strong synthetic fiber like capron, parlon, nylon etc. The rubber cover protects the belt from moisture, mechanical damage, and abrasive and impact action of material. Number of plies can be assessed using the following relation.

Number of plies $\geq (R\, S_{max}) / B\, K_t$ (7.20)

Whereas: $S_{max.}$ – Maximum belt tension in kg
R – Factor of safety in the range of 8–10
B – Belt width in cm
K_t – Ultimate tensile strength per cm width per ply in kg/cm.

The plies are laid in different ways. Prominent amongst them are: cut ply, folded ply, spiral folded ply, stepped ply, cut ply over lain by a layer of rubberized textile belt with steel wires armouring. Amongst the various types of belting available, the prominent are PVC (polyvinyl chloride) and Neoprene or SBR™ rubber. PVC single ply solid woven-carcass belting and multi-ply elting using SBR™ compound or Neoprene is very much in use. Cable belt and steel cable belt are the other two types of belting, which are gaining popularity day by day. The steel-cable belt (fig. 7.6(i)) has steel cables embedded in the carcass to increase the belt's tensile strength. It is used for belts on steep gradients.

Rubber belts can meet some of the requirements such as resistance to hygroscopicity, high strength, low self weight, smaller specific elongation, high flexibility, high resistance to ply separation and long service life.

Take-ups: Take up is an integral part of the belt conveyor system that is incorporated with it to keep the belt tight so that sag is at a minimum. This tension is provided by the use of any of these devices: a screw tension roller (fig. 7.6(e)), take up loop tension carriage (fig. 7.6(f)) or a vertically suspended gravity tensioning device (fig. 7.6(g)). The first two devices are located at either of belt terminal points and the third one some where in between these two terminals. This mechanism (fig. 7.6(g)) allows shrinkage of belt to stitch it or join one segment of belt to other.

These conveyers are simple in design and occupy a limited space. This continuous system of transportation requires very few workers for its operation and maintenance. Belt conveyors are available with a capacity range of less than 100 t/hr. to 14,000 t/hr. (or less than 500 m³/hr to 5000 m³/hr). Its width ranges between 0.5 m to 3 m. For less than 1 m wide belts, the speed limit is 1–1.5 m/sec; whereas for belts more than 1 m wide the speed ranges from 4–5 m/sec. The length of such installation is practically unlimited; as one such unit can be up to 1.6 km or more. Additional units can be put in series if hauling distance exceeds than this. There is hardly any flexibility once the system is installed. It is a sequential operation; hence, breakdowns are also cumulative. It does not allow overloading.

While installing this system the roads should be checked for their uniform gradient and clearance from all sides. Installation should be checked for proper alignment, fastening, erection and tensioning. Use of correct size of belts, drums, idlers, rollers, and motors should not be ignored. The safety devices which are incorporated with the system include hold backs, side guards, protection switches, shock absorbers, fault indicators for slips, sequential controllers etc. The sequential control allows the conveyors to start in sequence from out bye to in bye end and stop in the reverse order i.e. first of all the in bye conveyor stops and the out bye conveyor last. Signaling arrangement runs parallel to its length all along the roadway of its installation and so does the fencing. No wood or inflammable material should come in contact with any of its components to avoid the risk of fire. Provision for over-bridges, pass-over etc. must be made to allow safe travel for the workers. When used in underground coalmines, special precautions are taken to prevent fire and explosion. This requires provision of stone dust barriers and spray of stone dust and water in the roadways where these units operate.

7.4.1.1   Conveyor calculations[12]

Power/H.P. required to run a belt conveyor system = (power reqd. to run the moving parts) + (power reqd. to move mineral) +/- (power required to negotiate gradient)

$$= ((QCVL/75) + (TCLx1000/75 \times 3600) +/- (HT \times 1000/75 \times 3600))K/D$$

> Where: Q – weight of moving parts/m of run in kg. (This includes wt. of belt/m run, wt. of idlers. Belt width and idler spacing effects this factor.)
> C – idler's friction factor (its value varies from 0.018 to 0.046 depending upon belt width).
> V – speed of conveyor in m/sec.
> L – length of conveyor in m.
> H – height of lift or drop in m.
> T – hourly output or production in tons.
> D – overall efficiency of the system.
> K – a coefficient which takes into account resistance at terminal drums.
> K – 1 if L > 30 m, else 1.2 if L < 30 m.

Euler's Formula: No slip occurs between belt and the drum if the following condition is satisfied, i.e.

$$S_t / S_l <= e^{\mu\theta} \quad \text{(for up hill)   Or}   \quad S_l / S_t <= e^{\mu\theta} \quad \text{(for down hill)} \qquad (7.22)$$

Where: S_t – tension in tight side in kg; S_l = tension in loose or slacken side in kg;
   μ – coefficient of friction between belt and drum;
   θ – angle of contact between drum and belt in radians;
   V – conveyor speed in m/sec.

Based on this criteria the *H.P. required* = $(S_t\text{-}S_l)V/(75 \times \eta)$ \qquad (7.23)

Carrying Capacity of Belt Conveyors (Hourly Output) $T = a\, b\, V \times 3600$ \qquad (7.24)

Where: T is hourly production in tons.

        a – average cross-section of material loaded on conveyor, and it is given: (Depending upon type of material), by $W^2/10$ to $W^2/12$

        W – is the width of belt in m

        b – bulk density of material loaded in tons/cum.

        V – belt conveyor speed in m/sec.

1. A single driving drum of a trunk conveyor is 1 m in diameter. The tension in the carrying belt is 6000 kg and that in the return belt is 2500 kg if the driving drum revolves at 50 rpm. What should be the H.P. of the motor to drive this conveyor, assuming gear efficiency to be 90% and the motor efficiency as 85%?
(Hint – speed = angled turned/sec. = 2 × 3.14 × .5 m × 50 r.p.m/60)

$$\text{B.H.P} = (S_t - S_l)V/(75 \times \eta)$$
$$= ((6000 - 2500) \times (2 \times 3.14 \times .5 \times 50))/(60 \times 75 \times .9 \times .85) = 159.6$$

2. A belt 0.9 m wide conveys ore of bulk density 1.35 t/cum. at a speed of 1.75 m/sec. Calculate its carrying capacity. (Hint: Use eq. 7.24)

The value of S_t can be increased either of these methods or combination there of

1. By increasing coefficient of friction 'μ', this can be achieved by lagging the driving drum with a suitable rubber-like material but it is done rarely in practice.
2. By increasing angle of wrap (θ), as shown in figure.
3. By increasing value of S_l: This can be achieved by pre-tensioning the belt by a *Screw Tightening Device (fig. 7.6(e))*, which is put at the tail end of the belt, or by *Loop Take Up*, as shown in figure 7.6(f). For higher-powered conveyors, gravity operated tensioning devices may be used. i.e. it can be either by steel cable/rope on loop take-up carriage or by a heavy roller mounted in a frame as shown in figure 7.6(g).

7.4.2 Cable belt conveyors[12]

In this type of conveyor system the belt is relieved from the tractive force which is transferred to the ropes, which are attached at both sides of the belt all along its length. The belt conveyor needs to carry and bear a load of the material that is to be conveyed. A cable belt conveyor system consists of:

- Two endless ropes, which are supported on the rollers and pass over the drum or pulleys at both ends.
- Spring steel strips embedded in the body of the belt. To these strips are mounted the steel shoes which grip the ropes.
- The belt conveyor itself, which could be of a single ply construction.

Introduction of the cable belt conveyor system has resulted in the use of cheap quality belt with less number of plies. It could be of longer length. The only problem is the complicated construction at both the ends of this installation.

7.4.3 Scraper chain conveyors[12]

These types of conveyors have been designed to withstand rigorous mining conditions particularly at a working face, which is always advanced ahead. These conveyors can negotiate gradients of more than 18° i.e. more than the belt conveyors usually negotiate. These conveyors are easily extendible without dismantling but their length is limited and usually not more than 200 m. A scraper chain conveyor consists of the following parts (fig. 7.6(j)): troughs; endless chain/chains (which may be one or two depending upon the design); sprocket wheel; drive head having motors, gears, tensioning devices and other parts.

Classification:
Mainly there are two types of scraper chain conveyors:

1. Conveyors that are conveying material only, these are usually the conventional chain conveyors (fig. 7.6(j)). These conveyors are either rigid, which means they need to be dismantled before their advance; else they could be of snaking or flexible types which can be shifted without dismantling.
2. Conveyors that are conveying material + cutter loaders; the armored chain conveyor falls under this category. They are also used along with coal ploughs. These conveyors form part of the self-advancing type of supports and are the mechanism that is used on longwall faces. These conveyors usually have two chains. The main features of these conveyors include:
   • Large weight, longer length in the range of 120–150 m
   • High capacity and power.
   • Installed adjacent to face and pushed forward without dismantling
   • Capacity could be up to more than 200 tons/hr.

Chains are the important part of the scraper chain conveyors. Chains should be of high breaking strength and longer life. Also their design should be such that damaged links can be easily replaced. The following type of chain designs are in use:

• Simple flat link and pin type
• Detachable chains with punched links
• Modern conveyors with round steel link chains.

These conveyors can be pushed forward by one of the following devices:

1. Hand operated jacks
2. Compressed air or hydraulic cylinders
3. Using a steel wedge
4. Special advancing devices, which are pulled along the face by winches.

7.5   HOISTING OR WINDING SYSTEM[4,10,11,14]

Hoisting or winding operations are carried out in the vertical or the inclined shafts in the mines. Thus, they establish a link between the surface and underground horizons, and are used to transport men, material, equipment, ore, and waste rocks. Practically this is a lifeline for the miners. Also the shaft, hoist and the fittings, when combined together, constitutes the major capital expenditure of the total investment made for a mine. Earlier the hoist engine used to be steam driven but now this is obsolete and only the electrically driven hoists are operating. These are basically of two types: Drum and Koepe (friction). In order to make these systems operational, the shaft needs to be designed, driven (excavated, sunk) and equipped accordingly. A modern

Table 7.4 Main components of a hoisting plant.

Locale →	Surface	Within shaft	Underground

Equipping details →	(a) Hoist Room equipped with:
	  – Hoist drum or sheaves;
	  – Mechanical & electrical fittings;
	  – Hoist ropes, safety devices etc.
	(b) Head frame or gear equipped with:
	  – Cage landing arrangements and/or
	  – Skip dumping mechanism;
	  – Storage bin, etc.
		Shaft equipped with:
		  – Cage or skip or both
		  – Rope guides
			– Shaft insets or landings
			– Storage bins, crusher chamber
			– Skip loading pockets
			– Handling facilities for man, material, more waste etc.

hoisting plant, which is to be used for services and production purposes, consists of the elements shown in table 7.4.

7.5.1 Head-frame or head-gear[14]

Head frames are constructed over a shaft to support the sheave, which is used to suspend the suspension gear. The suspension gear includes the ropes, conveyances (cage or skip), capels, hooks etc. This construction is also required to allow dumping of the hoisted material above ground. These headgears can be constructed using steel and concrete, as shown in figure 7.7. The prevalent types in steel structure are: 'A' shaped (fig. 7.7(a)), a frame with four or six stands (fig. 7.7((b)) and tent type (fig. 7.7(c)). The Koepe winders (fig. 7.7(d), fig. 7.7(e)) usually use concrete towers.

7.5.2 Shaft conveyances

The common shaft conveyances are skips or cages, which could be used as balancing or with a counter weight. The skips are of three designs; Overturning, with swing out body and with fixed body. The second one has been shown in figure 7.7(h). The cages are available in different sizes and could be single or double deck (fig. 7.7(i)). For production hoisting skips are better whereas for waste hoisting, man riding and material conveyance cages are better. In table 7.5, a comparison between these two systems has been outlined.

7.5.3 Rope equipment[14]

Ropes that are used in a vertical shaft for the purpose of hoisting have been illustrated in figure 7.7(f). The rope equipment consists of rope guides, balance ropes, rope clips, rope tensioning weights, tensioning frame, guide or shoe, conveyance, loading and unloading stages.

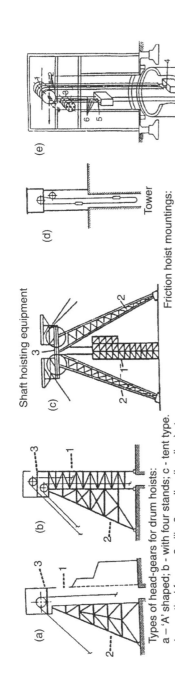

Shaft hoisting equipment

Types of head-gears for drum hoists:
a – 'A' shaped; b - with four stands; c - tent type.
1 – vertical frame; 2 - jib; 3 - pulley (landing) stage.

Friction hoist mountings:
d – Koepe pulley - tower mounted;
e – Koepe multi - rope tower mounted.
1 – drive pulley, 2 - nondeflected ropes,
3 – deflection pulley, 4,5 conveyances
(skip or cage), 6 - deflected ropes.

Tower

Rope equipment in
a vertical shaft:
1 – rope guide
2 – balance rope
3 – rope clips
4 – rope tensioning weights
5 – tensioning frame
6 – guide or shoe conveyance
7,8 – loading & unloading stages

Fleet angle between drum face and hoist
sheave is the angle the rope makes with the
drum as it deviates from the perpendicular
when winding across the drum. The angle
should never exceed 1.5⁰. For multi stage
winding, the minimum is 0.5⁰.

Head-gears of different types and rope suspension schemes.

Skip in filling, discharging
position, a double deck cage.

Figure 7.7 Headgears of different types and rope suspension schemes.

Table 7.5   Comparison of shaft hoisting conveyance systems.

Parameters	Skip winding system	Cage winding system
Shaft size; its space occupied to accommodate system's fittings	Small size shaft can be used, Lesser space required	Large sized shafts essential. More space is occupied by the cages
Arrangements at pit top and bottom	Top – for skip unloading; Bottom – Skip loading pockets, measuring hopper, bin etc.	Both at surface and u/g layouts to handle mine cars required
Rolling stock required	Less mine cars required in circulation, mine car size independent of shaft and skip sizes. Trackless mining possible	More mine cars required in circulation, Mine car size dependent on shaft and cage sizes. Track mining is essential
Handling from multi levels	Possible if arrangement made	No problem
Utility as up or down cast shafts	Dust generation favors installation in up-cast shafts	Can be installed in any shaft
Transfer of waste to u/g	Requires separate arrangement, hence, not feasible	Sand, waste rock etc. can be transferred through cages
Suitability for man-winding	In inclined shafts possible but not very convenient	Very good to handle man, machine and material
Cost	Initial cost may be high but running costs are low	More running costs
General suitability	Useful for production hoisting in coal and non-coal mines	Useful for man winding, mine services and waste handling
Productivity	It is almost continuous, hence better productivity	Considerable time is spent in handling the rolling stock

7.5.4   Classification of hoisting system[4,10,11]

Drum and friction hoists are the two types of hoists that are in operation in the mines. Their main features have been compared in table 7.6, and their sub-classification has been presented by way of a line diagram shown in figure 7.8.

Singled drum hoist with counter weight: It can be used as service or production hoist with the use of cage or skip (fig. 7.9(b)). The balance weight allows its use for multi level hoisting since the position of counter weight at any time is not important.

Single drum hoist with skips in balance: This arrangement is best suited for production hoisting from single level. If it is to be used as multi-level hoisting, then skip resetting is required which reduces the efficiency.

Double drum hoist – one drum clutched: As a service hoist and counter weight, this hoist can serve several levels efficiently, the clutch facilitating quick adjustment of ropes to compensate for initial stretch. This hoist is also used occasionally as a production hoist with skips in balance for one level hoisting. In both the above cases, the selection of this hoist over the single drum hoist would be justified only when savings would offset the added expenses of second drum and clutch in rope adjustment time.

Table 7.6   Main features of a hoisting system.

Features	Drum hoist	Friction sheave hoist	Multi rope friction drum hoist
Relation w.r.t. rope	It stores rope	Rope not stored but extends in the shaft	Use multiple ropes which are not stored but extends in the shaft
Important features	Single rope; Multi level; Medium depth; Widely used	Multi rope; Single level; Limited depth; High output; Efficient	Used for great depths
Max. skip capacity	25 tons.	75 tons	50 tons
Max. output tons/hr	800	2500	1600
Optimum depth	<1800 m	<900 m	>1800 m
Mounting	Head frame	Tower	Head frame

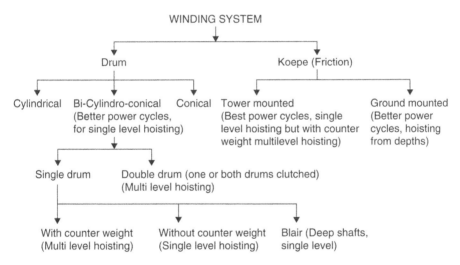

Figure 7.8   Comparison of different hoisting systems.

Double drum hoist, both drums clutched: The main advantage claimed by this type of hoist is that if something happens in one of the two compartments, the hoist can operate in the other compartment to raise and lower men and supplies. This hoist arrangement is practically favored if there is only one shaft entrance to the mine.

7.5.4.1   *Multi-rope friction winding system*[4,10,11]

This is an improvement over the Koepe system using a pulley. This is also known as friction drum winder. The friction drum, which replaces the Koepe pulley, has several grooves on it to accommodate more than one rope. Ropes can be many, but in general,

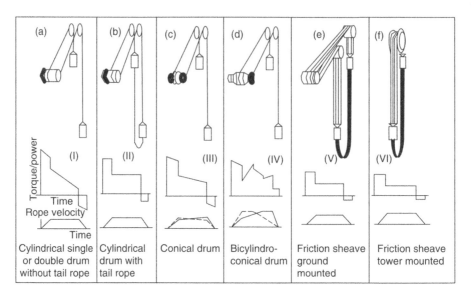

Figure 7.9   Hoisting equipment for shafts. Hoisting schemes with their duty cycles.

these are limited to four. The ropes run parallel to each other and fit into the grooves, about 30 cm apart, and at least as deep as the rope dia. In this system a counter weight is used. These winders could be ground (fig. 7.9(e)) or tower (fig. 7.9(f)) mounted. The conveyance attached to ropes could be a skip or cage (in balance i.e. two in number with tail rope). With the use of multiple ropes, the load can be shared by more than one rope, for example, if 4 ropes are used, then these ropes can have a dia. = $1/\sqrt{4}$; i.e. half the rope dia., of a pulley mounted Koepe winding system. Thus, this system in turn requires small size drum and winding engine. The space required to accommodate this equipment in the shaft as well as that of winding room is reduced. These features allow this system to be used in deep mines. The number of motors varies with the horsepower required. When the gear size for one motor is excessive two motors are used. Generally for a gear ratio up to 1:12 or less and h.p. 1500 or less, one motor is used. Utility of ground and tower mounted friction winders can be understood by considering the following slippage relation:

$$(T_1/T_2) \le e^{\mu\theta} \tag{7.25a}$$

Where: T_1, T_2 – tensions on loaded and empty sides respectively.

　　　　　e – natural logarithmic base;

　　　　　μ – coefficient of friction = 0.45 to 0.5, it can be increased by changing lining or using a suitable lubricant.

　　　　　θ – angle of wrap between rope and the drum, this is 180° for tower mounted and 240° for the ground mounted system. To increase this angle a deflection sheave can be used.

Substituting the value for tower mounted and ground mounted hoists, one can find the limiting ratio in the first case is 1.5 to 1.6, whereas, in the second case it is 1.8–1.9.

Factors need consideration while selecting a hoist:

- Production rate
- Depth of shaft
- Number of levels to be served.

The performance characteristics (fig. 7.9) include consideration of the following:

1. The main consideration is the power consumption/cycle that is related to the cost.
2. Duty cycle is the plot:
   - Power required v/s time
   - Hoist speed v/s time.

Attractive amongst them is the hoisting system, which can be operated at high speed, requires low power/cycle and which can give high output by single level hoisting. A friction hoist meets all these requirements.

7.5.5   Hoisting cycle[4,10,11]

It consists of acceleration time t_a, constant speed time t_v, retardation time t_r and change over time i.e. load or dump time t_d (figs 7.10(a) and (b)). Thus cycle time t_t (in sec.) can be numerically expressed as:

$$\textit{Cycle time: } t_t = 2(t_a + t_v + t_r + t_d) \tag{7.25b}$$

$$\textit{Acceleration time: } t_a = (V/a) \tag{7.25c}$$

$$\textit{Retardation time: } t_r = (V/r) \tag{7.25d}$$

$$\textit{Acceleration distance: } h_a = 0.5 \, at_a^{\,2} \tag{7.25e}$$

$$\textit{Retardation distance: } h_r = 0.5 \, rt_r^{\,2} \tag{7.25f}$$

Where: V – is hoisting speed;
   a and r – are the acceleration and retardation rates;
   h_a – acceleration distance;
   h_r – retardation distance;
   h_v – constant velocity distance;
   h_t – total hoisting distance from loading pocket to head frame bin.

7.5.6   Calculations of suspended load during hoisting

$$\textit{Weight of complete rope: } W_r = w_r \, (h_t + h_h) \tag{7.26a}$$

Where: W_r – weight of complete rope;
   w_r – rope weight/m
   h_h – distance from bin to idler or drive sheave at the apex of head frame if multiple ropes are used, multiply by the number.

$$\textit{Total weight of load: } W_l = W_r + W_s + W_o \tag{7.26b}$$

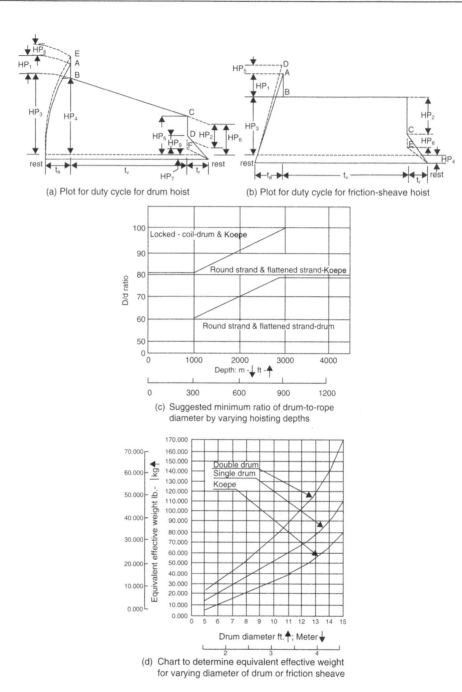

(a) Plot for duty cycle for drum hoist

(b) Plot for duty cycle for friction-sheave hoist

(c) Suggested minimum ratio of drum-to-rope
diameter by varying hoisting depths

(d) Chart to determine equivalent effective weight
for varying diameter of drum or friction sheave

Figure 7.10   Hoisting equipment – some details.

Where: W_l – total weight;

   W_s – weight of skip or cage i.e. dead weight;

   W_o – weight of live load in skip or cage.

Design load: $L = F_s W_l$ Whereas F_s - factor of safety   (7.26c)

Total suspended load: $W = W_e + W_o + 2W_s + 2W_r$ (7.26d)

Where: W – is total suspended load;

   W_e – equivalent effective weight which takes into account the balancing system and other rotating equipment.

Procedure to draw a Duty Cycle Plot of a hoisting system:

1. Plot $(HP_1 + HP_3 + HP_4 = HP_A)$v/s t_a
2. Plot point B by going down by HP_1 at point t_a (i.e. end of acceleration)
3. Plot $(HP_4 + HP_3)$ for t_v period
4. Project down from B by HP_2 amount to get point C
5. Join C to end of retardation period
6. Project up point A up to D by amount HP_5. Join D to the starting point
7. Project downward from C up to amount HP_6. Join this point to the last point
8. Plot HP_4 for all the three periods $(t_a + t_v + t_r)$.

Key Points of Duty Cycles of a hoisting system and H.P. required to perform its various stages: (The subscript refers the corresponding points on duty cycle diagram, figure 7.10(b))

$$HP_1 = (WV^2) / (550\ g\ t_a) = (WV^2) / (17700\ t_a) \qquad (7.27a)$$

Where: HP_1 is the H.P. required to produce acceleration in the system;

   W – Total suspended load in lbs or kg.

   V – Speed of hoisting in ft/sec or m/sec.

   t_a – Acceleration time in sec.

   g – Acceleration due to gravity = 32.2 ft/sec^2 (9.81 m/sec^2)

Similarly; $HP_2 = - (WV^2) / (17700\ t_r)$ (7.27b)

Where: HP_2 is the H.P. gained during retardation in the system;

   t_r – Retardation time in sec.

$$HP_3 = (W_0 V) / 550 \qquad (7.27c)$$

Where: HP_3 is the H.P. required to hoist the live load

$$HP_4 = HP_3 (1 - \eta) / \eta \qquad (7.27d)$$

Where: η is the efficiency of the system

$$HP_A = HP_1 + HP_3 + HP_4 \qquad (7.27e)$$

Where HP_A is H.P. required during acceleration period.

$$HP_B = HP_3 + HP_4 \qquad (7.27f)$$

Where HP_B is H.P. required during constant speed period.

$$HP_C = HP_2 + HP_3 + HP_4 \qquad (7.27g)$$

Where HP_C is H.P. required during retardation period.

Note: To change over from one state of motion to another some H.P. is required. This is denoted by HP_5

$$HP_5 = (0.9 \, HP_A) / t_a \qquad (7.27h)$$

Similarly for the retardation period

$$HP_6 = -(0.9 \, HP_A) / t_r \qquad (7.27i)$$

$$HP_D = HP_A + HP_5 \qquad (7.27j)$$

Where HP_D is the Peak H.P. required during acceleration period.

$$HP_E = HP_C + HP_6 \qquad (7.27k)$$

Where HP_E is the Net H.P. required during retardation period.

$$HP_{rms} = \sqrt{[\{ (HP_D)^2 \, t_a + (HP_B)^2 \, t_v + (HP_E)^2 \, t_r \} / \{ 0.5 \, t_a + t_v + 0.5 t_r + 0.25 \, t_d \}]} \qquad (7.28)$$

Where: HP_{rms} – is the root mean square H.P. of the motor;
t_v – Constant speed period in sec;
t_d – is loading or dumping time in sec.

$$E = \{ 0.7457 \, HP_B \, (t_a + t_v) \} / (3600 \, \eta) \quad \text{in } K_w \text{-} h_r / \text{trip} \qquad (7.29)$$

Where: E is the approximate energy required/trip or cycle.

7.5.7 Use of safety devices with a hoisting system

Within the shaft:

- Signaling arrangement – electrical as well as mechanical
- Separate ladder-ways in the shaft
- Guides – rail or rope

With hoisting engine/hoist:

- Depth indicator, level indicator
- Brakes, emergency breaks
- Over speed controller (such as Lilly controller)
- Over wind controller
- Slow banking devices with separate setting for man winding and ore/rock hoisting.

7.6 AERIAL ROPEWAY[12]

This is a suitable mode of transportation in hilly and difficult terrain where rail and road transports are difficult to operate. In conjunction with underground mining it is used to transfer ore from one mine to another, or, from surface of an underground mine

Table 7.7   A comparison between mono and bi-cable aerial ropeways.

Item	Bi-cable (B.C.)	Mono cable
Max. carrying capacity	Up to 500 t/hr	Up to 250 t/hr.
Max. allowable angle	Any angle	Up to 19°
Weight & cost, e.g. for a 8 km long aerial ropeway	It weighed 210 tons so its cost is high. The structures need to be of heavy duty i.e. trestles etc.	It weighed 70 tons so it is cheaper
Rope changing task	Difficult	Not so difficult
Reliability	More safe and reliable	Less than B.C
Rope life	More	Needs frequent change
Carrying passengers	Safely	Suitable for material transport
Flexibility	There is hardly any scope to change once a system is installed	There is hardly any scope to change once a system is installed

to the processing plants. Sometimes, it is also used to bring backfills to the mine from the river beaches or sand gathering stations. In this system the moving rope carries the buckets, which are fixed to it through the clamps. The rope is endless and is driven either from the loading station or unloading station. The aerial ropeways are of two types:

- Mono-cable
- Bi-cable
- Mono-cable – In this system an endless rope is used. The rope used is so strong that it can carry the loaded and empty buckets/cars when it moves. The cars are attached to the ropeway by means of brackets that could be attached or detached when needed. When the cars are detached from the rope they travel on the rails at the terminal stations. These cars are provided with wheels that run on the rails at the loading, unloading and intermediate stations when cars detached from the continuously running haulage rope.
- Bi-cable – aerial rope way (fig. 7.11): In this system the cars or buckets under circulation travel on a fixed rope of large diameter called tack rope, and these are hauled by a continuously running low diameter rope called a traction rope. The cars are attached to the traction rope by means of a lock or grip. One-track rope is laid each side i.e. one on return side and one on the loaded side, but the traction rope is endless that runs continuously.

7.6.1   Aerial ropeway calculations

Number of carriers in circulation/hr. nc = 2 × Hourly production in tons × Reserve Coefficient/Capacity of carriers in tons.

$$nc = 2 \times Q \times K / G \tag{7.30a}$$

Time interval between two adjacent carriers, in Secs. t = 3600/nc;

$$t = 3600 \times G / 2 \times Q \times K \tag{7.30b}$$

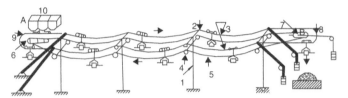

Details of a bi-cable aerial ropeway

A, B Terminal stations

1 – Trestles	6 – Driving sheaves for endless traction ropes
2 – Rocking saddles	7 – Pulley at tensioning station
3 – Track rope	8 – Rails at discharge end
4 – Traction sheaves	9 – Shunt rails
5 – Traction rope	10 – Loading chute

Figure 7.11   A bi-cable aerial ropeway.

Distance between two adjacent carriers (a) = speed × time; a = v × t

$$a = v \ x \ 3600 \ x \ G \ / \ 2 \ x \ Q \ x \ K \tag{7.30c}$$

Total number of carriers in circulation = Length of ropeway/distance between two adjacent carriers

$$Nc = L/a = 2 \ x \ L \ x \ Q \ x \ K/ \ (v \ x \ 3600 \ x \ G) \tag{7.31}$$

It is proposed to transport the copper ore from the pit top of an underground mine to the main concentrator plant 10 km. away from this mine, using an aerial ropeway system. The system is to be designed for an hourly output of 100 tons of ore. The carriers for the system to be used are of 1 ton. capacity. The system is likely to be operated at the speed of 2 m/sec. Calculate the total numbers of carriers you shall put in circulation.

7.7   ROPES[12]

Use of ropes in mines for haulage and hoisting operations is indispensable. If a rope is not selected properly it may result the transportation operation to be unsafe, nonproductive and a costly affair. The following section gives a brief account of types of ropes used in mines.

Stranded Ropes: A strand (fig. 7.12) consists of number of wires laid in a particular direction, whose dia. may or may not be the same. These wires can be laid to form a round or flattened (triangular) cross section. The ropes consist of several strands, laid either in Lang lay or ordinary lay (defined below) fashion, are known as stranded ropes. The central portion of a stranded rope is made of a core, which can be of fiber or metal (fig. 7.12). Fiber core is usually made up of a Manila hemp, Sisal hemp or Russian hemp. The metal cores are made of softer and ductile steel than that

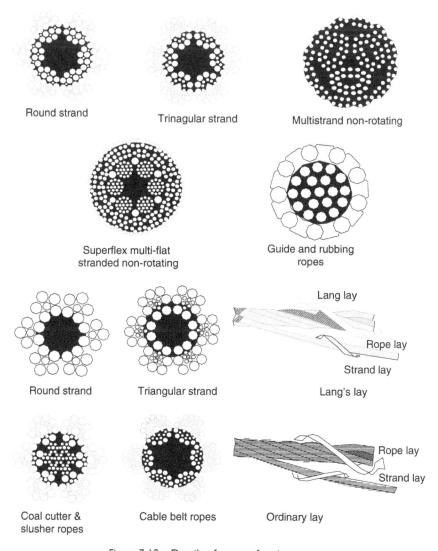

Round strand

Trinagular strand

Multistrand non-rotating

Superflex multi-flat
stranded non-rotating

Guide and rubbing
ropes

Round strand

Triangular strand

Lang lay

Rope lay

Strand lay

Lang's lay

Coal cutter &
slusher ropes

Cable belt ropes

Rope lay

Strand lay

Ordinary lay

Figure 7.12   Details of ropes of various types.

of wires in a strand. These ropes can be classified as round, flattened and multi stranded ropes. In *Lang's lay* wires and strands are laid in the same direction, whereas in the *Ordinary lay* wires and strands are laid in opposite directions (fig. 7.12).

There can be different ways to classify ropes, as shown by way of a line diagram shown in figures 7.13(a) and (b). In this figure materials used to manufacture ropes have been also shown. The cross sectional view of ropes of various kinds that are useful for haulage and hoisting operations have been illustrated in figure 7.12.

Round Stranded Rope: If a core of wire is covered by six wires, it forms a seven wire strand, and six such strands around a central fiber core makes a round stranded

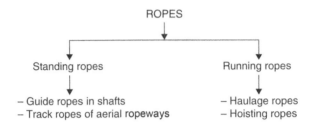

Material used for the construction of ropes: It is high carbon steel having the following composition, in general: C = 0.5 to 0.85%; Si = 0.11%; Mn = 0.48%; S = 0.033%; FE = remainder.

(a)

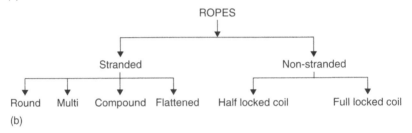

(b)

Figure 7.13   Mine Ropes Classification (a) based on use, (b) based on construction.

Table 7.8   Rope calculations.

Rope type	Weight in kg/cm²	Breaking strength in tons.
Round stranded	$0.04C^2$	$0.7C^2$
Flattened stranded	$0.045C^2$	$0.77C^2$
Locked coil	$0.06C^2$	$0.85C^2$

C – circumference of rope in cm.

rope which is a very common haulage rope (fig. 7.12). If the same strand is covered with a second layer of 12 wires, and using such strands when rope is prepared, this again is a round stranded rope that has its application in winding. To this a further layer of 18 wires will result a strand of 37 wires. 1-6-12-18 can represent this, and rope made by 6 strands is represented as 6(1-6-12-18). Use of ropes stranded in this manner, is common in mines. The gauge of all these wires may or may not be equal. When gauge of wires in a strand differs, it is also known as compound stranded. For example, the 'Warrington construction' is made up of wires of three gauges; but in 'Seale construction', around a central wire higher gauge, the wires of smaller gauge are laid out, and then again around these wires, a layer of wires of the same gauge as that of central wire, is laid out. Similarly in a rope known as 'Filler construction', in between the wires of higher gauges, the wires of smaller gauges are laid out to fill up the gap. All these designs are manufactured to obtain varying degree of flexibility to suit a particular operation.

Flattened Stranded Rope: It has wires in each strand laid around a triangular core made either of a single steel wire (fig. 7.12) or, formed by laying wires together to form

Table 7.9   Rope selection based on its use.

Operation	Type of ropes
Haulage	Round stranded, flattened stranded
Hoisting shallow depth	Round stranded, flattened stranded
Hoisting great depth	Multi stranded, locked coil
Aerial ropeways	Traction rope: flattened stranded; Track rope: locked coil
Guide rope in shafts	Rope with large diameter wires of round or flattened construction, sometimes half-locked coil
Suspension of pumps during sinking, crane and capstones ropes	Compound stranded rope
During shaft sinking*	Ordinary lay round stranded

*Except shaft sinking in all other operations usually Lang's lay rope is used.

almost a triangular shape; it results an outer flattened surface and hence the name flattened stranded rope. This type of rope gives better durability, strength, resistance to crushing and less tendency to twist and stretch.

Multi Stranded Rope: In this rope there are two or more layers of strands (fig. 7.12). The outer layer is laid in the opposite direction to the inner layer. The strand may be either round or flattened. Use of such ropes is wider for haulage, hoisting and as standing ropes.

Non Stranded Ropes – Locked Coil Ropes: This rope is not stranded but made by laying concentric sheaths of wires in one direction around a central wire, one or more layers of special wires are laid over it in the opposite direction and ultimately the outer wires are so shaped that they inter-lock each other. Maximum strength, no tendency to rotate and smooth wearing surface are some of its advantages, whereas low flexibility, difficult to detect broken wires, non-splicing and higher costs are some of its limitations.

Guide Ropes: These ropes are usually made of wires or rods (fig. 7.12) of large dia. some times even up to half an inch or more. They may be made of simple strand, compound strand or locked coil. The idea is to provide a maximum wearing surface.

Desirable qualities that a mine rope should have include: flexibility to fit into sheaves and pulleys, resistant to internal and external wear, weight/unit length, load bearing capacity to withstand tension, torsion and bending.

7.7.1   Rope calculations

Diameter of rope, $d = \sqrt{[(XL) / \{(F/K) -(D/850)\}]}$ (7.32)

Where: L – suspended load in tons
       F – ultimate tensile strength of steel in tons/cm^2
       K – factor of safety (8–10)
       D – depth in meters
       X – a constant which is 3, 2.6 and 2 for multi stranded, flattened stranded and locked coil ropes respectively.

7.8   TRACK AND MINE CARS

7.8.1   Track[3,9,12]

While adapting a track mine system, properly laid track and suitable types of mine cars are mandatory to achieve smooth and trouble free transportation of ore, rocks and material in the mine. Selection of a track depends upon the weight of locomotive or the weight on a wheel in case of rope haulage system. In figure/table 7.14(c) suggested rail weights given by Bethlehem catalog 2314 can be used as guide. Grosvenor[9] gave a thumb rule to choose rail weight in pounds/yard to be 10 lb./yard for each ton. of weight on wheels. Track lying is a skilled job. A track laying operation involves laying of ballast, sleepers, track with fish-plates and bolts, switches and crossings. Ballast provides an elastic bed for the track, distributes the applied load over a large area and holds the sleepers in place. The ballast should be crushed rock, which is not affected by water and produces minimum dust when subjected to heavy weights as it has got adverse effect on locomotive parts and the rolling stock. Its size should be in between 12–50 mm (0.5″ –2″). The sleepers could be of wood, concrete or steel. Spacing (L) between them should not exceed 850 mm (2′9″), or the following relation[1] can determine it:

Spacing between sleepers: $L = 5000 \times W/P$ (7.33)

Value of W = 37, 56, 82, 109 for rails of 14, 17.5, 24.8 and 30 kg/m respectively. P is the wheel pressure in kg.

The sleeper length should be gauge +0.6 m (2′). At each rail joint two sleepers must be laid. Suitable crossovers and crossings (fig. 7.14(a)) and turnouts (fig. 7.14(b)) should be used. A proper super elevation as per the gauge, speed and radius of the curve must be provided at each turning. Roadways should be as straight as possible. The track gauge of 600 mm is known as narrow gauge and this is suitable for rope haulage. The 750, 900 or 1000 mm gauges are known as meter gauges that are suitable for locomotive haulage. 1000 mm – meter gauge and 1524 mm broad gauge are common for the surface locomotives.

7.8.2   Mine cars[3,9,12]

Mine cars, bundies or tubs are the different names given to mine cars, which are made of a steel body; and are available with up to four axles and with carrying capacity up to 100 tons.[9] Mine cars up to 15 tons. capacity are very common. In coalmines mine cars of small size (<1, 1, 2, 3, tons.) are used where as in metal mines big size mine cars (3, 5, 7, 10 or more) are common. These cars available in various designs particularly as per their mode of discharging muck they hold, and prominent amongst them are: side discharge (fig. 7.15(a)), end discharge (fig. 7.15(b)), bottom discharge (fig. 7.15(c)), side discharge Granby cars, and automatic bottom discharge cars. Bottom discharge cars have the advantage of getting rid of muck sticking at their bottoms and also require less height of workings during muck discharge operation. Figure 7.15(d) illustrates the process of muck discharge from the bottom discharged cars as per Swedish practice.

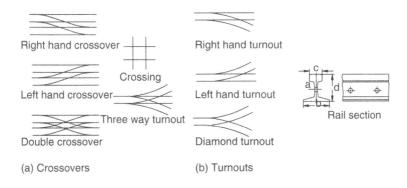

(a) Crossovers   (b) Turnouts

Locomotive Weight	Weight of rail per Yard		
Tons.	4 – Wheel Loco	6 – Wheel Loco	8 – Wheel Loco
2	20	-	-
4	25	-	-
6	30	-	-
8	30	-	-
10	40	-	-
13	60	30	-
15	60	40	-
20	60	40	-
27	80	60	-
37	100	70	60
50	-	85	70
		100	90

(c) Table: Suggested rail weights for the mine locomotives. Source: Bethlehem Catalog 2314.

Figure 7.14   Details of track system. Rail sections, crossovers and turnouts. Rail weight corresponding to locomotive weight for mines/tunnels shown.

7.9   THE WAY FORWARD

Deployment of a fleet of giant sized equipment, such as off-highway trucks of 320 tons capacity (fig. 7.16) has already begun to handle bulk quantities of minerals speedily. As detailed in sec. 19.5.8, in order to reduce wastage and losses which are key to reducing costs, *automation in mining* has begun and it is predicted that during the next generation, mine development with the use of equipment and practices, as shown in the illustration in figure 19.14 would be undertaken. The critical path to full automation in underground mining is shown in figure 19.15(d).

But one should certainly do nothing which leads to more unemployment as it has a booming effect on the all the work that we do. It is no good just copying what is being done in the highly industrialized countries. Certainly everything should be judged in the context of conditions that prevail in your own country, basically accepting the fact that better techniques have to be always employed, wherever feasible. In fact, if we make rock fragmentation and its disposal efficient, the entire mining and tunneling operations become efficient.

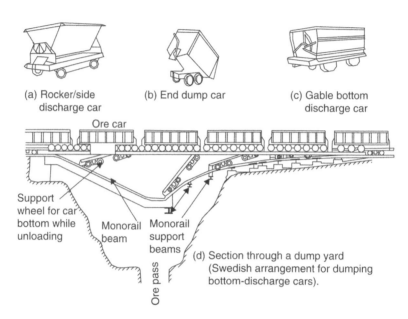

(a) Rocker/side
    discharge car

(b) End dump car

(c) Gable bottom
    discharge car

Ore car

Support wheel for car bottom while unloading

Monorail beam

Monorail support beams

Ore pass

(d) Section through a dump yard
    (Swedish arrangement for dumping
    bottom-discharge cars).

Figure 7.15   Mine cars and their unloading arrangement.

Figure 7.16   Giant sized 320 tons capacity trucks working at one of the surface coal mines in Australia.

QUESTIONS

1. A belt 0.9 m wide conveys ore of bulk density 1.35 t/cum. at a speed of 1.75 m/sec. Calculate its carrying capacity in tons/hour.
2. A hoisting cycle consists of the following elements: Acceleration period = 6 sec. Retardation period = 6 sec. Constant speed period = 80 sec. Detention period = 8 sec. The mine operates 3 shifts/day and effective shift time is 7.5 hours. If a skip

of 11 tons capacity has been installed with this balanced hoisting system, calculate the daily production from the mine.

3. A single driving drum of a trunk conveyor is 1 m in diameter. The tension in the carrying belt is 6000 kg and that in the return belt is 2500 kg. If the driving drum revolves at 50 rpm, what should be the H.P. of the motor to drive this conveyor, assuming gear efficiency to be 90% and the motor efficiency as 85%. (Hint – speed = angled turned/sec = 2 × 3.14 × .5 m × 50 r.p.m/60) B.H.P = (St – Sl)V/(75 × overall efficiency) = (6000 − 2500) × 2 × 3.14 × .5 × 50))/(60 × 75 × .9 × .85).

4. Calculate the chief dimensions of direct rope haulage to haul 350 tons of coal/shift through an incline having a gradient of 1 in 12 and length equal to 1000 m. The weight of a mine car to be deployed with this system is 250 kg and its carrying capacity is 500 kg of coal. Consider the speed of the haulage to be 12 km/hr. Assume the frictional resistance between mine car and track as 1/50, and the frictional resistance between rope and the friction rollers to be 1/10 of the weight of the rope. The effective working hours/shift at this mine is 7. Take changeover time/trip = 5 min. Haulage efficiency of the system to be 90%.

5. Classify the transportation system that is used in mines. Illustrate it by a line diagram. Give application as per the locale (i.e. place of use of each one of them) and limitations of each of these systems based on hauling distance and gradient (inclination).

6. Compare the important features of mono- and bi-cable aerial ropeways. Mention the suitability of each one of them.

7. Define the terms: track, trackless, haulage, hoisting.

8. Give details of these in conjunction with haulage: selection of rail sections, types of crossovers and turnouts and places of their applications. Rail weight corresponding to locomotive weight for use in mines and tunnels.

9. Derive a relation to find out the total power required to drive a belt conveyor system. Give Euler's relation. How can contact angle be increased? Illustrate it. Also write and illustrate different types of take ups used with this system.

10. Give a classification of mine ropes based on their construction. Draw the construction of each of these ropes describing their lay. Suggest the type of rope for the following operations: haulage underground, hoisting at great depth, guide rope in shafts.

11. Give a relation to find the hourly output of a belt conveyor system.

12. Given the following data calculate the diameter of rope, for each of the rope types: Depth of shaft 300 m; Total suspended load: 25 tons; Ultimate strength of rope = 100 tons/sq. cm; Factor of safety: 9.

13. How is performance of a hoisting system judged? Sketch different hoisting systems and draw their duty cycles. Compare the duty cycle performance and work out the utility of each of these hoisting systems. Mention the main components of the hoisting cycle.

14. It is proposed to transport chromite ore from an open pit mine to the main concentrator plant, 5 km away from the mine, using an aerial ropeway system. The system is to be designed for an hourly output of 50 tons of ore. The carriers for the system to be used are of 1 ton capacity. The system is likely to be operated at

the speed of 2.5 m/sec. Take the reserve coefficient to be 1.2. Calculate the total numbers of carriers you propose to put in circulation.

15. List the main components of a hoisting plant. Describe and illustrate it. What factors you will consider while selecting a hoisting system? Compare the main features of the prevalent hoisting systems.

16. List the type of locomotives that are used in underground mines. Compare their main features. Mention under which circumstances each one of them will be a better selection.

17. List types of mine cars known to you and describe unloading arrangements in each case.

18. List various types of conveyors that are used in mines. Give a line diagram to illustrate the various parts/elements of a belt conveyor system. Give the range with regard to the following parameters for which belt conveyors are available: width, single stretched length, gradient, capacity.

19. Mention the different rope haulage systems commonly used in mines. In which specific situation will you recommend each one of them? Give their main features and layouts. Compare their main features.

20. Mention the types of head gears/hoisting towers commonly used in mines. What is a fleet angle? Mention its limitations.

21. Name the conveyances that can be used with a hoisting system. How can hoists be mounted? What does a hoisting arrangement consist of?

22. Name types of surface mines known to you and suggest mucking and transportation equipment for each of these mines. What are off-highway trucks? Mention the range of capacities in which they are available for use in surface mines.

23. Outline the key points on the duty cycle of a hoist and give relations to find out the H.P. at each of these points. Give a relation to find out the RMS (root mean square) H.P. of the hoist motor. Also mention the relation to find the approximate power consumption/hoisting cycle.

24. Prepare a line diagram to describe and compare batch and continuous systems of transportation.

25. Using the following data, calculate the B.H.P. of a locomotive engine: Weight of locomotive = 25 tons. Weight of train = 60 tons. Gradient = 1 in 100 up. Acceleration = 0.005 m/sec. sq. Acceleration due to gravity = 9.81 m/sec. sq. Speed = 20 km/hr. Value of coefficient of frictional resistance = 10 kg/ton. Engine efficiency = 85%.

26. What is winding? List types of winders or hoists that are available for their use in mines. Mention the depth and output range known to you from these types of winders.

27. What material is commonly used to construct the belt of a belt conveyor? Write down the different types of belts that are available based on their 'PLY' layout. Give a relation to find out number of plies in a belt.

28. Where you will recommend the use of a shuttle car? How many motors are fitted in it and what is the function of each one of them? Give the specification of any two models of shuttle cars known to you.

29. Work out/give the operating dimensions of a low profile truck for its use in underground mines. Where you will deploy such trucks?

30. Write down the main features of a cable belt conveyor system. Also write down the application of shaker conveyors in mines.
31. Write down the main features, constructional details, and field of application and classification of the chain conveyor system used in mines. What are the precautions that should be observed while installing the conveyors underground? Give a list of the safety appliances/devices that are incorporated with a belt conveyor installation.
32. Calculate the motor B.H.P, rope diameter and diameter of the surge wheel for an endless rope haulage system, given the following operating parameters:
    – Output/shift, taking 7 effective hrs/shift = 350 tons
    – Length of haul where it is installed = 1000 m
    – Gradient of haulage road = 1 in 12 up
    – Weight of empty mine car = 250 kg
    – Carrying capacity of car = 750 kg
    – Rope speed = 4 km/hr, haulage efficiency = 90%
    – Coefficient of frictional resistance between mine tub' wheels and track = 1/55, frictional resistance of rope = 1/10 of rope weight
    – Ratio of rope dia. to surge wheel dia. = 70.
33. The following data have been obtained from a Koepe winding installation: – Duty cycle operations' details:

    Acceleration period = 10 sec. Retardation period = 8 sec. Constant speed period = 39.75 sec.
    Detention period = 10 sec. Hoisting overall efficiency = 90%
    H.P. required during acceleration period (P1) = 193
    H.P. generated during retardation period (P2) = −241
    H.P. required to hoist live load (P3) = 364

    Calculate the following, using the notations as indicated:

    P4, H.P. required to compensate efficiency losses
    PA, H.P. required from starting to the end of acceleration period
    PB, average H.P. required during constant speed period
    PC, H.P. required during retardation to rest period
    P5, H.P. required during the changeover from rest to acceleration
    P6, H.P. required for the changeover state during retardation period
    PD, Max. H.P. required during acceleration period
    PE, Max. H.P. required during retardation period
    Plot the duty cycle i.e. time v/s H.P., on graph paper.
34. Estimate the cycle time and hourly production from 138,000 lbs (62,583 kg) GVW off highway truck with 80,000 lbs (3628 kg) on its rear wheels when loaded to its rated capacity. It is to be operated on a 3000 ft (915 m) road having a gradient of 1 in 5 (up). The road flexes under load, is rutted and has little maintenance. Assume the following:

    Loading time = 1.30 min. Maneuvering and dumping time = 0.75 min. Truck capacity = 35 tons
    Job efficiency = 50 min/60 min. Coefficient of traction = 0.4.

Also determine whether this route will be suitable for this truck or not. You should use the characteristic curves and tables (given in sec. 7.3.4), else supplied by the manufacturer for these calculations.

35. Compare skip v/s cage winding systems. Work out in which situation each one of them would be most suitable. What are main operations included in a winding cycle? Calculate the daily production of a shaft equipped with a balanced friction sheave hoisting system, given the following:

Shift time = 7.2 hours. Shifts/day = 3. Skip capacity = 11 tons Cycle time = 85 sec./trip.

Also calculate the approximate energy consumption/skip hoisted if the average power consumption is 975 H.P. The hoist efficiency is 85%; the acceleration time is 6.5 sec and constant velocity time is 63.5 sec.

36. Given the following data work out the design parameters of a friction sheave hoist:

Two balanced skips, tower mounted hoist
Production rate 6300 tons/day
Working time = 3 shifts/day; 7 hr/shift
Shaft depth = 686 m
Head frame height = 30 m
Bin height = 15 m hoist rope 4 stranded 19 mm dia.; weight 1.41 kg/m
Skip load = 0.67 live load
Load time = dump time = 6 sec
Hoisting speed = 11.43 m/sec
Acceleration = retardation = 2.29 m/sq. sec.

Calculate the following:

a. Components and total cycle time (1 round trip) in second
b. Skip capacity
c. Rope slippage
d. Sheave diameter.

37. The following data are available for a friction hoist installed at a mine:

Two balanced skips; tower mounted hoist; working 3 shifts/day; 7 hr/shift.
Shaft depth = 1000 ft (305 m). Hoist rope 4 flattened stranded 1 inch (25.4 mm) dia.; wt 1.8 lbs/ft (2.68 kg/m)
Skip live load = 5 tons (4.5 tonnes). Sheave dia. = 10.3 ft (3.14 m). Skip dead load = 1.2 live load.
Load time = dump time = 6 sec. Hoisting speed = 20 ft/sec (6.1 m/sec.) Acceleration = 10 sec.
Retardation = 8 sec. Constant velocity time = 39.75 sec. Halt time = 10 sec. Hoisting efficiency = 90%.
Effective equivalent weight = 3700 lbs (1678 kg) (from table).

Check rope spillage and determine the RMS rating of an AC electric motor to drive the hoist and also find the approximate energy consumption/trip.

REFERENCES

1. Atlas Copco, leaflets and literature.
2. Bise, C.J.: *Mining engineering analysis*, SME, Littleton, Co, 1986. pp. 120–30.
3. Brantner, J.W.: Mine haulage locomotive calculations. In: Cummins and Given (eds.): *SME Mining Engineering Handbook*. AIME, New York, 1973, pp. 14, 17–18.
4. Brucker, D.S.: Faster and deeper – The sign of the times in hoisting, Rept. ASEA, *Swedish Trade Commision*, Los Angeles, 1975, pp. 29.
5. Buckeridge, R.M.; Carey, W.T. et al.: Rail haulage system. In: *Underground Mining Methods Handbook*. W.A. Hustrulid (edt.), SME-AIME, New York, 1982, pp. 1227–45.
6. Caterpillar – Performance handbook, 2000. pp. 10, 22–26.
7. Connel, J.P.: Truck Haulage. In: Cummins and Given (eds.): *SME Mining Engineering Handbook*. AIME, New York, 1973, pp. 18, 16–22.
8. G.I.A. Industries, Sweden, *leaflets*.
9. Grosvenor, N.E.: Mine cars and track. In: Cummins and Given (eds.): *SME Mining Engineering Handbook*. AIME, New York, 1973, pp. 14, 3–7.
10. Harmon, J.H.: Hoists and hoisting systems. In: Cummins & Given (eds.): *SME Mining Engineering Handbook*. AIME, New York, 1973, pp. 15, 2–69.
11. Hartman, H.L.: *Introductory Mining Engineering*. John Wiley & Sons, New York, 1987, pp. 317–31.
12. Karelin, N.T.: *Mine Transport*. Orient Longmans, Bombay, 1967, pp. 15–37, 94–96.
13. Mann, C.D.: Swedish-type car dumps. In: Cummins & Given (eds.): *SME Mining Engineering Handbook*. AIME, New York, 1973, pp. 14, 41–43.
14. Borshch-Komponiets, V.; Navitny, A. and Knysh, G.: *Mine surveying*. Mir Publishers, Moscow, 1985, pp. 196–210.

Chapter 8

Supports

Keeping disturbance to natural ground settings to a minimum could be considered as directly proportional to cost reduction and minimizing problems encountered during ground excavation and mining.

8.1 INTRODUCTION – NECESSITY OF SUPPORTS[1,4,5,6,7,10]

Basic factors determining the physical and mechanical properties of the rocks include:

- The depth of deposit
- The local geological structure (tectonics)
- The stratigraphy and geological age of the rock
- The weathering
- The presence of water and its condition.

The rock pressure depends upon the geologic factors (such as: the physical and mechanical properties of rock, bedding conditions, presence of water), the dimensions of excavation, and also the method of driving and the care with which the excavation has been made.

The basis of rock pressure is weight, because the upper layers of the rock press on those below them. Usually these forces are in balance when the ground is stable.

After the mass of rock has been penetrated by a mine roadway or tunnel, the loss of balance results in a very short interval of time. Therefore, in those roadways and tunnels where the rock does not deform beyond its elastic limit, the support does not experience any pressure and the roadway can stand for a long time without support. For example wide excavations driven in strong unfissured rocks such as granite etc. or even a small excavation driven in dense clay can stand unsupported for an unlimited time without noticeable change.

However, since all rocks are not so strong and also the excavations can be large, the elastic deformations increase to plastic ones; the continuity of the rock is disrupted and it begins to break. The outward sign of this stage of deformation is usually deflection of roof and cracking in it, at first barely noticeable and later on steadily increasing. As the cracks widen, the beds separate and the rocks drop off in sizes which vary with the rock type and its fissuring. Further fractures can cause roof falls. To maintain the size and shape of the roadway, it becomes necessary to support to resist the pressure of the surrounding ground i.e. the rock pressure.

In designing the support system it is necessary to know the direction and magnitude of rock pressure, and also, the strength of the rock around the roadway or tunnel. The change of stress caused by the deformation of surrounding rock can be regarded as bounded by so-called spheres of influence of the roadway or tunnel. In the roof this sphere of influence (fig. 8.9) includes the zone of breakage, or zone of progressive fissuring, and the active zone, both of which apply direct pressure on the roadway.[5] Outside these zones the rock is not subjected to the effects of excavation but the zones may extend later.

The strength of rock is measures by the specimen collected, which are taken after the redistribution of stresses and from the neighborhood of the excavation. The established factors that contribute determination of rock pressure are summarized below:

- The stressed state of rock mass and the mechanical properties of the rock;
- The shape, dimensions and location of the excavation;
- The duration of the exposure of the rock excavation; and,
- The depth of excavation.

8.2 CLASSIFICATION OF SUPPORTS

Proper selection of support is very vital to mines and tunnels. In mines it determines safety of work, ore production cost, losses and dilution, intensity of mining and productivity of the mine. Mining operations disturb the *stressed equilibrium state* of rock that is found in solid. A stress field called rock pressure develops around the workings. It acts upon the surrounding rock, pillars and supports. Thus, all these elements constitute a support network in any mine.

Similarly in civil tunnels support determines safety during its drivage and afterwards, tunneling rates and costs. A line diagram given in figure 8.1(b) represents supports of various kinds with their sub-classification.

8.3 SELF SUPPORT BY IN-PLACE (IN-SITU) ROCK

Use of the in-situ rock to support the rock is the best way of designing supporting system wherever feasible. *A competent rock is defined as the rock, which, because of its physical and geological characteristics, is capable of sustaining openings without any heavy structural supports.*[12] Rock mechanics tests are performed to evaluate the structural properties of the in situ rocks. If the rock is to support effectively it must not be allowed to loosen. This, in turn, requires a careful blasting and selection of properly shaped openings. The size should be kept as minimum as possible. Adherence to these practices could prove a useful guideline to minimize the need of artificial supports.

8.3.1 Support by the use of natural pillars

Pillars of different kinds practically form a near-rigid type of supports. In some methods they form the integral part of the stope design e.g. board and pillar; room and pillar; stope and pillar; post and pillar etc. whereas in other they are used

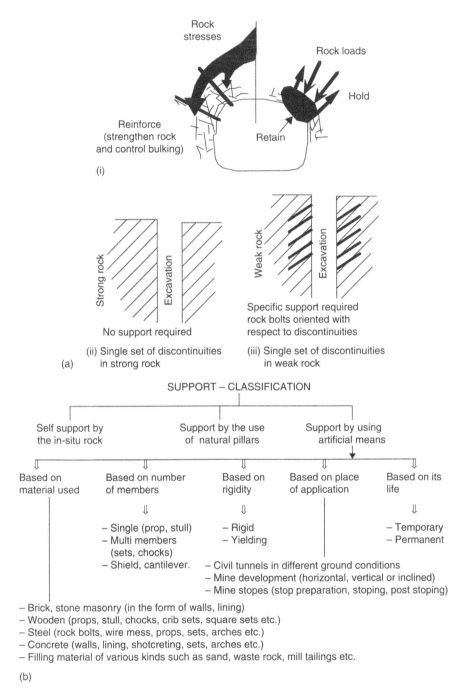

Rock
stresses

Rock loads

Hold

Reinforce
(strengthen rock
and control bulking)

Retain

(i)

Strong rock

Excavation

No support required

(ii) Single set of discontinuities
in strong rock

Weak rock

Excavation

Specific support required
rock bolts oriented with
respect to discontinuities

(iii) Single set of discontinuities
in weak rock

(a)

SUPPORT – CLASSIFICATION

Self support by
the in-situ rock

Support by the use
of natural pillars

Support by using
artificial means

Based on
material used

Based on number
of members

Based on
rigidity

Based on place
of application

Based on its
life

– Single (prop, stull)
– Multi members
(sets, chocks)
– Shield, cantilever.

– Rigid
– Yielding

– Temporary
– Permanent

– Civil tunnels in different ground conditions
– Mine development (horizontal, vertical or inclined)
– Mine stopes (stop preparation, stoping, post stoping)

– Brick, stone masonry (in the form of walls, lining)
– Wooden (props, stull, chocks, crib sets, square sets etc.)
– Steel (rock bolts, wire mess, props, sets, arches etc.)
– Concrete (walls, lining, shotcreting, sets, arches etc.)
– Filling material of various kinds such as sand, waste rock, mill tailings etc.

(b)

Figure 8.1   (a) Supports – some concepts. (b) Support – classification.

to maintain stability between the stopes (all stoping methods for steeply dipping deposits). Ore pillars are left either forever or for the duration of working of a given section. Depending upon the purpose and arrangement, the pillars can be classified as under:

- *Protective pillars*: These pillars are required to preclude caving of shafts or a particular structure.
- *Level pillars*: These are the pillars left above and under the workings of main horizons of the levels/sections to support them. Crown pillar and sill pillars belong to this category.
- *Rib/block/side pillars*: These pillars are left between two adjacent stopes or blocks.

The support with ore pillars is a simple and economic method. However, it is not practicable in the areas of high-grade deposits, ore of high values and also in the situations where the grade is not very much above the cutoff grade.

8.3.2   Use of artificial supports

An artificial support is needed to maintain stability in the development and exploitation (stoping) openings, and systemic ground control throughout the mine. Application of artificial support is made when caving or a self-supporting system cannot be exercised. As illustrated in figure 8.1(a);[11b] any support component performs one of the three functions: (i) to hold loose rock, key blocks, and other support in place; (ii) reinforce the rock-mass and control bulking; and (iii) retain broken or unstable rock between the holding and reinforcing element to form a stratified arch.

The prevalent types of artificial supports can be classified based on various criteria. Prominent amongst them are the material of its construction, the life it requires to serve, its characteristics in terms of rigidity or yield it can provide to the superincumbent load (as in certain circumstances some yielding is acceptable and preferred) and a few others. Adapting these criteria, the classification has been outlined in figure 8.1(b).

8.3.2.1   *Brick and stone masonry*[7]

Material such as local stones and bricks, which can be dressed and sized properly and easily, available at the low cost are used for the mine roadways' and tunnels' lining and forming the arches (fig. 8.6(a)). A wall by itself does not form a mine support but girders or bars are placed over it. Arching can be constructed using suitable stones and bricks particularly at the mine portals.

8.3.2.2   *Wooden (timber) supports*[4,7]

Wooden supports have been used in underground mining and tunneling operations since their inception. Different types of timber such as red wood, Sitka spruce, Douglas and white fir, and many other types, which can be locally available, are used to construct the wooden supports. The wooden supports are light in weight. Wood can be easily cut, manipulated, transported and put in the form of a support. It can be reused and gives indication before its failure. It is the best suited for temporary support works. In any situation if used it should be able to bear the load safely and its consumption should be minimum to economize on its material and erection costs. Its strength depends upon its fibrous structure.[4] Presence of knots, non-concentric layers, fiber inclination, outside and inside cracks are the common defects, which are observed in the timber used for this purpose (fig. 8.2(i)). These defects further weaken it. Humidity also affects its strength as many funguses that live in humid conditions affect it. It is a combustible material and needs due precaution against the outbreak of fire. The wood when cut from the forests is wet and needs seasoning i.e. allowing for its natural drying. Wood is considered wet if moisture is >30% and dry when it is below 20%.[7]

To make timber resistant to fungi, bores and insects it should be treated with preservatives, which can be organic or inorganic types. Tar and creosote oil are the common organic compounds, which are used; whereas salts such as zinc chloride, copper sulfate, iron sulfate and lime wash are the common inorganic compounds to treat the timber. These preservatives can be applied by various techniques and prominent amongst them are: spraying or brushing, cold dipping, hot and cold open tank treatment and pressure tank treatment. In some mines special treatment to the wood is given to make it fire resistant.

The wood is used as props (fig. 8.2(a) & (b))[7], stulls (fig. 8.2(c))[7], chocks (fig. 8.2(d)[7] & (e)), cogs, square-sets (fig. 8.2(f))[7], bars (fig. 8.2(g)), multi-member sets (fig. 8.2(h))[7] and as lagging between support and the ground. A new development that has been incorporated in some of the South African mines is the use of timber filled pipes props, called pipe sticks, in which the timber protrudes from each end of steel tube. The timber acts as the axial load-bearing member, and the steel pipe provides lateral constraint, thereby greatly improving the yieldability of the prop. The pipe stick is a 150–200 mm diameter wooden prop encased in 3–4 mm thick steel pipe.[16]

Besides the advantages as outlined in the preceding paragraphs, timber has the advantage of flexibility and can accommodate large strains before it breaks. Crushing of timber becomes apparent long before its failure. One limitation of timber used as lagging between steel supports is that it tends to rot.

The erection and use of timber in different forms has been illustrated in figure 8.2.[7] Use of timber in the form of reinforced sets (fig. 8.3(a)),[7] junction support (fig. 8.3(b))[7] and in the process of fore polling (fig. 8.3(c))[7] has been illustrated in figure 8.3. Sylvester, monkey winch and power driven winches are the devices used to withdraw props.[7] A chock-releasing device is used to withdraw chocks.

8.3.2.2.1   Calculations with regard to wooden supports[4,15]
As given by Saxena and Singh (1969):

Load bearing capacity of a prop:[15] $P = 47.2 - 1.5\ h/d$ (8.1)

Where: P – load bearing capacity of prop in tons;
       h – height of prop in mm;
       d – dia. of prop in mm.

Buckling Strength of a prop:[4] $\sigma = (\Pi^2 E / \lambda^2)$ *for* $\lambda > 100$ (8.2a)

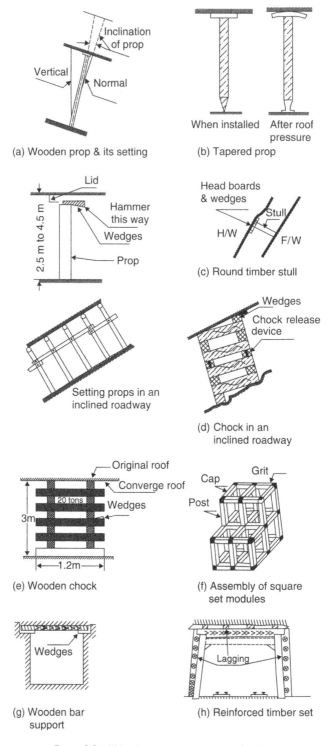

Figure 8.2　Wooden supports – some details.

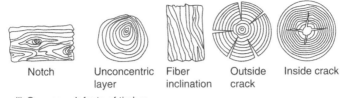

Notch   Unconcentric   Fiber   Outside   Inside crack
layer   inclination   crack

(i) Common defects of timber

Figure 8.2 (Continued).

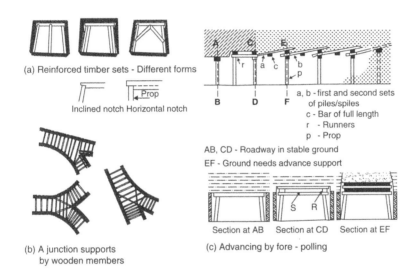

(a) Reinforced timber sets - Different forms

Inclined notch Horizontal notch

(b) A junction supports
by wooden members

a, b - first and second sets
of piles/spiles
c - Bar of full length
r - Runners
p - Prop

AB, CD - Roadway in stable ground
EF - Ground needs advance support

Section at AB   Section at CD   Section at EF

(c) Advancing by fore - polling

Figure 8.3 Wooden supports – some details of their use.

$$\sigma = \sigma_c \ (1 - a\lambda + b\lambda^2) \quad for \ \lambda < 100$$
Slenderness ratio $\lambda = 4l/d$

(8.2b)

Where: l – is length of prop;
    d – dia. of prop;
    E – elasticity modules of timber;
    σ – buckling strength of timber;
    σ_c – crushing strength;
    a, b quality constants for mine timbers a = 0, b = 2.
Bending strength of timber:[4]

Bending strength or modules of rupture $\sigma_b = M_{max}/W$

(8.3a)

$$M_{max} = P_K \, l/4$$

(8.3b)

$$W = bh^2/6$$

(8.3c)

$$\sigma_b = (8.3b) / (8.3c) = (3P_k l) / (2b\ h^2) \qquad (8.3d)$$

Where: σ_b – bending strength i.e. modules of rupture
     M_{max} – maximum bending movement
      W   – section modules
      l   – span, length of beam
     P_K – breaking load.
To calculate load on a wooden gallery:

Earling's formula:[4]

$$h = \alpha\, L_a \qquad (8.4a)$$

Pressure on support in ton./m^2 $\sigma_t = h\gamma$ (8.4b)

Load per unit length, t/m $q_t = \sigma_t\, a$ (8.4c)

Total load produced by parabolic dome $P_t = \alpha\, La^2\, a\, \gamma$ (8.4d)

Where: h = height of load in meters,
    α = loadings factor; depends upon rock formations under normal conditions 0.25–0.5; for bad roof with cracks, it may be 1–2.
    L_a= span of set at the roof/back in meters. or length of cap on wooden set.
    a = distance between two adjacent sets.
    γ = rock density in tons/m^3.

Protodyakonov formula:[4]

Load height (parabola height) in meters, $h = l/f$ (8.5a)

$$f = \sigma_c /100 \qquad (8.5b)$$

Pressure on support in ton./m^2, $\sigma_t = h\gamma,$ (8.5c)

Load per unit length, t/m, $q_t = \sigma_t\, a$ (8.5d)

Total load produced by parabolic dome, $P_t = (4/3)\, l\, h\, a\, \gamma$ (8.5e)

Whereas: f = Protodyakonov constant, may be taken as 0.01 of compressive strength of the rock in which gallery is driven.
    σ_c= compressive strength of rock in kg/cm^2
    2l = gallery width in m.
Note: other nomenclatures are the same as mentioned above.

8.3.2.3   Steel supports

Steel is an expensive material but it is widely used to manufacture mine and tunnel supports due to the following facts:

- It is free from natural defects
- It has high Young's modules of elasticity [up to the order of 2 millions kg/cm^2][4]
- It is least affected by temperature and humidity
- It can be reused.

Table 8.1   Common steel sections[4] used for manufacturing mine supports.

Parameters	Steel sections		
	Rail	Clement	Toussiant Heinzmann
Unit weight, kg/m	33.5	14	21
Rankin Ratio*	1.5	5.3	1.3

*Ratio of compressive strength to buckling strength in a beam of 2 m length.

Various types of steel sections can be used to manufacture the mine supports; prominent amongst them are listed in table 8.1 (particularly to manufacture beams or sets): Using steel the following types of mine supports are manufactured:

1. Steel props: 1 – Friction (fig. 8.4(c) & (d)); 2 – Hydraulic (fig. 8.4(a))
2. Steel chock (hydraulic) (fig. 8.4(b)), Cantilevers i.e. powered supports (fig. 8.4(b)).
3. Steel beams and sets
4. Steel arches: 1 – Rigid; 2 – Yielding
5. Shield support (fig. 8.4(b))[14]
6. Steel tubing
7. Wire-mess, roof truss and rock bolts.

8.3.2.3.1   Steel props, powered and shield supports

There are two types of steel props: friction and hydraulic, *having yielding characteristics* which is a desirable feature for their use in the mines particularly at the longwall faces. The former work on the principle of friction and the latter on the hydraulic. The construction details of these props have been illustrated in figure 8.4. Characteristic curves as shown in figure 8.4(c)[7] and figure 8.4(a)[7] for friction and hydraulic props respectively, depict their working behavior under the roof pressure. Friction props suffer the disadvantage of aging of friction surfaces and human errors in pre-loading the props. Hydraulic props work better than friction props with easy setting and withdrawal mechanisms. Figure 8.4(e) illustrates a friction props installation scheme at a longwall face.

Self advancing or power operated support (fig. 8.4(b)) *(powered support)*[14] (figs 16.24(e) and (f)) system consisting of an assembly of hydraulically operated steel hydraulic support units which are moved forward by hydraulic rams coupled or connected by pin or other means to the face conveyor. This type of support provides unobstructed room for plough or shearer and flexible conveyor equipment with roof-beams cantilevered from behind the working face.

Shield support: When considering shield support (figure 8.4(b))[14] in conjunction with drivage or tunneling work, different types of shields are employed to support different types of ground. They may cover the entire rock surface, including the face, or just the curved surface, or they may give partial shielding in the crown only. They can be self propelled as part of a tunnel borer or independently propelled for use with a partial face TBM or other form of a mechanical miner. With the advent of tunnel boring, now

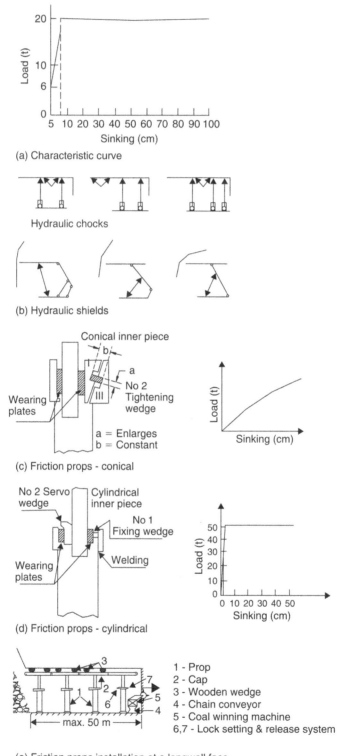

(a) Characteristic curve

Hydraulic chocks

(b) Hydraulic shields

Conical inner piece

Wearing plates

No 2 Tightening wedge

a = Enlarges
b = Constant

(c) Friction props - conical

No 2 Servo wedge   Cylindrical inner piece

No 1 Fixing wedge

Wearing plates   Welding

(d) Friction props - cylindrical

1 - Prop
2 - Cap
3 - Wooden wedge
4 - Chain conveyor
5 - Coal winning machine
6,7 - Lock setting & release system

max. 50 m

(e) Friction props installation at a longwall face

Figure 8.4 Hydraulic and friction props – some details. Hydraulic chocks and shields.

the shield can be carried, similar to the shell of a tortoise, as a part of a self-propelled boring machine. A finger shield consists of parallel steel strips separated by gaps, through which the rock bolts can be installed and inspection can be carried out.

Its other application is mainly in coalmines. In any design a shield support consists of a canopy, a base, hydraulic legs and controls system. To cope up with easily caving faces during longwall mining these supports have been developed.

Steel sets: Steel sets constructed of 'H' sections are used to prepare supports of different curvature (fig. 9.20). The advantages include ease and speed of installation, a relatively reliable and high load carrying capacity and little maintenance required. High cost, and difficulty in adapting to the varying ground conditions, requirement of more space than Shotcreting and rock bolting are some of their limitations. However, they used as a permanent support in many mines.

Steel Arches: Basically there are two designs of steel arches: *Rigid* and *Yielding*.

Rigid arches: These are used as permanent support and are popular in mines to support the permanent workings (fig. 9.20). Two, three or four-segment arches forming a near semicircular design are mostly used. The arch shape is a more efficient use of the steel sections than flat cross bars. With an arch, the steel member is in compression instead of bending.

Yielding arches are composed of three sections. The top section slides between two side elements. After a regular interval (perhaps a fortnight or so), the tightening elements are loosened; and the arches slide, converge, and thus relieve stresses accumulated on them. This step eliminates their deformation. Toussaint Heinzmann patented the first yielding arches. The profile is shown in figure 9.20. More designs were brought in later on. In some cases the yielding can be provided with insertion of wooden pieces between the steel elements.

Steel tubing: Cast iron steel tubing is used to permanently line the shaft walls during its sinking when the other methods of lining cannot work effectively, particularly where the freezing method to treat the ground is applicable. These are technically known as English and German tubing (figs 14.7(c) and (d)). In both cases, the tubing is built up of cast-iron rings, each of which comprises a number of flanged segments shaped to suit the curvature of the shaft.

8.3.2.3.2   Rock bolting

The use of rock-bolts in underground mines and civil excavations is increasing rapidly since its first use in 1918[8] in the underground mines of Poland. It has very largely replaced timber, made the excavation safer, released space previously obstructed by timber, and gave improved ventilation. Today in all types of mines, caverns and tunnels its use is extensive and at an increasing trend. The paragraphs below outline the theory and concept to help understand its functioning.

If L is the width of an opening and a uniform load q is applied to it, then maximum bending will be in the center of this opening and the magnitude of this bending stress can be calculated using the following relation:[4]

$$\sigma = (0.75 \, q \, L^2) / (bh_1^2 + bh_2^2) \qquad (8.6a)$$

Where: h_1 and h_2 are the thickness of two rock layers and b is their width.

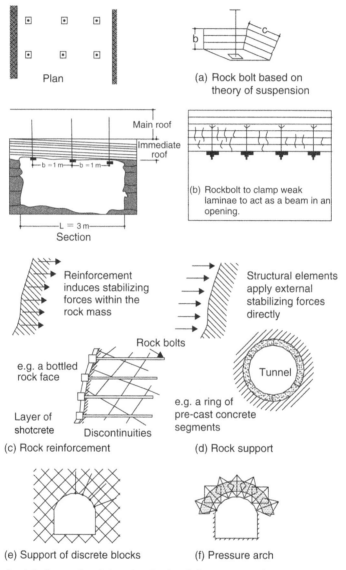

Figure 8.5 Rock bolts – related theories. Rock reinforcement and support – some concepts.

If these two layers are tied together, the bending stress can be expressed as:[4]

$$\sigma' = (0.75\, q\, L^2) / b(h_1 + h_2)^2 \qquad (8.6b)$$

If $h_1 = h_2 = h_0$.

Dividing eq. (8.6a) by eq. (8.6b), we get: $\sigma'/\sigma = 1/2$ i.e. the bending stress becomes just half.

This explains the concept of beam theory that works for layered or stratified deposits when rock bolts are used to support them. The concept is that by bolting, the immediate roof acts as a beam to support the overlying strata. This has been illustrated in figure 8.5(b). The other concept behind roof bolting is the theory of suspension that states that by using roof bolts the immediate roof is suspended to the main roof which is stable and strong, as illustrated in figure 8.5(a). Rock bolts also reinforce the rock by pressure arch[9b] and support of discrete blocks, as shown in figure 8.5. Use of rock bolts in mines is extensive. It can be used as permanent support to support the roof and sides of the main roadways, roadway junctions and wide chambers. In the stoping areas it finds wide applications to support brows of the draw points and other openings that require immediate and temporary supports.

To support the roadway junctions (2-way staggered, 3-way or 4-way) and galleries the usual pattern adopted in the mines have been illustrated in figures 8.6(f) and (g).[6] The number of bolts per square meter is called 'bolt density'. The spacing between the rows of roof bolts, and within a row, can be calculated using the guidelines outlined below. The dia. and length range:[6] 5/8" dia. 36–72" length (in 65% cases); 3/4" dia. 60–120" (30% cases); 1" and more dia. 60" and longer (5% cases).

Rock bolt calculations[4]

$$Length\ of\ rock\ bolt\ (l) = Thickness\ of\ immediate\ roof + 0.5 \tag{8.7a}$$

$$Number\ of\ bolts,\ m \geq (L\ h\ c\ \gamma)/R \tag{8.7b}$$

$$Allowable\ axial\ force,\ r \geq (0.785\ d^2\ \sigma_a)/n \tag{8.7c}$$

$$Bolt\ density\ m_o = m/(L\ c) \tag{8.7d}$$

$$Bolt\ spacing,\ b = L/m \tag{8.7e}$$

Where: h – thickness of immediate roof, in m
　　　l – length of rock bolt, in m
　　　m – number of rock bolts
　　　L – Gallery width, in m
　　　c – distance between rows of bolts, in m
　　　γ – immediate roof rock density in t/m^3
　　　n – factor of safety
　　　σ_a– yielding strength of steel in tons/m^2
　　　d – diameter of bolt in m
　　　R – allowable axial force in tons.

To calculate the anchorage force to keep the bolt in position[4]

$$\mu = k\ q \tag{8.8a}$$

$$P = F_t\ q\ (sin\alpha + \mu\ cos\alpha),\ in\ kg. \tag{8.8b}$$

Where: P – anchorage force to keep the bolt in place, in kg
　　　F_t – area of anchorage, in cm^2
　　　q – bearing capacity of rock, in kg/cm^2
　　　α – conical angle of the wedge
　　　k – coefficient = 0.0014
　　　μ – coefficient of friction between roof rock and bolt steel.

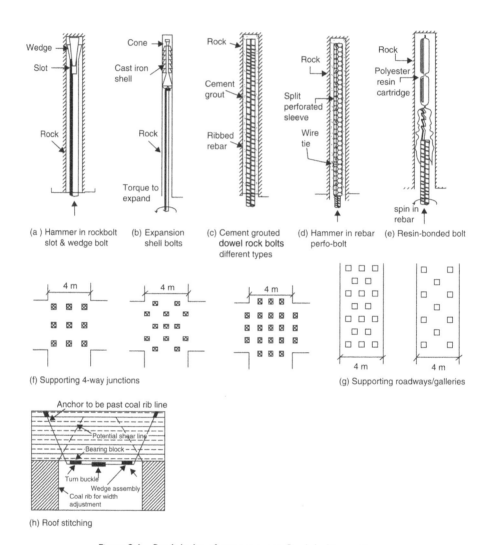

Figure 8.6   Rock bolts of various types. Rock bolting patterns.

Classification of rock bolts[4,7,8,9,13]   The following are rock bolts of various types. They have been also illustrated in figure 8.6.

1. Slot and wedge
2. Expansion shell
3. Bolts with distributed anchorage: 1 – Grouted dowels. 2 – Cable bolts; 3 – Perfo types.
4. Special types of bolts – such as resin bolts.

Slot and wedge bolts: These bolts were the earliest, although they are not the best, but still continue to be used for some temporary support applications. They consist of a steel bar with a slot cut at one end, which contains a steel wedge (fig. 8.6(a)).

The wedge end is placed in the hole first and the wedge is driven into the slot by hammering the exposed end of the bolt. Expanding halves grip the hole, allowing the bolt to be tensioned and carry load. Blast vibrations and ground movement easily loosen bolts of this type. Holes for the bolts are first drilled using a suitable drill. The hole length is 5–7 cm less than the bolt length.

Expansion shell bolts: It contains toothed blades of malleable cast iron with a conical wedge at one or both ends (fig. 8.6(b)). One or both of the cones are internally threaded onto the rock bolt so that when the bolt is rotated by a wrench, the cones are forced into the blades to press them against the walls of the drillhole. The grip increases as the tension increases. These are the least expensive and very widely used for short-term support in underground mines. They are most effective in hard rocks but in soft rocks they tend to slip and loosen. In some mines this loosening is avoided by introducing a cement grout through a plastic tube running alongside of the bolt. A few days after its installation, the nut should be once again tightened, as it gets loose during the initial few days after installation due to active workings in the nearby areas.

Grouted dowels: A dowel is a fully grouted rock bolt without a mechanical anchor, usually consisting of a ribbed reinforcing bar, installed in a drillhole and bonded to the rock over its full length (fig. 8.6(c)). Dowels are self-tensioning when the rock starts to move and dilate.[8] They should therefore be installed as soon as possible after excavation, before the rock has started to move, and before it has lost its interlocking and shear strength.

The normal grout mix consists of sand cement ratio as 50/50 or 60/40. Water-to-cement ratio should not be greater than 0.4 by weight.[8] Grout injection particularly in the upholes requires care to ensure its complete filling. Sometimes air pockets may be left in the hole, which are difficult to detect. Pneumatically operated grout pumps/loaders are used to fill the holes, to which the bar is driven. The dowel is retained in upholes either by a cheap form of end anchor, or by packing the drill hole collar with cotton waste, steel wool or wooden wedges.

Cable bolts: Grouted cables, called cable bolts can be used in the stoping areas to support the back to prevent its fall or caving, also to stabilize and prevent caving of the hanging wall by installing these bolts horizontally or at a certain angle. In some of the Australian mines cable bolts up to 18 m length have been used successfully. Upward cable bolting of open stope crown pillars can provide improved support within it. This allows an increase in stope span, and reduction in pillar dimensions and can eliminate need for any other type of support within it.

Used and old haulage or hoist ropes can be used for this purpose after removing grease and washing them. Almost the same technique is used to stitch a weak flat roof or back. This technique is known as roof stitching and similar to this is a roof truss as shown in figure 8.6(h).[6]

Perfo-bolts:[4] In this system (fig. 8.6(d)) instead of pumping the grout into the hole, it is trawled into the two halves of a split-perforated sleeve. The halves are placed together and bound at the ends with soft iron wire, and the tube full of cement is inserted in the hole. The dowel is then driven by sledgehammer into the sleeve, forcing grout out through the perforations and into contact with the rock. This practice is very popular in Scandinavian countries for its application in all types of rocks.

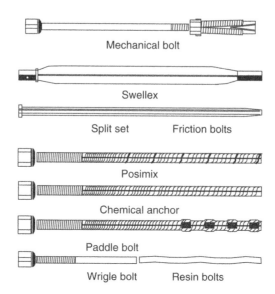

Figure 8.7(a)   Rock bolts – some prevalent designs.[9b]

Installation of reinforced concrete roof bolt consists of two operations: filling the hole with grout and placing a roof bolt into the hole. In the absence of compressed air, the grout in some cases is injected into the hole with the aid of a hand-grouting gun.

Resin grouted bolts:[4] Where high and quick strength is required, the resin bolts (fig. 8.6(e), 8.7) although costlier, find applications. In this practice a ribbed reinforcing rod is cemented into the drillhole by a polyester resin, which in a few minutes changes into a thick liquid to a high strength solid by a process of catalyst initiated polymerization.

In comparison to cement grouts, resin has the advantages of quick setting and reaching its full strength quickly i.e. within 2–4 hours. The bond strength is much stronger than the cement grout. Complete grouting combined with tensioning can be achieved by inserting several slow-setting cartridges behind the fast ones. If the resin starts to set before installation is complete, the bolt is left sticking out of the hole and is practically ineffective.

Optimum bolt inclination: The most common cause of tension loss in bolt installations is an angle less than 80° to the rock face as shown in fig. 8.7(b). Spherical washers, however, can tolerate some variations. The bolt should also be arranged to intersect any joints at an angle of more than 45°. A series of examples meeting this requirement has been shown in fig. 8.7(c).

Wire mesh: This is also known as screen. This is available in different wire gauge thickness and mess apertures. Its main purpose is to support the rock between bolts, which is particularly necessary when the rock is closely jointed and the bolts are moderately to widely spaced. It can also serve as reinforcement for shotcrete.

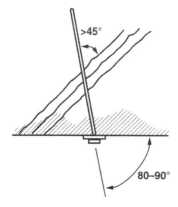

Figure 8.7(b)   Optimum bolt inclination.[17]

Figure 8.7(c)   Systematic bolting patterns in jointed ground, after Choquest (1991)[17] (www.smenet.org).

In Figures 8.8(a) to (d), application of split set dome plates, split sets (rock bolts), sheet mesh and cables (strand Graford) at Mount Isa Mines for good and poor ground conditions, used for the long-term and short-term accesses, have been illustrated. These hard rocks mines have attained a depth up to 1.8 km.

8.3.2.4   *Concrete supports*

High compressive strength, easy to erect and manufacture, fire resistant, smooth finished surface and suitable under adverse mining conditions including presence of abnormal make of water, are some of the advantages a concrete support commands

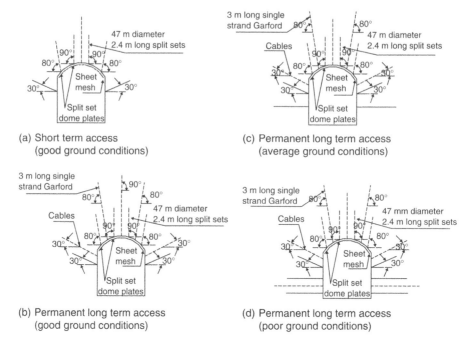

Figure 8.8   Application of split set dome plates, split set (rock bolts), sheet mesh and cables (strand Graford) at Mount Isa Mines for good and poor ground conditions prevalent for the long term and short term access. Excavation size 4 m × 4.7 m. Max. bolt spacing = 1.2 m; Ring spacing = 1.4–1.5 m. Cable bolt ring spacing 2.5 m; Tension 3–5 tons. Sheet mesh = (100 mm × 100 mm × 5 mm). Courtesy: Mount Isa Mines, Australia.

over the steel and wooden supports. Low tensile strength, failure without warning and requirement of curing time for its setting are some of its limitations. Concrete finds its application for the following types of mine supports:

- Shaft lining
- Mine roadways lining
- Concrete arches (fig. 9.20(c))
- As shotcrete or gunite.

Placement of concrete is achieved in the following manner:

- Monolithic concrete
- Concrete blocks
- Shotcreting or guniting.

Monolithic concrete (cast-in-place): A monolithic concrete is mass concrete instead of the concrete with blocks. This involves placing a 40–60 cm (Cemal et al. 1983) wall around an opening's roof and sides. In order to cast it, first the shuttering/folds are built and then concrete mixture which could be in the ratio of 1:2:4 to 1:1:2 (cement:sand: coarse aggregates), depending upon the strength required. It is allowed to cure for 2–4 weeks. It can be further strengthened by incorporating steel tie rods, straps, angles etc., and then it is known as reinforced cement concrete (R.C.C.). Monolithic concrete finds application is shaft lining, filling the gap between the steel arches and the ground (sides and back of mine roadways/openings). This concrete lining is very widely used in permanent workings such as shaft, chambers, and pit bottom openings including the shaft insets.

Shaft lining:[3,4] use of monolithic concrete is made to prepare the shaft lining. Mathematical relations used to determine the thickness (t) of concrete lining in the shafts have been worked out by some of the authors as outlined below:
   According to Protodyakonov:

$$t = [(Pr)/\{(\sigma_b/F) - P\}] + \{150/(\sigma_b/F)\} \tag{8.9a}$$

$$t = \{0.007 \sqrt{(2rH)}\} + 14 \tag{8.9b}$$

$$P = \text{Horizontal stress to the shaft lining}$$
$$P = (\gamma H)/(m-1) \tag{8.9c}$$

According to Brinkhaus:

$$t = (2r/10) + 12 \tag{8.10}$$

r = radius of shaft in cm.
   According to Heber:

$$t = [\sqrt{\{(\sigma_b/F)/(\sigma_b/F) - 2p\} - 1]} \, r \qquad \text{if depth} < 400\,m \tag{8.11a}$$

$$t = [\sqrt{\{(\sigma_b/F)/(\sigma_b/F) - \sqrt{3}p\} - 1]} \, r \qquad \text{if depth} > 400\,m \tag{8.11b}$$

Where: t = thickness of lining in cm.
         P = side pressure on lining in kg/cm^2.
         σ_b = fstrength of concrete (after 28 days) in kg/cm^2.
         F = factor of safety, usually 2.
         H = depth of shaft in cm.
         γ = rock density in kg/cm^3.
         m = Poisson's ratio.

   Using these relations,[3,4] given the following data, calculate the thickness of concrete lining.
Depth of shaft = 300 m; Radius of the shaft = 2.5 m; Factor of safety = 2;
Formation = sandstone; Poisson's ratio = 5; Density = 2.5 t/m^3.
Concrete details: compressive strength = 225 kg/m^2; density = 2.4 t/m^3

Concrete blocks: Sometimes using the concrete blocks arches or sets are built. In order to provide yieldability to them, the wooden pieces/blocks are inserted while carrying out the masonry work to build these sets.

Shotcreting and guniting:[2,8] Mortar or concrete conveyed through a hose and pneumatically projected with high velocity on a surface, is known as shotcreting. Similar to this is guniting in which only cement mortar is applied and it contains no coarse aggregate. It has limited applications due to its higher cost and lower effectiveness.

There are two techniques that are used to apply shotcrete: 1 – Dry mix process – The mixture of cement and damp sand is conveyed through a delivery hose to a nozzle where the remainder of the mixing water is added; 2 – Wet mix process: all the ingredients are mixed before they enter the delivery hose. A dry process equipment has components such as a gun, compressor, hoses, nozzles and water pump (in some designs), whereas in a wet mixer there is a mixing chamber to which compressed air is fed from the mains or through the separate supply.

It requires a skilled operator. It is applied in a limited space; as such the working atmosphere becomes tedious and requires good ventilation and illumination. For its application the surface should be clean and free from dripping water or running water. Safety types of couplings secured with chains should be used to avoid any accident in the event of improper fastening of couplings.

1. General use: It finds its application in shafts, adits, haulage-ways and service chambers to acts as:
   - Primary support.
   - Final lining.
   - Protective covering for excavated surfaces that are altered when exposed to air.
   - Protective covering for steel or wooden supports, rock bolt plates etc.
   - As lagging material in place of timber, steel or concrete in between steel or wooden supports.
2. Use as rock sealant: This application of shotcrete can prevent slaking of shale or other rocks and thereby their weathering can be prevented.
3. Use as safety measure: When an opening is required to be rock bolted, the shotcreting can be applied before and after the rock bolts are installed. When blocky ground tends to produce fragments, it may be applied with wire mesh. The shotcreting is usually applied after mesh is bolted to the surface.
4. Use as structural support: 1 – Non-reinforced with the thickness in the range of 2–4″ (52–104 mm) 2 – Reinforced with fiberglass, wires etc. to the same thickness.
5. For repairing stonework, brickwork and concrete lining. In making the bulkheads airtight. Sometimes as a temporary support during shaft sinking.
6. Hoek et al. (1995) provided a table of recommended shotcrete applications in underground mining for different rock mass conditions (table 8.2).[18] This table provides a simple link between rock mass description and behavior and recommendations for shotcrete. It can also serve to check designs obtained by other means. The support recommendations cover the whole spectrum of anticipated excavation behavior including wedge type instability in low stressed rock to moderately violent rock fracturing during rock bursts. Support recommendations for burst-prone ground are also covered.

8.3.2.5 Support by filling[7]

In effectiveness next to natural pillars is the use of back-fills as support. It has almost 100% ability to support the superincumbent load without yielding. The concept of filling, as described in section 16.3.2, is to pack a worked out area with a fill, which could be waste rock, mill tailings, sand etc. The method has got applications while mining all types of ores. It is a reliable means of support though costlier than most of other means, but allows almost a full recovery of ores, decrease in the use of other type of supports

Table 8.2   Summary of shotcrete recommendations in hard rock underground mining for different rock mass conditions (after Hoek et al. 1995)[18] (www.smenet.org).

Rock mass description	Rock mass behaviour	Support requirements	Shotcrete application
Massive metamorphic or igneous rock. Low stress conditions.	No spalling, slabbing or failure.	None.	None.
Massive sedimentary rock. Low stress conditions	Surfaces of some shales, siltstones, or claystones may slake as a result of moisture content change.	Sealing surface to prevent slaking.	Apply 25 mm thickness of plain shotcrete to permanent surfaces as soon as possible after excavation. Repair shotcrete damage due to blasting.
Massive rock with single wide fault or shear zone.	Fault gouge may be weak and erodible and may cause stability problems in adjacent jointed rock.	Provision of support and surface sealing in vicinity of weak fault of shear zone.	Remove weak material to a depth equal to width of fault or shear zone and grout rebar into adjacent sound rock. Weldmesh can be used if required to provide temporary rockfall support. Fill void with plain shotcrete. Extend steel fibre reinforced shotcrete laterally for at least width of gouge zone.
Massive metamorphic or igneous rock. Moderate to high stress conditions.	Surface slabbing, spalling and possible rockburst damage (see also last two classes).	Retention of broken rock and control of rock mass bulking.	Apply 50 mm shotcrete over weldmesh anchored behind bolt faceplates, or apply 50 mm of steel fibre reinforced shotcrete on rock and install rockbolts with faceplates; then apply second 25 mm shotcrete layer. Extend shotcrete and bolt application down sidewalls where required.
Massive sedimentary rock. High stress conditions.	Surface slabbing, spalling and possible squeezing in shales and soft rocks.	Retention of broken rock and control of squeezing.	Apply 75 mm layer of fibre-reinforced shotcrete directly on clean rock. Rockbolts or dowels are also needed for additional support.
Metamorphic or igneous rock with a few widely spaced joints. Low stress conditions.	Potential for wedges or blocks to fall or slide due to gravity loading.	Provision of support in addition to that available from rockbolts or cables.	Apply 50 mm of steel-fibre-reinforced shotcrete to clean rock surfaces on which joint traces are exposed.
Sedimentary rock with a few widely spaced bedding planes and joints. Low stress conditions.	Potential for wedges or blocks to fall or slide due to gravity loading. Bedding plane exposures may deteriorate in time.	Provision of support in addition to that available from rockbolts or cables. Sealing of weak bedding plane exposures.	Apply 50 mm of steel-fibre-reinforced shotcrete on clean rock surface on which discontinuity traces are exposed, with particular attention to bedding plane traces.
Jointed metamorphic or igneous rock. High stress conditions.	Combined structural and stress controlled failures around opening boundary.	Retention of broken rock and control of rock mass dilation.	Apply 75 mm plain shotcrete over weldmesh anchored behind bolt faceplates or apply 75 mm of steel fibre reinforced shotcrete on rock, install rockbolts with faceplates and then apply second 25 mm shotcrete layer. Thicker shotcrete layers may be required at high stress concentrations or, alternatively, intentionally weak zones in shotcrete may be created (slots) and sprayed later after deformations have stabilized.
Bedded and jointed weak sedimentary rock. High stress conditions.	Slabbing, spalling and possibly squeezing.	Control of rock mass failure and squeezing.	Apply 75 mm of steel fibre reinforced shotcrete to clean rock surfaces as soon as possible, install rockbolts, with faceplates, through shotcrete, apply second 75 mm shotcrete layer.
Highly jointed metamorphic or igneous rock. Low stress conditions.	Ravelling of small wedges and blocks defined by intersecting joints.	Prevention of progressive ravelling.	Apply 50 mm of steel fibre reinforced shotcrete on clean rock surface in roof of excavation. Rockbolts or dowels may be needed for additional support for large blocks.
Highly jointed and bedded sedimentary rock. Low stress conditions.	Bed separation in wide span excavations and ravelling of bedding traces in inclined faces.	Control of bed separation and ravelling.	Rockbolts or dowels required to control bed separation. Apply 75 mm of fibre-reinforced shotcrete to bedding plane traces before bolting.
Mild rockburst conditions in massive to moderately jointed rock subjected to high stress conditions.	Localized spalling and slabbing to depth of about 0.25 m.	Retention of broken rock and control of rock mass degradation. Little energy dissipation capacity.	Apply 50 to 75 mm of steel-fibre-reinforced shotcrete or shotcrete over mesh or cable lacing that is firmly attached to the rock surface by means of (mechanical) rockbolts or cablebolts.
Moderate rockburst conditions in massive to moderately jointed rock subjected to high stress conditions.	Localized spalling and rock fracturing to depth of about 0.75 m.	Retention of broken rock and control of rock mass bulking. Moderate energy dissipation capacity many be needed.	Apply 50 to 100 mm of shotcrete over mesh or cable lacing which is firmly attached to the rock surface by means of rockbolts or cablebolts. Use short grouted rebar in combination with mechanical or cable bolts is recommended. Bolting through at least part of shotcrete is preferred.
Severe rockburst conditions in massive to moderately jointed rock subjected to high stress conditions.	Wide spread spalling and rock fracturing to depth of less than 1.5 m.	Retention of broken rock, control of rock mass bulking, and ability to yield. Moderate to high energy dissipation capacity.	Apply 75 to >100 mm of shotcrete over mesh or cable lacing which is firmly attached to the rock surface by means of yielding/frictional rockbolts or plain cablebolts. Use short grouted rebar in combination to control bulking. Prevent stress raisers by introducing longitudinal slots or thin mesh-protected zones of weak shotcrete. (If shotcrete spalling is encountered, mesh over shotcrete may have to be added.)

such as timber etc., and, improved fire safety and ventilation. Degree of packing depends upon the type of fill used. There are two main classes of fill – *Dry* and *Wet*.

The wet fill is mainly referred to as a *hydraulic fill* that is a mixture of filling material with water. After mixing, a slurry is obtained which is transported through pipes to the void to be filled. In the *hardening type of hydraulic fill* together with water, sand or mill tailings some hardening ingredient is mixed which allows cementing of the filled massif. In *pneumatic fill* dry filling material is moved via pipes by compressed air. *The ores suitable to work as support temporarily should not cake, ignite or oxidize in its loose state.* Given below is a comparison made to cover the important features of these back-filling practices.

8.4 SELECTION OF SUPPORT

Support is very vital for the mines to exploit the ores safely and for civil tunnels to drive and maintain them safely for their day-to-day use. Its improper selection may not only jeopardize the safety of the mine, but its overall productivity, economy and recoveries. Experience and skill of the mineworkers and supervisors play an important role to judiciously select it. Unwanted supports increase costs and inefficiency, and an inadequate support network can cause ground stability problems, thereby creating unsafe conditions. Supports are needed at all the stages of mining i.e. during development, stoping and post stoping (in some cases). Influence of supports on the drivage costs of different kinds of development entries has been illustrated in the figure 16.34(a).[9b] Costliest are the stoping practices, as shown in figure 16.34(b),[9b] which require use of artificial supports in the form of timber sets or backfills.

Proper stability in a mine can be achieved by following some of the guidelines outlined below. These measures can save the support costs considerably.

8.4.1 Measures to preserve the stability of the stoped out workings or to minimize problems of ground stability[1,12]

- *Limiting exposure size (i.e. volume or span)*: An excessive exposure of roof and wall leads to partial or mass caving of the roof and walls, development of excessive rock pressure, destruction/damage of supports and the need for their restoration, ore losses and ore dilution, unsafe working conditions and accidents. Therefore, keeping a proper span of the worked out area is the primary measure to retain the stability of the workings. The permissible exposure is achieved by allowing a hanging wall to cave in some stoping practices. Arching of the roof of the room (the drive or working area) increases the stability of the working area.
- *Reduction in duration of stoping/exposure time of the excavated area*: The rock strength decreases with time; pressure from the surrounding solid makes the rock crack; and a previously latent crack opens and penetrates into the rock. In addition, the exposed rock when weathered gets weakened. Low advance rate/intensity of stoping operations increases expenses on support, leads to ore losses and dilution and reduction in the safety and productivity.
- *Direction of stoping*: The direction of joints, planes of weakening or lamination in the ore should be taken into account while selecting the stoping direction to facilitate breaking and ensure stability of the roof. If the exposed surface of the rock is parallel to joints or laminations, the roof will exfoliate readily; cave, and high strength support will be needed. Thus, roof stability can be improved by changing the direction of stoping.
- *Reduction in seismic effects due to blasting*: The larger the blasting charges (explosive), the higher is its seismic effect on the surrounding solid rock mass. This causes cracking of the roof and wall rocks. This is the reason that blasthole diameter and length are reduced at deeper levels.

8.5 EFFECT OF ORE EXTRACTION UPON DISPLACEMENT OF COUNTRY ROCK AND SURFACE[1]

The magnitude and nature of pressure depends upon a number of parameters that can be termed as model parameters and design parameters. The *model parameters* include depth of working; deposit size, shape and orientation (dip); physical and mechanical properties of the enclosing rocks and ores to be mined. These parameters can be determined based on their natural occurrence. The *design parameters* include: method of support, face advance rate, method of rock fragmentation (dislodging or breaking), duration of working a particular block etc.

As the mine deepens, the stoping methods and ore winning procedures need change. Deeper levels warrant minimum hanging wall exposure, discontinuation of stoping methods (that were applied at shallow depth) and large sized pillars.

But beyond a certain depth, say, 700–800 m,[1] the support by pillars becomes impossible due to problems of rock bursts. This is due to the fact that at these pillars because of continuous working in and around them, a heavy concentration of stresses is usually noticed/experienced. The outburst of these pillars is accompanied by ejection of ore lumps from face and walls of the workings. In strong rock the pillars' failure may be sudden and severe, resulting in shock waves.

Rock displacement:[1] The void created by extraction of ore gets filled with caving rock in due course of time resulting in deformation and subsidence of the overlying strata. This is called displacement of the rock. Displacement causes a slow and smooth subsidence of earth's surface without rupture, or abrupt subsidence with considerable movements, caving and collapses.

The rock displacement zone includes a caving zone, as shown in figure 8.9. The earth's surface which experiences displacement is called a *'trough'*. This trough area covers subsidence over 10 mm.[1] Rock displaces along a curvilinear surface, but for graphical representation (fig. 8.9) they are assumed to be planes forming with a horizontal boundary angle γ_0, Displacement angle γ' and Caving (fault) angle γ''.

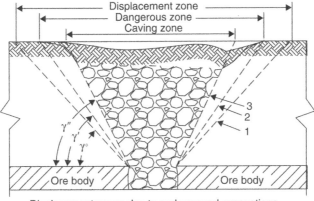

Displacement zones due to underground excavations

Figure 8.9 Ground disturbance above the excavation area influencing the above lying strata right up to the surface datum.

The boundary angle defines the entire area of rock displacement. The caving (fault) plane extends through the extreme outer cracks on the earth's surface; the displacement angle determines the zone of dangerous displacement for the surface and underground engineering structures. The value of displacement angle[1] depends upon the physical and mechanical properties of rock, nature of rock, water permeability, angle of dip of the deposit and mining depth. For massive rocks it ranges from 45 to 70°, whereas for laminated rocks it is between 30 to 65°. The surface structures to be protected should be positioned outside the displacement zone else a protective pillar below them should be left.

Safety factor:[1] The mining depth at which stoping of ore does not cause Earth surface displacement is called the safe mining depth. The ratio of minimum safe depth to deposit thickness is called the safety factor. The safety factor depends upon the physical and mechanical properties of the rock and it is about 200 for mining without filling, 80 for complete dry filling and 30 for mining with wet filling.

8.6   THE WAY FORWARD

Table 19.2 details the application of powered supports during longwall mining; it also includes automation in mining of thin seams with the application of supports having a height as low as 0.5 m (fig. 19.13(b)) and as high as 7 m; a remarkable breakthrough in the mining of extremely low as well as thick coal seams. Equally impressive is the application of shotcreting for hard rock underground mining for different rock mass conditions as detailed in table 8.2.

Research carried on rock mass quality as detailed in sec. 3.5 and figure 16.30(f), provides very useful guidelines which take into account cavability, subsidence angles, failure zones, fragmentation, cave front orientation, undercutting sequence, overall mining sequence and support design. Sec. 16.3.2.9.1B which describes 'top slicing' under a concrete slab; ore recovery of 100% has been achieved; it is also an impressive work to mine out difficult deposits particularly of high valued ores such as uranium, diamond, platinum and gold.

QUESTIONS

1. Classify 'mine supports' using different criteria.
2. Classify different rock bolts; also give a sketch of each type of rock bolt. Give a relation to calculate the anchorage force to keep the bolt in place. Illustrate the use of rock bolts in mines.
3. Classify natural pillars. Mention their function.
4. Define 'safety factor and safe mining depth'; rock displacement; artificial support; self-support; competent rock.
5. Define slenderness ratio. Calculate the slenderness ratio λ (lambda) of a wooden prop if its diameter is 16 cm and length = 2 m. Calculate load bearing capacity of a wooden prop if its height-diameter ratio is (i) 7, (ii) 12 and (iii) 19.
6. 'Keeping disturbance to natural ground settings to a minimum could be considered as directly proportional to cost reduction and minimizing problems encountered

during ground excavation and mining'. Do you agree with this statement? How could least disturbance to the natural setting, in practice, be achieved?

7. Does any excavation created at the subsurface disturb the above lying strata right up to the surface datum?

8 .Draw steel arches: (i) yielding type, (ii) rigid type. Write the characteristic features of: hydraulic prop and friction prop.

9. Give Earling's and Protodyakonov relations to calculate the total load produced by a parabolic dome.

10. Given relations to determine the bearing capacity of a timber prop and also its buckling strength and the bending strength.

11. How do you support the following, in an underground mine: A junction (three way) – give different ways. A goaf edge. A weak (roof) gallery. A roof fall area. Two walls of a steeply dipping vein. A small patch of rock in the roof/back of mine roadway.

12. How will you test the roof in underground mines? Where do you apply advance timbering while driving an inclined or horizontal drive? What is this technique called?

13. In what form are the following materials used for mine support: Wood, Steel, Bricks, Concrete and Sand or mill tailings.

14. List measures to preserve the stability of the stoped out workings or to minimize problems of ground stability.

15. List the problems of deep mining.

16. List the factors upon which rock pressure depends.

17. Mention the steel sections commonly used for the purpose of mine support.

18. Name the appliance used to withdraw props.

19. Name the steel props that are commonly used as mine supports. Draw their characteristic curves to compare their features.

20. What is a 'trough'? What does the trough area cover? Draw a trough showing various zones within it.

21. What is shotcreting? How does it differ from guniting? Write down the application of this support in mines.

22. Where (i.e. at what location) in an underground mine will you use mill tailings as support? In an underground coal mine what type of fill can be used?

23. Why is timber the most suitable to use as a temporary support? Mention the natural defects found in timber to be used as a support. What way is the timber treated before it can be used as support in underground mines? When would you call timber dry, based on its moisture content?

24. Design a rock bolting pattern (i.e. length of bolt, allowable axial force, bolt density and spacing between the bolts in each row), for a gallery in an underground coal mine using the following data:

Immediate roof – fractured rock, 1.8 m thick

Width of gallery $= 3$ m

Spacing between rows of bolts $= 1$ m

Rock density of immediate roof $= 2.5$ t/m^3

Diameter of steel bolt to be used $= 2.5$ cm

Yielding strength of steel used $= 24,000$ t/m^2

Factor of safety $= 2$.

25. Suggest the specific places in mines where use of following types of supports can be justified:
    a. Power supports
    b. Square set timbering
    c. Mill tailings
    d. German and English tubing
    e. Stulls
    f. R.C.C.

26. In an underground coal mine, 3 member wooden sets are to be erected in a gallery. Use following data:
    – width of gallery = 2 m
    – length of wooden cap = 2 m
    – loading factor = 0.25
    – rock density = 2.5 t/m^3
    – compressive strength of rock = 300 kg/cm^2
    – distance between two adjacent sets = 1 m.
    Calculate total load produced by the parabolic zone using Protodyakonov or Earling's relations.

27. Answer the following:
    a. List the forms in which concrete is used as mine support.
    b. Where do you apply advancing timbering while driving an inclined or horizontal drive. What is this technique called?
    c. Name the techniques available to apply shotcreting.
    d. Which steel section is used to manufacture yielding type steel arches?
    e. Where do you use power supports and shield supports?
    f. Name a quick setting rock bolt.
    g. Name the theories of rock bolting.

28. Draw suitable sketches to illustrate the following:
    (i) Prop erected in a horizontal drive.
    (ii) A chock in a horizontal working.
    (iii) Supporting a steeply dipping gallery using wooden sets.
    (iv) Supporting a goaf edge.

29. A circular shaft 5.5 m in dia. is to be sunk to a depth of 300 m from the surface. Calculate the thickness of the concrete lining using a relation given by the various authors. The relevant data are given below:
    Factor of safety = 2
    Formation = sandstone
    Poisson's ratio = 5
    Density = 2.5 t/m^3
    Concrete details:
    Compressive strength = 225 kg/cm^2
    Density = 2.4 t/m^3.

REFERENCES

1. Agoshkov, M.; Borisov, S. and Boyarsky, V.: *Mining of Ores and Non-metallic Minerals.* Mir Publishers, Moscow, 1988, pp. 42–44.

2. American Concrete Institute, *Shotcreting, publication SP-14 Committee 506*, Detroit, Michigan, 1996.
3. Auld, F.: Design of shaft lining. In: *Proc. Inst. Civ. Eng;* part 2, no. 67, 1979, London.
4. Biron, C. and Arioglu, E.: *Design of supports in mines.* John Wiley & Sons, 1983, pp. 89–101, 105–11, 126–29.
5. Boky, B.: *Mining.* Mir Publishers, Moscow, 1967, pp. 300–309.
6. Chlumecky, N.: Artificial supports. In: Cummins and Given (eds.): *SME Mining Engineering Handbook.* AIME, New York, 1973, pp. 13, 150–55.
7. Deshmukh, R.T.: Winning and working coal, 1965, pp. 60–120.
8. Franklin, J.A. and Dusseault, M.B.: *Rock Engineering.* McGraw-Hill, New York, 1989, pp. 556.
9. Hadjigeorgious, J. and Charette, F.: Rock bolting for underground excavations. In: Hustrulid and Bullock (eds.): *SME Underground Mining Methods.* Colorado, 2001, pp. 547.
9b. Hartman, H.L.: *Introductory Mining Engineering.* John Wiley & Sons, New York, 1987, pp. 311–15.
10. Hudson, J.A. and Harrison, J.P.: *Engineering Rock Mechanics.* Pergamon, 1997, pp. 270–72, 284.
11. Holland, C.T.: Coal mines bumps: Some aspects of occurrance, causes and control. *Bull 35, 1954, USBM.*
11b. Kaiser and Tannant, D.D.: The role of shotcrete in hard rock mines. In: Hustrulid and Bullock (eds.): *SME Underground Mining Methods.* Colorado, 2001, pp. 583.
12. Obert, L.; Duvall, W.L. and Merill, R.H.: Design of underground openings in competent ground rock. *Bull 587, 1960, USBM.*
13. Panek, L.A. and McCormic, J.A.: Roof/Rock bolting. In: Cummins and Given (eds.): *SME Mining Engineering Handbook.* AIME, New York, 1973, pp. 13, 126.
14. Peng, S.S. and Chiang, H.S.: *Longwall Mining.* Wiley (Inter science), New York, 1984, pp. 708.
15. Saxena, N.C. and Singh, B.: Props in longwall workings, *J. Mines, Metal & Fuels,* 1969.
16. South African Chamber of mines, 1977, *News Bulletin.*
17. Hadjigeorgiou, J. and Cjarette, F.: Rock bolting for underground excavations. In: Hustrulid and Bullock (eds.): *SME Underground Mining Methods.* Colorado, 2001, pp. 547–554.
18. Kaiser and Tannant, D.D.: The role of shotcrete in hard rock mines. In: Hustrulid and Bullock (eds.): *SME Underground Mining Methods.* Colorado, 2001, pp. 583.

Chapter 9

Drives and tunnels (conventional methods)

Mine development involving drivage and tunneling operations is the toughest task as it has to encounter new sets of conditions concerning ground, water and gases every moment.

9.1 INTRODUCTION – FUNCTION OF DRIVES AND TUNNELS

This chapter covers the methods to create, construct or make openings of different shapes and sizes with their inclination either almost horizontal or inclined. In mining these openings are blind ended and have been designated by different names based on their purpose, utility or orientation with respect to a deposit for whose exploitation purpose they are driven or constructed. Adits, inclines, declines/ramps/slopes, cross-cuts, levels, sub-levels etc. come under this category. But the openings of similar configuration having both the ends exposed to the atmosphere are known as tunnels. Tunnels are driven to provide passage to rails, roads, navigation, pedestrians etc. and also for conveyance of water and to serve sewerage. The difference between civil tunnels and mine openings has been described in chapter 1, section 1.8.

9.2 DRIVAGE TECHNIQUES (FOR DRIVES AND TUNNELS)

Here the meaning of drivage is to construct, drive or make the openings as referred in the preceding paragraphs. There has been a consistent development with regard to techniques, methods and equipment to be deployed while undertaking drivage work in underground mines and tunnels. Choice of a particular parameter (i.e. technique, method or equipment) depends upon the types of ground/deposit (in terms of its strength, presence of geological disturbances, water, gas etc.) through which the drive (or tunnel) needs to be driven; its size, shape, inclination, disposition w.r.t. the deposit or a particular reference point; speed of drivage and availability of resources in terms of capital required.

Figure 9.1 classifies various techniques that are available to drive openings in mines and tunnels in civil engineering tasks. Except cut and cover, and submerged tubes (tunneling) methods (fig. 9.1); the rest of the methods are common in both the disciplines – mining and civil, and hence, the description given below on methods of driving these opening is common. However, special attention is required while driving tunnels which will be described separately.

In drivage work rock fragmentation (primary breaking) is the first operation, which can be carried out with or without the aid of explosives. Fragmentation using

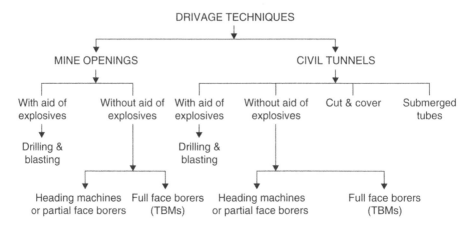

Figure 9.1   An account of available techniques to drive openings for mining and civil engineering works.

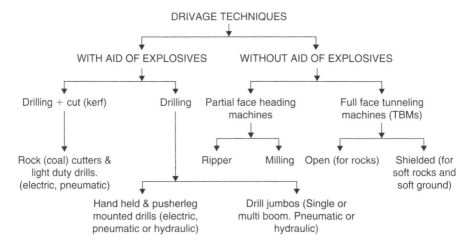

Figure 9.2   A detailed breakdown of driving techniques used for the mine openings and tunnels.

explosive requires drilling and blasting but if it is to be carried out without explosives then rock cutting machines, which are known as heading and tunneling machines, are used. Figure 9.2 gives a detailed breakdown of methods to construct mine opening and tunnels.

9.3   DRIVAGE TECHNIQUES WITH THE AID OF EXPLOSIVES

9.3.1   Pattern of holes

The terms development heading, workings, drivage work used here refer to mine openings and civil tunnels. In drivage work placement of holes properly, while designing a pattern of holes, is of prime importance due to the following reasons:

• To obtain a desired shape, size, orientation and gradient of the mine openings and tunnels.

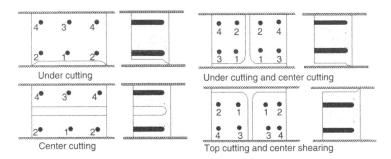

Figure 9.3   Coal/soft rock fragmentation using 'Kerf' techniques (Not to scale, only schematic).

- To achieve an accurate contour of the tunnels or mine openings, with least over break and smooth surface of the floor and face.
- For compact heaping of the blasted muck after blasting at the working face.

The following two techniques are available to accomplish this task:

9.3.1.1   *Mechanized-cut kerf* [6,11,17]

In this technique use of rock cutting machines (fig. 9.3) is made to provide a cut/kerf/cavity, which can be put at the bottom, middle, top or at any other desirable position[6] of a mine opening or tunnel face. This acts as an initial free face towards which blasting of the holes drilled in the face is directed. This free face reduces the amount of drilling and explosives considerably; but cutting the kerf is practicable in soft and medium hard rocks such as coal, salt, potash etc. Hence, its use is usually restricted to coal mines only. Rock cutting machines together with their accessories and power packs are, thus, the additional items required in comparison to the technique – blasting off the solid, described below.

9.3.1.2   *Blasting off the solid*

This is a universal technique applicable for any type of strata, tunnel or mine. A particular pattern is followed to position the holes, as shown in figure 9.4.[9] Types of pattern of holes mainly differ in the arrangement of breaking in holes (known as cut holes i.e. the holes, which are used to create an initial free face, towards which the blast is subsequently directed), easers, trimmers and line (side) holes (fig. 9.8). Broadly, these patterns can be classified as:

1. Parallel hole cuts
2. Angled cuts.

9.3.1.2.1   Parallel hole cuts

If breaking in holes are put at right angle, or parallel to the direction of the working face, these types of pattern of holes are known as parallel hole cuts. Burn cut, cylindrical cut and cormorant cut, fall in this category (fig. 9.4).[9] A cluster of parallel shot holes,

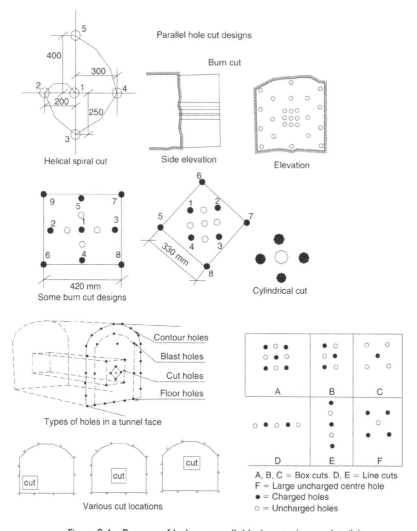

Figure 9.4   Pattern of holes – parallel hole cuts (some details).

known as 'cut'; is drilled almost parallel to the intended direction of face to blastoff a cavity in the center of the heading. Some of the holes are charged while others are kept empty. When the shock waves are reflected at these empty holes, the rock is shattered and subsequently blown out by the escaping gases. There is a specific geometrical relationship between the diameter of empty holes and the spacing between the empty and charged holes, in a given rock, which gives essential conditions of breakage. Figure 9.5(a)[13] shows this geometrical relationship. With spacing less than 1.5 times the diameter of an empty hole, the blasting is expected to be clean while with higher spacing the breakage would tend to be uneven. With spacing in excess of twice the diameter of an empty hole, there is a likelihood of plastic deformation and burning of rock. For achieving the greater advance, it is therefore necessary to increase the diameter of the empty hole, which in

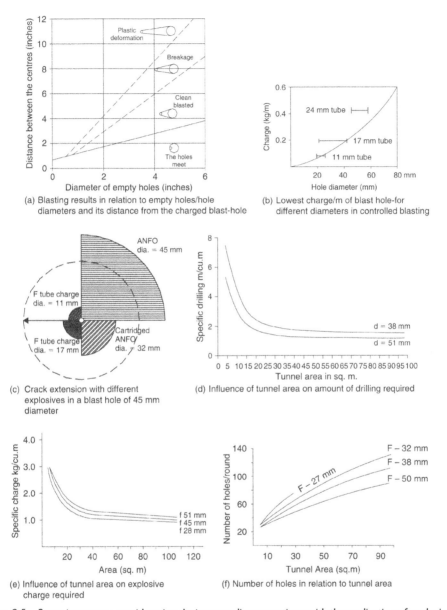

(a) Blasting results in relation to empty holes/hole diameters and its distance from the charged blast-hole

(b) Lowest charge/m of blast hole-for different diameters in controlled blasting

(c) Crack extension with different explosives in a blast hole of 45 mm diameter

(d) Influence of tunnel area on amount of drilling required

(e) Influence of tunnel area on explosive charge required

(f) Number of holes in relation to tunnel area

Figure 9.5 Some important consideration during tunneling operations with the application of explosives.

turn gives better scope of for increasing spacing between the holes without jeopardizing the condition of free breakage. Even with proper burden if the charge concentration in the hole is too high, a miss-function of cut may result due to rock impact and sintering, preventing generation of the planned free face. There are variations in shot hole patterns employing large diameter empty holes or providing empty space by not blasting a few shot holes. Various patterns employed at the cut are shown in figure 9.4.[9]

It has been found that at some Russian mines introduction of this pattern reduced the scattering of rock at the face from 15–23 m to 9–10 m; which in turn increased the loader efficiency by about 15%.[16] These patterns are suitable for hard, brittle and homogeneous rocks. These patterns are also known as fragmentation or fracture cuts.

Number of holes in a pattern The number of blastholes in a pattern is a function of the variables, as given in equation 9.1.[16] The number of blast holes (N) in a round can be divided into two groups: Breaking in or cut-holes, easers and bottom holes – N_b; and Line (peripheral) holes – N_l

$$N = N_b + N_l = [(q\ S/\gamma) + (\{c\ \sqrt{S} - B\}/b) + 1]\ (1 - \{\gamma_0/\gamma\}) \qquad (9.1)$$

Where: q – powder factor (kg/m³)
γ – explosive consumption/m length of hole (kg)
c – a coefficient depending upon shape of workings, for square section c = 4; trapezoidal c = 4.2; arch c = 3.86.
S – cross sectional area of a working (m²)
B – width of working
b – average spacing of line holes which in practice taken as 0.75 to 0.8 m irrespective of rock properties and the working's cross-section.
γ_0 – explosive consumption/m length of line holes (kg)

The consumption of explosive per cubic meter of rock i.e. powder factor q, as given in equation 9.2[16] is

$$q = q_1 f_1\ v\ e\ d_c \qquad (9.2)$$

Where: q_1 – a factor that characterizes the properties of the rock being broken by blasting. Varies from 1.5 to 0.15 depending as per very tough to jointed loose rocks.
f_1 – coefficient allowing for the structure and texture of the rock
v – a factor that takes into account the additional resistance offered to blast by the surrounding rock mass. For workings of limited cross sectional area and with one free face $v = 6.5/\sqrt{S}$
S – cross sectional area of a working (m²)
e – a factor that is indicative of the power of explosives. It varies from 0.8 to 1.17 (from least to most powerful explosives).
d_c – diameter of explosive cartridge.

In different situations different relations to calculate the number of holes are used. Given below are some of the empirical relations that can be used as a guide. However, a factor can be applied to suit the local conditions that differ from one tunnel or mine to mine to another. The relation given below has been used in some of the Swedish mines and tunnels for medium hard to hard strata.

Swedish relation: $N = (30.9 + W \times H_t)\ (44/d)$ or $(30.9 + S)\ (44/d)$ \qquad (9.3)

Where: N = number of holes
W = width of drive (m), H_t = height of drive (m)
d = hole diameter (mm), S = cross-sectional area (m²)

Willber[21] gave the following relation for the tunneling work in the USA.

$$N = 0.124A + 10 \text{ for soft or highly fractured rocks.} \tag{9.4a}$$

$$N = 0.158A + 28 \text{ for hard or massive rocks} \tag{9.4b}$$

Where: A – cross sectional area (ft^2)

Hole diameter:[16] It is function of explosive cartridges to be charged. The difference between the diameter of cartridge and that of the hole should be small, as an air gap reduces the effect of the explosive. The optimal ratio is 1.2.[16] The effect of decoupling is also considerable. The number of blastholes in a round should decrease in proportion to the increase in the diameter of holes [equation (9.4a)[16] and fig. 9.5(f)[12]] i.e.

$$(N_1 / N_2) = (\gamma_2 / \gamma_1) \quad \text{or} \tag{9.5a}$$

$$(N_1 / N_2) = (d_2 / d_1) \tag{9.5b}$$

Where: γ, d and N are the explosive consumption in kg/m, diameter of hole and number of holes respectively in a round.

The cut: As described above, the location of cut in figure 9.4 and its design within a drilling pattern meant for driving tunnels with the aid of explosives is vital. For good results consideration for these parameters must be given:

• Diameter of large empty hole/holes
• The burden
• The charge concentration.
• Drilling precision and charging skill.

The larger the empty hole diameter the better, and the depth of round could be deeper.[15] A slight deviation in drilling can jeopardize the complete blast. Too large a burden will cause breakage only, or plastic deformation in the cut area. But the right burden will result in a clean blast up to the planned length (depth) of round, and thereby better advance per round. This is how optimum utilization of the input resources could be achieved. As shown in figure 9.6(a),[13,15] the burden between the large empty hole and shot hole is the distance between their centers. For best results it is taken as 1.5 D_2; where D_2 is the diameter of the empty hole. Where several empty holes are used, a fictitious diameter, D, could be calculated using the following relation.[15]

$$D = D_2 \sqrt{n} \tag{9.6a}$$

D_2 is diameter of empty hole and n = number of holes.

In order to calculate the first square in the cut area, this D is used.

$$B_1 = 1.5 D_2 \tag{9.6b}$$

B_1 is the center-to-center distance between the large hole and the shot hole. In the case of several large holes;

$$B_1 = 1.5D \tag{9.6c}$$

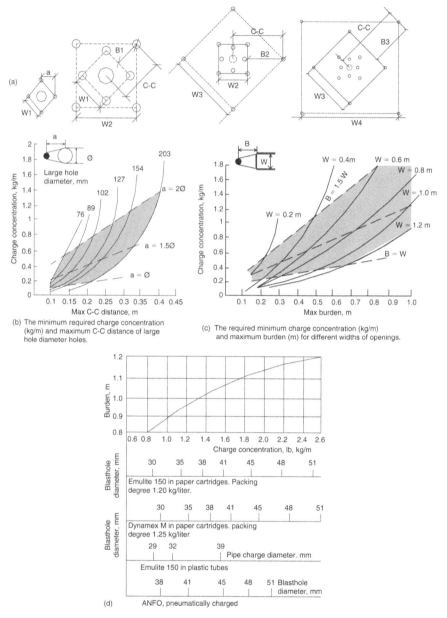

Figure 9.6 (a) Calculation of width of the resultant squares/rectangles in a cylindrical cut pattern. (b) Graph to determine minimum charge concentration (kg/m) in the cut-holes to be charged with explosive; and maximum centre to centre distance (m) for different large empty hole/relieving hole diameters. (76 mm, 89 mm, –, 203 mm are the usual relieving/empty hole diameters). (c) Determination of charge concentration based on the width of resultant openings (square/rectangles) by blasting cutholes. (d) Graph to calculate burden, spacing, charge concentration, for different types of explosives.

Charging the holes in the first square
The holes closest to the empty hole must be charged very carefully. Too low a concentration of charge may not break the rock; while too high a concentration may recompact the rock, usually known as 'freezing' and thereby not allowing the rock to blow out through the large hole. The charge concentration can be found by the graphs given in figure 9.6(c).[15]

The calculation of the remaining squares of the cut[12,15]
The calculation method for the remaining squares has a difference that the breakage is towards a rectangular opening instead of a circular. Normally, burden (B) for the remaining squares of the cut is equal to the width W of the opening:

$$B = W \qquad (9.7a)$$

From figure 9.6, the charge concentration in kg/m of hole can be obtained for the calculated burden.

$$Q = l_c \, (H - h_0) \qquad (9.7b)$$

Where: Q = charge quantity (kg);
l_c = charge kg/m from table;
H = hole length (m);
h_0 = stemming length = 0.5B
The number of squares in the cut is limited by the fact that the burden in the last square must not exceed the burden of the stoping holes for a given charge concentration in the hole. The cut holes occupy approximately $2 \, m^2$ (Small tunnels, as a matter of fact, consists only of cut holes and contour holes[15]). The given numerical example (for an empty hole of 127 mm dia.) demonstrates as how cut holes can be designed.

The charge concentration of the explosives into the holes of the first and remaining three squares can be assessed using graphs, shown in figure 9.6.

After calculations for the cut area, the details of rest of the tunnel round may be worked out. For this purpose the rest of the area of the tunnel other than cut holes is divided into following sections/zones:[12,15]

- Floor holes
- Wall holes
- Roof holes
- Stoping – upwards & horizontal holes
- Stoping – downwards holes.

To calculate burden, spacing and charge concentration, the graph shown in figure 9.6(d) may be used. In this graph the use of explosives such as Emulite-150 in paper cartridges; Dynmex in paper cartridges, Emulite-150 in plastic tubes and ANFO pneumatically charged have been shown. By projecting cartridge diameter of any of the explosives mentioned above, the burden B of the floor holes and charge concentration in kg/m can be obtained and then using table 9.3,[12,15] the charging pattern can be assessed.
Lifters:

$$B = 0.9 \sqrt{\frac{q_l \, x PRP_{ANFO}}{(S \, / B \,) c_0 f}} \qquad (9.8)$$

Table 9.1 Calculation of the cut section dimensions of a tunnel round in parallel hole cut hole pattern.[12]

Section of cut	Burden value (center to center)	Side of section
First	$B_1 = 1.5 D_2$	$W_1 = B_1 \sqrt{2}$
Second	$B_2 = 1.5 W_1$	$W_2 = W_1 \sqrt{2}$
Third##	$B_3 = 1.5 W_2$	$W_3 = W_2 \sqrt{2}$
Fourth##	$B_4 = 1.5 W_3$	$W_4 = W_3 \sqrt{2}$

– Stop calculation if burden exceeds that of the stoping holes (refer table 9.3).

Table 9.2 Charge calculation for the different zones of tunnel face. L = shot-hole/blast-hole length.[12]

Section of cut	Burden value	Side of section
First	$B_1 = 1.5 \times 127 = 191$	$W_1 = B_1 \sqrt{2} = 270$
Second	$B_2 = 1.5 W_1 = 404$	$W_2 = W_1 \sqrt{2} = 560$
Third####	$B_3 = 1.5 W_2 = 840$	$W_3 = W_2 \sqrt{2} = 1180$
Fourth####	$B_4 = 1.5 W_3 = 1770$####	$W_4 = W_3 \sqrt{2} = 2495$

– The calculations of third and fourth squares may not be required as the burden may exceed, the one (i.e. burden) at the stoping holes (refer table 9.3).
Note: These values are applicable if 38 mm shot-holes are used; for larger diameter shot holes which can accommodate more explosives, values can be adjusted.

B = Lifters' Burden[12] (m);
q_l = explosive linear charge concentration (kg/m);
PRP_{ANFO} = relative weight strength of the explosive with respect to ANFO
f = fixation factor, generally 1.45 is taken to consider gravitational effect and delay timing between blast holes
S/B = spacing and burden ratio, which is usually 1
c_0 = corrected blastability factor (kg/m^3); $c_0 = c + 0.05$ for $B = 1.4$ to 1.5 m but $c_0 = c + 0.07/B$ for $B < 1.4$ m
c is the rock constant whose values varies from 0.2 to 0.4 depending upon type of rock;[18] for brittle rocks 0.2, and rest all other rocks it is 0.3 to 0.4.
The number of blast holes[12] (lifters)

$$NB = \text{Integer of} \left[\frac{AT + 2LX \sin \gamma}{B} + 2 \right] \tag{9.9}$$

Where: AT = tunnel width (m);
$\quad$ L = hole depth (m);
$\quad$ γ = lookout angle.

For the rest of the calculation, table 9.3, should be referred. Burden should comply with the following condition:[12] $B \leqslant 0.6 L$ i.e. burden should not exceed 60% of the hole depth.

Table 9.3   Designing of drilling and blasting pattern in tunnels and mine drives.

Part of round	Burden (B) (m)	Spacing (m)	Length of bottom charge (m)	Charge concentration Bottom (kg/m)	Column (kg/m)	Stemming (m)
Floor (lifters)	B	1.1 B	L/3	I_b	I_b	0.2B
Wall*	0.9 B	1.1 B	L/3	I_b	0.4 I_b	0.5B
Roof*	0.9 B	1.1 B	L/3	I_b	0.36 I_b	0.5B
Stoping						
Upwards & horizontal	B	1.1 B	L/3	I_b	0.5 I_b	0.5B
Downwards	B	1.2 B	L/3	I_b	0.5 I_b	0.5B

I_b = charge concentration in bottom of hole = $7.85 \ 10^{-4} \ d^2 \ \rho$; d = cartridge diameter (mm);
ρ is explosive density (gms/cc); B = burden in stoping area = $0.88 \ I_b^{0.35}$
L = hole depth in the round.
* – in some cases smooth blasting is essential and these relation are not applicable.

With regard to bottom and column charge; the column charge could be up to 70% of the bottom charge but in practice it is difficult to obey such rules. As such usually the same concentration at both the sections is kept. Stemming depth is usually 10 times the hole diameter

Fixation factor   For different situations different fixation factors are used[18] for calculating burden [Roger Homberg1982]. During bench blasting when holes are vertical, it is 1; and when holes are inclined and it becomes easier to loosen toe, it is considered less than 1. In tunnel blasting a number of holes are blasted at a time. Sometimes the holes have to loosen the burden upward or sometimes downward. Different fixation factors are used to include the effect of multiple holes, and that of gravity.

In locating lifters one must consider lookout angle, which depends upon the drilling equipment and hole depth. For an advance of about 3 m a lookout angle (γ) equal to 0.05 rad (3°) (corresponding to -5 cm/m)[18] should be enough to provide room for the next round.

Angled cuts

If breaking in holes are put at an angle to the axis of the working face (drive/tunnel), the patterns of holes are known as angled cut. The tunnel face is utilized as free face toward which the initial blasting power is directed. This results in a cavity, towards which the subsequent blasting is directed. The angled cuts have some limitations, such as:

• The width of tunnel and size of drill with its mountings, as shown in figure 9.7(f),[16] limit the angle relative to the axis of tunnel at which the holes can be drilled.
• Accuracy of drilling: in some angled cut patterns, every pair of holes drilled should meet as close as possible to promote flash over. In practice it is difficult to achieve, thereby, less pull is resulted.
• Dependence of blasthole depth on the width of working as holes require sloping (inclination). Difficult to collar and drill holes accurately in the desired direction.

In angled cut use of orientation of the rock beds, available joints and cracks, jointing pattern, lamination etc. is made. Apart from the soft rocks, these patterns are also drilled in hard and tough rocks. These patterns result in fly rocks and high consumption of explosives. The following are the common angled cut patterns:

Wedge cut/v cut In a 'V' cut two holes are drilled at a horizon (1–1.5 m above the floor) almost at the center of the face (width-wise). Ideally both holes should meet at their apex, but in practice it is difficult to achieve and thereby instead of 'V' a wedge is resulted, and hence the name 'wedge cut' (fig. 9.7(b), (e)).[9] The angles of subsequent holes drilled at the same horizon are increased in such a way that the round holes (at the sides) are at 90–95° to the face. In harder strata (formations) a double or triple 'V or wedge' can be drilled. While blasting, 'V' holes are given the initial delay and to the subsequent holes delays are given in an increasing order. Thus, a wedge of rock that is pulled out first is ultimately converted into a slot/kerf/slit that has been generated by way of blasting. The rest of the holes of the pattern are drilled parallel to the axis of the drive/tunnel and blasted in a sequential order using delay detonators and taking advantage of the initial free face created. This pattern is suitable for medium hard to hard strata. Theoretically, the advance that can be obtained by drilling these rounds is given by:[16]

$$A = B\,CotV \tag{9.10a}$$

$$B = (W/2) - E\,SinV \tag{9.10b}$$

Where: (as shown in figure 9.7(f)); A = advance in m;
W – width of face; V = 34.5 − 18.5°; E from 1.5 to 3 m.
  The following guidelines[15] for the V cuts could be used:

- An advance of 45 to 50% of tunnel width is achievable.
- The angle of cut should not be too acute and should not be less than 60°. A more acute angle requires higher charge concentration in the holes. The cut usually consists of two Vs; but for deeper round these Vs could be three or four.
- Holes within each 'V' should be given same delay number and there should be delay interval of 50 ms between the consecutive Vs to allow time for broken rock displacement and swelling. The graphs shown in figure 9.9(b)[15] could be used to determine burdens B_1 and B_2 for the cut and height of cut C.
- Charge concentration in the bottom of hole l_b can be found from the graph (fig. 9.9(b)).
- Height of bottom charge h_b for all cut holes = H/3; H is hole depth.

Concentration of bottom charge (l_c) = 0.3 to 0.5 of l_b
Stemming in the cut-holes = $0.3B_1$ and in the rest of holes = $0.5B_2$

Fan cut In this pattern (fig. 9.7(d)) holes are drilled in a fan like fashion. Holes are drilled at a horizon, 1–1.5 m above the floor, setting them at different angles in an increasing order, starting from one side of the face, so that last hole (at the other side of the face) is at 90–95°. Sometimes a second fan is drilled at another horizon within 0.3 m above or below this horizon. Holes are fired in a sequential order starting from the first hole (the one having least angle at one side of face). This results an initial ditch/kerf towards which the blasting of the rest of the holes of the round can be

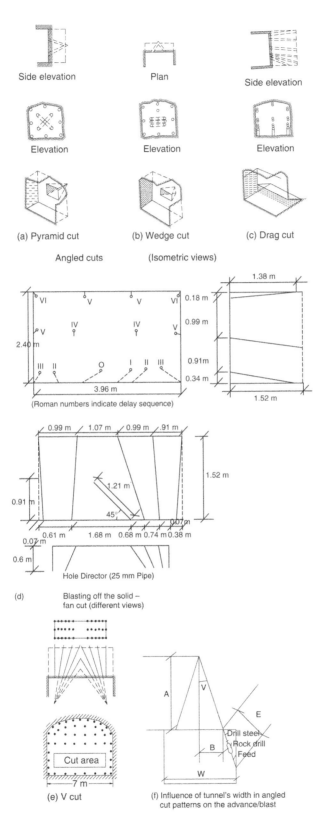

Side elevation

Plan

Side elevation

Elevation

Elevation

Elevation

(a) Pyramid cut

(b) Wedge cut

(c) Drag cut

Angled cuts   (Isometric views)

VI   V   V   VI   0.18 m

0.99 m

V   IV   IV   V

2.40 m

III II   O   I II III

3.96 m

(Roman numbers indicate delay sequence)

1.38 m

0.91m

0.34 m

1.52 m

0.99 m   1.07 m   0.99 m   .91 m

1.52 m

1.21 m

0.91 m

45°

0.07 m

0.61 m   1.68 m   0.68 m 0.74 m 0.38 m

0.07 m

0.6 m

Hole Director (25 mm Pipe)

(d)   Blasting off the solid –
           fan cut (different views)

Cut area

7 m

(e) V cut

A   V

E

Drill steel
Rock drill
B   Feed

W

(f) Influence of tunnel's width in angled
cut patterns on the advance/blast

Figure 9.7   Angled cuts – Some details.

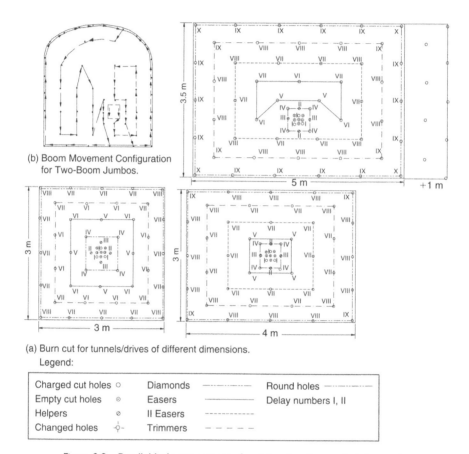

(b) Boom Movement Configuration for Two-Boom Jumbos.

(a) Burn cut for tunnels/drives of different dimensions.

Legend:

Charged cut holes	○	Diamonds	·············	Round holes	··············
Empty cut holes	⊙	Easers	———	Delay numbers I, II	
Helpers	⊘	II Easers	- - - - - - - -		
Changed holes	-⊘-	Trimmers	– – – –		

Figure 9.8   Parallel hole cut patterns for drives in mines and civil tunnels.

directed by the use of delay detonators. This pattern is suitable for soft to medium hard strata. A hole director, some sort of template, is used during manual drilling to achieve better accuracy of drilling.[11]

Drag cut   In this pattern holes are drilled like a fan cut pattern but in a vertical plane as shown in figure 9.7(c). When these breaking in holes i.e. cut holes are blasted, an undercut (slot/ditch) at the face is created. The drilling and blasting for the rest of the holes are directed towards this cavity. This pattern is suitable for soft to medium hard strata.

Pyramid cut   This pattern (fig. 9.7(a)) can be drilled for any drivage work – horizontal, up or downward. A cluster of holes (having 4–6 holes) is drilled in the center of the face directing towards a common apex so that a pyramid is formed after blasting. The angles of subsequent holes drilled around the cut holes are increased in such a way that the round holes (at the sides) are 90–95° to the face. This pattern is popular during shaft sinking and raising operations and suitable for any type of strata.

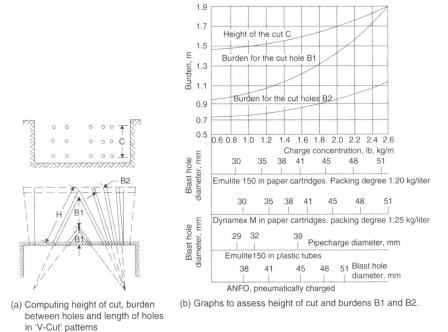

(a) Computing height of cut, burden between holes and length of holes in 'V-Cut' patterns

(b) Graphs to assess height of cut and burdens B1 and B2.

Figure 9.9   V cut and related parameters.

9.3.1.2.2   Verification of pattern of holes

The performance curves drawn in figures[12] 9.5 (d), (e) and (f) could be referred, as guide. Figure 9.5(e) illustrates powder factor as a function of tunnel area. Drilling required with respect to tunnel area has been shown in figure 9.5(d). It can be seen that the larger the tunnel area the less will be the powder factor and total drilling meters. The number of holes per unit area also decreases as the tunnel cross-section increases (fig. 9.5(f)).

Drilling operation   It is evident that blast holes of smaller diameter are easy to drill by the handhold and pusher-leg mounted drills, whereas, large diameter holes should be drilled using the rig or jumbo mounted drills/drifters. In small diameter holes cartridges of 25–28 or 30–34 mm diameter are used, whereas, for large diameter holes ANFO, slurry explosives or cartridge of large diameters could be used. Large diameter holes should be drilled in large sized drives and tunnels to reduce the number of holes and time of drilling.

In chapter 5 drills of different kinds have been described and their selection depends upon size of face, desired advanced per unit time, and the prevalent face conditions. As stated, the choice varies from a handheld drill to heavy-duty drifters mounted on multi-boom jumbos. In figure 9.8(b), the usual pattern followed for the movement of booms while undertaking the drilling tasks with boom jumbos has been illustrated.[14]

9.3.2   Charging and blasting the rounds

9.3.2.1   Placement of primer

Blastholes are usually charged with a continuous column of explosive cartridges. The effectiveness of blast in this practice is greatly affected by the location of the primer, and the stemming material and its length. Preparation of the primer has been illustrated in figure 5.6(b). The primer can be placed at the blast hole collar – *direct initiation*, or at the blast hole bottom – *inverse initiation*, as shown in figures 5.6(c) and (d). In the inverse initiation the blast energy is utilized to a greater extent than in the direct initiation due to prolonged action of explosion product on the enclosing rock. The effect of inverse initiation increases with increase of hole depth.

9.3.2.2   Stemming[17]

The amount of *stemming material* is specified by the safety regulations in some countries. In practice it is usually in the range of 0.6–0.8 m during drift/tunnel blasting. Usually a mixture of clay and sand in the ratio of 1:4 is used. Water stemming (polyethylene tubes filled with water) contributes to adsorption of toxic gases and suppression of dust.

9.3.2.3   Depth of round/hole

There is no empirical formula as such to determine the depth of round but as a guideline,[16] particularly in the workings of limited cross-section, the depth of hole $= 0.5\sqrt{S}$ for angled cuts and for parallel hole cuts, it is equal to $0.75\sqrt{S}$, (whereas, S is the cross sectional area of the drive).

9.3.2.4   Charge density in cut-holes and rest of the face area[12,13,14,15,16,17]

Charge density i.e. explosive/m length in the cut-hole area is a function of diameter of empty hole, its distance from the charge hole and explosive density. However, since the distance between the cut-holes is so short and also the volume of rock likely to be blasted by the cluster of holes (i.e. in cut-hole area) is small, hence the required charge concentration/m length should be low. This means either explosives of low density; or higher density explosives with loose confinement, or with the use of some inert material like wood, as spacers, should be used. Except ammonium nitrate (AN) based explosives, all other commercial explosives have a higher density; that ranges 1.2–1.7 gms/c.c. Hence, in the cut-holes the explosive cartridges must be just pushed in without any tight/hard tamping. The cut area after its blast creates a sufficient void/cavity. But the rest of the holes, which are drilled at an increased burden and spacing, and ultimately yield a sufficient amount of rock, should be charged thoroughly with tight tamping. At the line holes (peripheral holes) the spacing between them should be reduced and also the charge concentration by the use of either low density explosive or otherwise, to obtain smooth face profile with minimum over-break.

In chapter 5 different types of explosives, blasting accessories and firing techniques have been described. Selection of matching explosive for a particular hole diameter is vital; for example for a hole diameter in the range of 32–40 mm the explosive's VOD in the range of 2000–3000 m/sec is suitable. Similarly, large diameter holes require explosives of higher VOD. Cross-sectional area influences explosive consumption considerably, as illustrated in figure 9.5(e). For smaller tunnels higher powder factor is required and it gets reduced as their size increase.

9.3.3   Smooth blasting[12,14,15,16,17]

In practice over-break along the planned shape, contour or configuration results after blasting a round. This result in higher costs on supports, more mucking, ore dilution, generation of cracks and exceptionally into roof falls. To have a better confirmation of the actual cross-section with the designed one, a technique known as smooth blasting, is used. In this technique extra care is taken while drilling and blasting the line (peripheral holes). Increasing the number of line holes and reducing spacing between them could achieve this (fig. 9.10(a)).[14] Practically, the spacing of lines holes for burden of 0.7–0.9 m is taken as 0.5 to 0.6 m.[16] The line holes should be located as close to the working (tunnel walls) as the drill machine permits. In smooth blasting, in the outer profile drilling must be increased to about 1.5 times the normal. Drilling quality by handhold machines is usually poor so far as accuracy is concerned, whereas, drill jumbos can provide accurate drilling.

To charge the line holes special explosive is used in different countries. For example Gurit is a special explosive manufactured for this purpose by Nitro Nobel, Sweden.

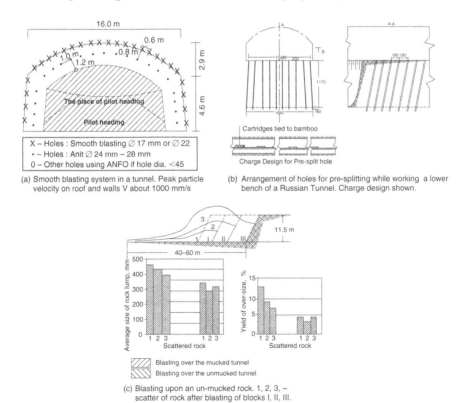

(a) Smooth blasting system in a tunnel. Peak particle velocity on roof and walls V about 1000 mm/s

X – Holes : Smooth blasting ⌀ 17 mm or ⌀ 22
• – Holes : Anit ⌀ 24 mm – 28 mm
0 – Other holes using ANFO if hole dia. <45

(b) Arrangement of holes for pre-splitting while working a lower bench of a Russian Tunnel. Charge design shown.

Cartridges tied to bamboo

Charge Design for Pre-split hole

(c) Blasting upon an un-mucked rock. 1, 2, 3, – scatter of rock after blasting of blocks I, II, III.

Blasting over the mucked tunnel
Blasting over the unmucked tunnel

Figure 9.10   Controlled blasting techniques in large sized tunnels and underground mine openings.

It is supplied in the form of small diameter rigid plastic pipes. Results of a tunnel blast-ing with and without use of Gurit (F-tube) have been illustrated in figure 9.5(c).[14,18] In Italy,[2] Profile-X, available in cartridges of 17–25 mm diameter and with special cen-tering devices to provide air cushion are used. In Australia use of ANFO and poly-styrene beads has been used for this purpose since 1975. In Germany detonating cord of varying strengths such as 40, 80 or 100 g/m are used for stone drifting.[9] Extension of the cracks with different explosives has been illustrated, as shown in figure 9.5(c).[17,18]

The importance of smooth blasting in drives and tunnels of large size is more important than the workings of smaller cross-sections. Smooth blasting lowers the consumption of concrete for lining and promotes a wider use of shotcrete lining which reduces the roughness of tunnel surfaces; a desirable feature from the ventilation point of view in mines; and water flow point of view in hydraulic tunnels.

Given table 9.4,[2] is the inter-relation between some parameters while drilling the perimeter holes to achieve a smooth profile. Also ref. tables 9.11 and 9.12.

In large sized tunnel another technique that is popular is the pre-splitting. The method consists of drilling accurately, as much as possible, with respect to the contour of a tunnel, a set of blast holes arranged in a single plane, or in line. Figure 9.10(b)[16] illustrates pre-splitting blast-holes for breaking the bottom bench of a Russian tunnel. The technique is also popular during open-pit and open cast mining operations to reduce vibrations, as described in chapter 17.

9.3.3.1 *Charging and blasting procedure*

- Before charging, prepare primers and get ready with the required quantity of explosive and detonators.
- Keep ready, and in proper condition the necessary blasting and charging tools, such as: scraper, stemming rod, circuit tester, connecting wires, blasting cable, warning display boards, explosive charging device (if any), exploder, etc.
- Check for the correct numbers, disposition and length of the shot holes to be charged.
- Clean the shot holes by blowing them so that sludge and water, if any, is flushed out.
- While charging, follow the standard procedure to charge the shot holes. It could be with the use of cartridge loader in large tunnels and as described in chapter 5.
- Tamp the explosive cartridges as required based on the location of the shot holes with respect to the cut.
- Make a tight and neat connections of the lead wires.

Table 9.4 Burden and spacing during smooth blasting in tunnel and mine openings.[2,17]

Practiced or as proposed by	Hole dia. (mm)	Charge dia. (mm)	Loading density (kg/m)	Spacing (m)	Burden (m)
By practice	32	17	0.220	0.4–0.6	0.55–0.75
By practice	51	25	0.500	0.65–0.90	0.8–1.20
DuPont	52	–	0.18–0.38	0.60	0.90
USBM	51	–	0.18–0.38	0.75	1.05

- Properly earth the charging equipment if charging is to be carried out pneumatically.
- Before blasting lay the blasting cables properly and test the circuit for its correctness.
- Post the guards at the appropriate locations. Display the warning display boards, if required. Ringing a siren or hooter can perform this task.
- Take shelter at the two right angles from the blasting face, wherever practicable.
- Reverse the ventilating current to act as exhaust, if so planned.
- Make sure that all the precautions have been taken prior to turning the key of exploder.
- After blasting allow sufficient time before re-approaching the face. Check for the misfires, if any.
- Follow the standard procedure to deal with the misfired shots.

9.3.3.2 Use of ANFO in drives and tunnels

Holes of less than 40 mm dia. charged with ANFO may not give proper blasting results as per the studies conducted by USBM, but when their dia. exceeds 40 mm, it has been established that use of ANFO explosive works out to be cheaper and productive. ANFO is charged pneumatically using usually ejector type loaders and anti-static detonators such as Anodets, to avoid the risk of static charge that is produced during the pneumatic charging. Anodets are costlier than electric detonators and in many countries they are not manufactured. Use of safety fuse with plastic igniting cord (PIC) and connectors can be an alternative to Anodets in such pneumatically charged tunnels and drives. It has been practiced at underground copper mines in India. The practice has been successfully adapted at some of the Australian and Canadian metal mines and tunnels.

9.4 MUCK DISPOSAL AND HANDLING (MUCKING AND TRANSPORTATION)

In chapters 6 and 7 on mucking and transportation; the units available to carry out these operations have been dealt in detail. After blasting, once the fumes are cleared and the face is inspected by the supervisor and the blaster, and declared free from any misfires. Water spraying then follows to suppress the dust. Once the face has been loose dressed (scaled), it is ready for mucking. Selection of mucking and transportation units differs from mine to mine (or from one tunnel to another) and even in the same mine sometimes. The line diagram given in figure 9.11 and 6.2, can provide a useful guide to match a specific situation.

Referring to figure 9.10(c)[16] that shows that blasting upon the un-mucked rock reduces the average lump size and the yield of oversize rock by 10–15% and 20–25% respectively comparing with blasting over the mucked tunnel face. This also results into lesser throw of rock and a compact muck pile. In a compact muck pile mucking efficiency could be increased up to 30%. Better fragmentation is due to proper utilization of energy, which otherwise is wasted in throwing the muck and creating noise and vibrations.

During muck disposal from tunnels and drives a proper match of these units can avoid delays and waiting time. This would allow for proper decision while buying these units, and money spent judiciously. Next important consideration is the layout

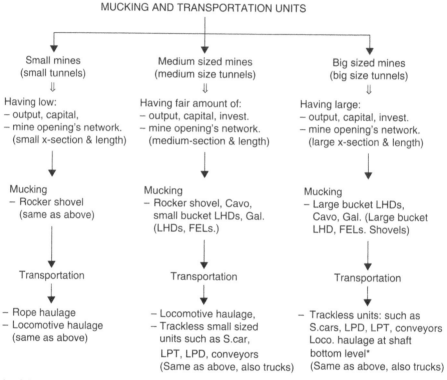

Figure 9.11   Selection of mucking and transportation units in underground mines and tunnels.

to load the muck from the face and then transfer it to transporting units (unloading). Various schemes are available for track and trackless systems as illustrated in figure 9.12.[16] For single-track system, the arrangement as shown in figures 9.12(a) to (c) include:

- Superimposed parting
- Bye pass system
- Traverser system.

With double track some of the layouts have been shown in figures 9.12(d) and (e) are:

I. With the use of a portable switch
II. With use of shunting locomotives.

For large output and to achieve almost a continuous mucking from the face, the second layout (fig. 9.12(e)) can be used in medium to large mines and tunnels. Selection

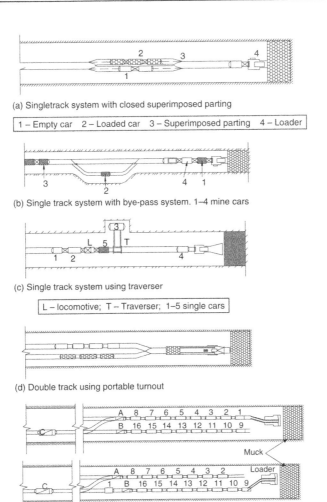

(a) Singletrack system with closed superimposed parting

1 – Empty car   2 – Loaded car   3 – Superimposed parting   4 – Loader

(b) Single track system with bye-pass system. 1–4 mine cars

(c) Single track system using traverser

L – locomotive;   T – Traverser;   1–5 single cars

(d) Double track using portable turnout

Muck

Loader

(e) Double track using a pair of shunting locomotives. 1–8 mine cars (first train);
9–16 mine cars (second train); A & B – Shunting locomotives; C – Main locomotive.

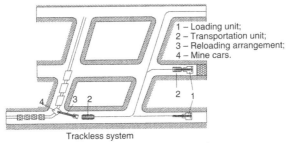

1 – Loading unit;
2 – Transportation unit;
3 – Reloading arrangement;
4 – Mine cars.

Trackless system

(f) Muck handling by trackless loading and transportation
units from single or multiple headings.

Figure 9.12   Some muck handling layouts while driving tunnels. Top: Single-track system with super
imposed parting. Bottom: Double track with locomotives.

of any one of these systems depends upon the local conditions that can differ from mine to mine or from one tunnel to another.

While driving blind headings or tunnels muck disposal arrangement in trackless system requires some space for turning of the loading and transportation units which should be provided at regular intervals in the intended haulage route. At the muck discharging (unloading) point of the mucking unit or muck receiving point of the transportation unit, more height than that of the tunnel is required to facilitate muck unloading operations. At least 0.3 m to 0.5 m clearance from the roof or back should be kept for smooth operation. In some mines the trackless loading units discharge muck directly into the mine cars running on track, as shown in figure 9.12(f).

Note: *Apart from the guidelines given in the text above; for selection of equipment and services please refer sections 6.18 and 15.10.*

9.5 VENTILATION[28]

In any enclosed area, the air can become fouled for various reasons; it must be removed and replaced by fresh air; this process is called ventilation. Underground workings are accessed by few openings such as shafts, inclines, declines etc., and air is circulated through a network of mine openings. In this process the air undergoes a number of physical and chemical changes. On the one hand, the oxygen content diminishes and the carbon dioxide content increases because of breathing, rotting timber, exhaust of diesel vehicles etc; and, on the other hand, various harmful gases and dust pass into the air from the rock and from the use of explosives. The amount of changes in the air resulting from these processes depends on the gas content of the deposit which is being driven through, the air speed through the mine openings, the extent of mine workings and tendency of the mineral (ore) and rock to absorb oxygen, and to oxidize. Other conditions being equal the underground air will be fouler, if the ventilation current is slow and it has a long distance to travel.

Gases of various types are emitted in the mines, as detailed in table 16.1(a); prominent amongst them are: methane CH_4 (usually in coal mines), carbon dioxide – CO_2, hydrogen sulfide – H_2S, sulfur dioxide – SO_2, and a number of others. When explosives are used carbon monoxide – CO and oxides of nitrogen – NO_x, are formed. While blasting 100 kg of explosive it can yield CO = 10–27 m^3; CO_2 = 1.2–4 m^3; oxides of nitrogen = 0.6 to 4.4 m^3 and ammonia = 0.03 to 0.3 m^3 (Vutukuri and Lama, 1986).[28] The diesel exhaust emission emits CO_2, SO_2, hydrocarbons, CO, particulates and oxides of nitrogen.

Mine air differs from the atmospheric air by its humidity, temperature, pressure and density. The function of ventilation, apart from providing the air to breathe in, is to maintain normal temperature and humidity. In general, the quality of air that is warranted is as outlined below:

Quality of air: O_2 not less than 20%; CO_2 not more than 0.5%, and temperature, not more than 20°C. Safe limits of the mine gases: CO – 0.0006%; Nitrogen oxides – 0.0002%; Sulfurous gases – 0.0007%; H_2S – 0.00066%.

Quantity of air required:
Quantity of air literally means that air must sweep all and not just some active areas in the mine in order to drive out the contaminants injurious to health of the

miner. It means also that air may be used to dilute some contaminants to such an extent that they are rendered harmless. If this is so then the composition of underground air should be maintained as close as possible to that of outside air on the surface. Except in coal mines in which the emission of methane gas decides the quantity of air required, in all other cases, the criteria followed for the purpose is outlined below; however it could differ from one country to another based on the prevalent regulations.

1. Based on the manpower employed:

   Q = 6 × N,

   where Q is the quantity required in m³/min.

   N – number of workers/shift; consider the shift in which maximum numbers of workers are employed.

2. Based on the population i.e. number of diesel driven units deployed during the shift:

   Q = Nv × 2–4 m³/B.H.P./min.

   where Q is the quantity required in m³/min.

   Nv number of diesel operated units in operation/shift; if a radioactive mine then take 5–6 m³ in place of 2–4 m³ in all other cases.

3. Based on explosive consumption during the shift:

   Q = (Ab/1000) × (100/nt) = 12.5 Ab/t, m³/min.

   where Q is the quantity required in m³/min.

   A = explosive consumption/shift in kg.

   b = volume of CO released by blasting 1 kg of explosives; it should not exceed 0.04 m³ as per safety regulations.

   n = maximum allowable concentration of CO in return air circuit = 0.008%.

   t = duration of fume clearance in min; it should not exceed 30 min.

Take the maximum of the 3 values of 'Q' so obtained.

9.5.1 Mine opening ventilation

Mine openings during their construction can be ventilated by any of the following ways:

9.5.1.1 Using general air flow

This practice can be made applicable if the length or distance to drive an opening is small. Air is directed towards the face by the use of some of the devices such as: baffles, ducts, doors and other contrivances. But this system proves to be not very effective and ventilation at the blind end of the face remains sluggish.

9.5.1.2 Using auxiliary fans: forcing, exhaust or contra rotating

Auxiliary fans, which could be of forcing, exhaust or contra rotating type, are used for this purpose. The fan and ducting are laid out in the drives as per the schemes illustrated by the line diagrams (fig. 9.13).[16]

Forcing or blowing ventilation: In this system[16] the fresh air is discharged from the ventilation ducting towards the working face, that mixes intensively with the foul gases and quickly dilutes them (fig. 9.13(b)). To avoid re-circulation of air the fan should be located not nearer than 10 m from the blind face. The effectiveness of this system is governed to a considerable extent by the distance between the face and the ducting pipe.

This distance L ≤ 4 √S meters from the face (9.11a)

Where: S is the cross-sectional area of the face in m².

Exhaust system: In this system (fig. 9.13(c)) the ducting should not be placed very close to the face, as it may get damaged due to blasting.

The maximum distance (L) between face and ducting ≤ 3 √S meters. (9.11b)

The effectiveness of the exhaust ventilation can be particularly poor in the workings of large cross sectional area where stagnation zones are liable to occur.

The major disadvantage of both these systems can be avoided by using a *combined technique*, or, with the use of a contra-rotating fan. In the combine system a blowing fan and an exhaust fan simultaneously ventilate the working face. The exhaust fan is the main air supplier whereas the blowing fan serves solely to accelerate the ventilation of remote areas of workings by forcing the contaminated air out of stagnation zones and moving it towards suction ducting outlets. A layout illustrating this system is shown in figure 9.13(d).

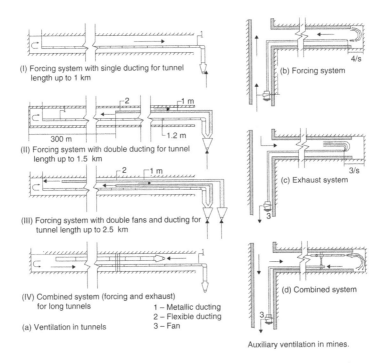

(I) Forcing system with single ducting for tunnel length up to 1 km

(II) Forcing system with double ducting for tunnel length up to 1.5 km

(III) Forcing system with double fans and ducting for tunnel length up to 2.5 km

(IV) Combined system (forcing and exhaust) for long tunnels

(a) Ventilation in tunnels

1 – Metallic ducting
2 – Flexible ducting
3 – Fan

(b) Forcing system

(c) Exhaust system

(d) Combined system

Auxiliary ventilation in mines.

Figure 9.13 Various schemes of ventilation in tunnels and mine workings/openings.

The amount of air or fan capacity can be calculated as per the workers employed, explosive consumption, and number of diesel units deployed.

9.5.2 Ventilation during civil tunneling

This scheme differs from the one used for the blind ended mine openings due to the fact that a fan exclusively for the tunnel is installed at its portal. Usually a fan drift, or adit is put within 30 m from the portal to deliver the air current to the tunnel. In case of mine openings the air is driven from a mainstream or ventilating current flowing through the nearby mine roadways. As per the length of a tunnel, the schemes shown in figure 9.13(a) can be adopted. In figure 9.13(a-I), a blowing or forcing system has been shown for tunnel lengths up to 1 km. The distance of metallic ducting from the face can be kept up to 50–80 m. Flexible ducting can be added to it, to make the air current more effective, if need arise.

For tunnels up to 1.5 km long, a single fan using two ducting pipes – metallic (1.2 m dia.) and flexible (1 m dia.) can supply air (fig. 9.13(a-II)). The flexible ducting can lag behind the metallic one by a distance up to 300 m. The provision for second ducting allows better ventilation to the spots where most of the equipment and work goes on. Fresh air supply can be made uniform along the tunnel by providing ports in the air duct every 80–100 m and regulating the discharge through them by dampers. For tunnels up to 2.5 km length an arrangement as shown in figure (fig. 9.13(a-III)), using two fans and three ducting can be made. In very long tunnels the number of fans can be installed at regular intervals in the metallic ducting. In figure 9.13(a-IV) a combination of exhaust and forcing fans has been shown, and to avoid mixing of foul and fresh air a barrier (sometimes created by mist generators) or ventilation doors are installed.

In table 9.5[19] ducts of different types with their important features have been shown. These ducts are used in mines and tunnels to carry the ventilation current.

Table 9.5 Ventilation ducts with their important features.[19]

	Type of ventilation duct			
	Hard line plastic/f-glass	Hard line (metal)	Smooth bag (plastic fabric)	Spiral bag (plastic fabric)
Typical resistance (K factor) $\times 10^{-10}$	13	15	20	60
Max. design velocity (fpm)	4000	3750	3350	2250
Air flow (cfm)	Maximum Duct Dia. (Inches) [Standard sizes: 18, 24, 30, 36, 48, 60, 72″]			
5000	15	16	17	20
10,000	21	22	23	28
15,000	26	27	29	34
20,000	30	31	33	40
40,000	43	44	47	56
50,000	48	49	52	62
75,000	59	61	64	78
100,000	68	70	74	90

Table 9.6 Main operations in a working cycle while driving mine openings and tunnels.

Steps	Operations
1	Allocation & reaching to the working face
2	Make the working face safe after scaling, water spraying and resuming ventilation if need arise.
3	Drilling the face – ensure the following prior to it: marking the centerline, grade line and drilling pattern. Connecting drills to water and compressed air, or to electric supply if hydraulic drills are used.
4	Blasting: check for correct drilling; charge the shot holes; connect detonators' leads and check the circuit. Blast the face after taking due precautions. Allow sufficient time for fume clearance. Repeat step 2 to make the working face safe. Install temporary support if need arise
5	Check the face for misfire or the residue of explosive, if any, in any of the buts. Blow (clean) the buts with water and compressed air.
6	Muck handling: bring the mucking and transportation units into operation.
7	Erect permanent support, if need arise.
8	Extend the service lines i.e. utilities such as compressed air and water pipes, electric cables, ventilation ducts, drainage, tracks etc.
9	The face is ready for drilling for the next round, repeat steps 1 to 8. Steps 1 & 2 will be needed whenever shift changes.

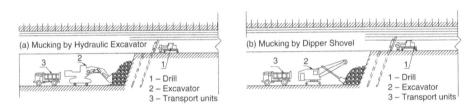

Figure 9.14 Application of dipper shovel; and hydraulic excavators in large caverns and tunnels.

9.6 WORKING CYCLE (INCLUDING AUXILIARY OPERATIONS)

In an underground mine while undertaking the task of mine development without the application of a continuous miner, the working cycle to complete a round (i.e. advance/ blast) consists of number of unit operations together with the auxiliary operations (services) as outlined in table 9.6.

Figures 9.14 illustrates application of a dipper shovel (right) and hydraulic shovel (left) for large sized tunnels which are driven by the benching method (section 9.7).

Figure 9.15(a)[14] depicts the necessary operations while carrying out tunneling works using explosives. The tasks included are surveying, drilling, charging, blasting, fume clearance, mucking, transportation and scaling. When cutting machines are used to fragment the rock in tunnels all other operations than drilling and blasting are necessary but in this work can go smoothly without disturbing workers and the surrounding strata.

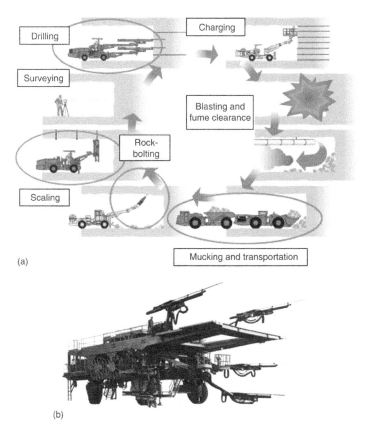

(a)

(b)

Figure 9.15 (a) Cycle of operations for tunneling and mine openings with the aid of explosives in a trackless environment. (b) A full face multi-boom drilling jumbo.

9.7 DRIVING LARGE SIZED DRIVES/TUNNELS IN TOUGH ROCKS[14,16,20]

The preceding sections describe the usual procedure that is followed to construct tunnels and mine openings of small to normal sizes, but when they are large sized; the special techniques available to drive them are outlined below:

- Full face driving/tunneling
- Heading and benching
- Pilot heading.

9.7.1 Full-face driving/tunneling

In tough rocks it is feasible to drive the full face of large sized tunnels/openings without use of any temporary support. Use of permanent support, however, can be made as the face advances. The technique involves carrying out all the unit operations in their sequential order for the full face. If the height of working exceeds 2.5 m and use of pusher-leg mounted jackhammer is made for the purpose of drilling then it is

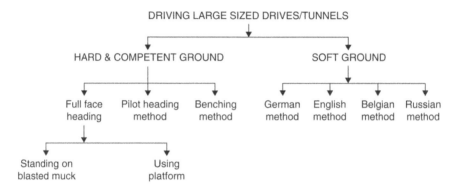

Figure 9.16 Methods of driving large sized drives/tunnels underground.

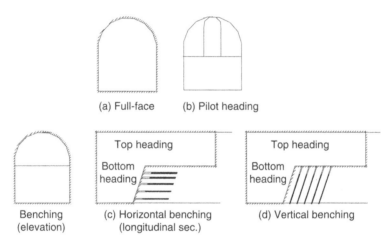

Figure 9.17 Different schemes/techniques to drive large sized tunnels and mine openings.

essential to drill the upper portion of the face either by standing on the blasted muck of the previous blast, or by erecting a platform. This practice has been almost replaced with the advent of multi boom jumbos (fig. 9.15(b), 9.17(a)) that are available to cope up a face of cross section[16] up to $110 \, m^2$.

A large sized face enables use of high capacity equipment to carry out the unit operations such as: drilling, blasting, mucking and transportation independently. This, in turn, results in faster rates and productivity of the operation. The shortcomings of the method include the requirements of high capital to buy equipment, difficulty in scaling the roof and the sides, and the problem of erecting supports.

9.7.2 Pilot heading technique

There are two ways to carry out the driving operation by this technique:

• Pilot heading in the bottom or top portion of the drive/tunnel (fig. 9.17(b)) or
• Pilot heading in the center of the drive.

In the first method a heading equal to 0.35–0.4 times the cross-section of the drive/tunnel is driven through the full length or as it is being cut. The method has proved useful for the drives up to 50 m^2 cross-sectional area.[16] Advance information regarding the type rock to be encountered can be obtained by this method. The short-comings of the method include the slow rate of driving the pilot heading and its subsequent widening. This technique is widely used.

In the second technique a central heading is driven first, then by radial drilling it is widened. In absence of proper drilling and blasting, an uneven profile of the tunnel/drive including at its floor can be resulted.

9.7.3   Heading and bench method

In this method the face is divided into two parts – top and bottom. Driving the top portion first and the bottom afterwards or vice-versa can be adopted. Type of rock, total cross-section area and the type of equipment available govern the ratio of top to bottom excavation. The ratio between top and bottom excavation varies from 0.75 to 1. The top bench is driven similar to the full-face method. If support is required it is erected simultaneously. The bottom bench is excavated under the protection of the previously placed permanent lining of the top bench. Since this bench has two free faces, that gives better drilling and blasting performances. In this technique due to the smaller size of the top heading, design and erection of temporary and permanents supports is simplified. High productivity while driving the bottom bench can be achieved. The shortcomings include longer overall time to complete the total drivage operation due its sequential working. If the rock is very stable then the reveres process, i.e. driving bottom heading first and then heightening it, can be followed. Figure 9.17 illustrates the schemes i.e. one with benching by vertical holes (fig. 9.17(d)) and the other by benching by putting horizontal holes (fig. 9.17(c)).

These techniques have been illustrated in figure 9.17.[16] Amongst the various techniques outlined above, the practice of full-face heading is widely used due to economy of the operation. Heading and bench is the next in line. In the mines and tunnels where jumbos are not available the pilot heading method can provide better results.

9.8   CONVENTIONAL TUNNELING METHODS: TUNNELING THROUGH SOFT GROUND AND SOFT ROCKS

Driving through soft rocks or ground is not an easy task, as it requires controlling the ground from collapse and subsidence. Soft ground or rocks can be described as ground which when dug is not self-supporting and it cannot withstand without support beyond a very little period. This period could be few minutes to several hours or days. In some circumstances advance timbering by the method known as fore-polling is necessary (refer fig. 8.3(c)). In a situation like this an explosive is practically not used to fragment the ground; and the conventional tools and appliances such as picks, spades, wedges, chisels, shovels and rippers or their equivalent moderns; which are meant for ground dislodging, digging and excavation are used. Ground needs to be excavated in a sequence and setting up the temporary supports goes side by side. Once the muck has been disposed off, the temporary support is replace by the permanent one. Presence of water may pose additional problems. It may result in mud and other unconsolidated material inflow conditions, which requires additional arrangements. Different countries, such as Germany, Belgium, England, Austria, Russia, and others use different ways or sequence to excavate the ground as shown in figure 9.18.[20] The numbers in each figure indicate the sequence of excavation. The practices followed at some of the Russian mines and tunnels are described below.

Driving tunnels with initial opening of the roof part of the cross-section; There can be two alternatives to follow this practice:[16]

1. A single heading, or double heading method can achieve this. As shown in figure 9.18(e), in the single heading procedure work is begun by driving top heading 1, next calotte 2 is enlarged and a permanent roof lining 3 is erected. Once the concrete hardens, first the middle part of bench 4, then its side portions 5 are dug, lining wall being finally built up underneath roof skew (slanting, rough) backs 6. The procedure is safer, economical and efficient. But due to lining works taken together some delay is resulted.

2. In *double heading* method, as shown in figure 9.18(f), first a bottom heading 1, then a top heading 2 are driven; calotte 3 is worked next and roof-lining 4 is finally erected. After concrete get sufficient strength, the middle part of the bench 5, then side portions 6 are excavated. Once this is completed, the wall lining is led up against roof skewbacks 7. In this method independent muck handing from the top and bottom headings can be carried out. The top heading is used only for the enlargement purpose. This method is widely used.

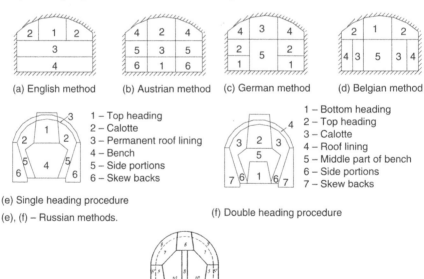

(a) English method (b) Austrian method (c) German method (d) Belgian method

1 – Top heading
2 – Calotte
3 – Permanent roof lining
4 – Bench
5 – Side portions
6 – Skew backs

1 – Bottom heading
2 – Top heading
3 – Calotte
4 – Roof lining
5 – Middle part of bench
6 – Side portions
7 – Skew backs

(e) Single heading procedure

(e), (f) – Russian methods.

(f) Double heading procedure

(g)

Figure 9.18 Different schemes/systems to drive large sized tunnels and mine openings in soft, unstable and weak ground.

3. *Driving tunnels/large excavations by initial opening of its section along the perimeter:* The procedure followed in this technique is that initially, as shown in figure 9.18(g), two side headings 1, and a central heading 2 are driven. These are connected by crosscuts 3. After excavating heading 1, lining 1' is put once the work at the first tier is completed. Similarly the second tier is worked by excavating and lining as shown by 4, 5, 4' and 5'. A top heading 6 is then driven. The muck of this heading is transferred through the opening 8. Lining 9 is then erected and sufficient time is allowed to cure it. Finally the left out portion 10 is taken. The method is simple and temporary supports suiting the rock conditions can be chosen and erected. But working in small sections may affect the quality of lining. The procedure is widely used for chambers and drives of large cross-section in inhomogeneous structure and soft to medium tough rocks. The stages of this method, which include the temporary support work, is divided into four sections to complete the task.

The main problem in tunneling through such a ground is that it weakens and tends to sink into the opening – a phenomenon called 'decompression' occurs. New and more advanced methods involve techniques to overcome the problem of such decompression or ground fall. Techniques applied are:

• Advance timbering or fore-polling using steel or concrete piles
• Ground improvement or consolidation (figs 11.21(c) to (f))
• Use of shields (section 11.10).

9.9   SUPPORTS FOR TUNNELS AND MINE OPENINGS

While considering the requirements of supports for the mine openings and tunnels, one should understand the basic difference between these two structures. Tunnel service life is practically unlimited and it can exceed even 100 years in many cases and their utility is round the clock, and as such repair of any kind if not impossible it is impracticable, and it can cause great disturbance to its users. Its supporting system should be waterproof (no seepage of water), smooth with even surface and aesthetic finish. Beside the primary requirement that the support should be strong enough to sustain the calculated load with a factor of safety not less than 2. Mine openings are not as shallow seated as the tunnels and except openings such as shafts, main levels and pit bottoms, the rest of the openings don't have a life exceeding a few years and usually in the range of 1–5 years. The on-going excavation works in the neighboring or adjacent workings also disturb the stability of these openings whereas in tunnels a situation of this kind hardly arises. In mine openings rough surfaces, make of water (seepage) and its disposal can be tolerated. Supports are subjected to vibrations but can be inspected and necessary repair and erection can be undertaken as and when required without many disturbances to the routine activities. A limited number of personnel, and that too for limited duration, accesses them. The span of mine working is also kept limited due to stability problems and it goes on reducing as the depth of working increases. Tunnels for traffic and transport purposes have large dimensions. Acceptable levels of deformation and deterioration of mine openings are higher than tunnels as such factor of safety in the range of 1.2–1.5 is acceptable.[20]

9.9.1   Classification[20]

The support used for mines and tunnels can be classified as:

1. Temporary
2. Permanent or primary lining
3. Secondary lining.

While driving tunnels and mine openings, in some situations exposing them without support even for a few minutes to a few hours can cause their collapse and in such cases use of temporary support is essential. In some circumstances even in advance or before digging the ground ahead of tunnel sight support is essential. This is achieved either by the method known as fore-polling with the use of timber or steel piles or with the use of shield supports. Use of wooden props, bars and sets, and rock bolting is made to support the site temporarily.

Permanent support or lining by some artificial means in soft ground, and the rocks, which could be soft to medium hard, is mandatory. Even a hard and competent ground needs support depending upon the life of mine openings or tunnels. The time gap and the span of the unsupported or temporarily supported ground differ from place to place and it could be a few hours to a few days (fig. 9.19(b)). Different types of supports, which are in vogue, are listed below:

• Natural (self support)
• Rock reinforcement using: Rock bolts, rock dowels and rock anchors
• Segmental supports: Tubings made of cast iron, steel or reinforced concrete
• Steel sets or rolled steel joist (RSJ) supports
• Concrete supports: monolithic (cast-in-place), prefabricated segments or blocks, shotcrete
• Wooden supports.

A description of these supports has been given in chapter 8 on supports. Also a brief reference is given in chapter 14 on shaft sinking. A comparative study with regard to application of supports of various types in civil and mining engineering tunnels and openings has been presented in table 9.7.

Secondary lining does not take any load but they are used to provide aesthetic finish and certain desired shapes to the primary lining. For example smooth circular profile is mandatory for sewer and water tunnels. They can be used as sealant, fire resistant coatings or protective covering to the temporary and permanent supports.

Apart from supporting temporarily or permanently the tunnels and mine roadways, the rock reinforcement can be achieved by making use of rock bolts and their variants in different manners as shown in figures[8] 8.5 to 8.8. These figures illustrate that this technique can be used to suspend the individual blocks, increase resistance to sliding of individual blocks and to prevent the progressive failure of blocks in the tunnels and mine openings.

Use of gray iron and cast iron tubing is made in the manufacturing of tunnel supports. The arches made of steel can be yielding or rigid types, as shown in figures 9.20(b) and (d) to (f). They can be assembled using two or three members. They can be circular, horseshoe or arched shapes.

Table 9.7 A comparative study with regard to application of supports of various types in civil and mining engineering tunnels and openings.[20,21]

Support type	Application in tunneling	Application for mine openings
Natural (Self support)	Good quality competent rocks in low stressed condition w.r.t. rock strength.	Good quality competent rocks in low stressed condition w.r.t. rock strength.
Rock reinforcement using: Rock bolts Rock dowels Rock anchors (figs 8.6 to 8.8)	Rock bolts or rock reinforcement techniques are used as temporarily support during tunneling operations.	During mine tunneling use of rock bolts of different kinds are increasing day by day. There will be hardly a mine opening where bolts are not used. Rock bolts are very widely used in good quality rock conditions as a permanent support. Tensioned bolts improve effectiveness. In weak ground cement and resin anchors are suitable. Mechanical anchoring requires competent ground for suitable anchorage. Bolts are used as a temporary support and to hold wire mesh in friable ground.
Segmental supports: including tubing made of cast iron, steel or reinforced concrete (table 11.4; fig. 14.7(c) and (d)).	Use of steel tubing was first started in 18th century to support shafts during their sinking. English and German tubing are famous designs (Ref. Chapters 8 and 14). It has been used for a number of underground tunnels in UK, USA and all over the world in last 100 years or more particularly in metro tunnels. It has excellent waterproofing, resistance to corrosion and tight fitting features. Since the fifties, shortage of cast iron has given birth to pre-cast concrete liners, which found to be cheaper than cast iron tubing. They find applications in weak and soft grounds, and ground with heavy water makes. They are usually circular or the shape that has been resulted by the borer machines. Cast iron tubing has good water sealing properties compared with the concrete. Concrete tubing is weak in tension compared to compression but has the cost advantage over the others. In general, these liners find application in tunnels driven by borers.	Except during shaft sinking, this support hardly finds any application for mine tunnels or openings.

(Continued)

Table 9.7 Continued.

Support type	Application in tunneling	Application for mine openings
Steel sets or rolled steel joist (RSJ) supports (fig. 9.20(b), (e) to (g))	Where rock mass is fractured and inherently weak to the extent that rock bolts cannot function effectively, and therefore, at all such locations during tunneling RSJ supports can be used as temporary supports to be replaced or followed by erection of the primary lining or permanent supports.	In mine tunnels or openings in a situation where rock mass is fractured and inherently weak to the extent that rock bolts cannot function effectively and also at the depths even in the good rock conditions, RSJ supports are used as primary supports. Due to timber scarcity and cost, its strength limitations and liability to fire and decaying fungi, steel sets have replaced it very widely. Steel sets could be of rigid as well as yielding characteristics. The sets of the desired shapes and size can be fabricated or supplied by the manufactures. A set could be of 3 to 5 members. Usually 'H' sections are used for manufacturing rigid arches and 'V' sections to construct yielding arches. The arches are placed at distances from 0.5 m to 1.25 m from one another. Lagging behind them could be that of wood, R.C.C concrete slabs of various shape or metallic screen or glass fiber etc.
Concrete supports	Monolithic concrete is very widely used particularly for water or hydraulic tunneling projects. Most suitable conditions for their application are the tunnels around which the strata movement is negligible.	For the lining of shafts, chambers or large excavations at pit bottoms, shafts insets and mine portals concrete can be used as monolithic, R.C.C, prefabricated blocks or prefabricated segments. Prefabricated arches find applications in mine tunnels at the main levels. But steel arches or sets are more popular due to their high strength, low per unit weight, installation and manufacturing characteristics.
Shotcrete	It is used as a temporary support before installing concrete lining as a primary or permanent support. In some situations it is used to cover the surface to prevent spalling and slabbing of the weak and loose ground or rock mass.	Used as surface coating to prevent rocks from further deterioration due to ventilation air currents. Used as covering to steel or wooden supports. As lagging between excavated rock surface and the support that has been erected.
Wooden supports (figs 8.2 and 8.3)	During tunneling in soft and weak ground its use, as a temporary support is almost mandatory.	Used as a temporary support in coalmine openings but in metal mines very limited use as a temporary support during the drivage work.

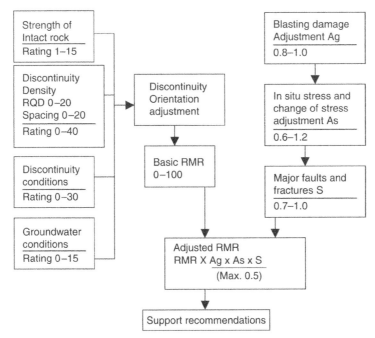

(a) Adjustments for RMR system for application in mines and tunnels.

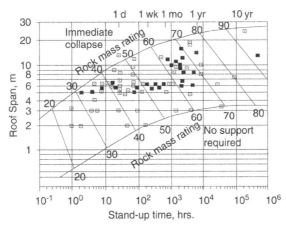

(b) Relationship between standup time and roof span.

Figure 9.19   Relationship between standup time and roof span of mine openings and tunnels for rock mass classes as per RMR system. Symbols: Block squares – roof fall in mines; Open squares – Roof fall in tunnels. Contour lines are limits of applicability.

9.9.2   Selection of supports

Deer et al.[7] recommended support requirement based on the RQD concept for the tunnels having a dia. in the range of 6–12 m and driven by TBM or using the conventional methods. Bieniawski[3,4] also suggested a support network for 10 m wide horseshoe-shaped

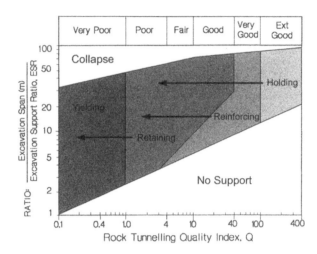

Figure 9.19   (d) Support functions as a function of rock mass quality[24] (For description of Q refer table 3.5 sec. 3.6.3) (www.smenet.org).

tunnels driven by conventional methods using RMR concept. Combining these two concepts, in the table 9.8, a guideline for supports requirements, in general, has been evolved.

In table 9.8, it may be noted that wire mesh requirement with rock bolting may be zero in good rock to 100% in very poor ground conditions, similarly lagging behind the steel sets may be 25% in excellent rock to 100% in very poor rock.

While selecting support network span of the workings, standing time required and types of ground based on the rock mass rating should be taken into consideration. Based on this logic[4] and analyzing field data, the results are shown in figure 9.19(b).[3,4] This figure shows that when roof span is high and rock mass rating is low, immediate collapse may occur and on the contrary when roof span is less and even the rock mass is poor, support may not be required for a considerable time. Collapse and roof falls have occurred in the mine openings and tunnels even when the rock mass rating is good but the span is high, after certain time which could be few hours to few years.

Bieniawski[3,4] also proposed that adjustment in the RMR should be done taking into account parameters such as: blasting damage, change in in-situ stress and presence of any major faults or fractures while determining RMR. But the value of these factors when multiplied should not exceed 0.5. The over value of this factor is multiplied by the RMR determined based on the rock strength (rating 0–15), discontinuity density (rating 0–40), discontinuity condition (rating 0–30) and ground water condition (rating 0–15). This has been illustrated in figure 9.19(a),[4] by a flow diagram.

General applications of the Rock Mass Classification schemes: These schemes in tunneling and drivage work not only provide the quantitative empirical guide to the support requirement but also significant benefits as described by Whittaker and Frith:[21]

- They allow subdivision of tunnel routing requiring different supports.
- They initiate the systematic collection and recording of the geological data.
- They provide an estimate of unsupported span of ground and stand-up time. Thereby phasing of support requirement can be made.

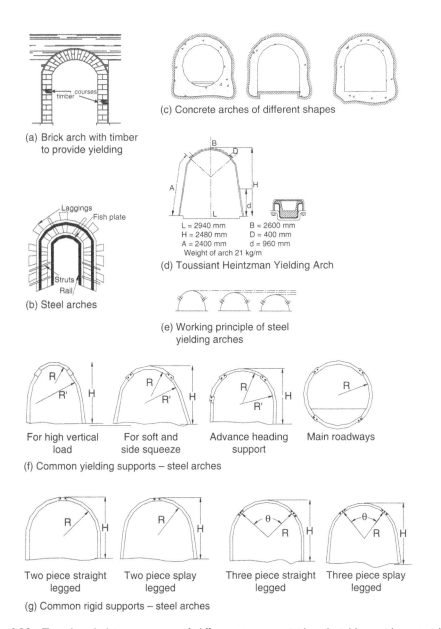

(a) Brick arch with timber to provide yielding

(c) Concrete arches of different shapes

(b) Steel arches

L = 2940 mm   B = 2600 mm
H = 2480 mm   D = 400 mm
A = 2400 mm   d = 960 mm
Weight of arch 21 kg/m

(d) Toussiant Heintzman Yielding Arch

(e) Working principle of steel yielding arches

For high vertical load   For soft and side squeeze   Advance heading support   Main roadways

(f) Common yielding supports – steel arches

Two piece straight legged   Two piece splay legged   Three piece straight legged   Three piece splay legged

(g) Common rigid supports – steel arches

Figure 9.20   Tunnel and drive supports of different types – rigid and yielding with material of construction bricks, concrete and steel.

However, Bieniawski[4] is of the following view:

- They should not be used as rigid guidelines
- Alternative schemes should also be considered
- Application should be judged on case to case basis

Table 9.8 Support requirements while driving tunnels through different ground conditions.[3,4,7]

Class number	Rock condition	RMR	RQD	Types of supports
I	Excellent	81–100	90–100	Steel sets, rock bolts and shotcrete requirement none to occasional i.e. as and when required.
II	Good	61–80	75–90	Steel sets, rock bolts and shotcrete requirement none to occasional i.e. as and when required. But the requirement could be more than in class I.
III	Fair	41–60	50–75	Light to medium duty sets of steel or concrete; systematic rock bolting; 50–100 mm thick shotcreting in the back or crown of the tunnel.
IV	Poor	21–40	25–50	Medium duty sets of steel or concrete; systematic rock bolting; 100–150 mm thick shotcreting in the back or crown, sides and over the rock bolts.
V	Very poor	<20	<25	Medium to heavy duty sets of steel or concrete including tubing; systematic rock bolting; 150 mm or thicker shotcreting to the whole section.

- At least two classifications must be applied
- Q and RMR system have been found to be superior to others.

Each support component is intended to perform one of three functions as illustrated in fig. 8.1(a)(i): (1) to hold loose rock, key block and other support in place; (2) reinforce the rock mass and control bulking; and (3) retain broken or unstable rock between the holding and reinforcement element to form a stabilized arch. In good rock, holding and reinforcement are the most important, whereas in fair to poor rock, all the three functions must be integrated, as shown in fig 9.19(d)[24]. In addition, in very poor ground (Q < 1, Barton et al. 1974), large deformation must be anticipate, and support components must retain their functionality over a large displacement range (i.e., must yield).

9.10 DRIVING WITHOUT AID OF EXPLOSIVES

Driving the mine openings and tunnels without aid of explosives could be accomplished by any of the following methods/techniques:

- Using conventional digging and excavation tools and appliances (section 9.8)
- Using heading machines (Chapter 10)
- Using full face tunnel borers (Chapter 11).

Details of these methods have been given in the sections and chapters as indicated above.

9.11 PRE-CURSOR OR PRIOR TO DRIVING CIVIL TUNNELS

The preceding sections described the common features between the mine openings and tunnels for civil engineering purposes. At the beginning of this chapter a distinction between tunnels and mine openings has been also made. Apart from these, there are some additional basic differences between the two, which will be dealt in the following paragraphs.

9.11.1 Site investigations

The success of any tunneling project lies on reliable forecasts of soil, rock, ground water and ground stress conditions. As without such forecast even the up to date methods and designs may be of little use, and may fail. It may result in unexpected problems, disputes with contractors or other agencies, cost overruns and delays in completing the tunneling program. If latent adverse geological features remain undetected during the design and construction phases, the potential of failure during operation remains in the years to come.

Another basic difference between mine openings and tunnels is that while driving mine openings sufficient information about the type of ground to be encountered is already available. This information is established during exploration, prospecting and feasibility stages and even in more detail while deciding a mining method. When a mine is opened or developed, additional information on ground conditions also proves a useful guide while driving the mine openings.

Before starting a tunneling project, information about the proposed site needs to established, as described in sections 3.4 and 3.5.

9.11.2 Location of tunnels

Location for the mine openings are governed by the deposits for which they are driven and their relative positions w.r.t. them cannot be changed. But tunnel locations can be altered to a great extent in favor of benefits that may be available to follow a particular route. The locations could be mountains or hilly terrain, below water-bodies or they may pass through the urban areas. While driving through hilly terrain and mountains, tough rocks are usually encountered which may be self-supporting to those requiring some support. While driving below water bodies prefabricated support is mandatory as strata are usually sebaceous. In urban areas usually soft ground is encountered as most of the cities are located near the rivers, and away from hilly terrains.

Another consideration with regard to tunnel location is its datum i.e. whether it is going to be below the valley level or above it. Keeping the tunnel portal at least 5 m above the highest flood level in the area will prevent water inflow in the tunnel during the rainy season.

Location with respect to depth is also an important consideration as in the case of urban tunnels a minimum capping or overburden is necessary for their stability, else, cut and cover method should be adopted which allows positioning them at shallow depths.

9.11.3 Rocks and ground characterization

Useful guidelines in this regard have been given in section 3.5 and 3.6. Similarly type of ground that can be encountered has been classified in table 9.9.

Table 9.9   Tunnel-man's Ground Classification.[26,29]

Classification	Behavior	Typical soil types
Firm	Heading (tunnel face) can advance without initial support, and final lining can be constructed before ground starts to move.	Loess above water-table, hard clay, marl, cemented sand and gravel when not highly over stressed.
Ravelling Slow Ravelling Fast Ravelling	Chunks or flakes of material begin to drop out of the arch or walls sometimes after the ground has been exposed, due to loosening, or to over stress and 'brittle fracture' (ground separate or breaks along distinct surfaces, opposed to squeezing ground). In fast ravelling ground, the process starts within a few minutes; otherwise the ground is slow ravelling.	Residual soils or sand with small amounts of binder may be fast ravelling below the water table and slow ravelling above. Stiff fissured clays may be slow or fast ravelling depending upon degree of over stress.
Squeezing	Ground squeezes or extrudes lastically into tunnel, without visible fracturing or loss of continuity, and without perceptible increase in water content. Ductile, plastic yield and flow due to over stress.	Ground with low frictional strength. Rate of squeezing depends upon degree of over-stress. Occurs at shallow to medium depth in clay of very soft to medium consistency. Stiff to hard clay under high cover may move in combination with ravelling at execution surface and squeezing at depth behind face.
Running Cohesive Running	Granular materials without cohesion are unstable at a slope greater than their angle of repose (+/− 30–35°). When exposed at steeper slopes they run like granulated sugar or dune sand until the slope flattens to the angle of repose.	Clean, dry granular materials. Apparent cohesion in moist sand, or weak cementation in any granular soil, may allow the material to stand for a brief period of ravelling before it breaks down and runs. Such behaviour is cohesive running.
Flowing	A mixture of soil and water flows into tunnel like a viscous fluid. The material can enter the tunnel from the invert as well as from the face, crown, and wall, and can flow for great distances, completely filling the tunnel in a few cases.	Below water table in silt, sand or gavel without enough clay content to give significant cohesion and plasticity. May also occur in highly sensitive clay when such material is disturbed.
Swelling	Ground absorbs water, increases in volume, and expands slowly into the tunnel.	Highly pre-consolidated clay with plastic index in excess of about 30, generally containing a significant percentage of montmorillonite.

9.11.4 Size, shape, length and orientation (route) of tunnels

The size of a tunnel depends upon its purpose. While designing it, consideration of the vehicles or equipment of largest dimension plus the clearance from both sides and roof plus thickness of support work, are considered. Allowance for space for pedestrians, drainage and other facilities should also be taken into consideration. The cross-sectional area should be verified by the ventilation requirements in terms of adequate (quantity) circulation of fresh air that should flow through a tunnel, within allowable velocity range, based on the local environmental laws. Details with regard to shapes are dealt in section 12.9.

For tunnels driven by using borers, circular or elliptical shapes cannot be avoided and they offer disadvantages to effectively utilize the space. However, they offer better stability to tunnels.

The length of a tunnel could be a few meters to 50 km or more. Tunnel length dictates equipment selection; as short length tunnels are mostly driven using conventional methods while for longer tunnels use of borers and modern technology proves to be advantageous.

Orientation of tunnel or the route through which it should pass is an important consideration and many a times characteristics of the ground through which it is to be driven, dictates it. Passing though the difficult ground conditions, sometimes, can jeopardize a tunnel not only during its construction phase but also later on during its regular use. Shortest route with minimum support work is an ideal situation.

9.11.5 Preparatory work required

Apart from the proper design details w.r.t. location, orientation, gradient (inclination), size, shape, support types and position of tunnel portals; there are many other facilities that need to be established. The prominent amongst them are the access roads; warehouses; stack yards; shunting yards; provision for power, potable water, telephone, maintenance facilities, first aid, waste disposal, offices, canteen, lamp room, rest shelter, magazine, hoist room, compressed air, drilling water, waste disposal arrangement etc. Most of these installations are temporary and can be removed after the completion of the tunneling project.

Tunneling appliances, equipment and services: Some of these equipment and appliances can be hired but if this task is contracted, then the contractor brings them. Special items needed for this purpose are: haulage equipment, tunnel surveying devices, tunnel ventilators with rigid and flexible ducting, face and main pumps with suction and delivery pipe ranges, compressed air and water pipelines, portable pneumatic lights, concrete mixers and delivery range, blasting cables, winches and few others. The services to be provided include power supply, water supply, transport, stores, repairs, refreshment, housing, social life etc.

9.12 PAST, PRESENT AND FUTURE OF TUNNELING TECHNOLOGY[22,23]

Hartwig, S. et al. (Ref. 23) tried to compare the tunnelling scenario that prevailed 25 years before, with the one that prevailed in 1998. The authors also forecasted a likely scenario after 25 years until 2023, in tunneling technology. The main operations considered for the purpose of comparison have been tabulated in table 9.10.

Table 9.10 Comparison of unit operations time for a round in a face of 70 m² cross section.[23]

Operation	1973	1998	2023
Drilling	3.75 hours Mechanized pneumatic drilling began. Concept of hydraulics came into limelight.	2.75 hours for longer round up to 5 m. Hydraulic drills proved faster and efficient, use of computer for monitoring and operation begun. But there have been some ergonomic-based problems.	1.25 hours for longer rounds of 7 m or more. Multi boom, hydraulic rock drills with faster drilling up to 11 m/min of 50–65 mm dia. blast-holes. controlled. Bit changing remote facilities would be incorporated. Equipped with charging mechanisms.
Charging	0.25 hours Mostly NG-based explosives with electric detonators in vogue.	1.25 hours ANFO widely used. Emulsions are also used. Non-electric detonators in use.	0.25 hours** Use of emulsion of varying strength suiting even the contour hole for smooth profile. Use of electronic detonators for ideal delay mechanism. Use of explosive with reduced toxic emissions.
Ventilation	0.50 hours	0.50 hours	0.25 hours** Reverse ventilation after blasting.
Scaling	0.50 hours (manual)	1.50 hours (manual as well as mechanized but time consuming).	0.50 hours. Seismic and water pressure type scaling equipment would be in operation.
Mucking	3.50 hours	3.50 hours, Electric powered LHDs (loaders) more environmentally friendly and faster.	1.75 hours. Faster mucking would be there but not perfectly matching the drilling and other operations.
Shotcrete supports	1.25 hours	1.00 hours, Application of better quality shotcrete mixtures efficient than in case I.	1.00 hours** Use of steel fibered reinforced shotcreting.
Bolting 15 bolts	1.00 hours	1.00 hours. Time consuming but anchored bolts are replaced by untensioned grouted bards.	1.00 hours Drilling rig would carry out automatic bolting with prefixed design.
Surveying	0.50 hours	0.00 hours. Automatic positioning system in operation.	0.00 hours. Automatic positioning system in operation.
Time lost	0.75 hours	1.50 hours to account for unplanned delays.	1.50 hours to account for unplanned delays.
Sum	12.00 hours	13.00 hours	7.50 hours
Pull	3.7 m	4.9 m	6.9 m or even more. Pl. refer also to sec. 9.14 .
Advance/hour	0.31 m	0.38 m	0.92 m
Performance Increase	Base case (100%)	122%	300%

Table 9.10 shows that within a span of 25 years (1973 to 1998), the increase in performance hardly exceeded 1% every year but the one forecasted for next 25 years (1998–2023) is 4%/year. ** – time required could be more than forecasted.

The analysis shows that there is hardly any improvement in most of the unit operations, except that there are longer rounds with faster drilling. But scaling time has increased compared to previously. In the past it used to be undertaken by the working crews standing on a muck pile. It is likely to be faster in the near future with the application of pressurized water jets as described in sec. 9.13.

9.13   OVER-BREAK AND SCALING – SOME INNOVATIONS[27]

Deviation from the planned profile of an opening after blasting often results in its under-break or over-break. There are a number of parameters that could be held responsible for this, and they include: improper drilling, improper charging and explosive that has been used, quality of rock mass, and encountering abnormal structures or features within the rock mass, such as fault zones, dykes, joints etc. Sometimes abnormal make of water (it is a common term used in mining which means water that comes out as a result of porosity and permeability from the strata/rock-mass) also adversely affects this aspect and ultimately results in uneven floor and sides of tunnels and openings. It requires more scaling than usual. There have been innovations in scaling techniques. Conventional scaling manually using crowbar and scaling-rods has been replaced by mechanized scaling. These include:

- Light mechanical scaling.
- Hydraulic jet scaling.
- Road-header scaling equipment.

Light mechanical scaling equipment consists of a very small hydraulic hammer attached to a crane arm and mounted on a platform. It was developed to avoid manual scaling during sub-level caving operations. It could be deployed for scaling tunnels to save time but when the blast is not perfect, comparatively more time would be required.

Hydraulic jet scaling – this is similar in operation to that of hydraulicking, which is an aqueous extraction method used for underground coal mining and surface mining operations (sec. 1.6). But for scaling purposes the water-pressure is kept to not exceeding 100 bars. This technique has also been reported to work satisfactorily in properly blasted rounds and even giving better performance than mechanical scaling, but during poor blasting this too also requires more scaling time.

Road-header scaling equipment – This method also has proved to be faster and better than the previous two techniques.

9.14   LONGER ROUNDS – SOME TRIALS[25]

During the 1980s at LKAB in Malmberet tests for longer rounds indicated that a longer round, up to depth of 7.4 m, could be undertaken. Based on these project results, LKAB has introduced the predrilled large diameter cut hole method in the Malmberger mine. Following are the conclusions drawn from the Malmberger mine project work.

Results: advance were measured before and after scaling, and compared with standard 64 mm dia. blasthole long drift rounds. Scaling was performed using a

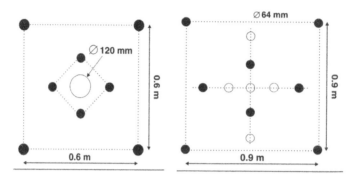

Figure 9.21   (Left):The standard Kiruna parallel hole cut. (Right):The cut with 64 mm hole diameter. The reference or the standard practice that was prevalent before introduction of longer rounds.[25]

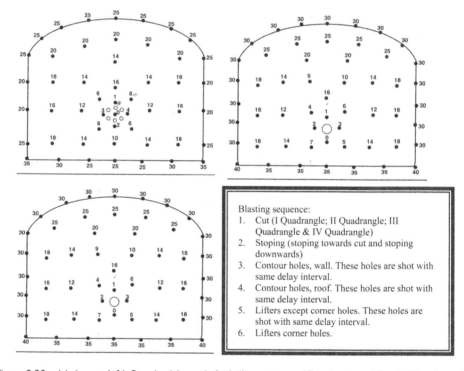

Blasting sequence:
1.   Cut (I Quadrangle; II Quadrangle; III Quadrangle & IV Quadrangle)
2.   Stoping (stoping towards cut and stoping downwards)
3.   Contour holes, wall. These holes are shot with same delay interval.
4.   Contour holes, roof. These holes are shot with same delay interval.
5.   Lifters except corner holes. These holes are shot with same delay interval.
6.   Lifters corner holes.

Figure 9.22   (a) (upper-left) Standard long drift drilling pattern 57 holes in a 6.5 m × 35 m face. (b) (right) Drilling plan with central hole of 250 or 300 mm dia. (c) Cork screw drilling/ignition pattern (hole placement based on their rock removal capability) and central hole of 250 or 300 mm dia.[25]

Mountabert BRP 30 hydraulic hammer and with water at 100 bars pressure. As shown in figure 9.23(a), rounds with larger dia. 300 mm (central cuthole) holes were better and even before scale the advance was 97%. Use of electronic detonators at the contour holes gave best results w.r.t. radial cracks beyond the excavation. Different

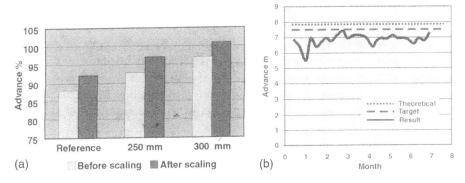

Figure 9.23   (a) Advance achieved with a large diameter central hole. (b) Advance for long rounds; depicting progressive increase in the advance/round.[25]

Table 9.11   Tested rounds for optimal blasting plan.[25]

Contour charging method	Rounds with 250 mm dia. large hole	Rounds with 300 mm dia. large hole
Cord 40 g/m or 80 g/m	6	4
String loaded emulsion	12	3
Cord 40 g/m or 80 g/m + EDS	3	
String loaded emulsion + EDS		4
Total	21	11

Table 9.12   Data on contour blasting explosives.[25]

Explosive	Density (kg/l)	VOD (m/s)	Gas volume STP (l/kg)	Energy (MJ/kg)	Linear charge conc. (kg/m)
Cord 40 g/m	1.05	6500	780	5.95	0.04
Cord 80 g/m	1.0	6500	780	5.95	0.08
Kimulux R	1.21	5500	906	2.94	3.86
String loaded Kimulux R	1.21	5500	906	2.94	0.55

explosives and initiation combinations were used as shown in tables 9.11 and 9.12. Contour blasting with a decoupled string of emulsion initiated instantaneously with electronic detonators resulted in no blast-initiated cracks. Contour holes fully charged with emulsion ended with radial cracks of at least 0.5 m; it also showed that when electronic detonators are used, the type of explosive at the contour holes has a minor influence on the results. At the end of the project in 1995, the advance achieved was 99.5% (fig. 9.23(b)) and need for scaling was cut by 50% when use was madeof a large diameter central hole. Based on these project results, LKAB has introduced the predrilled large diameter cut hole method in the Malmberger mine.

Conclusions from Malmberger mine project work are as below:

- Short round in the range of 4–4.5 m; use of 64 mm diameter cut holes work well.
- Longer rounds 7.8 m long were found to be economically viable.
- By precise delayed intervals at the contour, the quality of contour profile can be improved. Contour tests also showed that standard explosive to suit this diameter is not manufactured, and that is why results of using 48 mm dia. contour holes were better than those of 60 mm contour holes.
- Precise laser reference for alignment and accurate marking of holes are very important factors not only to keep over-break low but also to keep the drift-heading at the right direction and right level. Lifter with lock-out angle of 3° is essential for a clean floor.
- The amount of over-break depends upon many factors such as: the alignment of drilling rig, the drilling accuracy, the method of scaling and the geology. The over-break varied, but it was sometimes exceeding 15% but at the end it has been on average 12%.

To minimize damage to the walls – the techniques used are as below.

The intended holes should be precisely placed as per design and when perimeter holes are shot simultaneously. Experiments have shown that if adjacent holes are separated in time more than 1 ms, the result deteriorates. Such precise timing will require use of electronic detonators. Many techniques are used to reduce linear charge in the contour row and in the buffer row; for example:

- Decoupled plastic pipe charges.
- Detonating cord.
- String-loaded bulk emulsion.
- Low density/strength bulk explosives (e.g. ANFO or emulsion with polystyrene).
- Notched holes together with a very light charge.

It is not very unusual for the blaster to fail to consider the effect of the charges in the rows adjacent to the often well planned smooth blasted contour row. Charging the adjacent rows with a heavy charge results in cracks spreading further into the remaining rock than would result from the smooth blasted contour holes/rows.

9.15 THE WAY FORWARD

The mining industry today uses widely diversified types of tools, appliances and equipment. This poses problems to both the manufacturers and the users. As discussed in sec. 19.5.9, amongst various best practices and systems, standardization of equipment, operations, procedures and practices would mean following the laid out norms, guidelines and instructions. Standards are laid out taking into consideration the best performances and industrial benchmarks. They are based on experience, scientific studies, debates and consensus arrived at amongst those involved and concerned. For carrying out mine development activities and driving tunnels this initiative would prove to be a right approach in achieving excellence.

QUESTIONS

1. A development heading of 4 m × 2 m is to be driven using an electric drill to drill the shot holes and a coal cutting machine to provide the undercut of 1.65 m depth, at the bottom of the face. Find the length of shot hole you propose to drill. If the powder factor to be achieved is: (i) 0.15 kg/ton. (ii) 0.2 kg/ton. (iii) 0.3 kg/ton. (iv) 0.5 kg/ton, calculate the number of shot holes that should be drilled in each case. Assume that each shot hole is to be charged about 54% of its length, with permitted type of explosive using cartridge of 200 mm length and 150 gms (weight). Take bulk density of coal to be 1.3 gms/c.c.

2. A drive of 3 m × 3 m in an underground lead and zinc mine to be developed on a track-system of mining. Give any two layouts known to you for mucking and transportation from the development headings.

3. An adit/level drive/cross cut of 3 m × 3 m is to be driven in a hard and competent ground. Name the type of pattern you shall recommend in this situation. Draw the pattern of holes (using Swedish Relation to compute numbers of holes, else using any other method known to you) if the average diameter of the shot holes to be drilled is 35 mm. (a) If the dimension of any of these drives is changed in the following manner, draw the pattern of holes: (i) 4 m × 3 m; (ii) 5 m × 4 m; (iii) 2.4 m × 2.4 m. Use graph paper to draw these patterns of holes assuming a suitable scale. Also mark the blasting sequence using Roman letters in each case. (b) If the size of round to be drilled in each case is 2.4 m and the bulk density of rock is 3 ton/m^3 calculate the drill factor in each case. (c) If the shot holes driven in each case (except some of the cut holes) are to be charged to about two thirds of their lengths, with dynamite/gelatin explosive having cartridges of 200 mm (length) and 150 gm (weight), calculate the powder factor in each case.

4. Classify the pattern of holes giving application of each of these patterns. Prepare suitable sketches for each of these patterns. Suggest suitable drilling patterns for the following: (a) An underground coal mine. (b) Drivage work in sandstone, shale etc. i.e. where the strata are medium hard. (c) In an underground metal mine where the stratum is hard.

5. Classify the supports used for mines and tunnels.

6. Compare the available techniques (with aid of a line diagram) to drive openings for mining and civil engineering works.

7. Compare the technique "blasting off the solid" with the one in which the kerf is cut to provide the initial free face for the blasting operation to go ahead in the development heading of an underground coal mine. Illustrate the various ways to cut the kerfs (mention the positions). What types of machines are deployed to cut these kerfs?

8. Define the following terms: adit, shaft, winze, raise, drive, tunnel, incline, decline, ramp, crosscut, level, sump, ore pass, waste pass, mine portal, level drive, drift, stope, stoping, shot hole, blast hole, rock drill, kerf, pattern of holes, drill factor, powder factor, unit operations.

9. Draw a scheme to illustrate a mine ventilation system. What types of fans can be installed to ventilate mines? How can tunnels be ventilated? What type of harmful gases can you find in underground mines? How you can boost ventilation in tunnels?

10. Draw sketches to illustrate the prevalent schemes/systems to drive large sized tunnels and mine openings in soft, unstable and weak ground. List the governing factor to determine the location for the mine openings.

11. Give any two layouts known to you for mucking and transportation from the development headings.

12. Give schemes of ventilation of a detached face or blind heading in a mine. What type of ventilation systems can be used for such headings?

13. How do you assess support requirements while driving a tunnel through different ground conditions? Review the prevalent practices in this regard.

14. How is drivage work without the aid of explosives carried out?

15. How is the ventilation requirement in a mine calculated? What type of ventilation ducts are commonly used in mines?

16. Illustrate the cycle of operations that will be necessary while driving a tunnel through a strong ground.

17. Illustrate the cycle of operations while driving a tunnel with and without using explosives.

18. In an underground radioactive uranium mine a development heading of $4.5\,m \times 3.5\,m$ is to be ventilated by an auxiliary ventilation arrangement. Calculate the quantity of air required and its velocity in meters/sec. The mine is operated 3 shifts with labor strength of 45, 65 and 80 persons in I, II and III shifts respectively. The explosive consumed during the shift is $1550\,kg$ and the fumes should be cleared within 10 minutes after blasting. Maximum permissible concentration of CO is 0.008%. The volume of CO released by blasting 1 kg of explosive should not exceed $0.04\,m^3$ as per safety regulations. In the mine, 5 items of diesel-operated equipment, 3 shuttle cars of 60 B.H.P. each and 2 LHDs of 55 B.H.P. each, are to be deployed.

19. In an underground copper mine a ramp of $5\,m \times 3.5\,m$ is required to be driven at a gradient of 1 in 10 to connect two levels 60 m apart. Calculate the length of ramp that will be needed to drive. If a monthly progress of 100 m is to be achieved, propose a suitable scheme including the equipment that will be needed for the various operations to achieve this progress. You can assume your own data wherever needed but mention them.

20. It is proposed drive a sub-level of a sublevel stope of size $4\,m \times 3\,m$ in an underground chromite mine. An ore pass/waste pass is available for muck dumping. Suggest the type of equipment, matching each other, to carry out various unit operations. Consider the ground to be competent. Name the type of drill hole pattern you shall recommend in this situation. If the average diameter of shot holes to be drilled is 34 mm, calculate the number of holes and mark the pattern on a graph sheet using a suitable scale. Also mark the blasting sequence using Roman letters. What type of explosive will be suitable for this face?

21. Prepare a comparative statement to describe the application of supports of various types in civil and mining engineering tunnels and openings.

22. What are innovations that have taken place for scaling operations?

23. What is 'Tunnel-man's Ground Classification' and how it is useful?

24. What is the meaning of precursor to driving civil tunnels? What exactly is required to be done prior to starting a tunneling project? List the steps required to be taken.

25. What is the past, present and future of tunneling technology? Elaborate it. You could make a literature survey to answer this question.

26. Where you can deploy the dipper shovel and hydraulic excavators?
27. Why is mine development involving drivage and tunneling operations considered the toughest task?
28. Why is mine development necessary? Classify it.
29. Propose a suitable match of the equipment to be used to complete the cycle of operations for undertaking the drivage works of the following kinds:
    a. Driving sublevel drives of a sublevel stope of size 3 m × 2.8 m in an underground copper mine. An ore pass/waste pass is available for muck dumping.
    b. Driving a main level of 5 m × 3 m size for a distance of more than 1.5 km at a gradient of 1 in 200 in black rock. The rock produced is to be dumped outside the mine through an adit which connects this drive at a distance of 100 m from its starting point.
    c. Driving a level roadway in a underground coal mine of 4 m × 2.5 m for a length of 1 km without aid of explosives.
    d. A stone drift at a gradient of 1 in 7 to connect two coal seams, 25 m apart.
    e. Galleries/drives in underground coal mines of 3.6 m × 2.0 m.
    f. A drive of 3 m × 3 m in an underground lead and zinc mine to be developed on a track system of mining.
    g. A tunnel of 7 m × 4.5 m in a hilly terrain for a distance of 7 km in total.
30. A working cycle for a development face in an underground gold mine is as below:
    - Allocation 30 minutes.
    - Reaching to the working face – 30 minutes.
    - Face preparation – 1 hour (this includes water spraying, scaling, face checking, marking the face as well as the drilling pattern, hooking up the m/cs etc.)
    - Drilling the face, with 53 shot holes with the average rate of drilling of 0.3/minute/jack hammer drill. Two j/h drills are deployed at the face.
    - Charging the face – 1 hour.
    - Blasting and fume clearance – 1 hour.
    - Reproaching the face and mucking out of 30 ton/hr., consider size of pull obtained to be 2 m and bulk density of rock as 3 ton/m^3. The rock transportation time has been included.
    - Extension of services – 2 hr.
    - Draw this cycle of operation on a graph sheet to scale. Propose the required manpower for this task. Assume the data and conditions wherever needed but mention them.
31. Give schemes to transport from the development headings in case of a mine having:
    a. Track mining with single track.
    b. Track mining with double track.
    c. Trackless system of mining.

REFERENCES

1. *Atlas Copco manual and leaflets.*
2. Berta, G.: *Explosives – An engineering tool.* Italesplosivi, Milano, Italy, 1990.
3. Bieniawski, Z.T.: Ground control. In: *SME Mining Engineering Handbook*, Hartman (edt.). SMME, Colorado, 1992, pp. 897–911.

4. Bieniawski, Z.T.: *Rock mechanics design in mining and tunneling*. A.A. Balkema, 1984, pp. 97–132.

5. *British Steel*: Common yielding steel support configurations for tunnels and roadways.

6. Chironis, N.P.: Shooting coal pay off. *Coal Age*, Vol. 88, No. 9, 1983, pp. 86–91.

7. Deere, D.U et al.: Design of tunnel support systems. *Highway Research Record*, no. 339, 1970, pp. 26–33.

8. Douglas, T.H. and Arthur, L.J.: A guide to the use of rock reinforcement in underground excavations, *CIRA Report No. 101*, 1983, pp. 74.

9. Gregory, C.E.: *Explosives for North American Engineers*. Trans. Tech. Publ. Rockport, M.A., 1984, pp. 314.

10. Hoek, E. and Wood, D.: Rock support. *World tunneling*, No. 2, 1989, pp. 131–136.

11. IEL Ltd. Explosive division. *Leaflets and literature*.

12. Jimeno, C.L.: Jimeno, E.L. and Carcedo, F.J.A.: *Drilling and Blasting of Rock*, A.A. Balkema, Netherlands, 1997, pp. 217–225.

13. Langefors, U.: Fragmentation in rock blasting. Mine and Mineral Engg. 1966, Vol. 2, No. 9, pp. 339.

14. Matti, H.: *Rock excavation handbook*. Sandvik – Tamrock, 1999, pp. 214–233; and *Tamrock leaflets and literature*.

15. Olofsson, S.: *Applied Explosives Technology for Construction and Mining*. Applex, Sweden, 1997, pp. 131–155, 180.

16. Pokrovsky, N.M.: *Driving horizontal workings and tunnels*. Mir Publishers, Moscow, 1988, pp. 12–36, 66, 105–106, 268–273, 291.

17. Pradhan, G.K.: *Explosives and Blasting Techniques*. Mintech Publications, Buuvenswar, 1996, pp. 214, 368.

18. Roger, H.: Blasting. In: W.A. Hustrulid (edt.), *Underground Mining Methods Handbook*. SME-AIME, New York, 1982, pp.1581–86.

19. Vergne, J.N.: Hard Rock Miner's Handbook. McIntosh Redpath Engineering, 2000, pp. 236.

20. Whittaker, B.N. and Frith, R.C.: *Tunneling – Design, Stability and Construction*. IMM Publication; 1990, pp. 3–5, 270–80.

21. Wilber, L.D.: Rock tunnels. In: J.O. Bichel and T.R. Kuesel (edts.): *Tunnel Engineering Handbook*.Van Nostrand-Reinhold, New York, 1982, pp. 123–207.

22. GIA Industry AB, Sweden – leaflets and literature.

23. Hartwig, S. and Nord, G.: Atlas Copco, *Underground construction in Modern Infrastructures*, Frazein, Bergdahl & Nordmark (eds.) 1998, pp. 335–341. Balkema publishers.

24. Kaiser and Tannant, D.D.: The role of shotcrete in hard rock mines. In: Hustrulid and Bullock (eds.): *SME Underground Mining Methods*. Colorado, 2001, pp. 583.

25. Holmberg, R.; Hustrulid, H. and Cunningham, C.: Blast design for underground mining applications. In: Hustrulid and Bullock (eds.): *SME Underground Mining Methods*. Colorado, 2001, pp. 635–661.

26. Leonard, R.J.: Flowing and Raveling clays. In: *Proc. RETC*, AIME, 1987, pp. 242 (www.smenet.org).

27. Rockmore – leaflets and literature (soft and hard copies).

28. Vutukuri, V.S. and Lama, R.D.: *Environment Engineering in Mines*. Cambridge University Press, 1986. pp. 87–100, 244, 264.

29. Whittaker, B.N. and Frith, R.C.: *Tunneling – Design, Stability and Construction*. IMM Publication; 1990, pp. 1–17.

Chapter 10

Tunneling by roadheaders and impact hammers

Roadheaders are a viable alternative to full face TBMs, and to create openings of any configuration and size with minimum disturbance to the surroundings.

10.1 TUNNELING BY BOOM-MOUNTED ROADHEADERS[1,5,8,12]

In the 1960s use of roadheaders for tunneling began and by the end of the 1970s it had gained considerable acceptance worldwide. The roadheader, or continuous miner, is a heavy equipment, which uses a pick-laced cutter-head, much smaller in diameter than the tunnel itself. The cutterhead is mounted on the end of a boom that can swing up and down, left or right. The boom is most frequently tread mounted but can also be mounted within a shield.

When working in more massive formations, where all rock must be cut, the roadheader has efficiency in the range of 15–20 HP hr/ton.[1] In rocks with poor bonding between striations, such as shale, the roadheader is plunged into the face near the bottom of the heading and rips upward. The rock slabs off in large chunks. Under these conditions the mass of rock cut per unit of energy improves dramatically.

Basically a roadheader consists of a cutting unit, a gathering unit and a delivery unit (figs 10.1, 10.2, 10.3, 10.4). This equipment is very mobile and versatile compared with a full face TBM. It can cut a variety of cross sections; change diameters at will, change directions quickly, and move to and from a face under its own power. The machine usually incorporates gathering arms and a conveyor system to move the material cut from the face to a loading point at the rear of the machine. From this point, the muck may be handled by variety of methods available; and that includes shuttle trains, conveyors, or trucks. A roadheader cutting boom is usually mounted on crawler track but increasingly the booms are being mounted on other machines such as: hydraulic breakers, trucks, traveling gentries and inside the shields.

Modern roadheaders are equipped with electronic/hydraulic-controlled systems linked to microprocessor-based guidance and profile control systems.[4,5] (This is referred as a ZED Miner). Any deviation from the desired position and orientation can be detected by the laser system and required corrections are automatically applied. Boom hydraulics is controlled electronically to ensure perfect profile of the cut. Mechanical and hydraulic components are monitored electronically which ultimately results in proper preventive maintenance of the equipment. Application of

(a)   (b)

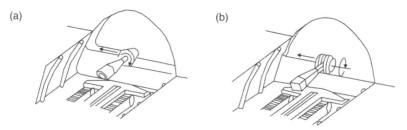

Comparison of two cutting systems:
(a) Milling (auger) type cutter head can mine narrow bands and lenses selectively
(b) Ripper type cutter head having 25–30% higher rate of production

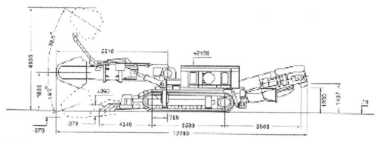

(c) Total weight 70 t. Total installed power 353 kW. Length – 12.28 m. Min. Height – 2.15 m.
Cutting thrust up/down 100 kN. Cutting power 115/230 kW. (Courtesy: Paurat, Germany)

Figure 10.1   Roadheaders – working principle.

CAD (Computer Aided Design) is made in designing cutter heads to obtain a proper design, which ultimately helps in reducing pick consumption and resultant vibrations.

Switch gears on roadheaders were incorporated while developing the Alpine Miner AM 105. Via switch gear, the advantages of variable cutting speed, previously achieved only through pole changing motors and thus only at reduced power available at lower speed, can be now utilized without drop of available power.[5]

All modern roadheaders utilize the gathering arm loading system and chain conveyor in the center of the machine. The other loading mechanisms that are in use are spinner loader and swinging loading beams. These roadheaders are mounted on the crawler track assembly, tramming at a speed exceeding 15 m/min.[5] Roadheaders employed in rock cutting use conical self-sharpening picks.

This equipment can handle small boulders by breaking them loose from their matrix and picking them up with the muck. Large boulders are difficult to handle in two respects (Asche & Cooper, 2002): (i) picks can break when they suddenly strike a large boulder, and (ii) the muck handling system is usually unable to deal with large boulders and can get jammed.

Figure 10.1(c) presents a light duty and small sized roadheader manufactures by Paurat, Germany;[8] the same company also manufactures the large and medium sized roadheaders.

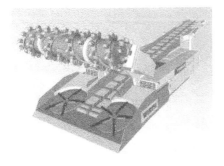
(a) Bar type

(b) Ripper type

(c) Milling type

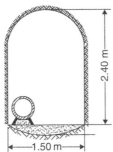

2.40 m

1.50 m

(d) Opening shape obtained
    by a partial face tunnel borer

Figure 10.2   Roadheaders – some models.

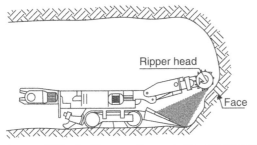

Ripper head

Face

Approx. 80% of cuttings are thrown to the gathering head

Figure 10.3   Ripper or transverse type roadheader.

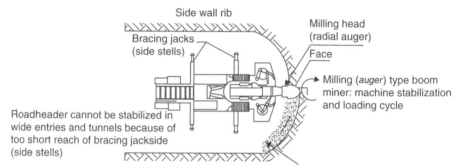

Side wall rib

Bracing jacks
(side stells)

Milling head
(radial auger)

Face

Milling (*auger*) type boom
miner: machine stabilization
and loading cycle

Roadheader cannot be stabilized in
wide entries and tunnels because of
too short reach of bracing jackside
(side stells)

Cuttings (muck) are thrown sideways on floor and
must be picked up in unproductive additional
loading cycle

Figure 10.4   Milling/longitudinal roadheader.

10.2   CLASSIFICATION OF BOOM-MOUNTED ROADHEADERS

Based on the cutting principle employed, the roadheaders, or continuous miners can be classified[3] as listed below:

- Ripper or transverse type: Bar and Disk
- Milling or longitudinal (Auger) type
- Borer type.

10.2.1   Ripper (transverse) type roadheaders – Cutter heads with rotation perpendicular to the boom axis

In the ripper type the full weight of the machine acts as the counter reaction for the cutter head. The rock is ripped off the face and thrown on the gathering head. The miner shown in figure 10.3 utilizes the weight of machine as the reaction force for cutting in a better way than the milling type of roadheaders, and no bracing jacks 'stells' are required.[3] These machines have 20–35% lower weight (which means lower price) than boom type roadheaders of equivalent capacity. In the USA 75% of the road headers are of ripper type.[3] Ripper type roadheaders can be further classified as: Bar & Disc types (fig. 10.2(a) & (b)).

10.2.1.1   Bar type

Its cutting element consists of a ripper bar, or cutting head that tears the coal/rock from the face. The cut rock/coal is carried by the moving chains and discharged into an intermediate conveyor. The rock/coal that falls on the floor is picked by gathering arms and loaded on the conveyor. The cutting head consists of five to seven cutting chains with picks which runs in guides all around the ripper bar (fig. 10.2(a)). The ripper bar is hinged at the rear end, which permits the front end to raise or lower. The chains are driven by electric motors. The whole unit i.e. the cutting chains, the motors and the conveyors can be telescoped forward to cut the rock/coal.

To give a cut, the machine is positioned in the center of the face with cutting head retracted. The machine's cutting head is swung to one side of the face and lowered down to the floor. With the chain in motion the ripper bar is hydraulically pushed forward in the rock/coal (usually 460 mm). After it has been fully advanced the ripper bar is gradually raised until it reaches up to the roof and this completes one strip. With a meter wide ripper bar, five such slices are required to complete a 5 m wide gallery (tunnel). After full face has been cut, the machine is moved forward.

10.2.1.2   Disc type

In this type of ripper unit, there are two cutting heads, each consisting of two vertical discs laced with tipped bits. The cutting heads are carried in the front end of an extension boom. In this machine the cutting heads are rotated and at the same time a to-and-fro movement of about 25 times/minute is imparted to the gear case and the cutting heads. The resulting motion, therefore, of each pick is spiral which results in low consumption of power and less degradation of rock/coal. In one of the designs, for example, the cutting head cuts a kerf of 2.5 m wide and 0.6 m deep. The discs rotate at the speed of 75 r.p.m. The cutting head can be lowered or raised hydraulically. Apart from the cutting head, the machine has got an apron with gathering arms, a conveyor extending up to the rear of the machine and crawler tracks. These types of machines can be operated for coal seams having a thickness range of 1–3 m.

There are different combinations of ripper type boom headers with different muck loading units, including backhoe excavator loading the muck into trucks; cutter boom dischar ging muck on to a chain conveyor unit for its onward transportation into trucks and the integral unit gathering arms of a ripper-heading machine removes the muck from the face as shown in figure 10.3.

Using these heading machines it is possible to drive or construct large sized chambers or even tunnels of large size can be driven in two lifts. The method is known as 'benching'. First an upper bench is advanced in the upper half portion of the face and then the lower follows it.

10.3   MILLING OR LONGITUDINAL (AUGER) ROADHEADERS

In milling type roadheaders a cylindrical or cone shaped cutter head rotates in line with the axis of the cutter boom (fig. 10.1(a); fig. 10.4).[3] The cutting force is exerted mainly sideways, which prevents utilization of full weight as counter force. When cutting harder rock the machine is braced against the sidewalls with hydraulic jacks ('stelling'). This consumes time and the bracing jacks, which protrude side ways, make the machine inflexible in narrow headings. For wider and high tunnel faces, particularly in hard rock, these types of headers are unsuitable because their bracing jacks (stells) cannot reach both sidewalls (ribs) and the roof to stabilize the roadheader (fig. 10.4).[3]

These headers rip the rock from face and throw it sideways on the floor. These headers have small diameters than ripper heads and are better suited for the selective mining of thin ore bands or lenses of high grade. At present the roadheaders with inter-changeable cutter heads either ripper or milling (augers) are available. In the UK[3,4] 65% of the roadheaders are of milling type. A two-boom milling type road-header is also available. In some designs it has a very distinct feature that its heads are exchangeable with ripper type heads.

The two systems (fig. 10.1) could be compared[5] as outlined below:

- Ripping (transverse) cutter heads cut in the direction of face, and therefore, they are more stable than milling heads of the same weight and power. They are less affected by changing rock conditions including hard rocks or bands, if encountered. This feature enables them to be used for a wider range of applications. These cutter heads always cause certain over-break regardless of machine position.
- Milling (longitudinal) heads has lower cutting speeds, which result into lower pick consumption. Pick array on these heads is easier because both cutting and slewing motions go in the same direction.

10.3.1 Borer type roadheaders

These types of roadheaders have been designed to cut the core of coal from the face by boring large diameter holes. A large amount of coal is obtained without being actually cut, therefore, the proportion of large sized coal is high. Numbers of designs of this type of headers are available with a variation in the design of cutting heads.

In the roadheader manufactured by one of the companies, the machine consists of two rotors and a cutting chain as the cutting element. Each rotor has a central core barrel and two or three cutting arms, laced with picks, make two, three or more concentric kerfs in the face and the central core breaks the core of coal barrel. The relative position of cutting arms can be altered or in some designs one or more of them can be eliminated to suit different seam thickness and to prove the most effective pattern of cut. The outermost arms can be hydraulically adjusted to vary the diameter of cut. An adjustable outer chain gives a desired shape of gallery.

Coal from the face drops to the bottom and is pushed on a scraper chain conveyor by rotating arms, which are provided with ploughs. The scraper chain conveyor runs centrally through the machine and carries the coal to the rear end. The rear end can be swung on either side by 40° or can be raised or lowered hydraulically. The entire cutting unit is supported on hydraulic cylinders, which can tilt the cutting unit by 4 degrees in the horizontal or vertical plane. This allows the borer to negotiate the irregularities of the seam. This machine can drive a gallery of 1.8 to 2.3 m high and 3.2 to 3.8 m wide while advancing at the rate of 0.3 to 0.75 m/minute. Likewise there are few other designs that are available and used for mining coal seams.

Table 10.1   Classification of roadheaders.[5] RH – Roadheader.

Roadheader	Weight range (t)	Cutter head power (kW)	RH with standard cutting range		RH with extended cutting range	
			Max. Section (m²)	Max. UCS (MPa)	Max. Section (m²)	Max. UCS (MPa)
Light	8–40	50–170	~25	60–80	~40	20–40
Medium	40–70	160–230	~30	80–100	~60	40–60
Heavy	70–110	250–300	~40	100–120	~70	50–70
Extra heavy	>100	350–400	~45	120–140	~80	80–110

10.4 CLASSIFICATION BASED ON WEIGHT

The roadheaders based on machine weight, cutter head power could be classified[5] as light to heavy duty units, which are capable of covering face size up to 45 m²; and they could cope up with rocks of compressive strength in the range of 20–140 MPa. The details are shown in table 10.1.[5] Figure 10.5(a)[5] also is a useful guide to select a roadheader based on uniaxial compressive strength, tunnel cross section, and required weight and power of the machine. Figure 10.5(b)[5] illustrates roadheaders with different features and their corresponding weight and power (kW). Tamrock-Sandvik also advocates that roadheaders' range of applications be extended into harder formations.

10.5 ADVANTAGES OF ROADHEADERS[3,4,5]

Versatility and Mobility: TBM generates circular openings whereas this unit can generate a variety of sections. Faces can be accessed, and by retracting a roadheader from the face, all required measures for rock protection (support) can be performed without significant shutdown; and this feature enables its use in changing rock conditions. Large tunnels and openings can be subdivided and excavation activities can be carried simultaneously in these sections/divisions.

Low Investment: Compared to the TBM, for the similar size of cross-section; the investment costs for roadheaders amount to approximately 0.15 (large sections) to 0.3 (small sections). Since they are available on a rent basis, they can be used even for small projects.

Quick and Easy Mobility: Comprehensive assembled equipment and chambers are not required. Delivery time is usually 3–6 months, and immediately after the delivery they can be brought into operation.

10.6 IMPORTANT DEVELOPMENTS

Alpine Miner Tunneller AMT 70 has been specially developed for NATM.[11] NATM uses the techniques and equipment that can produce high quality excavations in soft ground and adverse conditions. This unit has special features to excavate tunnels of the size 60 m² or more.

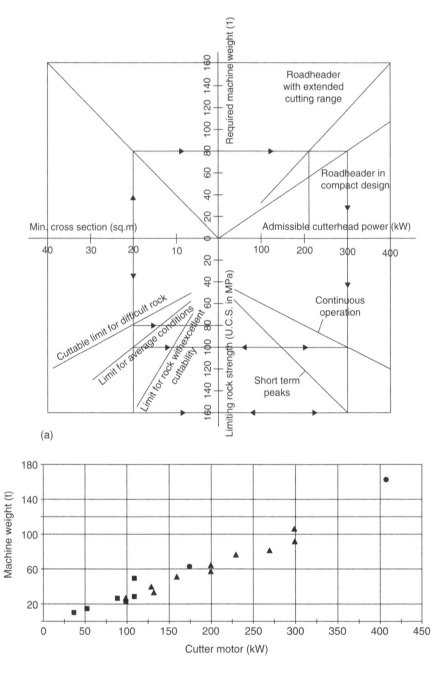

Figure 10.5   (a): Indicative diagram for roadheader selection. Inter-relationship of weight of machine, its power (kW), rock strength and operating environment. (b): Relation between weight and power (kW) of roadheaders. A useful guide for roadheader selection.

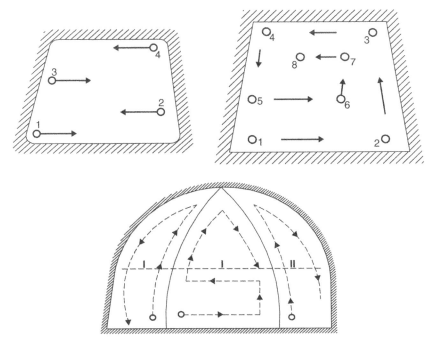

Figure 10.6   Working sequence of roadheaders. Left: Attacking from bottom to top. Right: Attacking perimeter. Bottom: Attacking middle of face first then sides. Figure 10.5

Use of a water jet operating at pressures up to 700 bar can make significant reductions in pick cutting forces.[13] This reduces the specific energy required for cutting and make of dust.

10.7   PROCEDURE OF DRIVING BY HEADING MACHINES

Least delays are the key for success of the operation. In order to achieve better utilization, 20–25% time during the day can be allocated for the maintenance. The heading machines also operate in a cycle, whose duration is a function of length of unsupported roof, or the roof that can be allowed unsupported. It varies from face to face. There are different ways to excavate, as shown in figure 10.6.[9]

- Excavating the perimeter first and then the central part, if the strata are friable.
- Ripping bottom first then the upper part to take advantage of undercutting. It gives better advance/output rates.
- If different bands or layers of the strata are encountered, then soft band can be attacked first and the other ones afterwards.
- If the roof rock is weak it is better to attack the central portion first, and then to rip the sides of the face/working.

10.8   AUXILIARY OPERATIONS

Generation of dust and heat while performing cutting operations by the roadheader are obvious. Water-assisted cutting suppresses the dust but results in humidity. Proper ventilation could control these problems.

10.8.1   Ground support

It is an established fact that roadheaders produce smooth, proper sized and shaped openings compared with conventional drilling and blasting techniques of tunneling. Blasting damages the surroundings but with this technique there could be a saving of 10–15% in support costs for the excavation requiring temporary support; or in suitable ground conditions, the need of support may be eliminated completely.[6] The amount of support required at the face, in fact, determines the roadheader's utilization,[3] as outlined below:

Support Type	% of cutting time available during the working cycle at the face
None:	60–80
Rock bolts, or Shotcrete:	40–50
Shotcrete and rock bolts; or steel sets:	30–35
Steel sets with full lagging:	20–25

When ground conditions require, a roadheader can be mounted inside shields (section 10.14.1, fig. 10.10), or it can be advanced within self-advancing powered supports. The roadheader can move independently within a shield; and the operations such as: excavation, mucking and erection of segmental lining can proceed concurrently.

10.9   HYDRAULIC IMPACT HAMMER TUNNELING

Use of hydraulic impact hammers for full face tunneling (fig. 10.7) began during the 1960s. In Italy a number of tunnels have been driven using this technique;[13] and it has proven to be economical in Asia and Mediterranean countries.[5] In this technique a 3000 kg (usual range 2000–3500 kg, or even over)[5] hydraulic hammer with impact energy of around 6000 Joules (the range is 2000 to 12,000 Joules (2740 to 8760 ft. lbs.)) is used for tunneling purposes (Giovene 1990). The impact hammer usually has a chisel tool of self-sharpening design.

Rock Conditions: This system is best suited for fissured, jointed and well layered or foliated rock mass. Massive rock conditions usually slow down the progress but strong rocks with pronounced joints and bedding planes and weak bonding favor this technique.

10.10   EXCAVATION PROCEDURE AND CYCLE OF OPERATIONS[5]

This technique is suitable for the tunnels exceeding cross section 30 m² but in smaller tunnels than this will pose operational problems due to the restricted space. In narrow

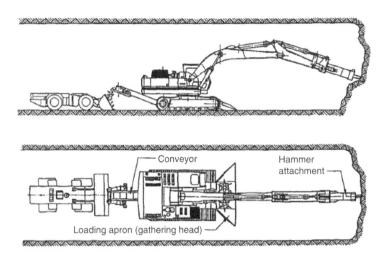

Figure 10.7   Hammer tunneler with loading and conveying system for excavation of narrow tunnels.

tunnels (width less than 8 m); only one set of 'Hammer and Excavator' can work at the face. The work is usually divided into following unit operations:

• Excavation by hammer
• Muck handling by the excavator
• Scaling and handling the scaled muck
• Supporting the tunnel using type of support as applicable.

If the tunnel is of more than 70 m² due to large sized face, hammering and muck handling operations can be carried out simultaneously. In this situation the work progresses in this manner:

• Excavation by hammer and scaling; Muck handling by the excavator
• Supporting the tunnel using types of support as applicable.

For the longer tunnels; if the tunnel can be driven in two opposite directions from its middle position; it gives added advantages of the best use of the resources – men, machines and services. For tunnels exceeding 7 m heights, the work can be divided into two benches.[5]

10.10.1   Hammer's working cycle

Excavation at the tunnel face is first made at the height of 1 to 1.5 m above the floor at the center of tunnel. This cut is made to a depth of 1 to 2 m. This small ditch, or slot is then extended towards the sides and floor of the tunnel. Once this big slot is created, then the hammer works to break ground in slices, till the final shape and size of the tunnel are dug. The cycle is then repeated to advance further. The procedure has been illustrated in figure 10.8(a).[5] If the rock mass is jointed, natural planes of weaknesses are used to achieve maximum gain, or yield of the hammer's impacts. This has been illustrated in figure 10.8(b).[5]

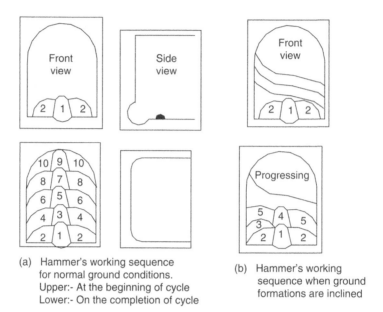

(a) Hammer's working sequence for normal ground conditions.
Upper:- At the beginning of cycle
Lower:- On the completion of cycle

(b) Hammer's working sequence when ground formations are inclined

Figure 10.8 Hydraulic hammer working sequence.

AEC Alpine Company developed loading apron and chain conveyor attachments, which fit onto standard excavators permitting excavator-mounted hammers to work in narrow tunnels.

10.11 MERIT AND LIMITATIONS[5,7,13]

- Comparing convention drilling and blasting methods, low over-break, less vibrations (5–10% of blasting) and smooth tunnel profile can be achieved. Low vibrations favor their use in urban areas. Large sized lumps can be handled effectively at the face itself, thereby no problem during the subsequent muck-handling route.
- In large sized tunnels simultaneous mucking and rock fragmentation are achieved that enables effective utilization of the resources.
- Compared TBM tunneling, investment costs are much less, and tunnel profile is not restricted to a particular shape. However, the rock type, tunnel's overall dimensions (size, shape, length), location and utilization of the equipment governs the economy. During tunneling by hammering availability is extremely high (60–80% of excavator time compared to 30–50% in primary breaking).[5]
- Smaller tunnels than 30 m^2 cross-section, and tough rocks without presence of natural planes of weakness, the performance of this technique may not be satisfactory.

10.12 PARTIAL FACE ROTARY ROCK TUNNELING MACHINES[2]

In partial face either a pilot hole is reamed to achieve the desired tunnel size (circular), or an oscillating head is employed for this purpose.[2] These tunneling machines work on the principle of undercutting. During up-stroke cutting is achieved, while during down stroke

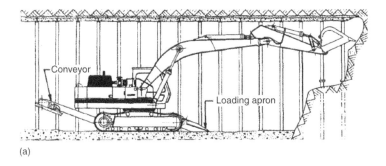

(a)

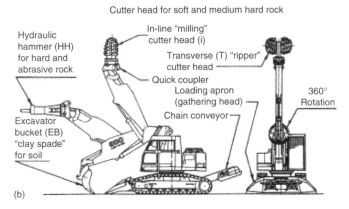

(b)

Figure 10.9   (a): Soft Ground excavation with standard bucket. (b): Excavator with attachments for multiple tasks.

the rotating cutter head draws the muck on the panzer type conveyor for its transfer to the rear. The resultant shape of drive/tunnel using such a unit is shown in figure 10.2(d).[2]

10.13   EXCAVATORS

Use of excavators for tunneling operations for ground excavation, particularly in soft formations, has been in vogue long since. These excavators are used in following two ways:

- Excavator buckets and excavator with multiple attachments.
- Excavator mounted within a shield.
- Excavator-Mounted cutter booms (partial face machines for New Austrian Tunneling Method (NATM)).

10.13.1   Excavators mounted within shield[1,6,7,13]

10.13.1.1   Excavator buckets

Buckets (fig. 10.9) of excavators are being used for excavation of soft soil such as clay, silt and sand and for mucking out of blasted rock. Buckets can be mounted on roadheader chassis, portals (gantries) and shields (described later).

10.14 EXCAVATOR WITH MULTIPLE TOOL MINER (MTM) ATTACHMENTS

During the last decade this has been considered as a dramatic revolution in the tunneling industry. The MTM is the Swiss Army Knife of construction equipment. It provides three machines for the price of one. The ALPINE Multi-Tool Miner (MTM) system includes the following attachments:

1. Cutter Head (milling as well as transverse (ripper)). For excavating of soft and medium hard and abrasive rocks.
2. Hydraulic Hammer. For breaking of large and hard boulders and for excavation of extremely hard seams and intrusions.
3. Bucket and Clay-Spade. For excavation and mucking out of soft soil.
4. Drill Mast and Roof Bolter. For drilling of exploratory holes ahead of the tunneling machine and for installation of rock bolts in exceptional conditions/circumstances.
5. Shotcrete Manipulator. The boom can be equipped with a shotcreting nozzle.
6. Grout and Cement Injector. In exceptional cases, grout and cement can be injected in the ground.
7. Breasting Plate. In an emergency the robotic boom can be used to press a breasting plate against the face thus closing the open tunneling machine.
8. Man Basket. The robotic boom can be equipped with a basket.
9. Crane. The robotic boom can be converted into a crane and tool carrier.

Introduction of quick-change couplers takes less than a minute (as claimed by the manufacturers) to change from a bucket to a hammer or a cutter head, and the excavator's operator does not have to leave the cab. Some prominent tunnels in Europe have used this equipment.

10.14.1 Excavator mounted within a shield

The excavator contains a powerful hoe, a pick, or combination tool within a shield (fig. 10.10). Pushing off the tunnel's lining thrusts the shield forwards. It operates like a single shield, alternately thrusting forward and setting lining rings. The hoe is either tripod or swivel mounted and also serves the function to move the material from the face to the conveyor pickup point. The excavators are generally limited to reasonably competent, but soft ground. When truly hydraulic ground conditions are encountered, however, the excavators become unsatisfactory. Despite the excavator's limited range of use, it is a popular choice due to its relatively low cost, the ability to handle large boulders, and visibility at the face.[1] It is a high advance rate machine for compact talus, alluvium, or ancient flood plains, the types of soil on which the majority of urban areas have been developed.

The shields with partial face excavation could achieve admirable tunnelling performance in a wide variety of geological conditions. It is possible to implement widely differing tunnel profiles with these machines. The shields with partial-face excavation can be controlled directly. Additional advantages include:

- simple handling
- variable deployment opportunities
- uncomplicated installation
- economical method of operation.

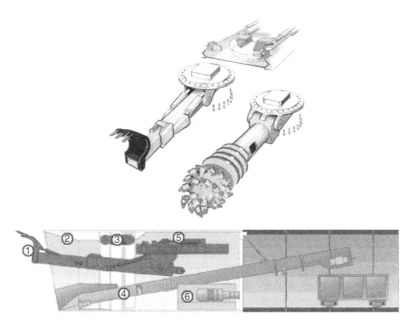

Figure 10.10 The shields with partial face excavation. 1. excavator. 2. shield. 3. steering cylinder. 4. conveyor belt. 5. machine pipe. 6. hydraulic power pack. (Courtesy: Herrenknecht, Germany).

10.14.2 Excavator-mounted cutter booms (Partial face machines for NATM)

Excavator-mounted cutter booms developed by Alpine Equipment Corporation (AEC Alpine), augment the flexibility and cost–effectiveness (\$ per m^3 of excavation) of NATM.[10] As NATM prevails over heavy and expensive ground support, these innovative machines are more flexible and cost–effective than heavy and expensive NATM roadheaders.

The cutter boom, which is mounted on the excavator (hoe or shovel) covers a wider range of excavation.[10] It can excavate deep trenches and pits and cut cross-sections of any shape and size. Tunnels, stations and caverns of great height can be driven in one single pass.

Cutter booms are available with hydraulic drives and with electric motors upto 450 kW (600 hp).[10] For most tunneling applications, electrical cutter head drives of 75 to 123 kW (100 to 165 hp) proved to be practical. For the excavation of stations, halls and wide tunnels, electrical cutter motors in the range of 123 to 160 kW (165 to 215 hp) are recommended. In addition to the cutter boom attachment, the excavator is also equipped with a loading apron and an under-slung conveyor.

When comparison made between cutter boom with roadheaders, the following features are found to be in favor of these excavator mounted cutter booms:[10]

- In soft rock up to approximately 62 MPa (9000 psi) unconfined compressive strength (UCS), an excavator-mounted cutter boom delivers the same production rate as a roadheader with identical cutter horsepower.

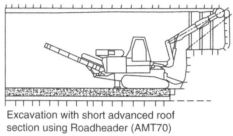

Excavation with short advanced roof
section using Roadheader (AMT70)

Figure 10.11   Application of roadheader AMT 70 with NATM.

- It has a far greater cutting range, and a significantly higher tramming speed. It can negotiate steeper gradients than a roadheader. This feature is of importance while constructing ramps and slopes. Excavators can have both diesel or electrical drives. If frequent long distance tramming is required, a diesel engine should be selected.
- A typical cutter boom attachment costs about one-fifth of the price of a roadheader with identical cutter power. Operating and maintenance costs of excavator-mounted cutter booms are normally considerably lower than with roadheaders because spare parts for standard excavators (hoes) cost less and contractors' mechanics are familiar with excavators. In figure 10.11 a NATM (sec. 11.13) roadheader has been shown. The cutter-loader performs the same function as a NATM roadheader for about one-third of its price.
- At most tunneling projects the contractor does not require to buy an additional excavator because these popular machines are already required for site preparation and construction of tunnel portals. Furthermore, excavators are widely available at reasonable lease rates.
- For excavation of wide tunnels and stations, the excavator-mounted cutter boom excavates the face on one side and the muck is stored on the floor. Then the machine moves to the other side and mines (worksout) the face.

10.15 THE WAY FORWARD

Road-headers are proving to be a viable alternative to full face TBMs, particularly to create openings of any configuration and size with minimum disturbance to the surroundings. An example of recent work is shown in figure 19.14(c): several Australian mining companies are working on 'Oscillating Disc Cutting' – a novel technology for cutting hard rock at high rates with low cutter forces. It is a dramatic shift away from drill-and-blast. Hopefully, it would boost the utility of the road-header's community.

QUESTIONS

1. "Road-headers are a viable alternative to full face TBMs to create openings of any configuration and size with minimum disturbance to the surroundings". Is this statement true? Justify your answer.

2. Classify the road headers, or continuous miners, based on the cutting principle employed.
3. Classify the road-headers based on machine weight and cutter head power.
4. Compare milling (longitudinal) heads with ripping (transverse) cutter heads.
5. Describe a milling or longitudinal (augur) road header.
6. Describe the working principle of road headers. List the units with which they are equipped.
7. How are the dust and heat generated while a road-header is performing cutting operations taken care of?
8. How old is the use of hydraulic impact hammers for full face tunneling? How has their performance been in Italy, Asia and Mediterranean countries? How does this equipment for loading and conveying systems work for the excavation of narrow tunnels? Describe its excavation procedure and cycle of operations.
9. Illustrate the procedure of driving by heading machines. What is the key consideration for the smooth running of this equipment? What is its working sequence (i.e. attacking front)?
10. In what type of formations have excavators been made use of for tunnelling operations? Give two ways in which these excavators are used.
11. List the components of a shield with a partial face excavation mechanism.
12. List the different types of road headers/continuous miners. Name their main units. Give specifications of some of the road headers known to you.
13. What is an excavator with multiple tool miner (MTM) attachments? Why is the MTM considered as the "Swiss army knife" of construction equipment? List the attachments of an ALPINE-MTM system.
14. What does an indicative diagram depict? How does it provide a useful guidance in the selection of a road-header?
15. What are shields with partial face excavation mechanisms? Where will you find them in use? What are their advantages?
16. What are the borer type road headers used for? Where do they find an application? Describe them briefly.
17. Why are excavator-mounted cutter booms considered to be better when compared with cutter booms with road-headers?

REFERENCES

1. Friant, James, E. and Ozdemir, Levent.: Tunnel boring technology – present and future. In: *Proc. RETC, AIME, 1993*, pp. 869–88.
2. Handwith, H.J. and Dahmen, N.J.: Tunnelling machines. In: *Underground Mining Methods Handbook*, W.A. Hustrulid (edt.) SME-AIME, New York, 1982, pp. 1107–10.
3. Kogelmann, W.J. and Schenck, G.K.: Recent North American advances in boom-type tunneling machines. In: *Tunneling'82*, IMM Publication, Michael, J.J. Jones (Edt.), 1982, pp. 205–210.
4. Kogelmann, W.J.: Roadheader application and selection criteria; *Alpine Equip. Corpo.* State College, PA, 1988.
5. Matti, H.: *Rock excavation Handbook. Sandvik* – Tamrock, 1999, pp. 254–272.
6. McFeat Smith, I. and Fowell, R.J.: The selection and application of roadheaders for rock tunneling. In: *Proc. RETC, AIME, Atlanta, 1979*, pp. 261–79.

7. Morris, A.H.: Practical results of cutting harder rocks with picks in United Kingdom coal mine tunnels. In: *Tunneling'85*, IMM, M.J. Jones (Edt.), 1985, pp. 173–77.

8. *Paurat – leaflets*.

9. Pokrovsky, N.M.: *Driving horizontal workings and tunnels*. Mir Publishers, Moscow, 1988, pp. 239.

10. Randy, F; George, H.K. and Ignacy Puszkiewicz, Two firsts in tunneling in Canada: *Proc. RETC, AIME, 1993*, pp. 273–93.

11. Sandtner, A. and Gehring, K.H.: Development of roadheading equipment for tunneling by NATM. In: *Tunneling'88*, IMM Publication, 1988, pp. 275–88.

12. Singh, J.: *Heavy Constructions – Planning, Equipment and Methods*. Oxford & IBH Pub. New Delhi, 1993, pp. 531.

13. Whittaker, B.N. and Frith, R.C.: *Tunneling – Design, Stability and Construction*. IMM Publication; 1990. pp. 141–45.

Chapter 11

Full-face tunnel borers (TBMs) & special methods

Tunneling is opening up a future. It is narrowing gaps/distances by passing through the most difficult ground and hazardous conditions. Present technology could meet this challenge.

11.1   INTRODUCTION

There has been tremendous growth and development with regard to methods, techniques and equipment to construct civil tunnels as well as underground mine openings. This is due to consistent efforts from the last 150 years[30] in this particular field. Today there is hardly any locale, site, or set of conditions where tunnels can be driven without aid of tunnel borers. Not only horizontal tunnels but also ground/rocks can be bored in any directions using these machines (fig. 1.6(b)). Hard-rock boring that used to be a challenging task has now become a routine job. Remote controlled tunneling projects are on their way and performing very well. This chapter deals with application of tunnel borers for driving small- and large-sized tunnels in soft as well as hard rocks.

A brief history describing the relationship between underground mining and civil tunneling has been dealt with in section 1.8. In this modern era rapid excavation has become almost mandatory to develop infrastructures. Tunnels play an important role for transportation, conveyance, storage, defense and underground mining operations. Underground tunnels are amongst the critical development activities for every nation, particularly in the urban areas, where surface land is less available, and at these locales use of conventional drilling and blasting methods to construct tunnels or any underground opening is practically prohibited.

11.1.1   Improved understanding[26]

Based on the experience from many recent TBM-projects, following are the two issues that need mention:[26]

1. Rock anisotropy, such as foliation and/or bedding, can have a significant effect on TBM performance. The Brazilian tensile strength test can provide a reliable indication of the degree and extent of rock anisotropy. This information together with punch penetration and/or fracture toughness tests can be used to develop an assessment of anisotropy influence on TBM performance.
2. The importance of emphasizing the most relevant factors in petrographic (thin section) analysis. For TBM performance evaluation, emphasis should not be placed on

the traditional mineralogical and petrographic study only, or on discussing the rock name or origin, as commonly found in most geotechnical reports, but rather on factors of key importance for evaluation of TBM performance and cutter wear, such as:

- Grain suturing/interlocking.
- Micro fractures.
- Orientation, directional properties.
- Grain size/shape/elongation.
- Content of particular hard minerals (such as quartz, garnet and epidote).
- Any other unusual microscopic features.

However, some factors perhaps not properly investigated or often underestimated still cause the basis for claims[26] in TBM projects, as follows:

- Ground water, which in cases of excessive inflow, can cause great problems and considerable extra cost.
- Rock stresses, which in the worst case may cause the TBM to get stuck, particularly in squeezing ground.
- Adverse ground conditions, such as running or swelling ground.
- Geologic features, such as joints, fractures, bedding/foliation, which can have a significant impact on TBM performance.
- Microscopic features of the rock such as grain suturing/interlocking, which can increase the difficulty of excavation. Based on experience, particular attention for future projects should be paid to these factors, although the others are also important, and definitely should not be omitted.

11.2   TUNNELING METHODS AND PROCEDURES[8,13,25,30,39]

Tunneling operations, which used to be an art in the past, have now become an engineering task. Transition through art to engineering has involved development of new methods, techniques and equipment. In the following paragraphs these aspects have been briefly dealt with. Figure 11.1 outlines the tunneling methods that are in use under different situations.

Tunneling with the use of explosives has been dealt with in chapter 9. This chapter will be confined to all other tunneling methods except those with the use of roadheaders and impact hammers dealt with in chapter 10.

Use of tunneling machines in civil construction works is worldwide. It finds its applications in mining too, for driving the mine roadways. Driving circular tunnels with the use of tunnel boring machines (TBMs) in the diameter range of 1.75 m to 11.0 m is common in civil engineering, whereas in mines, this range is from 1.75 m to 6 m. Use of tunnel borers in rocks eliminates conventional cyclic operations of drilling and blasting. Today borers are available for not only the horizontal drivage work but also for vertically up, down and inclined openings of different shapes and sizes, as shown in figure 1.6(b). The tunneling machines can be classified in the following three categories:

- Full-face tunneling machines – open and shielded
- Partial face rotary rock tunneling machines
- Boom type rotary excavator tunneling machines (dealt with in chapter 10).

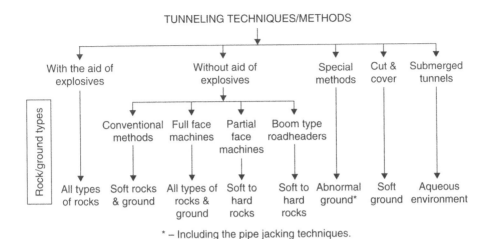

TUNNELING TECHNIQUES/METHODS

| With the aid of explosives | Without aid of explosives | Special methods | Cut & cover | Submerged tunnels |

Conventional methods   Full face machines   Partial face machines   Boom type roadheaders

Rock/ground types

All types of rocks   Soft rocks & ground   All types of rocks & ground   Soft to hard rocks   Soft to hard rocks   Abnormal ground*   Soft ground   Aqueous environment

* – Including the pipe jacking techniques.

Figure 11.1   A general classification of tunneling methods.

11.3   FULL-FACE TUNNELING MACHINES

Full-face tunneling machines are available for practically any situation: from soft to hard rocks; soft to consolidated ground; stable to unstable ground; ground saturated with water; geologically disturbed ground and a few other conditions. A classification covering all these aspects has been presented in figure 11.2. In this figure the following definitions are applicable to describe the terms used.

I – Non-mechanized and mechanized: if ground excavation is with the use of conventional tools and manual (as described in section 9.8, fig. 9.18) it is termed as non-mechanized and if excavation is by use of some cutting tools such disc cutter etc. it is known as a mechanized. A mechanized shield is equipped with an integrated unit that can excavate and load the ground beyond the shield and erect lining, apart from the protective casing and jacks for the movement of the shield which are common for both types of shields (mechanized and non-mechanized). In a non-mechanized shield partitions are there to carry out different operations.

II – Open and closed; if the forward end is open and without any cover it is an open shield but when it is covered or closed, it is known as a closed shield. Open shields are employed in competent rocks or ground and closed, in unstable rocks or ground. In open shields counteracting earth and groundwater pressure at the face is not required. These shields are distinguished based on the ground excavation mechanism, which could be either manual, partial face excavation or full-face excavation. These shields may have cross-sections other than circular. They could be rectangular or semi-circular also. Their older versions are with manual excavation. In countries with cheap labor these still find applications but where labor is costlier these version are not used. Manual excavation suffers with the disadvantage of slow progress.

Blind face (partly closed requiring support): this method calls for dividing the forward end of the tunnel (face) into a number of segments. The face is supported using wooden members, and is dug manually from top towards the bottom. Along with the

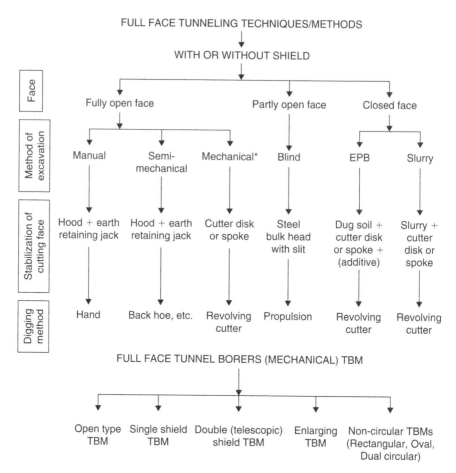

Figure 11.2   Classification of tunneling machines based on Method of excavation, Mechanism of stabilization the cutting face and digging method applied[20] with modification. * – A further classification of this type of tunneling machines is given in the line diagram below.

face digging, the support erection goes simultaneously. However, the method is slow and tedious. Sometimes the breast plates could be pushed hydraulically to the face.

Semi-mechanized and mechanized support: When full-face rotating disc cutters are used it is known as fully mechanized. The details are discussed in section 11.3.1.

Partial face extraction units: In these units ground excavators, which could be shovel, bucket teeth or hydraulic hammer, are installed within the shield, as described in chapter 10.

11.3.1   Full-face tunnel borers (mechanical) TBM – open and shielded[8,30,36]

In 1856 Herman Haupt in the U.S.A. at the Hoosac Tunnel (Western Massachusetts) made the earliest attempt at a true TBM for hard rocks. This machine had a full

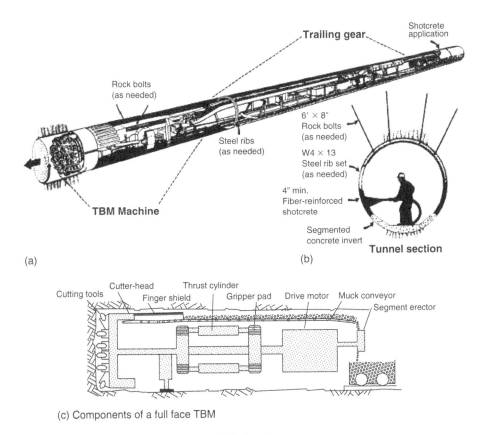

(a)

(b)

Trailing gear

Shotcrete
application

Rock bolts
(as needed)

6′ × 8″
Rock bolts
(as needed)

Steel ribs
(as needed)

W4 × 13
Steel rib set
(as needed)

4″ min.
Fiber-reinforced
shotcrete

TBM Machine

Segmented
concrete invert

Tunnel section

Cutting tools

Cutter-head

Finger shield

Thrust cylinder

Gripper pad

Drive motor

Muck conveyor

Segment erector

(c) Components of a full face TBM

Figure 11.3   TBM system.

diameter rotating cutter-head and picks arranged to cut concentric circles. It had grippers, thrust mechanisms and a conveyor for muck removal. He completed a 10-foot advance before he was fired for innovation. The concept languished for the next 100 years.[8] TBMs with drag picks for cutting tools, were applied in soft grounds and coal, but each attempt to use these machines in rock failed.

In 1956, James Robbins applied the idea of using a rolling disc cutter, rather than a drag pick, on a sewer project in Toronto, and a best day advance record of 115 feet was achieved. This project was the first to demonstrate the economic feasibility of the full-face TBM over a wide range of soft to moderately hard rocks.[8] From this project through trials and errors and collaboration with research institutes and universities, manufacturers of TBMs were improving machine performance allowing the users to venture into harder and harder rocks. And since then the Robbins Company alone has bored 3500 km of tunnel in more than 700 projects worldwide with machines ranging from 2.0 m diameter to 11.87 m diameter.[30]

These units consist of a rotating head fitted with the rock cutting tools (figs 11.3(a) and (b)).[14] This head is forced into the tunnel face. A single pass is sufficient to create a round or elliptical (oval) hole (i.e. full face). The cuttings are removed by the

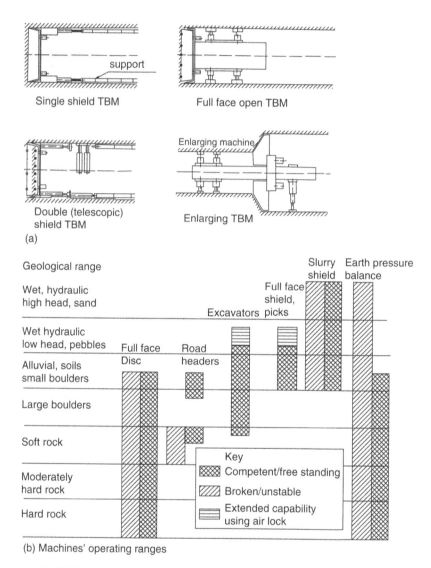

Figure 11.4 (a) TBM types. (b) Suitability of full-face TBM, roadheaders, shield machines and excavators under different ground conditions/geologic ranges.

cutter-head buckets or scoop that transfers them to a conveyor belt (fig. 11.3(c)). After completing a boring stroke, the tunneling machine is advanced by hydraulically pulling the gripping mechanism in from the tunnel walls, and then stroking forward and resetting the gripper to new forward position on walls (fig. 11.3(c)).[6] The unit in this way is set for the new position.

These machines are available as 'open' or 'shielded' (fig. 11.4(a)). The open machines allow the ground support to be as near as possible (fig. 11.6). These types of TBMs find their applications for driving through rocks of compressive strength up to $1300\,kg/cm^2$. But their use to drive through the rocks such as shale, limestone, dolomites etc. having compressive strength in the range $600–800\,kg/cm^2$, is common.[13]

After nearly 150 years of development, the TBM has come to the point where mechanical excavation can now be considered for virtually any tunneling conditions[12] (fig. 11.4(b))[8]. A machine type is available to construct a tunnel under circumstances where formerly only hand tunneling or drill and blast could be considered. In fact, in many environments today, only a TBM will meet the requirements and accomplish the job.

Figure 11.4(b)[8] shows the type of TBMs for various rock and soil conditions in which they are extensively used. The following systems are used to excavate large cross-sections (see also Figure 11.4(a)):

- Mechanical excavation of the full cross-section with open type machines.
- Mechanical excavation of the full cross-section with shield machines or telescope type shield machines.
- Mechanical excavation of a pilot tunnel with subsequent mechanical enlarging.

11.3.2   Mechanical excavation of the full cross-section with open type machines

Worldwide, the majority of bored tunnels lie in the diameter range of 3–5 m, with a peak of 4.0 m. Relatively few rock tunnels have been bored exceeding diameter 6.5 m and to date none with diameter large than 12 m.[37] Nearly 140 km of tunnels over 9 m diameter have been driven in the rocks to date.[12]

11.3.2.1   *Open main beam machines*

Robbins main beam TBMs are basically for hard formations.[30] The basic design consists of a cutter-head mounted on a robust support and main beam (fig. 11.8(a)). As the machine advances a floating gripper assembly on the main beam transmits thrust to the sidewalls. A conveyor installed inside the main beam transfers muck from the face to the rear of the machine. The boring cycle of this design is given in figure 11.7(a).[30]

Wirth's TBMs (fig. 11.6, 11.8(c))[39] consist of a basic unit with the outer and inner Kelly and trailer system for backup equipment. The inner Kelly takes the form of a solid square section construction. It carries the cutter head, the main bearing and the drive, and is mounted axially, and movable in the outer Kelly. The outer Kelly is clamped to the tunnel walls in two planes. In its clamped position, it serves as the guide for the inner Kelly and simultaneously acts as the thrust pedestal for the thrust cylinders, which press the cutter head against the rock face during boring. The cutter head is equipped with a scraper and bucket mechanism that collects the cuttings from the tunnel's invert and transmits it to the machine's belt conveyor, which is installed in the inner Kelly or top of the machine, depending on the machine's diameter and specific tunneling conditions.

Open main beam machines allow the most versatile rock support. Ring beams, lagging straps, rock bolts and mesh, invert segments; these all can be installed just behind the cutter-head. It is an excellent design to accommodate blocky or squeezing ground.

In 1990, using conventional open beam TPM a roadway of 10.80 m diameter was driven successfully at the German coal mine, Lohberg.

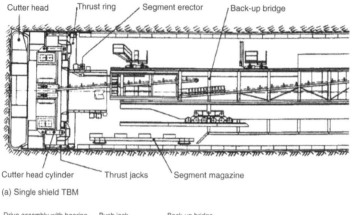

(a) Single shield TBM

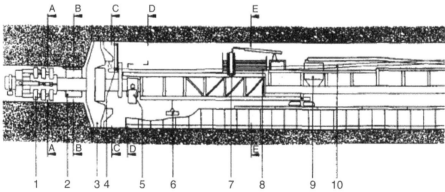

(b) Double shield (Telescopic) TBM

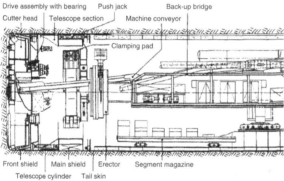

(c) Enlarging TBM

1 – Gripper system and thrust cylinder. 2 – Support system for pilot hole. 3 – Cutter head with drive assembly. 4 – Rock bolting equipment. 5 – Rear support. 6 – Invert segment. 7 – Erector for ring beam and wire mesh. 8 – Work platform. 9 – concrete silo. 10 – Muck conveyor. 11 – Concrete transfer.

Figure 11.5   Details of single, double and enlarging TBMs.

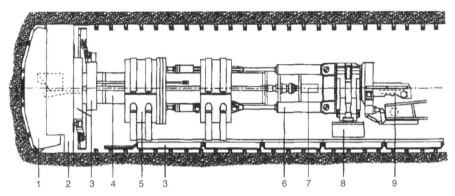

1 – Cutter head. 2 – Cutter head shield with support cylinder. 3 – Lining installation and transporting system. 4 – Inner Kelly. 5 – Cutter Kelly with gripper shields and adjustment cylinders. 6 – Advance cylinder. 7 – Cutter head drive. 8 – Rear support. 9 – Belt conveyor.

Figure 11.6   Wirth's TBM and its components.

11.3.2.2   *Single shield*

This type of machine (fig. 11.5(a))[12] features a complete circular shield, and is used when a tunnel is to be completely lined. Either ring beams and lagging or segments are placed directly behind or within the tail of the machine. It develops its thrust by pushing off the previously set lining. Cutter arrangement, cutter-head, and cutter-head support structure of this type of machine can be built very similarly to a main beam machine. Steering, however, is quite different. Boring cycle of Robbins's design is illustrated in figure 11.7(a).[30] The Bozberg Tunnel in Switzerland was bored with a diameter of 11.67 m using this type of machine[30] (Robbins, Herrenknecht Joint Venture).

11.3.2.3   *Double shield*[8,30]

This machine (fig. 11.5(b), 11.9(a))[12] is capable of working in two modes, as a single shield, or by using a set of grippers in a second tail shield, which telescopes into the head shield. Its advantage comes when operating in better ground conditions where lining placement in the stationary tail shield and boring ahead by pushing off grippers, can take place simultaneously. The boring cycle of Robbins's design is illustrated in figure 11.7(b).[8]

A major breakthrough occurred[30] in the early 1980s, on a double shield machine when a shielded cutter face, superior bucket designs and back loading or recessed cutters were introduced to provide a smooth cutter-head. Cutters and buckets were protected from damage when operating in broken rock.

In addition, discs with wide spacing were used with spectacular results in soft, (as little as 1000–2000 psi) non-welded tuff. Thus, the full-face disc machine can be used from the hardest rock, whether massive or fractured, to soft rock and even in self-supporting compact soils. Both the single and double shield types now share this technology.

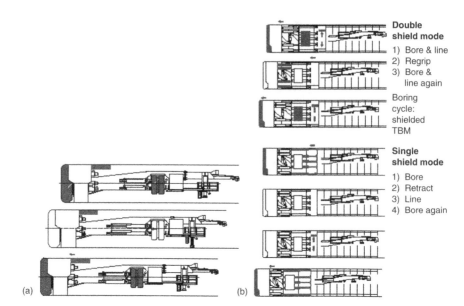

Figure 11.7 (a) Robin's Single shield TBM working Cycle. Top – Initial position; Middle – Re-grip; Bottom – Bore again. (b) Robin's Double shield TBMs boring cycle (Courtesy: Robbins).[24]

The Robbins's TBMs were used at the UK and France sides of the Channel Tunnel. The Gripper TBMs: Herrenknecht, Germany[16] manufactures single (fig. 11.8(b)) and double gripper machines (fig. 11.9(b)). The gripper principle is simple and suitable for boring in solid rock formations. The rocks could be stabilized at the nearest and earliest possible point from the front face. In these units substantial automation has been incorporated. The cutterhead is equipped with cutters (disks). The rotating cutterhead presses the disks against the tunnel face applying high pressure. The disks perform rolling movements on the tunnel face causing the loosening of the rock.

The TBM has a gripper system, which extends radially against the tunnel walls. Hydraulic cylinders push the cutterhead against the tunnel face so that another section of tunnel can be excavated. The maximum stroke depends on the piston length of the thrust cylinders. After completion of a stroke, excavation is interrupted and the machine is repositioned. An additional support system stabilizes the gripper TBM during the repositioning cycle. The single gripper machine (fig. 11.8(b)) braces itself at the back with two gripper plates against the rock. It has merit of making available a spacious working area for installation of rock support. The double gripper machine (fig. 11.9(b)) has a total of four hydraulically operated gripper plates. In comparison to the single gripper, however, it has less free space for the rock support.

11.3.2.4 *Enlarging TBM* [8,12,30]

Wirth developed the enlarging method for the excavation of inclined shafts and large tunnels. Characteristic for the enlarging method is the use of a TBM to drive a small

(a) 8 m diameter TBM
    (Robin's main beam)
    designed for hard granite.
    (Courtesy: Robbins).[24]

(b) Single Gripper TBM Gripp
    1. Cutterhead. 2. Gripper shield.
    3. Finger shield. 4. Ring erector.
    5. Anchor drill. 6. Work cage with
    safety roof. 7. Wire-mesh erector.
    8. Gripper plates. L1 – Area for
    rock support. In Switzerland for
    the longest railway tunnels in the
    world, the Gotthard (57 km) and
    the Loetschberg base tunnel
    (34 km) are under construction
    using six single gripper-TBM
    type, equipped with rock support
    mechanism. (Courtesy:
    Herrenknecht).[9]

(c) Wirth's open type hard
    rock TBM used at Quinling,
    China.

Figure 11.8   TBMs manufactured by various manufacturers – some models and types.

pilot tunnel in the center of the final cross-section and utilize this tunnel to guide the enlarging machine (fig. 11.5(c)).[12] The gripper assembly of a full-face machine grips the tunnel behind the cutter head, whereas the enlarging machine grips the pilot bore ahead of the cutter head (fig. 11.5(c)).

The enlarging or reaming technique was applied for the first time in 1970 by selecting a two-stage Wirth machine to excavate the Sonnenberg Tunnel in Lucerne. The previously mechanically excavated pilot tunnel with a diameter of 3.5 m was enlarged in two steps, first to 7.7 m and then to the final diameter of 10.46 m. The same machine was used again to build two more highway projects in Switzerland.

The initial observation might imply, that the full face method is superior to the enlarging method, since a full-face machine has to traverse the tunnel route only once. However, experience has shown, that this disadvantage can be compensated for by

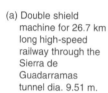

(a) Double shield
machine for 26.7 km
long high-speed
railway through the
Sierra de
Guadarramas
tunnel dia. 9.51 m.

(b) A gripper machine for small diameter
(3000 mm) was used at. **Zurich,
Switzerland.** This means that it doesn't
have a shield, but is braced in the tunnel
for drilling. Equipped for deployment in
mountains with gas hazards. The cutting
head consists of a center plate fitted with
external rollers. Nineteen chisel rollers,
each with a diameter of 14 inches, are the
drilling tool for the machine. Four electrical
motors, each with 160 kW, drive the cutting
head. For the drilling the cutting head is
braced horizontally in the tunnel by means
of four gripper plates. All of the operation
elements of this unit are located in a
closed cabin. (Courtesy: Herrenknecht)

(c) In Sörenberg (Switzerland) a hard
rock shield (Ø 4.52 m) excavated a 5.3 km
transit tunnel for a new gas pipeline from
the Netherlands to Italy. The best weekly
performance of 200 m and best daily
performance of 38 m in 18 hours were
recorded. Special features: In case of
methane emissions, the 115 m long hard
rock TBM was equipped with methane
detectors. During an alarm, the machine was
automatically stopped and only ventilation,
emergency lighting and emergency
telephone continued to be active.

Figure 11.9   Some typical TBMs used around the globe.

other factors. There are many projects in existence where the enlarging method was
preferred over the full-face method.[12] Tunnel reaming process makes possible:

- Detailed geological survey through pilot tunnel
- Implementation of safety provisions through the pilot tunnel
- Reliable drainage through the pilot tunnel
- Early access to critical tunnel sections and ventilation shafts

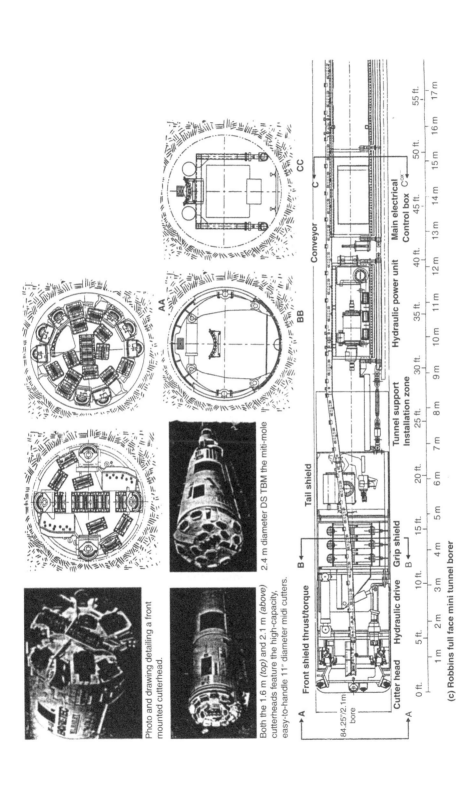

Photo and drawing detailing a front mounted cutterhead.

2.4 m diameter DS TBM the miti-mole

Both the 1.6 m (top) and 2.1 m (above) cutterheads feature the high-capacity, easy-to-handle 11″ diameter midi cutters.

AA

BB

CC

Cutter head

Front shield thrust/torque

Hydraulic drive

Grip shield

Tail shield

Tunnel support Installation zone

Hydraulic power unit

Main electrical Control box

Conveyor

84.25″/2.1 m bore

0 ft.   5 ft.   10 ft.   15 ft.   20 ft.   25 ft.   30 ft.   35 ft.   40 ft.   45 ft.   50 ft.   55 ft.

1 m   2 m   3 m   4 m   5 m   6 m   7 m   8 m   9 m   10 m   11 m   12 m   13 m   14 m   15 m   16 m   17 m

(c) Robbins full face mini tunnel borer

Figure 11.10   Robbins full-face mini TBMs. Cutter heads used are shown.

- Fast trouble-free completion of support work immediately behind the cutter head
- Easy variation of bore diameter during tunneling.

11.4 MINI TUNNEL BORERS[30]

Tunnels around 3.2 m diameters or less are commonly referred to as 'small diameter' tunnels. Such tunnels are used for sewer and water conveyance. Lee Tunneling method developed in recent past; use small diameter tunnel to provide an initial opening, which is then widened to large sized tunnels. Small sized tunnels give restrictive workspace and can pose problems[38] with regard to:

- Muck handling – mucking & transportation. Only single-track layout is possible.
- Ventilation – needs to be highly efficient due to restricted space.
- Cutter head access and operator's comfort
- Rock bolting, shotcreting and supporting in difficult ground conditions.

In figure 11.10 Robbins mini tunnel borers have been shown. Typical 1.6 m and 2.1 diameters borers have been also shown. In addition, there are a few models of Herrenknecht, Germany,[17] that are suitable for a mini tunnel in varying ground conditions. One of them has been illustrated in figure 11.9(b).

In figure 11.14 an EPB machines with muck pump in homogeneous soft geological conditions has been shown. This version is available for small to large sized tunnels.

In figure 11.13, a tunnel borer for non-homogeneous and alternating geological conditions (loose and medium-hard rocks) has been shown. This machine can be changed from Mixshield to slurry mode underground. This version is available for small to large sized tunnels.

11.5 BORING SYSTEM

The fundamental principle governing rock fragmentation efficiency of a hard rock TBM (or any excavator for that matter) is that system performance improves with the increasing size of cuttings produced. This means that the cutting tools should be arranged and used in the manner which produces the largest size cuttings. In a hard rock TBM this is accomplished by increasing individual cutter loads to attain deeper penetrations into the rock. Deeper penetration, in turn, allows wider cutter spacing. The combination of deep penetration and wide concentric cuts or kerfs produces the largest average chip size. With proper design, a hard rock TBM type can achieve an excavation efficiency of 3–6 HP–hr/ton (2.24 kw to 4.8 kw-hr/ton).[8]

The boring or excavation system is the most important part of a TBM and is responsible for its effectiveness. It consists mainly of the *cutter head with the cutting tools, the cutter head drive, and the thrust system*. The rock is removed from the face by means of disk cutters, which roll with an applied load in concentric kerfs over the face of the tunnel.

The *machine related, and geologic factors* as listed in table 11.1, influence TBM's advance rate.[12]

Table 11.1 Parameters influencing the advance rates of TBMs.[12]

Machine parameters	Geologic parameters**
Cutter head rpm	Rock strength
Cutter head thrust	Hardness
Cutter head torque	Abrasiveness
Disk geometry	Jointing
Disk wear	Bedding
Disk diameter	Schistosity
Cutter arrangement	Orientation relative to the tunnel axis

** Of the rock properties, the uniaxial compressive strength, tensile strength, and point load index correlate well to the penetration rate. These properties are the basis for various prediction models and should be available from the geologic investigations. In addition to the above mentioned properties the prediction model should take into account considerable influence of the schistosity and jointing of the rock.

The net advance rate is a linear function of cutter head rpm and the penetration of a disk per cutter head revolution. An analysis has been made[35] for the machine speed starting from 1980 to post 1990, and it shows that for the tunnels' diameter in the range of 3–6 m; the usual speed is 5–15 rpm; and those exceeding 6 m diameters it is in the range of 4–8 rpm. During 1980–1990 the cutter thrust have been 150–250 kN; whereas during post 1990 it was in the range of 175–250 and it has gone up to a maximum of 300 kN. For sedimentary rocks it was in the range of 140 to 200 kN, and for hard rocks in the range of 150–300 kN.

11.6 ROCK CUTTING TOOLS AND THEIR TYPES

The cutting tools are the essential ingredients in the process of cutting the rocks. The type of cutter that is to be used with a roadheader machine and TBM depends upon the type of rock for which this equipment is to be used. The very soft rock requires very high torque and low thrust. A soft to medium hard rock needs very high thrust and medium torque. For hard rocks high thrust and torque are required. Given in table 11.2 are the different types of cutting tools/bits or picks that are commonly used depending upon the type of rocks.

In figures 11.11(d) to (f)[13] different types of cutters have been shown. Figure 11.11(d) illustrates the drag pick – for very soft to soft rocks applying very high torque and low thrust and figure 11.11(e) disk cutters for medium hard to hard ground applying high thrust and medium torque. In figure 11.11(f) button cutters for hard rocks applying very high thrust and high torque have been shown. Also see section 19.5.7 and figure 19.14(c).

11.6.1 Cutting head configuration

In soft ground conditions drag cutters are usually used throughout the cutting head face but for other rocks various combinations of cutter types and their layouts are

Table 11.2   Cutting tools used in conjunction with various tunneling machines.[13]

Rock type	Comp. Strength, Mpa (psi)	Cutting tool	Type of attack
Very soft to soft	0–124 (0–18,000)	Drag, chisel, picks	Point i.e. applying force parallel to rock surface
Soft to hard	140–180 (20,160–25,920)	Disk cutter	Small surface area of contact, cutting force normal to rock surface (indenture tools)
Hard	>180 (25,920–36,000 or more)	Roller studded with buttons	Large surface area of contact, cutting force normal to rock surface (indenture tools)

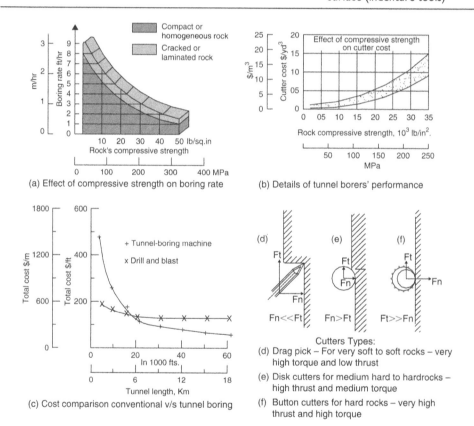

(a) Effect of compressive strength on boring rate

(b) Details of tunnel borers' performance

(c) Cost comparison conventional v/s tunnel boring

Cutters Types:

(d) Drag pick – For very soft to soft rocks – very high torque and low thrust

(e) Disk cutters for medium hard to hardrocks – high thrust and medium torque

(f) Button cutters for hard rocks – very high thrust and high torque

Figure 11.11   Effect of rock strength on TBM's performance. Cost considerations – cutter cost and length of tunnel. Application of different types of cutting tools/cutters based on rock types.

prevalent. In TBMs the cutting head configuration takes three distinct zones; namely Center, the Face, and the Outer gauge cutters.[36] In some designs, the center, the cutters are arranged in the form of a tricone in order to facilitate rock breakage. Disk or roller cutters, depending upon hardness of rock (and including drag cutters for the soft rocks) usually excavate the main face area. The gauge cutters are located at the outside edge of the cutting head to excavate the opening of the desired size.

11.7   TBM PERFORMANCE

Overall performance[27] by TBMs by considering advance on a m/h basis has been made. This analysis was based on 65 tunnels driven in USA, Europe, Australia and South Africa (table 11.3). This shows that best performance can be achieved in the tunnel diameter in the range of 4–6 m and not in very small or very large sized tunnels.

Type of rock is responsible for the rate of progress and cost of cutters to a great extent. Compressive strength is the useful criterion to determine the boring rate and cutters' cost. The advance rate based on rock strength of compact and laminated rocks, has been shown in figure 11.11(a).[14] Rock strength also influences the cutter cost, and hence, the ultimate driving costs by TBMs as shown in figure 11.11(b).[14]

Experience indicates that *Percentage availability and utilization of TBM units* has remarkable influence on overall performance. The percentage utilization[14] at present is in the range of 35%–50%, which of course is improving but this performance is not impressive. The reasons for the delays are mechanical breakdowns, interruptions due to haulage and auxiliary operations such as: ground control, ventilation etc.

11.7.1   Economical aspects

The length of tunnel, its cross-section, the type of strata, rate of water inflow at the face and amount of support work govern drivage costs. Generally tunnel diameter for mining purposes should be in the dia. range 2.1 to 6 m; in a rock having unconfined compressive strength below 40,000 psi (276,000 kpa) with a length not below 2 km to get a reasonable good results on economical grounds.[13] An increase in the heading/tunnel length lowers the erection, commissioning and dismantling costs/m (on overall basis). By obtaining the tunnel's profile, almost the same as planned, without any over-break and undue strain to the surrounding strata often results in reduction in costs. This saves in support costs and helps to preserve the stability of the workings.

Apart from the machine's availability, the rock types determine the rate of penetration and hence, the *Rate of advance and its unit cost*. Besides the cost of capital and energy, the bit or cutter cost is the major cost element while determining the overall drivage cost by these units.

The consumption of power and cutting tools is substantial and it is governed by the toughness of the rock and its other properties (content of abrasive components,

Table 11.3   Influence of TBM diameter on advance rates.[27]

Diameter (m)	TBM best advance rates (m/h)	Rock category	TBM best advance rates (m/h)
2	1.0	Hard rocks	2.5
4	1.9	Soft rocks	5.1
5.5	1.55		
8.5	1.0		
9	0.95		
11.5	0.7		

coarseness of grains and their cohesiveness). The life of bits ranges from 200–300 hrs based on the rock toughness.

11.8 SIZE OF UNIT AND ITS OVERALL LENGTH INCLUDING ITS TRAILING GEAR

The TBMs is a large sized equipment, its length could be 15 m or so, and including its trailing gear meant for ground support, ventilation and haulage purposes; the length of the complete unit may range from 100–135 m (fig. 11.3(a)). This warrants for longer tunnels or drives which are almost horizontally driven. In figure 11.11(c),[14] the cross-over point comparing the cost of driving by TBMs with the conventional (i.e. the one with the aid of explosives) is at 6.7 kms. In fact, tunnels or drives of this length in mining are very rare but in civil they are common.

It may be noted that for the mining jobs, if the tunnel is round in shape then it provides a disadvantage on account of effective utilization of its cross-sectional area. The bottom (floor) needs to be leveled by filling rock material and compacting it, enabling its use as a mine roadway for haulage, and travel ways.

Apart from the above, the merits and limitations of the rapid drivage work with the application of TBMs are summarized below.

11.8.1 Advantages

- The process is continuous and it includes all the unit operations such as rock fragmentation, its handling and disposal including the auxiliary operations that are necessary.
- If operated under the optimal conditions a faster rate of advance is possible (3–10 times the conventional, based on the prevalent conditions[13]). Advance rate, ranging from 150 ft (45m/day) to record day rates exceeding 400 ft (180 m/day), are not uncommon.[8]
- Better wall configuration and smooth walls, which in turn require reduced support work and lesser air resistance to the airflow during ventilation.
- Better safety, low manpower requirement and high productivity under suitable working environments.
- Fewer vibrations to disturb sensitive existing structures. Smooth walls provide superior fluid flow in water or sewer tunnels. Any lined tunnel requires less roof support and concrete or grout.

11.8.2 Disadvantages

- Inflexible and restricted mobility.
- Limited to soft to medium hard rocks, although already entered into the hard rocks regime but cutters' costs increases substantially.
- High capital and running costs.
- Encountering abnormal make of water, gas or strata jeopardize the progress badly.
- Longer tunnels may justify its economic viability but not the shorter ones, particularly those with lots of turnings and undulations.
- The compressive strength of rock is also a limiting factor, as it governs the energy required for breaking, consumption of cutting tools, mass of tunnel borers and thrust against the tunnel face and a few other parameters.

- Other factors include: bulkiness (length and weight), high power (energy) required, high consumption of cutting tools, poor space utilization factor, if bored round shape, requirement of the elaborate arrangements for dust suppression and air cooling.

11.9   BACKUP SYSTEM/ACTIVITIES[30]

Based on 3 shifts working/day using two new Robbins's TBMs that were used to drive two water tunnels of 5 m dia. at Lesotho Highlands Water Project; a boring cycle essentially composed of: Boring, Support, Change of cutters, Regripping, Probe drilling, Maintenance, Rail/ventilation advance and others (includes events such as: electrical outage, much train absence, shift change, dealing with water inflow etc.).

A cycle of TBM comprises of *rock breaking, transportation and loading of rock and erection of supports.* The support is erected simultaneously by keeping a lag between the face and maximum unsupported area; a fixed distance (at some places equal to length of the unit including its trailing units for muck handling), to prevent any collapse from roof or sides. Based on rock types and condition of the strata all types of supports such as rock bolting, steel arches and concrete are used for this purpose.

11.9.1   Muck disposal

As shown in figure 11.12(a) various options are available to handle the muck generated during the boring cycle. Any of the following alternatives can be adapted based on a merit of one over another for a particular situation.

- Muck removal by train:
  - Single track
  - Double track
- By truck
- Continuous conveyor system
- By muck pump.

11.9.2   Single track

When the tunnel is small, or the equipment assembly area is confined, a single track backup is generally the best solution. This system, which is meant for shorter small sized tunnels, is simple to operate and maintain.

11.9.3   Double track

In a situation where the tunnel is large enough and long, and if the company already has the rolling stock, this system proves to be suitable. With this system usually a California double track switch is incorporated.

11.9.4   Continuous conveyor system

This system has proved to provide the highest performance in terms of productivity together with the other advantages as listed below.

- Few tunnel workers required
- Lighter rolling stocks can be used

- Reduced ventilation requirements
- Less tunnel congestion
- Improved tunnel safety
- Adaptable to tunnel inclines and declines.

The TBM trailing conveyors were introduced during the early 1980s. This system allows high production TBM's to continuously advance with minimum delays. In a study made by Perini (1994, unpublished) to compare train haulage system with conveyor system revealed that TBM could advance at its maximum rate of 5.5 m/hr (18 ft/hr) up to tunnel length of 4.572 km (15,000 ft) Train haulage would require five locomotives with 14, 15.3 m³ (20 cu yard) muck cars per train with switches located at maximum distance of 1.5 km apart. As tunnel length exceeds 4.5 km, the TBM advance rate would decrease due to train haulage availability. The TBM advance rate would have been approximately 2.7 m/hr when the TBM approaches the end of the tunnel about 15 km; whereas the conveyor system would allow advancing at its maximum rate with minimum delays.

Comparing the two systems (unpublished literature), for a tunnel length up to about 7 km (20,000 ft); both systems are almost are at par, but beyond that the conveyor system performs well and improves productivity and reduces the cost. The conveyor would permit the TBM to complete the project in about half the time of train/rail haulage when reaching up to the entire length.

Once the conveyor system is started the speed of the trailing conveyor can be varied by the TBM operator manually or automatically based on TBM production level. This production level is monitored via the TBM primary conveyor, which is equipped with a belt scale.

11.9.5   Other back-ups include[30]

Ventilation

- Dust suppression (wet or dry)
- Fresh air flexible ventilation ducts cassettes
- Booster fans for fresh air.

Electrical

- High voltage cable reels
- Emergency generator
- Emergency lighting.

Water and Compressed air

- Backup mounted compressors (or hose reels)
- Compressed air distribution system
- Water systems (including re-circulation system).

Ground Support

- Shotcreting equipment, decks and robots
- Concrete/steel segment transport and storage
- Back filling system for concrete segment lining
- Drill systems for rock bolt drilling patterns.

Other Equipment

- Muck train moving devices (car movers)
- Closed circuit television and remote control devices including PCA/data logging system (fig. 11.12(b))
- Signal lighting systems for trains and other equipment
- Communication system for TBM and backups; TBM to portal.

Dust suppression: The dust generation[28] may range 200–500 mg/m^3 (Porkovsky 1980) of air depending on the cutting rate. It is prevented from propagating by equipping the tunneling machines, in most of the cases, with spray nozzles on the face cutter head. Various types of devices collect dust.

Center-line and grade line to the drive are given by a laser beam which can be obtained from a laser source that is fastened to the tunnel walls behind the machine, and shinning on a pair of targets mounted on the tunneling machine itself.

11.10   TBMS FOR SOFT GROUND/FORMATIONS

Use of shields: the first time use of a shield was made was during driving a tunnel under the Thames River in London (1824–69). A shield is a mobile metallic support, round in configuration (figs 11.8, 11.9, 11.13), preventing the face area of a tunnel from collapsing and affording protection for rock excavation and erection of permanent lining. Hydraulic jacks shove the shield forward. The shielded machines have been designed to allow placing the pre-cast concrete segments. The shielded machines protect the machine and man at the heading. These type of TBMs are useful for driving the permanent roadways and tunnels in the geological disturbed areas, soft and loose formations, aqueous strata with considerable water inflow, and particularly, at the sites where no ground settling (subsidence) can be tolerated. They find application to construct tunnels meant for underground railway, irrigation and water supply. At these locations geological conditions usually vary through out the tunnel. They can be classified in the following manner.

11.10.1   Full-face shield with picks

This TBM is very similar to a hard rock, single shield machine, except that the cutter-head is dressed with drag picks rather than disc cutters. Almost by definition, picks limit the machine to softer, non-abrasive soils. Usual features of these machines are the muck controlling gates, and a semi-sealed bulkhead. By controlling the gate's opening and cutter-head speed, inflow of material is metered.

This machine is not suitable for use in truly hydraulic soils unless specially equipped. With the addition of tail seals and replacement of the belt conveyor with a screw conveyor, some manufacturers have devised a way to convert this type of machine into an Earth Pressure Balance machine (EPB). Thus, it is capable of going back and forth between the operating modes.

A significant operating factor in these machines, as well as with slurry or EPB machines is to meter the removal of material in proportion to the forward advance of the machine. These machines are always used at relatively shallow depths and if

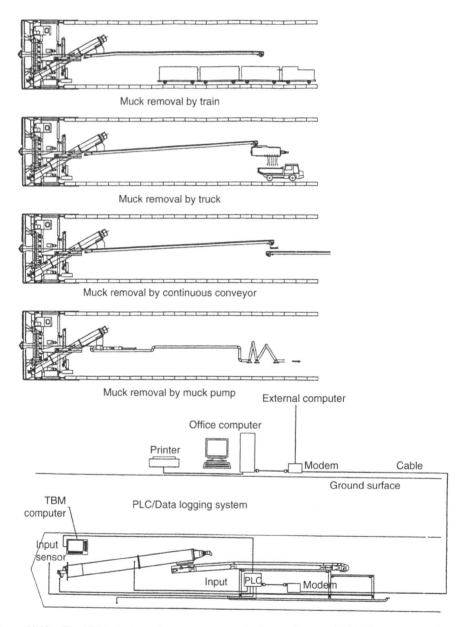

Muck removal by train

Muck removal by truck

Muck removal by continuous conveyor

Muck removal by muck pump

Figure 11.12   Top (a): Muck disposal arrangements and schemes. Bottom (b):TBM's monitoring through PLC/Data logging System (Courtesy: Lovat).

mining continues without forward motion, a funnel can form all the way to the surface. This can and has caused devastating accidents on the surface; a road collapses or a building foundation shifts.

Classification of shield tunneling machines is shown in figure 11.2.[20] This classification is based on the face support method and cutting method. The face support

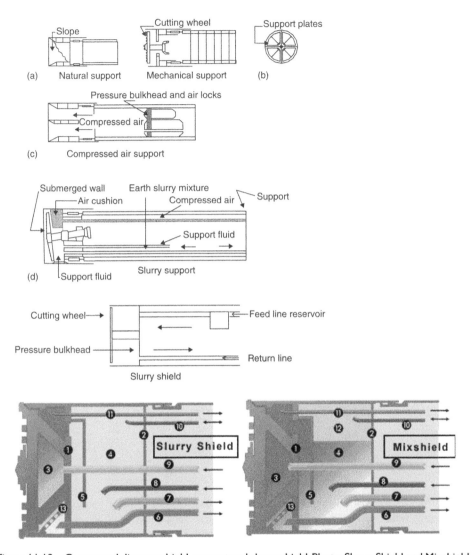

Figure 11.13   Conceptual diagram shield support and slurry shield. Photo: Slurry Shield and Mixshield.
1. Submerged wall. 2. Pressure wall. 3. Extraction chamber. 4. Pressure chamber.
5. Communicating pipes. 6. Slurry conduction line. 7. Pressure chamber conveyor pipe.
8. Pressure chamber supply pipe. 9. Extraction chamber supply pipe. 10. Compressed air
supply and outlet. 11. Extraction chamber ventilation. 12. Compressed air buffer. 13.
Suction screen (Courtesy: Herrenknecht).[16]

methods and cutting methods of most typical shields have been illustrated in figures
11.13[20] and 11.14.

11.10.2   Compressed air shields[8]

Holding water pressure under control pneumatically is the concept that has come to
limelight in early nineties. This works on the principle (fig. 11.13(c)) that water pressure

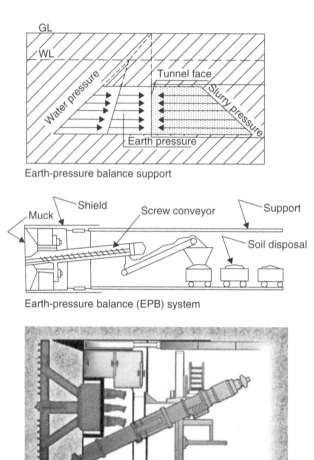

Figure 11.14   Conceptual diagrams for Earth Pressure Balance (EPB) Shields. Method to encounter water and earth pressures and mechanical components and arrangements in EPB concept shown. Photo: EPB Tunneling Machine. This machine is a shield machine with an earth pressure balanced working face. The soil extraction takes place via the cutting wheel. The tunnel is supported with lining segments (moulded segments of steel reinforced concrete) (Courtesy: Herrenknecht).[16]

at the working face increases linearly with its distance from the ground water level (GL) above (in the presence of confined water such as that of Artesian ground water). In order to avoid water ingress, the compressed air pressure should be higher or equal to the highest water pressure at the tunnel face. At the tunnel face at its lowest point (i.e. at the invert), the water pressure is the highest. Thus, if air pressure inside the tunnel is adjusted exactly to the water pressure at the invert, no water will enter into the void. But in practice the air pressure inside the tunnel remains the same at any point. This means the air pressure at the crown area of the tunnel is higher than water pressure and it would allow air to be released in this area. Where there is little cover (over burden) there is a danger that due to flow phenomenon, the soil particles loose their balance that could lead to a blow-out.[8] Especially in porous ground it should not be used to counteract the ground pressure.[8]

11.10.3   Slurry shield

The slurry shields were developed in the 1960s and the Earth Pressure Balanced shields (EPB) in 1970s. These are classified as closed face type shields.

It is defined as the shield equipped with an excavation mechanism to dig the ground, an agitating mechanism for the dug soil, a slurry feed mechanism to circulate slurry, a slurry processing mechanism to process the slurry transported after initial excavation and a slurry adjustment mechanism to feed the face with slurry of predetermined properties.

The slurry shield is designed for operating in true hydraulic soils or a mixed situation where the tunnel alignment goes in and out of these conditions. The design is also a full shield, developing thrust by pushing against the liners and with a variable rpm, full-face cutter-head. Picks are generally used, but sometimes in combination with disc cutters. The discs are placed slightly forward of the picks with the concept that should boulders be encountered, the discs will break them up and minimize damage to the picks. The machine differs from the full-face pick shield in that the slurry shield always operates with a pressurized bulkhead.

A slurry fluid, frequently bentonite, is pumped into the space (void) between the face and the bulkhead. The slurry is mixed with the in-situ material as it is scraped from the face. The combined or 'pregnant' slurry is then removed by slurry pump from an outlet hole, usually near the bottom of the bulkhead. The slurry is generally pumped to surface where a separation plant removes the solids. Cleansed slurry is then returned to the face.

The critical operating parameters are: (i) not to over pressurize, and thus not to cause any bubble at the surface, and (ii) to carefully meter output to avoid creating a funnel at the surface. This type machine is used at shallow depths and may be pressurized up to about 3 bars.[8]

This machine is extensively used in both Japan and Western Europe where it is essential to pressurize the excavation face to prevent collapse. In general use of closed types shield is increasing, for example, in Japan these account for up to 90% of the shield tunneling projects.

11.10.4   Earth pressure balance[16,17,20,23]

The EPB shields are intended to secure a stable face by means of applying a constant pressure to the earth to be dug. It is defined as a shield equipped with an excavation mechanism to dig the ground, a mixing mechanism to agitate the dug soil, a discharge mechanism to discharge the dug soil and a control mechanism to give a certain degree of binding strength to the dug soil. In some cases, this method uses soil additives to increase the plasticity of dug soil.

Like the slurry shield, the EPB machine (fig. 11.14) is designed to seal and pressurize the face cavity to control water or ground inflows. These machines have been designed to operate under pressures as high as 10 bar.[8] The operational objective is to allow the pressure in the cutter-head and face cavity to build up naturally by the pressure of the ground itself and the accompanying ground water. These machines operate best at ground moisture contents of 10–15% or less. Water, or a mud, is sometimes pumped into the bulkhead to maintain a desired moisture content. A gated outlet, however, controls face pressure. Most frequently, a screw conveyor is used by itself or in conjunction with a piston discharger. The EPB may be designed to operate on a wide range of rock and ground conditions, ranging from very hard (with discs) to soft ground (with picks). The use of slurry machines is declining in favor of the newer EPB concept because of its simpler control and versatility.

The machine is frequently built to allow operation in multiple modes.[17,23] It may operate either as a single shield or as a double shield, switching back and forth as the need arises. It can also switch back and forth between a closed (sealed and pressurized mode) and, an open (atmospheric pressure) depending upon the competency of the rock. These machines have achieved some outstanding results, as demonstrated in the English Channel Tunnel,[30] but are too expensive a design to use unless absolutely necessary.

In fact years ago the company Wayss & Freytag AG and German company Herrenknecht GmbH developed the concept of *Mixshield, TBM* together.[16,23] The machine type was conceived for use in very changeable geology. The original idea incorporated some of the technologies used in German slurry shield, EPB shield, and the Pressure shield. Experience has also shown that in using this type of shield, some features of hard rock TBM have to be integrated to the system.[16,23]

Some of the Herrenknecht's EPB and Mixshield TBMs used at various projects have been presented in figures 11.15[17, 23] and 11.16.[17] The grading curve (fig. 11.16(a)) could be used as a guide to select a tunneling method based on type of ground.

In figure 11.16(b)[16,17] a Seismic Surveying System; the Sonic Soft Ground Probing System has been shown. The system is integrated in the cutter head structure and provides continuous information about the ground in front of the shield that can be transmitted to site offices so that any erratic feature of the ground can be predicted.

11.10.4.1   *Segments*

The shield tunneling method uses two types of linings; primary and secondary. The primary lining is made of the blocks, called segments, which are assembled into a ring inside the shield of the tail section. The secondary lining is made in a ring shape with a cast in place concrete. The materials, sectional form and the type of joints used have been shown in table 11.4.

11.10.4.2   *Back filling*[20]

The back filling is mainly done to achieve the uniformity of the working external force (the earth pressure) into the void behind the lining segments. A proper back filling is essential to avoid any water leakage. The type of back filling system from shield concurrently with digging has been shown in figure 11.17.

(a) The Mixshield (Ø 9,785 m) used for the Sophiaspoor Tunnel in the Netherlands having capability of continuous tunneling. Installation of the segment does not interrupt the excavation process, as it would be the case with conventional procedure. (Courtesy: Herrenknecht)

(b) Earth Pressure Balance shield - The soft Asian ground is an ideal area for it. During excavations for the subway tunnel in Bangkok, the tunnel line ran through sand and clay. Two EPB-shield machines with a 6.46 m diameter were chosen for this project. Equipped with 66 cutters and 8 overcutters per turn direction as well as a center rake with 4 cutters, the two machines excavated a total of 8.25 km of tunnel. Various times, the best daily performance of 38.4 m and best weekly performance of 200.4 m were accomplished. (Courtesy: Herrenknecht)

(c) World's largest tunnel boring machine (diameter of 14.2 m) for the excavation of the 4th Elbe River Tunnel, Germany manufactured by Herrenknecht, in 1997. Special feature that was incorporated in the design was an active center cutter with a diameter of 3 m integrated in the middle of the cutterhead could drill independently ahead up to 0.6 m. This prevented material cloggings on the complete cutterhead. Other innovative safety measures helped to take the project to the breakthrough.

Figure 11.15   Application of EPB & Mixshield at some tunneling projects.

11.10.4.3   *Auxiliary construction measures*[20]

With shield tunneling use of auxiliary construction measures or special ground treatment methods are essential to strengthen the ground in and around the intended excavation. The techniques used and their application to the characteristics of the ground are summarized in table 11.5. This shows that chemical grouting is common with all methods and type of ground. This treatment is needed at the entry, departure and other sections where working conditions are very severe.

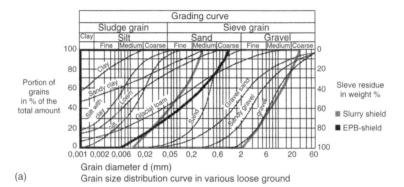

Portion of
grains
in % of the
total amount

(a)   Grain size distribution curve in various loose ground

(b) Millions of years of Earth history of the earth hold surprises in tunnel excavations. To improve safety, a seismic probing ahead procedure (SSP) "Sonic softground probing" developed. It probes ahead virtually and warns the TBM operator ahead of time, when significant geological changes are to be expected.

Specially adapted loudspeakers and microphones are installed in the cutterhead. The loudspeakers send sound waves into the ground every second. The microphones receive reflective signals, which are recorded simultaneously. A data giant located on the backup system evaluates the measured and statistic data and visualizes significant geological changes up to 40 m ahead of the cutterhead. SSP has already delivered precise predictions. (Courtesy: Herrenknecht AG)

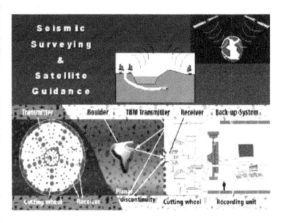

Figure 11.16   (a) Grading curves[16] to identify application of Slurry and EPB. Note the common zone where both shields could be applicable (Courtesy: Herrenknecht).[16] (b) Some renovations in TBMs.

11.10.5   Developments

Circular shields are common but in recent years different shaped shields have been developed.[20] These are: Rectangular (fig. 11.18(d)), Oval shaped (fig. 11.18(c)), and Dual Circular shaped (fig. 11.18(b)). The enlarging tunnel shield (figs 11.5(c) and 11.18(a)) has been developed to enlarge the tunnels.

The newly developed 'Extruded concrete lining' (ECL) construction method which does not use segments for lining is some where between the shield method and NATM method and is lately being given much attention as construction method possessing the advantages of both the shield and NATM methods.[20] Although the ECL method still has problems with high water pressure and sharply curved sections. It is more economical than the shield tunneling method that uses segments. Its technology

Table 11.4   Segments type with material of construction.[20]

Material	Sectional form	Joint
Steel	Box type (core type)	Straight bolt joint
Spherical Granite Cast Iron	Plate type	Curved bolt joint
Reinforced Concrete (RC)	Corrugated type	Pin mortice joint
Composite (Steel + Concrete);		Hinge (knuckle) joint
(Cast Iron + Concrete)		

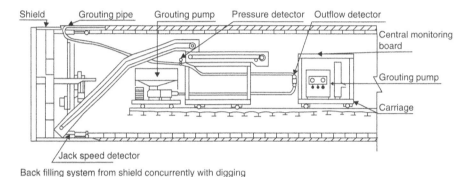

Back filling system from shield concurrently with digging

Figure 11.17   Back filling system from shield concurrently with digging.

Table 11.5   Types and characteristics of auxiliary construction measures.[20]

Construction measures	Characteristics
Compressed air	• Prevention of seepage of water at the face (sandy soil, cohesive soil) • Soil retaining at the face (cohesive soil) • Increase of ground strength by means of dewatering
Groundwater lowering	• Lowering water level (sandy soil)
Chemical grouting	• Prevention of seepage water at the face (sandy soil) • Increased adhesion of soil (sandy soil) • Reduced air permeability of ground (sandy soil) • Increased strength of ground (sandy soil, cohesive soil)
Freezing	• Prevention of seepage water at the face • Increased strength of ground

is also highly regarded as vis-à-vis the possibility of containing the adverse effects on environment, including ground subsidence.

The Lee Tunneling Method (LTM) developed in the recent past[20(b)] use a small diameter tunnel to provide the initial opening, which is then widened to a large sized tunnel using the conventional drilling and blasting techniques (fig. 11.18(e)).

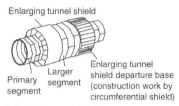

Enlarging tunnel shield

Primary segment | Larger segment | Enlarging tunnel shield departure base (construction work by circumferential shield)

(a) Enlarging tunnel shield

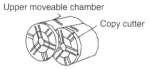

Upper moveable chamber

Copy cutter

(b) MF tunnel shielding method

(c) Oval shield tunneling method

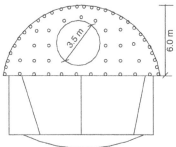
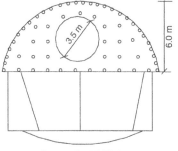

(d) Rectangular shield

3.5 m

6.0 m

(e) Lee tunneling method (LTM) – a combination of mini TBM with conventional drilling and blasting method of tunneling

(f) A tunnel portal

(g)
A modern tunnel is equipped with pipes. It has a concrete floor, emergency exits, drainage and fire extinguishing equipment. Afterwards, the soundproofed, vibration-free, 'fixed rails' for the long distance and regional trains which will travel through the tunnel at a speed of 120 km/h are installed and, finally, the electrical signal and safety equipment which is connected to the controlling computer Railway Station. The train engines and railcars obtain their electrical power from an electricity rail attached to the top of the tunnel.

Figure 11.18 Latest development in tunneling techniques/shields.

11.11   PHASES OF TUNNELING PROJECT

Usually a civil tunneling project has to pass through the following three segments:

- Construction of portal
- Tunneling through soft ground
- Tunneling through rocks.

11.11.1   Tunnel portal

Tunneling work starts with the construction of its portal (fig. 11.18(f)). The ground encountered at this site is usually weak and weathered and may not require drilling and blasting. Sometimes this portion is too weak so that complete ground including the cover i.e. its overburden is also excavated. Situations are similar to tunneling by cut and cover method (sec. 11.15). This means first the ground is excavated and then prefabricated concrete segments or steel tubing, whatever has been chosen as the lining material, are erected in place and covered by the overburden. For abnormal ground conditions any of the special methods (sec. 11.15) as per the prevalent site conditions can be adopted. Once the rock is touched and a few rounds are advanced then permanent lining is erected to avoid any damage due to blasting. But where tunnel borers, heading machines or driving in soft ground using conventional tools is to be applied, permanent lining follows the tunnel face. The lag between temporary and permanent lining depends upon the ground conditions. The cross-sectional area of the portal is usually larger than the tunnel crosses sectional area.

11.11.2   Phases of a TBM project

Driving tunnels/headings by the TBMs can be divided into three steps:

- Preparatory work
- Heading proper and
- Dismantling the equipment.

Preparatory work includes transportation of the TBM's components to the site, driving of an assembly chamber that is larger in dimensions than the actual size of the drive or tunnel. This assembly chamber is equipped with lifting facilities such as over-head crane etc. and also with some repair facilities. The dissembled parts are assembled here.

To begin with, initial few rounds of the tunnel (of the size and shape that almost corresponds to that of the intended configuration) are driven using the conventional methods. This face provides a footing for the TBM cycles to begin with. Sometimes it requires shifting this machine from one site/face to another, particularly the continuous miners or the heading machines. In such a situation most of its components are left intact and only peripheral hardware are removed. But in any case dismantling and shifting of this unit requires substantial time and workforce.

11.12   FUTURE TECHNOLOGY[8]

The future of Tunnel Boring Technology as described by James E. Friant and Levent Ozdemir[8] could be summarized in the following paragraphs.

The use of the mechanical mole for tunnel construction in all types of rock and ground conditions has not seen the end of its development. TBMs will see increasing use in even more challenging environments.

11.12.1 Hard rock TBMs

The basic disc cutter and full-face cutter heads will be with us for some time yet. In the laboratory, on a 2.0 m diameter machine, a penetration rate of over 36.5 m/hr was achieved using a cutter spacing of 236 mm. An amazing efficiency of 1.2 HP-hr/ton was achieved. This approaches the energy efficiency of explosives and is more than double the best of today's equipment.

Equally big improvements in overall advance rates will come from improving utilization. While a TBM is said to be 'continuous', the fact is they are seldom utilized more than 50% of shift time. The record (so far as the authors know) is 63%. Non-boring time is re-gripping, cutter changes, repairs, and roof support under adverse conditions. There are some human factors too, but we are probably stuck with these for a few generations yet.

Sequential gripper TBMs to permit continuous boring are coming to the market. The objective is to improve overall machine utilization by eliminating the downtime to reset grippers.

A problem, which is beginning to surface with the advent of high power, high thrust TBMs is the creation of very high stresses in the cutting disc. Plastic deformation and/or macro fracturing of the edge have both been observed. In fact, the cutter ring composition and metallurgy is now the limiting factor to transferring more power into the rock. As a result, a great deal of attention needs to be focused on seeking and developing new materials with the capability to sustain higher stress and acceptable wear resistance while reacting to the high cutter loads which the TBMs of tomorrow will generate.

11.12.2 Soft ground machines

The full face pick machine, slurry machine and EPB machines will improve their efficiency primarily through automation. Automation in operation, monitoring of pressures, amount of muck being removed, thrust control, and perhaps most significantly by automatic lining installation.

The authors' only controversial prediction for the future involves the above three machine types. The machines will merge into a common more versatile combination machine, a sort of an EPB which can use an injected slurry where needed but which could also completely re-pressurize when in competent ground. The pure slurry machine is a more tedious beast to control as it requires careful monitoring of slurry pressure, control of advance to match the amount of material removed and the necessary remote separation plants. The basic EPB unit or such a combination unit is likely to erode the pure slurry machine market.

In general – modern TBMs are being fitted with more electronic systems to provide early detection of impending component failures and wear on critical components. One of the more exciting new developments is the automatic steering of machines. Technology is now available to directly interface the laser guidance system with the machine steering circuit to enable fully automatic steering of the TBM. Electronic systems are also being installed on TBMs to provide automatic optimization of machine performance in response to changes in rock and ground conditions (fig. 11.16(b)). In particular with the use of variable frequency AC drive systems, the cutter-head RPM and the thrust pressure can be continuously varied to allow maximum penetration at all times during the advance cycle. Electronic systems are also aiding in the scheduling and control of various operations at the tunnel heading and the back-up/transport system.

Not to be overlooked or minimized in their importance are improvements to TBM backup systems. Recent improvements in TBM performance can be attributed to the use of continuously advancing conveyor systems in place of the traditional rail haulage for muck removal. The mechanical reliability of the conveyor systems have improved dramatically over the last several years, as well as the distance over which they can be used effectively. Several tunnels are currently being bored with conveyor haulage systems extending over distances of several miles. For shaft hoisting of TBM muck, conveyors can also readily interface with the vertical belt hoisting systems, resulting in the most continuous muck handling system yet devised for TBMs.

In table 11.6, comparison of important parameters while driving tunnels and mine opening with conventional (drilling and blasting method), roadheaders and tunneling machines have been shown.

Table 11.6   Comparison of different techniques of tunneling (Courtesy: Tamrock).

Parameters	Drilling & Blasting	TBM	Roadheaders
Configuration:			
• Size	Any	1.75–11 m dia. in civil, 1.75–8 m dia. in mining	Boom height governs it, but it can be any height
• Shape	Any	Any	Arch and rectangular
• Length	Shorter lengths up to 3 km	Lengths more than 3 km	Upto 3 km; longer can be tried
• Gradient	Not exceeding 18°	Not exceeding 6°	Not exceeding 6°
• Turning radius	Any	30–60°	30–60°
Rock strength:			
• Uniaxial compressive strength (UCS)	Any	Upto 220 MPa	Upto 70 MPa by light duty & 150 MPa by heavy duty; beyond that performance not guaranteed
• RQD	All ranges	Not good if it is between 25–45%	Good for all RQD
Geological conditions:			
• Running ground	Not suitable unless pre-grouted	Specially designed m/cs	Not suitable
• Squeezing ground	Some difficulty	Some difficulty	Some difficulty

(Continued)

Table 11.6 Continued.

Parameters	Drilling & Blasting	TBM	Roadheaders
• Boulder & glacial till	Drilling difficult	Difficult for boulders but okay for till	Boulders not that difficult; till okay
• Faults	Precautions required, ground need to be supported but excavation not difficult.	Faults difficult to handle; beyond 10 m wide faults can not be handled	Medium difficulty
Operational details:			
• Air blasts and slaps	Yes; by delay blasting can be reduced	None	None
• Dust generation	Very dusty after blasting	Very much	Some dust
• Noise level	High due to drilling & blasting	Not that much	Medium level
• Multi drift excavation	Possible	Not possible	Not usually used
• Partial face excavation	Possible	Not possible	Possible
• Working schedule	Cyclic	Continuous	Continuous
• Muck removal	Flexible using track or trackless equipment	Conveyor belt discharge into rails or trucks	Collecting arms and conveyor belt discharge into rails or trucks
• Versatility and	Maximum Mobility	TBM practically confined to circular cross-section	Face is accessible. Without significant shut down support work can be done
Performance & costs:			
• Progress rate	5–40 m/week	Faster (50–200 /week)	About 15–90 m/week
• Equipment utilization	35%; Higher in multiple faces	40%	60%
• Initial cost	Not high	Very high	Medium (0.15 to 0.3 times TBM)*
• Lead time	Very less	3–18 months to get a TBM	Not more than 3–6 months
• Renting option	Usually not	Usually not	Usually rented, if small project

11.13 NEW AUSTRIAN TUNNELING METHOD (NATM)

11.13.1 NATM design philosophy and typical features

New Austrian Tunneling Method (NATM) was developed soon after the Second World War but achieved worldwide recognition in 1964 in conjunction with the application of shotcreting in the Schwaikheim Tunnel, which was designed under the guidance of Mueller and Rabcewicz. In fact Mueller, Rabcewicz, Brunner and Pacher have contributed significantly in developing this method in Austria.[38] NATM has been applied in a variety of ground conditions ranging from hard to soft rocks, soft stable ground to weak, friable and unstable grounds. It has been successfully used in rural as well as urban areas particularly under some major cities. For mining applications (shafts and tunnels particularly in coal mines in FRG), as well as for civil engineering applications involving road, railway and water tunnels. The main advantage[32] of the NATM over conventional drill and blast techniques, TBMs and shields is its outstanding flexibility. The excavation sequence and suitability of NATM for soils and rocks have been summarized in figure 11.19(a).[31]

11.13.2 Ground categories and tunneling procedures[31]

NATM can be applied to soil and rocks having uniaxial compressive strength up to 40 MPa and tunnels' cross-section up to 60 m^2 or more. The driving methods include:

- Tunnel Heading by manual mining using conventional tools and appliances
- Tunnel Heading with the aid of explosives (drilling and blasting)
- Tunnel Heading using Excavator, Boom Mounted Excavator, or Roadheader.

11.13.2.1 Excavation sequence

The excavation sequence used minimizes ground disturbance in order to preserve the inherent strength of the ground. Excavated area is kept small and timely installation of initial support is required. Typically, the full range of excavation sequence consists of the following items:

- Crown excavation
- Bench excavation
- Excavation of invert.

Figure 11.19(b) presents a typical example for mechanization of NATM with application of Alpine miner. The procedure to divide the working face of the tunnel into different sections has been shown.

11.13.3 Semi-mechanized methods

Pre Vault method: This method was developed in 1970's. It has wide applications in the recent years for rocks as well as soft ground tunneling projects. It allows advance or pre-support of the section to be excavated by:

- Sawing a small slit (or slot) along the outer line of the section to be excavated; 15–20 cm thick and 3–4 m long (fig. 11.20))[4,6]
- Shotcreting of slit
- Excavation under the protection of thin vault.

This method, initially applied usually to the upper half section of tunnels, was progressively used in full section, even for very large tunnels up to 150 m^2.

The following merits of precut method while using hard rocks have been advocated:[34]

- Use of explosive creates many adverse impacts such as: ground vibrations, noise and over excavations. These impacts are particularly unsuitable in the urban areas.

Excavation sequence	Soil cross section						Rock (UCS ~ 40 MPa)					
	<40 m²		40–60 m²		>60 m²		<40 m²		40–60 m²		>60 m²	
Excavating roof section (including mucking)	9%	•	9.0%	•	8.0%	•	10.5%	•	11.0%	•	12.0%	•
Placing of roof arch	4.5%	■	4.5%	■	5.5%	■	4.5%	■	5.0%	■	5.5%	■
Placing of wire mesh in roof section	6.5%	○	6.0%	○	6.5%	○	5.5%	○	5.5%	○	5.5%	○
Shotcreting in roof section	13.0%	■	12.0%	■	12%	■	10.5%	■	11%	■	12.0%	■
Excavation of bench (including mucking)	9.0%	•	9.0%	•	9.5%	•	10.0%	•	11%	•	11.0%	•
Placing of steel arch (lateral)	6.5%	○	5%	○	4.0%	○	4.0%	○	3.0%	○	2.5%	○
Placing of wiremesh (lateral)	4.5%	○	3.5%	○	3.0%	○	4.0%	○	3.5%	○	3.0%	○
Shotcreting of side wall	13.0%	•	12.5%	•	12%	•	11.0%	•	10%	•	9.0%	•
Excavating of floor arch (including mucking)	4.5%	■	5.5%	■	6%	■	7.5%	■	8.0%	■	8.5%	■
Placing of floor segment (steel arch)	6.5%	○	7.0%	○	7%	○	6.0%	○	6.5%	○	6.5%	○
Concreting of floor arch	6.5%	○	7.0%	○	7%	○	6.5%	○	6.5%	○	7.0%	○
Refill of floor arch	4.0%	○	4.5%	○	4.5%	○	4.0%	○	4.0%	○	3.5%	○
Driving of steel piles	12.5%	○	14.5%	○	15%	○						
Placing of rock bolts							16.0%	■	15.0%	■	14.0%	■
% of operation time with AMT 70 – system	66.4%		67.0%		56%		70%		71%		58.0%	
INDEX▶	First column: approx. % time of overall time of one cycle of advance. Second column: • – Basic function. ■ – Already existing, or planned additional function. ○ – Not planned for integration.											

Figure 11.19(a)   Application of NATM to cover soil and rocks up to UCS of 40 MPa; and cross section up to 60 m². Percentage of the operation time covered by NATM using AMT70 Roadheader has been also shown.

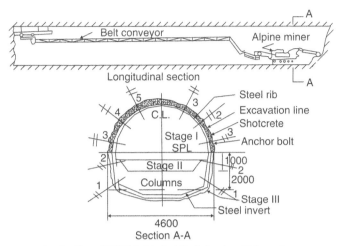

Application of NATM at Lokat (India) using Alpine miner

Figure 11.19(b)   Application of NATM using Alpine miner.

Table 11.7   Comparison of major criteria for Shotcrete tunneling methods and TBMs.[11]

Phase	Assessment criteria	Shotcrete tunneling methods	TBM
Construction Phase	• Supporting agent in face zone	Variable	Safer
	• Lining thickness	Variable	Constant
	• Safety of crew	Lower	Higher
	• Working & health protection	Lower	Higher
	• Degree of mechanization	Limited	Higher
	• Degree of standardization	Conditional	High
	• Danger of break	Higher	Lower
	• Construction time – short tunnels	Shorter	Longer
	• Construction time – long tunnels	Longer	Shorter
	• Construction cost – short tunnels	Lower	Higher
	• Construction cost – long tunnels	Higher	Lower
Operational Phase	• Tunnel cross-section	Variable	Constant
	• Cross-section form & its effective utilization	As desired; Generally higher	Usually circular Generally low

- This method is also superior to the well-known controlled blasting techniques such as pre-splitting that requires a higher amount of drilling and thereby resulting in higher costs. The ground vibrations and risk of cracking rock mass are not completely eliminated by them.
- In pre-cut techniques the rock to be blasted is quickly and cleanly separated from the rock mass with no significant vibrations, cracks, nor irregular profile and the surrounding rock remain intact and stable.

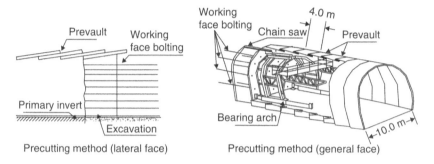

Figure 11.20 Special tunneling method – Precutting.

- The void of the cut acts as a barrier and helps to stop transmission of vibrations to the surroundings and ultimately to the surface.
- Due to presence of precut, during tunneling the explosive consumption and its related adverse impacts are minimized.

11.14 TUNNELING THROUGH ABNORMAL OR DIFFICULT GROUND USING SPECIAL METHODS

11.14.1 Ground treatment

Difficult and troublesome ground includes the ones which cause difficulties during or after construction of tunnel. The type of ground that can be encountered includes:

- Weathered ground/rocks
- Soft ground conditions
- Presence of intense jointing
- Encountering fault, fold or any other geological structure
- Heavy inflow of water
- Variable hardness – mixed face conditions
- Hard and abrasive rocks.

Soft ground conditions[38] include ground which could be firm, raveling, squeezing, running, flowing, or swelling. It could be either of these or their combination and it needs treatment.

Ground can be treated before, during, or after driving tunnels. Different alternatives to treat the ground are available and the selection of any one of them depends upon the magnitude of problem, site conditions, judgment and experience of tunnel engineers. The following two ground treatment techniques are available:

- Reinforcement
- Treatment that tackles the problems arising from the presence of water.

11.14.1.1 Reinforcement

Bolting, anchoring and surface coating can reinforce rock. Rock bolting is the established practice to reinforce the rocks. Lining the ground by spraying concrete – shotcreting, guniting or use of prefabricated concrete blocks in the form of supports of different kinds is the usual method. The application and other details of these techniques have been dealt in chapter 8 on supports.

11.14.1.2   *Treatment that tackles the problems arising due to the presence of water*

Water can be a major cause of instability. It can delay the progress and, sometimes, makes use of explosives and equipment unsafe. Proper drainage and dewatering can be direct and effective ways to stabilize rocks and treat the ground. There can be following four ways to control water:

1. Lowering the water table or ground water by:
   a. Well points
   b. Deep well pumping
   c. Gravity drainage or under-drainage below tunnel
2. Use of compressed air to hold back water
3. Grouting
4. Freezing.

11.14.1.3   *Lowering water table/ground water[25]*

Lowering of water table[25] (fig. 11.21(a))[10] in granular soils can be effective by using techniques such as digging well points, pumping from deep wells or by allowing the water to drain out to a lower horizon than the tunnel's datum.[25] Well points are the tubes that are sunk to a maximum depth of 6 m (the limit of effective suction head) and water is pumped using suction pumps. These are spaced at 1 m or so and the tubes are perforated at their bottoms and fitted with a strainer to exclude sand. Continuous pumping for several weeks lowers the water level within the cone of depression. But the method is effective for depths not exceeding 6 m. Using deep-water wells and submergible pumps, water can be pumped effectively from sufficient depth; and the water table can be lowered after a continuous pumping from these wells that surround the tunneling site. Digging an array of wells around the tunnel as it progresses and continuous pumping from them is a costly affair, which can limit application of this technique. Sometimes it is practicable to drive a pilot tunnel below the main tunnel. Arrangement is made to drain off the water, that surrounds the main tunnel, by gravity into a sump which is located at the pilot tunnel and from there it can be pumped to the surface through a shaft or some other openings. This practice is practicable and applied in mines while driving drives and openings at the upper levels or horizons.

11.14.1.4   *Use of compressed air to hold back water*

The technique of holding water-borne ground or strata using compressed air was developed somewhere at the end of nineteenth century and since then it has been used considerably during the tunneling operations. It is used in water bearing silts, sands, gravels and fissured material to counteract the pore water at the face and sides including the back of the tunnel.[25] It also provides support to inadequately thick cover of impermeable clay that may be present in some cases. In this technique compressed air

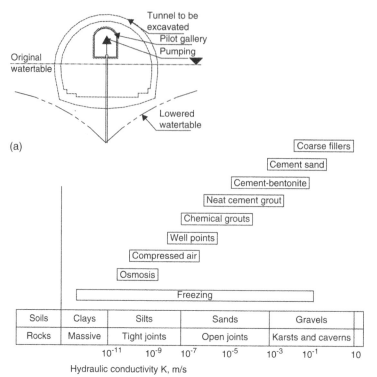

(a)

(b) Grout materials to suit different classes
of ground (Johnson, 1958)

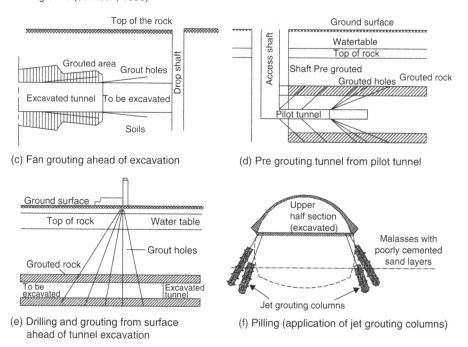

(c) Fan grouting ahead of excavation

(d) Pre grouting tunnel from pilot tunnel

(e) Drilling and grouting from surface
ahead of tunnel excavation

(f) Pilling (application of jet grouting columns)

Figure 11.21 Special methods to lower water-table (a); and unconsolidated ground. Classification of
grout material (b). Techniques of ground consolidation – grouting (c to f).

is retained by airtight bulkheads, through which men, material or machine and tools can pass through air locks. Two aspects that should be looked into are magnitude of air pressure and physiological effects on the workers. Water pressure (hydrostatic pressure) increases as the depth increases so air pressure is set based on the local conditions and the experience of the engineer. Excessive pressure can cause high air losses and decomposition of ground and lengthening of decompression time to workers. Lower pressure increases water flow and possibly ground handling and supports works at the face.

11.14.1.5 *Grouting*

Grouting means injecting the grout, or sealant to seal off the cavities, fissures, cracks or small apertures or pores through which the water can access the intended opening, or tunnel and thereby reducing the make of water to a minimum. Grouts of various types are available and prominent amongst them are the grouts of following two types:

1. Suspension of particles which could be that of cement, clay, or bentonite
2. Liquids which are usually colloidal solutions and which set into gel.

Grout's stability, settling times, viscosity, rigidity and overall cost of its application are the essential qualities a grout should posses. Pure cement is unstable and as such clay or bentonite is added while using it. Use of bentonite clay improves penetration and reduces permeability. Bentonite has thixotropic properties, whereby it becomes fluid when stirred but sets to gel when undisturbed. It takes a few hours to days to settle down.

Chemical grouts such as sodium silicates can set quickly and are cheapest amongst the other chemical grouts. Other chemical grouts include chrome lignin, resins and polymers. Resins and polymers are the costliest but form a solid mass.[7] For fine grained sand, chemical grouts of low viscosity can be used and for very finest sands expensive resins of very low viscosity may be used but spacing between holes must be low and injection rates also should be slow.[25] Chemical grouts reduce permeability by void filling and strengthen the ground formation. Drilling and grouting from the surface as shown in figure 11.21(e) can achieve grouting ahead of tunnel. Drilling and grouting can also be done from a pilot tunnel as shown in figure 11.21(d). Drilling fans ahead of excavation from a full-face tunnel has been shown in figure 11.21(c). Jet grouting (fig. 11.21(f)) is used to reinforce the ground by pumping in cement paste into the soil at high pressure. This results in a reinforced ground which is harder, more watertight and stronger than the native soil. These are some of the prevalent practices to apply the grout and the selection of any particular depends upon the local conditions. In fact, grouts are applied by drilling the holes radially in a similar manner, as described in chapter 14 during shaft sinking but during tunneling holes are drilled radially at a little inclination with the horizontal.

It may be noted that grouting is an expensive way of reducing inflow of water during tunneling operations. It also delays the work. It should, therefore, be used only when drainage and pumping which are cheaper and faster; are impractical; where seepage must be reduced substantially in the long term as well as short term; or when the strength as well as the water tightness of the rock mass needs improvement.[18]

Grouting also can be beneficial in underground mines where water inflows would otherwise lead to difficult and unsafe mining conditions.

11.14.1.6 Freezing

Details of this technique has been given in chapter 14 on shaft sinking and as with shaft sinking and tunneling the ground can be frozen around the tunnel to seal off the difficult ground which otherwise is difficult to deal with by other techniques. In this technique freezing the ground, which contains water, and the channels, through which water can pass or percolate, creates a cylinder of about 1 m thickness. Thus, freezing prevents the inflow of water and provides the cohesive strength to the ground. Drilling an array of boreholes to which the tubes are fitted effects freezing. Through these tubes refrigerant such as brine or liquid nitrogen are circulated to extract the heat from the ground and achieve freezing. The layout of such an arrangement is similar to the one shown in figures 14.10 and 14.11. Pipe arrays for the freezing may be in the same pattern as described for the grouting process, i.e. vertical or inclined from the surface, radially from the pilot tunnel or ahead of the tunnel face. The process is longer and requires elaborate arrangement. The time required could be weeks together and this increases further if water is saline. Special explosives are required to drive through the frozen ground. Slowly thawing the frozen ground is equally important. All these requirements make the process the costliest.

Grout materials to suit different classes of ground have been illustrated by a bar chart as shown in figure 11.21(b).[19] This figure also illustrates that freezing is effective in almost all the conditions.

11.15 CUT AND COVER METHOD OF TUNNELING[25]

For shallow depth tunnels, this is the fastest method. In this method a trench at the tunnel site is created with the use of conventional methods of rock fragmentation and ground digging. In soft ground sides are supported using piles. The whole width of tunnel can be excavated first and then support of sides and roof is undertaken. In some situation first the sides are dug, and concrete or some other form of support is built at both sides and then it is covered with roof slab. Excavation of ground enclosed between two sides then follows. The back filling on top of roof slab is the next step to resume the surface, which can be utilized again in the same manner or even better manner than previously. This method is also utilized to construct portals, particularly, in a hilly terrain.

11.16 SUBMERGED TUBES/TUNNELS[25]

There are two methods to build tunnel under water bodies. In the first method, leaving a sufficient cover of ground between the floor of water body and the back of the tunnel, the tunnel is driven by the use of borers. In the second method a trench is prepared below the floor of a water-body. The tunnel is built by assembling the prefabricated steel tubing or R.C.C. segments. The section of the tunnel could be circular or rectangular. The segments manufactured elsewhere are brought to the site and sunk into tunnel trench using the pontoons or other floating crafts. In this trench the floor is laid with sand and gravel. Care is taken to perfectly join the segments of the tunnel to ensure water tightness. The sides are back filled with suitable sand and gravel, and sand is also pumped in water over the site where the tunnel has been built.

11.17 THE WAY FORWARD

- To improve safety, a seismic probing ahead procedure "Sonic Softground Probing" (SSP) has been developed. It probes ahead virtually and warns the TBM operator ahead of time, when significant geological changes are to be expected. SSP has already delivered precise predictions and will continue to be beneficial in future as well.
- As described in sec. 19.5.1; Geological vision (Radar) – ability to see through rock to locate an orebody in space, grade and geological structures (CRC-Australia) as shown in figure 19.12; it could be a useful technique in the days to come and have great potential in bringing precision to mining operations.
- In fact, as shown in figure 19.15(a), precision in ore exploitation techniques results in enormous knock-on benefits.

QUESTIONS

1. " 'Tunneling technology' is opening the future's opportunities. For example: the Euro Tunnel between London and Paris (France) which was not there in the 1980s is now a reality. It is narrowing gaps and distances by passing through the most difficult ground and hazardous conditions. Present technology could meet this challenge." Today there is hardly any locale, site, or set of conditions where tunnels cannot be driven without aid of tunnel borers. Do you agree with these statements? Describe how TBM have developed rapidly in the recent past by undertaking a literature survey using different sources including websites.
2. A modern tunnel is equipped with what type of services and fittings? List them.
3. Briefly summarize 'The future of Tunnel Boring Technology as described by James E. Friant and Levent Ozdemir'.
4. Circular shield tunnels are common but in recent years irregular-shaped shield tunnels have been developed – name what they are.
5. Classify full face TBMs based on method of excavation, mechanism of stabilizing the cutting face and the digging method applied. You should answer by drawing a suitable line diagram.
6. Classify the different types of borers. Name some of the full-face boring machines.
7. Classify the different types of rock cutting tools. Suggest appropriate rock cutting tools for: (a) soft to medium rock, (b) medium to hard rock, and (c) very hard rock.
8. Classify the machines that are available for driving the headings without the aid of explosives. Prepare a comparative statement to compare their main features.
9. Compare the important features of different types of tunneling machines.
10. Define: rock anisotropy.

11. Describe a shield. When and where was one used for the first time?
12. Describe a TBM System.
13. Describe the important features and field of applications of: slurry shield, compressed air shields and full face shield with picks.
14. Detail these (i.e. write notes on): submerged tubes/tunnels, grouting, cut and cover method of tunneling, freezing.
15. For TBM performance evaluation, emphasis should not be placed on the traditional mineralogical and petrography study only, but rather on factors of key importance for evaluation of TBM performance and cutter wear. List those factors.
16. Give a line diagram to depict a general classification of tunneling methods that are used under different situations including varying ground conditions.
17. Give a line diagram to illustrate the suitability of full face TBM, road headers, shield machines and excavators as per ground conditions/geologic ranges.
18. Give a conceptual diagram of the Slurry Shield and Mixshield giving their main components.
19. Give conceptual diagrams for earth pressure balance (EPB) shields. Cite some tunneling projects where these (EPB & Mixshield) have been used.
20. Give details of single, double and enlarging TBMs.
21. How does rock strength affect TBM performance?
22. List the options that are available to handle the muck generated during a boring cycle. Briefly describe each one of them.
23. List the various types of coal cutters.
24. List the TBMs manufactured by various manufacturers – some models and types known to you. Also include some typical TBMs used around the globe.
25. List the parameters influencing the advance rates of TBMs.
26. List the special methods to lower a water-table. Illustrate them giving suitable sketches.
27. List the appropriate TBMs for soft ground/formations. How do they differ from those meant for hard ground and rocks?
28. List what you consider as the difficult and troublesome types of ground (which cause difficulties during or after construction of a tunnel) that can be encountered.
29. Make a comparison of major criteria for shotcrete tunneling methods with those of TBMs.
30. Mention the latest developments in tunneling techniques/shields known to you.
31. Millions of years of the history of the earth hold surprises in tunnel excavations. To improve safety, a seismic probing ahead procedure "Sonic Soft-ground Probing" (SSP) has been developed. What does it do? Describe it.
32. New Austrian Tunnelling Method (NATM) – what is this concept? Write down briefly its important features and field of application. Up to what UCS can soil and rocks be covered by this technique?
33. On which size range (diameters of tunnels) could best performance be achieved, based on the experience gained in this regard so far?
34. Prepare a comparative statement comparing various features of the TBM and road headers.
35. Some factors whether not properly investigated or often underestimated still cause the basis for claims in TBM projects; List these factors.

36. Tabulate a comparison of different techniques of tunneling.
37. TBMs are large-sized equipment – mention their usual length including the trailing gear. List the backups that are essential to running this system.
38. Usually a civil tunneling project has to pass through three segments/phases. What are they? List them.
39. Water can be major cause of instability. It can delay the progress and, sometimes, makes use of explosives and equipment unsafe. Proper drainage and dewatering can be direct and effective ways to stabilize rocks and treat the ground. List the four ways to control water.
40. What is a boring cycle essentially composed of?
41. What are 'Mini Tunnel Borers? Where do they find their applications? What problem could small-sized tunnels pose? List them.
42. What are the methods to drive large-size tunnels/drives in varying ground conditions? Illustrate the various methods used.
43. What are these concepts, describe them: (a) Mechanical excavation of the full cross-section with open type machines. (b) Mechanical excavation of the full cross-section with shield machines or telescope type shield machines. (c) Mechanical excavation of a pilot tunnel with subsequent mechanical enlarging.
44. What is a tunnel portal? How it is driven? What problems are usually encountered during its construction?
45. What is the Pre Vault method? Where does it find application? List how (by steps) it allows advance or pre-support of the section to be excavated.
46. What is a boring or excavation system? Mention its importance for TBM effectiveness. What does it consists of?
47. Where was the world's largest tunnel boring machine (diameter of 14.2 m) used? Mention a special feature that was incorporated in its design.
48. With shield tunnelling use of auxiliary construction measures or special ground treatment methods are essential to strengthen the ground in and around the intended excavation. List them together with their characteristics.

REFERENCES

1. Babenderede, S.: Application of NATM for metro construction in the FRG, *Euro-tunnel 80*, IMM, London, 1980, pp. 54–58.
2. Bieniawski, Z.T.: *Rock mechanics design in mining and tunneling*. A.A. Balkema, 1984, pp. 272.
3. Breeds, C.D. and Conway, J.J.: Rapid excavation. In: *SME Mining Engineering Handbook*, Hartman (edt.). SMME, Colorado, pp. 1871–78, 1905–07.
4. Cazenave, B. and Le Goer, Y.: Mechanical precutting. *North American Tunneling*, 1996.
5. Deacon, W.G. and Hughes, J.F.: Application of NATM at Barrow-upon-Soar gypsum mine to construct two surface drifts. In: *Tunneling '88* IMM Publication, 1988, pp. 69–77.
6. Dias Daniel and Kastner, R.: Effects of pre-lining on tunnel design. In: *Underground construction in modern infrastructure*, Franzen, Bergdahl and Nordmark (eds.). A.A. Balkema, 1998, pp. 391–397.
7. Franklin, J.A. and Dusseault, M.B.: Rock Engineering. McGraw-Hill, New York, 1989, pp. 563–85.

8. Friant, James. E. and Ozdemir, Levent: Tunnel boring technology – present and future. In: *Proc. RETC, AIME, 1993*, pp. 869–888.

9. Graham, P.C.: Rock exploration for machine manufacturers. In: *Proce. Sympo. on Exploration for rock engineering, Johannesburg, 1976*, pp. 173–180.

10. Guilloux, A.: French national report on tunneling in soft ground. In: *Underground construction in soft ground*, Fujita and Kusakabe (eds.). A.A. Balkema, 1995, pp. 97–100.

11. Hack, A.: TBM Tunneling, Hagenberg/Austria. *Proceedings of International Lecture Series 5*, 1996.

12. Hamburger, Harmann and Weber, Walter: Tunnel boring of large cross sections with full face and enlarging machines in hard rock. In: *Proc. RETC, AIME, 1993*, pp. 811–831.

13. Handwith, H.J. and Dahmen, N.J.: Tunnelling machines. In: *Underground Mining Methods Handbook*, W.A. Hustrulid (edt.). SME-AIME, New York, 1982, pp. 1107–10.

14. Hartman, H.L.: *Introductory Mining Engineering.* John Wiley & Sons, New York, 1987, Tunneling – pp. 498–503.

15. Heflin, L.H. and Irshad, Mohammad: Soft ground NATM tunnel design. In: *Proc. RETC, AIME, 1987*, pp. 112–129.

16. Herrenknecht GmbH, Schwanau, Germany: *Literature and leaflets.*

17. Herrenknecht, M. and Burger, W.: The new generaions of soft ground tunneling machines. Proc. RETC 1999, AIME, 647–662.

18. Johnson, G.D.: Thorough grouting can reduce lining costs in tunnels. In: *Proc. Conf. Grouting in Geo-tech. Eng.* Am. Soc. Civ. Eng; New York, 1982, pp. 892–906.

19. Johnson, S.J.: Cement and clay grouting of foundations; Grouting with clay-cement grouts. *J. Soil Mech. and Found.* Div. ASCE, 1958, paper 1545, pp. 1–12.

20. Kurihara, K.; Kawata, H. and Konishi, J.: Current practices of shield tunneling methods – A survey on Japanese shield tunneling. In: *Underground construction in soft ground.* Fujita and Kusakabe (eds.). A.A. Balkema, 1995, pp. 329–336.

20b. LTM Corporation, Seoul, Korea – Leaflets and literature (soft and hard copies).

21. Lislerud, A.: Hard rock tunnel boring – prognosis and costs. In: *Tunneling and underground space technology*, 1988, vol. 3, pp. 1, 9–17.

22. Lislerud, A. et al.: Hard rock tunnel boring. Norwaegian Institute of Technology, Engineering project report 1–83, 1983, pp. 159.

23. Maidl, B. Herrenknecht, M. and Anheuser, L.: *Mechanised Shield Tunneling.* Ernst & Sohn, Berlin, 1996, pp. 4, 35–36, 275–277.

24. Martin, Herrenknecht Ing. E.h. and Burger, Werner: The new generations of soft ground tunneling machines. In: *Proc. RETC, AIME, 1999*, pp. 647–662.

25. Megaw, T.M. and Bartlett, J.V.: *Tunnels: Planning, Design, Construction.* Ellis Horwood Limited, Chichestre, 1983, Cut and Cover: pp. 11–19.

26. Nilsen, B. and Ozdemir, L.: Recent development in site investigations and testing for hard TBM projects; *Proc. RETC, AIME*, 1999, pp. 715–731.

27. Parkes, D.B.: The performance of tunnel boring machines in rock. In: *CIRIA Special Publication no. 62, 1988*, pp. 56.

28. Pokrovsky, N.M.: *Driving horizontal workings and tunnels.* Mir Publishers, Moscow, 1988, pp. 328–332, 348–350.

29. Randy, Fulton; Geogrge, H.K. and Ignacy, Puszkiewicz: Two firsts in tunneling in Canada. In: *Proc. RETC, AIME, 1993*, pp. 273–293.

30. Robbins TBMs, Robbins Company, Kent, USA.: *Literature and leaflets.*

31. Sandtner, A. and Gehring, K.H.: Development of roadheading equipment for tunneling by NATM. *Tunneling '88*, IMM Pub. pp. 275–288.

32. Sauer, G. and Gold, H.: NATM ground support concepts and their effect on contracting practices. In: *RETC Proc. 1989*, pp. 67–86.

33. Sauer, G.: When an invention is some thing new: from practice to theory in tunneling, *Trans. IMM*, London, No. 97, 1988, pp. A94–108. Tab. 15.1.

34. Singh, J.: *Heavy Constructions – Planning, Equipment and Methods*. Oxford & Ibh Pub. New Delhi, 1993, pp. 528.

35. Stevenson Garry, W.: Empirical estimates of TBM performance in hard rock; *Proc. RETC, AIME, 1999*, pp. 993–1006.

36. Thon, J.G.: Tunnel boring machines. In: *Tunnel engineering handbook*, J.O. Bickel and T.R. Kuesel (eds.), 1983, pp. 670.

37. Toolanen, B.; Hatwig, S. and Janzon, H.: Design considerations for large hard rocks TBMs when used in bad ground. In: *Proc. RETC, AIME, 1993*, pp. 853–868.

38. Whittaker, B.N. and Frith, R.C.: Tunneling – Design, Stability and Construction. IMM Publication; 1990, pp. 69–91, 199–227, 351–360.

39. Writh's TBMs, Wirth Company, Erkelenz, Germany: *Literature and leaflets*.

Chapter 12

Planning

It is the person behind the machine that matters. Hence quality of human resources must be ensured. Imparting proper education and practical training to the students; organizing vocational training, refresher courses, symposia and seminars on a regular basis for the working crews, could achieve this.

12.1 ECONOMIC STUDIES[4,5]

A feasibility study is undertaken to establish whether a mineral deposit under consideration is technically feasible and economically viable to mine out or not. Referring to the definition of the term 'ore', which can be defined as a metal bearing mineral or aggregate of such minerals together with barren material (termed as 'gangue'), that can be mined profitably. This means a feasibility study is undertaken to determine how much proportion of a mineral deposit constitute ore and how much waste, or sub-grade material from the economics point of view.

Thus, a feasibility study of a mineral deposit refers to a process of examining the technical details together with economic analysis, to exploit it. A feasibility study goes parallel with geological and technical investigations, with the following aims to achieve:

- To *ascertain* what exists? In terms of geology, mineralogy, geography and infrastructures.
- To *determine* what can be done technically? In terms of mining, processing and handling the materials produced.
- To *investigate* whether the proposed product can be sold? In terms of quantity and rate (i.e. price/unit).
- To *estimate* the project costs: In terms of cost of construction (capital) and operating.
- To *calculate* the revenues after meeting expenses to fulfill the financial goals (i.e. profit margin) set by the sponsoring organization/company.

12.1.1 Phases or stages in economic studies

A feasibility study has to pass through three distinct stages:

- Preliminary studies or valuation
- Intermediate economic study or pre-feasibility study
- Feasibility study.

12.1.1.1 Preliminary studies or valuation

This study is the first step of feasibility studies when practically very little information is available on the *quantity and quality* of the ore, and other parameters pertaining to it. During this study very little money and time are spent; as the return is distant and indirect. Based on the available information and data, this study aims to investigate: What may be? What is known to be? What may be worth looking for? This study is carried out to find whether further money should be spent on exploration or/and the prospecting tasks, or it should be dropped then and there.

12.1.1.2 Intermediate economic study or pre-feasibility study

Based on the green signal given by the preliminary studies, this study is carried to get more and more data and information. However, many assumptions are required to be made during this phase. This data set is further analyzed to give further signals for spending resources in terms of time and money on various parameters relating to these studies.

12.1.1.3 Feasibility study

Finally comes the stage of feasibility study. This could be a task of several months and write-up of several volumes involving various agencies and experienced manpower. A feasibility study should be able to achieve the following functions:

- It should be reliable
- It should possess sufficient supporting documents, drawings, analysis results etc. that can be presented to any financial institute such as a bank etc.
- A firm conceptual plan must be ready at this stage which would not be radically changed and which could be used as the basis for the preparation of a Detailed Project Report (D.P.R) (section 12.2).

Parameters need to be considered while undertaking feasibility studies for metal, nonmetal or fuel mineral deposits: Given below is a guideline to gather and compile information/data on the various parameters need to be included while undertaking the feasibility studies for different types of mineral deposits.

12.1.1.3.1 Information on the deposit

A. Geology: type of mineralization and its grade, rock types, geological structure, extent of leached or oxidized zones, if any.
B. Geometry of the deposit: size, shape, and altitude above ground level, depth extension and continuity.
C. Geography: its location and access from the main population areas, surface topography, climatic conditions, type of land, and the political boundaries. Rank & chemical analysis in case of coal.
D. Exploration & Reserves details: exploration history, if any, current and future programs. Mineral inventory details, Grade-tonnage curves. Sampling and assaying procedures.

12.1.1.3.2   Information on general project economics

A. Markets: likely purchasers, marketable form of the product, expected marketing costs and selling rate.

B. Transportation: type of transportation (road, rail & air) available; for public and goods, its cost and other details.

C. Utilities: such as electric power, natural gas or any other form of energy – their availabilities, costs etc.

D. Communication: such as post office, telephone, internet and other means of communication and information technology – availability & costs.

E. Ownership rights for surface land, water & mineral bearing zones: ownership details & acquisition procedures, costs and rents, legal requirements etc.

F. Potable (drinking) water: its quality, availability, cost etc., also mine water details with regard to quality, quantity, treatment and disposal. Hydrology in case of coals i.e. permeability, porosity, etc. for the coal and over burden strata.

G. Labor: types of labor available – skilled, unskilled or semi-skilled. Wages pattern, Labor organization. Availability of housing & transport facilities for them.

H. Social & welfare facilities: such as schools, hospitals, recreation means, play grounds, churches, mosques, banks, police, rail & bus stations.

I. Political & Government considerations: Taxation pattern, type of subsidy or depletion allowance, if available. Mining & environmental laws – existing & proposed.

J. Financing: Sources and its alternatives if need arise, repayment terms, interest rates and other conditions.

12.1.1.3.3   Mining method selection

A. Physical controls: Strength of ore and enclosing rocks, geology, geometry, subsidence considerations etc.

B. Exploitation constraints: Expected recovery, dilution, waste rock production – its quantity & handling costs.

C. Pre-production – Development layouts, time schedule and capital sum required.

D. Production: Rate, requirement of labor, machines & equipment. Capital required. Production schedule.

E. Selectivity: is a technical feasible mining method available, if yes, selecting the one after comparing various design alternatives. If not, then specify the technical reasons for its rejection.

12.1.1.3.4   Processing methods

A. Type of product required: specification, treatment arrangements as per metallurgical, chemical & physical properties of the ore.

B. Layout of the plants required.

C. Requirement of the labor, machines, equipment, infrastructures and capital money.

D. Disposal of the finished product.

12.1.1.3.5   Ecology

This includes assessment of likely impact of mining and processing on the environment, its remedial measures and management plan.

12.1.1.3.6 Capital and operating costs estimates

A. Capital investment required:
- Capital requirement for mining: For acquisition of surface and mineral rights, exploration, pre-production, development, production, consultancy, working capital, etc.
- Capital required for process plants: Land, building, plants & equipment, disposal of waste etc.
- Capital required for social benefits: Accommodation, recreation facilities, welfare, social overheads, etc.

B. Operating costs:
- Mining – Wages with fringes, supplies & material, fuel & energy, maintenance.
- Processing – Wages with fringes, supplies & material, fuel & energy, maintenance.
- Miscellaneous – Administrative & supervisory, cost of capital, depreciation, amortization etc.

12.1.1.3.7 Project cost & rates of return

- Total costs: Mining through process plants including social overheads, taxation.
- Total revenues: Likely to be received by selling the finished product and also from the by-products, if any.
- Profit/loss – Annual profit/cash flow and rate of return on the investment.

12.1.1.3.8 Comments

Assess *Economic Viability & Technical Feasibility* of the project including the social benefits.

12.1.2 Conceptual mine planning and detailed project reports

Engineering studies and models: Based on the feasibility studies a decision is taken whether to mine a particular deposit or not. For the deposit selected for mining a basic planning model is prepared by involving the following engineering studies and the framework.

1. Conceptual studies/model
2. Engineering studies/model
3. Detailed studies/model.

12.1.2.1 Conceptual studies/models

For any deposit during the feasibility studies based on the data, information and the analytical work carried out concerning the parameters such as: economics, geology, geography and geo-mechanical and others; a suitable mining method is selected keeping in view a certain rate of production or annual output from the mine.

For the method so selected a conceptual framework or model is then prepared (fig. 12.1(a)). For example, if an underground mining method is selected, first the drawings regarding the profile of the orebody are prepared. Based on the shape, size, location, geometry of the orebody a general scheme of the stopes, pillars and levels' disposition is then drawn; similarly the general schemes with regard to mining sequence, ore fragmentation and extraction systems, ore and waste handling systems, mine services schemes are also outlined. These schemes form the basis to calculate and estimate the

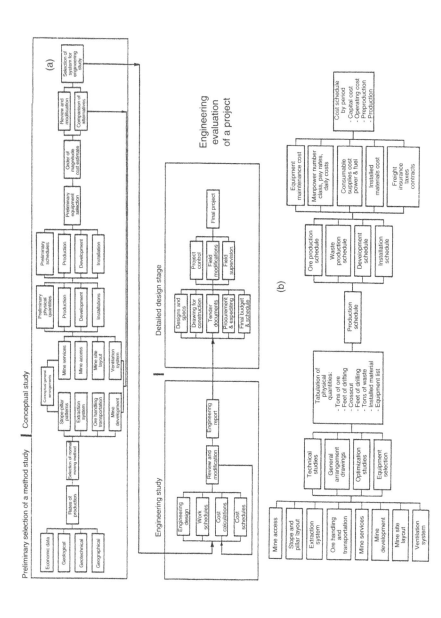

Figure 12.1 Conceptual model (a); and engineering evaluation (b) of an underground mining project. The same logic could be applied for surface mines or a tunneling project by including what else needed, or excluding whatever is not applicable.

physical quantities with regard to ore, waste, and installation activities. Equipment required, time schedule and budget (i.e. money required) are also assessed. This complete exercise is first carried out for one alternative, if other alternatives are also possible then such exercise is undertaken for each of these alternatives, and amongst them the best one is selected. For the alternative so chosen, the engineering model is then built. Figure 12.1(a) illustrates a conceptual model for an underground mine with sublevel stoping as mining method and figure 12.1(b) the engineering evaluation. These models could be applied for tunnels and surface mines too but with certain modifications that might be warranted; i.e. by including the specific features and excluding those not required.

12.1.2.2 Engineering studies

During the engineering studies, the details of the concepts developed are further studied, drawn and detailed (figs 12.1(b), 12.2(a)). The drawings related to mine access; method design, layouts and mine services are prepared. The specifications for the different sets of equipment and various installations are worked out. Based on these technical studies; the physical quantities with regard to ore tons, waste tons, amount of development work and equipment numbers are assessed. A schedule with regard to construction, development, production and resources requirement in terms of manpower, material, energy, equipment and money is prepared. Cost of production and revenues to be received are assessed.

12.1.2.3 Models and detailed design

From the above framework, the construction drawings, specifications of various machines and equipment, tender documents, budget forecasts, procurement and recruitment schedules can be prepared. This task, in fact, is known as preparation of *Detailed Project Report (D.P.R.)*. This is the main document for the project, from which information regarding anything related to the project can be drawn. In order to assess costs, prepare schedules for production, development, construction and installation activities certain norms regarding performance of the equipment to be deployed, manpower to be engaged are required. The planners or consultants using their expertise and the experience decide these norms. Sometimes they can be obtained from the other mines or projects having the similar working conditions or set up, else these can be calculated (figs 12.2(b) and (c)). However, for any estimate to be reliable and precise (within the practical limits) a correction factor can be applied based on the local conditions, and the judgment and experience of the planner.

12.2 MINE DESIGN ELEMENTS

The following constitutes the mine design elements:

- Mineral resources and reserves i.e. mineral inventory.
- Cutoff grade and ore reserves.
- Production rate and mine life.
- Price of mineral/ore.

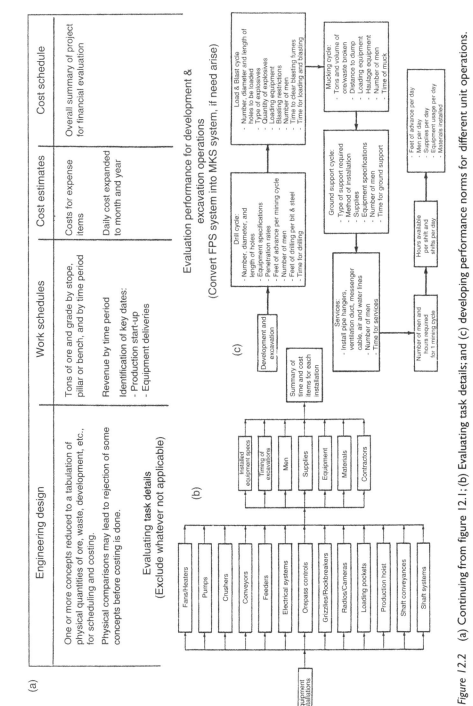

Figure 12.2 (a) Continuing from figure 12.1; (b) Evaluating task details; and (c) developing performance norms for different unit operations.

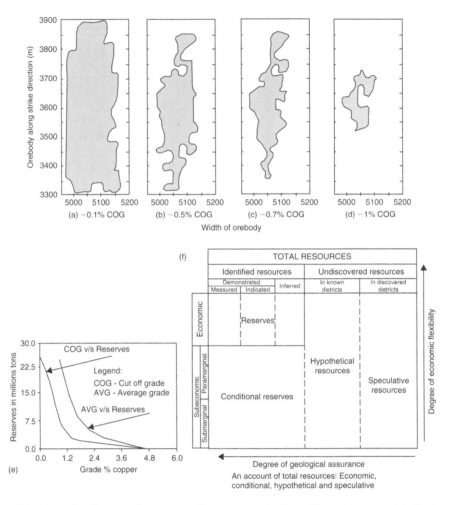

Figure 12.3   (a to d): Influence of raising cutoff grade on orebody profile and reserves. (e): Grade –
Tonnage curve. (f): An account of total resources.

12.2.1   Mineral resources and reserves

The difference between the terms mineral resources and reserves has been illustrated
in figure 12.3(f), prepared by the U.S. Geological Survey. One axis of the diagram represents the degree of geologic assurance about the existence of the resource and the
other axis represents the economic feasibility of the resource recovery. Out of the
known mineral resources the portion of it, which is economical to mine, is referred as
ore reserves. The ore reserves have been further divided as measured (proved) indicated (probable) and inferred (possible) reserves. The first two types of the reserves
are compiled by the exposed faces and drill hole results. Certain geological evidences
project the possible reserves. The sub-economic reserves are not economical at present
but they could prove to be ore reserves with the change of time and technological

scenario. The hypothetical resources are those that have escaped discovery in the known mining districts. Speculative resources are those that could occur in untested places where favorable conditions are known or suspected. In any deposit all the ore reserves cannot be mined and, therefore, a further categorization is made to assess the net reserves that will be available at the pit top for final dispatch.

Geological reserves:[1] The ore reserves that have been estimated by the geologist within the defined geological boundaries and norms, are known as geological reserves.

Workable or mineable ore reserves:[1] Due to certain natural problems such as low thickness, abnormal pinching and swelling, very irregular ore boundaries, presence of abnormal gases, water or geological disturbance which compel to leave certain portion of total geological reserves. This portion of the reserves, which cannot be recovered, is known as non-workable reserves. Deducting these reserves from the geological reserves gives the workable or mineable reserves (fig. 12.5(a) & equation 12.1a)).

Commercial ore reserves:[1] Based on the mineable ore reserves and the boundaries, a mining method is designed. During stoping operations due to certain operational losses the recovery is not full. Deducting these operational losses from the mineable reserves gives the commercial reserves (figs. 12.5(a), 12.5(d) & equation 12.1b).

These operational losses are due to ore blocked in shaft pillars and mine or district barriers, poor recovery from the stopes' pillars, not full recovery from the stopes during stoping (may be due to improper drilling or blasting or both), leaving broken ore in worked out stopes beyond the reach of the mucking equipment etc. Thus, the commercial reserves are the reserves, which are available at the pit top for the purpose of its commercial use. The following relations can be used to express various types of reserves numerically.

Workable or Mineable Reserves, $Q_w = Q_g - N_w$ $\qquad\qquad\qquad\qquad$ (12.1a)

Commercial Reserves, $Q_C = Q_w - O_L$ $\qquad\qquad\qquad\qquad\qquad$ (12.1b)

Whereas: Q_g – Geological reserves.
$\qquad\qquad$ Q_w – Mineable or workable reserves.
$\qquad\qquad$ Q_C – Commercial reserves.
$\qquad\qquad$ N_w – Non-workable reserves.
$\qquad\qquad$ O_L – Total operational losses = $L_1 + L_2 + L_3$
$\qquad\qquad$ L_1 – Losses in pillars, barriers, boundaries etc.
$\qquad\qquad$ L_2 – Losses due to poor breakage, blasted ore losses in worked out stopes, etc.
$\qquad\qquad$ L_3 – Losses due to abnormal adverse working conditions such as excessive
$\qquad\qquad\qquad\qquad$ ground pressure, heat, humidity, gases, water etc.

Mineral Inventory:[1] For any mineral deposit accountability of its reserves grade-wise is known as the mineral inventory assessment for that deposit. Thus, there is difference between mineral inventory and ore reserves. In ore reserves estimation the technique, method, equipment to mine out the deposit together with its cost of mining and selling price comes into picture, whereas, in mineral inventory estimation these considerations are not required to be made.

12.2.2   Cutoff grade[7,8,10,11,12]

As discussed earlier, in any mineral deposit only the mineralization that can be exploited commercially to yield profit is classified as ore. To determine and map an orebody, it is necessary to establish a cutoff grade that represents the lowest grade at which the mineralized rock qualifies as ore. Evaluation of a deposit indicates that the grade distribution is not uniform. There are areas of high mineral concentration and low mineral concentration within it. Intermix areas with high and low grades are also encountered. In a situation like this some kind of selectivity is employed to define geographically and quantitatively the potential ore limits. *Cutoff grade* could be defined as "any grade that for any specific reasons, is used to separate two courses of action, e.g. to mine or to dump"[11]. Where the grade of the mineralized material is less than the cutoff grade it is classified as waste and where it is equal to or above cutoff grade it is classified as ore.

The reasons for continuing interest in cutoff grades are obvious. Too high a grade can reduce the mineral recovered and possibly the life of the deposit (figs 12.3(a) to (d)). Too low a cutoff would reduce the average grade (and hence profit) below an acceptable level. In project evaluation it is important to determine a cutoff grade, which is normally set to achieve the financial objectives for the project.

Studies on cutoff grade theory may fall into two basic categories. The fixed cutoff grade concept assumes a static cutoff for the life of the mine, while the variable cutoff grade concept assumes a dynamic cutoff maximizing the mine net present value. Lane[7] outlined three distinct stages in a mining operation: ore generation (mining), concentration (milling) and refining. He demonstrated that in establishing cutoff grades, consideration of costs, capacities, waste: ore ratios and average grade of different increments of ore of the orebody as well as the present values of annual cash flows are essential. For each stage, there is a grade at which the cost of extracting the recoverable metal equals the revenue from the metal. This is commonly known as the *break-even grade*.

If the capacity of an operation is limited by one stage only, the break-even grade for that stage will be the optimum cutoff grade. Where an operation is constrained by more than one stage, the optimum cutoff grade may not necessarily be a break-even grade. In such a case the "balancing" cutoff grade for each pair of stages needs to be considered as well. For calculation of break-even grade assuming that output of the mining operation is constrained only by its capacity to handle ore, the following formulae could be used depending upon the policy of a company.[8]

Constant price and cost
Simple formula $(s-r) \, y \, g = c$ (12.2a)

Maximum total profit
$(s-r) \, y \, g = c + f/C$ (12.2b)

Maximum present value
$(s-r) \, y \, g = c + (f + d \times v)/C$
Varying prices and costs   (12.2c)

Maximum present value
$(s-r) \, y \, g = c + (f + d \times v + \delta v)/C$ (12.2d)

Where: s – price of metal/mineral
       r – smelting and marketing costs
       y – recovery of mineral
       c – marginal (variable) processing costs
       f – annual fixed cost
       d – discount rate
       C – capacity: units of ore p.a.
       g – cutoff grade
       v – present value
       δv – the decline in present value over a year.

Only the variable costs[4] are used in cutoff grade analysis because the inclusion of other operating costs (fixed, direct etc.) reduces the ore reserves by an amount that would pay for these other costs. The total costs should be used in the optimization analysis with regard to net present value and fixing production rates.

While using the simple formula (Eq. 12.2a), it may result in an average grade that does not provide sufficient revenues to cover all the fixed costs of the operation. In such a situation, it would be necessary to consider formula (Eq. 12.2b) which would result in a higher cutoff grade. Alternately an intermediate value might be selected.

A decision on cutoff grade is a matter of the policy of an enterprise based on its financial or other goals and consequently different formulae/relations are used for computing this parameter. However, a decision to achieve an optimum cutoff grade (i.e. maximizing present values) throughout the life of a mine, would mean designing the cutoff grade, average grade and cash flow to decline from some high starting level.

Parameters influencing cutoff grade: In underground mines, even for the same deposit, the cutoff dependent costs vary significantly according to stoping methods, orebody thickness, degree of mechanisation and working efficiencies of the men and machines.

Different stoping methods require a different cutoff grade. The sublevel stoping and its variants can be operated at a lower cutoff grade than cut and fill and its variants (fig. 12.4a(i)).

Thinner orebodies adversely affect the cutoff grades but once a certain minimum thickness is attained, the cutoff grades become almost stable (fig. 12.4a(iii)).

The cutoff grade for the same stoping method, in the same mine, can vary if operated using different sets of machines (fig. 12.4a(ii)). Labor-intensive methods need higher cutoff grades.

An allowance for working efficiency, that is the productivity, should be made when calculating cutoff grades (figs 12.4a(i) and (ii)).

There is a sharp decline in cutoff grades with an increase of process recoveries (fig. 12.4a(iv)).

All these parameters (whichever are appropriate) are also influencing cutoff grade while mining the deposit by any of the surface mining methods.

Grade-tonnage calculations and plotting the curves:[10] While evaluating any deposit establishing the grade-wise reserves should be the prime objective, as compilation of this information will mean knowing the deposit fully. It is a basic foundation for any mine upon which it can be developed, constructed and run for the purpose of obtaining the desired production. For this purpose through the exploration program, the

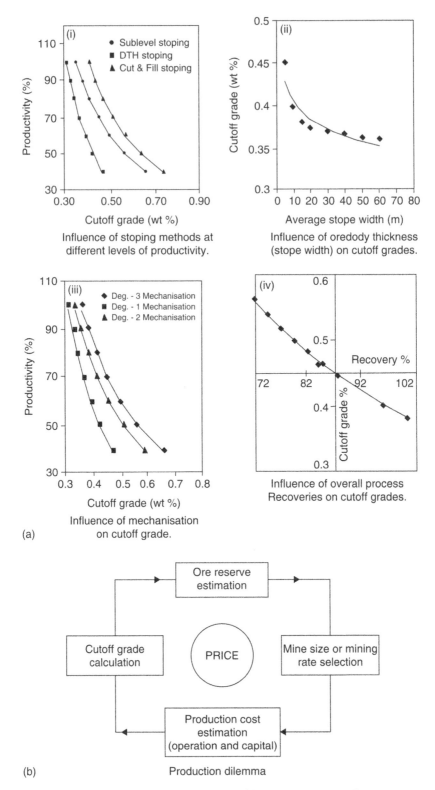

Figure 12.4 (a): Parameter influencing cutoff grades. (b): Inter-relationship of various parameters.

Table 12.1   Calculation of mineral inventory for establishing Grade-Tonnage details. Also for plotting
the G-T curves: plot col. I v/s col. 4 and col. 7 v/s col. 4. (C^I, —Column number)[10]

1 Cutoff Grade %	*2* Average Grade %	*3* Quantity Tons	*4* Cum. Res. tons	*5* Metal tons ($C_2 \times C_3$)/100	*6* Cum. Metal Tons	*7* Avg. grade, %, ($C_6 \times$ 100)/C_4
			ΣC_3		ΣC_5	
			$\Sigma C_3 - R_1 C_3$		$\Sigma C_5 - R_1 C_5$	
			Likewise—		Likewise—	
		ΣC_3		ΣC_5		

data are grouped following a certain class interval and average grade and tonnage for
each class is determined. Following the procedure illustrated in Table 12.1 the
tonnage at each of the cutoff grade considered can be estimated. As shown in the table
each cutoff grade has its average grade. These data can be plotted to obtain two
curves; firstly – cutoff grade v/s tonnage and secondly, average grade v/s tonnage
(fig. 12.3(e)). Using these curves one can find out tonnage and average grade at any of
the cutoff grades, which is an important parameter for the purpose of mine planning.

12.2.2.1   Mining & process plant input-output calculations[10] (for a copper mining complex)

Calculation for amount of CONCENTRATES generated from the CONCENTRATOR:

$INPUT = OUTPUT$

$A_{ORE} \times G_{ORE} = A_{CON} \times G_{CON} \times RF_{CON}$

or

$A_{CON} = (A_{ORE} \times G_{ORE}) / (G_{CON} \times RF_{CON})$

Where: A_{ORE} – Amount of ore input to concentrator
G_{ORE} – Grade of ore feed to concentrator in %
A_{CON} – Amount of concentrates generated (output) from the concentrator
G_{CON} – Grade of concentrates in %
RF_{CON} – Recovery factor of the concentrator

Calculation for amount of ANODES generated from the SMELTER:

$A_{CON} \times G_{CON} = A_{ANOD} \times G_{ANOD} \times RF_{ANOD}$

OR

$A_{ANOD} = (A_{CON} \times G_{CON}) / (G_{ANOD} \times RF_{ANOD})$

Where: A_{ANOD} – Amount of anodes generated (output) from the smelter.
G_{ANOD} – Grade of anodes produced from smelter in %
RF_{SMT} – Recovery factor of the smelter

Calculation for amount of CATHODES generated from the REFINERY:

$A_{CATHOD} = (A_{ANOD} \times G_{ANOD}) / (G_{CATHOD} \times RF_{REF})$

Where: $A_{CATHODE}$ – Amount of cathodes generated (output) from the refinery.

G_{CATHOD} – Grade of cathodes produced from the refinery in %

RF_{REF} – Recovery factor of the refinery.

12.2.2.2 Cutoff grade calculations:

$$(g/100) \times RF_{CON} \times RF_{SMET} \times RF_{REF} \times (P_{MET} + P_{BY-P}) = C_{MIN} + C_{MIL} + \\ \{(g \times C_{SMET})/(G_{CON} \times RF_{CON})\} + \\ \{g \times C_{REF})/(G_{ANOD} \times RF_{CON} \times RF_{SMET})\}$$

$$(12.2e)$$

Where*: g – is cutoff grade of ore in %

G_{CON} – Grade of concentrates in %

G_{ANOD} – Grade of anodes in %

RF_{CON} – Recovery factor of concentrator

RF_{SMET} – Recovery factor of smelter

RF_{REF} – Recovery factor of refinery

P_{MET} – Metal price/ton of metal

P_{BY-P} – By-product price/ton

C_{MIN} – Mining cost/t of ore

C_{MIL} – Milling cost/t of ore

C_{SMET} – Smelting cost/t of concentrates

C_{REF} – Refining cost/t of anodes

* – Not to be used if not appropriate based on the processes involved.

12.2.3   Interrelationship amongst the mine design elements

Mineral inventory, cutoff grade and ore reserves: as already defined, mineral inventory is an accountability of grade-wise reserves and cutoff grade is the minimum grade at which if mining is carried out, it will be in profit. The reserves out of the total mineral inventory, at and above cutoff grade, are called ore reserves.

Production rate, cost of mining and mine life: To examine the relation between production rate and cost of mining, the analysis carried out indicates that lower the production rate the higher are the total costs, which include operating costs plus the capital investment (fig. 12.5(b)). And these costs get lower as the production rate increases till a certain rate and after that it again starts getting higher and higher[4]. Based on this logic the curves are drawn, production rate v/s mining costs, at their turning point i.e. the point or range at which the trend changes from lower to higher costs.

There is another concept, in which a curve is drawn between the production rate v/s net present value of the revenues to be received after accounting for the revenues to be received and costs to be incurred. The second concept takes into account the time value of money and therefore, it should be preferred, when a decision in this regard is to be taken.

Too short a mine life requires higher overall costs as the capital investment increases substantially, whereas a longer mine-life has higher operating costs due to low production rates. Hence, the approach should be to operate at an optimum production rate range where the costs of mining per unit are the lowest.

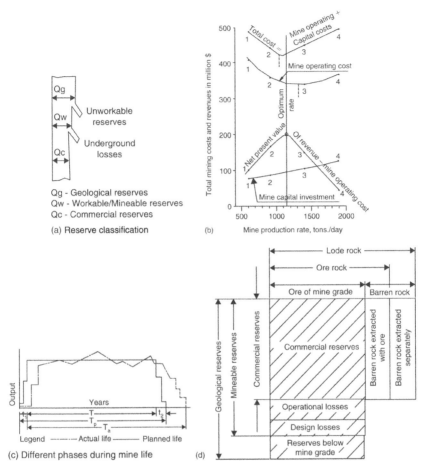

Figure 12.5   Mine design elements. (a): Commercial reserves. (b): Concepts to determine optimum production rates. (c): Different phases during mine life. (d): Losses of various kinds.

Price of mineral and expected rate of returns i.e. profit: The focal point of this dilemma is the price of the mineral and the rate of return an investor wishes to achieve. High priced minerals allow deep seated and difficult deposits to be mined out out. For example, it is the price of gold which is allowing mining its deep-seated deposits up to depths of 3 km or so. Low priced minerals can be mined up to shallow depths only and warrants cheaper and bulk mining methods to exploit them. This dilemma of inter-relationship of various parameters has been presented in figure 12.4(b).

12.2.4   Mine life

Mine Life, $T = Q_C / A_Y$
A_Y - Production rate/year.

(12.3a)

Mine Life using Taylor's relations:

Mine Life (T) in Years = 0.2 (Expected Reserves In Tons.) $^{0.25}$ (12.3b)

Mine Life (T) in years = 6.5 (Reserves In Million Tons.) $^{0.25}$ X (1 +/- 0.2) (12.3c)

This relation is a rough guide to assess the expected mine life during the feasibility or pre-feasibility period.

12.2.4.1 Phases or stages during mine life

Once the decision is taken, based on the feasibility studies to mine a particular deposit, there are three stages or phases of the mine life (fig. 12.5(c)):

1. Construction or pre-production phase
2. Rated or regular production phase
3. Liquidation phase.

Construction or Pre-production phase: This is the initial phase of the mine life (fig. 12.5(c)). During this phase activities such as mine development, construction, installation and exploration (to keep the exploration program an ongoing activity) are undertaken. The infrastructure facilities such as power, water, communication, transport, accommodation, community social & welfare, health & safety etc. are established. Recruitment of manpower, and procurement of material, consumables, machines and equipment also take place simultaneously. While construction or driving the development entries, sometimes, ore is also recovered. In underground mines stope preparation activities are also included in this phase so that the mine starts producing partially. Once a few stopes are ready for production, the regular production from the mine can be obtained.

Rated or Regular Production Phase: During this phase while few stopes or benches (in surface mines) yield the rated production of the mine, the other stopes or benches are developed to sustain this production rate. New horizons or levels are developed simultaneously to provide production when the working stopes or districts or benches have been exhausted. Exploration work to look for the new areas also goes on side by side during this phase.

Fluctuations in the production rate are experienced during this phase but efforts are made to sustain the rated output from the mine.

Liquidation Phase: This period aims to close down the mine and it starts when the ore reserves have been almost exhausted. The recovery of ore from pillars of various types is carried out during this phase (sec. 16.6). The production rate shows a declining trend. The resources in terms of labor, machines, equipment, facilities etc. are wound up during this period. Before closing, certain safety and legislation requirements need to be fulfilled. The duration of each of these phases can differ from the ones that have been planned due to changes in working environment/scenario from time to time.

Gestation or Pre-production Period: The period that is required to bring the mine into production stage from the initial stage of preliminary studies, is referred to as the pre-production or gestation period. It differs from project to project and also on the size of deposit and availability of basic resources. Comparing deep-seated deposits likely

Table 12.2   An account of time period required for various activities during the gestation period.

Activities	Duration-range
*Preliminary studies and investigations and sanction for pre-feasibility studies	1 year
**Detailed exploration and carrying out the pre-feasibility & feasibility studies	2–5 years
Preparation and submission of feasibility report	1 year
Hunting and appointment of the consultant	1 year
Mine construction and development period and start of partial production	2–5 years
Total	7–13 years

* – During this phase prospecting for the mineral deposit is carried out by the application of direct search techniques such as physical or visual examination, geologic study and mapping, sampling, and this can be supplemented by indirect methods such as geophysical (air borne or/and ground), geo-chemical and geo-botanical (section 3.2.2).

** – This phase is for sampling and evaluation. The samples may be collected by way of trenching, rotary drilling, core drilling, and by driving exploratory shafts, adits, tunnels or other excavations. Based on the assay values thus obtained, the mineral inventory of the deposit is assessed with the application of any of the reserve estimation techniques available.

to be mined by the u/g mining systems and the shallow deposits to be mined by surface mining methods; less gestation period is required in the later case. Given in table 12.2, is an approximate estimate of the time required for each of the activities involved to bring a deposit into production stage.

12.3   DIVIDING PROPERTY FOR THE PURPOSE OF UNDERGROUND MINING[1,2,3,9,13]

The basic concept of mining a deposit is to work from whole to part i.e. the deposit in totality is considered and then it is divided into workable mining units if the deposit is very extensive. Professor Popov (1964) proposed the following relations/equations to calculate the dimensions of 'take' i.e. extent along strike or across strike directions (fig. 12.6(a)).

$$S/H = 7\sqrt{\sin\alpha} + 1 \tag{12.4a}$$

Since; $SH = Q_w/\Sigma P$; $Q_c = Q_w \times C$ and $T = Q_c/A_y$ substituting these values we can get

$$S H = (T A_Y) / C\Sigma P \tag{12.4b}$$

Solving these equations:

$$H = \sqrt{\{(T A_Y) / C\Sigma P (7 \sqrt{\sin\alpha} + 1)\}}, m \tag{12.4c}$$

$$S = \sqrt{\{(T A_Y) \times (7 \sqrt{\sin\alpha} + 1)/ C\Sigma P\}}, m \tag{12.4d}$$

Unit content of the seam or deposit $P = \gamma \times m$, $tons/m^2$
   When there are number of such seams or layers

$$\Sigma P = \gamma (m_1 + m_2 + m_3 + m_4 + m_5 ----- m_n), tons/m^2$$

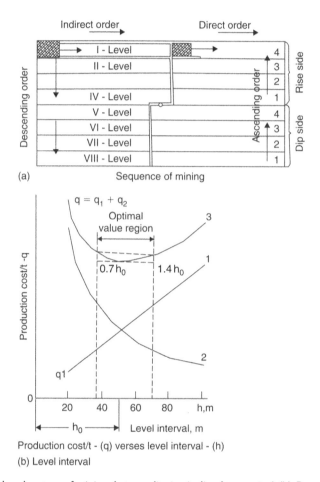

Figure 12.6 (a): Level system of mining that are dipping inclined to vertical. (b): Procedure/Technique to determine level interval.

S – dimension of take along strike direction in meters
H – dimension of take along dip direction in meters
α – the angle of dip in degrees
Q_W – workable or mineable reserves in tons
Q_c – commercial reserves in tons
A_y – annual production in tons
T – life of mine in years
C – overall recovery factor
P – unit content of ore or coal in tons/m²
γ – specific gravity of ore or coal in tons/m³
m – thickness of orebody or coal seam in m
m_1–m_n – thickness of each of the orebodies or coal seams in m, if they are more than one.

Now, the deposit within a mining unit can be further divided into two systems that are in practice:

1. Panel system
2. Level system.

12.3.1 Panel system[2,3,9,13]

If the dip of the deposit is flat to 25° or so, the deposit is divided into rectangles known as panels. The size of each panel is a function of incubation period (in case of coalmines; incubation period is the duration between the dislodging coal to begin with in a panel and appearance of fire in it due to self oxidation), output, degree of mechanization and productivity to be achieved. In coal mines its usual dimension is 800–1500 m along the strike direction and 800–1000 m across it. In some specific situations such as: extension of the deposit along the strike within 3–3.5 km and if area is opened by inclined shafts, in place of division by the panel system, the level system (described below) of mining can be recommended. Usual underground mining/stoping methods adopting the panel system are room and pillar, board and pillar and longwall mining. The panel system provides an intensive way to mine a deposit per unit time, thereby, better productivity can be achieved; but the development work required for developing the property increases considerably in the dip-rise direction comparing this with the level system. Access to the deposit is usually through the inclines if the deposit is at the shallow depths else it is by shafts for deep-seated deposits. The layout of a mine for panel system of mining is shown in figure 12.7. Thus, in this system the mine is divided into workable panels of appropriate size, separated by the pillars that are left in between them. Isolating a panel in this manner offers advantages of mining it; as an isolated portion of the deposit which can be provided with fresh ventilation and all unit operations can be undertaken independently to obtain the desired rate of production from it. In the event of outbreak of fire, explosion, inundation or any other hazardous conditions it can be isolated from the rest of the panels.

12.3.2 Level system[1,2,3,6,13]

For the deposit having dip exceeding 25° to vertical, it is divided into levels spaced usually at the height of stopes of a stoping method that is likely to be adopted and it ranges from 30 to 100 m (fig. 12.6(a)). However, the level interval depends upon a number of factors as discussed below. Levels are usually worked in the descending order, starting from the upper horizon and advancing towards the depth (fig. 12.6(a)). But in the mines where the level system of mining has been adopted, particularly if the dip of deposit is in the range of 25–40° and the make of water in the mine is excessive, the levels can be worked in the ascending order (fig. 12.6(a)). In coalmines if there is a problem due to presence of blackdamp, the ascending system proves to be useful. The shaft or access to the deposit is usually put in the center of the deposit, then the direct system of mining in which the deposit is won from the center towards the mine boundaries can be adopted to get production from the mine at the earliest. This is also termed as the advancing system. The reverse i.e. the retreating system can be also followed to avoid maintaining of the roadways passing through the worked out areas

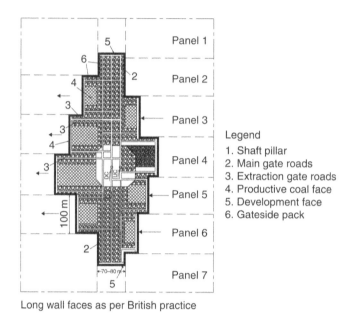

Long wall faces as per British practice

Figure 12.7   Panel system of mining.

and if advance information about the deposit (up to the mine boundaries) is warranted. Figure 12.6(a) also illustrates the sequence of mining.

12.3.3   Level interval[1,6,13]

The level interval is a vertical distance, when projected on a vertical plane, between two consecutive levels. In mines the usual interval is in the range of 30–100 m, rarely exceeding this. It depends upon geological, mining and economical factors. The geological factors include thickness, dip, and strength of rock and ore. The mining system includes factors such as method of stoping, degree of mechanization for carrying out the unit operations (such as drilling, blasting, mucking and transportation), ventilation and mine services arrangements. The economical factors comprise: I – expenses on delivering personnel, materials and equipment to the stopes, cost of repair and maintenance of the workings, cost of services in terms of ventilation, illumination, drainage etc. Let this be designated by q_1. The cost of fittings in the level such as: track, service lines, cables pipes etc. is inversely proportion to the level interval. II – the cost of driving shaft sidings, crosscuts and other main horizon workings etc. Let this be designated by q_2.

It is evident that expenses q_1/ton, increases as the level interval increases, whereas expenses q_2/t follow a trend as shown in figure 12.6(b), by curve 2. Total expenses/t of ore (q); have been represented by curve 3. This curve indicates that as the level interval increases, the value of q first decreases progressively (the curve goes down) and then, after reaching the minimum, starts increasing (the curve goes up). The lowest point in the curve 3 is represented by h_0. It has been observed that decreasing level interval to $0.7 h_0$ or increasing it to $1.4 h_0$ does not exert any noticeable influence on the economic indices.

Table 12.3 Level interval based on the practices of past for the different stoping methods.

Stoping method	Level interval, in m	Remark
Sublevel/blasthole stoping	50–80	
Shrinkage stoping	30–75	
Breast/longwall system	30–60	
Cut & fill	50–70 or more	
Squareset stoping	30–50	
Sublevel caving	50–80	
Block caving	60–100 or more	

Thus, this economic calculation can indicate the region where the lowest cost/t of mining will be obtained. But this needs to be verified further from the practical data based on the level interval that has been kept for the various stoping methods, as per the practices known to be safe. The level interval of the prevalent stoping methods is illustrated in table 12.3.

When selecting the level interval within the limits indicated in the above table, care must be taken for the prevalent conditions e.g. maximum level interval should be adopted for thin, steep, regular, stable (rock and ore), high valued orebodies. The small level interval corresponds to the opposite conditions. The modern trend is to increase level interval from 50–60 to 75–100 m due to availability of large size and highly productive equipment. Some mines are equipped with elevators for hoisting of personnel, material and equipment allowing greater level interval. In some circumstances, for example, the increasing depth or increase of rock pressure in the stoping space limits the level interval.

Some of the stoping methods require that level pillars be left between adjacent levels and adjacent stopes. Extraction of these pillars involves greater ore losses and higher expense than extracting the other part of a stoping block. Since the ore reserves blocked in pillars are usually constant at any level interval, increasing the latter decreases the percentage of ore loss, and the costs involved for robbing the pillars.

12.4 MINE PLANNING DURATION

In order to carry out the task of mine planning effectively, a mining plan should be prepared. This plan could be for a shorter or longer duration, as illustrated in figure 12.8. The shorter duration mine planning is often referred as the Short Term or Micro Planning, whereas the long duration mine planning is referred as Long Term or Macro Planning.

Micro Planning: This includes day-to-day, weekly, fortnightly monthly, quarterly, half-yearly and yearly planning. The planning for a period up to 5 years could be considered in this class. The routine maintenance tasks, daily-weekly-fortnightly-monthly production, development and installations schedules are planned. Planning for the major overhauls shutdowns and annual budgets fall in this category.

MINE PLANNING
⇓

MICRO PLANNING	MACRO PLANNING
OR	OR
SHORT TERM PLANNING	LONG TERM PLANNING
(For a period upto 5 years)	(For a period of 5–20 years)
Day to day, Weekly, Fortnightly, Monthly, Quarterly, Half yearly, Annual or specified interval Activities: Mine development, Construction, Installation, Production, Equipment maintenance, Equipment replacement, Major overhauls, Annual shutdown, Material procurement, Annual budgeting, etc.	Setting company's objectives, goals & policy. Decision on mining methods, Equipment & techniques. Sequencing exploration, Mine development & exploitation activities. Setting targets with regard to exploration, Production, Development, Procurement, Vocational training, Recruitment, Training, expansion of existing mine & plants. Search & feasibility studies for new deposits.

Figure 12.8 Types of mine planning based on duration.

Macro Planning: This planning is carried out as per the policy of the company which is usually decided taking into consideration the *National Goals and Policies* with regard to exploration, development and exploitation of the mineral resources. Overall mine planning (which includes selection of stoping method, stope and pillar layouts, sequence of mining), expansion programs and opening up the new deposits come under this category. This planning is usually for a period of 5–20 years.

12.5 MINE DEVELOPMENT – INTRODUCTION[1,2,9]

Mining is a process for extraction of any mineral or precisely speaking any ore, which can be defined as the portion of a mineral deposit that can be extracted economically (i.e. with profit). The mine is the place where this process is carried out. Before carrying out any mining operation the mine needs to be provided with the basic infrastructure facilities such as water, power, means of transport, communication etc. It needs to be equipped with facilities such office buildings, warehouses, workshops, first aid and rescue stations, basic welfare amenities, mine services; mineral handling, transportation and processing plants.

At the mines, access to the deposit is the first mining operation. If a deposit is to be mined by underground mining, then driving openings/entries, which can be horizontal, inclined and vertical, or their combination can access it. Once the deposit is accessed, to exploit it, similar types of mine openings are needed. This network of mining openings, to open and finally extract a deposit, is referred as 'Mine Development'. Development required to access the deposit is often termed as Main Mine Development or Primary Development, and the development work required for the final ore exploitation is termed as Secondary Development or Stope Preparation. Development work can also be classified as Vertical and Horizontal/Inclined development as shown in figure 12.9. The task of mine development requires a great amount of skill and experience. It is a difficult task amongst all mining operations, as it is tedious, costly and time consuming. The development is tedious due to the fact that the

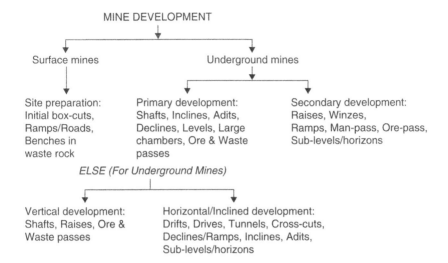

Figure 12.9 Type of excavation work required during development phase of a mine.

development openings are often driven through the strata about which no or very little advance information is available.

Similarly if a deposit is to be mined by any of the surface mining methods it needs to be developed first (see 17.5.4). The development work in this case includes site preparation, putting of initial box-cuts, driving of ramps/roads and benches in the waste rock to strip or uncover the deposit.

12.6 ACCESS TO DEPOSIT OR MEANS OF MINE ACCESS[1,2,3,6,9,13]

A deposit to be mined by underground methods can be accessed by any of the following types of mine openings or their combination:
- Adit
- Incline
- Decline/Ramp
- Inclined shaft
- Vertical shaft.

Adits can be driven across (as shown in fig. 12.10), or along the strike direction of the deposit.

Inclines (fig. 12.10) can be driven in the overlying strata of a flatly dipping shallow deposit. An inclined deposit of low thickness commencing from a shallow depth can be accessed by an incline driven from the surface and passing through the deposit itself. It can also be driven in the f/w side as illustrated in figure 12.12. Multiple seams can be accessed by an incline driven in the f/w most (bottom most) seam and connecting it by the cross measure drifts, as shown in figure 12.10.

Declines/Ramps can also be driven to join different horizons in underground mines. Shallow flatly dipping deposits or steeply dipping deposits extending from shallow

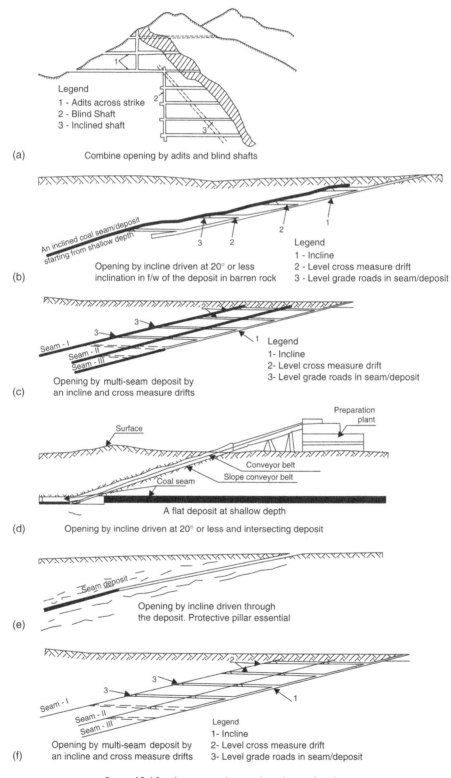

(a)   Combine opening by adits and blind shafts

Legend
1 - Adits across strike
2 - Blind Shaft
3 - Inclined shaft

(b)   Opening by incline driven at 20° or less
       inclination in f/w of the deposit in barren rock

Legend
1 - Incline
2 - Level cross measure drift
3 - Level grade roads in seam/deposit

(c)   Opening by multi-seam deposit by
      an incline and cross measure drifts

Legend
1- Incline
2- Level cross measure drift
3- Level grade roads in seam/deposit

(d)   Opening by incline driven at 20° or less and intersecting deposit

A flat deposit at shallow depth

(e)   Opening by incline driven through
      the deposit. Protective pillar essential

(f)   Opening by multi-seam deposit by
      an incline and cross measure drifts

Legend
1- Incline
2- Level cross measure drift
3- Level grade roads in seam/deposit

Figure 12.10 Access to deposit by adits and inclines.

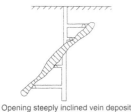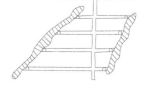

Level drive

Crosscut

| Opening steeply inclined vein deposit by vertical shaft driven through it. Keeping protection pillar is essential. | Opening steeply inclined vein deposit by putting vertical shaft in f/w side. Joininig shaft by cross cuts driven in barren rock is essential | Opening steeply inclined vein deposit by putting vertical shaft in between them. Joining shaft by crosscuts driven in barren rock is essential |

Figure 12.11   Access to deposit by vertical shaft.

depth to a considerable depth can be accessed, at the earliest by declines, as shown in figure 12.12.

Shafts serve deeper levels, which could be tracked or trackless. For steeply inclined, almost vertical or deep-seated flat deposits, access by a vertical shaft is an obvious choice as shown in figure 12.11. If the shaft is allowed to pass through the deposit, a protective pillar is required to be left in the orebody. Positioning the shaft in the f/w side requires drifting the crosscuts. Accessing a steeply inclined vein type deposit by an inclined shaft gives advantage of short cross-cutting in the barren rock, as shown in figure 12.12.

In any mine there can be more than one type of openings driven to access the deposit and provide different services. Table 12.4 compares important features of different modes of entering into deposits and provides a basis to select the one that suits a specific situation.

12.7 SYSTEM – OPENING UP A DEPOSIT[1]

Mainly the following two approaches are used to mine-out a deposit with the application of underground methods.

12.7.1 Opening deposit in parts

In this approach the initial part of the deposit or whatever portions that have been explored, are opened by driving a shaft at one or two level intervals. First the levels are developed and the stoping process proceeds there after. Simultaneously the shaft is sunk through the deeper horizons. This approach is suitable if the deposit is not fully explored and the exploration activity is to be kept as an on-going process along with the stoping at the upper levels. It is practicable if the life of a level exceeds five years or so[1], which means if the orebodies have wider extension along and across their strike direction.

12.7.2 Opening up the whole deposit[1]

In this approach the shaft is sunk through the whole deposit, intersecting or passing through all the levels planned (fig. 12.12). This may take several years. To exploit the

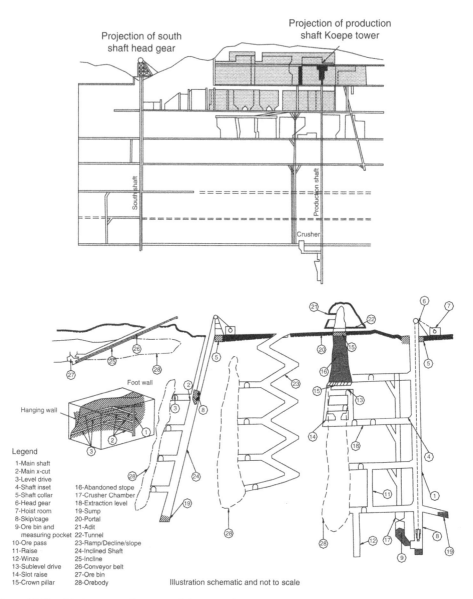

Projection of south
shaft head gear

Projection of production
shaft Koepe tower

South shaft

Production shaft

Crusher

Foot wall

Hanging wall

Legend
1-Main shaft
2-Main x-cut
3-Level drive
4-Shaft inset   16-Abandoned stope
5-Shaft collar   17-Crusher Chamber
6-Head gear   18-Extraction level
7-Hoist room   19-Sump
8-Skip/cage   20-Portal
9-Ore bin and   21-Adit
  measuring pocket   22-Tunnel
10-Ore pass   23-Ramp/Decline/slope
11-Raise   24-Inclined Shaft
12-Winze   25-Incline
13-Sublevel drive   26-Conveyor belt
14-Slot raise   27-Ore bin
15-Crown pillar   28-Orebody

Illustration schematic and not to scale

Figure 12.12   (a): Access to deposit and division of mining property into levels (longitudinal section).
(b): Access to deposit – combination of various modes of entires.

deposit at an early stage, the deposit is often accessed by other modes of entries such as adit, incline, decline or their combination so that development, stope preparation, and stoping activities can be carried out to exploit or mine-out its shallow seated portion. This approach lowers the expenses on sinking by 30–35% (Agoshkov et al., 1988) and avoids the process of shaft deepening. In this approach several development workings can be driven simultaneously at several horizons/levels.

Table 12.4 Modes of entering into a deposit and their selection.

MODES OF ENTERING INTO A DEPOSIT

PARAMETERS	ADIT (figs 12.10, 12.12)	INCLINE (figs 12.10, 12.12)	DECLINE/RAMP (fig. 12.12)	SHAFTS – Vertical or Inclined (figs 12.11, 12.12)
Definition & its suitability w.r.t haulage or hoisting systems	Almost horizontal passage of limited cross-section driven from surface to access orebody & /or provide mine services. It could be tracked or trackless	An inclined passage of limited cross-section driven from surface to access orebody & /or provide mine services. Suitable for conveyor, rope haulage/hoisting.	A passage of limited x-section driven in zigzag fashion from surface giving access to orebody & facilitating use. of trackless haulage	A vertical or steeply inclined passage of limited x-section driven from surface or u/g giving access to orebody. It serves deeper levels which could be tracked or trackless
Opening's inclination limit	Almost flat	Up to 20°	Up to 8°	>20° to vertical
Opening's shape	Rectangular, trapezoidal or arched	Rectangular or arched.	Rectangular or arched.	Rectangular, circular or elliptical.
Depth limitation	Driven at or above the valley level	Not exceeding 150 m	Not exceeding 250 m	Depth exceeding 100 m or so.
Usual rock-types through which an entry driven	Mostly in waste or black rock	Can be driven in waste rock as well as orebody	Mostly in waste or black rock	Mostly in waste or black rock
Positioning w.r.t. deposit & surface datum	F/W or H/W based on its purpose of driving. At least 5 m above highest flood level (HFL) recorded in the area	Along (within) deposit or in F/W side in waste rock. At least 5 m above HFL recorded in the area.	Preferably in F/W side of the deposit. At least 5 m above HFL recorded in the area.	For flat deposits in overlying strata but for a steep deposit in F/W. At least 5 m above HFL recorded in the area.
Principal purpose	Early access to the deposits which are extending above the valley level for carrying out u/g exploration, development & auxiliary operations	Early access to the shallow deposits to develop & produce ore at the earliest. Also equipped with mine services & serve as man-way access.	Early access to the shallow deposits to develop & produce ore at the earliest using trackless equipment. Also equipped with mine services & serve as man-way access.	Access to any deposit to develop & produce ore on a regular basis. Usually serve as permanent mine entry. Also equipped with mine services and serve as man-way access.
Other utilities details	For hauling waste rock. As ventilation & drainage outlets. Laying power cables, compressed air lines, water pipes etc. Also serves as travel roadways.	For hauling waste rock. As ventilation & drainage outlets. Laying power cables, compressed air lines, water pipes etc. Also serves as travel roadways.	For hauling waste rock. As ventilation & drainage outlets. Laying power cables, compressed air lines, water pipes etc. Also serves as travel roadways.	For hoisting waste rock. As ventilation & drainage outlets. Laying power cables, compressed air lines, water pipes etc. Also for hoisting personnel, materials & equipment.
Driving rate	Fastest	Faster	Fast	Slow
Construction cost	Least	Low	High	Highest

Note: The definitions and other important features are included in this comparison.

Shaft stations (including shaft insets) are made during shaft sinking to avoid damage to its lining and delays. The cross entries (connection) between intake and return or production or service shaft are not made at each of the levels but at an interval of two or three (some times even more) level's interval to reduce the costs. At the main haulage level which usually, is the pit bottom, together with cross entries (crosscuts) to join the two shafts, chambers/excavations are made for providing facilities such as first aid, fire fighting, electric sub-station, sumps fitted with pumps, garage, repair shops, battery charging station etc. (fig. 12.15(b) to (d)). Ore bins, crusher chambers, loading pockets etc. are the important structures, which are driven and equipped with necessary equipment and fittings.

Provision is made to deliver material (ore, waste or supplies) and movement of the man, machine, equipment and mine services, between levels by driving/connecting them through orepasses, waste passes, raises, winzes, inclines or ramps.

The shaft stations are equipped and connected with mine workings as per the specific purpose for which a particular shaft station has been planned. Under appropriate conditions only, increasing the level interval can reduce the volume of openings (i.e. development entries).

12.8 POSITIONING AND DEVELOPING THE MAIN HAULAGE LEVELS[1]

The orebody's orientation, dip, thickness, depth and properties of country rock and the ore govern this. In addition, type of haulage and stoping method also dictate its design. In thin orebodies only one drift is sufficient and w.r.t. orebody it can be positioned as shown in figure 12.13(a) in which its one of the corners is passing through the orebody, or as shown in figure 12.13(b) keeping the orebody in the middle. Lifespan of level and stability of the country rock determine the option of positioning the level entry w.r.t. h/w or f/w.

For orebodies up to 10–20 m thickness a haulage level can be positioned either in the middle (fig. 12.13(c) and (e)) or of in the f/w side (fig. 12.13(d)). In thicker orebodies several drifts in the ore and one drift in the country rock, serving as the main level drive, can be put as shown in figure 12.13(f), which was very common for the scraper haulage but in the case of trackless equipment the arrangement shown in figure 12.13(g) is more common. In this arrangement the main level can be put either in f/w or h/w side in the country rock. The crosscuts leading to the orebody can be interconnected in the form of loops, as shown in figure 12.13(h), or they can be dead-ended (fig. 12.13(g)).

Figure 12.13(i) illustrates that sharp corners should be avoided when deciding the shape of openings. Similarly Figure 12.13(j) advocates the preferred orientation of development entries. Figure 12.13(k) provides guidelines for working in geological disturbances such as faults and folds. Figure 12.13(l) details stable and unstable breakaways – useful guidelines while planning the mine development network.

12.8.1 Selecting development in ore or rock (country rock)[1]

Positioning the levels/level-drives in the rock or ore is governed mainly by the parameters such as: orebody thickness, direction of stoping, method of ventilation etc.

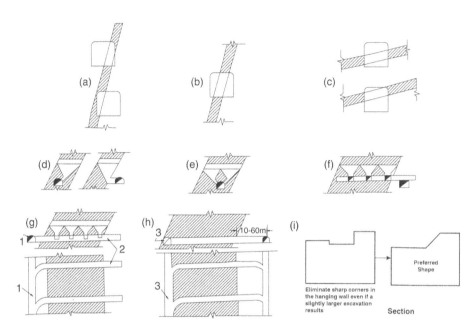

Figure 12.13   Positioning drives/levels, x-cuts and other openings with respect to ore-body. (a): Intersecting ore-body at corners; (b): Intersecting ore-body at middle of drive. (c) & (e): single drive within ore-body. (d): Single drive in f/w side. (f): for thick ore-bodies several drives in ore with a common crosscut to draw ore to the haulage level becomes essential. (g): for very thick ore-bodies putting cross entries, which could be dead ended, are essential; else it could be loop system as shown in figure (h). 1 – double track or two-lane roadway. 2 – cross entries/roadways. 3 – single track or one lane roadway. (i): Preferred opening shapes. Shape of openings should be kept as regular (without sharp corners) as possible[14] (www.smenet.org).

Positioning these drives in the country rock is almost mandatory in thick orebodies. In general, it offers some of these advantages:

- Decreased losses of ore in the form of level pillars.
- Insignificant cost of maintaining these drives as they are away from the stopes.
- Pillar recovery may be carried out without disturbing the ventilation circuits.
- Ventilation schemes become simple and practicable.
- In thin and curved (irregular boundaries) orebodies, the level drive if driven, it will not be suitable for locomotive haulage, and in case of trackless haulage system the chances of accidents increases.

The disadvantages/limitations include their driving costs compared with the one in ore in which substantial cost is paid off by ore recovery. However, the selection between the two is mainly governed by the consideration of ventilation and stoping schemes.

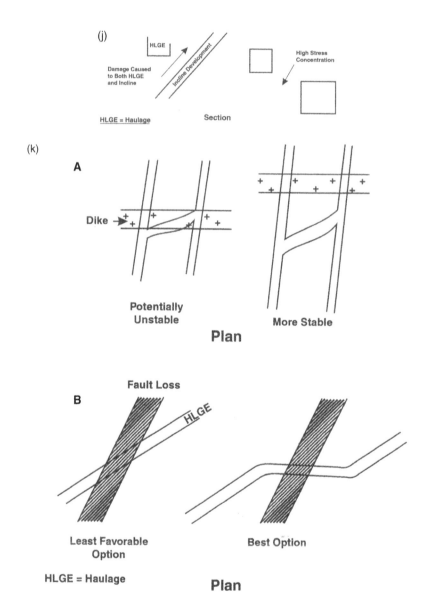

Figure 12.13   (j): The preferred orientation of the development workings. The most highly stressed
part of an excavation is its corner, and therefore, positioning the development entries
in such areas is not a good practice, and should be avoided[14] (www.smenet.org). (k):
Preferred approach to an intersection with (A) dikes and (B) faults. In permanent
structures such as hoist chambers, shaft insets, sumps etc., the excavations should be
kept away from the dykes wherever possible. The breakaways should not be sited in
dikes, even at the expense of extra development, as shown in (A). Haulages should not
be positioned in fault losses, and development should not made alongside a fault. A fault
should always be intersected at an angle as near to normal as possible, as shown in (B)[14]
(www.smenet.org).

(I)

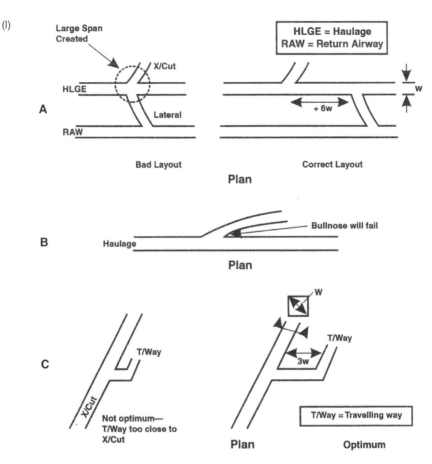

Figure 12.13 (I): Stable and unstable breakaways. A – Offsetting breakaways. B – Angle of breakaways. C – Distance between breakaways and large crosscut. Multiple breakaways should be avoided to reduce large hanging wall (roof) spans. The right design in spacing the breakaway is at six times the width of the excavation between successive tangent points, as shown in (A). Acute breakaways (i.e. at less than 45°) should be avoided since these result in 'pointed' bull noses, as shown in (B). The breakaways for inclines are often brought too close to the connecting crosscut. The length of connection between an incline and the flat should be three times the diagonal dimension of the flat end, as shown in (C)[14] (www.smenet.org).

12.8.2 Vertical development in the form of raises[1]

In general, two types of raises are put in a mine. The first type of raises are driven prior to development of a stope. They could be located within rib pillars, outside the orebody, partly in the ore and partly in country rocks. They can be vertical or steeply inclined. In a vertical raise transportation of man and material is convenient and shorter but driving them vertically without departing from the orebody is only feasible in thick orebodies. Positioning of these raises w.r.t. an orebody has been illustrated in figure 13.1(ii). If these raises are to serve a number of stopes then they should be located in country rock away from the stopes. Connecting them to a particular horizon is accomplished by driving crosscuts, as shown in figure 13.1(ii)(e). These service raises are put in most of stoping methods and used as personnel, material and ventilation outlets. These raises are equipped with ladder-ways, compressed air pipes, water lines, cables etc. These are also used to open new sub-horizons between two or more main levels, where from a particular horizon can be developed. The size of these raises depends upon the purpose they need to serve, and accordingly, they consist of one or multi-compartments.

The second type of raises are driven as the stope progresses from a lower horizon towards upper ones. These raises are used to transfer ore, waste, material or serve as man-ways. According to their purpose these are termed as: ore-pass or man-ways. Their application is limited to some stoping methods only. In addition, in some stoping methods raises are driven following the orebody profile (usually half in ore and half in waste rock in the extreme hanging wall side) to provide the initial free face in a stope. These are usually termed as 'slot raises' and ultimately converted into 'slots' (an excavation to provide initial free face for the stoping operations to start with). (Please refer to section 16.2.3).

12.8.3   Connecting main levels by ramps/declines/slopes

With the application of self-propelled trackless haulage and mucking equipment such as low profile trucks, dumper, shuttle cars and LHDs, the haulage layouts have been changed considerably. Use of ramps/declines or slopes is made to develop and carry out production activities at several levels simultaneously. This system has resulted in faster rate of development and stoping with reduced costs and better productivity (figs 1.7(a), 16.16).

12.8.4   Determination of optimal load concentration point[2,9]

12.8.4.1   Analytical method

1	2	3	4	n	m
q_1	q_2	q_3	q_4	q_n	q_m

Let us consider loads q_1, q_2, q_3,—q_n—q_m are concentrated along a certain route having distances l_1, l_2, —l_n—l_m between them.

Aim: To find the OPTIMAL LOCATION POINT, which means to find a point where all other loads should be hauled, to minimize tons-km performance.

Let us designate this point 'O', which may be located at: either of the terminals 1 or m, or, at any of the intermediate points, say, 2, 3, 4, —n etc. Or, somewhere in the section between these points.

(1) Let us consider 'O' lies at one of the terminal points, say, m.
This can be optimal, only if the total of the loads concentrated at all other points is less than what it is at point m, mathematically:

$$Q - q_m < q_m$$
$$\text{Or} \quad Q < 2q_m$$

Or $q_m > Q/2$

(2) Let us consider that point 'O' coincides with any of the points, say, n
This will be optimal if the sum total of the loads concentrated at this point and the load coming from one direction is less than what is coming from the opposite direction, mathematically:

$\Sigma\ q_{left} < q_n + \Sigma\ q_{right},$
similarly: $\Sigma\ q_{right} < q_n + \Sigma\ q_{left}$ (12.5a)
But $q_n + \Sigma\ q_{right} = Q - \Sigma\ q_{left}$

Substituting the value value of $(q_n + \Sigma\ q_{right})$, into eq. (12.5a), $\Sigma\ q_{left} = Q - \Sigma\ q_{left}$

Or $2(\Sigma\ q_{left}) < Q$
Or $\Sigma\ q_{left} < Q/2,$ (12.5b)
Similarly, $\Sigma\ q_{right} < Q/2$

Thus, any point on the route will be an optimal point if the load brought from any one direction (left or right) is less than half the total load.

12.8.4.2   *Graphical method: funicular diagram (fig. 12.14(a))*

- Use the X axis for marking distances at which various loads are located. Use the Y axis for work done in ton-km or ton-m
- Project by dotted lines the ordinates at each of the points marked on the X-axis.
- Start plotting the points in the following manner:

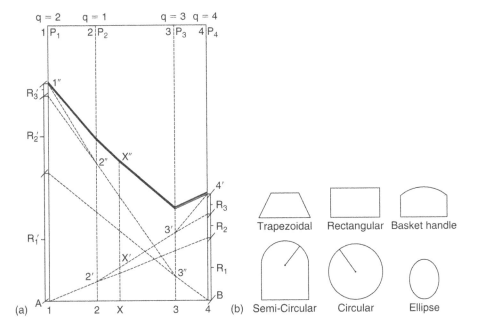

Figure 12.14   (a): Optimum positioning of production outlets/openings. (b): Openings of various shapes.

(1) Hauling loads to any of the terminals
While bringing loads to A, refer to the figure:
- Calculate work required to be done by B to haul the load to A, plot it on the ordinate of point A.
- Join it with B, it is intersecting the ordinate of point 3 at 3″.
- Extend the ordinate marked at A by the amount equal to the work required to be done by point 3 in bringing the load to A, join this to 3″. This line is intersecting the ordinate of point 2 at 2″. Again extend the ordinate marked at A by the amount equal to work required to be done by point 2 in bringing the load to A, join this to 2″.

(2) Similarly plot the ordinate at another terminal B.

(3) To plot the performances at intermediate points on the route, say point 2 or 3:
- Extend the ordinate 2-2″ by the amount of work required to be done by bringing the load from A to point 2 i.e. equal to 2-2′.
- Similarly extend the ordinate 3-3′ by 3-3″, to obtain total work done at 3.
- Join these final points by a firm-line.

(4) To find out amount work required to be done at any point x
- Project the ordinate at this location.

12.9 SIZE AND SHAPE OF MINE OPENINGS AND TUNNELS[2,3,13]

To determine the size, which means width and height, of a mine opening the following guidelines can be adopted:

Width: For trackless mine roadways, the minimum width to be kept:

- *One lane traffic: width of a unit* + 2 m* $\qquad$ (12.6a)

- *Two lane traffic: width of 2 units + 1.5 m* $\qquad$ (12.6b)

* The one which is the wider most.

For tracked mine roadways, the *minimum width* to be kept:

- Single Track = Clearance from the travel side not less than 750 mm + width up to sleepers or tubs + clearance of 0.3 m other side. (12.7a)

- Double Track = Clearance from the travel side not less than 750 mm + width up to sleepers or tubs with a minimum clearance of 0.3 m between in-bye and out-bye tubs or locomotives + clearance of 0.3 m other side. (12.7b)

- Locomotive haulage – the width of the roadway should be considered at a height 1.5 m above the railhead in the case of a trolley wire locomotive system. For battery loco-haulage this with is considered at a height of 1.420 m above the railhead. In the case of mine cars up to 2 tons capacity, with rope haulage, this width is considered at 1.3 m above the rail-head; else it should be considered at a height of 1.3 m above the rail-head for mine cars of bigger capacities.   (12.7c)

In track laid roadways: Ballast thickness = 180–200 mm
Height from ballast surface to rail head = 150–180 mm
Thus, from floor the height occupied by these items = 330–380 mm

> To calculate the width of excavation = Width as calculated above + Thickness of support/sets + Gap behind these sets to accommodate lagging, at least, 50 mm both sides.   (12.8)

> *Height of roadways – the minimum height to be kept*:   (12.9)

- Haulage roadways not less than 1.9 m
- Roadways with trolley wire locomotives not less than 2.2 m
- Height of auxiliary workings should not be less than 1.8 m

> *Other points to note*:   (12.10)

- In arched roadways the height should be considered excluding the arched portion.
- For trackless or any other equipment there should be a minimum clearance of 0.3 m from the roof.
- The post of the frame of a trapezoidal timber/steel set should be kept at 80°
- Allowance for roof subsidence up to 100 mm, should also be made.

Ventilation requirements: The size of opening should be such that:

- Velocity of air should not exceed 8 m/sec in ventilation and main airways.
- In all other workings it should not exceed 6 m/sec.

This may be noted that civil tunnels will have different criteria to determine their size and shapes. It would be mainly governed by their purpose, utilities and life.
> *Shape of the openings and tunnels depend upon:*

- Type of rock/strata
- Depth of working, or planned depth of tunnel
- Life of opening (or tunnel) and its utility
- Stability of opening/tunnel based on its shape
- Presence of geological disturbance, if any
- The available useful cross-sectional area
- Its construction/driving, and maintenance costs.

From the stability point of view, the various shapes available in increasing order of stability are (fig. 12.14(b)):

- Trapezoidal
- Rectangular
- Wider arch
- Narrow arch

- Circular
- Pearl/pentagon
- Hexagonal with vertical apex.

The ratio between the whole cross-sectional area and useful cross-section:

- Rectangular – 1:1
- Arched sides both ways – 1.22:1
- Elliptical – 1.27:1
- Circular – 1.30:1.

12.10 PIT TOP LAYOUTS[2,3,9,13]

Mine pits are the lifelines of any mine. Positioning them judiciously in relation to the deposit to be mined is of prime importance, and so are the arrangements that need to be made around them at the surface as well as underground at the landing stations and pit bottoms. Pit top and pit bottom are the terminal stations for the vertical transport system and as such they should be equipped with all the necessary facilities to handle personnel, machines, equipment, materials, ore and waste rocks. The layout at any station including the terminals should be compact, tidy and well illuminated. In order to handle output from the mines effectively, several types of pit top layouts or designs, as described below, are available.

1. *Run Round Type – Pit Top Layout*: This layout is suitable for high output and requires a large surface space. Handling of ore of different grades can be achieved. The waste rock can be handled separately.
2. *Shunt Back Type – Pit Top Layout*: It is a cheap, simple and effective arrangement for reversing the mine cars. This is best suited for the long wheelbase mine cars. It can handle an output up to 2000 ton/day. A similar design is applicable for pit tops also but it can be spread over a larger area comparing the one at any of the underground shaft stations.
3. *Turn Table Type – Pit Top Layout*: This ensures continuous feed of mine cars. The reversing of cars is achieved within a restricted space. Electric power operated turn tables are used if the output is more than 500 ton/day.
4. *Traverser Type – Pit Top Layout*: It is a very compact circuit and once installed cannot be changed. Where limited space at the surface is available this arrangement is better. This circuit can handle output of 45–60 winds per hour.

12.11 PIT BOTTOM LAYOUTS[2,3,9,13]

The pit bottom is a link between vertical hoisting and horizontal transportation and it must ensure full utilization of both the systems. There are two types of *shaft station intersections* – single and double, based on whether the shaft has one or two outlets at the pit bottom. The first type is naturally simple in design and less costly. But its main disadvantage, when considered in conjunction with track system, is that before a loaded car can be pushed into the hoisting cage the empty one has to be pulled out of it in the direction opposite to the first. Hence, considerable time and labor is required. With the

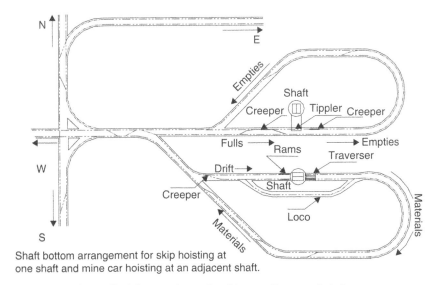

Shaft bottom arrangement for skip hoisting at
one shaft and mine car hoisting at an adjacent shaft.

Figure 12.15   Shaft bottom layout for skip as well as cage hoisting systems.

double stations, the mine cars are loaded and unloaded from the cage in one direction (fig. 12.15). This takes less time and so this operation can be mechanized with the use of mine car pushers. Hence the single intersection shafts have very limited applications such as for the purpose of exploration, low output mines, short service life or with no hoisting plants or with auxiliary plants operating irregularly. To secure direct communication between two sides of the station, a passageway for men is usually provided near the shaft under the ladder compartment of the shaft or by a by-pass. At its intersection the shaft inset needs to heightened, as per the maximum length of material to be hoisted/ lowered, e.g. at least 4.5 m with arched ceiling and at least 3.5 m with flat back to facilitate the handling of longer pipes, timber, equipment etc. to be received from the pit top.

Figure 12.15 illustrates a shaft station, which could be a pit bottom layout or any of the stations serving a particular horizon of an underground mine. The types of facilities and arrangements which need to be made have been shown. In the case of track mining for proper handling of mine cars, the layouts with suitable designs are essential. In the following paragraphs pit bottom layouts of different types have been described.

12.11.1   Types of pit bottom layouts[3,5b]

1. *Shunt Back System*: This layout avoids loops and brings empties to the full side of the cage with the help of a traverser, turntable or shunt back arrangements/ switches incorporated in the circuit. This reduces a long run round, and also keeps the travel time to a minimum. When the shaft axis is in line with the main haulage axis or at a right angle to it, there can be one or two reversing switches as shown in figures 12.16(a) and (d) respectively. There can be a combination of loop and reversing switch, as shown in figure 12.16(f). Capacity to handle mine cars in such an arrangement is limited.

Pit bottom layouts

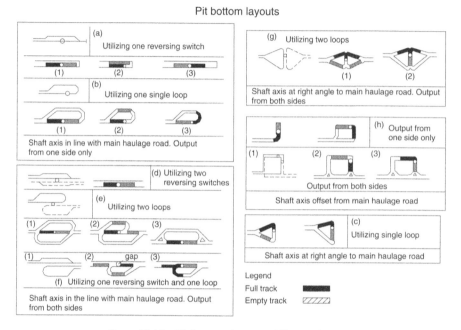

Figure 12.16   Pit bottom layout – different designs.

2. *Loop System*: In this type of layout a loop is provided for bringing load on one side of the shaft and taking empties to the districts. A large loop can provide space for keeping the empty mine cars. There could be two loops (figs 12.16(e), (g)) and the output can be received from two opposite directions of the pit bottom. For high output and large mines this pattern can be used. With single loop (fig. 12.16(b) and (c)) moderate output can be achieved. When the shaft axis is in line with the main haulage axis or at a right angle to it, there can be single or double loops, as shown in figure 12.16.

3. *Blind pit bottom*: This pattern is adopted at the peripheral shafts (the shafts at the terminal ends of a deposit to be mined) including the staple shafts, where small capacity hoisting installations are in use and there is very little scope for mechanization. Here usually a cage system of hoisting is adopted and the transport axis is in line with the hoisting axis (fig. 12.16(a)).

In practice many patterns of shaft bottom layouts are available. In general, the following relation can represent it:

$$X_n - N - Y - Z - \alpha° - L$$

Whereas: X_n – Mines as the per daily output

N – Number of shafts at the pit bottom

Y – Number of skip hoisting installations

Z – Number of cage hoisting installations

$\alpha°$ – Angle between the main shaft axis and main haulage road (in line with, perpendicular or at an angle)

L – Max. number of haulage tracks in the main haulage roadways at the shaft bottom horizon.

Thus, a pit bottom layout is a function of the parameters such as number of shafts, their orientation in relation to the haulage roadways – tracked or trackless, number of tracks/ lanes, output required, type of conveyances used – cage or skip or both, type of mine cars – their size and shapes, provision for handling waste, ore and/or mine services etc.

The layout of the pit bottom should ensure reasonable capacity and safety. It should be simple in switching operations with the requirement of minimum labor force. The volume of excavation work should also be minimum. To secure traffic safety and reduce the number of people working at the shaft stations, wide use of signaling and automatic devices should be made.

The shaft layouts can also be referred to as: I – Shaft stations with cage hoisting (main hoisting is carried by cages). II – Single shaft stations with combined skip and cage hoisting. III – Double shaft stations – skip and cage hoisting are done through the separate shafts as shown in figure 12.15. The shaft can be at an angle with the main axis and the output can be obtained from one or both sides.

12.12   STRUCTURES CONCERNING PIT BOTTOM LAYOUTS[2,3,9]

Skip loading pockets
In case of skip hoisting (figs 12.12, 12.15, 1.7) special pockets have to be provided to load the skips. These pockets have the capacity, usually, equal to the payload of a skip. These measuring pockets are fed with ore through the various routes or mechanisms.

Surge pockets/bins
In the skip hoisting system storage bins are driven to store the ore equal to production of a few shifts to a few days, as per the planning made in this regard. These bins receive the crushed ore; from the crusher chamber wherever crushing is essential else the ore is screened through the grizzlies is fed into it. In some mines ores of different grades are stored in the separate compartment of the same bin. The reinforced concrete partition is made to create separation. In large mines *ore passes* also serve a good means to store the ore.

Underground crushing
Sometimes it becomes necessary to undertake primary crushing of ore underground. To have this facility a crusher chamber equipped with a crusher and its fittings is installed. Usually gyratory or jaw crushers are suitable for hard ore crushing. The ore after crushing can be fed into a storage or surge bin from where through a suitable gate or belt conveyor (in some layouts) the ore is fed to the measuring pockets, for its ultimate discharge into the skips. This ore is then hoisted up to the surface to feed the concentrator plant or a stockpile.

Figures 12.12 and 15.9 depict a section through the ore crushing installation and orebin, which feeds the ore into the skip via a measuring pocket.

12.13   THE WAY FORWARD

As shown in figure 18.21 and described in sec. 18.6.6 more than 300 mining companies are using software popularly known as '*ERP – Entrepreneur Resources Planning*' which has the potential to revolutionize processes and optimize resources. Such standardization and integration strategies should be considered by every mine owner including use of available software for the various modules related to mine planning and design.

QUESTIONS

1. "It is the person behind the machine that matters." Is this a true statement? How could the quality of human resources be ensured/improved?
2. A feasibility study goes parallel with geological and technical investigations, with certain aims to achieve; list those aims.
3. A feasibility study has to pass through three distinct stages, list them.
4. A lead and zinc deposit, having geological reserves of 20 million tons is to be mined by underground methods. 1.5 m million tons of these reserves have thickness less than 1 m and are also scattered in the form of small lenses and patches, and cannot be mined economically at present. Out of the workable reserves 25% are likely to be blocked in the form of pillars of various kinds, and recovery during the pillar recovery phase is likely to be 80%. The expected recovery from the stopes is 90%. Calculate the commercial reserves and overall recovery factor. If the optimum rate of production for this mine is 1600 tons per day, considering 300 working days/annum, calculate the life of the mine.
5. The cut-off grade of a mine is increased from 0.5% to 1% Cu; is this going to result in an increase or decrease in ore reserves?
6. Define: ore, gangue, break-even grade, balancing cut-off grade, cut-off grade, fixed cut-off grade, variable cut-off grade.
7. Describe and illustrate the parameters that are relevant in the production dilemma to win a deposit.
8. Distinguish between commercial reserves and mineable (workable) reserves.
9. Distinguish between macro-planning and micro-planning used in conjunction with mine planning.
10. For the deposit selected for mining, a basic planning model is prepared by involving specific engineering studies and framework – list what they are and describe them briefly.
11. From a stability point of view, list the various shapes available in the order of their increasing stability.
12. Give an account of gestation period, mentioning its duration.
13. Give Popov's relation for dividing a deposit into working units i.e. mines. Also derive this relation to calculate the dimensions of the mine along and across the dip of the deposit. Prepare a table showing the H/S ratio taking dip angles 0 to 90 at 10 degree intervals.
14. Given below is a guideline to gather and compile information/data on the various parameters which need to be included while undertaking the feasibility studies for different types of mineral deposits – list the information and data which need to be gathered.
15. "Grade-tonnage computation (curve) is a basic foundation for any mine upon which it can be developed, constructed and run for the purpose of obtaining the desired production". Is this true? Write down in detail how is it established.

16. How are shafts or mine openings optimally located? Four ore shoots A, B, C and D have reserves of 8, 4, 12 and 18 million ton respectively. These reserves have been planned to come to a common crosscut during their mining for the purpose of hoisting them to surface. The ore shoot A is located to the extreme west side of the deposit, and from this ore shoots B, C and D are located at a distance of 0.2 km, 0.5 km and 0.65 km respectively. Suggest the optimum position of the shaft to be put in.

17. How shall you decide the level interval? Answer to the point and also illustrate it graphically. Also mention the parameters you shall consider to decide the size of a coal panel.

18. How shall you divide the deposit within a mine? Illustrate your answer if the deposit is: (i) having a dip less than 20 degrees, (ii) having a dip greater than 20 degrees. How are the mines developed in each case? Also mention how the exploitation process goes in each case.

19. Has the use of ramps/declines resulted in a faster rate of development and stoping with reduced costs and better productivity?

20. How would you take care to position drives/levels, crosscuts and other openings with respect to an orebody while designing underground mines?

21. If 1 ton of copper having an average grade of 1.5% Cu is fed to a mill to produce concentrates of 18% Cu, calculate the amount of concentrates, if recovery from the mill is 90%.

22. If copper that is produced at a mining company is sold at the rate of $1400 per ton, and during its processing the recovery is 80%, calculate the cut-off grade of the copper ore if its cost of mining and processing is $6 per ton of ore.

23. Illustrate the concept of production dilemma. What is the focal point of this strategy? How do you determine the optimum rate of production of a mine? Mention which concept is better and why?

24. In conjunction with pit bottom layout, detail out these systems: shunt back system, loop system and blind pit bottom.

25. In general, two types of raises are put in a mine. What are they? Work out the details.

26. In the process of planning an underground metalliferous mine, once the preliminary selection of mining method/methods has been made, how shall you undertake its engineering evaluation to cover the following aspects: conceptual model/studies; engineering model/studies; detailed design. Illustrate your answer by way of suitable line diagrams. On a similar logic prepare these models if the deposit is to be mined by open pit mining methods.

27. In the process of planning an underground metalliferous mine, the preliminary selection indicated mining by the application of sublevel/blasthole stoping. Prepare a line diagram to illustrate the conceptual aspect/model of the engineering studies/evaluation. Name other two models that should also be prepared to cover the engineering evaluation of the method selected.

28. In underground mines, even for the same deposit, the cut-off dependent costs vary significantly according to what? List those parameters.

29. "In underground mines, even for the same deposit, the cut-off dependent costs vary significantly according to stoping methods, orebody thickness, degree of mechanisation and working efficiencies of the men and machines". Is this true?

30. List two approaches that are used to mine-out a deposit with the application of underground methods.
31. List mine design elements. List parameters influencing cut-off grades and describe the inter-relationship amongst them.
32. List the level intervals based on practices of the past for different stoping methods.
33. List the types of pit bottom layouts available to handle the mine cars at a pit (shaft) bottom. Draw any two such layouts.
34. List the types of pit top layouts available to handle the mine cars from a production winding shaft. Draw any two such pit top layouts.
35. List the various shapes of mine openings on an increasing order of stability. Also mention the ratio of whole cross-section to useful cross-section for any two of them. Design any four shapes known to you for the type of roadway specified below: A trackless roadway with two lane traffic, to allow use of following equipment: LHD with a height of 2.4 m (up to canopy), width = 1.9 m; truck with a height of 2.5 m (up to driver's cabin), width = 2.5 m. Use a graph paper to draw them on a suitable scale.
36. List patterns of shaft bottom layouts that are available. How would you represent them? Write the nomenclature used to represent them.
37. List the range of economic studies that are usually done prior to declaring a deposit suitable for mining.
38. List the structures that are encompassed within a pit bottom layout describing the function and utility of each one of them and what design considerations should be taken care of.
39. Mainly two different approaches are used to mine-out a deposit with the application of underground methods. Describe them.
40. Make a distinction between the following: (i) Mineral inventory and ore reserves. (ii) Cut-off grade and break-even grade. (iii) Commercial reserves and workable reserves. (iv) Macro- and micro-mine planning.
41. Name the different phases of the life of a mine.
42. Outline the salient features that need to be considered while undertaking feasibility studies and preparing a report for metallic, non-metallic and fuel deposits.
43. Positioning the levels/level-drives in the rock or ore is governed mainly by the same parameters; list what they are. List the advantages of positioning these drives in the country rock.
44. Propose the size of a mining unit for a coal deposit which may measure 10 km × 2 km. The average dip of the deposit is 40 degrees. Consider the life of each unit to be 30 years. The overall recovery factor during mining is expected to be 0.9. The production of 1.5 million ton/annum is planned for each of these underground mines. Take average thickness of coal to be 11 m (including all the seams) and specific gravity of coal as 1.3 ton/cum.
45. What are the formulae which could be used, depending upon the policy of a company, to calculate cut-off grade? List/describe them.
46. There are chromite deposits scattered along a hilly terrain within a span of 50 km in Oman. The distance between each of these deposits, on average is 1.9 km. The mineable reserves at these locations, in thousands of tons, are as follows: 40, 50, 60, 70, 80, 90, 100, 100, 90, 80, 70, 60, 50, 40, 30, 20, 10, 120, 65, 35. Propose an optimum location of a crushing station so that ore from all these locations/mines

can be brought (to this location) for the purpose of crushing and then to be dispatched to UAE harbor. Write down the rule you applied.

47. For a trackless roadway with one lane traffic, to allow use of following equipment: LHD with a height of 2.5 m (up to canopy), width = 1.9 m; truck with a height of 2.5 m (up to driver's cabin), width = 2.5 m; calculate the height and width of the rectangular roadway.

48. What do you understand by the following: geological reserves, mineable reserves and commercial reserves? Losses of various kinds/types.

49. What do you understand by the optimum rate of production? Give different logic/ways to find it out, and mention which logic is the best.

50. What is DPR, what is its purpose?

51. What is a gestation period? What is its usual range (write in years)?

52. What is mine planning? Classify it. If the life of a mine is more than 15 years, mention what agencies/organizations should be consulted during its planning.

53. What will be the optimal concentration point of a route having loads of 5, 10, 20, 30, 50, and 60 million tons, each being at a distance 1 km apart.

54. When selecting the level interval what precautions should be taken for best results?

55. Why should only the variable costs be used in cut-off grade calculation?

56. Write guidelines to determine the size (which means width and height) of a mine opening.

57. List the modes of entering deposits suitable for underground mining. Suggest a mode of entering for the following situations:
   a. Deep-seated flat coal deposit.
   b. A steeply dipping deposit blanketed by a hill which is 50 m above the valley level.
   c. A shallow coal deposit dipping at 10°.
   d. A steeply dipping deposit between depths of 50 m to 200 m from the surface. The mode should be suitable for the use of trackless haulage equipment.

58. List the different modes of entering a deposit to be worked by underground methods. Compare them with regard to the following:
   a. Openings inclination limit.
   b. Depth limitation.
   c. Driving rate.
   d. Construction cost.
   Illustrate each one by way of suitable sketches.

59. Distinguish between mineral inventory and ore reserves. Define cut-off grade, break-even grade and average grade/mill grade. Draw a grade tonnage curve using the following data. Also mention its significance in the mine planning process.

Cutoff grade (% Cu)	Average grade (% Cu)	Quantity in million tons
0.1	0.25	20
0	0.60	6
0.5	0.80	5
0.7	1.00	4
1.0	1.20	4
1.3	2.0	16

Find the ore reserves at 0.5% cut-off grade. Also calculate the metal (in tons) at this cut-off grade if the overall recovery factor is 0.85. Also mention its significance in the mine planning process.

1	2	3	4	5	6	7
Cutoff Grade %	Average Grade %	Quantity tons	Cum. reserve tons	Metal tons $(C_2 \times C_3)/100$	Cum. metal tons	Average grade %, $(C_6 \times 100)/C_4$
			ΣC_3		ΣC_5	
		ΣC_3		ΣC_5		

Use the above drawn table for the calculation of the mineral inventory, and for assessing and plotting data for GT Curve. Note: for GT Curve: plot Col. 1 v/s Col. 4 and Col. 7 v/s Col. 4.

60. Given the following input data for a copper mining complex, calculate the cut-off grade.
Cost data:
Variable mining cost/t of ore = $4.10
Variable milling cost/t of ore = $1.8
Variable smelting cost/t of concentrates = $33
Variable refining cost/t of anodes = $25
Price data:
Copper price/t = $1700
By-product price/t = $225
Grade and Recovery data:

Plant	Grade (% Cu)	Recovery (%)
Concentrator	16.81	86
Smelter	99.80	96.5
Refinery	99.99	99

61. Draw the shapes of mine roadways for each of the following situations, allowing their use for purposes such as: haulage, mine services and travel ways.
   a. Single track with 1 m gauge, allowing use of mine cars of 1.6 m width and 1.5 m height (above the track head) – trapezoidal. Show position of mine services and drain.
   b. Double track with 1 m gauge, allowing use of mine cars of 1.6 m width and 1.5 m height (above the track head) – rectangular. Allow supports with lagging of 150 mm all the three sides.
   c. Trackless roadway with one lane traffic, to allow use of following equipment: LHD with a height of 2.5 m (up to canopy), width = 1.9 m; truck with a height of 2.8 m (up to driver's cabin), width = 2.5 m; – Arch with two centres.

d. Trackless roadway with two lane traffic, to allow use of following equipment: LHD with a height of 2.6 m (up to canopy), width $= 2.0$ m; truck with a height of 2.6 m (up to driver's cabin), width $= 2.5$m; – Arch with three centres.

Use graph paper to draw them on a suitable scale.

REFERENCES

1. Agoshkov, M.; Borisov, S. and Boyarsky,V.: *Mining of Ores and Non-metallic Minerals.* Mir Publishers, Moscow, 1988, pp. 13–14, 33–35, 45–55, 63–68, 245.
2. Boky, B.: *Mining.* Mir Publishers, Moscow, 1988, pp. 254–58, 338–49.
3. Desmukh, R.T. and Deshmukh, D.J.: *Winning and working coal.* Mitra Press, Calcutta, 1967, pp. 236–269.
4. Dowis, J.E.: Detailed cost estimating for a mining venture. In: *Mining Industry Costs,* J.R. Hoskin (edt.). Northwest Mining Asso. Spokane, W.A. 1982, pp. 193–212.
5. Folinsbee, J.C. and Clarke, R.W.: In: *Design and operation of caving and sublevel stoping mines,* D.R. Steward (edt.). SME-AIME, New York, 1981, pp. 55–65.
5b. Fritzsche, C.H. and Potis E.L.J.: *horizon mining.* George Allen & Unwin Ltd., London, 1954, pp. 510–515.
6. Hartman, H.L.: Introductory Mining Engineering. Johan Wiley & Sons, New York, 1987, pp. 288–98.
7. Lane, K.F.: Choosing the optimum cutoff grade. *Colorado School of Mines Quarterly,* 59, 1964, No. 4.
8. Lane, K.F.: Commercial aspects of choosing cutoff grades, *16th APCOM Symposium,* 1979, 280–5.
9. Shevyakov, L.: *Mining of Mineral Deposits.* Foreign Language Publishing House, Moscow, 1988, pp. 109–144.
10. Tatiya, R.R.: *PhD Thesis,* Imperial College of Science and Technology, London University, 1987, London.
11. Taylor, H.K.: General background theory of cutoff grade. *Trans. Inst. Min. Metal.* London (Section A: Mineral Industry), 1972, A160-79.
12. Taylor, H.K.: Cutoff grades – some further reflections, *Trans. Inst. Min. Metal.* London (Section A: Mineral Indst), 1985, A204-16.
13. Vorobjev, B.M. and Deshmukh, R.T.: Advanced coal mining. Mitra Press, Calcutta, 1964, pp. 83–193.
14. Bullock, R.L.: General planning of the non-coal underground mine. In: Hustrulid and Bullock (eds.): *SME Underground Mining Methods.* Colorado, 2001, pp. 15–22.

Chapter 13

Excavations in upward direction – raising

Raising operations for raises exceeding 10 m in length used to have the highest accident rate in the past but advancements in technology for driving raises exceeding 1000 m in length was once a dream but is now a reality.

13.1 INTRODUCTION

In an underground situation one of the important openings is the raise, which is driven in the upward direction. It can be vertical or steeply inclined. Opposite to raise is a winze, whose driving mechanism is just the reverse. While driving raises a crew has to approach the non-scaled back after blasting a round whereas in a winze there is no problem of this kind although their feet, in most of the cases, are in water. During raising gravity assists in drilling and mucking, thereby making the process faster and cheaper; but in winzing it slows down the drilling speeds and the blasted muck needs hoisting. Thus, driving a winze is a slow, tedious and costly affair but provides better safety to working crews than raising. Earlier raising used to be considered as one of the most hazardous mining operations but with the advent of new techniques the process has become more safe and economical than winzing. However, winzing or sinking is an indispensable operation to give an access to the deeper horizons, and to join lower horizons (levels) to the upper ones raising is an established practice.

13.2 RAISE APPLICATIONS IN CIVIL AND CONSTRUCTION INDUSTRIES

Raises are one of the important structures in many civil and construction projects (fig. 13.1(i))[2] as detailed below:[13]

1) Hydroelectric projects
   (a) Surge chamber
   (b) Ventilation shaft
   (c) Elevator shaft
   (d) Pressure shaft
   (e) Cable shaft
2) Water supply
   (a) Access or service shaft
   (b) Ventilation

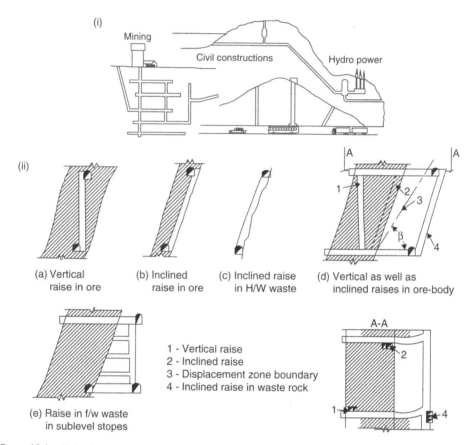

Figure 13.1   (i) Application of raises in various industries. (ii) Positioning raises in relation to orebody.

 (c) Supply riser
 (d) Uptake or down-take shaft
3) Waste water shafts
 (a) Drop shafts
4) Tunnel projects
 (a) Ventilation
 (b) Accelerators housings
 (c) Access ways.

13.3   CLASSIFICATION – TYPES OF RAISES FOR MINES

The raises can be grouped into two categories (ref. Sec. 12.8.2 also), the one which is driven prior to stoping an ore block i.e. the stope, and the other one, which is driven as the stoping process progresses upward in a fill or loose ore. In the former group also there are two types of raises, the one which is used to have an access within the stope to the personnel, machine, material, air, fills, utilities such as compressed air, water, electric cables etc. This is known as a service raise or man-way. The other one is used to provide an initial free face for the stoping operations to start with in some of the stoping systems. This is known as a slot raise. The dimensions of a service raise depend upon its utility. It can be divided into a number of compartments. One of them is usually a stepladder man-way and others are intended for discharge of ore to the lower haulage horizon, lowering of fill from the upper horizon, delivery of material, and ventilation. Vertical raises are more expedient because of small wear to the sides, more convenient transportation of men and shorter length. The raises driven should be able to provide a safe access to the stopes and minimum cost of their driving and repair, and good ventilation of the stoping faces. Raise placement in relation to the orebody is governed by the purpose it needs to serve and based on this logic it could be placed within the orebody, following the profile of the orebody, half in the ore and half in the contact rock and away from the orebody (usually in f/w side), as shown in figure 13.1(ii).[1] Because raises need to be preserved for a longer duration they are usually located in the country rock on f/w side. Raises meant to serve as an ore pass should have inclination of at least 60°. Apart from the stoping sections, the raises are nowadays driven to act as ore passes, waste passes, backfill inlets, ventilation inlets or returns and emergency access to the mines from surface to underground or from one horizon to others within a mine. To summarize, in underground mines the applications include:

1. Ventilation access (for the whole mine)
2. In stopes: ventilation, services (to accommodate supply lines, ladder ways, material hoisting etc.)
3. In some stoping methods as a slot raise it provides initial free face and others for creating funnels, and also as fill pass, drainage tower, ore pass and man-pass.
4. Waste and ore passes for transferring waste rocks and ores from one horizon to other.
5. Emergency escape
6. Pilot for shaft sinking.

Raises could be used as ore passes to haul up the production from deep-seated open pit mines, as shown in figure 17.16(e)[1].

13.4   RAISE DRIVING TECHNIQUES

The raises, for the purpose of their driving, can be classified as blind and the raises that have two levels available to access them. The former is more difficult to drive than the latter one. The other classification can be based upon with and without use of explosives while driving them. The line diagrams, in figures 13.2 and 13.3, presents the classification based on both these criterion. The techniques in use have been described in the following sections. This may be mentioned that use of stoper or parallel raise feed drills is made while driving all types of raise except those driven by the use of blasthole drills and raise borers. In these raises use of conventional explosives is also made to blast them.

13.5   CONVENTIONAL RAISING METHOD: OPEN RAISING

This is the oldest and simplest method of driving raises of very short length particularly in the competent ground. The process consists of drilling and blasting the initial

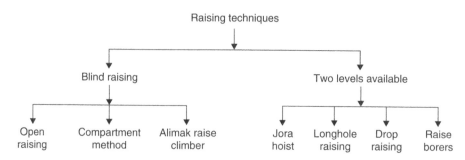

Figure 13.2   Classification of raising methods/techniques based on the availability of access to the intended raise site at the time of its drivage.

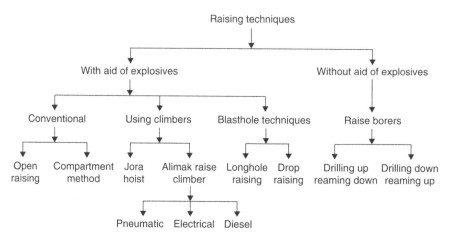

Figure 13.3   Classification of vertical/steeply inclined raise driving techniques based on the rock fragmentation mechanism.

2–3 rounds of 1.5 to 2 m, using the blasted muck to stand on. But after this stulls or stage bars are used to prepare a stage (platform), as shown in figure 13.4(a). This stage is used to carry out drilling and blasting operations. Before blasting any round; holes are drilled for the next platform, fixing a pulley (sheave) block, and to hang a rope ladder. The pulley block is used for supplying material at the face and the rope ladder for accessing the face. Before blasting a round the planks from the working stage are removed, and only the stulls are kept in place. Thus, the operation is not without risk and difficulties. This, in turn, limits application of this method for driving raises usually not exceeding 10 m.

13.6   CONVENTIONAL RAISING METHOD: RAISING BY COMPARTMENT

This technique is an improvement over the open raising method. The method involves dividing the raise into two compartments, as shown in figure 13.4(b). One of the compartments, known as the man-way compartment, is used to install service lines such as water, compressed air, ventilation ducts, pulley block for material hoisting and the ladders. The other one is used to accommodate the blasted muck. The parting is built by fixing the wooden logs/sleepers skin to skin. After every blast the muck is drawn from the bottom so that its level is maintained at the same height as that of the man-way compartment. Before blasting any round the man-way compartment is covered using an inclined bulkhead to divert the blast's fragments towards the muck compartment and thereby avoiding any damage to its fittings. However, while re-approaching the face after blasting due precaution is necessary. The method is slow and tedious but allows raises of longer lengths than the open raising method. In place of two compartments it could be equipped with three compartments to drive raises of large cross-section. Open raising and this method are practically useful for small mines with low output. Earlier these methods were very popular and even today these are almost a mandatory for driving the blind raises of short lengths. However, the raises that have accesses at their both ends can be driven quickly and economically with the advent of modern methods, as discussed below.

13.7   RAISING BY THE USE OF MECHANICAL CLIMBERS: JORA HOIST[4,10]

In a situation where two levels are available a method known as Jora hoist was developed in the past.[10] The method consists of drilling a large diameter hole at the center of the intended raise to get through into the lower level (fig. 13.4(c)).[4,10] From the upper level a cage is suspended using a steel rope that can be hoisted up and down using a winze. This arrangement was known as Jora hoist. The cage has got a flat surface at its top, which is used as the working platform to carry out drilling and blasting operations. With the jack mounted in the sides, the cage can be fixed against the raise sides. While drilling a round the parallel holes are drilled around the central hole, which acts as a free face. Before blasting, the hoist is lowered down in an access specially driven to hold the hoist. This practice suffered with a number of disadvantages, such as: requirement of accesses at both ends of the raise, necessity of a large capacity drill to drill the central large dia. hole, damage to the rope during blasting, slow and tedious hoisting operations, etc. Hence the practice was discontinued, particularly when the Alimak raise climber, described below, was brought into commercial use in the mines.

13.8   RAISING BY MECHANICAL CLIMBERS: ALIMAK RAISE CLIMBER[2,14]

The Alimak Company,[2] Sweden, introduced this technique in 1957 and for driving the blind raises of longer lengths and even today it is indispensable. The Alimak raise climber is designed by keeping sufficient safety margins with regard to the material used for its manufacturing. The drive gear is operated with an air-operated brake that automatically actuates when motors are connected or disconnected. There is a safety device, which comes into operation automatically at over-speed. Also the brake regulates speed while descending by gravity. Using the gravity the cage can be brought down to the bottom of the raise in case of cutoff of air supply. The men travel in the

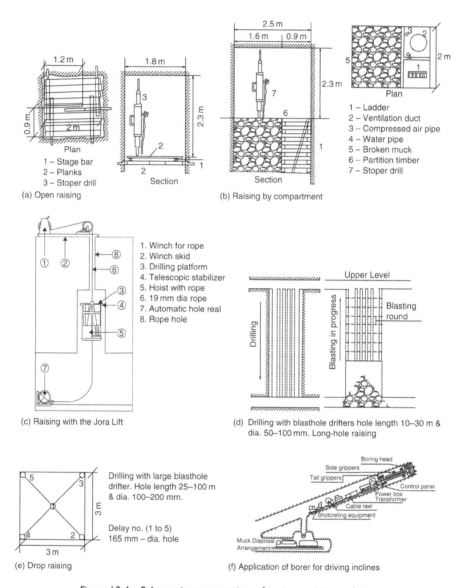

Figure 13.4   Schematic presentation of various raising techniques.

cage up to the face while the material is transported on the platform. The hoist is driven by air, and it climbs along a pin rack, which is bolted to a guiderail. The guiderail is composed of pipes for air and water, as shown in figure[2,4] 13.5(c). The guiderail can be extended as the driving progresses. Each guiderail section is bolted to the rock wall (side wall) using expansion bolts. This method has the following features:

- It makes it possible to drive very long raises, vertical (fig. 13.5(c)) or inclined (fig. 13.6(ii)), straight or curved and mostly rectangular in shape. For driving blind raises of these features, even today, this method is almost mandatory.

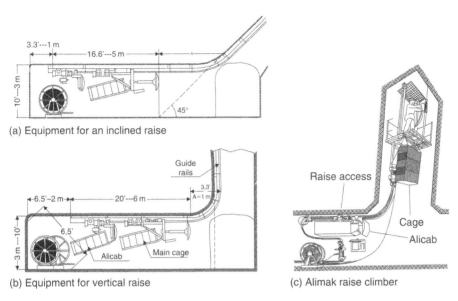

(a) Equipment for an inclined raise

(b) Equipment for vertical raise

(c) Alimak raise climber

(d)

Figure 13.5   Alimak raise climber – some details.

- Using guiderails the raise climber can be driven to a safe position. The guiderail curves also offer the possibility to arrange quick communication between the bottom and the work platform by a special service hoist, known as Alitrolley or Alicab, which is ready for operation on the guiderail all the time.

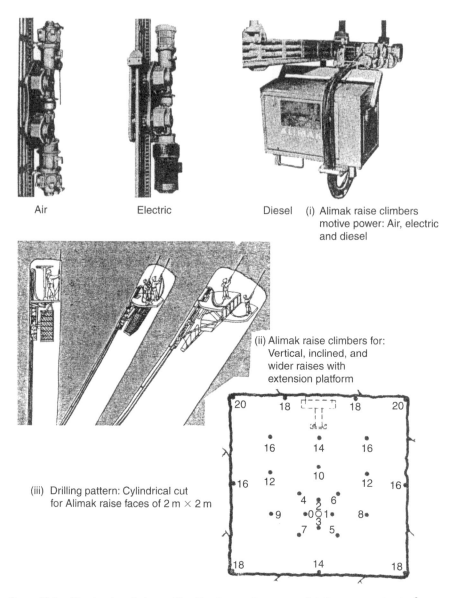

Air   Electric   Diesel   (i) Alimak raise climbers
                                            motive power: Air, electric
                                            and diesel

(ii) Alimak raise climbers for:
     Vertical, inclined, and
     wider raises with
     extension platform

(iii) Drilling pattern: Cylindrical cut
      for Alimak raise faces of 2 m × 2 m

Figure 13.6   Alimak raise climber – Classification, applications and drilling pattern for 4 m² raise.

- All work is performed from the platform, which is easily adjusted for height and angle.
- Because of its design features for blowing air and water at the face after blasting, risks of foul gases are eliminated and the time required for ventilation is reduced.
- The men travel in the cage under the platform when ascending to the face or descending. All open exposure below the blasted face is thus eliminated.
- Connecting an additional extension piece to the platform; it may be used for large areas, thereby, raises of large cross sectional areas, or shafts can be driven. To achieve a large area two parallel or opposite climbers can be used.

13.8.1   Preparatory work and fittings

A horizontal cutout (also known as raise access) as shown in figures[14] 13.5 (a) & (b), of 9 m (length) × 4 m (width) × 3 m (height) is required to accommodate a raise climber with Alicab, but without Alicab its length could be 7 m. For vertical raises the curved guiderails used are: (8°, 25°, 25°, 25°, 8° i.e. the sum should be around 90°). First a vertical raise by conventional method is driven for a length of 5 m. To install the curve the brow is slashed at 45° (about 1 m from the corner), then the guiderail curve is fixed by lifting it with the help of pulley – block.

The manufacturer can supply platforms of 1.6 m × 1.6 m or 2.4 m × 2.4 m or any other size. Unit has safety devices for over-speed control, and to guard against air supply failure. It has got a steel umbrella and fencing attached with platform for the safety purposes.

13.8.2   Ignition and telephone systems

To eliminate the need of a separate cable for the blasting operation, a steel wire having strong insulation, is pulled through one of the guiderail pipes. When firing a round, the current goes through the closed circuit formed by the wire and through the guiderail itself. The same cable system is used for providing alarm and telephone communication between the platform and the base. While drilling the face a header is fitted to the topmost guiderail but on completion of the drilling operation this is replaced by the header-plate having nozzles fitted to it; so that it can be utilized to blow air and water effectively after blasting operations at the face to clear the fumes, dust and ventilate the face effectively. It is also possible to adjust water and air supply through the central hole; however, for the raises of longer lengths installation of a booster pump for achieving required water pressure at the time of drilling, may be necessary.

There are four pipes in the guiderail, two for compressed air, one for water and the fourth one meant for the remote control operation of the air and water valves.

The central is fixed in the Alimak raise access, to this incoming and outgoing connection for air and water is made. Air supply to the hose reel is also given from this. While extending a guiderail, a guiderail of required length, bracket, expansion bolts, O-ring, U-washers, bolts and nuts, spacers and header plate with cocks, are required.

13.8.3   Cycle of operations

Drilling is done from the working platform. A safety belt must be used if the gap (distance between side walls and the platform) exceeds 15 cm. The drilling patterns usually drilled are: burn cut or pyramid cut, based upon the situation and the type of the rock. In case of burn cut, the cut holes, should be positioned either at the center or opposite side of the guiderail to protect them from the direct rock hit.

On completion of drilling the face is charged with explosives and the header is replace by a header-plate. Before blasting the cage is brought to the raise inset. After blasting, the air and water jets from the header plate clear the fumes and then the face is re-approached under the safety roof – umbrella. The face is scaled and after extension of the guiderail the cycle is repeated. The cycle of operation has been illustrated in figure 13.5(d).

13.8.4 Performance

This is the safest method to drive blind raises of longer lengths. The economy of the operation lies in compensating the cost of driving the raise access which is required to install this unit i.e. for raises of longer lengths this cost will be very nominal if raising cost/m length is calculated, but for the raises of shorter lengths this will be a substantial amount. However, its use for shorter lengths than 40 m can be made, but it would result a higher cost/m length. With regard to performance, often a two-person crew can achieve installation of the climber. Initially this crew can complete a round of 1.8 to 2.4 m/shift and later on when raise length increases considerably these rounds of 1.8 to 2.4 m are taken in the alternate shifts.

13.8.5 Design variants

This unit is available in three basic drives operated by compressed air (e.g. STH-5L), electricity and diesel-hydraulic.[2] In figure 13.7, the economical range[14] and the maximum possible ranges with respect to their driving lengths have been illustrated. In figure 13.6(i) a diesel-operated unit has been shown. Figure 13.7 shows cost[14] of guiderail and mounting hardware in each case for the purpose of comparison. Given below is the brief description of these units.

13.8.6 Air-driven unit

The compressed air comes through the hose and the reel winds it when the climber descends. These units are normally recommended for raises upto 200 m lengths but the system has been used for the raises up to 320 m.

13.8.7 Electrically driven unit

For this unit (such as STH-5E) through a specially designed cable (weight = 1–1.6 t/1000 m) the current is supplied to the electric motors. This unit is capable of driving

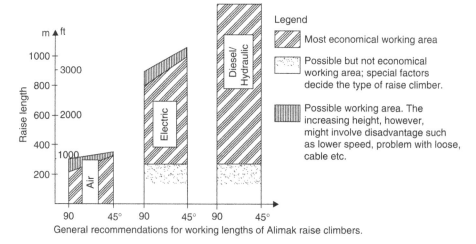

Figure 13.7 Alimak raise climber – range of applicability of different models.

raises up to 1000 m at a stretch. The longest raise driven by this unit is 950 m at 45° of 4 m² cross-section at Denison mine in Canada.

13.8.8   Diesel-hydraulic unit

This unit can drive raises of more than a kilometer length. This unit is self-contained (fig. 13.7) and it does not require any cable or hose. Since the air can be blown through the header, when this unit ascends up extra air is not required to dilute the fumes of the diesel motor. During its descent the motor is not run and use of gravity is made. Specially designed brakes control the speed of the unit.

13.9   BLASTHOLE RAISING METHOD: LONG-HOLE RAISING

This method[15] consists of drilling long parallel holes in a cylindrical or burn cut pattern. The hole length and inclination is kept the same as that of the intended raise to be driven (fig. 13.4(d)). In order to adopt this technique two levels (i.e. the top and bottom of the raise) must be available. From the upper level down holes are drilled to get through into the lower level. On completion of drilling, the blasting is undertaken in stages as shown in figure 13.4(d). Raises driven by this technique are having inclinations exceeding 50° to vertical.

In order to carry out drilling the pneumatic or hydraulic drifters mounted on the pre-fabricated rigs or vertical columns and horizontal bar structures are used. The former type of mounting requires an extra space all around the raise configuration equal to 0.75 m to 1 m to accommodate the equipment. When a drifter is mounted on a jack type vertical column and horizontal bar structure, it gives accuracy and flexibility while drilling the holes in any position and angle. This type of structure also needs a clear space of 0.5 m from all sides of the raise at the drilling site. All the components being lighter in weight than rig mounting, shifting them from one site to another is quicker, hence, use of this type of mounting is very common. The holes drilled are from 50 mm to 100 mm dia. In practice, the steps outlined below are followed.

13.9.1   Marking the raise

To begin with, the surveyor marks the center point, the raise boundary and its intended direction at the drilling site (locale or face). Then the site should be inspected by the driller and the supervisor to check for its suitability to install the drilling equipment with respect to its height and necessary clearance required for the machine to run effectively. The site is then cleaned, the floor is checked by the blaster, the service lines are brought up to the site and the drill pattern is marked at the back of the raise face and not at the floor to avoid its obliteration.

13.9.2   Equipment installation

At the site the drilling rig is brought and installed. If column and bar mounting is used the jacks must be perfectly tightened against the roof.

13.9.3   Drilling

Drifter drills, with the mountings, as discussed in the preceding paragraphs, are used to drill the holes up to 100 mm dia. The holes are drilled either in burn cut or cylindrical cut patterns (fig. 13.8, also refer fig. 9.4). In the cylindrical cut pattern, a few cut holes (may be from 1 to 3) of large dia. (upto 100 mm or so) are required to be drilled. This can be achieved by reaming the normal size holes. But in burn cut all hole diameters is kept the same. Placement of the holes in any of these patterns should be such that there is minimum time required while shifting the machine from one hole to another, as shown in figure 13.8. The total number of holes shown in this figure is for a mine in which this practice was new but these numbers can be reduced, and any of the designs to place the cut holes can be adopted. During drilling the drill machine is set to drill at the intended angle at which the raise is to be driven. For measuring angles an instrument such as "Brunton" can be used. The central hole is usually drilled first. Each hole is collared for its initial 0.5 m using a little large dia. bit than the normal one for fixing the PVC stand pipe e.g. for a 57 mm dia. hole, a 65 mm dia. bit can be used to insert the standpipe. It is important to run the machine slowly while drilling these down holes and flushing them after every 0.5 m of drilling, to avoid rod and bit jamming. Once the hole/s get through to the lower level then it assists in draining out the drilling sludge to the lower level, and thereby keeping the drilling site neat and tidy. The success of drilling lies in drilling the holes accurately without deviation. To achieve this, machine and its components including various clamps should be tightened before and during the drilling operation.

The drilling accessories that are required at the site includes: bits for normal drilling (e.g. 57 mm), collaring (e.g. 64 mm dia.) and reaming (e.g. 104 mm dia.); coupling and adopters; extension drill steels; rod and bit spanners; lubricant; PVC stand pipes; tapered wooden plugs; crow bars; spades; picks etc.

13.9.4   Raise correlation

Before carrying out blasting operations and shifting the drill to another site, the holes should be surveyed for their accuracy, and if deviation exceeds the tolerable limits (not exceeding 1–2%), then additional drilling may be necessary to replace the deviated holes.

13.9.5   Blowing and plugging the holes

Since all the holes get through to the next lower level, their plugging is necessary before charging them with the explosive. This is achieved by using wooden plugs, or sometimes polythene bags, or Hessian cloths, or old used jute bags (fig. 13.9(c)). To achieve a perfect plugging, first of all the plugging material is tied by a wire or rope and lowered down the hole till it gets through to the lower level. This plugging material suspended in this manner is then pulled up slowly while rock cuttings are also poured from the top into the hole simultaneously so that the bottom part of the hole gets plugged. However, the rope/wire should be pulled tight and tied to the anchor bolt fixed at the raise top to ensure proper plugging of the hole. This is a skilled operation and should be carried out by trained personnel. The techniques used to plug the holes for drop raising, described in the following sections, can be also used.

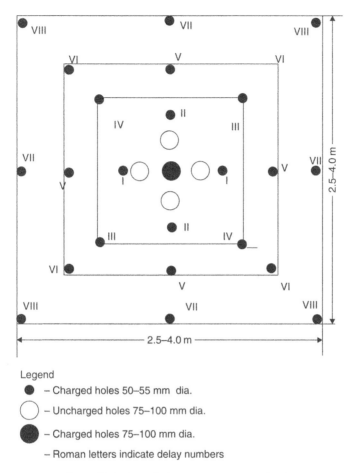

Legend

● – Charged holes 50–55 mm dia.

◯ – Uncharged holes 75–100 mm dia.

⬤ – Charged holes 75–100 mm dia.

– Roman letters indicate delay numbers

Figure 13.8 Long hole raising – pattern of holes.

13.9.6 Charging and blasting

After plugging, the holes are charged with explosive, equal to the length of round that is usually, 2.5 to 3 m, while keeping rest of the hole empty. Of course to stem it over the explosive column; the drilling rock cuttings/chips, or some stemming material can be poured into the hole. In non-watery conditions use of explosive ANFO with suitable primer/booster is usually made; as its use gives advantages of ease in charging and low cost. Use of delay detonators is made to achieve sequential blasting. Usually ANFO is poured manually without use of pneumatic loader otherwise electric detonators will have to be replaced by the antistatic detonators Anodets. This is to note that right from the first round itself all the holes should be kept equally empty and at the same level (horizon). Before charging the next round, all the holes should be thoroughly cleaned and blown using compressed air. If some of the holes get jammed due to previous blasting, it is important to get them through, sometimes, by blasting their neighboring holes using a mild charge. Once all the holes are through, the same procedure for blasting the subsequent rounds follows. Before blasting, the workers at both levels should be warned and approaches to the raise site (at both levels) should be well guarded. Special care should be taken when the parting to get through into the upper level is 5 m or less. This portion should be taken in one round only to avoid damage to the holes and formation of excessive loose rock around the raise collar. In this technique, this is how the charging operation is conducted from the upper level, while the blasting progresses upward from lower level towards the upper one.

13.9.7 Limitations

This technique can be applied only if raise site can be accessed from both the levels. Blind raising is not feasible. Raises up to 40 m lengths and 45° inclinations can be driven. Accurate drilling and proper blasting is the key to the success of this method. Disturbed ground with joints, fissures etc. may result in frequent jamming of drill rods and bits.

13.9.8 Advantages

Safety: This is a valuable technique by which no-one is required to enter into the raise during its drivage as both drilling and blasting operations are conducted from the upper level only.

Productivity: Operation is not cyclic, thereby, better productivity and faster rate of drivage can be achieved. The raise can be drilled in advance and it can be blasted as and when required. Muck removal is not essential after every blast. While drilling is going on at one site, preparatory arrangements including shifting of all fittings other than the drilling equipment can be undertaken at another site.

Better working conditions: Workers are required to work at the levels where better working conditions are available comparing the same if they are required to work within the raise in a limited space and under the same arduous conditions that prevail with the other raising methods. This technique is specially beneficial in the areas of bad ground and also at depths where excessive pressure and temperature prevail.

Better raise configuration: A smoother raise configuration is obtained which helps in equipping it in a proper manner. For driving the service and slot raises this technique finds wide application.

Flexibility and simplicity: This method does not require any elaborate arrangement to accommodate the equipment as is required in case of raise borers and Alimak raise climbers (e.g. raise access of 9 m × 4 m × 3 m). The people, machines and equipment that are meant for blasthole drilling in the stoping operations can be utilized to undertake this operation.

Economical: For the raises of small lengths this is the most economical method as better productivity can be achieved by utilizing the same resources meant for stoping operations in a mine. In addition, to fit this equipment no extra preparation and excavation of any kind are required. It allows the use of cheaper explosive like ANFO, contrary to the costly conventional high-density explosives used in other raising methods.

For the raises up to 25 m lengths this method is almost mandatory particularly when two levels are available, and blasthole drills are used in the stoping operations.

13.10   BLASTHOLE RAISING METHOD: DROP RAISING

The advanced version of the long-hole raising technique is the "Drop Raising" in which large diameter and longer holes are used to drive the raises (figs 13.4(e) and 13.9(a)). This technique is basically based on the vertical crater retreat (VCR) concept, discussed below.

VCR concept: The term 'crater' in blasting terminology is applied to creation of a surface cavity in a rock mass as a result of detonating an explosive charge into it. This blasting concept was initially used as a tool to evaluate the capability of an explosive. It gained importance in surface blasting operations, and in the recent past, in underground blasting operations too.

Based on the research work carried out, the explosive charges used in crater theory are spherical or its geometric equivalent. In blasting practice the spherical charge is defined as the one which has a length to diameter (L: D) ratio of 1:4 or less, and up to, but not exceeding a L:D = 6:1. Thus, for holes of 165 mm dia. a charge of 990 mm length would constitute a spherical charge.

Crater blasting when used for research purposes, the charge is fired in the upward direction, enabling a crater to form towards a horizontal free face. But in an underground situation when a spherical charge is blasted in the downward direction towards a free face, which could be back of an opening or ceiling of an excavation, an entirely new concept of crater formation has emerged. In this the case crater is formed in a downward direction. Adverse effects of gravity and friction do not affect the results. To the contrary, the gravity enlarges the crater dimension by removing the entire ruptured zone, as shown in figure 16.14(c).

Once the excavation of an underground opening disturbs equilibrium of the mass, a stressed zone of elliptical shape is formed above this opening. Depending upon the stability of the rock the material within the stress zone caves in sooner or later, if not supported by some artificial means. Depending upon the rock properties and structural geology the total height of the cavity may exceed the optimum distance of spherical charge from the back many times. Thus, the cratering characteristics of the rock mass to be blasted are studied, and use of this concept is made to carry out the blasting operations in the stopes or raises by blasting vertical or steeply inclined holes of large diameter (165 mm) in the upwards direction and retreating towards the top cut or drilled horizon. The blasting operation when carried out in this manner, the method is known as Vertical Crater Retreat (VCR).

Formulae Used:[10] A crater is consists of 5 holes, one at the center and the other four at the corners of a raise. Crater theory is valid for the central hole only, and for the rest of the holes, the charge depth increases from 10–20 cm between each hole, as shown in figure 13.9(b).[10] The charge depth can be determined by using following formulae.

Charge length, $l = 6 \times d$ d is big-blast-hole diameter in mm   (13.1)

Optimum charge depth is 50% of the critical depth; $L_{opt} = 0.5 \times L_{crit}$ (13.2)

$L_{crit} = S \times Q^{1/3}$ (13.3)

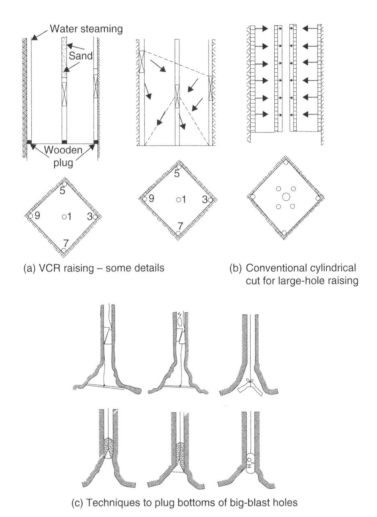

(a) VCR raising – some details

(b) Conventional cylindrical
cut for large-hole raising

(c) Techniques to plug bottoms of big-blast holes

Figure 13.9   Drop raising – some details.

S is the Strain Energy Factor usually 1.5 but depends upon explosive and type of rock.

Charge weigh $Q = 3 \times d^3 \times \pi \times \rho /2$ (in kg);   (13.4)
ρ explosive density in gm/cc

Thus, $L_{opt} = 0.5 \times S \times Q^{1/3}$ (13.5)

Drilling: The term down-the-hole (DTH) is used as a generic name given to all those drills which are referred as or known by their trade names such as down-hole drills or in-the-hole (ITH) drills. These drills differ from the conventional drills by virtue of placement of the drill itself in the drill string. The DTH drill is placed immediately after the bit, which always remains in the hole. Thus, no energy is dissipated through the steel or coupling and penetration rate is almost constant regardless of depth of hole. It is pneumatically operated and flushing is done by the compressed air with water mist injection. In some cases dry drilling with dust collectors becomes essential to avoid use of water as this drill can tolerate only a very small amount of water for the purpose of flushing. Most DTH drills operate up to air pressure of 250 p.s.i. (1724 kPa). Mission Megamatic is one of such units,[9] which have been tried in some of the mines in Canada, India and a few other places.

In this technique use of these drills is made to drill the parallel holes in the intended direction of the raise to be driven. All holes are drilled to get through into the next lower level up to which the raise is to be driven. With careful drilling hole deviation should not exceed 1–2%. Thus, this technique is similar to the longhole raising (fig. 13.9(b)) but in place of drilling holes of 50–60 mm dia. following a burn or cylindrical cut pattern, five to six holes (one or two in the center and the rest at all the corners), as shown in figure 13.9(a), of dia. 100 mm or more are drilled. An extra hole in the center is purposely drilled to take care of any abnormal hole deviation and damage of the central hole during raise blasting. However, on completion of drilling holes a survey for their deviation is undertaken. Raises of longer lengths up to 150 m can be drilled with the application of the drills used for this method.

Blocking the blastholes: As in the conventional crater formation the charge covers the bottom of hole for some length and rest of the hole length remains empty. Now, think of inverting this figure or the scenario, it will reveal that the charge should be placed at a certain height (which can yield the desirable results) from the free face, and hence, blocking the hole at a certain height above the free face is essential.

This process involves securing two wedges at the desired location near the bottom of the blasthole i.e. from the free face. Explosive is charged on top of the blocking. Angles of holes determine where the hole is to be blocked. Given below are generally accepted values for blocking heights:[3] Hole angle 80–90°: 1.2 m; 57–79°: 1.5 m; 50–56°: 1.8 m, less than 50°: 2.1 m. The process of blocking the holes has been shown in figure 13.9(c).[8] For the purpose of blocking the holes, conical or rectangular wooden blocks (fig. 13.9(c)), which can be suspended from the top sill by 5–6 mm dia. polypropylene rope, are usually used. In the recent past at some mines, the rope has also been replaced by the primacord (40 grain). This process could reduce the hole loading time by 20% or so.

Hole blocking begins by tying one wedge block onto the 40-grain primacord and lowering the block up to the pre-determined blocking height. After this the second wedge block is lowered down so that it hits the first one. The primacord is tugged to ensure that the blocks wedge together against the wall of the hole. About one foot of small rock cuttings is poured into the hole to ensure proper blocking of the hole. The holes can be blocked even following different techniques. Figure 13.9(c) illustrates some of these practices. This may be noted that any extra hole which is not charged, should be filled with some stemming material to avoid its damage and jamming at the time of blasting the raise round.

Blasting: The hole is then loaded with the explosive. Its amount depends upon its density and ratio of hole diameter to length so that a spherical charge can be obtained. For example for a 165 mm dia. hole, the charge amount of an explosive of 1.40 gm/cm^3 density works out to be 27.2 kg; Hence, as per this calculation half the weight of the explosive is first dropped or charged, then booster primer with proper delay is lowered down. The rest of the explosive i.e. the other half is then charged. The hole is then stemmed, for a length of 1–2 m or so, using suitable material. The same procedure is repeated while charging the rest of the holes. It may be noted that the low density explosives such as ANFO are not suitable for this technique, and therefore usually slurry explosives, which can also be used in the watery holes, are used. Use of double deck or multi deck charging has been tried successfully in some mines but during stoping operations its application is common.

Performance: Since drilling and blasting are the independent operations so better productivity can be achieved by this technique. While drilling performance depends upon the usual parameters such as rock factor (type of strata), operating factors (drill power, blow energy, speed and flushing mechanism), drill hole factors (hole dia., length and inclination) and service factors (working conditions, skill of operator etc.). But after completion of drilling, by a trained crew, a blasting round of 3–4 m/shift or at least a round in the alternate shifts can be achieved.

Scope, advantages and limitations: This technique has got the same scope of its application, advantages and limitations as the long hole raising practice, described in the preceding section. In addition, there is scope to reduce the cost further as a better drill factor, powder factor and productivity can be achieved by this technique. Longer raises up to 80 m or so have been driven successfully, however, the only limitation is the requirement to use the same drilling equipment in the stoping operations too, in order to justify the investment made.

13.11   RAISING BY THE APPLICATION OF RAISE BORERS[11]

This is another technique (fig. 13.4(f))[11] that can be applied to drive a raise between two levels. Using this technique raises up to 910 m length and 30° to vertical (usually steeper than 45°) of 0.9 m to 3.7 m cross section can be driven. This is the usual range but Robbins – Atlas Copco raise borers have even crossed this range (table 13.2). Raises have been drilled successfully even in a relatively poor ground. A circular configuration is obtained by this technique without application of drilling and blasting. The machine is set up at the top and a pilot hole of 225 to 250 mm dia. is first drilled down to get through into the lower level, as shown in figure 13.10. Then a large reamer bit is put on at the bottom of the drill rod and the raise is reamed to the desired diameter up to the upper level. The pilot hole also provides information about the type of strata to be encountered and helps in driving the raise accurately. In the case of large deviation, the pilot hole can be abandoned. The reverse procedure can be also adopted i.e. first driving the pilot hole upward from the lower level towards the upper level and then reaming it from the upper level towards the lower one. This technique is less popular. However, machines are available to drive either way. The space and facility available at each end of a raise mainly govern the choice of method.

Raise borers are available for driving in soft as well as hard ground. This unit can disassemble into its various components that can be transported to the raise site and can be assembled again. The unit is available with crawler, wheel and skid mountings. Usually an operator with a helper can operate this machine. Trained personnel are required to operate and maintain it.

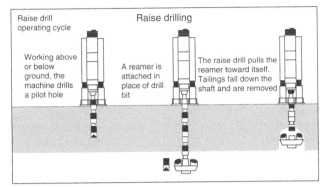

Figure 13.10   Driving vertical raise using raise borers. Most popular and widely used model Robbins 73RM. Considered as a reliable workhorse for virtually any raise boring application. It is a medium size raise drill, ranging from 1.8 to 3.1 m (6–10 ft) in diameter (Courtesy: Atals Copco – Robins).[3]

When drilling stabilizers are used, a stabilizer follows the pilot bit; and then drill rods are inter-spaced with stabilizers, the spacing of which varies with type of rock. On completion of the drilling up to the targeted end, the reamer replaces the bit. This operation is then carried in the reverse direction to withdraw the drill rods. Removal of cuttings from a down-drilled pilot hole is done by air or sometimes by water – which deposits them around the hole collar from where they can be removed by hand shoveling. During up-reaming, the cuttings fall by gravity. Where the raise borer is located at the lower level of the raise, the cuttings from both operations (drilling and reaming) drop by gravity into a hopper and then via a chute or pipeline to the ore conveyance (i.e. any transportation unit). In general, it is cheaper to drill down and ream up. However, a decision on drilling manner requires consideration of several factors. Important amongst them are: the availability of access to the raise site, ease in the transportation and installation of the raise borer, speed, energy required and overall economy of the operation.

The rotational speed during pilot hole drilling varies from 35–72 r.p.m and the pressure on the pilot bit from 30,000 (13608 kg) [for 9.8 in. (250 mm) bit] to 125,000 lb. (56700 kg) for a 15 in. (381 mm) bit[11] [Paul et al.]. The r.p.m. during reaming is in the range of 10–20 and the pressure on the reaming head in the range of 20,000 lb. (9072 kg) for 48 in. (1220 mm) to 36,000 lb. (16330 kg) for 60 inches (1524 mm).

Since the early 1960s, the name Robbins has been synonymous with raise drilling. This has been recently taken over by Atlas Copco Rock Drilling Division. Since the development of the first production raise borer, the Robbins 41R, more than 35 different models of raise borers have been designed and built to suit the needs of many different applications. Based on past experience the models developed by Robbins Co; the range of some of the specification[7] have been summarized in table 13.1. Table 13.2[3] displays working range of some of the Atlas Copco/Robins very recent (2002) models. In table 13.3 field data of some 17 raises bored by borers in different rock types have been presented.

To site an example of its use in mines, LKAB Iron Ore Mines, Sweden some 16.5 km of sinking and raising have been done.[3,4] This mine has been divided into

Table 13.1   Robbins raise borers – usual range of some of the parameters.[7]

Robbins raise borers – important parameters	Usual range
Models designation (with max. raise dia. and length capabilities in meters)	23R (0.9, 120), 32R (1.2, 180), 52R (1.5, 90), 61R (1.8, 240), 71R (2.1, 460), 72R (2.1, 240), 82R (2.4, 300), 85R (3.7, 300) and 121R (3.7, 910).
H. P. Range	100 to 400
Drive speed	0–117 r.p.m
Full load torque	116,000 to 2000,000 N-m
Machine weight	4.1 t to 112 t
Type of drive	Hydraulic except in two models
Recommended raise dia.	0.9 to 3.7 m
Length range	90 m to 910 m
Accessories:	
Tri-cone blasthole bits	225 to 375 mm dia. i.e. 9 to 15 in. dia.
Drill rods	8–10 in. dia. and 4–5′ length
Stabilizers and reaming heads are some of the drilling accessories	

Table 13.2   Some of latest raise borer models by Atlas Copco/Robins with their working ranges.[3]

Model	Nominal diameter, m	Diameter range, m	Nominal length, m	Maximum length, m
34 H	1.2	0.6–1.8	340	610
43 RM	1.2	1.0–1.8	250	610
53 RE/RH	1.5	1.2–2.4	490	650
73 RM	2.4	1.8–3.1	550	700
83 RM	4.0	2.4–5.0	500	1010
97 RL	4.0	2.4–5.0	600	1010
123 RM	5.0	3.1–6.0	920	1100

eight production areas, each containing its own group of ore passes and ventilation systems. At this mine a total of 32 ore passes between the 775 and 1045 m levels were driven using Tamrock and Robbins raise borers and rest of the task has been by some other companies. This technique offers several advantages such as:

- Safety: Personnel are not required to enter into the raise during its drivage.
- Stability: Circular raise configuration is stable and allows smooth flow of air. The raise sides are obtained without impact of drilling and blasting. Even in weaker grounds it has been driven successfully. Sides can be rock bolted and lined in the incompetent formations.
- Productivity: The process in not cyclic but a continuous one, thereby, in the ground suitable to this technique a faster rate of raising can be achieved.
- Economical: If enough sites are made available so that utilization of this unit is maximum, it proves to be economical.

Due to the above listed merits, this method is gaining popularity and contractors are using it for their civil and construction projects, as outlined in section 13.1.

Table 13.3   Performance of raise borers in different rock-types, diameter and length.

Raise dia., ft	Length, ft	Inclination degrees	Advance rate ft/hr	r.p.m.	Purpose	Rock type
4	511	90	2.2	14	Shaft pilot	Quartzite
4	879	90	17.7	32	Ventilation	Sandstone, siltstone
4	1677	90	11.3	32	Ventilation	Limestone
5	289	90	1.3	18	Ore pass	Quartzite
5	391	89	7.4	14	Ventilation	Limestone
5	776	85	3.1	15	Waste pass	Sulfide, rhyolite
5	1078	90	6.8	22	Ventilation	Limestone, dolomite
6	447	78	4.4	18	Ore pass	Diopside, quartz
6	938	90	1.3	14	Shaft pilot	Diorite, gneiss
6	1114	90	2.3	22	Ventilation	Graywacke
7	764	90	1.6	18	Ventilation	Gabbro, diorite
7	750	90	1.5	18	Pilot vent	Quartz, Quartzite
7*	1140	90	11	10	Ventilation	Limestone
7*	1200	90	4.9	10	Ventilation	Shale, sandstone
9	145	90	3.3	7	Ventilation	Limestone, dolomite
12	440	90	2.7	8	Ventilation	Shale, sandstone
12	678	90	3.6	8	Ventilation	Shale, sandstone

*Using Robbins Raise Borer. Rest using Dresser Raise Borers[13] (1 ft = 0.3048 m)

13.12   RAISE BORING IN A PACKAGE – BORPAK[6]

It is a recent addition to the raising techniques.[6] This unit is self driven and mounted on crawler track. It is used for blind raising. This unit is set up underneath the intended raise (fig. 13.11),[6] and starts boring upward through a launching tube. After the head has penetrated a few meters into the rock, grippers hold the body while the head rotates and bores the rock similar to the tunnel-boring machine. It can bore a blind raise of 1.2 to 2.5 dia., up to 300 m lengths.

13.13   ORE PASS/WASTE ROCK PASS[5,12,16]

Passes are openings that are driven in the rock massif for conveyance of waste rock, ore, filling material from the upper levels (elevations) to the lower horizons (elevations). They are underground conduits for the gravity transport of the broken rock, ore, waste or backfill; inclination varies from vertical to 30°. Historically cross-sections are less than $5.6 \, m^2$ (60 sq. ft); but at present there are ore passes (OPs) which can accommodate boulders up to 1.5 m size. Their length varies from 18 to 180 m, or shorter or longer OPs exist. Transport and storage are the two important purposes that are served by OPs, as shown in figure 13.12(g) illustrating a dumping mechanism by LHDs in an ore pass.[17] Gravity assists the flow of material. Such openings are an integrated part of the ore and waste rock movements in the modern mines. These

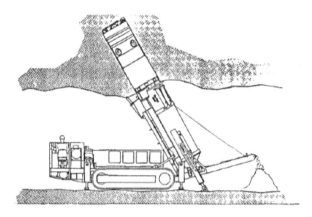

Figure 13.11   BorPak – raise boring machine.

passes are also useful for deep seated open pit mines where the rocks are hauled up or down through such openings for their onward disposal (fig. 17.16(e)).

13.13.1   Size and shape

They can be driven with the application of raising methods, with or without aid of explosives (fig. 13.2). However, the method selected should meet the objective. Square, rectangular and circular are the usual shapes.

Size of Ore Pass:[16] Apart from rule of thumb, and adopting the size based on experience of other mines having almost the same environment; some empirical relations have been developed based on experience gained at Czechoslovakian mines. In specifying these formulae lump size, size distribution pattern, gradation, stickiness, etc. have been considered.

Square cross section of side length, $L = 4.6\sqrt{(d^2 k)}$ (13.6a)

Rectangular cross section of width $W = 4.6\sqrt{(d^2 k)}$ (13.6b)

Circular cross section of diameter $D = 5.2\sqrt{(d^2 k)}$ (13.6c)

d is the size of largest lump. Value of k is determined from the nomograph. For typical hard rock mines, given below are some of values of "k" under different set of conditions:

k = 0.6 when the content of sticky fines = 0%.
k = 1.0 when the content of sticky fines = 5%.
k = 1.4 when the content of sticky fines = 10%.

In hard rock mines all very fine material (passing 200 mesh) is considered sticky due to water sprayed on it to suppress the dust. A borer-driven circular ore pass is more stable than a square or rectangular ore pass. The latter invites arching on the sides and stress concentration at the corners. A circular ore pass made with a borer suffers with the problem of placing ground support in the raise and such raises with a smooth circular configuration are more prone to hang-ups than a rectangular raise that has been drilled and blasted. In highly stressed, burst-prone ground, an ore pass usually has to be raise-bored for safety reasons. The same reasons make placing ground support in the raise problematic.[16]

13.13.2   Ore pass lining

In bad ground, ore passes are often lined with concrete. In many cases, the concrete is faced with high strength steel liner that also provides the formwork that is required to pour the concrete. The steel lined ore pass has proven its suitability worldwide; however, the high cost and time required often make this design impractical.[14] Different means have been developed for placing a concrete lining to reduce the cost.

It is generally accepted that the resistance to wear of concrete and shotcrete is mainly dependent on the aggregate employed. It has been proposed that the most economical procedure is to select an aggregate material that has a relatively high abrasion resistance and toughness, such as: basalt, andesite, diabase, diorite, etc. or some metamorphosed trap rocks. The method has proven effective in South Africa and at some locations in North America; however, the high price of the special aggregate makes it cost-prohibitive for most applications.

In hard rock, a long glory hole is not normally lined. Instead, two glory holes are provided near the same location for the following reasons.

• Lining a long glory hole properly is more expensive than drilling the hole.
• A second raise provides the required relief to exhaust the air blast (when both raises are inter-connected).
• If one raise becomes inoperable, the second one is available.

13.13.3   Design consideration of rock pass/ore pass[5,12]

In figures 13.12(c) to (f) orepasses of different designs have been presented. Figure 13.12(c) illustrates the ore pass with midway knuckle and figure 13.12(d) an orepass with knuckle at the bottom. In figure 13.12(e) an inclined orepass with inspection raises and drifts have been shown. The orepass shown in figure 13.12(f) is a vertical and having dia. of 9 m. These ore passes are fitted with chute gates of different types so that ore flow from them can be controlled and regulated to feed directly into the transportation units, which could be tracked or trackless.

While handling the ore underground, one of the problems is that with the ore passes and car loading, should the transfer be vertical or inclined? Is a knuckle desirable just ahead of the loading chute in order to break the momentum of fall? What are the optimum dimensions of the transfer? The decision involves a knowledge of bulk solids flow, particularly for the specific ore to be handled. Designs are made on the basis of past experience and model flow studies. The important factors[5] that must be considered while designing an ore pass are:

• The size distribution and size segregation of the particles (ore fragments) in the ore mass.
• The shear strength (s) of the ore mass, with $s = c + p \tan \phi$, where c is the cohesive strength, p – compressive loading on the material, $\tan \phi$ – coefficient of internal friction.

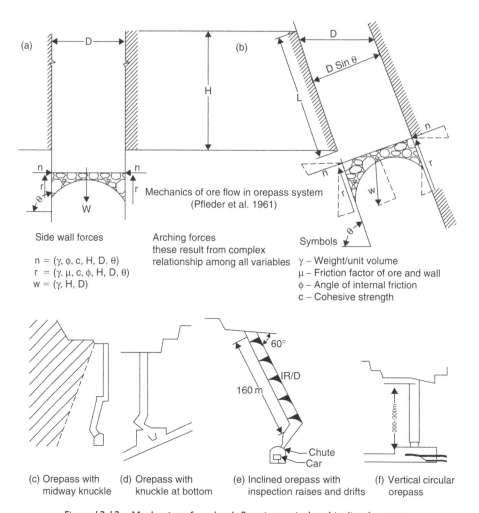

Side wall forces

n = (γ, φ, c, H, D, θ)
r = (γ, μ, c, φ, H, D, θ)
w = (γ, H, D)

Arching forces
these result from complex
relationship among all variables

Symbols

γ – Weight/unit volume
μ – Friction factor of ore and wall
φ – Angle of internal friction
c – Cohesive strength

Mechanics of ore flow in orepass system
(Pfleder et al. 1961)

(c) Orepass with midway knuckle   (d) Orepass with knuckle at bottom   (e) Inclined orepass with inspection raises and drifts   (f) Vertical circular orepass

Figure 13.12   Mechanics of ore/rock flow in vertical and inclined passes.

- The height of fall in the ore passes, as related to the tendency of ore to crush and to pack.
- The characteristics of the wall rock of the ore pass – its resistance to slabbing and abrasive wear, with a resultant friction factor (μ) between sliding ore mass and wall.
- The flow rate and storage capacity desired.
- The climatic conditions (heat, humidity, presence of water etc.)

The mechanics of the ore flow in the vertical and inclined passes have been compared in figures[5,12] 13.12(a) and (b). Some of the important aspects in this regard, as described by David, J.S. and Pfleider, E.P.[5] are discussed below:

- Arching phenomenon is of great importance. The forces and strengths developed in an arch are due to various reasons and amongst them are the tendencies of

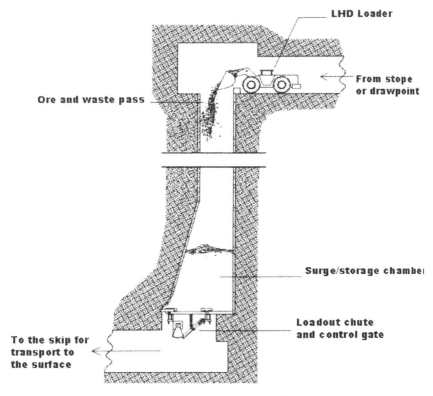

Figure 13.12   (g): Typical ore pass (OP) for LHDS[17] (www.smenet.org).

an ore to pack or segregate, which can have a profound effect in this respect. Experience has shown that ore passes generally will not "hang up" if the least cross-sectional dimension is four to six times in dia. of the largest ore pieces. However, of the broken ore has a large percentage of fines, which develop considerable shear strength when moist, the mass tends to form an arch. This can happen in cold climates during a prolonged shutdown, and generally calcium chloride is employed in such situations.

- It is felt that for the ore passes of equal dimensions, the ore flow is better in the inclined ore pass than the vertical one, for the reason that arch forming is more pronounced in vertical ore passes than in the inclined ones where the weight of flowing media keeps breaking the arch on the upper side of the ore pass. However, the ore has a greater tendency to pack and hold on the footwall side of an inclined ore pass. In order to minimize the damage of the orepass's walls (wear and slabbing) it should be put at normal to dip of the formations.
- The orepass collar's elevation should be such that maximum tonnage above this elevation are available.
- If there is choice between a vertical or inclined ore pass, the merits and limitations of each should be noted. For the same elevation difference a vertical ore pass is a shorter, less direct hit or impact to sided walls, easy to drive, slash (if required to

enlarge their capacity) and maintain. A vertical pass does not require knuckle and has none of the problems inherent with knuckles. Ore flow velocity is less in an inclined ore pass, therefore, there is less problem with entrained air, fragmentation and dust. For some conditions of ore and ore pass, arching is less likely in an inclined pass due to imbalance of normal forces.

Glory Hole Ore Passes:[16] For some mining applications at high altitude, very long vertical ore passes are required. These long passes are normally excavated with a raise-borer. They are similar to the long waste rock passes except that they are normally run empty while a waste rock pass is designed to be kept nearly full. The glory hole ore passes have all the problems of regular ore passes (as described in the preceding paragraphs), plus some of the following problems.

• Three design considerations should be made: Avoid these two malfunctions: (i) failure to flow which results in hang-ups; (ii) failure to flow in the entire cross-section of the ore pass, known as 'piping' or rat-holing; a third design consideration is the stability of the ore pass.

Air Blast: Because this type of ore pass is normally designed to feed directly into an underground bin, it is run empty. This means that the ore stream obtains very high velocities resulting in intermittent air blasts due to a piston effect that must be relieved by an underground connection to a relief airway. A second ore pass connected to the same bin underground may provide the required relief.

Ricochet: The high velocity of the ore stream produces tremendous impact at the bottom of the raise; therefore, the geometry is designed to provide an impact bed (rock box) at the bottom of the raise. In some cases, liners are required to take care of the ricochet (bounce) from the rock box. Another ricochet phenomenon occurs when the glory hole raise is fed with a conveyor. The horizontal motion of ore on the conveyor continues when the ore stream falls into the raise. The result is a first impact on the far side of the raise and subsequent ricochet to the near side. If nothing is done to mitigate this action, it produces wear in the upper portion of the raise.

Attrition: The loss of potential energy due to the drop of the ore stream is divided between friction and comminution of the ore. The total potential energy is simple to calculate. The portion of this energy that results in attrition is difficult to estimate in advance. The amount of attrition is important if the ore is to be treated in an autogenous mill. If the ore has a high work index, and a conservative fraction is assumed for comminution, the amount of reduction in lump size is not normally significant. In some cases, such as a limestone quarry or a rock fill quarry, the generation of fines by attrition may result in an unsatisfactory product.

13.14   THE WAY FORWARD

Raise borers offer several advantages such as: safety, stability, productivity and if enough working sites are made available so that utilization of this unit is maximum, it proves to be economical. Hopefully research and development in this area will make the shaft borers an attractive option for vertical development including driving ore and waste passes.

QUESTIONS

1. "Raising operations exceeding 10 m used to be highly accident – prone in the past, but the advancement in technology has made driving raises exceeding 1000 m not a dream but a reality". Briefly review the raising operations and how this technique has developed in the recent past. You could undertake a literature survey in this regard including making use of the relevant websites.
2. Alimak raise climber – list the different models that are available and work out their range of applicability.
3. Classify raising methods/techniques based on the availability of access to the intended raise site at the time of its drivage. Define blind raising.
4. Classify vertical/steeply inclined raise driving techniques. Illustrate these techniques (except the technique by Alimak) giving suitable sketches.
5. Describe and illustrate the mechanics of ore/rock flow in vertical and inclined passes.
6. Describe the BorPak raise boring machine, mentioning its applications.
7. Describe these aspects which are used in conjunction with ore-pass/waste-passes: attrition; ricochet; air blast. What is a glory hole ore pass?
8. Differentiate between a raise and a winze. Name different methods of raising. Mention the specific conditions/situations under which each one of them could be applied. Also mention the limitations of each method.
9. For driving vertical raise using raise borers, the most popular and widely used model is the Robbins 73RM, considered as a reliable workhorse for virtually any raise boring application. Describe this model or the main features of its equivalent models.
10. Give a line diagram to classify vertical/steeply inclined raise driving techniques based on the rock fragmentation mechanism.
11. Give sketches to illustrate the following raising methods: (i) open raising; (ii) raising by compartment method; (iii) longhole raising; (iv) drop raising; (v) raising by use of Alimak raise climber; (vi) raising by use of borers.
12. Give the drilling pattern for the raise driven by the following method, as per the sizes given here: Long hole raising 3 m × 3 m. Alimak raise climber 2.5 m × 2.5 m. Drop raising 3 m × 3 m. Open raising (2.5 m × 2.5 m).
13. How are raises positioned w.r.t. an orebody? Illustrate it with suitable sketches.
14. If there is choice between a vertical or inclined ore pass, the merits and limitations of each should be noted – list these aspects in each of these two versions.
15. Illustrate with the aid of sketches the working cycle of an Alimak raise climber. Mention the different units of Alimak raise climber giving their functions.
16. In bad ground, ore passes are often lined – mention the type of lining and its salient features. Different means have been developed for placing a concrete lining to reduce the cost; what are they?
17. List the application of raises in various industries.
18. List the important factors that must be considered while designing an ore pass.
19. Mention the usual range of raise borers in terms of raise length and diameter.
20. Mention the scope, advantages and limitations of drop and longhole raising techniques. Write in detail the blasting procedure to be followed in each case. Include logical steps that should be followed to achieve the best results of blasting operations.

21. The opposite to a raise is a winze, whose driving mechanism is just the reverse. How? Describe procedure to drive winzes.
22. Propose under what specific situations you would recommend the use of pneumatic, electrical and diesel operated Alimak raise climbers. Draw the cyclic unit operations using the Alimak raise climber.
23. Raise borers are available for driving in soft as well as hard ground. Give details of such units.
24. Raise placement w.r.t. an orebody is governed by their purpose – mention those purposes.
25. Raises are one of the important structures in many civil and construction projects; describe their utility in a hydroelectric power project.
26. Raising by the application of raise borers – describe details of this technique and up to what inclination and length of raise could this technique be used?
27. Robbins raise borers offer several advantages; list them.
28. Summarize (list) applications of raises in underground mines.
29. What is the concept: vertical crater retreat (VCR)? How it is made applicable during the drop raising operation? Give details of the formulae to determine the charge length, optimum charge depth, critical charge depth and charge weight, which are used to perform the requisite calculations enabling the blasting operation to be carried out in a correct manner.
30. Propose suitable raising method/methods, for each of the situations outlined below:
    a. An ore-pass 2.5 m × 2.5 m, 180 m long in competent ground. Two levels are available.
    b. A slot raise 2 m × 2 m, 55 m long at an angle of 70° to terminate before the upper level and not joining it.
    c. A 35 m long 3 m × 2.5 m raise when only one level is available.
    d. A raise of 2.5 m × 2 m, 10 m long to fit Alimak raise climber to start with.
    e. A vertical raise of 2.5 m diameter to connect two main levels 70 m apart. Both the main levels are available.

REFERENCES

1. Agoshkov, M.; Borisov, S. and Boyarsky, V.: *Mining of Ores and Non-metallic Minerals.* Mir Publishers, Moscow, 1988, pp. 67–69.
2. *Alimak: Raise Climber (leaflets).* Vein Mining, 1977.
3. *Atlas Copco – Robbins: Leaflets and website,* 2002.
4. *Atlas Copco:* Raise driving with Jora Hoist, 1970.
5. David, J.S. and Pfleider, E.P.: Ore-pass, Tunnels and shafts. In: *Surface Mining,* E.P. Pfleider (edt.) AIME, New York, 1964, pp. 637–642.
6. Hamrin, H.: *Guide to Underground Mining Methods and Application.* Atlas Copco, Sweden, 2002, pp. 12–13.
7. Home, L.W.: Raise Drills. In: *Underground Mining Method Handbook,* W.A. Hustrulid (edt.). SME (AIMMPE), New York, 1982, pp. 1093–1097.
8. Jimeno, C.L.; Jimeno, E.L. and Carcedo, F.J.A.: *Blasting and Drilling of Rocks.* A.A. Balkema, Netherlands, 1997, pp. 241.
9. Kolihan Copper Mine, India: *Technical Innovations – Long-hole Blasting,* 1986.
10. Olofsson, S.: *Applied Explosives Technology for Construction and Mining.* Applex, Sweden, 1997, pp. 160–170.

11. Paul, R.B. and Eric, G.B.: Drifting and Raising by Rotary Drilling. In: Cummins & Given (eds): *SME Mining Engineering Handbook*. AIME, New York, 1973, pp. 10: 89–97.
12. Selleck, D.J. and Pfleider, E.P.: Ore passes, Tunnels and Shafts. In: *Surface Mining*, Vol. II, 1994, pp. 6: 38–40.
13. Singh, J.: *Heavy Constructions – Planning, Equipment and Methods*. Oxford & Ibh Pub. New Delhi, 1993, pp. 554–560.
14. Svensson, H.: Rasie Climbers. In: *Underground Mining Method Handbook*, W.A. Hustrulid (edt.). SME (AIMMPE), New York, 1982, pp. 1051–1055.
15. Tatiya, R.R.: Longhole Raising, *Mining and Engineering Journal*, Vol. 18, 1979, pp. 1–8.
16. Vergne, J.N.: *Hard Rock Miner's Handbook*. McIntosh Redpath Engineering, 2000, pp. 324–329.
17. Beus, M.J. et al.: Design of ore passes. In: Hustrulid and Bullock (eds.): *SME Underground Mining Methods*. Colorado, 2001, pp. 627–34.

Chapter 14

Shaft sinking

Sinking shafts traditionally has been regarded as dangerous. This is no longer the case.[4] The size and relative stability of the sinking sector has generated a core of experienced people. The achievements of these professionals in the field of safety are more impressive than the product. There are several instances now on record where a shaft has been completed without fatality. Fatality-free runs in excess of 200,000 shifts are becoming commonplace.

14.1 INTRODUCTION

Shafts are required for the following purposes:[8]

- Mining the mineral deposits
- Temporary storage and treatment of sewage
- Bridge and other deep foundations
- Hydraulic lift pits
- Wells
- In conjunction with the tunneling system or network for the purpose of lifts, escalators, stair and ladder-ways, ventilation, conveyance of liquid, carrying pipes and cable in river crossings, drainage and pumping particularly from sub-aqueous tunnels.

Shaft sinking is a specialized operation, which requires trained and skilled crew. Amongst the different types of openings that are driven for mining and civil engineering purposes, the inclined or vertically down – 'sinking operation', is the costliest to drive, as this task is slow and tedious. Decision with regard to size, shape and its positioning are taken based on the purpose a shaft intends to serve. Circular shafts are preferred in almost all situations due to their stability characteristics. When strata are competent ones, such as in most of the metal mines, rectangular or elliptical shafts give the advantage of proper use of their cross-sectional areas.

14.2 LOCATION[2]

While selecting a shaft site the following points should be born in mind:

Positioning regarding deposit geometry: The main shaft which is meant for the production hoisting should be located in the geometrical work-load center of the deposit i.e. ore hauled, in terms of tons-km, from any side to the shaft should be almost equal (refer sec. 12.8.4). Choosing this proposition, for a flat deposit a protective pillar around the shaft in ore will have to be left; but if the deposit is steeply dipping, then the shaft should be positioned in the f/w side of the orebody in the country rock from its stability point of view. To achieve centralized services, the same location should be preferred and the shaft meant for this purpose is usually called – an auxiliary shaft. In a large mine apart from a main shaft, there could be a few more shafts at other locations to serve different purposes including the ventilation air outlets.

Positioning regarding surface topography: The shaft collar should be at least 5 m above the highest flood level recorded in the area. It should be away from the places of public utility and water bodies. It should be, preferably, within an easy access to the available infrastructure facilities in the area such as roads, rail, power and communication links. There should be enough space around it to establish necessary facilities.

Position regarding geological disturbances, water table and ground conditions: Through the exploratory borehole records or by drilling a borehole (in absence of any information) at a little distance away, perhaps 50–60 m, from the proposed site it is worthwhile to know the type of the strata the shaft will have to pass through. This borehole should not be drilled in the center of the shaft, because if there is water under pressure, it may rise up through it and flood the shaft during sinking. Passing through a highly geologically disturbed area with the presence of discontinuities such as faults, folds, dikes, washouts, joints, fractured and fissured zones, should be either avoided, or measures to deal with them should be planned pre-hand. Loose ground, water bearing strata, mud and running sand areas offer difficulties and require special treatment; hence, passing through such areas should be kept to a minimum.

14.3 PREPARATORY WORK REQUIRED[2,4,7]

Apart from the proper design details concerning location, orientation (inclination), size, shape, support types and position of shaft stations (i.e. the shaft insets and pit bottoms), there are many other facilities that need to be established. The most prominent amongst them are the access roads; warehouses; stack yards; shunting yards; provision for power, potable water, telephone, maintenance facilities, first aid, waste disposal, mine offices, canteen, lamp room, rest shelter, magazine, hoist room, compressed air, drilling water, etc. etc. Some of these installations are temporary and can be removed after the completion of the sinking operation.

14.4 SINKING APPLIANCES, EQUIPMENT AND SERVICES[2,4,7,13]

Some of these equipment and appliances can be hired but if this task is contracted, then the contractor brings them. Special items needed for this purpose are: hoist with head gear and suspension gear (fig. 14.1(b)), the scaffold (hanging platform or work stage) (fig. 14.1(b)), kibble or sinking buckets (fig. 14.2(e)), shaft centering devices (fig. 14.2(a)), folding door and muck disposal bins and chutes (fig. 14.1(b)), shaft ventilators with rigid and flexible ducting (fig. 14.1(f)), face and main pumps with suction and delivery pipe ranges (fig. 14.1(g)), compressed air and water pipelines, portable pneumatic lights, concrete mixers and delivery range, blasting cables, winches and few others. The services to be provided include power supply, water supply, transport, stores, repairs, refreshment, housing, social life etc. etc.

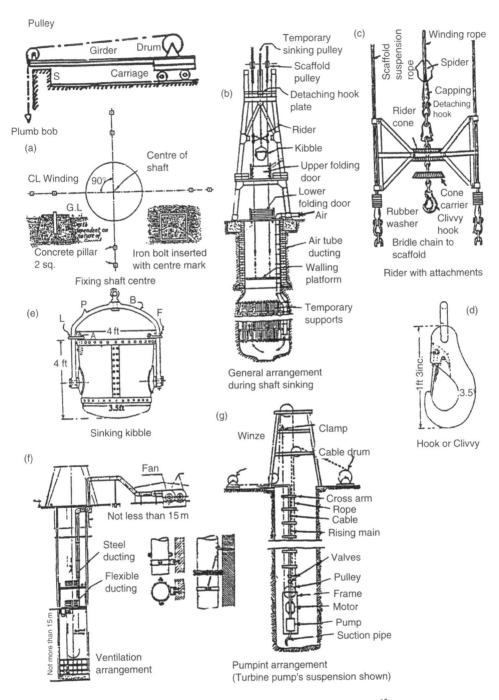

Figure 14.1 General arrangement during shaft sinking.[13]

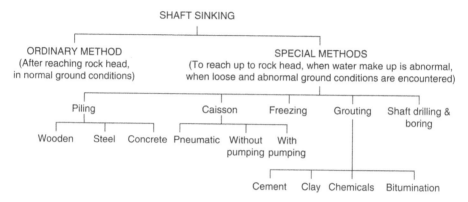

Figure 14.2 Line diagram classifying shaft sinking methods/techniques.

14.5 SINKING METHODS AND PROCEDURE[2]

Based on the techniques applied to sink a shaft, the methods have been classified by way of a line diagram shown in figure 14.2. However, this operation can be divided into three segments:

- Reaching up to the rock head
- Sinking through the rock
- Sinking through the abnormal or difficult ground, if any, using special methods.

14.6 REACHING UP TO THE ROCK HEAD[2,4,7]

Before the rock head at the shaft location is struck, there could be a presence of alluvial ground having sand, clay, gravel etc. or there can be an abnormal make of water and presence of running sand/ground. The thickness of this cover may vary from few meters to 30 m or more. This ground is excavated using ordinary excavating tools and appliances or by the use of mechanical excavators such as clamshells, backhoes etc. Cranes are used to hoist the muck if the cover is thick. If this procedure is not feasible then either the ground should be consolidated prior to carrying out any excavation, or a suitable special sinking method, as described in sec. 14.8, should be adopted. In this ground no blasting should be done to preserve its original strength. A larger area than the finished diameter of the shaft should be excavated taking into account the allowance for the thickness of temporary and permanent linings. When the rock-head has been struck, a few rounds in that are sunk though. Before advancing further, in the ground so excavated the shaft collar is built of concrete of required strength. Care is taken while designing the shaft collar so that it is keyed to the bedrock. Many times it becomes essential to extend the collar's concrete to the surface, so that it can provide a firm footing to the legs of the headgear to be installed. The whole idea is that the shaft collar and its surrounding should be keyed to the bed rock to have a sound foundation for the headgear and other installations. In figure 14.3(a) the procedure has been illustrated; this is as per the practices followed at the South African mines.

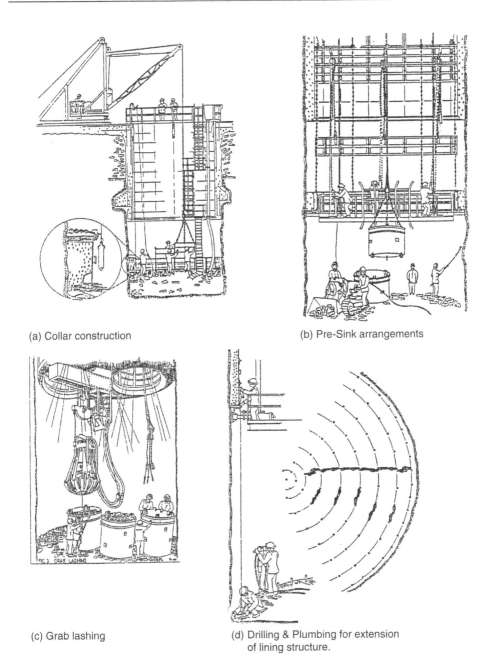

(a) Collar construction

(b) Pre-Sink arrangements

(c) Grab lashing

(d) Drilling & Plumbing for extension
of lining structure.

Figure 14.3   Procedure followed for collar construction; Pre-sink arrangements; Drilling, blasting and mucking arrangements during shaft sinking.

14.6.1   Pre-sink[4]

The objective[4] of pre-sinking (fig. 14.3(b)) is to construct a sufficient depth of shaft, to permit the assembly of the sinking stage and lashing unit (described later) in the shaft bottom. Another requirement is to open adequate clearance between the shaft bottom and the stage parking position to allow blasting without damaging the stage or, the more vulnerable lashing unit. A bottom to bank interval of say 90 m would be ideal. This would allow about 70 m between the shaft bottom and the underside of the garb's driver cab and about 20 m from the stage. Utopian conditions, however, rarely pertain and the pre-sink is usually not so deep. This situation then demands that shorter lightly charged rounds are pulled until a safe stage withdrawal height is obtained.

The main difference between collar construction and pre-sinking is that in pre-sinking the curb ring is suspended and lining can be placed some distance above the bottom, so that sinking and lining can go concurrently.

A stage is required for the shaft bottom protection and for access to the shutter. A specially prepared pre-sink stage is normally introduced to fulfill this need, or in some cases, the top two decks of the main sinking stage may be employed for this purpose.

Hand held drills (fig. 14.3(d)) and blast techniques are used for rock breaking and this section of the shaft serves as a useful training period for the shaft crews. Lashing is generally affected with crawler-mounted rocker shovels (fig. 14.3(b)). When the pre-sinking depth is attained, sinking is interrupted and the man sink stage assembled either at the bottom, or pre-erected at the surface alongside the shaft and lowered completely on to the bottom, with a large crane. The stage is roped up and raised, the lashing gear installed, commissioned, and the shaft stripped of pre-sink services and equipped with pipes and other services, and ready for the main sinking to commence.

The lead-time involved between the start of on site work on the project and this stage of maturity is at least six months,[4] during which the following should be ideally achieved:

1. Crew accommodation arrangement is established.
2. Services such as water, power, compressed air are established at the site.
3. Pre-sink is complete and this plant and equipment cleared from the site.
4. The main sink kibble and platform hoists are erected and commissioned.
5. Shaft concrete (lining) batch plant erected and commissioned.
6. Headgear, tipping and muck disposal arrangements, bank doors and bankman control cabin erected and in working order.
7. Offices, workshops, change-houses, garages, stores and all other site buildings built and occupied.
8. Stage, lashing unit and all in-shaft services ready to go.
9. Supplies of permanent and consumable materials secured, and deliveries scheduled.
10. The site adequately staffed and equipped for sinking to proceed.

14.7 SINKING THROUGH THE ROCK[2,3b,4,9,11]

A sinking cycle consists of the following unit operations:

1. Drilling
2. Blasting
3. Mucking and hoisting
4. Support or shaft lining

5. Auxiliary operations:
   a) Dewatering
   b) Ventilation
   c) Lighting or illumination
   d) Shaft centering

14.7.1   Drilling

Use of sinkers to drill holes of 32–38 mm. diameter and the shaft jumbos (equipped with number of drifters, (fig. 14.5) to drill holes of 40–55 mm. dia. is made. The hole length varies between 1.5 m and 3 m if the sinkers are used, and it can be up to 5 m in case of the shaft jumbos. Wedge cut, pyramid cut (figs 14.4(a), (b), (c)) and step cut (fig. 14.4(d)) are the common drilling patterns adopted. Wedge cut is more popular in the rectangular shafts whereas pyramid cut in the circular ones (fig. 14.5)). Step cut is adopted if the make of water is high and the shaft is of a large cross-section, so that the face can be divided into two portions to allow a continuous dewatering. The number of holes in a pattern is a function of hole diameter, shaft diameter and type of strata. The following formulae[11] for determining the number of holes, if drilling is with a shaft jumbo having hole diameter in the range of 45–55 mm could be used:

$$N = 0.234A + 22 \tag{14.1a}$$

$$N = 2.55A_1 + 22 \tag{14.1b}$$

where: A is cross sectional area in ft^2.
       A_1 in m^2.

The other method is the *powder factor method* in which, as per the type of rock, a suitable powder factor is selected based on the experience and the data available. To achieve this powder factor, the number of holes required, are calculated and arranged in a particular pattern.

In order to design a pattern in a circular shaft the holes are divided into a number of concentric circles, which can be 3 to 5, depending upon its cross-section. The ratio of hole numbers in a particular circle can be 1:2:3 for a three circled pattern and likewise 1:2:3:4:5 for a five circled pattern.[11] The diameter of these circles is a matter of shaft diameter Unrug, K.F.[11] suggested the following guidelines when explosive cartridge diameter to be charged is 32 mm:

Three circles, use 0.37, 0.66 and 0.93D (14.2a)

Four circles, use 0.35, 0.54, 0.7 and 0.93D (14.2b)

Five circles, use 0.27, 0.43, 0.6, 0.73 and 0.93D (14.2c)

where: D is the diameter of the shaft.

Use of drill jumbos, based on South African experience[4] has resulted in:

- Enhancement in overall productivity and thereby the sinking rates.
- Reduction in cost per meter of sinking.
- Improvement in overall environmental conditions and safety due to reduced human and machine populations at the working face.
- Small crews thereby better management.

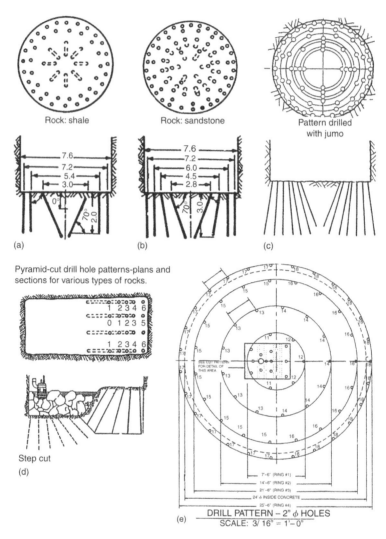

Figure 14.4 Drilling patterns during shaft sinking in different rock types using conventional sinkers (hole dia. 32–38 mm); (a) and (b) Pyramid-cut in a circular shaft (number of holes shown corresponds to medium hard rocks). (c) Pyramid cut pattern using shaft jumbos. (d) Step cut in watery conditions in a rectangular shaft. (e) Drilling pattern with shaft jumbo using 50 mm (2″) dia. holes to be charged with emulsion explosive.

The limitations include:

- Higher capital costs of rigs and their spares.
- Skilled maintenance and operating crews are almost mandatory.
- The kibble winder and opening through the working stage should able to cope up with Jumbo dimensions.

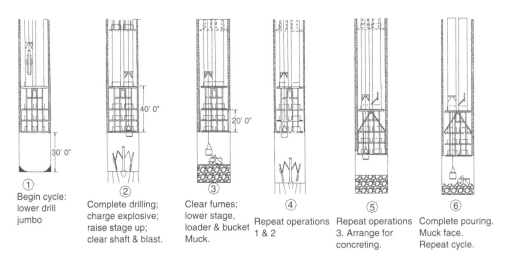

Figure 14.5 Sequence of operations while sinking through ordinary/normal ground. Relative positions of the shaft bottom shown.

The new shafts are being sunk with the use of hydraulic drills which are faster, less noisy and provides all the advantages as described in section 4.7.

14.7.2   Blasting[4,9]

In practice shaft bottoms during sinking are usually full of water; therefore, use of high-density water resistant explosives, such as nitro-glycerin based, is made to charge the holes. Use of water or sand-clay mixture can be made to act as a stemming material. Usually series-parallel connections are made to connect the detonators at the face and this circuit is then connected to the blasting cable suspended in the shaft and leading right up to the surface. Face is blasted after taking due precautions.

Aluminium-based water gel explosives (refer sec. 5.4.5) and high frequency electro- magnetically initiated detonators have been very successfully used in some of the South African shafts[4] and these have a promising future due to following facts:

1. Improved environmental conditions due to their low yield of nitrous fumes
2. Possible improved safety due to better shock sensitivity characteristics and immunity to stray currents
3. Ease of use of loose detonators and the simplicity of threading the toroid into the circuit
4. Initial tests have shown economic advantages.

Preceding the 1970s, NG-based explosives were very common. In the 1970s when 200 mm (8″) NG-based cartridges were replaced by 300 mm (12″) water-gel slurries, the first noticeable change was observed. In the later 1970s with introduction of jumbo drilling use of 1.5″ (dia.) × 48″ (long) detonator sensitive cartridges were made. These long cartridges reduced the explosive loading time. These long cartridges served the industry between the mid 1970s to the mid 1990s. Later on the application of hydraulic drill jumbos paved the path for large sized rounds of 12–14′ in shafts. The 'cut' of blasting patterns were changed and use of four 3.5″ dia. holes used as cut to provide initial free face for these 12–14′ rounds.

The latest development, as claimed by the Nitro-Nobel[9] is the use of emulsion explosive with booster and Nonel detonators. The rounds with the use of hydraulic drills are usually drilled with 50 mm (2″) diameter holes with a cut hole of 7.87″ diameter in the cylindrical cut (parallel hole cut) pattern. The round depth of 16.5′ has been successfully driven. Roach and Roy (1996) have listed following advantages of this practice:

- Less expensive than cartridge explosives
- Faster loading than cartridge explosives
- Provides full borehole coupling
- Reduce drilling as holes required are of larger diameter than those for cartridges
- Better fragmentation.

This pattern has been shown in figure 14.4(e).

14.7.3 Lashing and mucking[3b,4]

Lashing: It is the arrangement (fig. 14.3(b)) that is made for the loading of blasted muck into conveyance for its disposal. Thus, the Lasher is one who lashes it, and a lashing-unit is a mechanical device incorporating hoisting, slewing and radial traversing mechanism for the handling of the cactus grab (or any other mucking equipment), which completes the lashing system.

Presence of water, limited space and the time required to install mucking equipment makes this operation a time consuming activity. It occupies about 50–60% of the time of a sinking cycle. The mucking efficiency depends upon the size of rock fragments, hoisting depth, shaft cross-section and water inflow rate. Several types of shaft muckers are available; the prominent amongst them are listed below:

- Arm loaders such as riddle mucker, cryderman mucker, cactus grab (fig. 14.3(c)) and backhoe mucker.
- Rocker shovel such as Eimco-630.
- Scrapers – used for very large dimensioned shafts.

Details of these loaders have been outlined in chapter 6 on mucking.

In South African mines,[4] use of turret type lashing units and cactus grab (fig. 14.3(c)) is almost universal. The units are matched with $0.56 \, m^3$ grabs in shafts of 6–8 m diameters; larger units and $0.85 \, m^3$ grabs are used in the large shafts. Crawler mounted rocker shovels have been used in very large shallow depth shafts and in small shafts down to 5.5 m dia. Small $0.28 \, m^3$ remote mounted grabs have likewise been employed. In very small shafts hand lashing is sometimes considered but this is costly, rare and very unpopular nowadays and it is more usual to set the minimum shaft size, which will facilitate the mechanical cleaning system for reasons of efficiency and economy.

14.7.4 Hoisting

For hoisting/lowering of crew, material and muck two practices are prevalent i.e. by installing the permanent hoist and its attachments; else with the use of a temporary

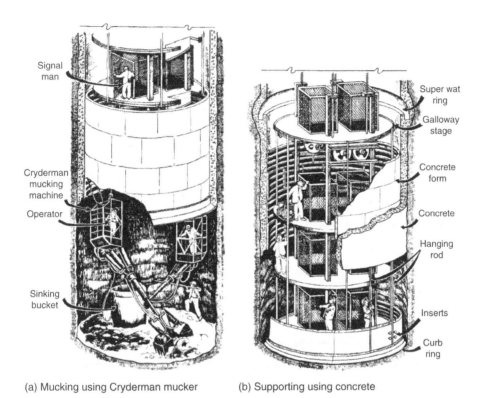

(a) Mucking using Cryderman mucker   (b) Supporting using concrete

Figure 14.6   Mucking during sinking and supporting system using concrete.

hoist, head-gear and other attachments. Usually the latter one is preferred. This is all the more essential if the sinking contract has been awarded, so that on completion of the sinking operation, the mine owner has no responsibility of caring for these items. This installation should be able to handle a load up to 150–200 tons. It should be compact to enable the permanent winding structures to be erected around it. The arrangement shown in figure 14.1, illustrates the use of various appliances that are required during this operation. The prominent amongst them are: *Head gear with pulleys* – as shown in figure 14.1(b), two pulleys are meant for winding the sinking kibbles and the other two are for the winding a scaffold i.e. work stage; *A Rider* – which enables the scaffold ropes to act as the guide ropes for the smooth run of the kibble from shaft top to the scaffold position in the shaft (fig. 14.1(c)); *Lower folding doors* – to keep the shaft top covered; *Top folding doors with kibble discharge mechanism* – for discharging the muck; *Kibbles* – a few kibbles (fig. 14.1(e)) are kept spare to speed up the mucking and for their use to lower the crew, material and even sometimes to hoist the water; *Work-stage or scaffold* – it is usually a multi-deck to carry out shaft lining and other works speedily. In addition, the air ducting, compressed air and water lines, cables etc. are required to provide the necessary services.

14.7.5 Support or shaft lining

Basically there are two types of lining: Temporary and Permanent. The make of water and strength of the strata through which the sinking operation is to be carried out govern the choice. In some situations temporary support is not required, whereas in others, it becomes essential to protect the crew and equipment from any side fall. Depending upon the conditions the length of temporary supports could range from 6 m to 40 m.[11] Once this length is covered by the temporary lining and before advancing further, the permanent lining is installed. Before installing the permanent lining if feasible the temporary lining can be removed else it is left in place. The permanent lining can be that of bricks, concrete blocks, monolithic concrete (figs 14.7(a), (b) and 14.12(c)), shotcrete and cast iron tubings (figs 14.7(c) and (d)). The bricks and concrete block were earlier used in the dry and shallow depth situations but at present the monolithic concreting of the desired strength is a common practice. The steel tubing is used in conjunction with the freezing method of sinking. The details of these linings have been dealt in chapter 8 on supports. The common types of shaft lining have been illustrated in figure 14.7.

14.7.6 Auxiliary operations

14.7.6.1 Dewatering

During sinking once the shaft has reached to the water table or beyond it, make of water is unavoidable. Even before, inflow of water is usual. Hence, one of the important auxiliary operations during sinking is dewatering. Arrangement has to be made as per the water inflow rate to be dealt with. The prevalent practices are as follows:

1. *Face pumps*: If the make of water is limited, this can be hoisted through the kibbles or water barrels. To fill these barrels, pneumatically operated membrane face pumps are most suitable, as they can deal with muddy, silted and dirty water.

2. *Sinking pumps*: If the make of water is beyond the handling capacity of the face pumps, then hanging pumps which can be suspended in the shaft together with the electric cables, motor, suction and delivery ranges, are used. The pumps used are of turbine type to which the impellers can be added as the water head increases. Adjusting the valve of the delivery side can also regulate the quantity. It can deal with dirty and gritty water. Being compact it can be readily raised or lowered. The arrangement has been shown in figure 14.1(g).

3. *Provision for the intermediate sump and pumps*: When the shaft depth increases and make of water is sufficient, it is always preferred, as per Boky[2] to have intermediate pump chambers with sumps at an interval not exceeding 250 m. To this sump, water from the face is delivered for its onward pumping to the surface. Keeping a standby pumping set is a normal practice during shaft sinking, as any moment an abnormal quantity of water inflow can be expected.

14.7.6.2 Ventilation

Fresh air, by a forcing fan installed at the surface is provided at the face through the rigid and flexible ducts, which are suspended at the side of the shaft. The rigid

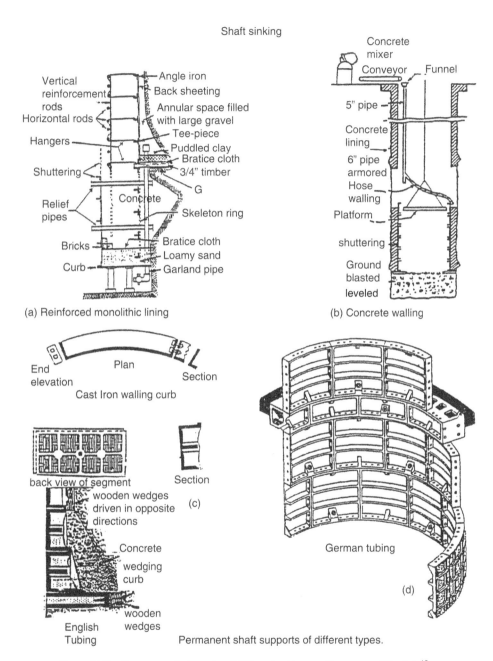

Shaft sinking

(a) Reinforced monolithic lining

Vertical reinforcement rods
Horizontal rods
Hangers
Shuttering
Relief pipes
Bricks
Curb

Angle iron
Back sheeting
Annular space filled with large gravel
Tee-piece
Puddled clay
Bratice cloth
3/4" timber
G
Skeleton ring
Bratice cloth
Loamy sand
Garland pipe

Concrete

(b) Concrete walling

Concrete mixer
Conveyor   Funnel
5" pipe
Concrete lining
6" pipe armored
Hose walling
Platform
shuttering
Ground blasted leveled

End elevation
Plan
Section
Cast Iron walling curb

back view of segment
wooden wedges driven in opposite directions
Concrete
wedging curb
wooden wedges
English Tubing

Section

(c)

German tubing

(d)

Permanent shaft supports of different types.

Figure 14.7 Permanent lining using RCC and tubings – German and English.[13]

ventilation duct range terminates at least 6 m above the shaft bottom to avoid its damage due to blasting. But thereafter to have the fresh air at the face flexible ducting of the canvas is joined to it. The whole shaft acts as return. But in many sinking projects now a days, the practice is to install a contra-rotating fan at the surface so that immediately after the blasting it is switched to act as an exhaust, and once the fumes are cleared it is re-switched to act as a forcing fan. Sufficient quantity of air with a water gauge up to 12 inch (0.3 m) is required to ventilate the face. In figure 14.1 (f) suspension of ventilation ducts to a shaft side has been shown.

14.7.6.3   Illumination

At the face a pneumatically operated light, consisting of a cluster of 4–6 bulbs fixed in a suitable water-tight fitting, is used to provide illumination at the working face during drilling, mucking, lining and other operations.

14.7.6.4   Shaft centering

Using the reference points, which are fixed before commencing the sinking operation to fix the shaft center, the shaft's center and inclination (i.e. verticality in case of a vertical shaft) are checked from time to time, by the use of a centering device (as shown in fig. 14.1(a)), installed at the surface. In figure 14.3(d), a survey plan to mark plumb line, and pop marks for installation of various steel structures (buntons and guide rails) during shaft equipping for a deep shaft at the Vall Reef shaft in South Africa, has been shown. (This shaft is having the following specifications: 10.6 m dia.; 2340 m deep and design for a production of 333,000 tons/month. 4 skip compartments; 4 double deck cage compartments; 2 service cage compartments).

14.7.6.5   Station construction and initial development
(figs 14.8(a), (b) and (c))[4]

Current practice is to establish the mining levels as and when the shaft bottom reaches the station floor elevation. Development of the station and tip crosscuts and in many cases the raise boring of rock passes and ventilation raises is erected during the sinking phases. This permits immediate commencement of development operations when equipping is complete; the shaft system can serve as a well-established nucleus of working mine.

The first cut of the station entries is drilled from the shaft bottom as it advances through the excavation. The shaft is usually taken about two rounds below the required finished floor of the station. This sump is allowed to fill with muck, which serves as a working floor while excavation on the level is in progress. The shaft lining is brought down to the station brow elevation. Depending upon the personal preference and ground conditions the station can be completely excavated, or, the excavation can be limited to that required to permit access for subsequent development.

Crawler-mounted rocker shovels are used to assist with cleaning. They back-lash into the shaft area where the grab re-handles the muck into a kibble for final disposal. These methods are applied until a safe parking area for LHD vehicle is created. The LHD machines of 1–2 m^3 size can pass through the stage kibble openings and then replace the rocker shovels for development cleaning and mucking. They likewise feed the grab in the shaft area. These methods yield high advance rates.

In instances where it is considered advantageous to carry out extensive development concurrently with shaft sinking, mid shaft loading arrangements can be considered. In these cases a section of the shaft would be equipped for skip hoisting from a loading box so that sinking and development operations could proceed independently to each other.

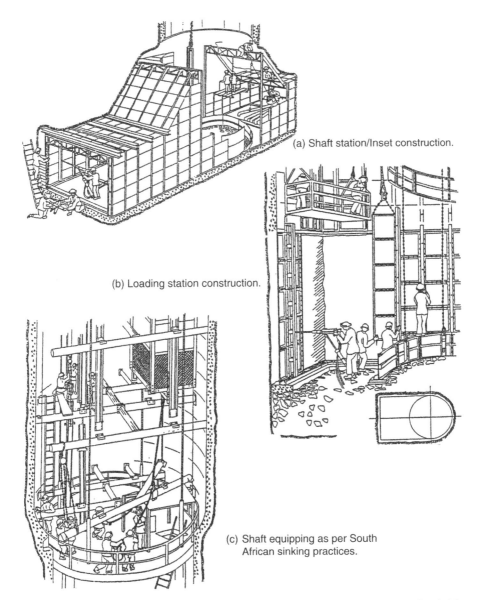

(a) Shaft station/Inset construction.

(b) Loading station construction.

(c) Shaft equipping as per South African sinking practices.

Figure 14.8   Construction details of a typical shaft inset and loading station based on South African practices. Shaft equipping also shown.

14.8   SPECIAL METHODS OF SHAFT SINKING[2,3,6,7,8,11,12,13]

In the process of shaft sinking, it becomes necessary to adopt a special method/ technique, if the ground through which shaft is to be sunk is loose or unstable such as sand, mud, gravel or alluvium, or when an excessive amount of water is encountered, which cannot be dealt with by the sinking pumps. Also when in some situations, both sets of these conditions are encountered. Following are the special methods that are used to deal with the situations, outlined above:

- Piling system
- Caisson methods
- Cementation
- Freezing method
- Shaft drilling & boring.

14.9   PILING SYSTEM

This method is suitable only for sinking through the loose ground near the surface. Wooden piles (fig. 14.9(a)) 2–5 m long, 50–70 mm thick & 150–200 mm wide, or steel piles (fig. 14.9(a)) are used. Steel piles are stronger those than of wood. Wooden piles are shod with iron at the bottom so as to pierce the ground. The piles are driven down by heavy mallets, and are placed edge to edge so as to form a complete circular lining. They are held in place by circular rings or curbs, placed at an interval of 0.8–1 m. After putting the first set of piles, another set of piles is then driven but before this the ground enclosed is dug out, to the extent that the first set of piles is about 0.6 m in the ground. In this manner a number of sets of piles are driven and ground is simultaneously dug, till driving through the loose and unstable ground is completed. This means sufficient extra ground all around the proposed site of the shaft needs to be dug and piled. Once the firm ground is encountered, the permanent lining which could be either that of bricks, steel tubing or concrete is built. The space between this lining and the piles is filled with some packing material.

In the recent past use of concrete pile wall is also becoming popular. This is accomplished either by driving the steel pipes of about 500 mm dia. using a pile driver, else the holes of 500 mm–1 m dia. are driven with a drilling rig and then concrete is poured into it. These pipes or holes are put all along the circumference of the shaft. If the sides of the holes drilled are liable to cave in, then it is treated with some mud or slurry.

14.10   CAISSON METHOD

This method is popularly known as Drop-Shaft and is common in civil engineering works while one has to sink through the riverbed. This method is suitable to sink through the running ground to a depth somewhat greater than the one, which can be negotiated by adopting the piling method. The method can be subdivided into three main classes, namely:

1. Sinking drum process
2. Forced-drop method
3. Pneumatic caisson method.

14.10.1   Sinking drum process

In this method (fig. 14.9(c)) basically a prefabricated lined shaft, like a drum with both the ends open and wall thickness equivalent to the thickness of lining, is forced down through the ground of the intended shaft site. The lining could be that of bricks, concrete or steel tubing, and it is fitted with a steel cutting shoe at its bottom.

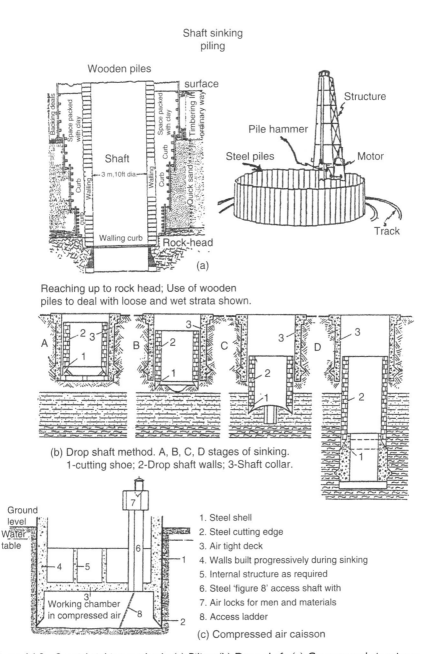

Shaft sinking
piling

Wooden piles

Reaching up to rock head; Use of wooden
piles to deal with loose and wet strata shown.

(b) Drop shaft method. A, B, C, D stages of sinking.
1-cutting shoe; 2-Drop shaft walls; 3-Shaft collar.

1. Steel shell
2. Steel cutting edge
3. Air tight deck
4. Walls built progressively during sinking
5. Internal structure as required
6. Steel 'figure 8' access shaft with
7. Air locks for men and materials
8. Access ladder

(c) Compressed air caisson

Figure 14.9   Special sinking methods: (a) Piling; (b) Drop-shaft; (c) Compressed air caisson.

The drum sinks gradually by its own weight and simultaneously the ground within the periphery of the drum is excavated manually or with the use of a mechanical excavator. As the drum pierces through the ground, at the top further lining structure is added to it. Care should be taken so that the drum sinks vertically down; sometimes additional weights are put at the top for smooth sinking.

Advantages:

- This process eliminates the temporary lining work, which in turn saves labor, material and time.
- The permanent lining too is built at the surface where it is easier, quicker and safer to build.
- The weight of sinking drum is sufficient to push aside the boulders whereas in the piling method it may pose problems.

Limitations:

- Sometimes it is difficult to keep the drum vertical.
- As the skin friction increases with the increase in depth, thereby sometimes it becomes difficult to sink further in spite of adding weights at the top.

14.10.2 Forced drop-shaft method

This method (fig. 14.9(b)) could be applied, if the sinking further by ordinary sinking drum process fails, or, in the strata where alternate layers of loose and tough ground are envisaged. In any case, first, in the upper part of the shaft it is essential to built the walling which may be of brickwork or concrete. This is called preliminary caisson. Through this structure, cast iron tubing drum having flanges in the inner side of this drum is used to sink through the ground. The drum is pushed downward with the help of hydraulic rams or jacks.

The lower end of the drum is provided with a cutting edge to ease the process of sinking. The ground within the tubing's drum is excavated either manually or with the use of a mechanical excavator. On completion of sinking by one segment, another segment is added from the top and the process is repeated till the sinking through loose and running ground finishes or the drum sticks to the ground, and it refuses to sink further.

This method has an advantage over the sinking-drum process, which works on the principal of gravity due to the fact that piercing through the ground is sure to a greater depth than the one which could be achieved by the ordinary sinking drum method. Sinking depth up to 60 m is the limitation of these methods.

14.10.3 Pneumatic caisson method[8]

A gentleman named Trigger invented this method (fig. 14.9(c)). In this method use of compressed air is made while sinking through the waterlogged quicksand or mud. It is a modification to sinking drum process where the quicksand or mud are kept to accumulate at minimum thickness at the shaft bottom face with the use of compressed air which is circulated about 2–3 m above it. A diaphragm or partition is fitted at this horizon to form the compressed air chamber. The compressed air's pressure is kept higher than that of the incoming water's pressure through the strata. An air lock is mounted on the top of the diaphragm to permit the passage of men and materials. The caisson sinks by gravity as in the ordinary sinking drum process. The method has been used in some of the western countries including Russia and is usually confined within a depth of 30 m. This method is costly. Working in the compressed air chamber causes health hazards. Hence, the method has its own limitations and is not very practicable.

14.11   SPECIAL METHODS BY TEMPORARY OR PERMANENT ISOLATION OF WATER[2,3,6,7,8,10,11,13]

14.11.1   Cementation

In this special method of shaft sinking (figs 14.10(a) and (b)) the liquid cement is injected through boreholes into the gullet strata in order to fill up any cracks, cavities, fissures and pores. The cement, in turn, strengthen the strata and ultimately make them impervious to water.

Thus, this method is applicable if the ground is firm but fissured. It is not suitable for running sand type ground conditions. The success of the method lies in to the fact that, at many locations, in the heavy water-bearing areas the pumps up to 10,000 g.p.m capacity failed but this method could succeed. The cement is injected at a pressure of 80 –4000 psi. Following steps are followed:

1. Boring/drilling
2. Cementation
3. Sinking and walling.

14.11.1.1   Boring/drilling

The long hole drilling drifters or diamond drills can be employed for this purpose. The number of holes to be drilled depends upon the porosity of the ground, if ground is more porous, more holes are drilled and vice-versa.

Holes are drilled in the fashion as illustrated in figures 14.10(a) and 14.10(b), all along the periphery of the shaft collar. The first series of holes is begun from the dry ground above water level. Preferably, if not all the holes but a few of them should be inclined radially or tangentially at 1 in 10 or 1 in 15, so as to ensure that these holes will intercept the fissures throughout the length of treatment and the area where these holes terminates i.e. the base, is away from the actual perimeter of excavation.

First, a 5 m long hole of about 70–80 mm diameter is drilled, and then it is fitted with the standpipe. The standpipe is projected about 0.15–0.3 m above the collar to fit a stop valve, so that water coming out from the hole during drilling can be kept under control. Down to 5 m, the hole of 35–45 mm. dia. is drilled up to a pre-determined length (may be 30–40 m) or up to the point where from the abnormal make of water is experienced. The drill rods are then withdrawn and in the hole using flexible hose, the liquid cement is injected. The same procedure is followed for each of the holes.

14.11.1.2   Cementation

The cementation plant includes: high capacity double acting ram pumps, cement mixing tanks, pipe range and other tanks to inject chemicals such as silicate of soda or sulfate of alumina, if pre-silication is adopted.

Initially 2.5% cement is injected, its quantity later on can be increased to 50% depending upon ground conditions and quantity of water. Total holes drilled are divided into three potions, say: A, B and C. Holes 'A' are used to seal off the main fissures. Holes 'B' are used to seal off the hair cracks. Holes 'C' are used to do the same function as holes 'B' but with reduced quantity of the cement injection. The selection

Table 14.1 Selection of grout density based on water absorbability of grouting hole.[11]

Absorbability, L/min.	Cement	Sand	Water	Proportion of cement to sand & water
0.0–0.001	1	0	12	1:12
0.001–0.01	1	0	10	1:10
0.01–0.1	1	0	8	1:8
0.1–0.5	1	0	6	1:6
0.5–1.0	1	0	4	1:4
1.0–3.0	1	0.5	3	1:2
3.0–5.0	1	1.0	2	1.1
5.0–10.0	1	2.0	1.5	1:5
Over 10.0	1	3.0	2.0	1:5

of grout (which is a mixture of cement, sand and water) density depends upon the water absorbability of the grouting hole. The table 14.1 illustrates this aspect.

To decrease absorbability and reduce the cost of operation, sometimes clay is mixed with cement in the ratio of 1:2 to 1:4. The grouting operation is considered successful when the control holes prove that water absorbability within the grouted rocks is less than 0.5 liters/min.

14.11.1.3 Sinking and walling

After cementation, for a ground column of certain length, the procedure of ordinary shaft sinking is carried out. However, the depth of blasting round should be limited to 1.5 m, and also, the explosive charge amount per shot-hole as well as the total charge/blast, should be kept at a minimum.

Depending upon the rocks through which shaft is to be sunk and the amount of water encountered while carrying out the process of cement injection and magnitude of the fissures (as the joints, fissures encountered are the potential source of water in future), the type of lining work that is carried out includes, the lining of concrete (R.C.C. or ordinary) or use of steel tubing. The former is cheaper and also very common.

14.12 THE FREEZING PROCESS

This method (figs 14.10(c) and (d)) is suitable for any kind of heavily watered strata including quicksand. It has proved its success even in most difficult ground conditions. The process consists of formation of a cylinder of frozen ground, in the center of which it is possible to sink a shaft, by following the ordinary method of sinking.

The freezing is accomplished by boring/drilling a ring of holes slightly outside, around the site selected, for the actual shaft to be sunk. In these holes through steel tubes brine solution is circulated. The brine solution, which absorbs the heat from the boreholes, progressively, causes the ground to freeze, and form the ice wall of sufficient thickness. This artificially created wall of ice prevents the inflow of water into the shaft being sunk. There are four distinct steps that are followed in this system, and these are:

- Drilling and lining of boreholes
- Formation and maintenance of the ice column

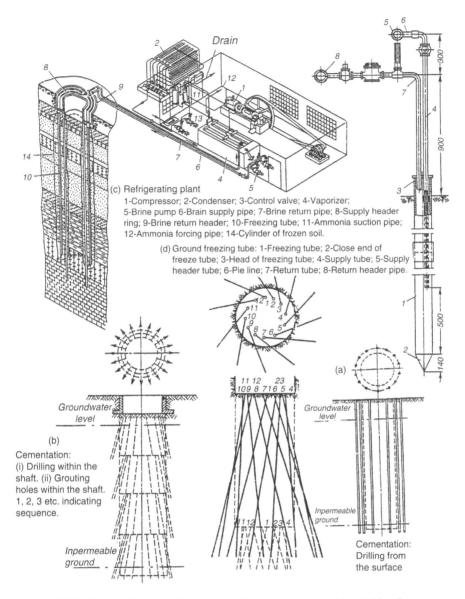

Figure 14.10 Special sinking methods: (a) and (b) – Cementation; (c) and (d) – Freezing.

- Actual sinking operations, and
- Thawing of ice-wall.

14.12.1 Drilling and lining of boreholes

To start with, vertical boreholes of diameter 150 mm, or more are drilled at some distance all along the circumference of the shaft site. All the holes are lined with steel tubes to prevent the caving from sides.

14.12.2   Formation and maintenance of the ice column

In the boreholes so drilled, two concentric freezing tubes are inserted. For a borehole of 150 mm. dia. an outer tube of 125 mm dia. and the inner tube of 50 mm dia. can be selected. The cold brine solution (at a temperature of –20°C) is pumped through the inner tube. It then ascends between the inner and outer tubes, extracting heat from the strata, and collected in the brine tank, which is fitted with a spiral shaped coil through which the refrigerant ammonia is circulated. This ammonia extracts heat from the brine solution and gets evaporated. The ammonia gas so formed is further pumped through the compressor to the water tank enabling it to reconvert in the liquid stage. This is how the cooling process goes on.

Thus, a plant consisting of the following items is installed at the surface for this freezing process to execute:

– Ammonia compressors
– Pumps for brine and water circulation purposes
– Pipe ranges to circulate water, brine and ammonia.

The circulation of the brine may be carried out in all the boreholes simultaneously; the ice wall then grows slowly around each freezing tube and ultimately joins with each other to form an ice cylinder. The time required to form the ice wall depends upon the size, depth, type of strata and the climatic conditions of the area where this process is being carried out. It may take to 2–6 months. For determining the thickness of ice cylinder, no formula as such is available, but it is kept as per the experience. The practice followed at German mines,[7] is described below. In figure 14.10(c), a typical plant layout has been shown.

Thermo-physical boundaries

Based on the German freezing shafts, the relation eq. 14.3[7] could be used as a useful guide to found the relation between freezing diameter (D) and excavation diameter (A) circle. This logically means that increasing frozen cylinder thickness as a function of increasing depth, as shown in figure 14.11(a).

$$D/A = [1.2 + (T/1000)] +/- 10\% \tag{14.3}$$

T is freezing depth in m.

The freeze wall thickness is most often calculated[12] using the Domke formula, which contain an appropriate factor of safety and produces dimension (S) equal to half the total thickness required.

$$\text{Domke's formula:}\ \ S/R = 0.95\ (P/K) + 7.54\ (P/K)^2 \tag{14.4}$$
$$\text{(Metric or Imperial units)}$$

where:

R – Radius of collar excavation (select any unit of length)
S – Freeze wall thickness inside the ring (same unit as that of R)
P – The ground pressure (select any unit of pressure)
K – Compressive strength of frozen ground (same unit as that of P) refer table 14.2 to get this value.

Table 14.2 An approximate unconfined compressive strength of frozen ground for different soil/ground types (Interpreted from the results of tests taken at various laboratories).[12]

Temperature °F	26		20		14		8	
Temperature °C	−3.3		−6.7		−10.0		−13.3	
	psi	MPa	psi	MPa	psi	MPa	psi	Mpa
Sand	1200	8.3	1500	10.3	1800	12.4	2000	13.8
Clayey Sand	700	4.8	1000	6.9	1150	7.9	1250	8.6
Sandy Clay	450	3.1	600	4.1	750	5.2	900	6.2
Clay	350	2.4	550	3.8	700	4.8	–	–
Silty Sand	–	–	–	–	1000	6.9	1200	8.3
Silty Clay	–	–	–	–	600	4.1	900	6.2
Silt	270	1.9	400	2.8	500	3.4	–	–
Ice	300	2.1	500	3.4	650	4.5	850	5.9

In table 14.2, an approximate unconfined compressive strength of frozen ground for different soil/ground types has been tabulated.

Example: Given following data, calculate S.
R = 3.75 m including an allowance for over-break
P = 1.38 MPa (Maximum ground pressure)
K = 6.9 MPa (from table)
Calculations: P/K = 0.2; S = 1.84 m; t = 2S = 3.68 m (12 feet).

In figure 14.11, performance of the ground freezing installations, as per the German practices, has been shown. The performance of the refrigeration plants is seen to reach up to 15 GJ/h. The pipe capacity per meter of installed freezing pipe averages 600 kJ/h, which is correspondence to approx. 143 kcal/h; (fig. 14.11(c)).

The distance between the freeze pipes (fig. 14.11(b)), as used by the German companies, which are expert and specialized in this task, states for the average freeze pipe spacing (f.p.s) to be:

$$f.p.s = 1.28 +/- 10\%, \text{ in m.} \tag{14.5}$$

14.12.3   Actual sinking operations

The sinking of shafts through this frozen ground is carried out by ordinary methods. Low freezing explosives are used for the purpose of blasting. Controlled blasting, sometimes, becomes necessary to prevent the damage to the brine circulating tubes.

In the sinking cycle, the shaft lining work is also carried out. This lining is usually of steel tubing, particularly at and below the water table level. The concrete walling to render it leakage proof backs the tubing lining.

14.12.4   Thawing of ice wall

Once the sinking through the difficult ground is completed, then it becomes essential to melt the ice cylinder. Circulating the hot brine through the tubes, thereby, melting the ice cylinder gradually, carries this process, known as thawing.

14.12.5   Freezing – shafts[8]

Where a shaft is entirely in water-bearing ground lacking cohesion, a plug at the bottom may need to be frozen, by means of further freezing pipes within the core. To avoid undesired and wasteful freezing of the core, these pipes can, with advantage, be insulated above the plug level. With mixed strata pipe can be insulated at level where freezing is not required.

Inclined shafts may be frozen either by use of inclined tubes parallel to the axis or by a carefully planned pattern of vertical tubes. Drilling of inclined holes requires specialized drilling equipment and skill to achieve the necessary accuracy. Deviations do occur as shown in figure 14.11(d).

The freeze must be maintained until the appropriate structural lining is completed and competent to carry the load. Any failure of refrigeration plant allows a thaw to be initiated, with progressive danger of inundation and collapse.

14.12.6   Ground freezing practices in Germany[7]

In areas of unstable overburden and water-bearing, the initial excavation is carried out using the ground freezing method. The excavation is first made within the protection of freeze circle, using an external concrete block lining down to a foundation level in the load bearing ground. The inner shaft casing, which rests on an annular footing just like a chimney is then constructed. The annular gap between outer lining and shaft casing is then filled with liquid asphalt so that any deformation of the strata due to mining activities does not damage the casing. When lining has been installed up to the shaft collar, the freeze plant is shut down and the plug of frozen ground around the shaft is allowed to thaw. A steel casing, with watertight welded joints, surrounds the inner lining to prevent the subsequent ingress of water (fig. 14.12(a)). These so-called sliding shafts act like stable pipes floating in a fluid. This type of lining system is called *Sliding Lining*.

This arrangement is costly, compared to the conventional practices, as much three times, as shown in figure 14.12(b). However, this system has the following advantages:

- Mining operations within the shaft pillar are possible
- The external welded steel membrane guarantees absolute water-tightness.

The following reasons[10] for lining damage in frozen ground have been pointed out.[10] These reasons are based on the literature survey, observations of 27 shaft linings and detailed investigation made of P-6 shaft at Legnica-Glogow Copper Basin (LGOM) in Poland.

1. Excessive shrinkage of concrete due to unsuitable quality and quantity.
2. Freezing of lining before cement has set. After thawing the cement does not reach its proper strength.
3. Structural changes in the concrete caused by freezing of excess water in saturated concrete.
4. Premature loading of lining before concrete has built up sufficient strength.

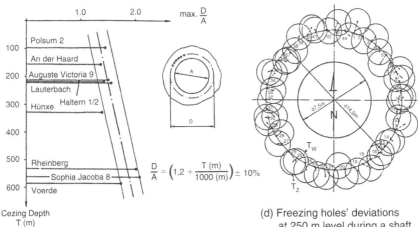

(a) Relation between freezing circle dia. D & excavation dia. A, as per German practices.

$$\frac{D}{A} = \left(1{,}2 + \frac{T\,(m)}{1000\,(m)}\right) \pm 10\%$$

(d) Freezing holes' deviations at 250 m level during a shaft sinking at a Poland mine.

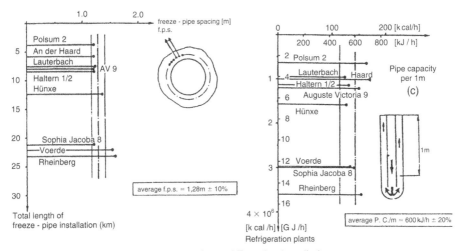

(b) Freeze pipe spacing and (c) Ground Freezing installations as per German shaft sinking practices.

Figure 14.11   Special sinking methods. Details of sinking by freezing as per German practices.

5. Local overloading of lining adjacent to plastic rock layers. Creep of the rocks occurs particularly during heat input from the concrete hydration.
6. Shut down in the freezing of mantle, which may result in excessive damage to the lining or even loss of the shaft.
7. Non-uniform loading, which may be due to an increase in volume of frozen mantle.
8. Point load due to high pressures during cementation.
9. Sudden increase in water pressure cause by collapse of frozen mantle during thawing.

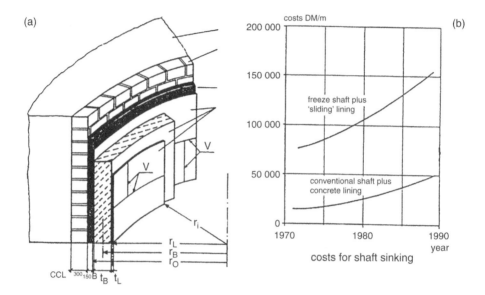

(a)

(b)

Sliding shaft construction (A German Practice)
1. Water bearing soil strata
2. Outer wall (concrete blocks)
3. Gap filled with asphalt
4. Inner wall: Steel membrane, concrete
   cylinder and liner.

r_O Outer radius
r_B radius of concrete bedding axis.
r_L radius of liner axis
t_B Concrete thickness
t_L Steel liner thickness

(b) Comparison of freeze shaft + sliding lining with
    conventional shaft + Concrete lining at German
    mines

(c) Concrete lining in a South African sinking shaft.

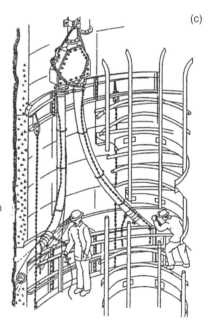

(c)

Figure 14.12   (a) and (b) Details of sinking by freezing as per German practices. (c) Details of lining
during sinking in South Aftrican mines.

The principal methods of handling water problems[8] during shaft sinking are:

1. Pumping, from sump in shaft kept ahead of main excavation
2. Ground water lowering either by well points or deep well pumping
3. Grouting of pores and fissures

4. Freezing
5. Compressed air working.

14.13   SHAFT DRILLING AND BORING

There are two methods: *Drilling and Boring*; which without aid of explosives can undertake shaft sinking operations. Sinking is the most hazardous work amongst all mining operations, and that too, while driving through the aqueous, cavable and soft ground. The drilling method gives the advantage of sinking a shaft without the entry of the crew into it during its drivage. Thus, the method is safe and proves economical in the conditions where the conventional methods may not prove viable.

14.13.1   Shaft drilling[2,3,6]

Use of rotary drilling has been made extensively for gas and oil wells. The same technique has been applied to sink shafts particularly through the profile of aquifers or caving formations that make the conventional shaft sinking techniques (including the special methods) economically impractical. Basically this technique is applied to drill the holes of large diameter; in the range of 64–300 inches (1.5 m–8 m), and up to a depth of 2000 m or so (fig. 14.13(c)).[6] Usually ventilation and emergency escape shafts can be sunk using this technique but in exceptional circumstances main shafts have been also sunk.

Basically this method uses a heavy oil rig to which a rotary drill with its drill string and bit are mounted. The bit is equipped with roller cutters with teeth that cut rock chips as the bit rotates at the bottom. The number and arrangement of cutters vary with the size of the hole. To keep control on hole deviation, stabilizers are used. Relatively low speeds are used while drilling large diameter holes. Soft formations require a few thousand kg of weight on the bit but for the hard formations, much more weight, to the order of 15 tons or so, may be required. The drilling mud or fluid (e.g. water-base bentonite-gel drilling mud) used in this technique supports the shaft walls, cools the drill bit and removes the cuttings. Due to the large diameter of shafts, cuttings cannot be removed by the conventional methods, and therefore, a double walled pipe with reverse circulation, as shown in figure 14.13(c), is employed for this purpose. At the surface, wire mesh screens separate cuttings from the drilling fluid, which is then recycled.

The method claims merits such as: carrying out all the operations from the surface and effectively dealing with the ground water, caving and soft formations. A smooth wall surface with fast penetration rates can be achieved by employing less labor. High capital cost and difficulties in drilling through the harder strata are some of its limitations.

14.13.2   Shaft boring[3]

Although the concept of shaft boring with the use of shaft borers (SBM), like tunnel borers (TBM) to drive horizontally, came during the sixties, it could not gain much popularity due to the fact that any difficult ground through which it needs to be driven, must be first treated or consolidated. Secondly, the problem of removal of the large volume of cuttings, which without a pilot hole leading to the lower accessing level, is a tedious task. The crew with the equipment has to travel on board.

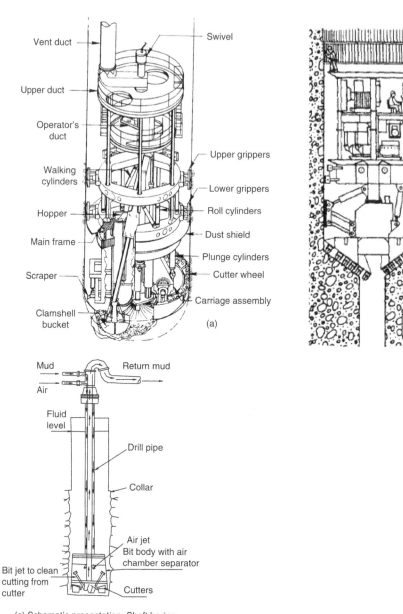

Vent duct

Swivel

Upper duct

Operator's duct

Upper grippers

Walking cylinders

Lower grippers

Hopper

Roll cylinders

Main frame

Dust shield

Scraper

Plunge cylinders

Cutter wheel

Clamshell bucket

Carriage assembly

(a)

(b)

Mud

Return mud

Air

Fluid level

Drill pipe

Collar

Air jet
Bit body with air chamber separator

Bit jet to clean cutting from cutter

Cutters

(c) Schematic presentation: Shaft boring.

Figure 14.13 Special sinking methods using shaft borers and drilling rigs. (a) Shaft boring using SBM[3]; (b) Using 'V' mole[3]; (c) Schematic presentation of shaft boring.

The system (fig. 14.13(a))[3] consists of a cutter wheel mounted on a carriage and a clam type mucking unit. The carriage is mounted on a slew structure that rotates about the vertical axis of the shaft. To fix this assembly into the shaft grippers are used. The rock cuttings are mucked into a hopper, which discharges the, muck into sinking kibbles/buckets for its discharge at the surface. The carboniferous rock is probably the most suitable formation for SBM. The strata, which give a large amount of water, must be sealed off by grout before the SBM begins boring.[11]

SBM includes the shaft lining and equipping facilities, laser beam and mechanical direction control devices, support installation facilities, water handling and ventilation systems, and simplified access for cutter changes and maintenance.

Innovation in this technique (fig. 14.13(b)) includes drilling of a pilot hole of about 1.2 m at the center of the intended shaft. This pilot hole provides the information about the type of strata going to be encountered and ease in subsequent reaming. In shaft boring techniques a drill string is not required and the precise verticality of the shaft can be maintained. To unstable ground immediate lining is possible by making use of the walling platform immediately above the machine installation. The muck generated goes through the pilot hole. This system is known as 'V mole'. The equipment shown in figure 14.13(b) consists of cutter-head, drive assembly, thrust and directional control cylinders, kelly, gripper assembly and working platform. This system was developed by a German company named Wirth and has been used successfully in some European mines.

14.14   SAFETY IN SINKING SHAFTS[4]

The harsh environment, the overwhelming noise and controlled violence of many of the operations combine with the force of gravity to create a climate in sinking shafts that had traditionally been regarded as dangerous.

This is no longer the case. The size and relative stability of the sinking sector has generated a core of experienced men who build carriers and spent their working lives in shaft construction. The achievements of these professionals in the field of safety are more impressive than the product, which they create and in which they take such a pride. There are several instances now on record where a shaft has been completed without fatality. Fatality free runs in excess of 200,000 shifts are becoming commonplace.[4] Indeed, accident rates in shafts are now comparable with the underground mining industry, in general. This is the result of vigilance, imagination and efforts exercised by every person involved in the great team efforts of these individual projects, which are coordinated into programmes which command the commitment of both management and crews. Basic elements for safety during shaft sinking operations:[4]

- Unyielding discipline, cooperation and mutual protection between individuals
- Use of standard procedures, records keeping, competition between the shifts and sites.
- Cooperation between the sinking organizations, vigilance and reaction to the observed hazards.
- Incentives, meticulous maintenance and control of equipment.
- Immediate investigation into accidents and taking the preventive measures for their reoccurrence.

14.14.1   Field tests and measurements[1]

A shaft is truly the 'life line' of an underground mine. Damage to shaft lining and guides as a result of ground movement can result in serious loss of production and jeopardizes safety and calls for extensive repairs. The Bureau of Mines in the USA and other countries for

Table 14.3 Laboratory test essential for shaft sinking.[1,5,10]

Test Type	Principal Area of use	Average Rock Type
Unconfined Compressive strength	It is strength index. Essential for lining and inset design, excavation characteristics and ground stability	All rock types
Unconfined Tensile strength	Very difficult test to measure an absolute value. Used mostly in the determination of excavation characteristics	All rock types
Tri-axial strength	Essential for determining the deformation characteristics of insets and lining design	All rock types. Critical for very weak rock types
Elastic properties; Young's modulus and Poisson Ratio	Essential for determining the deformation characteristics of insets and lining design	All rock types
Swelling strength and Slake stability	Important for argillaceous rock types – essential for excavation characteristics and for shaft drilling	Argillaceous rock types; i.e. mudstone.
Rebound hardness, abrasivity and specific energy	For determination of rock cutting and drilling characteristics	Stringer rock types
Permeability, porosity, Bulk density and Thermal tests	For ground treatment assessment – grouting, freezing, dewatering	Perméable strata
Moisture content	Important test to calibrate rock strength data	All samples

many years by way of regulations, monitoring, or research are following a testing approach to measure rock mass displacement, stress and strain in concrete lining and axial loads on shaft timbers and steel sets and rock bolts.[1] Data are collected and analyzed by a computerized system in the mine and downloaded via modem to the central office.

The instruments used are many but the following are the few that can be cited as examples: Multiple position borehole extensometer (MPBX) – to measure rock mass deformation; Pressure Cells (PC's) and strain gauges (SG's) – to measure tangential pressure in the lining; Flat Compression Load Cells (FCLC's) – measurement of loads on shaft sets; Thermistors and PRT's – to measure temperature in the range of 60–160°C.

In figure 14.14[1] a computerized data collection network (using sensors) for a deep shaft has been illustrated. The shaft sensors are connected to nearby (<75 m) computer that stores the data and transmits to the remote station. The computer may be located underground or at the shaft collar. It serves as intermediate data control and storage point and also provides preliminary data processing capability. For a copper mine in Poland, sensors installed have been shown in figures 14.14(a) and (c). Use

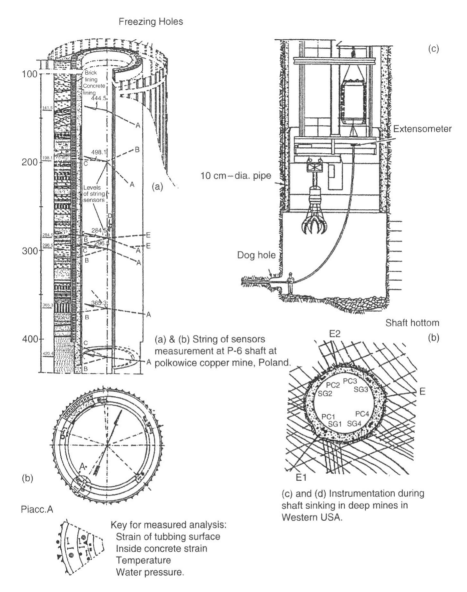

Freezing Holes

Brick lining
Concrete lining
444.5.

498.1

Levels of string sensors

(a) & (b) String of sensors measurement at P-6 shaft at polkowice copper mine, Poland.

10 cm – dia. pipe

Extensometer

Dog hole

Shaft hottom

(c) and (d) Instrumentation during shaft sinking in deep mines in Western USA.

Piacc.A

Key for measured analysis:
Strain of tubbing surface
Inside concrete strain
Temperature
Water pressure.

Figure 14.14   Instrumentation during sinking.

of various instruments for measurements of various types have been illustrated in figures 14.14(b), (c) and (d); while shaft sinking operations are in progress.

14.15   THE WAY FORWARD

Underground metalliferous mining challenges include: mining hotter, lower grade, highly stressed deposits at ultra depths. To deal with these abnormalities, the solution lies in the application of high speed shaft sinking and tunneling. The core of experienced crews engaged in the shaft sinking operations from recent decades, would hopefully, be able to meet this challenge. Research and innovations, however, would have to be at the forefront to maintain the tradition of fatality-free operations.

QUESTIONS

1. "Sinking shafts traditionally has been regarded as dangerous. This is no longer the case". Do you agree with this statement? Briefly review the shaft sinking operations with the application of the prevalent methods, techniques and equipment. You should go through a literature survey including use of the relevant websites.
2. A shaft of 5.5 m dia. is to be sunk through hard ground. Calculate the number of shot holes to be drilled for this purpose and also suggest the name of the pattern. Draw this pattern on a graph sheet. What type of explosive shall you recommend for this purpose?
3. Based on the techniques applied to sink a shaft, list the steps that are followed to carry out the sinking operations.
4. Classify shaft sinking methods.
5. Describe in detail the concept of shaft boring with the use of shaft borers (SBM). How does it differ from the TBM concept?
6. During shaft sinking at a site, the initial 20 m of the ground consists of a loose sand, gravels and clay. Hard rock has been encountered there after. Mention the special method of sinking you shall propose to deal with this ground, illustrate it with the aid of a suitable sketch.
7. For a shaft, assuming ground to be fairly competent, suggest a pattern of holes. Calculate number of holes and mark them on a graph sheet (plan and section). Use depth of round to be 2.2 m.
8. Give a schematic diagram to represent shaft drilling.
9. Give a complete scheme (brief description with suitable sketches) of a shaft sinking method in which the strata is good but due to lots cracks and fissures make of water is abnormal.
10. Give the different shapes of the shafts. What arrangements are required to be made prior to sinking a shaft? Prepare a surface layout illustrating a shaft sinking site.
11. Give the cycle of operations to be carried out while sinking a shaft from the surface.
12. How shall you deal with the following during shaft sinking: Water. Ventilation. Lighting and illumination. Shaft centering and alignment.
13. Illustrate Sequence of operations while sinking through ordinary/normal ground by clearly marking the relative positions of the shaft bottom.
14. List out the appliances, equipment and machines that are used during sinking operations.
15. List out the mucking equipment that can be used during shaft sinking.
16. List out the special methods of shaft sinking. Mention the conditions under which each one could be applied?
17. List purposes for which shafts are required.

18. List a sinking cycle's unit operations.
19. Mention the conditions under which you will prefer to use the following methods of shaft support: Brick lining. Concrete lining. R.C.C lining. Lining using German or English tubing. Shotcreting. Rock bolting any other type of support.
20. Mention the factors that you will take into consideration while selecting the site for a shaft.
21. Name any five sinking appliances known to you.
22. The harsh environment, the overwhelming noise and controlled violence of many of the operations combine with the force of gravity to create a climate in sinking shafts that had traditionally been regarded as dangerous. How would you ensure safety in such a dangerous operation?
23. Write down the unit operations that need to be carried out during shaft sinking, giving detailed procedure, equipment which can be used in each case.
24. What are these structures and what is utility of each of them: shaft inset, skip pocket, ore-pass, sump.
25. Suggest the shaft sinking methods under the following conditions:
    (i) Sinking through a river bed for a depth of 25 m.
    (ii) A firm ground having large make of water through pores, cavities and fissures.
    (iii) Loose and unstable ground for a initial depth of 15 m from the surface.
    (iv) Heavily watered strata with quick sand at a depth of 70 m from the surface.
26. Draw suitable sketches to illustrate the followings:
    a. Section of shaft showing sequence of cementation.
    b. Schematic presentation of shaft drilling technique.
    c. Step cut (drill pattern) used while sinking a rectangular shaft.
27. In conjunction with the Freezing method of shaft sinking, answer followings:
    a. Which liquid is circulated in the holes to form an ice column or cylinder?
    b. Name the process used to melt the frozen ground after sinking through the difficult ground is over.
    c. Write about the ground freezing practices that are followed at the West Germany.
    d. List reasons for lining damage in frozen ground.
    e. Type of lining suitable for this method.

REFERENCES

1. Beus, M.J.: An approach to field testing and design for deep mine shafts in Western USA. *Shaft Engineering Conference*, 1989, IMM, London, pp. 53–55.
2. Boky, B.: *Mining*. Mir Publishers, Moscow, 1988, pp. 228–42.
3. Breeds, C.D. and Conway, J.J.: Rapid excavation. In: Hartman (edt.): *SME Mining Engineering Handbook*, SMME, Colorado, 1992, pp. 1883–85.
3b. Bruce A. Mckinstry: Sinking the Silver Shaft. *Proc. RETC, AIME*, 1983, pp. 103–25 (www.smenet.org).
4. Douglas Alastair A.B. and Pfutzenreuter, Fred R.B.: Overview of current South African vertical circular shaft construction practice. *Shaft Engineering Conference*, 1989, IMM, London, pp. 137–54.
5. Falter, B.: Stability of liners in shaft design. *Shaft Engineering Conference*, 1989, IMM, London, pp. 169–70.

6. Franklin, J.A. and Dusseault, M.B.: *Rock Engineering*. McGraw-Hill, New York, 1989, pp. 563–80.
7. Klein: Shaft sinking by ground freezing in coal mining industry in the Federal Republic of Germany. *Shaft Engineering Conference*, 1989, IMM, London, pp. 276–78.
8. Megaw, T.M. and Bartlett, J.V.: *Tunnels: Planning, Design, Construction*. Ellis Horwood Limited, Chichestre, 1983, pp. 66–68, 117.
9. Nitro Nobel – *Website, literature and leaflets*.
10. Przygodzka, B.B. et al.: Extensometer investigations of frozen shaft lining. *Shaft Engineering Conference*, 1989, IMM, London, pp. 71.
11. Unrug, K.F.: Construction of development openings. In: Hartman (edt.): *SME Mining Engineering Handbook*. SMME, Colorado, 1992, pp. 1580–1620.
12. Vergne, J.N.: *Hard Rock Miner's Handbook*. McIntosh Redpath Engineering, 2000, pp. 152.
13. Universal Mining School (UMS), Cardiff University; Lessons-Distant education, 1950–55.

Chapter 15

Large sub-surface excavations

Health, Safety and Environment (HSE) are the three sides of an equilateral triangle but any imbalance may jeopardize financial and social goals of a company.

15.1 INTRODUCTION

The creation of large sized sub-surface excavations is on an increasing pace since the 1950s all over the world. In urban areas, the rising population has created a scarcity of surface land. During the last half century, the advances in rock mechanics to evaluate ground conditions together with the developments in ground consolidation and support techniques have enabled us to create large underground excavations. Method, techniques and equipment are available to excavate large volumes of rocks beneath the surface efficiently. This is the reason that these large excavations, which are known as 'Caverns', are created for many purposes, such as: civil works, storage facilities, defense installations, hydro-electric power plants, recreation facilities etc., as, illustrated in figure 15.1.

The second locale is the underground mines. The modern technology has allowed application of bulk mining methods for which large sized stopes and excavations meant for mine services and facilities are almost mandatory.

15.2 CAVERNS

In this chapter, all large sized openings other than underground mining stopes, have been designated as Caverns. Thus, caverns are the large sized underground openings driven for multiple purposes, as shown in figure 15.1. Following are some of their uses:

- Shelter for the people during war, and the same could be used as a recreation spot during peace.
- Defense installation for the utilities such as: Storage of arms, ammunitions, weapons and strategic commodities such as oil and other war fighting materials. A site for carrying out tests. Command control, communication and monitoring centers. Installation of radars.
- Powerhouses for the generation, transmission and storage of hydro-electric power.
- Storage of flammable oil and gases.
- Nuclear waste disposal – Repositories. Disposal of hazardous waste. In some of the countries such as Norway, environment regulations are so stringent that tailings or waste generated during mining has to be disposed of underground.[1]

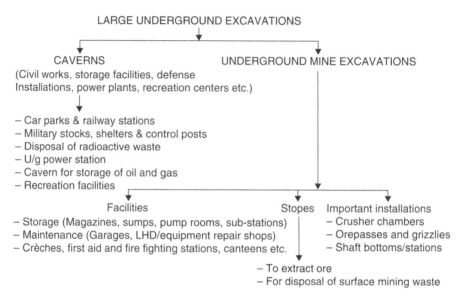

Figure 15.1   Classification of large underground excavations based on their function and utilities.

- Swimming pools, garages and parking lots, exhibition centers, markets and much other installation of public utilities.
- Warehouses and stores: Storage of goods and supplies for manufacturing units. In the this modern era, the variation in the supply and demand of various commodities including energy and fluctuation in their prices makes it necessary to store them; and match the supply with demand. Storage is also necessary to safeguard against crisis. The following factors favor underground storage:[5]
  a. Large volume of products can be stored
  b. Minimum surface area required
  c. Lower installation and maintenance costs
  d. High degree of safety
  e. Environment protection.

Thus, caverns are special large sized excavations that differ from the tunnels and other openings. These structures have the following unique features:

- These large sized openings have very long, or practically unlimited life; as such they should be built to take care all future forces that can influence their stability.
- Repair of such an opening if not impossible is a very difficult, time consuming and costly affair, which is seldom preferred.
- Their size, shape and location are adhered to designed specifications. This includes proper alignment, exact size and shape.
- Provisions for ventilation, emergency access, cross passages, illumination are the integral part of the design for such excavations.

- Practically, there is very little scope for the collapse and even repair of the tunnels support; hence, a support network with adequate safety factor is usually chosen. Infiltration of water, gases or any other liquid should not be there.
- Smooth walls are almost mandatory except in nuclear or hazardous waste repositories. In the earthquake prone areas, consideration to minimize this effect, should be given due importance during the design-phase itself.
- The containment created by underground structures protects the surface environment from the risks/disturbances inherent in certain types of activities.
- Underground space is opaque as any structure is only visible at the point(s) where it connects to the surface.

In underground mines the large excavations other than stopes, are also constructed almost on the same guidelines, as mentioned in the previous section while constructing the caverns. In fact stopes are also similar types of the openings, even larger than these caverns, but these are created to exploit the valuable minerals (ores) and there is a basic difference between the two. In stopes care is taken for their stability during the process of recovering ore from them and after that they are allowed to collapse. Nobody is allowed to enter into these worked out stopes. The configuration of these stopes is usually irregular and as per the outline (geometry) of the orebody. Life of a stope could range from a few months to a few years. The details are described in chapter 16.

However, mine openings have been used to store some commodities in North America and Europe. These openings could be used for oil, gas and their products. The merit of this option is that the cost of creating excavation is saved to a great extent, and hence, it has great potential in the years to come.

For proper site investigation and its selection the steps as described sec. 3.4 and 3.5, as appropriate, could be followed. This includes proper geological, geo-technical and hydrological investigations.

15.2.1   Constructional details – important aspects

Access: Caverns can be accessed by ramps, shafts or inclined tunnels; but these opening costs are a considerable part of the total project cost. After completion of construction the access tunnel is bulk-headed off, and shafts are used to have access during the operation phase (figs 15.3 and 15.4(a)).

Support: Due to varying geo-technical conditions the type of support varies from site to site. In quite competent ground it could be rockbolts, pre-stressed anchors, pre-stressed tendons to shotcreting, guniting and, full face reinforced concreting. In a fairly competent rock setup the bolt lengths in the roof could range from 0.15 to 0.30 times the cavern span, and in walls from 0.10 to 0.20 times the cavern height.[16]

Geo-Mechanical aspects: The study made by Khot et al.[9] could provide some useful information as listed below:

- The ratio of horizontal to vertical stress (lateral stress coefficient) at the location of the cavity is the most significant factor affecting the stability of a cavity/cavern. Since the mid roof and mid floor portion of such cavities are under tensile stresses, it is absolutely necessary to correctly estimate them at site.

- When a multiple cavern proposal is to be executed (such as in case of power houses), minimum spacing between two adjacent openings should be half the width of the larger opening, to ensure that compressive stresses are within the compressive strength of the enclosed rocks.
- The two dimensional photo-elastic technique can be effectively used for rapid and precise determination of boundary stresses around the cavity, while the finite element technique can be relied upon for determining the stresses inside the rock mass. The principal stress's plot indicates this.

In addition to the above, large caverns need to be given some special considerations as listed below:[2,7]

- Dimensions of caverns are limited by the support requirements in soft rocks/ grounds. These dimensions may also be limited by the permeability of rock mass.
- Cavern size may be restricted by the presence of major discontinuities. Joints with large aperture can also limit them. Geometry and size of caverns may be restricted by major inhomogeneity.
- Cavern shape may be restricted by rock mass structure. The ideal shape is controlled by in-situ stresses. In-situ stresses vary with depth. Large caverns need more support. They can destabilize a rock mass structure. As depth increases the cavern size decreases.
- Very large caverns can be created by drilling and blasting and that too if rock quality is better.
- Caverns created in hard rocks may require little or no support.

15.2.1.1 Construction procedure

As described in section 9.7, special techniques and procedures are followed to drive large sized tunnels and mine openings. In the similar way large caverns are constructed using the heading and bench method but these caverns are still higher and, therefore, they are worked in a number of lifts or benches (figs 15.2(a) and (b)).

The excavation proceeds from top to bottom by dividing the vertical span into a number of benches. This allows support of roof to begin with and then the sides as the excavation progresses downward; as illustrated in figures 15.2(a) and (b).[12] Following are the usual stages of such excavations:

- Top section by pilot heading and slashing the sides
- Horizontal benching
- Vertical benching.

The first step is to access the top of the cavern, which could be by way of raises, tunnels, or ramps. The accesses made are used for the transfer of the broken muck, crews, material and equipment during construction as well as operational phases.

The top section of a cavern can be driven using a multi-boom drilling jumbo, but in most of the cases it is unable to cover its entire width, as such first a pilot heading in the center of the cavern is drilled and blasted; and then its sides and roof are slashed. This is also known as side stoping, or slashing.

It is a common practice to take the next slice by horizontal benching so that use of the same drilling jumbo could be made. The height of this bench is governed by the

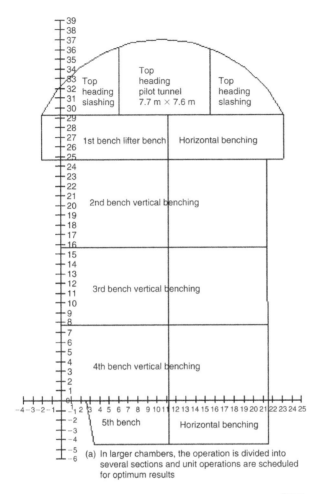

39
38
37
36
35
34
33 | Top heading slashing | Top heading pilot tunnel 7.7 m × 7.6 m | Top heading slashing
32
31
30
29
28
27 | 1st bench lifter bench | Horizontal benching
26
25
24
23
22
21
20 | 2nd bench vertical benching
19
18
17
16
15
14
13
12
11 | 3rd bench vertical benching
10
9
8
7
6
5
4
3 | 4th bench vertical benching
2
1
0
-4 -3 -2 -1 | 1 2 3 4 5 6 7 8 9 10 11 12 13 14 15 16 17 18 19 20 21 22 23 24 25
-2 | 5th bench | Horizontal benching
-3
-4
-5
-6

(a) In larger chambers, the operation is divided into several sections and unit operations are scheduled for optimum results

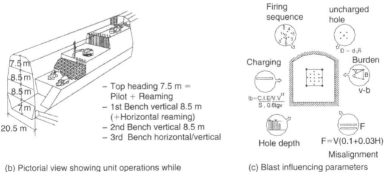

7.5 m
8.5 m
8.5 m
7 m
20.5 m

- Top heading 7.5 m = Pilot + Reaming
- 1st Bench vertical 8.5 m (+Horizontal reaming)
- 2nd Bench vertical 8.5 m
- 3rd Bench horizontal/vertical

Firing sequence   uncharged hole

Charging   Burden

$Ib = \dfrac{C.I.E/V.V^{\alpha}}{S, 0.6tgv}$ v-b

Hole depth $F = V(0.1 + 0.03H)$
Misalignment

$D = d\sqrt{n}$

(b) Pictorial view showing unit operations while creating large underground chambers/excavations

(c) Blast influencing parameters in caverns

Figure 15.2 Division of a large chamber into several horizons and scheduling unit operations for optimum utilization of resources.

capability of the drilling jumbo, which is usually within 5 m, and so it is necessary to decide the depth of round to be drilled; and usually it is within 4 m.

The next slice can be taken as a vertical bench and its height could be more but usually it is up to 12 m. The same logic is applied for the next few benches to arrive right up to the bottom of a carven.

In order to minimize over-break and achieve smooth configuration, smooth blasting of contour holes is almost mandatory in such excavations. For the top heading (upper most section) the smooth blasting as described in sec. 9.3.3; and for the benches pre-splitting (sec. 17.5.6), or even lining drilling could be adapted.

Drilling: Wagon drills or, DTH drills are used to undertake drilling at the benches. The diameter of these holes is usually in the range of 40 to 110 mm. It is a function of muck handling excavator. Equation 15.1 could be used:[13]

$$\text{Hole dia. } d = (0.07 \text{ to } 0.08) \, b_{ht} \tag{15.1}$$

Where: d – blast-hole dia. in mm;
   b_{ht} – bench height in cm.

In order to calculate spacing and burden for bench blasting patterns; the relations given in section 17.5.6, on bench blasting could be used. Over drilling of a bench should be also done to obtain the desired floor configuration. For this purpose, the same guidelines, as given in chapter 17, are applicable.

15.3   POWERHOUSE CAVERNS[2,4,12,16]

These caverns are made to house various units (equipment and facilities) of a powerhouse and include: Inlet valve (optional), turbine and generator (Mechanical hall), various mechanical and electrical sub-systems, transformers (optional) etc. The dimension of such excavations depends upon the number of units and power generating capacity. An analysis made[2] of the some of the world's largest spanned underground power units to generate power in the range of 40–475 MW provides the following data sets: the dimension varies as: width – 24 to 35 m; height – 19 to 57 m and length – 70 to 296 m. Typically a span in the range of 18–24 m is usual.[16] While determining the size of such caverns allowance for the movement of service equipment and crew should be accounted for.

In figure 15.3 excavation network within the structures in the form of large chambers, shafts, raises, winzes, tunnels and openings of a typical powerhouse complex has been shown. The figure illustrates a pictorial view along a powerhouse in North America. These complex structures require careful selection and installation of supports, which could be rock bolts, pre-stressed anchors and bars, concrete RCC or prefabricated arches, shotcreting etc.

15.4   OIL STORAGE CAVERNS[2,5,8]

The unique feature of an oil cavern is that the ground water level must be maintained above it.[16] The concept of unlined caverns in rocks to store oil is in vogue in Scandinavian and some other countries from the last 40 years, or more.[8]

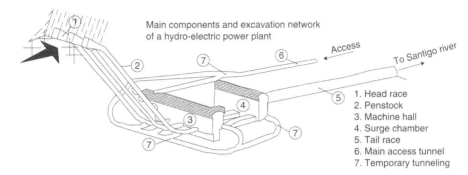

Main components and excavation network
of a hydro-electric power plant

Access

To Santigo river

1. Head race
2. Penstock
3. Machine hall
4. Surge chamber
5. Tail race
6. Main access tunnel
7. Temporary tunneling

Figure 15.3   Layout of a hydro-electric power house consisting of a network of vertical, inclined and horizontal openings.

The principle of the storage of crude oil in unlined u/g caverns utilizes the hydrostatic pressure of outside groundwater to contain the oil within the cavern. As the oil is lighter than water and being insoluble in it, it floats on water within the cavern. The roof of the cavern is located below ground water level in such a way that the pressure of water on the cavern walls would be higher than the pressure of oil stored within it. Since the liquid flows in the direction of falling pressure, the oil stored in the cavern cannot penetrate into the bedrock while the groundwater seeps continuously into the cavern through the fissure and cavities in the bedrock. The excess water entering the cavern is pumped out. The oil is thus stored on the waterbed in direct contact with the rock walls and therefore, usually, no lining is needed.

Uran caverns, a proposed site for the crude oil storage near Bombay, India consists of number of parallel caverns (fig. 15.4(c)),[8] which are unlined. The layout include the access tunnels, shafts, excavation to house submersible pumps to pump out oil and a pump installation to pump out water, which will be received through the leakage. An infiltration tunnel to augment of groundwater when ever required are some of the excavations that are included in this network. The caverns are 265 m (long) × 18 m (width) × 28 m (height). The caverns have an inverted U shape. The spacing between two adjacent caverns is minimum 20 m. The caverns have been proposed to be 30 m below water level and with a minimum overburden of 60 m. Seismic and electrical resistivity surveys have been proposed to select a site, which would be safe in the event of earthquake.

Dimensions: depending upon the capacity to store, these caverns may range in span from 10 to 21 m and height up to 30 m. Figure 15.4(a) illustrates a typical layout for large oil storage in Finland. Figure 15.4(b) depicts different shapes of oil storage caverns.[16]

15.5   REPOSITORY[10,11,15]

Radioactive waste is generated from various sources and quality and quantity also vary significantly. The sources are mining, milling, nuclear fuel fabrication, nuclear power plants, research reactors and reprocessing plants of spent fuels. Depending

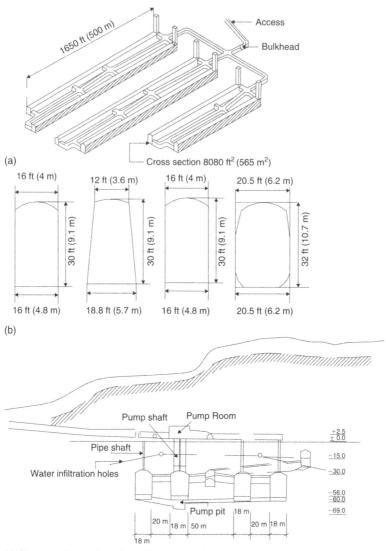

(a)

Cross section 8080 ft² (565 m²)

(b)

(c) Cross sectional view of an oil cavern system

Figure 15.4   Layout and shapes for the oil storage caverns. (a):A typical oil storage cavern in Finland.
(b):Typical shapes for oil storage caverns. (c): Section through a crude oil storage layout
consisting of unlined caverns near Bombay, India.

upon the radio nuclide content the wastes remain active for period ranging from a few
months to thousands of years. As such these structures are to be properly conditioned
and isolated from the human environment till they decay to the acceptable levels. To
achieve this, various types of u/g facilities are used for storage and dispose of these
wastes, after taking into consideration a suitable matrix. Some of these are shallow up
to a depth of 10 m such as: tile holes, trenches and storage vaults; while others such as

geological repository are as deep as 500–900 m.[11] Each of these installations has its own design, constructional and operational requirement to achieve isolation for a desired period of time.

It is a complex facility that must take into consideration a number of factors including those defined by the natural laws, regulatory authorities, and requirement established by political compromise.[15] A number of anonymous publications have been given during 1976 and afterwards, and also by others to formulate the concept for such an installation and to deal with various aspects. High-level radioactive waste is hazardous to man and the environment as it emits alpha, beta and gamma radiations for an extended period of time that can be of thousands of years. Many writers have proposed the conceptual designs for this complex structure. The life of a repository can be divided into four phases: Constructional, Operational, Sealing/Isolation and Post Isolation.

During the construction phase a network of excavation openings that includes shafts, ramps, tunnels, drifts etc. are built (fig. 1.10).[10] During the operational phase the horizontal and vertical bore holes in these emplacement drifts and tunnels will be drilled to place the specially designed nuclear waste canisters (fig. 15.5(a)).[10] These boreholes are then sealed and back-filled with specially designed rock mixture. After this the encompassing drifts and tunnels are backfilled and sealed. This also includes decommissioning of underground facilities that are installed during construction and installation phases, and including the artificial supports used, if any. Finally all the access ways including shafts are to be backfilled and sealed as per the regulatory requirement of any country.

Geo-technical aspects of waste disposal:[10] The desirable aspects of an ideal host medium for waste emplacement are:

- Good thermal conductivity
- High absorption capacity
- Low permeability
- Highly plastic and easy in mining
- Negligible mineral value (barren rock mass).

The technical, geological, and environmental factors to be considered:

- Formation depth; its vertical and horizontal span
- Permeability and porosity of rock and homogeneity of disposal horizon
- Tectonic and seismic potentiality
- Resource potential
- Hydrology and thermal properties
- Configuration on the surface
- Climate and population density; source of potable water supply and chances of surface impact (Environmental factors).

Figure 15.5(a) illustrates an overall concept of a typical repository; details of drill holes tunnel or room and the drill hole with the canister have been shown. The effect of this burial is estimated on the global as well as local basis (fig. 15.5(b)).[10] Global analysis undertakes adverse effects on the entire repository and the areas lying above, below and surroundings. The local model involves a limited area surrounding the

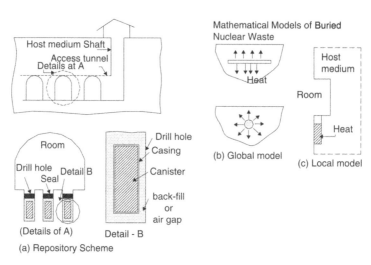

Figure 15.5 (a) Overall concept of a typical repository. Details of openings required for the purpose are shown. (b) & (c) – Mathematical models to access Local as well as Global impacts of the burried nuclear waste.

places of burial of nuclear waste. Thus, difference between these models is the scale on which various details are drawn.[10]

Various excavation and burial schemes[11] to isolate, dispose of and store the radioactive wastes are used. An earth trench is used for the disposal of solid wastes, which may have negligible, or likely radioactive substances. Reinforced cement concrete trenches have better containment integrity and are used for disposal of solid waste with radiation field up to 50 R/hr.[10]

A high-level waste storage tank *Farm* is the housing for large sized metallic tanks. These tanks are used to store highly active and corrosive liquid waste from a spent fuel reprocessing plant. For interim storage of high-level waste, '*Vitrified High Level Waste Storage Vaults*' are used.

In figure 15.6 some underground facilities that have been created by undertaking excavations of different kinds have been shown.

15.6 SALT CAVERN STORAGE[2,5]

Salt cavities are suitable for storing products, which do not react chemically with the salt. These cavities can be used for the storage of crude natural gas, kerosene and chemical products and few others. France has a number of such storage cavities. Formation of cavities in salt can be achieved by water leaching. Direct (fig. 15.7(a))[2] and Reverse (fig. 15.7(b))[2] leaching are the two techniques available to achieve this. Spherical, elliptical, or cylindrical are the common shapes. The latter is more common. Operation of such cavities is achieved by brine displacement. The product to be stored is injected in the cavity and thus water gets displaced. Similarly for de-storing, brine is injected to push the product out.

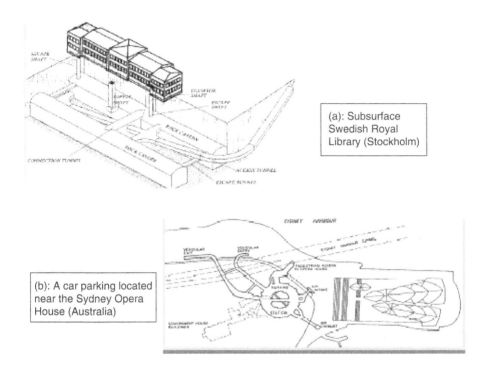

(a): Subsurface Swedish Royal Library (Stockholm)

(b): A car parking located near the Sydney Opera House (Australia)

(c): An underground storage facility at Kansas City (USA)

Figure 15.6   Some underground facilities – a network of excavations of various kinds.

15.7   AQUIFER STORAGE[2,5]

Aquifers are widely used in the US and Europe to store natural gas. The technique involves the injection of natural gas into the aquifer displacing the water and then creating an artificial reservoir for the natural gas (fig. 15.7(c)).[5]

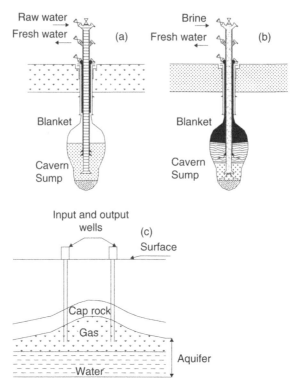

Figure 15.7   Storage for hydrocarbons – some typical techniques. Application of '*Salt Caverns*' for the storage of crude natural gas, kerosene, chemical products and a few other commodities. Formation of cavity in salt is undertaken by water leaching; (a): Direct method; (b): Indirect/Reserve method. (c): Application of aquifers in USA and Europe to store natural gas. The technique involves injection of natural gas in the aquifer displacing the water and then creating an artificial reservoir for the natural gas.

The cost of storage depends upon the factors such as characteristics of the product and its quality and quantity, geological and hydrological conditions of the site. The experience gained from operating such projects indicates that wherever feasible, it has proved to be a cheaper means of storing.

15.8   EXHIBITION HALL CAVERNS[14]

The use of underground space in urban areas is becoming most important due to scarcity of land in the densely populated/inhabited areas and environment concerns. Underground (u/g) shopping centers, exhibition centers, parking lots and sport facilities are being constructed all over the world.

A design flow diagram of a typical u/g cavern in Japan is shown in figure 15.8(a).[14] The earthquake condition were estimated based on past experience and

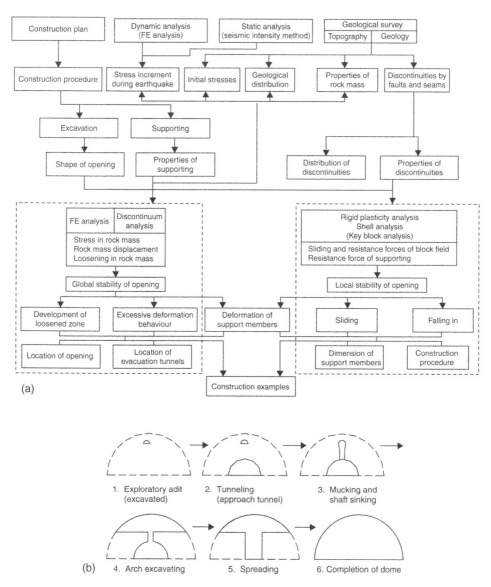

Figure 15.8   (a): Flow diagram showing design, excavation and construction consideration while undertaking large sized halls and facilities. (b): Excavation sequence of a dome-shaped exhibition hall.

used as input data for the static and dynamic analyses made. The occurrence of loose zones was estimated by means of finite element analysis. The stress concentrations at the various openings of the layout were specially investigated by stress-distribution analyses by using 2-D & 3-D finite element models.

The exploration tunnels driven for geological investigation purposes could be used as access tunnels to the main tunnels, and the halls that are driven subsequently

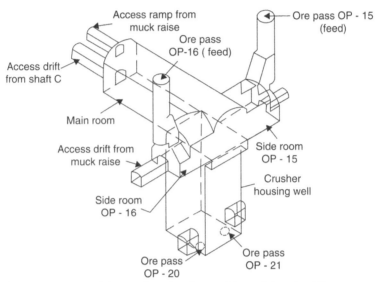

Schematic layout of the crusher chamber and its peripherals at the world's largest underground mine- El Teniente mine, Chile.

Figure 15.9   Crusher chamber, El Teniente mine, Chile.

in such a complex. Raises, winzes and chambers are some of the prominent structures that are driven with utmost precaution so that over-break is minimum. Application of smooth blasting is almost mandatory.

Another important feature of such construction is to install instruments for the measurement of stability parameters so that safety of workers during construction and safety of the general public even after hall (i.e. complex) is opened, can be ensured.

15.9   UNDERGROUND CHAMBERS IN MINES[6]

In underground mines, chambers of large size are required for various purposes and usually such chambers include: crusher room, magazine, crèches and rest room, first aid and fire fighting room, repair shop, garages and battery charging station, loco shop, canteen, electric sub-stations etc.

These openings can be made at any depth but due consideration is given with regard to their stability. The following procedure was followed while constructing a chamber (fig. 15.9) to house a crusher and its peripherals at the El Teniente mine in Chile, which is world's largest u/g mine:

1. Before designing this layout, a detailed investigation of rock types, which were within, and surroundings areas of this installation site was made. This included evaluation of stress conditions in the virgin rocks, geological – geotechnical exploration by diamond drilling and characterization of massive rock in-situ from geo-statistics on drill-core RQD data.[6]

2. During the excavation stage extensive instrumentation was made that included in-situ stress measurements (using over-coring and USBM deformation gages, extens-ometers, load cells). Additional instrumentation was made to monitor the subsequent behavior of this installation throughout its service life. Based on the data obtained on rock quality, application of finite element technique was made to develop two and three-dimensional models.

3. Based on the abovementioned engineering approach the installations, which included a few large chambers, were designed and constructed. This installation included the excavation of an underground room measuring 50 m (long) × 14 m (wide) × 14.5 m (high); two satellite chambers each measuring 14 m × 12 m × 12 m; and a crusher housing well of 33 m (depth) × 15 m × 18 m. The layout is shown in figure 15.9.[6]

15.10   EQUIPMENT AND SERVICES SELECTION

In order to select equipment for the operations required to drive or create civil as well as mining excavations, as described in this chapter; or those described in other chapters, the following guidelines could be used.

Table 15.1 details equipment, explosives, blasting accessories, service appliances and devices etc. For a particular excavation, matching sets of equipment should be selected, to carry out different unit operations. Some of the matches are given below for guidance but the one which could give optimum results should be chosen as per the local conditions, specific requirements and available resources in terms of personnel, machines, equipment, techniques, material and experience and exposure of the working crews.

I. Conventional Drifting or tunneling operations using rocks drills: This could be development drives, crosscuts, level drives, sublevel or small sized civil tunnels.
D1 + E1/E3 + B1 + M1/M3(a)/M3(b) + T1(a)/T2 + services as per table 15.1(b).

II. Mechanized Drifting or tunneling operations using one or two boom rock drills (pneumatic or hydraulic): this could be development drives, crosscuts, level drives, sublevel of medium size, or medium sized civil tunnels.
D3(a)/D3(b) + E1(as primers or in watery situations) + E2/E3 + E5(b)* + B2 + M3(b) + T1(a)/T2 + services as per table 15.1(b). * – if smooth blasting required.

III. Large sized tunnels, underground chambers and caverns using multi-boom hydraulic or pneumatic jumbos.
D3(a)/D3(b) + D2(c)** + E1(as primers or in watery situations) + E2/E3/E4 + E5(b)* + B2/B3# + M3(b)/M3(c) + T1 /T2 + services as per table 15.1(b). * – if smooth blasting required. ** – where benching is required. # – Practice in development stage.

IV. Drives and tunnels driven using roadheaders or full-face TBMs:
C2/C3 + M4 + T1(a)/T2 + services as per table 15.1(b).

V. Raising – vertical development in upward direction.
D4 + E1/E3 + B1 + M** + T** + services as per table 15.1(b). ** – as and when required using existing mucking and transportation units that are used for drifting and tunneling operations.

Table 15.1   Details of equipment, explosives, blasting accessories and services.

Symbol – Operation	Equipment with its suitability and locales of applications	
D – Drilling	1. Jack hammer with or without pusher leg (Pneumatic powered) for hole length upto 3.5 m and dia. 32–38 mm	Medium hard rocks to hard rocks, small output, small size horizontal mine openings and civil tunnels and chambers. Pin holes for services
	2. (a) Hand held jack-hammer, sinker for hole length upto 3.5 m and dia. 32–38 mm	(a) Shaft sinking, winzing in all and types of rocks. Funnel chamber excavations (downward)
	(b) Shaft Jumbo mounted with light duty drifters for drilling downward, hole length upto 5 m and dia. upto 50 mm; (for large dia. heavy duty drifter)	(b) Mechanized Shaft sinking, winzing in all types of rocks; Holes are suitable for large dia. explosives including emulsions
	(c) Rotary-percussive and DTH drills; wagon mounted. Hole dia. 50–100 mm or more; and length exceeding 5 m (All are pneumatic powered)	(c) Bench blasting for large u/g chambers, caverns, excavations and mine stopes. Allows bulk loading of explosives such as ANFO, slurries or emulsions
	3. (a) Single or multi boom drifting jumbos fitted with light duty pneumatic drifters capable of drilling inclined and horizontal holes upto 7 m length and 70 mm dia.	(a) Drifting and tunneling of almost all sizes in all types of rocks and also suitable for big sized chambers and caverns. Fast progress possible
	(b) Single or multi boom drifting jumbos fitted with light duty hydraulic drifters capable of drilling inclined and horizontal holes, length: upto 7 m; dia. upto 70 mm dia. (Note: for large dia. and longer holes use heavy duty drifters; logic applicable for (a) and (b))	(b) For fast drifting & tunneling operations, and also suitable for big sized chambers and caverns. Fast progress with better safety and quality possible
	(c) Ring and Fan drilling jumbos (pneumatic or hydraulic) in any direction; length: 5–40 m; dia. 50–100 mm	(c) For enlarging small sized tunnels and chambers into big ones. Mainly meant for stoping operation in mines
	4. Stoper, or parallel raise feed; for hole length upto 3 m and dia. 32–38 mm	Raising (vertically up and inclined upward) operations. Over-hand drilling in stopes
	5. Hand held electric drill for hole length upto 2 m and dia. 32–38 mm	Soft/weak rocks, light duty and small output, small size horizontal mine openings and civil tunnels. Pin holes for services

(Continued)

Table 15.1   Continued.

Symbol – Operation	Equipment with its suitability and locales of applications	
E – Explosives	1. NG based explosives and Dynamites in cartridge form. Dia. range 25 mm–50 mm; Length 200 mm or as supplied by the manufacturer	While driving tunnels, shafts, drifts, raises, winzes in watery conditions in medium to hard rocks. Strongest and Costliest
	2. Dry blasting agent ANFO for holes larger than 40 mm dia.; pneumatic charging in up holes and to avoid static hazards antistatic detonation system is mandatory	Dry rock conditions, suitable for all types of rocks, particularly with larger dia. and longer holes. For wet conditions heavy ANFO. Non-cap sensitive and needs booster charge for initiation. Cheapest
	3. Slurry explosives – wet blasting agents	While driving tunnels, shafts, drifts, raises, winzes in dry or watery conditions in medium hard to hard rocks. Cheaper than NG based explosives
	4. Emulsion explosives	Suitable for bulk loading in holes larger than 45 mm dia during drifting, tunneling, sinking and stoping
	5. Special explosives Permitted explosives Smooth blasting	To be used for specific conditions. In gassy coal mines and tunnels For charging perimeter holes of tunnels, shafts, caverns, chambers etc. and to reduce over break in weak and unstable ground conditions
	Seismic types	Exploration
B – Blasting	1. Electric detonators with multi shot exploder	With other than E2 explosives
	2. Antistatic detonators such as Anodets, or system such as Nonel, Hercudet	Where danger of electro-static charge generation exists; as in case of E_2 explosives
	3. Electronic detonators with computerized delay setting system	In tunnels and mines but at developing stage; commercial viability yet to be established
	4. Fuse blasting (use of detonating cord, connectors etc.)	In metal mines while using ANFO explosives
	5. Secondary breaking (pop or plaster shooting)	E_1 or E_3 types explosives. With detonating cord or electric detonators
C – Cutting	1. Cutting 'Kerf' using coal cutting machines while driving horizontal drives and tunnels in coal mines	In u/g coal mines

(Continued)

Table 15.1 Continued.

Symbol – Operation	Equipment with its suitability and locales of applications	
	2. Cutting rocks using heading machines (partial face boring)	Tunnels, Chambers, and horizontal Mine Openings of sufficient size and lengths
	3. Full face tunnel borers (TBMs)	Tunnels and horizontal Mine Openings of sufficient size and lengths
M – Mucking	1. (a) Over head loaders (Rocker shovels) track-bound, or trackless	(a) Small sized mine openings and tunnels. Crawler mounted EIMCO-630 during shaft sinking and winzing
	(b) Dipper shovels	(b) Large sized tunnels
	2. (a) Arm loaders – suspension types such as cactus, grab, cryderman etc.	(a) Shaft sinking and Winzing
	(b) Gathering arm loaders	(b) Mucking from coal mine openings (Development and stoping)
	(c) Digging arm loader (Hagg loader)	(c) Mucking in Tunnels and mine openings by discharging muck via conveyor into Hugg-haulers
	3. (a) Auto-loader Cavo	(a) Small sized tunnels and mine openings such as sub-levels
	(b) Integrated unit LHDs	(b) Mine Openings and Tunnels of $9\,m^2$ cross section or more. Mucking from stopes
	(c) Front-End-loaders (FELs)	(c) Large sized Tunnels
	4. Integral with cutting unit	4. Mine openings and tunnels driven using roadheaders, TBMs
T – Transportation	1. (a) Trackless Low-profile dumper, or Trucks	(a) Tunnels, large sized mine openings and stopes. Transportation from ore/waste chutes
	(b) Trackless – LHD	(b) Direct transportation from stopes to ore pass or waste passes (lead upto 300 m)
	(c) Hagg-Hauler	(c) Mucking through Hagg-loader for Tunneling and Mine openings
	(d) Large sized trucks	(d) Large sized tunnels
	2. Locomotives (battery, diesel, Trolley wire)	Tunnels and in mines for development and stoping operations
	3. Conveyors – Belt	From long wall mining faces at gate road and main (trunk) roadways in mines
	4. Rope haulage	Small to medium sized mines
	5. Hydraulic transportation	For transportation of mine fill in stopes

(Continued)

Table 15.1   Continued.

Symbol – Operation	Equipment with its suitability and locales of applications	
R – Ripping	Hand held rock breakers and rippers	In loose, soft and unstable ground during sinking and tunneling operations
H – Hoisting	1. Drum winder/hoist	Sinking, Regular mine production
	2. Koepe winder/hoist	Regular production from mine

Auxiliary Operations:

Symbol – Operation	System with Equipment/tools and appliances with their suitability and locales of applications	
V – Ventilation	1. Main Ventilation System using Forcing, or Exhaust fans installed at the surface and coursing air current using mine entries such as shafts, incline, declines etc.	Entire mine that includes the network of mine entries
	2. Main Ventilation System using Forcing, or Exhaust fans at the surface, and coursing air current by rigid and flexible ductings	Entire Tunnel, Cavern Network, Sinking shafts
	3. Auxiliary Ventilation using Forcing, Exhaust, or Contra-rotating fans and blowers with Rigid and/or Flexible ductings	For effective face ventilation by coursing the main air current by these means
	4. Spot coolers and Air Conditioning System	Deep mines
S – Support	1. Rock Reinforcement	To induce reinforcement forces within the rock mass:
	(a) Grouting by Bolts, Anchors, dowels, cables (b) Anchoring Rockbolts (c) Shot creting, Guniting	Single set of discontinuities in hard rocks, or Multiple discontinuities in soft rock
	2. Rock support (a) Single member – Props: Wooden, steel	To inhibit rock mass displacement: (a) individual blocks in tunnels, mines
	(b) Multi-member – sets and arches of wood, steel, or concrete	(b) continuous rock mass in tunnels, mine openings
	(c) RC (reinforced concrete) cast in place (d) Tubbings of various design	(c) & (d) continuous rock mass in tunnels and shafts

(Continued)

Auxiliary Operations: (*Continued*).

Symbol – *Operation*	*System with Equipment/tools and appliances with their suitability and* *locales of applications*	
P – Pumping	1. Portable Face pumps	To deal with muddy water at the face during shaft sinking, Tunneling, or driving an opening in mines
	2. Pumps mounted on trolley, or hanging platform To deal with muddy water at the face	During shaft sinking, Tunneling
	3. Main pumps installed at the sump to pump out water for its final disposal	To deal with water from tunnels, shafts and mines
I – Illumination	1. Portable face light (pneumatic)	While sinking shafts, driving tunnels and at the working faces in mines
	2. Fixed lighting arrangement	Tunnels, u/g stations, Caverns where no chance of getting damaged by day to day workings

VI. Winzing## and sinking operations – excavations in downward directions – shafts, winzes, inclined shafts in downward direction, inclines, ramps and declines.

D1*/D2(a)/D2(b) + E1/E4 + B1 + M2(a)/M1(a) + H1 + services as per table 15.1(b). * – If inclines, declines. ## – for winzing small sized units are used.

15.11 THE WAY FORWARD

- Explosive energy not used in the process of fragmentation and displacement, sometimes more than 85% of that developed in the blast, reduces the structural strength of the rock-mass outside the theoretical radius of action of excavation.
- New fractures and planes of weakness are created, and joints and bedding planes initially which were not critical, when opened, affect the rock mass cohesion. This is made evident by the over-break (as shown in fig. 5.15(f)), leaving the fractured rock-mass in a potential state of collapse.

The above mentioned observations were observed in practice by the author at a hydroelectric power project during it's construction phase, where excessive use of explosive during August and September (2010) (total quantity as well the powder factor) has resulted not only excessive over-break but an overall downward trend in progress during October and November (2010). This resulted in a radical shift in the rock mass quality from its Class III to Class IV, Rock Mass Rating (RMR) status due to excessive use of explosive attaining a very high PPV. Judicious use of explosive is the right approach while working on such projects; and those involved in these operations must be thoroughly trained.

QUESTIONS

1. Aquifers are widely used in US and Europe to store what commodity? Describe the technique used for this purpose. On what factors does the cost of storage depend? Have they proved to be a cheaper means of storage wherever they have been created?
2. Caverns are special large-sized excavations that differ from the tunnels and other openings. List the unique features of caverns.
3. Caverns are large-sized underground openings driven for multiple purposes; list them.
4. Classify large underground excavations based on their function and utilities.
5. Depending upon the radionuclide content radioactive wastes remain active for period ranging from a few months to thousands of years. Underground facilities are used for storage and disposal of these wastes. List the types of shallow as well as deep-seated storage facilities.
6. Draw a 'flow diagram' showing the design, excavation and construction considerations while undertaking large-sized halls and facilities including excavation sequence of a dome-shaped exhibition hall.
7. Draw the layout of a hydro-electric power house showing a network of vertical, inclined and horizontal openings. Mention the dimensions of such caverns known to you, if any.
8. For what type of commodities are salt cavities suitable? List some products. Does France have a number of such storage cavities? How can the formation of cavities in salt be achieved? List the two techniques available and illustrate these giving suitable sketches. What are the common shapes of such storage facilities?
9. Give a schematic layout of the crusher chamber, or any other structure known to you.
10. Give a design flow diagram of a typical underground cavern. Does it take into consideration the aspects concerning earthquakes, apart from those concerning its stability?
11. How can large-sized tunnels/drives be driven? Illustrate with the aid of sketches.
12. Illustrate an overall concept of a typical repository including the details of openings required.
13. Illustrate how division of a large chamber into several horizons is made to achieve the best output, deploy an equipment fleet and carryout unit operations efficiently.
14. In this modern era, the variation in the supply and demand of various commodities including energy and fluctuations in their prices makes it necessary to store them, and match the supply with demand. Storage is also necessary to safeguard against crisis. List the factors that favour underground storage.
15. In underground mines for what are chambers of large size required? List those structures which are usually required in large underground mines. These openings can be made at any depth: what consideration is given with regard to their stability?
16. Large caverns are constructed using the heading and bench method but these caverns are still higher and, therefore, they are worked in a number of lifts or benches; illustrate these by way of suitable sketch/sketches. The excavation proceeds from top to bottom by dividing the vertical span into a number of benches. This allows

support of the roof to begin with and then the sides as the excavation progresses downward; illustrate this concept. List the usual stages of such excavations.

17. Large caverns need to be given some special considerations, List them. Write down the construction procedure of caverns. Detail out these aspects also: geomechanical aspects; supports; access.

18. List desirable 'geo-technical' aspects of the host medium of radioactive waste emplacement. Also list the technical, geological, and environmental factors to be considered.

19. List some underground facilities which are a network of excavation of various kinds.

20. List the sources from where radioactive waste is generated. Does this waste remain active for thousands of years?

21. A repository is a complex facility that must take into consideration a number of factors; list them.

22. Why has the creation of large-sized sub-surface excavations been increasing in pace since the 1950s all over the world?

23. Why is the use of underground spaces in urban areas becoming most important? What type of facilities are being created in the subsurface the world over?

24. Write down the unique features of oil storage caverns. The concept of unlined caverns in rocks to store oil is in vogue in Scandinavia; what principle of the storage of crude oil in these caverns is utilized? How is the purpose of storage accomplished? Give a suitable layout and shapes for these oil storage caverns.

25. Equipment selection: Select equipment for the operations required to drive or create civil as well as mining excavations, as described in this chapter; or those described in other chapters, making use of table 15.1 which details equipment, explosives, blasting accessories, service appliances and devices etc. For a particular excavation, matching sets of equipment should be selected to carryout different unit operations. Using the nomenclature as given in table 15.1, work out a suitable match of equipment and services for the following operations:

a. Conventional drifting or tunneling operations using rocks drills

b. Mechanized drifting or tunneling operations using one or two boom rock drills (pneumatic or hydraulic): this could be development drives, crosscuts, level drives, sublevel of medium size, or medium sized civil tunnels.

c. Large-sized tunnels, underground chambers and caverns using multi-boom hydraulic or pneumatic jumbos; if smooth blasting is required.

d. Drives and tunnels driven using road headers or full-face TBMs.

e. Raising – vertical development in upward direction.

f. Winzing## and sinking operations – excavations in downward directions – shafts, winzes, inclined shafts in downward direction, inclines, ramps and declines. ## – for winzing, small sized units are used.

REFERENCES

1. Aarvoll, M. et al.: Storage of industrial waste in large caverns. *Int. Symp. nn large rock caverns*, 1986, vol. 1, pp. 759–70.

2. Anon.: Civil engineering guidelines for planning and designing. *Hydro-electric developments*. Vol. 3 Americal society of civil engineers, 1989a.

3. Azun, A.A., Chilingarian, G.V., Robertson, J.O. and Kumar, S.: *Surface operations in petroleum production, II. 1989*, Elsevier, New York, pp. 457–70.

4. Fidencio, M.: Aguampila underground power house complex excavation sequence. *Retc Pro.* 1993, pp. 1047–65.

5. Goyal, K.L., Kumar, M. and Mittal, A.K.: Underground storage of hydrocarbons. In: B. Singh (edt.). *Int. symp. on Underground Engineering, 1988*, Balkema, pp. 415–17.

6. Hector, A.N. and Guillermo, K.: Geotechnics for construction of chamber to house an underground crusher in Codelco – Chile's EI Teniente mine. *Tunneling '82*, IMM Publication; Editor Michael J.J. Jones. pp. 49–62.

7. Hudson, J.A. and Harrison, J.P.: *Engineering Rock Mechanics*. Pergamon, 1997, pp. 288, 90.

8. Iyengar, M., Soni, A.S., Malkani, S., Sharam, A.K. and Ramkrishna, K.S.: Proposed crude oil storage in underground rock caverns at Uran. In: B. Singh (edt.). *Int. symp. on Underground Engineering, 1988*, Balkema, pp. 441–48.

9. Khot, A.S., Vaid, D.K., Chaphalkar, S.G. and Mokhashi, S.L.: Stress distribution around machine hall cavity with special reference to Varahi underground power house. In: B. Singh (edt.). *Int. symp. on Underground Engineering*, 1988, Balkema, pp. 419–23.

10. Kumar, P. and Singh, B.: On the use of u/g spacefor burial of nucler waste. In: B. Singh (edt.). *Int. Symp. on Underground Engineering, 1988*, Balkema, pp. 523–25.

11. Mathur, R.K., Kasbekar, M.M., Natrajan, R. and Kumra, M.S.: Design, construction and operational aspects of underground facilities for storage and disposal of radio active wastes. In: B. Singh (edt.). *Int. Symp. on Underground Engineering, 1988*, Balkema, pp. 491–96.

12. Matti, H.: *Rock Excavation Handbook*. Sandvik – Tamrock, 1999, pp. 228–34

13. Pokrovsky, N.M.: *Driving horizontal workings and tunnels*. Mir Publishers, Moscow, 1988, pp. 66, 268–73.

14. Sakurai, S., Chikahisa, H., Kobayashi, K. and Tsutsui, M.: Design and construction of the Takyama festival underground art museum, Japan. Proced. of Stockholm Inter. Cong. on Underground construction in modern infrastructure; Franzen et al. editors, 1998. pp. 4–8.

15. Vieth, D.L. and Voegele, M.D.: Waste repositories. In: Hartman (edt.): *SME Mining Engineering Handbook*. SMME, Colorado, 1992, pp. 2138–49.

16. Willett, D.C.: Storage and power generation. In: Hartman (edt.): SME Mining Engineering Handbook. SMME, Colorado, 1992, pp. 2127–33.

Chapter 16

Underground mining/stoping methods & mine closure

Stopes are the blocks of hidden wealth surround by an environment that is risky to men, machines and equipment. Can we skillfully recover this wealth? Success lies in the selection of the proper stoping system and deploying matching methods, equipment and techniques.

16.1 INTRODUCTION

In underground mining situations selection of a mining method is very vital, as it has direct impact on safety, productivity, cost and recovery. As discussed in the following paragraphs, there are a number of parameters that should be studied in as much detail as practical. There are cases where due to selecting improper mining methods, mines had to close. While selecting a mining method, ground stability and nature of ore and enclosing rocks, must be thoroughly studied. Basically stopes can be classified into three major groups: open or self supported, supported by artificial means and caving stopes. In naturally supported stopes, which are also known as open stopes, the condition of ore and enclosing rocks permit large sized stopes where heavy blasting can be undertaken. In these stopes heavy-duty equipment can be deployed to deplete the reserves faster. In artificially supported stopes use of supports is mandatory, which in turn, reduce productivity and increase costs and deplete the reserves slowly. Weak and cavable ores instead of posing problems can prove to be helpful if stopes are properly designed. Production in bulk can be obtained from these caving stoping methods.

16.1.1 Factors governing choice of a mining method[1,9,10,37,38]

16.1.1.1 *Shape and size of the deposit*

Based on the *shape, the orebodies* can be divided as[1]:

- *Isometrical*: almost equal dimensions in all the three directions. Isometrical orebodies are of two types: (i) *Stock and Nests*: These are usually irregular in shape but having almost equal dimensions in all directions. Massive deposits fall in this category. Stocks can have dimensions even in kilometers, whereas nests are limited in size, within several meters. (ii) *Columnar*: extended in one direction downward. As the name suggests columnar deposits are like a column. Many diamond deposits are examples of columnar deposits. These are also known as pipe deposits, which are almost vertical and thin, and extending in depth.

- *Sheet*: extended in two directions. The sheet deposits have almost a constant thickness. Coal seams and ore veins are considered in this category. But the veins usually do not have uniform thickness. Tabular deposits fall in to this category. Lenses are considered to be a change from the first to third group, having irregular shape and unequal dimension in all the three directions.

In addition, there are numbers of other shapes, including saddle dome shaped, in which orebodies may occur.

A deposit usually consists of several orebodies separated from one another by barren rock. Many times they merge or separate out from each other. Mining becomes difficult with irregular shaped deposits.

Deposit contact with country rock: This aspect is also important as sharp and distinct contact of the deposit with country rocks makes the mining process easier. Massive ores (ore minerals combined with small amount of rock) have well defined contact with the country rock.

In other cases, such as impregnation ores (ore rocks with small ore minerals) have no sharp contact with country rocks. In some orebodies change from ore to barren-rock is gradual, and thereby, prediction of ore-barren boundary is difficult.

A regular shape of orebody favors any of the mining systems (methods) but irregular and intricate shapes preclude uses of some systems or lowers their efficiency. On the other hand some mining systems are most suitable for intricate orebody shapes.

16.1.1.2 *Thickness of deposit*

Orebody thickness is the normal distance between footwall and hangingwall (sec. 2.5.1 and fig. 2.3). If the distance is measured along the normal direction, then it is termed as true thickness. But when it is measured vertically or horizontally then it is vertical thickness or horizontal thickness. Vertical thickness is measured with flatly dipping orebodies, and the horizontal one with the steeply dipping. There is no standard classification with regard to thickness of ore, however, a thickness grouping could be as under:[1]

Very thin deposits: thickness less than 0.7 m.
Thin deposits: thickness range 0.7–2 m.
Medium thick deposit: thickness range 2–5 m.
Thick deposit: thickness range 5–20 m.
Very thick deposit: thickness exceeding 20 m.

Points to note:
- In very thin to thin deposits (up to 2 m thickness) in order to provide working space for the crew and machines, floor stripping in the country rocks becomes essential.
- In medium thick deposits, 5 m is the maximum length of prop, which can be fitted, if need arise.
- Steeply dipping thick deposits can be mined along the strike direction i.e. longitudinal stoping is possible.
- For very thick deposits ore mining from hanging wall towards footwall (transverse stoping) becomes essential.

16.1.1.3 *Dip of the deposit*

Dip of the deposit (sec. 2.51 and fig. 2.3) is one of the most important parameters, which governs a mining method/system. Usually the following classification holds good, with regard to dip of a deposit:

Flat dipping: 0° to below 20°
Inclined dipping: from 20° to below 50°
Steeply inclined dipping: exceeding 50°

The dip of the deposit has the decisive influence on the selection of a mining method and positioning of a stoping face. For flat deposit ore can be mined by breasting. Following are some of the important aspects regarding dip:

- Roof pressure decrease with increasing dip.
- Foothold of the workers deteriorates with the change of dip. Low dip provides firm foothold (figs 16.1(c) and (d)), further increase in dip does not give a firm foothold and needs a safety belt. Working in moderate dips needs a platform to stand on. For steep dip in over-hand stoping operations need an immediate filling of the worked out material, which may be sand, waste rock, mill tailings or accumulated broken ore to provide firm footing to the workers.
- With steeper dips and crumbly lode (orebody) there is a danger of large pieces of ore falling, or sometimes caving of the whole face (fig. 16.1(g)).
- Influence of dip during rock fragmentation: The process of rock fragmentation depends upon hardness, coherence and dip of the deposit. In overhand stoping gravity helps in varying degrees according to dip of deposit i.e. most in vertical deposits, and least in the flat once.
- While mining vertical and steeply inclined deposits gravity together with weight of the rock helps the rock breaking process, whereas this influence get diminished with decrease in dip (figs 16.1(a), (b)).
- In steeply inclined deposits the filling (with some foreign material) of worked out space is complete and settles well comparing the same with horizontal and low dip deposits (figs 16.1(e), (f)).
- In steeply dipping weak strata working by underhand stoping i.e. from upper level towards the lower level becomes essential. But underhand stoping is associated with the danger of rolling boulders, roof fall or its caving (figs 16.1(a), (b)).
- With increasing dip, the loading and transportation of the broken muck becomes easier (figs 16.1(a), (b)).
- Working under the non-coherent deposit is dangerous, as collapse of roof may occur any time but steep dips and non-coherentness of the deposits helps to design the caving methods of mining (fig. 16.1(g)).
- All these factors indicate that *to a considerable extent stoping methods change with dip.*

16.1.1.4 *Physical and mechanical characteristics of the ore and the enclosing rocks*

Rock strength: a set of mechanical and physical properties such as hardness, toughness, jointing, laminations, presence of foreign inclusions and intercalation determine it. The mechanical strength is measured as compressive, tensile, bending and shear strength.

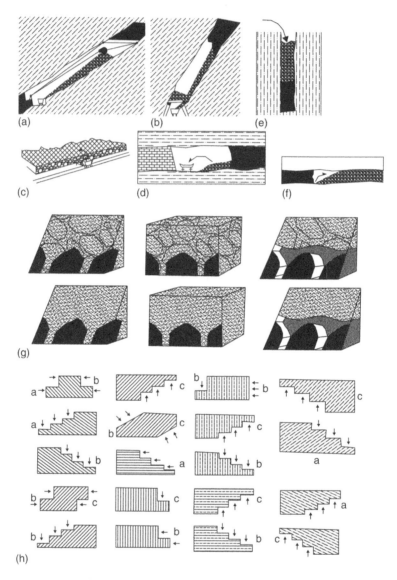

Figure 16.1   Influence of dip during stoping. With increase in dip loading and transportation of the broken muck becomes easier ((a) and (b)). The back-filling is perfect in steep dips (e), and not very tight in flat dips (f). (g) In non-coherent steeply dipping rock mining can be carried out by caving, and workers are not endangered by the type of deposit. (h) Influence of direction of cleats during mining. If mining is carried as marked *a* – this is not correct; if mined as marked *b* – it is fair; but when mining is carried as marked *c* – this is the best.

The strength of rocks has an influence on selection of a mining method as this feature has a direct bearing on selection of mining equipment and tools, also assessing the consumption pattern of materials (explosives, drilling accessories etc.), labor productivity and cost of extraction.[1]

In underground mining the stability of ore and country rock is equally as important as the rock strength. Stability is the ability of a massif, undermined (exposed) from beneath or sides, to resist caving for a certain period of time. This is a very vital characteristic, as based on this, the underground mining methods are classified.

Some rocks withstand exposure over a large area without caving for years and decades; others need to be supported only at few places; some cave immediately or shortly after their exposure over a small area. Finally, there are rocks which do not allow their exposure at all and must be supported immediately after undermining or exposure.

Apart from the physical and mechanical properties of the rock massif, stability is also affected by many other factors, and prominent amongst them are the depth of workings, their cross section – size and shape, and a few others.

With regard to stability of ores and country rocks, no index is available which determines the permissible value of the exposure – time and span. However, this may be divided into five groups as outlined below:[1]

- *Very unstable ores and rocks*: The rocks which do not allow any exposure of roof and walls and need advanced timbering; such rocks and ground are: quicksand, loose and water saturated strata etc.
- *Unstable ores and rocks*: In this type of strata the roof/back needs a strong support immediately after its exposure.
- *Medium stability ores and rocks*: This type of strata permit exposure of the roof over a comparatively large area and requires support if the roof exposure is to be allowed for a considerable time period.
- *Stable ore and rocks*: In these strata roof and side exposures can be allowed for a considerable time without support. However, some patches may require some support.
- *Very stable ores and rocks*: This type of strata can allow exposure for sufficient time and space without causing caving. Strata of this kind are rarely encountered comparing the same with the previous two groups.

To determine stability of the strata, it is important to know the nature of caving i.e.

- *Caving occurring at once over a large area*
- *Caving gradually in small sections by inrush of separate lumps or layers*
- *Caving after giving certain indications, or it occurs without warning/unexpectedly.*

Immediately after exposure the rock seldom shows signs of instability but in due course of time it starts losing stability due to rock pressure, and atmospheric conditions, and thereby it starts caving. Occasionally with time it develops a tendency to swell and bulge.

Degree of stability of ground plays an important role in evaluating its dilution factor, and deciding mode of supporting the excavated ground.

Roof Pressure: Roof pressure over a worked out space depends upon the texture of rock constituting the roof/back, its coherence, dip of the deposit, span, and rate of mining and duration of its exposure. The rock constituting the roof, and also the dip of the deposit cannot be changed but all other parameters can be controlled. The roof pressure can be counter-acted using supports, filling the worked out space or by controlled caving.

In some cases the roof is so firm that it stands for a longer duration and the worked-out room becomes too extensive. In such cases induced caving becomes essential to avoid large pressure building in and around such excavations. Excessive pressure built around these large excavations, if allowed, may result their sudden failure and that in turn may give way to air blast.

16.1.1.5 *Presence of geological disturbances and influence of the direction of cleats or partings*

Geological disturbance means presence of any one or a combination of more than one of structures such as fault, folds, joints, fissures, dykes etc. These structures usually require extra care in terms of strata stability, water seepage, gas leakage etc. making the mining process sometimes more tedious and slow. The presence of such structures usually result in higher costs and decline in productivity. Given a choice to select a geologically disturbed area with the one with minimum disturbances, one should prefer the latter one.

When choosing a mining method the strike direction and the direction in which the fissures are penetrating the deposit should be examined. The fissures running parallel to strike can be mined by overhand or underhand stoping. Of course, the ore transportation is difficult in underhand mining. If the transverse fissures, which run almost parallel to the dip, penetrate the deposit, it should be mined adopting breast stoping.

Figures 16.1(a) and (b) illustrate that nature of deposit to give way is more pronounced in steeply dipping deposits than the flat ones. Weaker deposits are worked in slices or strips. The working faces need support in such cases. After mining the strips or slices, the worked out space should be either backfilled, or caving of the roof (back) should be allowed.

In figure 16.1(h) the influence of direction of cleats on mining has been illustrated. If mining is carried out as shown in figures marked *a*, it is incorrect in relation to cleats' orientation. In figures marked *b*, cleats assist mining to a fair degree, but it is the best if mining can be carried out, as marked *c* in the figures.

This may be noted that *dip, thickness and strata stability* (ore and country rock) are the main geological and mining factors without which it might be impossible to select a safe and efficient mining system/method.

16.1.1.6 *Degree of mechanization and output required*

Mechanization means performing the underground operations using machines. The capacity of a machine is usually related to its size. Therefore, it is advantageous to select the largest units possible taking into account the aspects of flexibility, excavation and access size.

Use of higher bucket capacity LHDs, multi boom jumbos, large capacity dumpers and trucks in large underground mines is not uncommon. The types of equipment that are available in mines can be grouped in the following manner:

- Conventional pusherleg drills, rocker shovels, loco haulage and blasthole drills of 50–60 mm. dia. form degree-1 mechanization.
- Degree-2 mechanization means use of jumbos, trackless equipment such as LHDs (1 cu. yd. capacity or more), low profile dumpers and small capacity trucks. Drilling in stopes is by the same drills as in degree-1 mechanization.

- Degree-3 mechanization has the same set of machines as in degree-2, except that the drilling (for stoping) is by the down-the-hole drills capable of drilling holes of 150–200 mm. dia. of +40 m length.

In some situations the production requirements or the market demands select a mining method. Higher output warrants selection of bulk mining methods for which use of equipment of higher capacity and heavy duty becomes essential. The various sets of equipment used for this purpose are costly and require a huge sum of capital investment, and if not effectively utilized, lead to low productivity and higher overall costs. But when utilized properly, the cost of mining is substantially reduced comparing the same when small sized low capacity and conventional equipment are used.

There has been a tremendous growth so far of mechanization where the mining industry is concerned and the development that has taken place in the last half century is probably more than what had taken place in the previous five centuries. The reason is fast growing demand for minerals of all kinds due to rapid industrial growth and development in infrastructure sectors.

Higher output means quick depletion of a deposit and thereby a quick return of the investment made. To some extent it also helps in maintaining the stability of the strata, as there can be an open space which does not require an immediate support or support for a short duration, but if the same is kept open for longer it may require support. Large sized equipment also require more spacious drives and excavations which means more money should be spent on the development operations (i.e. more amortization fund). For small mines and low output, a lower degree of mechanization with small sized development entries and stopes should be preferred. For medium sized mines and output, degree-2 mechanization could prove beneficial, and for large sized mines and output, a high degree of mechanization with bulk mining methods becomes a usual choice.

Mechanization means carrying out the various unit operations such as drilling, charging and blasting, mucking, haulage and hoisting with the aid of suitable sets of equipment. As a matter of fact advent of equipment together with technology has brought about a drastic change in method design and its selection. This allowed designing bulk-mining methods. But a high degree of mechanization, if not effectively utilized, results in high capital costs and creates a problem of unemployment, whereas use of the primitive tools and machines lowers the productivity. One should certainly do nothing which leads to more unemployment as it has a booming effect on all the work that we do. It is no good to copy what is being done in the highly industrialized countries. Certainly everything should be judged in the context of conditions that prevail in any country, basically accepting the fact that better techniques have to be always employed, wherever feasible. In fact, if we make rock fragmentation and its disposal efficient, the entire mining becomes efficient.

Apart from the natural factors and the conditions that prevail with any deposit and influence significantly in selecting a mining method, there are several other factors that influence its selection, and the most prominent amongst them are the degree of mechanization, capital available and output required. For low output, use of conventional machines, equipment and methods can be used but large output mines require bulk-mining methods deploying large sized fast moving equipment. A balance is required to be made between the funds available to invest and the output rates to achieve optimum results.

Table 16.1   Proportion of valuable content in mineral deposit.[16]

Metal or mineral	Percentage/ppm of element	Percentage/ppm of mineral
Diamond	1/50 ppm (parts/million)	1/50 ppm (parts/million)
Gold	5 to 10 ppm	5 to 10 ppm
Tin	0.3 to 1%	0.5 to 1.5%
Copper	0.5 to 3%	1.5 to 10%
Nickel	1.0 to 4%	5.0 to 20%
Lead and zinc	10 to 20%	15 to 30%
Iron (high grade)	60 to 65%	85 to 93%

16.1.1.7   Ore grade and its distribution, and value of the product

Ore grade plays a vital role in selecting a mining method, as a low-grade deposit can be mined out profitably if bulk-mining methods are applied. High-grade deposits can be mined out by any of the mining methods and even up to a great depth. Also if grade distribution is not uniform it will be costly to mine out a deposit compared with one having a uniform grade distribution.

Table 16.1 presents the proportion of ore in a mineral inventory as a function of homogeneity. In general it can be stated that the more homogeneous a deposit, the less difficult it is to evaluate and mine out. A gold deposit can be considered to be more difficult to evaluate and win than a coal deposit. A copper deposit cannot be considered easy to evaluate and mine.

As the cost of mining remains practically the same for all types of rocks, which means a copper, lead, zinc or gold ore can be mined at almost the same cost, whereas the selling price of them differ significantly. This aspect imposes restrictions on mining a low valued deposit comparing the same with a high valued one. This is the reason the precious metals and stones such as gold, platinum, silver, diamond etc. are mined up to a great depth comparing the same with most of the other minerals and metals.

Ore value is the focal point while selecting a mining method. Based on the market value of the useful contents in the ore, cutoff grade of any mineral can be decided. The cutoff grade further decides ore reserves, as mineral reserves in any deposit below cutoff grade are considered waste and the one at and above cutoff grade as ore reserves. In high valued ores, costly mining methods, even at great depths, can be applied. Based on this logic, one will find that most of the gold deposits being mined currently, are at great depths.

16.1.1.8   Depth of the deposit

Deepening of workings below 600–800 m is often accompanied by a considerable rise in rock pressure, thus impeding the use of some systems. In addition, abrupt inrush of ore or rock burst from the stressed pillars is a phenomenon that has been encountered at very deep horizons. This warrants a change in stoping methods. Tracing the history of more than 120 years of mining at Kolar Gold Field, India, reveals that at these underground mines initially open stopings and timber supported stoping methods were in vogue. With the increase in frequency and severity of rock burst occurrence, use of granite masonry pack walls in cut and fill stopes was made. This did not help to improve the situation and subsequently deslimed mill tailings were used in the cut and fill stopes. But relief from the occurrence of rock bursts were felt when the filling material was replaced by concrete fill and by changing the stoping method to stope drive (sec. 16.3.2). Thus, with the change of depth stoping methods are altered. Cheaper methods need to be replaced by the costlier ones.

Apart from this, cost of ventilation, support, drainage, hoisting, transportation also increases. At greater depths apart from rock stability, the problems of heat and humidity equally arise.

16.1.1.9 *Presence of water*

Ores may be wet, dry or damp. Its moisture content depends upon the water inflow in the mine and ability to retain water by the ore itself. Ore moisture promotes caking and freezing during winter. Although presence of water does not have any direct impact its indirect impacts are many and lead to increased mining costs.

When make of water during driving and stoping is abnormal, it adversely affects safety and productivity of the mine. Pumping cost is increased. Acidic water adversely affects health and safety of the workers. It also damages the environment if it is discharged from the mines without a proper treatment. A proper drainage system means improvement in the haulage productivity and better life of mine track and roads. Presence of water during deep mining causes the problem of humidity.

16.1.1.10 *Presence of gases*

In a coal deposit presence of methane gas is common but for ore deposits (which means any deposit other than coal) it is rare but there might be some local pockets of methane. Other sources of methane in metal (also known as metalliferous) mines could be the old workings. In addition to this some other gases are emitted in mines; prominent amongst them are: carbon dioxide – CO_2, hydrogen sulfide – H_2S, sulfur dioxide – SO_2, and a few others. Presence of these gases is detailed in table 16.1(a). Special care is required while selecting and designing a mining method to win sulfur, uranium, pyrite and coal (as already mentioned).

16.1.1.11 *Ore & country rock susceptibility to caking and oxidation*

Caking: Some ore deposits are associated with small fractions of clays or some sticky material which causes problems after their fragmentation, as these clay or sticky materials after getting wet during their mining may result into cakes which are large immovable hard-to-loose solids. This phenomenon causes problems during ore drawing from a stope; due to ore mass bridging and plugging of ore passes.

Oxidation: Ore containing more than 40% sulfur in the form of pyrite or pyrhotite, etc. are liable to fires and are thus most hazardous. Fire may become intensive if ore fines and dust get mixed with some timber. Even in the absence of timber the pyrites ores after their fragmentation and prolonged storage, may cause intensive heating due to their self-oxidation. Apart from sulfide ores, any other ore after its fragmentation,

Table 16.1(a)   Common names and health effects of hazardous gases occurring in mines[47] (also ref. Sec. 9.5).

Gas	Common name	Health effects
Methane (CH_4)	Fire damp	Flammable, explosive; simple asphyxiation
Carbon monoxide (CO)	White damp	Chemical asphyxiation
Hydrogen sulfide (H_2S)	Stink damp	Eye, nose, throat irritation; acute respiratory depression
Oxygen deficiency	Black damp	Anoxia
Blasting by-products	After damp	Respiratory irritants
Diesel engine exhaust	After damp	Respiratory irritant; lung cancer

Table 16.1(b)   Heating of coal – hierarchy of temperatures.

	Temperature at which coal absorbs O_2 to form a complex and produce heat
30°C	Complex breaks down to produce CO/CO_2
45°C	True oxidation of coal to produce CO and CO_2
70°C	Cross-over temperature, heating accelerates
110°C	Moisture, H_2 and characteristic smell released
150°C	Desorbed CH_4, unsaturated hydrocarbons released
300°C	Cracked gases (e.g., H_2, CO, CH_4) released
400°C	Open flame

if stored for a prolonged period, may become oxidized and can cause problems during ore beneficiation/concentration.

Spontaneous Combustion: Spontaneous combustion is a process whereby a substance can ignite as a result of internal heat which arises spontaneously due to reactions liberating heat faster than it can be lost to the environment. The spontaneous heating of coal is usually slow until the temperature reaches about 70°C, referred to as the "crossover" temperature. Above this temperature, the reaction usually accelerates. At over 300°C, the volatiles, also called "coal gas" or "cracked gas", are given off. These gases (hydrogen, methane and carbon monoxide) will ignite spontaneously at temperatures of approximately 650°C (it has been reported that the presence of free radicals can result in the appearance of flame in the coal at about 400°C). The processes involved in a classic case of spontaneous combustion are presented in table 16.1(b) (different coals will produce varying pictures).

16.1.2   Desirable features of selecting a stoping method

1. *The facility it gives for rock fragmentation*: This may be examined by:
   - Ease it can impart during drilling operation
   - Effectiveness of the blast
   - Facility it can provide for the crew, machines and tools to be used for rock fragmentation purposes
   - Drilling is easy when workers get a sufficient and firm foothold, and there is no obstacle for installation, operation and shifting of the equipment used. Blasting

becomes more effective, if use of Earth's gravity can be made while sequencing the blast.

2. *Output*: Concentrating over few working faces with the application of the resources in terms of personnel, machines, equipment, services and supervision can yield better productivity i.e. output/person/shift. Also selecting of proper matching equipment for carrying out various unit operations is essential for better results.

3. *Rate of progress*: This aspect is important not only from the viewpoint of output but for the proper stability of ground. A fast rate of progress can reduce expenses on supports and mine services. Many methods have been designed for faster rate of development, stoping and production.

4. *Maximum recovery*: A good mining system should aim for maximum ore recovery in terms of tonnage and grade. Dilution should be minimum. The bulk mining methods may yield high output but often they are associated with higher dilution (sec. 16.4.2–3). Requirement of large number of pillars also results in poor overall recovery.

5. *Clean mining*: This means not contaminating the ore either with the filling waste or with the in-situ waste rock from the roof and sides of the stopes. Clean mining is possible if the method selected allows selective mining. Faster rate of progress can also yield a clean mining as in this case roof exposure will be minimum, and thereby, ore contamination by the falling fragments from the roof will also be very little.

6. *Easy access within stope*: This differs from one stoping method to another; however, mechanized mining can allow easy movement of personnel, machines, equipment and mine services.

7. *Energy & material consumption*: The methods which involve higher consumption of timber or other materials for the purpose of supporting the stope workings, are costlier. Another consumable item is energy/unit of output, which is required to carryout different unit operations and services such as ventilation, pumping, illumination etc.

8. *Ore handling*: Quick handling/removal of the broken muck from the stopes is essential for safety and productivity, and therefore, methods allowing this facility should be preferred.

9. *Proper filling/packing*: If mining method warrants back filling of the worked out space, method's adaptability to easy and effective filling should be looked into. A minimum void in the worked out space after filling should be aimed at.

10. *Ore winning postures and danger from roof fall/caving and muck rolling*: The ore can be extracted either by underhand, breasting or overhand stoping. Mining and transport are difficult in underhand stoping and easier in overhand stoping as shown in figures 16.1(a) & (b). In underhand stoping working below an exposed roof involves danger of roof fall or caving. To minimize this danger, roof bolting frequently becomes essential.

Some methods involve danger by rolling of the boulder or the broken muck (figs 16.1(a), (b)). This danger is enhanced if the floor is steep, smooth and fragments are round in shape. A danger of rolling boulders does not exist in overhand stoping. But with steeper dips overhand stoping becomes more difficult and dangerous. A worker has to hold feet on platform, broken ore or filling material.

In overhand stoping in steep and thick deposits there is a danger of whole mass running into workings. The big masses hang over the workings. Sometimes roof pressure becomes high, requiring efficient support system.

Flat deposits having large thickness can be mined by 'Breast and Benching'. In underhand stoping loading and transportation of broken muck is difficult.

Typical considerations to be weighted in selecting mining methods:[38]

- Maximize safety
- Minimize cost (bulk mining methods have lower operating cost than selective)
- Minimize the schedule required to achieve full production
- Optimize recovery (80% or more of the geological reserves)
- Minimize dilution (20% or less of waste rock that may or may not contain economic minerals)
- Minimize stope turn around (cycle time for various unit operations)
- Maximize mechanization
- Maximize automation (employment of remote controlled equipment)
- Minimize pre-production development (top down versus bottom up mining)
- Minimize stope development
- Maximize gravity assistance (underhand versus overhand)
- Maximize natural supports
- Minimize retention period (open stoping versus shrinkage)
- Maximize flexibility and adaptability:
  - Based on size, shape, and distribution of target mining areas
  - Based on distribution and variability of ore grade size, shape, and distribution of target mining areas
  - To sustain the mining rate for the mine life
  - Based on access requirement
  - Based on opening stability, ground support requirements, hydrology (ground water and surface runoffs), and surface subsidence.

16.1.3 Classification – stoping methods

As mentioned in the introduction to this chapter, stopes can be classified into three basic classes: Unsupported, Supported and Caving. This classification is based upon the characteristics of the ore and enclosing rocks. Each of these classes has different stoping methods as shown in line diagram (fig. 16.2). Each of these methods also has its variants, which have been dealt with in detail in various sections. The subdivision in each class is based on the thickness and dip of the deposit. In a few cases degree of mechanization also play role to design a stoping method. During the fifties very little mechanization was available and therefore, most of the mining operations used to be based on the strength and physique of workers. Method such as breast, room and pillar, shrinkage, cut and fill, square set and top slicing were in vogue. Consumption of timber used to be heavy. But with the advent of drills of different kinds and capabilities, explosives from the range of dynamite to ANFO and slurry, loading units from rock-shovel to LHDs and transportation units from battery and trolley wire locomotives to trackless units such as trucks and low profile dumpers of varying capacities and power, have changed the complete design of stoping methods. Now bulk mining

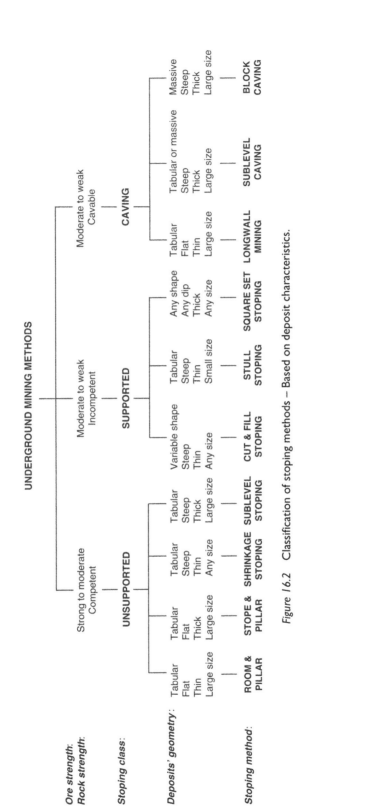

Figure 16.2 Classification of stoping methods – Based on deposit characteristics.

methods such as longwall mining, VCR and big blast stoping, mechanized cut and fill stoping, sublevel caving and block caving are available which can cope with the high production requirements.

16.2 OPEN STOPING METHODS

16.2.1 Open stoping method – room & pillar stoping

16.2.1.1 *Introduction*

In room & pillar mining the tabular or bedded thin to moderately thick orebodies dipping from flat to incline (up to 40° or so) are mined by driving rectangular or square shaped openings in the deposit. A pillar that separates two adjacent openings is left in between to provide natural support. If the deposit is continuous and the openings are driven in a systematic manner, then appearance in plan would be as if in a city the rectangular blocks are intersecting the streets.

The method is very popular for mining coal deposits (in U.S.A. 85% of underground coal production).[10] This term, room and pillar is also being applied when mining non-coal deposits. But when applied for non-coal deposits, the method in general, has been designated as Stope & Pillar mining. The method has been applied to mine out coal to a large extent and to lesser extent the non-metallic and metallic deposits such as lead, zinc, copper, potash, fluorspar, pyrite, limestone, dolomite, salt and many others.

Conditions & main elements: Following are the suitable conditions[3,9,10] for this method:

- Ore strength: weak to moderate (its variants can mine out strong orebodies also).
- Rock strength: moderate to strong.
- Deposit shape: tabular.
- Deposit dip: low (usually below 15°, but its variants can mine out orebodies up to 40° dip).
- Size and thickness: large extent, thickness below 5 m, if more benching will be required.
- Ore grade: moderate.
- Depth: Shallow to moderate (less than 500 m).

Main elements of the system/working parameters:[3,9,10,27]
- Dividing deposit in panels of the size = 600–1200 m × 120–240 m.
- Size of main and panel entries = 3–4.8 m.
- Number of entries = 2–12.
- Height of opening = deposit thickness.
- Span/width of rooms = In coal 6 m with rock bolting and may be up to 9 m with other supports; Range = 6–12 m.
- Minimum width of pillar in between = 3 m but usually its range: 6–12 m.
- Length/depth of room = 90–120 m (in some of its variants it could be less).

16.2.1.2 *Stope preparation*

The usual pattern to develop a mine to adopt this method of mining is to first develop the main haulage roadways. Simultaneously the mine is divided into panels (similar to

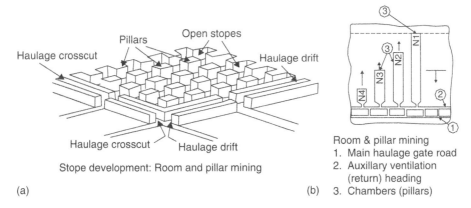

Figure 16.3   (a) Stope development during room and pillar mining. (b) Types of rooms, and nomenclature used for the system shown.

the one shown in fig. 12.7). The main roads are joined to panels by driving the panel haulage roads. Within a panel numbers of stopes are developed (fig. 16.3(a)).[30(b)] The rooms developed can have a neck[35] as illustrated in figure 16.3(b). It can be a single room with neck in the center; or single room with neck at one side of the room; or double room with two necks. Sometimes rooms without necks are also developed. As shown in figure 16.3(b), rooms are at a right angle to the longitudinal axis of the haulage roads but sometimes they can be put at slanting to facilitate the movement of the large sized equipment. It is interesting to note that during stope preparation about more than 20% of coal/ore is recovered.

16.2.1.3   Unit operations

This method when applied in coal has undergone many changes from time to time with respect to the unit-operations[2a,2b] which need to be carried out (fig. 16.4). The conventional cyclic operations are necessary in hard and tough ores that need fragmentation with the aid of explosive. Also when mines are gassy and seam thickness is varying. Continuous methods are suitable in conditions such as bad roof, limited number of working places and to produce fine product (coal). While mining coal or non-coal deposits there has been consistent change in the type of equipment being deployed with time. Based on the output, size of deposit and available resources the following sets of equipment, given as a degree of mechanization, can be deployed.

I. Stage mechanization: Coal cutters, drills and conventional explosives to fragment ore. Manual mucking into tubs. In non-coal deposits blasting off the solid by drilling and blasting is done to fragment the ore and no cutting machine is required.

II. Stage mechanization: Replacing hand loading by mechanical loading. This includes use of scrapers in inclined deposits and track or trackless loaders in flat deposits. Use of shuttle cars, low profile dumpers, trucks or conveyors as transportation units.

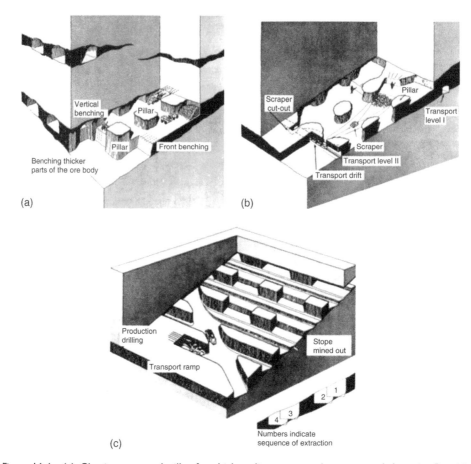

Figure 16.4   (a) Classic room and pillar for thick sedimentary coal or non-coal deposits. Benching shown to mine-out deposits in two lifts using trackless sets of equipment. (b) Room and pillar as applied for inclined deposits. (c) Step room and pillar mining for inclined deposits.

III. Stage mechanization: Use of cutter loaders (continuous miners) in conjunction with shuttle cars, chain conveyor or belt conveyors for winning the coal deposits. Use of Jumbo and blast-hole drills, ANFO explosives, LHD and trucks of higher capacities in non-coal deposits.

Figure 16.4(a),[9] gives a pictorial view of a stope and pillar's classic variant in a flat and thick deposit. Use of trackless equipment of various types has been shown. The deposit has been taken in two slices.

Figure 16.4(b)[9] depicts 'Post-Room and Pillar' layout, which is a combination of room and pillar and cut and fill stopings. This is suitable for orebodies having dip in the range of 20–55° and mined out space is back-filled. The fill ensures the stope's stability. A filled floor provides firm footings for trackless equipment. The recovery is claimed to be better than classic room and pillar mining.

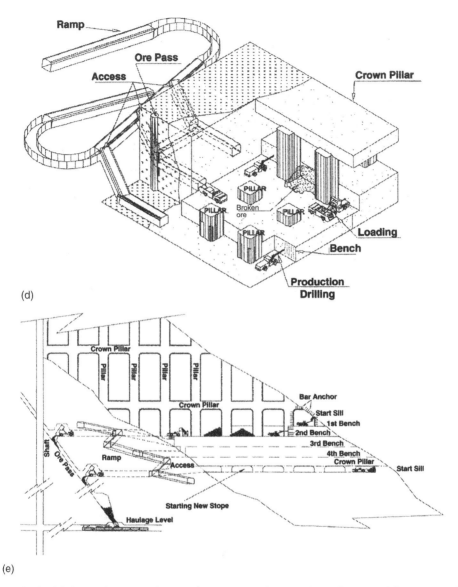

(d)

(e)

Figure 16.4   (d) Three dimensional view of mine layout for underhand Room and Pillar method[51] (www.smenet.org). (e) Cross section showing general layout for underhand Room and Pillar method[51] (www.smenet.org).

Figure 16.4(c)[9] depicts layout for 'Step Room and Pillar'. Its application can't be generalized but it is suitable for tabular orebodies with dip in the range of 25–30°. By orienting the stope at a certain angle across the dip, stope bottom assumes an angle that is suitable for trackless equipment to undertake various unit operations.

A special 'angle' orientation (it could be termed as apparent dip) of haulage drifts and stopes with respect to the dip, creates work areas with level bottoms. This feature enables trackless mucking and drilling equipment to operate in inclined ore bodies. Stopes are attacked from top to down side step by step as shown by sequence numbers in this layout.

In this layout access drifts are driven transversely across the dip at an angle suitable for the equipment movement. Ore extraction is made from a series of stope drifts that run horizontally following the strike of the orebody working from top to down. Pillars left are sufficiently narrow up-dip, thereby free movement of mucking equipment can extract broken ore efficiently. Stopes are cut successively down-dip, each stope slice having an approximately horizontal floor and being stepped in the middle of the second half of the stope. Crosscuts are also mined with horizontal floors for the movement of trackless equipment. This results in the footwall being stepped down-dip except where it is cut by equipment roadways.[11]

16.2.1.4   Stoping operations

Once the stopes in a panel have been developed, the stoping operations, which means final extraction to obtain production, then follows. It includes widening the rooms, pillar robbing and their final extraction. For extraction of pillars, based on the situation with regard to strata condition, sizes of workings and type of equipment to be used, different techniques are adopted as shown in figure 16.5. This includes different ways to split, stump or slab the pillars. While stumping a pocket is made in the pillar thus forming a stump (rib), then stump is cut across leaving a corner stump or fender. The width of pocket and the stump largely depends upon nature of roof and type of equipment used. In case of slabbing, a slab that is a part of a pillar is taken off. In all these methods mobile loaders can be deployed.

Use of temporary wooden and steel supports in a systematic manner is mandatory. The support material is recoverable to a great extent for its reuse. Same sets of

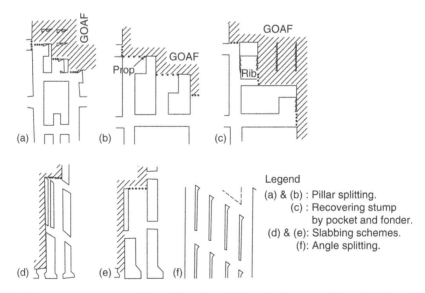

Figure 16.5   Schemes to extract coal/ore pillars from during development. (a) to (f) pillar splitting techniques/procedures in room and pillar system. (GOAF means already worked out area)

equipment, which are used during stope development, are deployed during its final extraction also. Thus the technology of mining the rooms is not different from that of mining the pillars. In the conventional system the technology of getting coal consists of drilling and blasting. The blasted coal is loaded manually into mine-cars, conveyors or gathering arm loader that discharges it to shuttlecars or conveyors. During stoping operations any of the sequence[35] listed below can be adopted but much depends upon the roof condition.

- Mining full advance (similar to as shown in fig. 16.24(c))
- Mining full retreat (similar to as shown in fig. 16.24(c))
- Mining half advance, half retreat.

If the roof is bad then the full advance method can be followed but for good roofs full retreat is usually adopted. For the roof conditions in between, the half advance and half retreat method can be followed. The layouts showing all these schemes, for longwall mining, have been illustrated in figures 16.24.

Variants – Room and Pillar Mining
- Room and pillar
    (i)   Classic room and pillar (fig. 16.4(a))
    (ii) Post room and pillar (fig. 16.4(b))
    (iii) Step room and pillar (fig. 16.4(c))
- Bord and pillar (fig. 16.6)
- Block system (fig. 16.7)
- Stope and pillar (fig. 16.8).

16.2.1.5   *Bord and pillar*[37]

In this room and pillar variant in the ore two sets of narrow headings, called galleries, stalls, rooms or bords; are driven in such a way that one set of headings is nearly at a right angle to another. Thus the ore deposit is divided into large number of rectangular or square blocks of ore, called pillars, and hence the name – Bord and Pillar. In figure 16.6 (a) layout of a bord and pillar within a panel has been illustrated.

This system is more suitable for working thicker coal seams than those suitable for room and pillar mining. The deposit may lie under valuable surface features, water bodies or waterlogged areas of another deposit – lying above, in such cases the pillars under any of these important features to be protected are left intact.

The cost of maintenance of haulage and ventilation roadways, formed by the network of pillars, is less. Where cheaper labor is available and capital to be invested has restrictions, this method is favored.

At great depth the system suffers disadvantages due to poor recovery, problem of strata control and complicated ventilation network.

Depillaring: Different procedures are followed to extract the pillars, as illustrated in the figure 16.6(b). The procedure followed mainly depends upon the type of roof and condition of the pillar being attacked.

Procedure shown in figure 16.6(A) is employed when roof condition is good. A stook is attacked from two sides. In the half moon method as shown in figure 16.6(B),

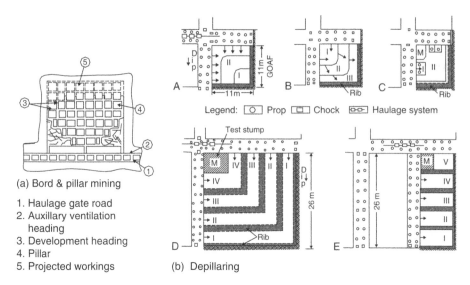

(a) Bord & pillar mining

1. Haulage gate road
2. Auxillary ventilation
   heading
3. Development heading
4. Pillar
5. Projected workings

Legend: [O] Prop [☐] Chock [⊙-⊙] Haulage system

(b) Depillaring

Figure 16.6 (a) Bord and pillar mining. (b) A to E depillaring techniques shown.

the ore is extracted in the steps shown. For a weak roof, the ore can be extracted in the manner shown in figure 16.6(C). Figure 16.6(D), illustrates another method where roof is weak. The pillar is extracted by driving the narrow headings, about 4 m wide. An indicator stump is left at the starting corner that indicates the roof condition. An ore rib, about 1 m thick, is left between these headings. In figure 16.6(E), another scheme to work in the weak roof areas, has been shown.

As stated in the preceding sections also that in all these methods use of temporary support by chocks, cogs, props (wooden, hydraulic or friction), crib sets and/or roof bolts is extensively done. Discussed below, in brief, are some of the variants of room and pillar method.

16.2.1.6   *Block system*[10,31]

In this modified method of room & pillar for mining coal deposits, instead of rectangular pillars nearly square pillars are formed. This system provides more than one working place in a pillar being robbed. Roof control is a little simpler and timber consumption is comparatively less than in the room & pillar system. In figure 16.7(a), a seven heading development plan has been shown. This layout can be developed by deploying a continuous miner or even by adopting cyclic unit operations. In figure 16.7(b), a scheme to mine out pillars, known as 'pocket and wing' has been shown. Figure 16.7(c) also shows how coal pillars are finally extracted by 'open end' method, while keeping a diagonal line of extraction. Figure 16.7(a) presents a scheme of developing the block system of mining. The manner in which seven headings can be driven using a continuous miner has been presented. Undertaking cyclic unit operations can also drive these headings. In figure 16.7(b)[10] in a panel the pillars are being won by the 'wing and pocket' method. Figure 16.7(c)[31] gives the layout of a 'block system' of

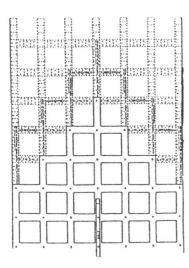

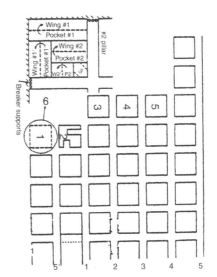

(a) Scheme of developing blocks by
     driving seven headings with or
     without using continuous miner.
     (Bullock, 1982)

(b) Winning blocks/pillars by
     wing and pocket method.
     (Hartman, 1982)

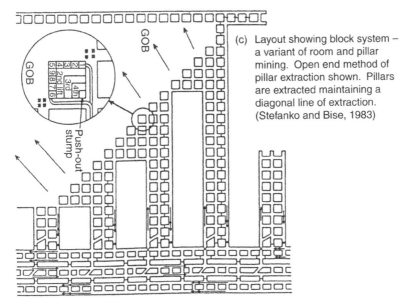

(c) Layout showing block system –
     a variant of room and pillar
     mining. Open end method of
     pillar extraction shown. Pillars
     are extracted maintaining a
     diagonal line of extraction.
     (Stefanko and Bise, 1983)

Layouts Block System of mining – a variant of room and
pillar mining.

Figure 16.7   Development and depillaring schemes for block system (a variant of room and pillar) for
               mining coal deposits. (GOB means already worked out area)

mining. 'Open end' method of pillar extraction has been shown. Pillars are being extracted by maintaining a diagonal line of extraction.

16.2.1.7 Stope and pillar

This method differs from the room and pillar in several ways listed below:

- The method is for mining the deposits other than coal, requiring natural support in the form of pillars.
- The pillars left are not systematic and can be at random. Also they may not be of the same dimensions all along.
- Adopting the advance system is almost a mandatory, whereas in room and pillar mining either system (advancing or retreating) or combination of both can be adopted.
- When a deposit is thick it can be mined by forming benches.

For the flat and thick deposits the orebody can be mined in benches while forming the pillars either in a regular interval (systematic pattern) or at random. The former is necessary when the ore grade is uniform and the latter one if grade is erratic so those portions are left as pillars where orebody has poor grade. The trackless mucking and transportation units, as shown in figures 16.4(a), could be deployed.[9] Thin orebodies do not require any benching but if they are thick it is essential; and can be mined in a manner shown in this figure. In this layout use of trackless units has been made, but if the deposit is inclined use of scraper (fig. 16.4(b)) could be made for mucking and transportation within a stope.

In figure 16.8, layout of a stope and pillar, which has been used at an Indian copper mine[28] has been shown. The deposit is dipping 25–40° and has a thickness in the range of 5–6 m. Keeping a level interval of 40 m, the stopes are developed and mined out by leaving a pillar 4–5 m thick (shown as rib pillar in the figure). Initially driving a room heading in the center of the stope joins both the levels, and then it is widened. Once the ore from hanging-wall side to a thickness of 2.5–3 m is stoped out in this manner, then the remaining footwall side ore is taken by benching. A dip of 35–40° requires ore to be scraped and hence scraper haulage is used for this purpose to muck out the ore into the ore chutes installed at the tramming level wherefrom it can be loaded to the transportation

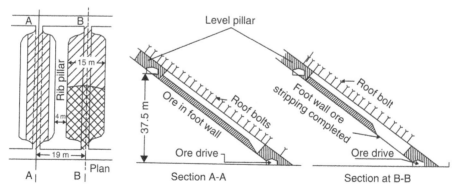

Figure 16.8 Layout of a stope and pillar stope to mine out an inclined 5–6 m thick deposit in two slices.

units which could be track or trackless. While mining the upper slice on the hanging wall side, the rock bolting is done through out the back of the stope for its stability. The pillars left between the stopes are finally recovered but ore recovery from them is usually poor.

Discussed below are the merits and limitations, in general, of room and pillar mining and its variants.

16.2.1.7.1 Advantages[3,10,31]

- Productivity: Moderate to high. There is a very good scope for mechanization in this method. But its variant bord and pillar is labor intensive.
- Range of OMS: Overall OMS 15–30 t/shift/man and face OMS 30–70 t/shift/man.
- Cost: Low to moderate (30% relative cost, when compared with square set stoping which is the costliest method – rating given 100%).
- Production rate: Moderate to high.
- Dilution: maximum up to 10–20%. Selective mining possible. Sometimes lean patches of orebody are left as pillars.
- Flexibility: unit operations can be carried simultaneously. Also change in layout, sets of equipment and number of working faces possible.
- Development cost and time: most of the development work is in ore, as such during development production up to 20–25% can be obtained. Quantum of development work is also not high, as such; bringing the stopes quickly into the regular production phase is possible.

16.2.1.7.2 Limitations[3,10,20,31]

- During stoping use of temporary supports is necessary. This lowers productivity and increases costs. Room and pillar and its variants are open stoping methods in which the strata remain intact during stoping operations and when the supports are withdrawn, caving starts. This results in subsidence of overlying strata in most of the cases.
- Recovery: without pillar extraction 40–50% and with pillar extraction 70–90%.
- Skill: requires special skill during development as well as stoping operations.
- Capital required: sufficient.
- Safety: working under an exposed roof is a potential source of accidents. In deposits thicker than 3 m working in benches under high roofs is dangerous.
- Ventilation in large open stopes is sluggish.

16.2.2 Open stoping method – shrinkage stoping

16.2.2.1 Introduction

As discussed in the preceding section, to mine out flat to inclined medium hard to strong orebodies with competent rock conditions (i.e. f/w & h/w) the prevalent method is either room & pillar or stope & pillar. But if the orebody's dip changes to steep and other conditions remain the same; the deposit can be mined by the method known as shrinkage stoping. In this method a stope is first undercut throughout its width and length at a height of about 10 m from the extraction level (i.e. haulage level). This horizon is known as the undercut level. The blasted muck so generated is discharged into the funnels developed below this level, and when these funnels are full of muck, the excess muck if any, is drawn from the draw points, which connect the funnels to the extraction drive. After undercutting the stope in this manner, the height of undercut back above the funnels filled with muck is about 2 m. Workers are allowed to stand on the blasted muck under the exposed back (roof), and when next slice, of say 2 m, is drilled and blasted in the back, the distance between this newly exposed back and the funnels is 4 m; but the space occupied by the blasted muck is about 2.6 m high and not 2 m. This is due to the fact that swelling in the blasted muck is 30–33% of the ore in-situ, and thus, the height that remains over the blasted muck is only 1.4 m, and this also means that a worker cannot work under this narrow height. Now this muck pile above the funnels is allowed to shrink, by allowing mucking from the draw points, by about 0.6 m, so that a working height is about 2 m. Thus, a mechanism of swelling in volume after blasting in-situ ore by about 33% and then allowing the blasted muck by the same amount to shrink is applied during stoping in this method, and hence, the name 'Shrinkage Stoping'.

Thus, the stope is always full of blasted muck, which works as a foothold for the men and machines. This muck also supports both the walls. When reaching to the full height of the stope by winning the ore in slices in this manner, the stope is full of muck and mucking from the draw points can be carried in a full swing. This method gives an opportunity to store the ore to safeguard against the fluctuation in the ore prices; which means that the ore can be stored when the market is down, and it can be drwan quickly (at the desired production rate) when the market is favorable. In pictorial view[9] shown in figure 16.9, some of these features have been shown.

Following are the *suitable conditions*,[9,10,19] in general, for the application of shrinkage stoping:

- Ore strength: strong (without caking, oxidizing & spontaneous heating).
- Rock strength: strong to fairly strong.
- Deposit shape: tabular or lenticular, regular dip and boundaries.
- Deposit dip: steep, (preferably 60–90°).

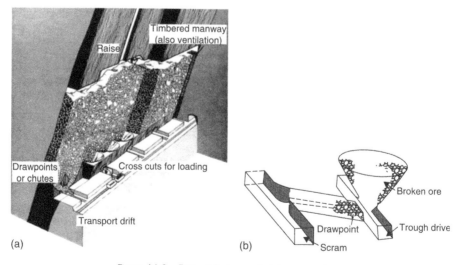

(a)

(b)

Figure 16.9 Pictorial view – shrinkage stoping.

- Size and thickness: Large extent, narrow to moderate thickness but not below 1 m and up to 30 m (up to 15 m is common. In fact rarely for the thick orebodies).
- Ore grade: fairly uniform and high.
- Depth: practiced up to 750 m.

In a continuous orebody a rib pillar of at least 10 m in between the two adjacent stopes separates them with each other. To support the level lying above the stope a crown pillar of 7–12 m is left. The vertical distance (i.e. height) between the undercut level and the extraction level, which forms the sill pillar, is usually 8–12 m.

Applications: Most popular hard rock mining method in the past. At present has limited applications. Ore types include that of copper, lead, zinc, iron, silver, nickel, gold and many others.

16.2.2.2 Stope preparation

The stope preparation work begins with connecting the stope's locale (the area where layout is going to be) with the main levels by drives and/or crosscut at both the levels. Then connecting these two levels by service raises driven at both ends of the stope. Development for the extraction layout, which includes driving of the extraction drive and draw points (a connection between funnel opening and the extraction drive), can be taken up simultaneously. The stope-accesses at shorter intervals starting from the service raises are driven to allow the access for the working crews at the different horizons when the stope advances in the upward direction. Driving of undercut level then follows. This level is stripped up to the full width of the stope i.e. the orebody. From the draw points raises of short length are driven to get through into the undercut level, from where these are converted into funnels. The following *dimensions*, in general, of the shrinkage stopes are adopted.

- Dividing deposit in levels with level interval in the range of 30–70 m all along strike extension.
- Size of entries at the main level and in stopes = 3–5 m.
- Height of openings = 2.7–3.5 m (Based on equipment height).
- Length of stope = 40–70 m (longitudinal).
- Minimum width of pillar in between stopes = 10 m.
- Height of stope = level interval, range: 30–70 m.

16.2.2.3 Unit operations

Drilling:

- Development – Use of jacklegs, stopers and jumbo drills for drivage work.
- Stoping – Use of jacklegs, stopers and wagon drills. Drifter mounted rigs for deep hole blasting.

Blasting:

- Use of NG based explosives, ANFO, slurries at the development headings.
- NG based explosives, ANFO, slurries in the stopes.
- Charging manually and with the use of pneumatic loaders (if ANFO is to be charged in up holes). Firing – electrically or by detonating fuse.

- Secondary breaking at the stopes: plaster or pop shooting, bamboo blasting at draw points.

Mucking and Transportation:

- Mucking: gravity flow of ore from stope to the extraction level horizon; mucking at the draw points with the application of LHD, FEL, rocker shovels. Also chute loading sometimes.
- Haulage: LHD, trucks or tracked haulage system using mine cars and locomotive.

16.2.2.4 Stoping operations

Once the stope development task, which includes the drivage work at the extraction level, driving undercut level and constructing funnels or any other means such as: trough-drawpoint system, slusher trench – millhole system etc. to collect muck from the stope, is complete; the stoping operation begins by taking the first slice of about 2–2.5 m above the undercut level. One-third volume after blasting each such slice is mucked at the extraction level.

16.2.2.5 Layouts

Some of the layouts as per the prevalent practices have been shown in figures 16.9(a) and (b). In figure 16.9(a),[9] a pictorial view of a shrinkage stope showing initial free face that need to be created for the stoping operations to start with, is by way of an undercut level. Drilling and charging shotholes have fragmented the ore in the stope. Muck handling is by way of funnels fitted with chute or by draw point loading.

Figure 16.30(a) shows layout with the use of slusher trench which is a drive connecting all the box-holes. A slusher is deployed to scrap muck into a millhole that discharges it into the mine cars. At the working horizon the back is kept flat but to provide an initial free face a stope raise of short length is driven within the stope at the working horizon.

In some of the Russian mines to mechanize the drilling operations, the deep hole/long hole drilling in the form of fans drilled in the horizontal plane, as well as in its upward and downward directions, are undertaken. This allows higher output from the stopes. The drilling is carried out in the chambers driven for the purpose at different horizons of the stope.

16.2.2.5.1 Winning the pillars
- Pillar recovery begins on completion of the stoping operations between two levels.
- Pillar recovery needs preparation in terms of driving accesses that are required to carry out the drilling and blasting operations. This includes recovery of sill, crown and rib pillars. Pillar recovery process has been dealt with in section 16.6.

16.2.2.5.2 Advantages[3,9,10,19]
- Gravity flow within the stope.
- Skill required – simple in operation, very useful for small mines.
- Capital required – low.
- Production rate – small to moderate.

- Recovery – during stoping 85–95%, during pillar extraction – 60–80%. Overall – up to 75%.
- Stope development – moderate.

16.2.2.5.3   Limitations[3,9,10,19]
- Productivity: low to moderate. Range of OMS = 5 to10 t/shift/man.
- Scope of mechanization – labor intensive, limited scope for mechanization.
- Mining cost – relative cost 50%.
- Safety – rough footing, working below the exposed roof.
- Ore withdrawal from stope – up to 35%, rest is tied up.
- It needs careful control of the broken ore surfaces in order to detect and eliminate hidden cavities or bridging, whose caving may lead to an accident.
- Sorting of ore of different grade within the stope is not practicable.
- No man is required to stay at the working horizon when the ore withdrawal from the stope is in progress.
- Presence of clay with the deposit or in the immediate h/w or f/w can form the cake of the blasted ore specially when it is wet. Formation of such cakes may cause problems while withdrawing muck from the draw points.
- Similarly, if sulfide ores after blasting are kept in the stope for a sufficient time due to ventilation current they may get oxidized. This oxidized ore poses problem during its concentration in the mill and requires extra reagents to overcome it.

16.2.3   Open stoping method – sublevel stoping[1,9,10,15,33(a)]

16.2.3.1   Introduction

As described, shrinkage stoping is one of the oldest methods but it was realized that it suffers with some bottlenecks during stoping. Undertaking the drilling and blasting operations by standing over the broken muck under the exposed roof/back and withdrawing only one-third of the blasted muck from the stope at a scheduled time are some of prominent bottlenecks amongst them. Hence, to mine out the steeply inclined fairly strong orebodies having competent wall rocks, and even if they are very thick, a method known as 'sublevel stoping' came into operation. In this method (fig. 16.10(a)) the orebody is vertically divided into levels, and between two levels the stopes of convenient size are formed. A rib pillar left in between them separates two adjacent stopes. Leaving a crown pillar at the top of the stope protects the level above, whereas lower level is used as haulage level to gather the ore from the stopes. Vertically the stope is divided into a number of horizons by suitably positioned 'drill drives', called sublevels, and hence the name 'sublevel stoping'. When the drills used for the purpose of stope drilling are the blasthole drills, as such, sometimes this method is also known as 'blasthole stoping'. Figure 16.10(b) displays layout of such a stope to mine out copper orebodies in seventies.

The method has gone through modifications with the advent of equipment available for drilling and innovations that have been brought forward in the blasting techniques. Based on this concept, sublevel stoping can be classified as under:

- Sublevel stoping with benching – with the use of conventional drills.
- Blasthole stoping – with the use of blasthole drills.
- Big/large blasthole stoping – with the use of large diameter drills such as DTH drills.

16.2.3.2 Sublevel stoping with benching

This is the earliest version of sublevel stoping in which ore between two consecutive sublevels used to be mined by drilling parallel holes using jackhammer drills from upper sublevel to lower sublevel through out the width of the orebody at the specified intervals in the entire length of the stope. The sublevel interval in this method is thus restricted by the drilling capability of the jack drills which are unable to drill holes longer than 6–8 m, and hence, by this method the sublevel interval cannot be kept larger than this height.

16.2.3.3 Blasthole stoping

With the advent of high capacity (length and diameter-wise) drifter drills capable of drilling holes of 50–100 mm. dia. and length of 30–40 m or more in any direction, the sublevel interval can be kept at 15–25 m or even more. Thus, using such drills the practice of drilling large diameter longer radial holes from the drill drives/sublevel came into force. Based on the orebody thickness, the blasthole stoping can be further classified as:

- Longitudinal sublevel stoping.
- Transverse sublevel stoping.

16.2.3.4 Longitudinal sublevel stoping

This method (figs 16.10(b) and 16.11), is essentially for orebodies having a thickness range of 5–20 m, but the stope height and length can be varied. Stope lengths up to 90 m and heights up to 120 m are not uncommon under suitable conditions.

Access to the stope is made through a service raise connecting two main haulage levels. Provision is made for three drill drives, commonly known as bottom, top and crown sublevels. The blast holes are drilled radially in the form of rings. A slot raise, ultimately to be converted into a slot is positioned at the wider-most end of the orebody within a stope. The blasted muck is collected at the extraction level, which is comprised of a trough, a number of draw points and an extraction drive. The length and orientation of the draw points vary depending on the dimensions of the equipment deployed for mucking, as shown in figures 16.10(b) and 16.11. The relation used to calculate draw-point length is given below:

$$\text{Draw point length} = \text{EL} + \text{SOM} + \text{CL}$$
$$\text{SOM} = \text{DPHT} \times \cot(\text{D})$$

where: EL – Equipment length;
  SOM – Distance occupied by muck;
  D – Angle of draw;
  CL – Clearance required for equipment movement;
  DPHT – Draw point height.

This method can be used for wider orebodies, up to 30 m or more, with certain modifications. In this case, provision for double drill drives (figs 16.10(b), 16.12(b) & (c)) at each of the drilling horizons and double troughs and extraction drives at the extraction level, are essential.

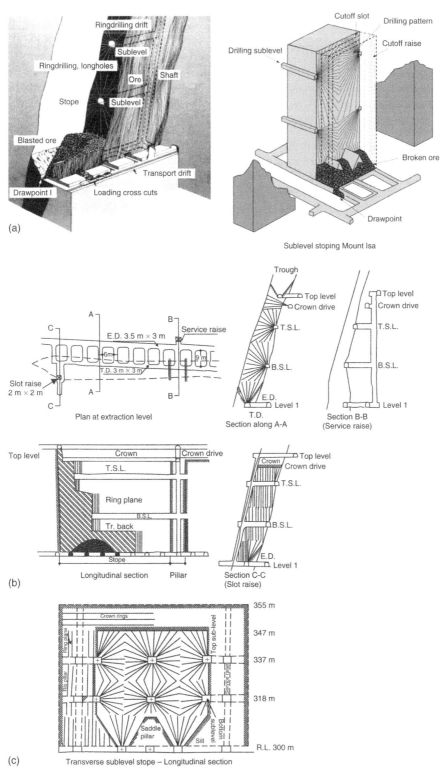

Figure 16.10 (a) Pictorial view – sublevel stoping. (b) Longitudinal sublevel stoping. (c) Transverse sublevel stoping.

16.2.3.5   *Transverse sublevel stoping*

This stoping method (fig. 16.10(c)) has the same features with regard to stope height, equipment deployment, dimensions and positions of various drilling horizons, and size of stope workings as that of a longitudinal sublevel stope. The following additional features can be incorporated in a transverse sublevel stope:

- The width of orebody should be exceeding 30 m. The stope length normally taken is 30 m but it can be more under suitable conditions.
- At the extraction level, double troughs and double extraction crosscuts are essential. Drilling crosscuts at various horizons should also be double.
- The slot follows the extreme hanging wall of the orebody. Stoping operation commences from the extreme hanging wall towards the footwall.
- The extraction crosscuts at the haulage level and the drilling crosscuts at the various horizons should be connected to a common footwall drive at each of the horizons. The footwall drives should be positioned at a minimum distance of 10 m from the extreme footwall (orebody) contact at each of the horizons.

16.2.3.6   *Blasthole drilling*

Blasthole drilling or longhole drilling are the names given to drill the holes of longer length (up to 40 m) and large diameter in the range of 45–75 mm. Drills are available (as discussed in chapter 4) to drill these holes in any direction and in any plane but they should be drilled following a specific design or pattern to fragment the ground in the desired manner. Given below are the guidelines to design such patterns. Ring design is the name given to a design in which blast holes can be marked to drill in any direction ranging from 0–360°. This includes design of fans, slot holes, rings for sublevel stoping and its variants; and also the blast hole patterns in the stopes for some other stoping methods such as sublevel caving, shrinkage stoping and block caving (to drill fans and caving rings sometimes). Figures 16.11 to 16.13 illustrate application of technique.

Ring Design: Given below are the related terms with such designs.

Ring burden: It is the shortest distance between the free face and the first ring, or between two consecutive rings; in other words, it is the perpendicular distance between two adjacent rings. Ring burden depends upon the type of rock, desired degree of fragmentation, explosive type and hole diameter. Undertaking few field trials and utilizing the experience usually decide it.

Toe spacing: This is the distance measured at a right angle between the adjacent holes of a ring at their toe. In order to design the rings, the author developed a relation to calculate the toe spacing:

$$TS = HL \times C_1 + C_2 \tag{16.1}$$

where: TS   – toe spacing in meters
      HL   – hole length in meters
      C_1 – constant-1 whose value can be altered based on hole dia., type of explosive & degree of fragmentation.
      C_2 – constant-2; whose value depends upon the type of rock to be blasted.

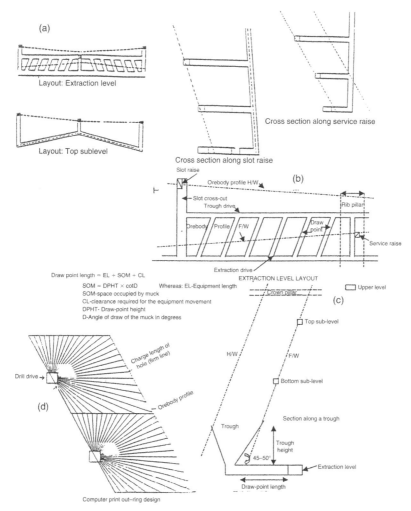

Figure 16.11   Design details of sublevel/blasthole stopes – computer printouts based on the algorithm developed.

By this relation giving values 0.05 and 0.75 for constants 1 & 2 respectively, for a hole of 1 m length, the toe spacing works out to be 0.8 m and it increases as hole length increases. Author derived this formula for its use at some copper mines. The constants were derived through the linear regression. However, their values can be altered to suit the rock characteristics.

Orebody thickness, interval between upper and lower limits of the area to be covered by the ring (measured along the dip of the orebody), toe-spacing (which is calculated using the relation mentioned above) and dip of the deposit form the input data to design a ring. When the above mentioned relation is used, it can be seen that the toe spacing is varying as per hole length (fig. 16.11(d)). An additional feature incorporated is the charging pattern of the rings using ANFO explosive. This can also be altered to achieve the desired powder factor and rational charging, if needed. ANFO is charged pneumatically using ANO loaders, shown in figure 5.2.

The designs shown in figures 16.11, 16.12 and 16.13 are self-explanatory and demonstrate the use of these designs to fragment any shape and size of the orebody. It also helps in the reduction of the stope development work.

16.2.4   Large blasthole stoping

This is the latest version of sublevel stoping in which use of Down-the-hole (DTH) drills (refer chapter 4) capable of drilling holes of 150–200 mm dia. and length of +40 m. With the use of such drills the blastholes can be drilled to a depth of 150 m, depending upon ground conditions and capability of the machine to retrieve the steel and drill. The stope height can, therefore, be varied accordingly.

Drilling is planned from only one horizon. Access to the drilling horizon is through a common decline/ramp serving a number of stopes and commencing from the immediate upper haulage level. The extraction layout for the stope mucking is the same as the one described for transverse sublevel stoping. The following are the two versions available for this method:

- Blasting in vertical slices by creating a slot (fig. 16.14(a)), or,
- Blasting in horizontal slices by adopting the crater blasting concept/theory – VCR method (figs 16.12(a to d); 16.14(a) and (b)).

Blasting in vertical slices by creating a slot: Similar to transverse sublevel stoping, a slot raise in the extreme hanging-wall is driven. By drilling and blasting the holes around the slot raise, it is widened and converted into a slot.

Close drilling to create the slot is essential, but stope drilling is designed with large spacing and burden. In this method holes drilled are either vertical or parallel to the hanging-wall of the orebody.

This method is essentially for wider orebodies with a stope length of normally 30 m or more under suitable conditions.

16.2.4.1   *Stope preparation (general procedure)*

In order to prepare a stope between two main (haulage) levels, first the access to the stope's location is made by driving the drives and/or crosscuts from these main levels. A raise, commonly known as service raise, connects these levels. At the lower level, the drivage work for the level layout may also be carried out simultaneously. This includes driving the extraction/haulage drive, a trough drive, and connecting these two drives at suitable spaced crosscuts or draw points (figs 16.10(a) and 16.11).

A slot crosscut is also driven at the wider most portion of the orebody within the stope. At the hanging wall contact (usually half in orebody and half in country rock) a slot raise is driven following the inclination of the orebody. This raise is usually known as slot, or cutout raise. This raise can be driven following any of the raise driving techniques based on the raise length and inclination (Chapter 13). But if longhole-raising technique is adopted, the same sets of equipment, manpower and services can be used for drilling the parallel slot holes and ring holes in the stopes.

On completion of a service raise, commencing from this raise at the different horizons sublevel/drill drives, of the size sufficient to accommodate the blast drills, are

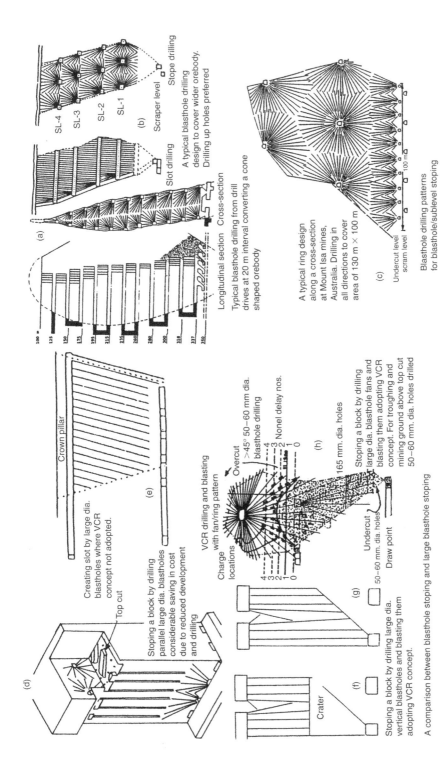

SL-4
SL-3
SL-2
SL-1

Scraper level   Stope drilling

(b)

A typical blasthole drilling
design to cover wider orebody.
Drilling up holes preferred

Slot drilling

(a)   Longitudinal section   Cross-section

Typical blasthole drilling from drill
drives at 20 m interval converting a cone
shaped orebody

A typical ring design
along a cross-section
at Mount Isa mines,
Australia. Drilling in
all directions to cover
area of 130 m × 100 m

(c)   Undercut level
       scram level

Blasthole drilling patterns
for blasthole/sublevel stoping

Crown pillar

Top cut

(e)

Creating slot by large dia.
blastholes where VCR
concept not adopted.

Stoping a block by drilling
parallel large dia. blastholes
considerable saving in cost
due to reduced development
and drilling

VCR drilling and blasting
with fan/ring pattern

Charge
locations

Overcut

>45° 50–60 mm dia.
      blasthole drilling

4
3
2
1
0   Nonel delay nos.

(h)   165 mm. dia. holes

Stoping a block by drilling
large dia. blasthole fans and
blasting them adopting VCR
concept. For troughing and
mining ground above top cut
50–60 mm. dia. holes drilled

4
3
2
1
0

(g)   50–60 mm. dia. holes

Undercut
50–60 mm. dia. holes

Draw point

(d)

(f)   Crater

Stoping a block by drilling
large dia. vertical blastholes and blasting them
adopting VCR concept.

A comparison between blasthole stoping and large blasthole stoping

Figure 16.12   Some typical designs for Sublevel/Blasthole stopes [(a) to (c)]. Some typical designs for Large Blasthole/VCR stopes [(d) to (f)]. Figures also
compare different features of the two stoping methods.

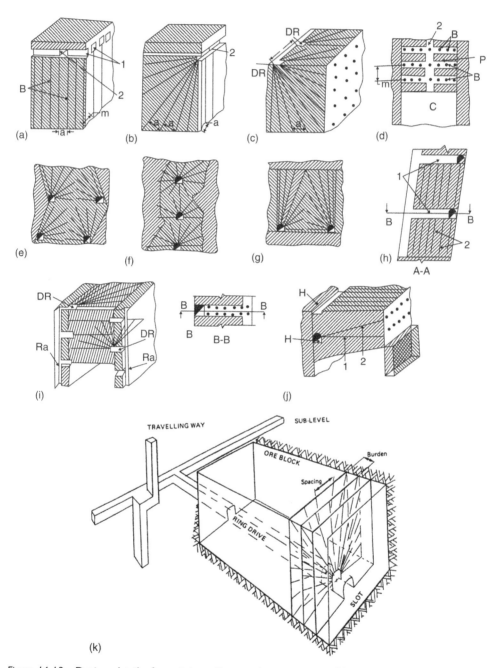

(a) (b) (c) (d)

(e) (f) (g) (h) A-A

(i) B-B (j)

(k)

Figure 16.13 Design details for mining pillars and stopes using blasthole drilling and blasting techniques. Drilling in different planes/horizons illustrated. (k)[45]: The three operations involved in ring blasting (AECL, 1980, a,b), involving three operations: driving drift/tunnel which is known as a 'ring drive'; creating a vertical slot at the end of the drill drive(s) to the full width of excavation (orebody); and drilling sets of radial holes which are known as a 'ring' (www.smenet.org).

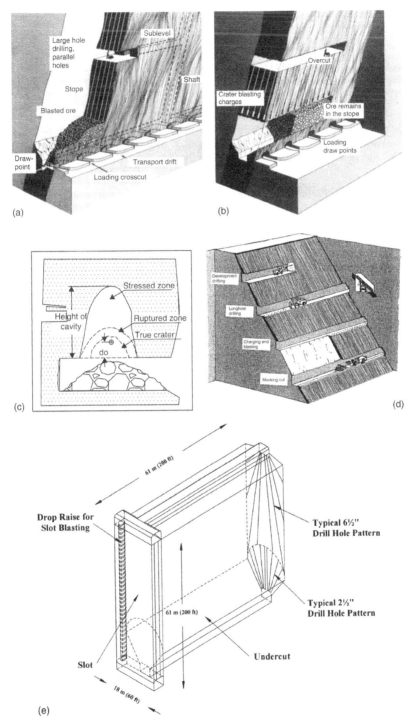

Figure 16.14 (a) Big-blasthole/DTH stoping – pictorial view. (b) VCR stoping – pictorial view. (c) Based on rock properties and structural geology, the total height of cavity may exceed the optimum distance of the spherical charge from the back many times. (d) Pictorial view showing mining of thin vein deposits using slim sized machines. Mechanised mining in 2.0 m wide sublevels shown. Note work distribution and deployment of different equipment within the stope. (e) Mechanized Sublevel (blast hole) Stope[50] (www.smenet.org).

driven either in f/w, center or in h/w contact of the orebody. These drives can be, in numbers, double (if put at the contact of both the walls), or triple (if put in the center and at the contact of both the walls), depending upon the thickness of the orebody (figs 16.12(b) and (c)).

The slot raise driven is then widened, depending upon raise inclination, by drilling vertical or inclined parallel holes. The slot holes are spaced at 1–2.5 m within same row and the burden kept to ranges between 1 to 2 m. If hole diameter is larger than 100 mm. the spacing and burden can be further increased. These holes are blasted against the slot raise; and ultimately the space so created up to the full height of the stope, width of orebody and for a length of 3–4 m, is known as a slot or kerf. Using this slot as a free face the rings drilled from the drill drives/sublevels are blasted.

16.2.4.2   VCR method[3,9,10,19,21]

VCR concept: For description of crater theory and VCR concept, please refer to section 13.10.

Suitable conditions,[3,9,10,19] in general, for the application of sublevel stoping:

- Ore strength: moderate to strong.
- Rock strength: fairly strong to strong.
- Deposit shape: tabular or lenticular, regular dip and boundaries.
- Deposit dip: steep, (preferably 60–90°).
- Size and thickness: large extent, thickness not below 5 m and up to 30 m or more.
- Ore grade: fairly uniform.
- Depth: practiced up to 1 km or even more.

Applications: Most popular hard rock mining method in underground metal mines to mine variety of ores. This includes copper, lead, zinc, iron, sulfur, nickel, gold and many others. Also this method is very popular as one of the bulk mining methods to mine large low-grade orebodies.

Main elements of the system/working parameters:[3,9,10,19]

- Diving deposit in levels with level interval in the range of 50–90 m all along strike extension.
- Size (width) of main level and stope entries = 3–7 m.
- Height of opening = 2.7–4 m (based on equipment height).
- Length of stope = 50–90 m (longitudinal); 25–40 m (transverse).
- Minimum width of pillar in between stopes = 10 m.
- Height of stope = level interval, range: 50–90 m.

16.2.4.3   Unit operations

- Drilling: Development – Use of jacklegs, stopers and jumbo drills for drivage work. Stoping – drifter mounted ring and fan drilling rigs. DTH or ITH drills for large blasthole drilling.
- Blasting: use of NG based explosives, ANFO, slurries at the development headings. ANFO and slurries during stoping. In VCR method use of spherical charge consisting of high-density explosives. Charging with the use of pneumatic loaders.

Firing – electrically or by detonating fuse. Secondary breaking at the stopes: plaster or pop shooting.

- Mucking: gravity flow of ore from stope to the extraction level horizon; mucking at the draw points with the application of LHD, FEL, rocker shovels.
- Haulage: LHD, trucks or tracked haulage system using mine cars and locomotive.

16.2.4.4 Layouts

In figure 16.10(a),[9] a pictorial view of a sublevel stope showing the initial free face that needs to be created for the stoping operations to start with, is by way of a slot. The ore in the stope has been fragmented from sublevels and it is handled at the extraction level by a troughing and draw point system.

Figure 16.10(b),[15] explains the various components/structures within a longitudinal sublevel stope where the rock fragmentation is achieved by the use of blasthole drills and explosive ANFO. A slot is driven to provide the initial free face and blasted muck is handled at the extraction level by a trough-draw point system.

In figure 16.10(c),[15] a transverse sublevel stope to mine out wider orebodies has been shown. The stope length is small and rib pillars are left at both the sides of the stope, and later on they are also recovered. Double trough and double extractions crosscuts are necessary at the extraction level. A slot is put in the extreme hanging wall covering the entire length of stope. Drilling is done from drill crosscuts located at the different horizons.

In figure 16.11(a),[33] stope layouts at extraction level and top sublevel horizons have been shown. Positioning of slot and service raise with respect to the orebody has been shown in the cross-sections drawn. In this figure a layout for two adjacent stopes has been shown. The author has drawn these layouts by computer using software developed by him.

In figures 16.19(a) (iv, v, vi) layouts[33] of six stopes of an orebody having continuity have been shown for transverse sublevel stopes using big-blasthole (DTH). Figure 16.19(a)(v) shows layout at the drilling horizon of all these stopes.

Figure 16.12[19,21] demonstrates design for the use of blasthole and big blasthole drilling patterns to create slots, troughs and to cover orebodies of varying configurations. In this figure a comparison between these two techniques can be seen. Large Application to large blastholes could be made to reduce amount of drilling and development task, thereby achieving reductions in costs and an increase in the overall productivity of rock fragmentation processes in the stopes.

This figure illustrates that ore can be sliced like bread with the application of blasthole drilling techniques. The necessary explanations have been given in the figure itself for better understanding.

Figures 16.12(e, f, g, h); 16.14(b)[9] and (c),[19] explain the main features of VCR stopes and mechanism of crater formation and the manner by which ore is blasted without creating any slot for the stoping operations to start with. Blasting progresses in the upward direction starting from the back of the trough. In figure 16.14(d), the pictorial view of a stope for mining steeply dipping thin vein deposits using slim sized equipment to carry out unit operations such as: drifting (2 m wide sub-levels), mucking and long-hole (Blasthole) drilling. This technique has a bright future for the thin and deep-seated deposits.

Figure 13.9 depicts different ways to plug the bottom of a large blasthole before charging explosive into it.

16.2.4.4.1 Advantages[3,9,10]

- Productivity: Moderate to high, not labor intensive. Very good scope for mechanization.
- Range of OMS: 15–30 t/shift/man.
- Cost: Moderate (40% relative cost), low fragmentation and handling costs.
- Production rate: Moderate to high.
- Recovery: during stoping 85–95%, during pillar extraction – 60–80% Overall – upto 75%.
- Dilution: maximum up to 20%.
- Flexibility: unit operations can be carried out simultaneously.
- Safety: little exposure to unsafe conditions, workers work under the protective roofs, easy to ventilate, thus better working conditions.

16.2.4.4.2 Limitations[3,9,10]

- Development cost and time: development cost fairly high. Stope preparation takes considerable time. Development and rock fragmentation costs during stoping increases when the deposit thickness is small and ore strength is high.
- Damage by heavy blasting: noise, vibrations, air blast and structural damages.
- Skill: requires special skill for accurate drilling otherwise hole deviation is common.
- Capital required: sufficient.

16.2.4.4.3 Winning the pillars

The same sets of equipment, which are used during stope preparation and stoping, are used to extract the pillars. The pillars are recovered using heavy blasting as described in sec. 16.6 on liquidation.

16.3 SUPPORTED STOPING METHODS

16.3.1 Supported stoping method – Stull stoping

16.3.1.1 *Introduction*

A stull is a prop, which is set between the two walls, h/w & f/w, of an inclined deposit (fig. 16.15(f)). For use as a stull, usually timbers longer than 4 m are not available. Taking into consideration all these parameters a method, known as stull stoping has been designed, that is to say, a stoping method, which is applicable for steeply inclined thin orebodies with weak walls requiring support in the form of stulls during stoping. Since the use of '*Stulls*' in this method is essential, hence the name, '*Stull stoping*'.

The open and supported stoping systems are applicable for mining the deposits of any shape, size, and thickness. But the supported system, without fill and with the use of stulls, is used to mine the deposits up to 4 m thick and seldom more. This is due to the fact that apart from non-availability of longer timbers, as mentioned above, timbering (i.e. erecting stulls) in the steeply dipping orebodies without fill, is a complicated process (figs 16.15(a), (b), (c), (d) & (f)). With the thickness increase, it becomes practically impossible particularly in the case of weak wall rocks.

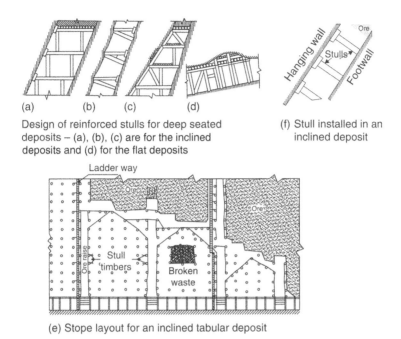

(a) (b) (c) (d)

Design of reinforced stulls for deep seated
deposits – (a), (b), (c) are for the inclined
deposits and (d) for the flat deposits

(f) Stull installed in an
inclined deposit

(e) Stope layout for an inclined tabular deposit

Figure 16.15   Stope layout – stull stoping. (a) to (d) Schemes for installing support network within a stope under varying dip and thickness of a deposit. (e) Longitudinal section of a typical stull stope.

Conditions: The system is characterized by the regular use of timbering in the stopes and also at the other openings made for the stope in the barren rock (fig. 16.15(e)). Following are the *suitable conditions*,[10,14,35] in general, for the application of stull stoping:

- Ore strength: fairly strong to strong (more competent than cut & fill).
- Rock strength: moderate to fairly weak.
- Deposit shape: approximately tabular; it can be irregular also.
- Deposit dip: usually steep but can be applied for flat dips also then it will be similar to longwall mining (sec. 16.4.1), or breast stoping.
- Size and thickness: small, thin not more than 4 m.
- Ore grade: fairly uniform and high.
- Depth: practiced up to 1 km.

Applications: In the past practiced at many mines but at present limited applications. Ores include that of copper, lead, zinc, silver, uranium and many others.

Main elements of the system/working parameters:[10,35]
- Dividing a deposit in levels with level interval up to 30 m all along strike extension.
- Size of main level and stope entries = 2.5–4 m.
- Height of opening = 2–3 m (based on equipment height).
- Length of stope = 40–50 m (longitudinal).
- Minimum width of pillar in between stopes = 10 m.
- Height of stope = level interval, ranges up to 30 m.

16.3.1.2   Unit operations

- Drilling: Use of jacklegs and stoppers to drill short length and small diameter holes.
- Blasting: Use of NG based explosives, ANFO, slurries. Charging manually. Firing – electrically or by detonating fuse.
- Mucking: gravity flow of ore from stope to the extraction level horizon through the orepasses built as the stope progresses upward. Chute loading into cars or trucks (fig. 16.15(e)).
- Haulage: trucks or tracked haulage system using mine cars and locomotive.

16.3.1.3   Auxiliary operations

- Most important task is the erection of stulls, immediately after blasting the ground. For steeply dipping deposits, interval (horizontal) = 1–2 m, the vertical spacing between rows = 1.8–2.5 m. Timbering also includes frames, chocks, cogs, reinforced stulls etc. The reinforced stulls usually used with this method have been illustrated in figure 16.15(a) to (d). The stull sets shown in figure 16.15(a), (b), (c) are for the inclined deposits and (d) for the flat deposits.
- Also preparation of the working platforms for the crews to work.
- Extension of the orepasses as the stope advances upward.

16.3.1.4   Stope preparation

This task starts with establishing the accesses from the main levels (both levels between which the stope is intended to form), by driving the drives and/or crosscuts. Service raises put at both the ends of the stope connect both the levels. Development for the extraction layout then follows; this includes driving of extraction drive and construction of chutes for the orepasses.

16.3.1.5   Stoping

This task starts by taking benches/slices of 2–2.5 m heights, covering the full width of the orebody for the stope length of 6–8 m. The slice can be taken up to the full stope length under favorable conditions. After drilling, blasting and mucking operations, the area is supported so that the next slice can be taken.

16.3.1.6   Layouts

In figure 16.15(e) a longitudinal section of a stull stope has been shown. The layout shows muck handling within the stope by making inclined passages that allow muck to fall into the vertical or inclined ore-passes which are fitted with a chute at their discharge end. The ore from these ore passes can be loaded directly into mine cars at the extraction level. Broken waste can be used to support the worked out space in the stope. The stope is worked upper hand, as shown in the figure 16.15(e). Raises with compartments, as shown in this layout, are driven, to accommodate muck in one compartment whereas the other compartment is used for services. The extraction drive is supported with wooden or steel sets in a systematic manner. The stope height and length up to 30 m is usually kept. It could be more in suitable conditions.

16.3.1.6.1   Variants

Mining of flat orebodies using this method is similar to longwall mining and differs from it only in using the regular/systematic timbering. The continuous (full) stope can advance along strike, down dip or up dip under favorable conditions.

16.3.1.6.2   Advantages[1,10,35]

- Simple in operation but requires skill, very useful for small mines even for irregular orebodies.
- Capital required – low.
- Recovery – good, if pillars are mined it could be up to 90%.
- Stope development – little.
- Possibility of mining orebodies under adverse geological conditions with better recoveries and low dilution (5–10%).

16.3.1.6.3   Limitations[1,10,35]

- Productivity: low, face productivity 0.5 to 2 m^3/shift.
- Production rate – low.
- Scope of mechanization – labor intensive, limited scope for mechanization.
- Mining cost – relative cost 70%.
- Safety – rough footing, working below the exposed roof but relatively safe in geologically disturbed areas.
- Heavy timber consumption, 0.1–0.2 m^3/m^3 of ore. Limited application at present.

16.3.2   Supported stoping method: cut & fill stoping

16.3.2.1   Introduction

As the name suggests, in this method the orebody is cut in slices and a fill of some kind replaces the void so created. Thus, a stope is worked in the upward direction starting from the level/horizon above the sill pillar. Initially two slices each of 2.5–3 m are taken. The first one is then replaced by some filling material, which becomes the foothold for the crew and equipment to mine out the subsequent slice. This procedure is repeated till stoping operation reaches up to the full height of the stope. During this operation the unit operations are carried out in a cyclic order i.e. drilling, blasting, mucking, transportation and filling within the stope.

Since stull stoping has got limited applications at present as the timber is getting scare and costlier day by day, and also due to other limitations of the method on account of production rate, productivity and safety aspects. But cut and fill stoping can be applied for not only the thin and steep orebodies with weaker walls (suitable conditions for the stull stoping) but also for the wider and even weaker orebodies than those suitable for the stull stoping. In fact, this method can be applied where the deposit cannot be mined by any of the open stoping methods, or in simple words, where open stoping fails, the substitute is the cut and fills stoping. This also means that the method is applicable for flat to steep deposits, however, the method is termed as long-wall mining with backfill when used to mine-out the flat deposits. For mining deep-seated deposits prone to rock bursts, application of this method is almost mandatory. This provides flexibility in terms of selective mining, degree of mechanization and for choosing the stope dimensions. Use of development waste-rocks and mill tailings as a backfill solves the problems of their handling and disposal, and thereby, minimizes the land degradation on this account.

Availability of a suitable fill, its effective placement and meeting its cost of application are the prerequisites for the success of this method. Development waste rock, crushed stone, sand, mill tailings or high density hardening material is the usual fill that replaces the in-situ ore. The fills can be placed manually, with the use of a mechanical stower, pneumatically or hydraulically. If the fill is not handled properly it can deteriorate the mine environment, and hence in the mines adopting this method, proper layout and skilled labor are essential.

Thus, this system is characterized by replacing the worked out area in slices by backfills of different kinds, and hence the name, 'Cut and Fill'. The method is almost mandatory in difficult ground conditions. It allows use of modern equipment to achieve moderate production rates.

Conditions and main elements of the system:[3,10,35] Following are the *suitable conditions*, in general, for the application of cut & fill stoping:

- Ore strength: moderate to strong.
- Rock strength: weak.
- Deposit shape: any, regular to irregular.
- Deposit dip: usually steep but can be applied for flat dips also then it will be similar to longwall mining.
- Size and thickness: fairly large extent, thin to thick (2–30 m).
- Ore grade: high but uniformity can be variable.
- Depth: practiced up to 2.5 km.

Applications: In the past as well as at present this method has had wide applications to mine out a variety of deposits. It has proved useful particularly to mine veins of non-ferrous, rare metals and gold due to flexibility in its application. It has applications in mining pillars and remnants. Barren rock can be left in the worked out space.

Main elements of the system or working parameters:[3,20,30,35] Mainly depends upon the degree of mechanization adopted.
- Dividing deposit in levels with level interval up to 45–90 m all along strike extension.
- Size of main level and stope entries = 2.5–7 m.
- Height of opening = 2–4 m (based on equipment height).
- Length of stope = 60–600 m.
- Minimum width of pillar in between stopes = 10 m.
- Height of stope = level interval, range: up to 90 m.
- Size of ore-pass/man-pass = 1.8–2.4 m², with spacing up to 60 m.

16.3.2.2 Stope preparation

Includes usually the tasks outlined below:

- Access from the main level to the stope by a drive or crosscut (at both the levels).
- Connecting the two levels by service raises at a proper interval for mine services including ventilation and conveyance of the filling material.

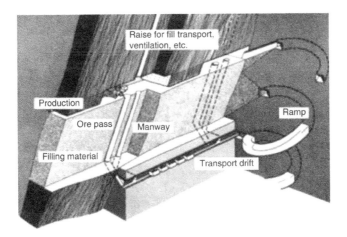

Figure 16.16 Pictorial view of cut and fill stope.

- Development for the extraction layout includes driving the extraction drive, and construction of chutes for the orepasses. Under cutting above the sill for the stoping operation to start with.
- If the stope is mined overhand, development commences at the sill and progress upward with mining the ore slices. The ore and man passes are built simultaneously using timber or tubing as the stope advances upward.
- To have an access to mobile equipment such as drill jumbos, LHDs, etc., a ramp sometimes becomes necessary (fig. 16.16); otherwise, maintenance of such sets of equipment sometimes may prove a bottleneck.
- If underhand stoping is used, mining begins just after the crown pillar in the downward direction in slices.

16.3.2.3 *Stoping*

The stoping operation involves mining slices, each of 2.5–3 m, and filling the void so created by a suitable material. This is a cyclic operation consisting of drilling, blasting, mucking and filling. The stope is usually divided into three segments/sections/panels to carry out these operations independently, and in any panel, only on completion of one operation, the other follows.

16.3.2.4 *Unit operations*

- Drilling: Use of jacklegs, and stopers for drilling short length and small diameter holes. Use of drifter drills mounted on jumbos for drilling holes of 45–76 mm. dia. and length upto 3 m or more. Holes drilled are inclined or horizontal.
- Blasting: Use of NG based explosives, ANFO, slurries. Charging manually as well as using pneumatic loaders. Firing – electrically or by detonating fuse.
- Mucking: at working horizon using LHDs, Cavo, rocker shovel or scraper up to ore pass; from ore pass through gravity up to the chute located at the haulage or extraction level. From chute loading into cars or trucks.
- Haulage: trucks or tracked haulage system using mine cars and locomotive.

16.3.2.5 *Auxiliary operations*

The auxiliary operations include some of these tasks:

- Back filling of the worked out area; preparatory work required to carryout the filling operations; drainage from the back fill and allowing it to set; preparation of the floor to carry out routine unit operations for the next slice (fig. 16.22).
- Extension of the ore and man passes as the stope advances upward.
- Undertaking support work, by other means if need arise. This includes rock or cable bolting, timbering, packs walling etc. The most important operation is ground control.

16.3.2.5.1 **Advantages**[1,3,9,10,35]
- Productivity: moderate 10–20 tons./man-shift, maximum up to 30–40 t/man/shift.
- Production rate – moderate. But can be applicable for small mines and even for irregular orebodies.
- Scope of mechanization – moderate.
- Safety – good safety records. Proved safer even at great depths and prevents occurrence of rock bursts.
- Flexibility: versatile, adaptive to a variety of conditions. If grade of ore is poor selective mining and ore sorting is possible.
- Ground conditions – suitable even for the worst ground conditions.
- Capital required – moderate.
- Stope development – little.
- Recovery – maximum if pillar mined, it could be up to 95% or more, and ore fines losses are eliminated. Low dilution.
- Stope development – little.
- Depth – proved vital for deep mining at high rock pressures.

16.3.2.5.2 **Limitations**[1,3,9,10,35]
- Mining cost – relative cost 60%.
- Cost of backfilling – up to 50% of the total mining cost.
- Operational skill – requires skilled labor. It is more labor intensive.
- Working atmosphere: at depth wet filling may create humidity problems.

16.3.2.5.3 **Variants**
- Cut and fill with flat back – (i) Conventional; (ii) Mechanized (fig. 16.17)
- Cut and fill with inclined slicing (fig. 16.20).
- Longwall cut and fill stoping.
- Post & pillar – cut and fill stoping (fig. 16.19).
- Stope-drive cut and fill stoping: (i) starting from upper level (fig. 16.20(e)); (ii) starting from lower level (fig. 16.20(f)).

Selection of a particular variant of cut and fill stoping would depend upon the strata conditions, as shown in figure 16.18.[17] Good ground conditions would allow conventional and mechanized cut and fill variants, whereas, for poor ground conditions post and pillar and stope drive methods would be a right choice.

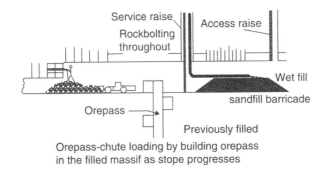

Service raise
Rockbolting
throughout
Access raise

Wet fill
sandfill barricade
Orepass
Previously filled
Orepass-chute loading by building orepass
in the filled massif as stope progresses

Figure 16.17   Mechanized cut and fill stoping.

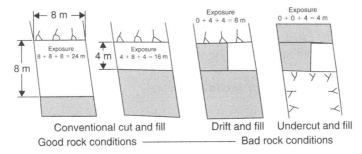

Figure 16.18   Relative exposure of unfilled space within a cut and fill stope. Good back/roof conditions allow conventional and mechanized methods; whereas bad back/roof conditions allow little unexposed space, and the costlier variants of cut and fill system are applicable.

16.3.2.6   *Cut and fill with flat back*[9,23,40]

This variant of cut and fill stoping can be applied for thin to medium thick steeply dipping orebodies with medium to high ore strength and unstable wall rocks. Irregular and valueless ore with barren rock inclusions, which could not be separated out, could be left in the stoping area. In practice these conditions may differ, for example, wall rock may be stable and ore boundaries could be regular without inclusions of the barren rock.

Stope development consists of driving a haulage or extraction drive, which could be doubled with interconnecting crosscuts if thickness is large. A raise either at the center or at one side of the stope is driven.

Stoping begins at the haulage drift roof level, or a sill pillar 3–4 m thick is left above the drift. The stope is mined by taking a slice 2–3 m thick. Increasing the slice thickness leads to better productivity but decreases safety. Drilling and blasting the vertical or horizontal holes causes rock fragmentation. The blasted muck is delivered to ore passes by scrapers, or loaders of different kinds including Cavos and LHDs (fig. 16.17).[9]

After stoping the first slice, a strong floor is placed on the haulage drift or pillar (as the case may be), and chutes are made at the locations intended to form ore passes.

After mining the second slice, the worked out space is filled with the filling material and ore passes are erected over the chutes. The ore passes are reinforced with timber, concrete or large diameter steel pipe (tubing 500–1200 mm. dia.) rings.

To avoid the losses of ore fines, the fill surface is covered with a strong wooden or metal floor. Sometimes, particularly when working high valued ores, used conveyor belts or canvas covers the floor. The best way is to concrete the floor, for a thickness of 15–20 cm. Such floors eliminate ingress of ore fines into fill, allow smooth running of equipment, improve ventilation and increase stability of the fill itself.

The stope is divided either into two or three compartments. In one (or two) of them rock fragmentation and mucking operations are carried out, whereas in the remaining compartment back filling, extension of ore passes and placement of hard floor or floor covering operations are undertaken.

There are two variations of flat back cut and fill stoping: *Conventional and Mechanized* (fig. 16.17). In the conventional method use of conventional drills and loading equipment are used and they are made captive. In the mechanized version high capacity self-propelled equipment is used during stope development and stoping. A ramp is put to facilitate equipment movement. The filling method suitable flat deposits is known as longwall mining with backfilling.

16.3.2.7   *Cut and fill with inclined slicing*

In this system ore is extracted in inclined slices having inclination in the range of 30–40° to the horizontal. The method is also termed as rill stoping (fig. 16.20(a)).[7] Incorporating this feature gives the advantage that ore after blasting is delivered directly to the ore passes. The fill is stowed in the stoped out area.

This method can be applicable for orebodies not thicker than 3–4 m because in wide inclined faces it is difficult to take care of the roof. In addition, since it is not practicable to sort out the ore on the inclined floor and leave the barren rock in the stopes; it cannot be used in orebodies thinner than 0.7–0.8 m. If inclination of orebody is less than 60°, the ore and fill is retained at the footwall, and need additional labor to transfer them and hence under such conditions the system should not be applied. The wall rock should be stable as it is not feasible with weak wall rocks.

16.3.2.8   *Post and pillar cut and fill stoping*

In this variant of cut and fill stoping, in addition to back filling the stope, additional thin ore pillars known as posts are left within a stope in a systematic manner (fig. 16.19(a) and (b)[33]). Rib pillars separate adjacent stopes, and hence, the name post and pillar cut and fill stoping. The posts can also be left at random in some cases, particularly when lean patches of ore (low grade ore pockets) are encountered within the orebody in a stope. The stope progresses in the upward direction. In wide orebodies with weak walls, leaving ore in the form of post and pillars becomes essential to prevent stope collapse. This reduces overall recovery from the stopes as the ore left in the form of posts cannot be recovered and recovery from the pillars is also poor. Due to wide orebodies and particularly when there are lots of fractures, cracks, fissures and geological disturbances of any kind, during stoping cable bolting becomes essential. In layout (fig. 16.19(a)), the orebody is 30–40 m wide with joints, cracks and fissures and as such cable bolting in the stope needs to be done. The layout shown in figure 16.19(b)

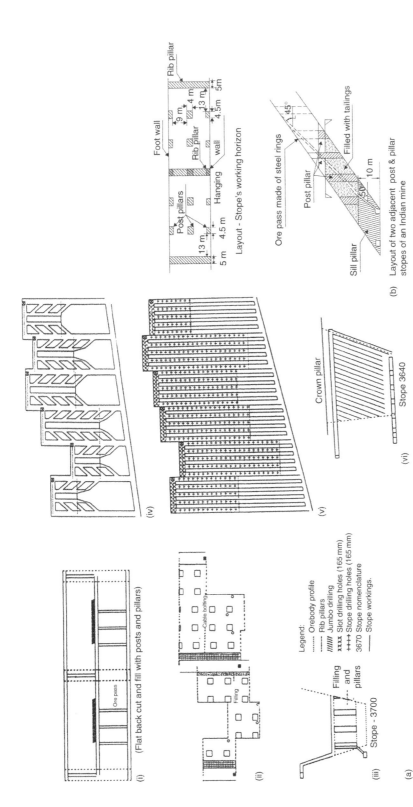

(a)

(i) (Flat back cut and fill with posts and pillars)

(ii)

Cable bolting

Filling

(iii)

Filling and pillars

Stope - 3700

Legend:

· · · · · Orebody profile
— — — Rib pillars
////// Jumbo drilling
xxxx Slot drilling holes (165 mm)
+++ Stope drilling holes (165 mm)
3670 Stope nomenclature
——— Stope workings.

(iv)

(v)

(vi)

Crown pillar

Stope 3640

Foot wall

Post pillars

13 m
5 m   4.5 m

Rib pillar

Hanging wall

9 m   4 m   13 m
4.5m

Rib pillar

5m

Layout - Stope's working horizon

Ore pass made of steel rings

Post pillar

45°

Filled with tailings

Sill pillar

50°   10 m

(b) Layout of two adjacent post & pillar stopes of an Indian mine

Figure 16.19 (a) Computer printout showing layout details of a post and pillar cut and fill stope for mining weak and wider Pb-Zn deposits. (b) Post and pillar layout of a copper deposit.

is for mining the wider copper orebodies that has been practiced at the Mosabani group of mines in India. Here posts of 4 m × 4 m at spacing of 9 m have been left systematically. The pillars left between two adjacent stopes are of 5 m thicknesses. In figure 16.19(a) a stope longitudinal section together with plan at the working horizon, has been shown. The stope can be accessed from the upper level through the service raises, which can be positioned at the rib pillars and to access the working horizon from the lower level; a man pass is built within the fill massif, as the stope progresses in the upward direction. Similarly the ore passes and the drainage raises are built in the fill. All these raises (or the passes) are extended as the stope progresses.

16.3.2.9   *Stope drive or undercut and fill stoping*

In this variation a slice of 1.8–5 m (based on strata conditions) is taken either from top to downward direction or vice-versa i.e. downward direction to upward, and each successive layer is filled with the cemented fill. Use of conventional equipment (fig. 16.20(e) and (f))[25] and also modern equipment such as LHDs can be made under suitable conditions. A ramp or man pass constructed in the filled material, as the stoping progresses can access the stope. Ore is dumped into an orepass driven between two haulage levels while preparing the stope. The orepass is equipped with a chute at the bottom.

This method is applied where either the wall rocks or vein is too unconsolidated/weak. The method is suitable for deposit of any dip and varying thickness. The method proved a success in the rock-burst prone areas and recovering the pillars.

If strata conditions are too bad or while working at great depths, a stope drive stating from the bottom is adopted. In this case the wall rock exposure during stoping is minimum. Since the back fill used in this case is high-density concrete fill, which is costlier, this method is suitable during deep mining to mine the valuable deposits such as gold or high-grade base metal deposits.

16.3.2.9.1   Filling methods during deep mining[25]

Kolar Gold Field (KGF) located in India, is one of the deepest gold bearing areas in the world, where mining up to a depth of about 3 km has been undertaken during the course of its mining which was for a duration of more than 120 years. Mining of gold started in this field in around 1880. The deposit in the form of steeply dipping thin veins was extending right from shallow depth to the depths exceeding 3 kms. Earlier the methods such as sublevel and shrinkage stoping were in vogue at the shallow depths. Later on use of granite was started to construct the pack walls in the stopes. As the mines started getting deeper and deeper the problem of rock burst began. Use of more effective filling material such as mill tailings was then started but this practice could not be continued more due to humidity problems that were arisen due to presence of water in the fill (i.e. hydraulic filling). The problem of rock burst still existed. The filling material was then replaced by hardening fill – concrete, and stoping methods were changed to the stope drive system (fig. 16.20(e) and (f)). Both variants of this system i.e. starting from upper level or starting from bottom level, were used. As illustrated in figure 16.20(e) the stope drive method starting from the upper level can be applied in bad rock conditions, and the one starting from the lower level upward can be applied in even the worst ground conditions. In the latter case at a time very little area

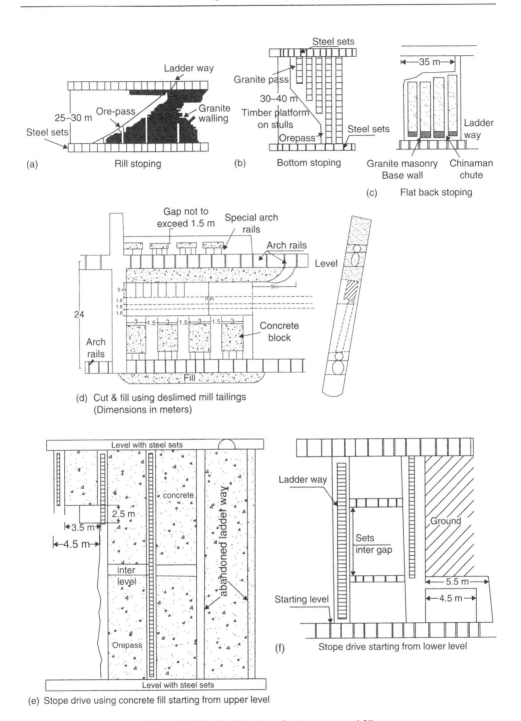

(a)   Rill stoping

(b)   Bottom stoping

(c)   Flat back stoping

(d) Cut & fill using deslimed mill tailings
     (Dimensions in meters)

(e) Stope drive using concrete fill starting from upper level

(f)   Stope drive starting from lower level

Figure 16.20   Application of various cut and fill.

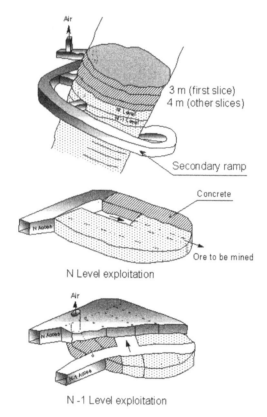

3 m (first slice)
4 m (other slices)

Secondary ramp

Concrete

N Access

Ore to be mined

N Level exploitation

Air

N Access

N-1 Access

N -1 Level exploitation

Figure 16.20   (g) Initial Top Slicing method[52] (www.smenet.org).

is kept open (without support) and this reduces the chances of rock bursts. The earlier methods used in this field have been also illustrated in this figure (fig. 16.20(a) to (d)).

16.3.2.9.2   Top slicing[52] (An undercut-and-fill method)
A case study: to mine out uranium pipe-like deposits; this was adapted during 1987–2001 at Jouac Mine, France. It is an illustration of using high cost concrete backfill to achieve 100% recovery to exploit valuable deposits such as uranium.

The general underground mine layout consisted of the following.

A primary 15% ramp with a $20\,m^2$ cross-section was driven in the granitic host rock from the surface to the various orebodies. A secondary ramp was driven around the various vertical columns. (The uranium ore body dipping at 45°–90°; appears like columns. These pipe-deposits had an average cross-section of $300\,m^2$ and height in the range of 20 to 700 m. There were about 10 pipes that were discovered at this mine). Ore was hauled using 20–40-ton trucks and ventilation was provided at $250\,m^3/sec$.

A brief description of the method is illustrated in figure 16.20(g).

- A ramp was driven in the consistent barren rock, winding around the pipe (deposit, as described above).

- A first slice was mined in the ore using a drill and blast method. Roof support is required because of poor ground conditions. Productivity obtained was about 9.5 tons/man-shift, which was low. Each drift making up the slice was backfilled as soon as it was completed. Various back filling ratios such as: 5/6; ¾ or ½ ratios were used depending upon the orebody geometry.
- The second slice was then extracted with the drift driving perpendicular to that used on the level above. Extra reinforcement was not required. Each drift was backfilled with concrete.
- The third slice was then undercut, and the process continued in this manner.

Advantages of this method include: the workers' safety is maximized as all mining is done under a concrete slab; ore recovery is 100%; dilution is low. Disadvantages include: high cost due to use of concrete backfill.

16.3.2.9.3   Filling materials

The types of backfills used in the cut and fill stopes includes sand, crushed rock (which may be mined from the mine or available at the surface), mill tailings, boiler and metallurgical slag. Following are some of the desirable qualities which a fill should possess:

- Its availability in large quantities at a cheaper rate,
- Easy to transport right from its place of origin/availability to the stoping areas to be backfilled,
- Non-caking (if not used during pillar recovery where caking is a desirable quality),
- Least shrinkage coefficient and
- Fire safe.

The fills are further distinguished as dry gravity, pneumatic, hydraulic and hardening fills based on the manner in which they are filled in the voids of the stopes. A comparison between different fills that are in use has been made in table 16.2. Gravity filling is achieved when fill is tipped from the trucks, mine-cars etc. in a void created by mining the deposits steeper than 25°. In order to place the fill in the voids created by the flat to inclined deposits either manual masonry work, or some mechanical stowers, such as skid mounted slinger belt, or truck mounted slinger belt[39] could be deployed. When material to be stowed is fed into a compressed air stream by a stowing machine for blowing into a stope by a pipeline, this is known as pneumatic stowing.[23] This practice requires control over the dust and noise generation. Power consumption and capital costs are also significant; thereby its use is limited to the places where hydraulic filling cannot be applied.

When sand or mill tailings are mixed with water and transported through the pipes to discharge this product in the stopes, the technique is known as hydraulic filling. This is a popular technique whose layout has been shown in figure 16.21.[25] For deep mines it gives a problem of humidity. Deep mines also have the problem of rock bursts which can be effectively tackled by the use of a fill which is made of a mixture of water, binder (such as cement) and aggregates; known as high density fill.[38]

The layout shown in figure 16.21 gives a general scheme of hydraulic fills using deslimed mill tailings in a cut and fill stope. The fill is sent down to the mine using a borehole or shaft. The borehole discharges the filling pulp into a pipeline, which is taken up to the stope where the fill is to be placed behind a bulkhead. The bulkhead

Table 16.2 Comparison of different types of fills.

Parameters	Back filling systems				
	Dry pack wall and gravity filling	Mechanical filling	Pneumatic filling	Hydraulic filling	Hardening or high density filling**
The system	This involves packing of filling material manually in the goafs of flat and shallow deposits. It is termed as gravity filling when fill is tipped by tubs or other means in goafs of deposits steeper than 25°.	The placement of back fill in the stopes is achieved by the use of equipment such as slushing, slinger belt placement or stower placement.	Material to be stowed is fed into a compressed air stream by stowing machine for blowing into a stope by a pipeline. Material is wetted by water before it discharges into goaf.	Sand/tailings or crushed material is transported u/g in the form of slurry through pipes. The material settles in the stope and water percolates to lower level and from there it is recycled.	A high-density mixture is placed in the stope. It hardens after placement to form a one-piece massif.
Material used	Crushed material In lumps, waste rock from mine, sand gravel, and mill tailings. Clay content should not exceed 15%.	Rock from development headings or waste fill available near mine. Lump size not exceeding 150–200 mm.	Not sticky. Crushed low abrasive rocks. This includes waste rock, tailings, slag etc. Size: 5–60 mm. (or 1/3 of pipe dia.) with clay addition up to 10–15%.	Coarse sand and mill tailings without silt. Solid liquid ratio – for sand 1:0.75 to 1:1. From 1:1.5–1:1.25 for coarser material.	Mixture of water, binder (slag or cement) & aggregate (which could be materials suitable for hydraulic or pneumatic fills).
Fill transportation	Vertical: via. rock pass, holes, pipes fitted in shafts, below pipes hopper and chute gate are installed. Horizontal: via cars, conveyors. In stopes by gravity and scrapers.	Material to be stowed is transported using transport system of the mine.	Material to be stowed is brought u/g using transport system of the mine. Then fill is transported via. pipes (not longer than 300 m) by the compressed air.	Mixing arrangement for fill with water made at surface. Slurry is transported through pipes, gravity assists the flow.	Due to high density of fill, a positive displacement concrete or mud pump is used to transport material by pipeline.

	20–25%	20–30%	10–15%	5–10%	Least.
Shrinkage coefficient					
Merits and limitations	System is simple in operation. Proved better for flat deposits where even hydraulic filling is less effective. Low capital and operational costs. Low productive requires considerable manpower. Relatively loose and uncompacted fill.	It requires bulky equipment and hence limited sphere of application. It strains transport capacity of mine and cannot be used during ore production shifts. Relatively loose and uncompacted fill.	In absence of water humid conditions not created. Very useful where hydraulic stowing not suitable. Relatively compact filling is achieved. Strain on transport system of mine significant. Capital required and power costs are the highest. Noise and dust generated should be controlled.	A flexible method effective to control strata and fires in stopes. In stopes provides good floors for equipment movement Preparation plant required. Excessive water and silt results in many problems. For deep mines not suitable due to humidity generated. Not very effective for flat deposits.	It minimizes mining losses for valuable ores. Minimize risks of bump, burst and fire. Effective protection of surface area as this is the strongest fill available. Preparation plant is more expensive and sophisticated there by filling is costlier.

Ice fill: Ice has been proposed as backfill in permafrost regions; however to date ice has only been used in Norway and CIS Russia.[38]

** Paste Fill: it is a hydraulic fill (total tailings without cycloning) mixed with cement. The distinction between paste fill and high-density fill is an item of concern. In general, a high-density fill has the properties of a fluid while paste fill has the physical properties of a semi-solid. Today paste fill is often a default selection when planning new projects and in number of instances, has been installed at old mines to replace or supplement an existing backfill system.[38]

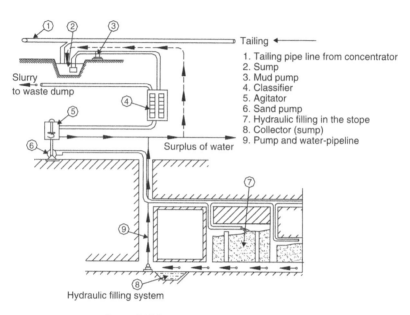

1. Tailing pipe line from concentrator
2. Sump
3. Mud pump
4. Classifier
5. Agitator
6. Sand pump
7. Hydraulic filling in the stope
8. Collector (sump)
9. Pump and water-pipeline

Hydraulic filling system

Figure 16.21   Hydraulic filling system.

is prepared within the stope before the fill is placed in a section or panel chosen for it. Different designs are available so far as the bulkhead is concerned, as shown in figures (16.22(a), (d) and (e)).[23] The purpose of bulkhead is to allow the fill to settle and water to drain off. Drainage tower made of wire mesh is also used. Other arrangements for the purpose of placing the hydraulic fills are also shown in figure 16.22. It is important to drain off water and pump back it to the surface, from where it can be reused. From the stopes the water is first allowed to collect into a settling sump, from where it is pumped to the surface as shown in figure 16.21.[1]

In many mines especially when binder is added to the slurry backfill, auxiliary drainage facilities (perforated piping) are installed in the stopes to accelerate the removal of free water as shown in fig. 16.22(b) (right), most mines cover the drainage pipes with a fine mesh sock to prevent the fines from entering the drainage system, allowing only water to be removed from the stope. Slurry transportation concepts should be made applicable while designing these backfill systems to avoid any transport problems enabling easy operation to achieve the targeted production.

Calculation for stowing material required

Relation to be used: $P' = (\gamma'/\gamma) \, K \, P$ (16.2)

where: P' – weight of stowing material to be used
P – weight of mineral (ore) to be extracted in tons.
γ' – density of coal/ore in tons/m^3
γ – density of stowing material in tons/m^3
K – factor of stowing (0.3–0.95) according back-filling/stowing system

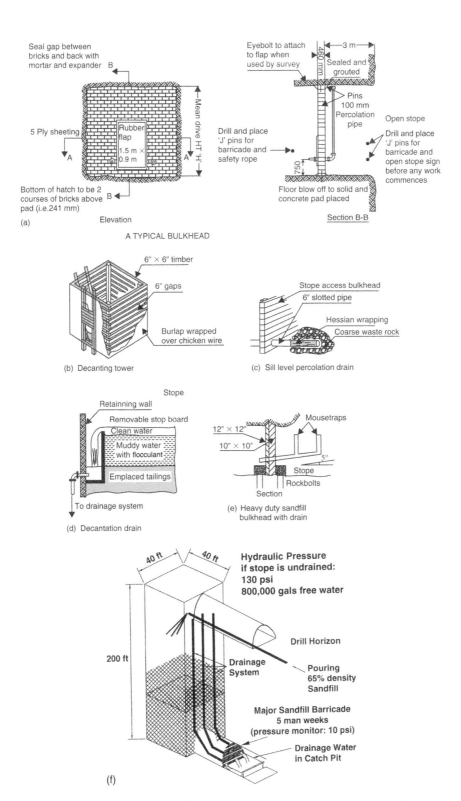

Figure 16.22 Details of various types of bulk-head; decantation/drainage towers used in conjunction with hydraulic backfills. (f) Hydraulic slurry backfill, auxiliary drainage[48] (www.smenet.org).

If a stowing material of 1.6 t/m³ is to be used in coal and metal mines, having densities of coal and metallic ore as 1.3 t/m³ and 3 t/m³ respectively, calculate the amount of stowing material that will be needed in terms (or multiple) of weight of mineral/production. Consider value of K factor for pneumatic stowing in coal mines as 0.8 and in metal mine as 0.7.

16.3.3 Supported stoping method – square set stoping

16.3.3.1 *Introduction*

Referring to different forms of supports that are used in mines, one of them is 'Square Set'. It consists of members known as grit, cap and posts (fig. 8.2). In these sets the cross-members (grits and caps) are usually 1.8 to 2.4 m in lengths whereas the posts (vertical members) are 2.4–3 m high. The important feature of this type of support is that it can be extended in any direction by adding the members to cover the complete excavation that needs immediate support after its exposure so that a support network like a nest is built within the stope. It is this feature that allows even the underhand stoping to be to carried out. The ground conditions that require immediate support are the deposits with weak orebody as well as the enclosing wall rocks. Hence, a method that is characterized by the regular use of timbering in the stope in the form of square sets or a method that warrants use of square sets is known as 'square set stoping'.

Following are the *suitable conditions,*[1,10,35] in general, for the application of square set stoping:

- Ore strength: weak to very weak.
- Rock strength: weak to very weak.
- Deposit shape: any, regular to irregular.
- Deposit dip: usually steep but can be applied for flat dips also, and then it will be similar to longwall mining.
- Size and thickness: size can be any but usually small, thin not more than 3.6 m.
- Ore grade: high but uniformity can be variable.
- Depth: practiced up to 2.5 km.

Applications: In the past practiced at many mines but at present limited applications. Ores includes that of copper, lead, zinc, silver, uranium, gold and many others. It has application in mining pillars and remnants.

Main elements of the system/working parameters:[1,10,35]
- Dividing deposit in levels with level interval up to 30 m all along strike extension.
- Size of main level and stope entries = 1.8 m–2.5 m.
- Height of opening = 2–3 m (based on equipment height).
- Length of stope = 30–40 m (longitudinal).
- Minimum width of pillar in between stopes = 10 m.
- Height of stope = level interval, range: up to 30 m.

16.3.3.2 *Stope preparation*

- Access from the main level to the stope by a drive or crosscut (at both the levels).
- Connecting the two levels by service raises at both ends of the stope.

- Development for the extraction layout, this includes, driving the extraction drive, construction of chutes for the orepasses. Under cutting and draw points not required.
- If the stope is mined overhand, development commences at the sill and progresses upward with mining. The ore passes are built simultaneously using the square sets structures. Each of the orepass terminates into a chute where from mine cars can be loaded directly.
- If underhand stoping is used mining begins at the crown.

16.3.3.3 Stoping

- Carried out by mining the rooms/blocks of square sets' sizes.
- In difficult ground conditions advance timbering – fore-polling or spilling becomes essential. After drilling, blasting (which should be kept minimum, or preferably avoided) and mucking operations, the area is supported by square sets, before taking the next slice/room/block. The timbered stopes are usually filled with waste rocks/backfills simultaneously.
- Man pass, ore pass and waste passes are also built simultaneously.

16.3.3.4 Unit operations

- Drilling: Use of jacklegs, and stopers, drilling of short length and small diameter holes.
- Blasting: Use of NG based explosives, ANFO, slurries. Charging manually. Firing – electrically or by detonating fuse. As mentioned above this operation should be kept to a minimum. Use of conventional digging and dislodging tools could be made to dislodge the ground.
- Mucking: gravity flow of ore from stope to the extraction level horizon through the orepasses built as the stope progresses upward. Chute loading into cars or trucks.
- Haulage: trucks or tracked haulage system using mine cars and locomotive.

16.3.3.5 Auxiliary operations

As in the stulll stoping, the most important operation is ground control.

- Most important is the erection of square sets, immediately after blasting the ground. A network of square set timbering sets is built as the stope progresses.
- Also preparation of the working platforms for the workers to work.
- Extension of the orepasses as the stope advances upward.

16.3.3.6 Layouts

In figure 16.23 square set stopes are shown. The stoping operations in these stopes start from lower level towards the upper one. Some sets are used to act as ore passes for transferring the muck generated within the stope to the haulage level below. The orepasses are fitted with the chutes to discharge the muck directly into transportation or haulage units at the extraction level of the stope.

Like a rill stoping, the sets within the stope can be installed. This allows ore flow within a stope by gravity. Sometimes ore and walls are very weak that in addition to square sets, use of a backfill becomes essential.

One of the important features of square set stoping is that it can be used for irregular ore boundaries as shown in figure 16.23 (left). Even the lean patches of ore can be left. The stoping operation can be commenced either overhand or underhand.

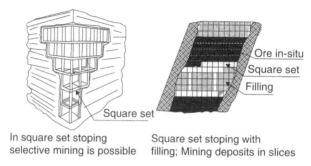

In square set stoping   Square set stoping with
selective mining is possible   filling; Mining deposits in slices

Figure 16.23   Details of square-set stoping.

16.3.3.6.1   Advantages[1,10,35]
- Flexibility: versatile, adaptive to variety of conditions. If grade of ore is poor selective mining and ore sorting is possible.
- Ground conditions: suitable even for the worst ground conditions.
- Skill required – simple in operation, very useful for small mines even for irregular orebodies.
- Capital required – low.
- Recovery = maximum if pillar mined up to 95% or more.
- Stope development – little.

16.3.3.6.2   Limitations[1,10,35]
- Productivity – low, productivity 1–3 t/man-shift.
- Production rate – low.
- Scope of mechanization – labor intensive, limited scope for mechanization.
- Mining cost – relative cost 100%.
- Safety – poor, fire hazards with sulfide ores. Rough foot holding, working below the exposed roof and poor ground conditions.
- Heavy timber consumption, more than $0.2\,m^3/m^3$ of ore. Limited application at present.

16.4   CAVING METHODS

16.4.1   Caving method – longwall mining[1,10,20,27,31,32,37]

16.4.1.1   Introduction

This is one of the oldest mining methods being practiced all over the world to mine thin and flat deposits, particularly the coal seams. The system is characterized by a long (around hundred or more, meters) face (known as wall) established across a panel between sets of entries and retreated or advanced by narrow cuts, allowing the hanging wall (roof) to cave in (fig. 16.24(c)). The width of working face is few meters and it is kept supported using yielding supports of different kinds – conventional or powered. As cut is taken along the length of face, the supports retract, advance and rearrange allowing the roof to cave behind. Since a long face characterizes the method, which look like a longwall with a cantilever-roof, hence the name 'Longwall mining'.

Following are the *suitable conditions*,[1,10,20,27] in general, for the application of long-wall mining:

- Ore strength: any, but in coal, which is weak the method is very well suited particularly to coal cut by continuous cutting equipment.
- Rock strength: weak to moderate but cavable.
- Deposit shape: tabular.
- Deposit dip: flat, low (<12°) and uniform.
- Size and thickness: large extent along and across the dip.
- Ore grade: moderate but uniform.
- Depth: 150 m–900 m in coal and up to 3.5 km. for non-coal deposits.

Applications: Widely in mining the coal all over the world and prominent amongst them are U.K.; U.S.; Germany. In metal and nonmetal deposits (i.e. hard rock mining) limited applications but the ores that have been mined includes that of copper, lead, zinc, silver, uranium, gold, potash and many others.

Main elements of the system or working parameters: The system is similar to Room and Pillar mining. Given below are prevalent dimensions[6]
- Diving deposit in panels of 900–2700 m (in length) × 150–300 m (in width).
- Height of opening = deposits thickness (0.9–4.5 m).
- Face width = 2.4–3.6 m.
- Face length = 90–230 m (U.S. 140–170 m, U.K. – up to 200 m; Germany up to 234 m).
- Depth of cut = 76–762 mm (for plough and shearer faces).

16.4.1.2 Unit operations

- Conventional (cyclic): Hard seams and deposits, gassy mines, variable seam thickness,
- Continuous: Bad roof, limited number of working places and to produce fine product (coal).

16.4.1.3 While mining coal

- Stage I mechanization: Coal cutters, drills and conventional explosives. Manual mucking into tubs.
- Stage II mechanization: Replacing coal fragmentation by earlier method by shearer (single or double drum types), and ploughs. Cutting and loading into armored chain and flight face conveyors. Stage loaders and conveyor, or track system for onward transmission/transportation.

16.4.1.4 Stope preparation

In a panel longwall faces are run. Developing the gate roads from the main haulage roads of the mine accesses the panel. A longwall face can be with single unit, or double unit (fig. 16.24(b)). The length of a single unit face is usually within 100 m but a double unit face is adopted where high degree of mechanization with faster progress is warranted. The length of double unit is double the length of a single unit face and usually it is within 200 m. The length could be determined by the relation given by Popov[37]:

$$L_{cl} = \frac{(T_{sp}-t_{pc})n}{\dfrac{1}{V_W}+t_R} \ldots \tag{16.3}$$

$$L_w = \text{Length of longwall face} = L_{cl} + N_s \times L_s \qquad (16.4)$$

where: L_{cl} – Length of longwall face excluding stables, m

T_{sp} – Duration of production shift, minutes

T_{pc} – Time needed for preparatory work, minutes

n – Number of production shifts during the day

V_w – Rate of advance of cutter loader, m/min

t_R – Time lost due to stoppage of the equipment, minutes/m of face length. This factor should also include time required for the replacement of cutting tools and other machine related delays

N_s – Number of stables

L_s – Length of stable along the longwall face, m.

To develop a longwall face at both the terminal ends of a longwall face, two blind ended faces known as stables (like pilot drive or tunnel faces), as shown in figure 16.24(b) are driven from the gate roads. On completion of a few rounds, a drive parallel to the longwall face, known as 'kerf drive' is driven. This, in fact, is the initial free face for a long wall face to start with. In double unit longwall face instead of two stables at the terminal points, a third stable at the center (fig. 16.24(b)) is driven. These stables i.e. the pilot faces, are always run ahead of the longwall face. Once the longwall face at the beginning of the stope is developed in this manner, after advancing it a few meters the face can be equipped with coal cutting and loading units, and also face supports are installed. In case of coal, use of armored or scraper chain conveyor at the face is common. The cutting units include plough, coal cutting machines or coal shearers (fig. 16.24(d)).[32]

The longwall faces can be driven along the strike, rise (i.e. opposite to dip direction), or along the dip directions of a deposit. The stoping sequence can be either advancing or retreating based on the same logic that holds good for Room and Pillar mining system (fig. 16.24(c)).[37]

While mining the ore deposits other than coal, ore fragmentation using drilling and blasting becomes essential, and therefore, the face is not equipped with a conveyor system of haulage. A scraper rope haulage system can be adopted.

16.4.1.5 Stoping operations

Stoping operations can be cyclic or even continuous depending upon the type of mechanization adopted at the face. The cyclic operations are carried out using the conventional equipment such as coal cutters and conveyors. In this system after cutting and loading operations, conveyors are shifted and the face is properly supported. Similar operations are carried out while mining the strong ore deposits but in place of cutting operations, drilling and blasting operations are conducted. Use of a self-advancing support system with shearers mounted on conveyors allows a continuous mining system (figs 16.24(d), (e)). This system is highly productive but requires high capital investment and a high degree of equipment maintenance facilities and skilled operators. Equipment down time should be minimum to get good results.

16.4.1.6 Layouts

- Long wall face single unit with backfilling, or caving.
- Longwall face double unit with backfilling, or caving.

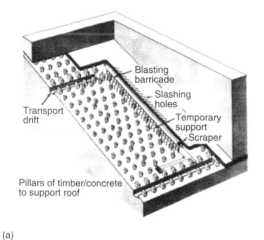

Blasting
barricade

Slashing
holes

Transport
drift

Temporary
support

Scraper

Pillars of timber/concrete
to support roof

(a)

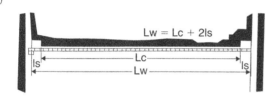

$Lw = Lc + 2ls$

Lc

Lw

ls

ls

Lc = Longwall face length
(without stable)
Lw = Long wall face length
(total)
ls = Length of stable
$L'c$ = Length of first unit
$L''c$ = Length of second unit

Single unit longwall face

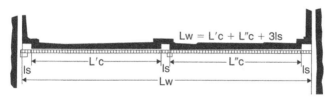

$Lw = L'c + L''c + 3ls$

ls

$L'c$

ls

$L''c$

ls

Lw

Double unit longwall face

(b)

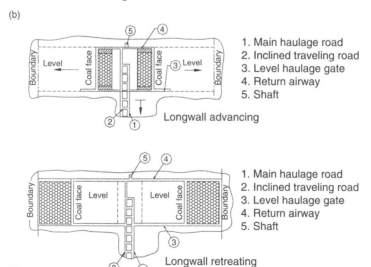

⑤ ④

Boundary

Level

Coal face

③ Level

Boundary

Coal face

② ①

Longwall advancing

1. Main haulage road
2. Inclined traveling road
3. Level haulage gate
4. Return airway
5. Shaft

⑤ ④

Boundary

Coal face

Level

Level

Coal face

Boundary

③

② ①

Longwall retreating

1. Main haulage road
2. Inclined traveling road
3. Level haulage gate
4. Return airway
5. Shaft

(c)

Figure 16.24   (a) Pictorial view – longwall mining. (b) Longwall faces with single and double units.
(c) Longwall faces – advancing and retreating.

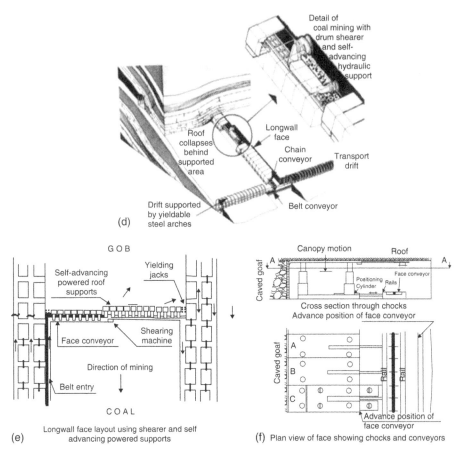

Figure 16.24 (d) Pictorial view – longwall mining using shearer. (e) Longwall mining layout using shearer with self advancing support system. (f) Relative position of supports and face conveyors shown.

These faces can be equipped with:

1. Conventional equipment (low degree mechanization) for hard coal in coal mines.
2. Conventional equipment for hard strata other than coal for rock fragmentation, mucking and transportation.
3. Mechanized faces equipped with plough and hydraulic support.
4. Mechanized faces with drum shearers and ploughs, and self-advancing powered supports.

Long wall face single unit with back filling or caving: Longwall mining is usually carried out allowing the overlying strata to cave in after the ore/coal is mined out. But in some circumstance, the surface structures need to be protected against caving. In such cases the worked out area is backfilled using fills such as pack walls made of waste rocks, hydraulic backfills using either sand or mill tailings. In figures 16.24(d) to (e), longwall faces with the use different types of face mechanization have been presented. Caving has been allowed in the layouts shown in these figures. These layouts could also be without caving when the stoped out area is suitably backfilled.

Use of conventional equipment for hard strata other than coal for rock fragmentation, mucking and transportation have been shown in layouts given in figure 16.24(a). Such layouts are popular in mining thin and fairly strong deposits of lead, zinc, copper, gold and many others.

Mechanized faces with drum shearers, may be single or double drum, and self advancing powered supports have been presented in figures 16.24(d) and (e). In these layout details of the self-advancing support system has been also presented.

In the layout shown in figure 16.24(d), a pictorial view of longwall retreating system[27] has been presented. In this layout the coal is cut by the use of double drum shearer that has been mounted on the face conveyor. This conveyor conveys the coal from the longwall face to the stage loader installed in the gate roadway for its onward transmission to the main conveyor. In figure 16.24(f) use of hydraulic chocks at the longwall face with respect to the goaf (the worked out area) has been shown.

16.4.1.6.1   Advantages[9,10,27,30,31,32s]

- Productivity: highest, comparable with bulk mining methods such as block caving. Max. up to 90–100/man-shift in U.S. mines.
- Production rate – medium to large.
- Mining cost – relative cost 20%.
- Scope of mechanization – sufficient. Suitable for remote control and automation.
- Ground conditions: suitable even for the worst ground conditions.
- Early production possible by adopting longwall advancing. Stope development – little and goes simultaneously in case of advance longwall mining. Retreat method helps in knowing the strata conditions and other parameters in advance.
- Recovery – fairly high, maximum if pillar mined up to 95% or more, without recovering pillars recovery ranges between 70–90%. Low dilution (10–20%).
- Safety – good safety records. Proved safer even at great depths.

16.4.1.6.2   Limitations[9,10,27,30,31,32]

- Longwall with caving causes subsidence over wide areas.
- Capital required – High, and so is the capital cost/unit of production.
- Flexibility – little flexibility once the layout is executed and equipment installed.
- Reliance on the face machinery is great, hence in the event of any break down, it effects the complete cycle. Any interruption proves costly.
- Operational skill – requires skilled labor.
- Working atmosphere – at depths if wet backfilling used, it may create humidity problems. Also heating of goaf/gob may create heat and humidity problems, if caving allowed.

16.4.1.7   Mining at ultra depths[53]

In addition to what has been described in sec. 8.4.1 the most significant recommendations that remain valid are:

- Formation of isolated areas of the reef – a vein of ore (i.e. remnants) should be avoided.

Table 16.2(a)   Expert design specifications for the various layouts[53] (www.smenet.org).

		Mining methods			
	(units)	**LSP**	**SGM**	**SDD**	**CSDP**
Generic macro design parameters					
Raise-to-raise spacing	m	n/a	200	100	180
Maximum length of back	m	240	240	240	175
Dip pillar's maximum deviation from true dip	degrees	n/a	45		
Dyke and fault bracket pillar width (on either side)	m	20	10	20	20
Preliminary vertical distance between levels	m	<120	92	70	68
Width of dip-stabilising pillar for depth range < 4 km	m		30		
Width of dip-stabilising pillar for depth range between 4 km and 5 km	m		40		
Width of dip-stabilising pillar up to a max. depth of 4.6 km	m			25	40
Width of strike-stabilising pillar for depth range between 3 km and 4 km	m	40			
Width of strike stabilising pillar for depth range between 4 km and 5 km	m	50			
Design of flat development infrastructure					
Minimum vertical distance of haulage / RAW below reef	m	90	120	50	90
Minimum distance to reef of replacement haulages	m	90			
RAW-HLGE connecting crosscut spacing	m	250		200	180
Footwall drives/Haulage and RAW tunnel dimensions (height × width)	m²	3.0 × 3.5	4.5 × 4.0	4.5 × 4.0	4.0 × 3.0
Replacement haulage and RAW tunnel dimensions (height × width)	m²	4.0 × 3.5			
Minimum distance from a HLGE - RAW connecting to reef X/cut breakaway	m		25	25	25
Minimum distance between opposite breakaways (longwall)	m	15			
Haulage and RAW spacing (centre to centre)	m	30	30	30	30
Minimum middling of follow-behinds from reef plane	m	20			
Maximum middling of follow-behinds from reef plane	m	40			
Minimum middling of incline man & material way from reef plane	m	60			
Follow-behind tunnel dimensions (height × width)	m²	3.0 × 3.5			
Crosscut-to-crosscut spacing (grid definition)	m	63	200	100	180
Crosscut to reef dimensions (height × width)	m²	3.0 × 3.5	3.5 × 3.5	3.5 × 3.5	3.0 × 3.0
Minimum crosscut distance below reef (perpendicular) at reef intersection	m	10	12		15
Design of stoping-related infrastructure					
Panel face length between 3 km and 4 km	m	40	30-40	46	25
Panel face length between 4 km and 5 km	m	35			
Maximum number of panels per longwall section	unit	6			
Maximum number of panels per raise	unit	n/a	7	2	7
Average stope width (VCR considered)	m	1.5	1.5	1.5	1.5
Raise dimensions (height × width)	m²	2.0 × 1.8	3.0 × 1.5	2	1.8 × 2.5
Strike gullies dimensions (height × width)	m²	1.6 × 2.0	1.8 × 2.8		1.8 × 2.0
Ledging distance from centre line (on either side of raise)	m		10	6	15
Maximum strike scraper distance	m	100	80		70
Maximum on-dip distance for continuous scrapping	m				150
Maximum length of winze	m		75		
Depth of wide raise (on either side from centre of raise)	m			6	
Design of vent control infrastructure					
Bulk-cooling loop dimensions (height × width)	m²	3.0 × 3.5			
Cooling car cubby (one side extra width, in excess of x/cut width)	m	4 + 4		5	
Cooling car cubby length	m	6	15	20	
No. of dedicated vent holes per x/cut	unit	1	1		
Design of ore-handling infrastructure					
Conventional stope box hole dimensions (height × width)	m²	2.0 × 1.8	1.8 × 1.4	1.4 × 1.5	1.8 × 2.5
Raise bore stope orepass diameter	m		1.4 m		2.1
Box hole stub length (extra width in excess of x/cut width)	m				5
Box hole stub length	m		12		20
Maximum raise bore length	m		120		
No. of conventional box holes/box fronts per X/cut	unit		2	2	2
No. of Y-leg box holes/box fronts per X/cut	unit		3		
Inclination of conventional stope box hole	degrees		55	55	65
Maximum inclination of raise bore orepass	degrees		45	15	65
Design of access-ways infrastructure					
Incline man and material way (longwall) (height × width)	m²	4.0 × 3.0			
Length of the station for the incline man and material way (longwall)	m	25			
Travelling way dimensions (height × width)	m²	2.8 × 2.0	2.4 × 2.4	2.4 × 2.4	2.0 × 2.5
Travelling way length	m		15		15
Travelling way inclination	degrees		34	34	34

(Continued)

Table 16.2(a)   Continued.

	(units)	Mining methods			
		LSP	SGM	SDD	CSDP
Design of materials-handling infrastructure					
Timber bay dimensions (height × width)	m²		3.0 × 4.0	3.0 × 3.0	3.0 × 2.0
Timber bay length	m		30	20	30
Timber bay distance from reef	m		20		10
Generic scheduling rates and mining constraints					
Normal haulage and RAW development	m/mth/end	35	30	35	35
High speed haulage and RAW development	m/mth/end	70	70	70	70
X/cut development	m/mth/end	35	30	35	35
Raise development	m/mth/end	30	25	12	30
Winze development	m/mth/end		15		
Box hole development (conventional)	m/mth/end	20	10	35	20
Drop-box hole development (conventional)	m/mth/end				30
Raise bore development	m/mth/end	35			35
Delay in stripping of services at a raise	months		1		
Delay attributed to stope equipping	months		6		
Lodging face advance	m/mth/face		25		
Stoping face advance	m/mth/face	12	15	15	15

- Stoping operations should be concentrated as far as practical.
- Working faces should be advanced rapidly and continuously.

Mining at ultra-depths (Iponeleng model orebody).

The Witwatersrand basin of South Africa has produced more than 50,000 tons of gold since 1986; the gold production reached at its peak at 1000 tons per annum in 1970 and 500 tons thereafter. With the increase in prices of gold and its further boom in the days to come, gold will be mined below 3 km depth, where the gold reserves are expected to be more than whatever has been mined so far from this field. As such there is great potential for mining of gold from this field but the challenge of ultra-depths such as high rock temperature and virgin rock stress are likely to further increase. In order to address this challenge the DEEPMINE programme with close liaison amongst government, mining companies, research institutes, universities and the mining community have been formulated to work out the Mining Plan to undertake mining between 3 to 5 km for which the term '*Ultra Deep*' has been used. From the last several decades, the tabular reefs of this region were mined using the longwall mining concept with strike pillars providing regional support. But during the recent past a new generation of layouts have been introduced based on the concept of scattered mining with dip pillars for regional support. In order to choose the most appropriate mining method for the ultra-depths, a study was undertaken to analyse various options and alternatives. The 1.5 m thick reef considered for this study is $24\,km^2$ between depths of 3800 to 5000 m below the surface, with an average dip of 23°. The geological information was used as input. The four variants of longwall mining, namely: longwalling with strike stability pillars (LSP), sequential grid mining (SGM), the sequential down dip (SDD), and closely spaced dip pillar (CSDP) methods were considered. Some operational assumptions were made. The mining rules, scheduling rates, and design specifications for each layout on both micro- and macro-scale were defined by the team experts, as given in table 16.2(a).

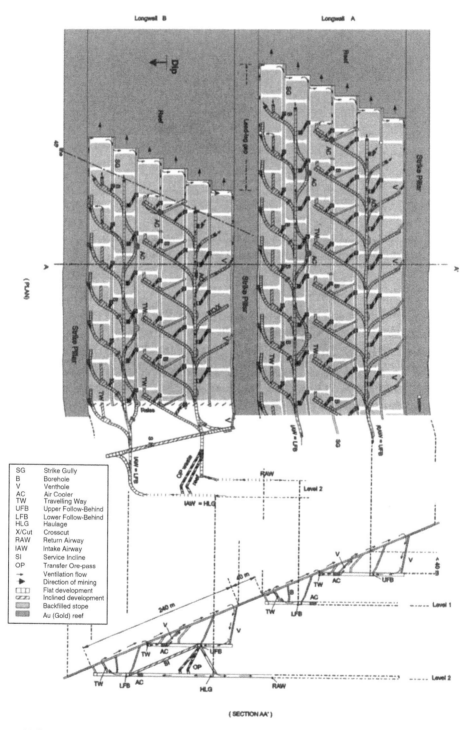

Figure 16.24(g)   Micro-concept of a longwall layout with strike stabilizing pillars. Longwall A is
depicted as mature longwall production unit. Longwall B is depicted with all footwall
infrastructure required for longwall establishement[53] (www.smenet.org).

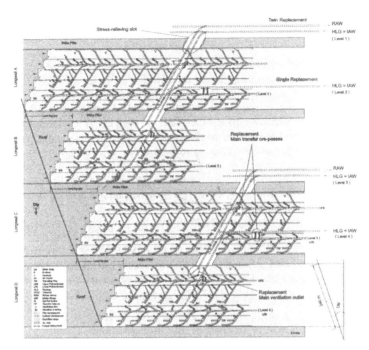

Figure 16.24(h)   Macro-concept of a longwall layout with strike stabilizing pillars. Three Longwall
units are depicted. Replacement haulages are shown, these being developed roughly
at every 750 m[53] (www.smenet.org).

A *Micro layout* consists of a plan and section of a raise connection and its
footwall infrastructures showing all the components such as timber bays, travelling
ways, box-holes, cooling loops and cubbies, vent holes etc. The design parameters
include: level spacing, back length, cross-cut spacing, pillar sizes, panel lengths, panel
shapes, mining directions, face orientation (overhand, underhand), and lead legs, as
shown in figure 16.24(g). A *Macro layout* expands the micro-mine design over four to
five mining levels, as shown in figure 16.24(h). For the project under consideration,
this enabled the designers to understand the practicability of the parameters that
were considered. This was done for a monthly target of 30,000 m² excavation
which is equivalent to production of 15,000 tons of ore/month. Table 16.2(a) details
the scope of each of these variants, for example, LSP for a depth between 3–5 km;
SGM – for depth between 4–5 km, SDD & CSDP – for a maximum depth up
to 4.6 km.

The study of the ultra-deep Iponeleng model orebody has identified some weakness and strengths. It suggested that longwall mining is ideal for a large mining block
with regular grade distribution, whereas other methods give flexibility in the presence
of geological disturbance such as dykes, faults and erratically grade-distributed orebodies.

16.4.2   Caving method – sublevel caving[1,5,9,10,18,35]

16.4.2.1   Introduction

In this mining method, similar to sublevel stoping, the steeply inclined strong orebody is divided into a number of sublevels (between two haulage levels) but mining of ore proceeds by blasting the fans drilled in the orebody, at the top most sublevel horizon and the blasted muck is removed immediately and dumped into the orepass meant for this purpose to deliver the ore at the main haulage level below. The void so created allows its h/w and cap rocks, which are weak, fractured and cavable (unlike sublevel stoping where wall and cap rocks are competent) to cave in. In a very wide orebody the stoping at this horizon proceeds transversely from its extreme h/w towards its extreme f/w contact all along the stope length (fig. 16.25(a)[9]), whereas, in wider to narrow orebodies (not less than 6 m thick) it proceeds from one end of the stope towards its other end longitudinally (fig. 16.25(b)). The same procedure is repeated at each of the sublevel horizons starting from the top most till it reaches to its bottom most. Since this stope is divided into a number of sublevels and caving of h/w is allowed simultaneously, hence the name 'sublevel caving'.

In order to ensure proper caving of h/w and cape rocks, and smooth flow of the blasted ore through the collecting crosscuts/drives; a proper study must be carried out to decide the size of various openings that are included within a stope layout. Prominent parameters amongst them are the dimensions of the crosscuts/drives, interval between sublevel horizons, spacing between the crosscuts, inclination of fans and burden between fans.

Thus, this method is characterized by blasting the fans drilled between two sublevels either longitudinally or transversely, and replacing the void created by the caved waste rock obtained by allowing the hanging wall and capping over it to cave in. Hence, a weak and cavable hanging wall, and acceptability of surface subsidence are the pre requisites for the success of this method.

Following are the *suitable conditions*,[5,18,35] in general, for the application of sublevel caving:

- Ore strength: medium hard to strong requiring fragmentation by drilling and blasting but less strong than the one suitable for the unsupported stoping methods.
- Rock strength: weak to moderate but fractured, jointed and cavable.
- Deposit shape: tabular or massive.
- Deposit dip: steep but can be applied to flat dips; in that case it will resemble to longwall mining with caving.
- Size and thickness: large extent along and across the dip, thickness >6 m.
- Ore grade: moderate but uniform as sorting is not possible.
- Depth: moderate, practiced up to a depth of 1.2 km.

Applications: Limited applications for mining the coal (anthracite). But widely applied to mine out the ores of metal and non-metal deposits such as iron, copper, lead, zinc, nickel and many others. One of the world's largest u/g mine – Kiruna Iron Ore Mine, LKAB, Kiruna, Sweden has adopted this method.

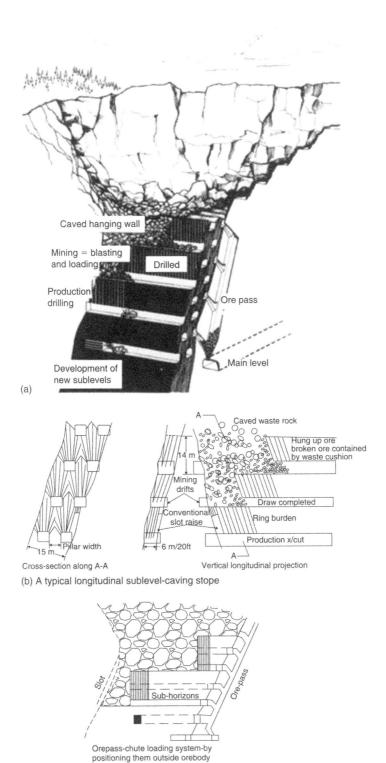

(a)

Caved hanging wall

Mining = blasting and loading

Drilled

Production drilling

Ore pass

Development of new sublevels

Main level

A — Caved waste rock

Hung up ore broken ore contained by waste cushion

14 m

Mining drifts

Draw completed

Conventional slot raise

Ring burden

6 m/20ft

Production x/cut

A —

15 m — Pillar width

Cross-section along A-A

Vertical longitudinal projection

(b) A typical longitudinal sublevel-caving stope

Slot

Sub-horizons

Ore-pass

Orepass-chute loading system-by positioning them outside orebody

(c) A transverse sublevel caving stope

Figure 16.25 (a) Pictorial view – sublevel caving. (b) Layout – longitudinal sublevel caving. (c) Transverse sublevel caving.

Main elements of the system/working parameters[5,35] – An optimum range of these parameters could be as under:

- Dividing deposit in levels spaced 60–80 m apart, all along the strike extension.
- Size of main level and stope entries = 3–5 m.
- Height of opening = 2.7–4 m (based on equipment height).
- Length of stope = 50–90 m (longitudinal), 50–60 m (transverse).
- Height of stope = level interval, range: 60–80 m.

Details of the sublevel entries:

- Distance between sublevels vertically = from 9–14 m to 20–32 m (fig. 16.26(d)).
- Spacing between crosscuts (if transverse stoping) = from 7.5–11 m to 23 m (center to center, fig. 16.26(d)).
- Dimensions of drill crosscuts (drives) – width 3 m–6 m × height 2.5 m–4 m.
- Fan inclination = vertical to 70–80° (figs 16.26(e) and (f)), Inclination of outer holes = 70–85° (fig. 16.26(d)).
- Ring burden = 1.2–1.8 m.

16.4.2.2 *Unit operations*

- Drilling: Development – Use of jacklegs, stopers and jumbo drills for drivage work. Stoping – drifter mounted single or twin boom fan drilling rigs.
- Blasting: use of NG based explosives, ANFO, slurries at the development headings. ANFO, slurries in the stopes.
- Charging with the use of pneumatic loaders to charge ANFO else conventional. Firing electrically.
- Mucking: Use of LHDs, FEL, rocker shovels and Cavos.
- Haulage: LHDs, trucks or shuttle cars at the sublevel horizons from the working faces to the ore passes. From the ore pass gravity flow and muck is discharged to the trucks or tracked haulage system using mine cars and locomotive.

In figures 16.25(a) to (b) most of these operations with the use of equipment of different types have been shown.

16.4.2.2.1 **Variants**
- Longitudinal sublevel caving (figs 16.25(b), 16.26(a)).
- Transverse sublevel caving (figs 16.25(a) and (c), 16.26(b)).
- Top slicing.

16.4.2.3 *Stope preparation (Transverse sublevel caving)*

In order to prepare a stope between two main (haulage) levels, first the access to the stope's location is made by driving drives and crosscuts from these main levels. A raise connects these levels; commonly known as a service raise or man-pass. To facilitate equipment movement within the stope, particularly in a highly mechanized mine, access by ramp to the stope becomes essential. The stope is then divided into number of sub-horizons, spaced at an interval of 9–14 m[5], when blastholes of 50–75 mm. dia. are used.

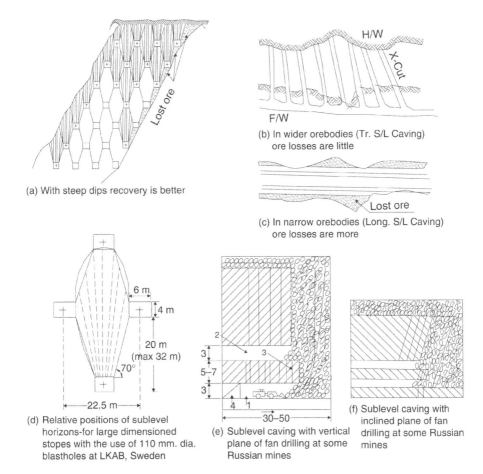

(a) With steep dips recovery is better

(b) In wider orebodies (Tr. S/L Caving) ore losses are little

(c) In narrow orebodies (Long. S/L Caving) ore losses are more

(d) Relative positions of sublevel horizons-for large dimensioned stopes with the use of 110 mm. dia. blastholes at LKAB, Sweden

(e) Sublevel caving with vertical plane of fan drilling at some Russian mines

(f) Sublevel caving with inclined plane of fan drilling at some Russian mines

Figure 16.26   Sublevel caving system – influence of dip and thickness. Relative position of sublevel horizons.

With the application of large diameter holes of 100–110 mm this spacing could be increased to 32 m, as per the prevalent practices at iron ore mines LKAB, Sweden (fig. 16.26(d)),[18] and some of the mines in Russia (figs 16.26(e) and (f))[1] and Canada.

The development work first of all commences at the top sublevel and it consists of driving the footwall drive throughout the length of the stope in waste rock at a location at least 10 m away from the footwall contact of the orebody. From this drive the drilling crosscuts, spaced at a predetermined interval are driven across the orebody up to its hanging wall contact. At hanging wall contact each of these crosscuts are connected and the resultant opening is termed as a hanging wall drive. At this hanging wall drive a slot raise is driven to ultimately convert into the slot by widening it by long parallel holes. Using this slot as a free face, the fans drilled (as described above) from the drill crosscuts are blasted. Similar development work is carried out at each of the sublevel's horizons. An ore pass is driven between two levels, and it is connected to each of these sublevel's horizons to facilitate the ore dumping into it.

16.4.2.4   Stope preparation (Sublevel caving – longitudinal)

The stope is developed on a similar pattern, as that of transverse sublevel caving but instead of drilling crosscuts across the deposit, the drill drives along the deposit are driven at each of the sublevel horizons, as shown in figure 16.25(b). Stope blasting commences from the slot created for the purpose at the wider most portion of the orebody within the stope.

16.4.2.5   Layouts

The layout shown in figure 16.25(a) is a prospective view of a transverse sublevel caving stope. In figure 16.25(b) longitudinal section along the strike direction of the deposit for a longitudinal sublevel caving stope has been shown. Details of the blasthole required to create a slot has also been shown.

This stoping method can be applied to any dip but overall recovery of the ore from a stope reduces as the deposit's dip decreases from vertical to the horizontal and flat dips (fig. 16.26(a)).

The influence of orebody thickness also plays important role so far as the overall recovery is concerned. Narrow orebodies using longitudinal sublevel caving results in lower recoveries comparing the same when wider orebodies are mined using the transverse sublevel caving (figs 16.26(b) and (c)). Sometimes ore is medium-hard which requires rock bolting during stoping operations.

An account of various unit operations that are undertaken at the different horizons of a sublevel caving stope for a steeply dipping deposit has been presented in figure 16.25(a). While mucking is in progress at the top most sublevel horizon of a stope, the fans are blasted at the immediate level below. The drilling operations are in progress at the next level below, and the drivage work including making a slot and putting in sublevel crosscuts, is in progress at the bottom most horizon.

The diagram shown in figure 16.26(d) presents the relative positions of sublevel horizons vertically. The horizontal distance between the adjacent crosscuts at the same horizon that need to be kept to allow a smooth flow of the blasted ore and the caved rocks have been also shown. This design is the development of several years of practice at some of the prominent mines using this method. Layouts shown in figures 16.26(d), (e) and (f) are latest designs with the application of large diameter holes, as described above.

The mechanics of the progress of gravity flow of ore and waste rocks within an extraction ellipsoid has been presented in figure 16.27.[18] As shown in figure 16.27, material mobility is a function of shape and eccentricity of the ellipsoids of extraction and loosening.

16.4.2.5.1   Advantages[1,10,18,35]

- Productivity: Fairly high, OMS in the range of 20–40 t/shift/man.
- Production rate – High. It can be considered as one of the bulk mining methods.
- Recovery – with pillar extraction 80–90%, but with dilution sometimes it (recovery) exceeds 100%.
- Commencement of regular production; at an early stage even during stope development 20–25% of the designed production rate can be achieved. Stope development and stoping activities go on simultaneously.

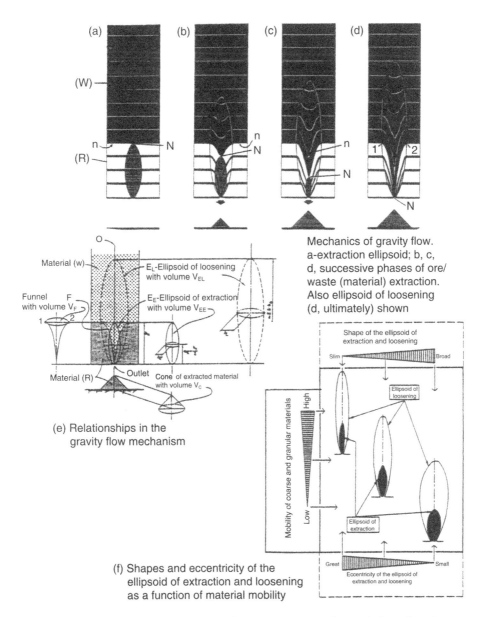

Mechanics of gravity flow.
a-extraction ellipsoid; b, c,
d, successive phases of ore/
waste (material) extraction.
Also ellipsoid of loosening
(d, ultimately) shown

(e) Relationships in the
gravity flow mechanism

(f) Shapes and eccentricity of the
ellipsoid of extraction and loosening
as a function of material mobility

Figure 16.27   Important parameters influencing operation of smooth flow of muck.

- Adaptability to mechanization: suitable for varying degrees of mechanization.
- Safety – little exposure to unsafe conditions, workers work under the protective roofs, easy to ventilate, thus better working conditions.

16.4.2.5.2   Limitations[1,10,18,35]
- Dilution is high, ranging from 10 to 35%, and that results in considerable ore losses in the stopes. This feature also limits its application to high valued ores.

- Subsidence and caving occur over wide areas causing land degradation.
- Cost is comparatively high (40–60% relative cost). Development cost and time are fairly high. Stope development is responsible for the high costs. Capital required – sufficient.
- Provision for the equipment access in the stope needs to be incorporated.

16.4.3   Caving method – block caving[1,4,9,10,13,24,36]

16.4.3.1   *Introduction*

This system is characterized by breaking the ore by the caving action initiated when support is withdrawn from a sizable area (which has been created first of all) under a column of ore (block), allowing it to cave in. At the same time a series of workings are cut along vertical plane boundaries of the block to weaken its bond to the solid. Gravity and rock pressure of the overburden make the undercut ore to cave in, thereby filling gradually, the undercutting space. Thus, unlike sublevel caving, not only the walls (h/w as well as f/w) and the capping rocks but also the ore itself must be weak, fractured and cavable. The caved ore is discharged through the ore passes, which are known by the structures such as: funnels, finger raises, bells, etc. connecting either to grizzly level for screening through it (in some layouts), or directly to troughs that are connected to the extraction drives/crosscuts. LHD mucking, or direct discharge into mine cars through chutes enables muck transfer from the stope for its onward disposal. Since the orebody is divided into large sized blocks, and the caving and mining of the whole block starts at one time, hence the name 'block caving'.

In order to ensure proper caving of walls and cap rocks, and smooth flow of the caved ore through the collecting crosscuts/drives; a proper study must be carried out to decide the size of various openings that are included within a stope layout. Prominent parameters amongst them are the dimensions of the funnels/finger raises/troughs, and spacing between them.

Following are the *suitable conditions,*[1,10,24,36] in general, for the application of block caving:

- Ore strength: weak, soft, friable, fractured and/or jointed. It should cave freely under its own weight when undercut. It should not be sticky if wet and not readily oxidized.
- Rock strength: weak to moderate but fractured, jointed and cavable. Almost similar characteristics as that of ore. Ore rock boundary should be distinct.
- Deposit shape: thick tabular or massive. Preferably regular.
- Deposit dip: steep but can be applied to flat dip if deposit is very thick.
- Size and thickness: large extent along and across the dip, thickness >30 m.
- Ore grade: can be low but uniform as sorting is not possible.
- Depth: moderate, practiced up to a depth of 1.2 km. If sufficient depth then strength of overburden could exceed that of rock strength (h/w rock), thereby surface subsidence can be reduced.

Applications: Widely applied to win the ores of metal and non-metal deposits such as iron, copper, lead, zinc, diamond (South Africa), molybdenum, nickel and many others. This is a bulk mining method that has been applied in the mines with high production.

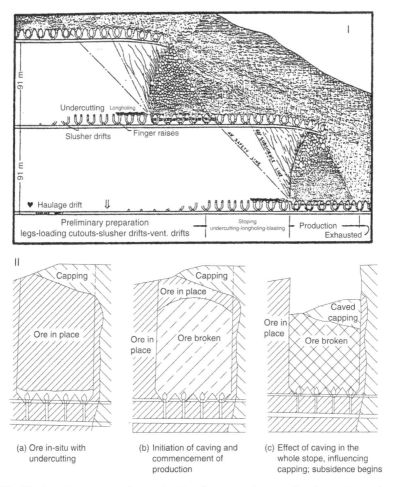

Figure 16.28   Block caving system. Some details of mass caving starting from the surface (Top). Progressive stages during caving shown (Bottom).

16.4.3.2   *Unit operations*

- Drilling: Development for undercutting – use of jacklegs and jumbo drills for drivage work. Stoping – drifter mounted single or twin boom fan drilling rigs to drill fans for undercutting or troughing. Also for drilling long holes in the form of fans or rings to induce caving in some cases.
- Blasting: use of NG based explosives, ANFO (if hole dia. larger than 40 mm), slurries at the development headings as well as during stoping.
- Charging with the use of pneumatic loaders to charge ANFO else conventional. Firing – electrically. Secondary breaking in stopes by plaster or pop shooting; bamboo blasting, use of dynamite bomb and/or application of impact hammer.
- Mucking (figs 16.29, 16.30): gravity flow through funnels/bells, finger raises equipped with chute. If trough-draw point system adopted then loading at the draw

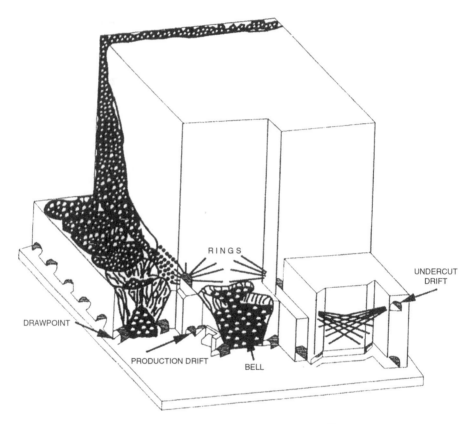

Figure 16.28   (d) Isometric view of panel caving operations[49] (www.smenet.org).

points with the use of LHDs, FEL or rocker shovels.[9] Use of scrapers made if mill hole system adopted.
- Haulage: at the main level use of LHDs, trucks or tracked haulage system using mine cars and locomotive.

16.4.3.2.1   Variants
Based on deposit's division:[1,10,24,36]
- Block caving: Regular square or rectangular areas (30 m × 30 m to 60 m × 60 m) are undercut and usually these are mined in alternating or diagonal order. The height of block is in the range of 40–100 m, seldom more. Ores that are weak, fractured and break fine, usually fall into this category. In order to minimize dilution and provide space for service and ventilation openings, blocks are separated by pillars (both sides). It is being applied in about 50% cases.
- Panel caving: When ore is unstable and caves readily, panels of 20 m to 60 m wide and 150–300 m long, are arranged along or across the strike of the orebody, and they are mined retreating. Caving in several panels can be performed simultaneously with a definite lead. In order to minimize dilution and provide a space for the

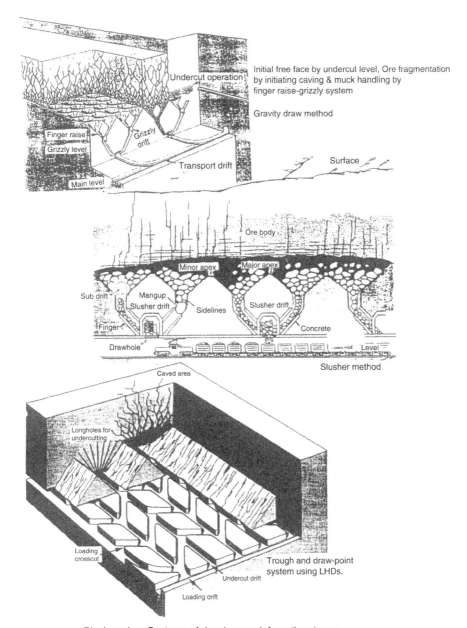

Initial free face by undercut level, Ore fragmentation by initiating caving & muck handling by finger raise-grizzly system

Gravity draw method

Undercut operation

Finger raise
Grizzly level
Grizzly drift
Transport drift
Main level
Surface

Ore body

Minor apex   Major apex

Sub drift   Mangup
Slusher drift   Sidelines   Slusher drift
Finger   Concrete
Drawhole   Level
Slusher method

Caved area

Longholes for undercutting

Loading crosscut
Undercut drift
Loading drift
Trough and draw-point system using LHDs.

Block caving: Systems of drawing muck from the stopes.

Figure 16.29   Block caving with grizzly and finger raise system.

service and ventilation openings; panels are separated by pillars (3–10 m wide). Adaptability of this variant is in 25% cases.

• Mass Caving: In this scheme no division of the area is made as in block or panel caving variants. Also no pillar is left. Irregularly sized prisms are mined as large as

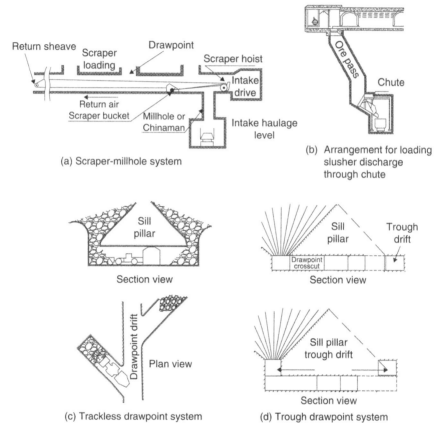

Figure 16.30 Different schemes for drawing muck from block caving stopes. Some of these draw systems are applicable for Sublevel and VCR stopes also.

consistent with caving properties of the ore and stress on the opening below. Undercutting is initiated on the retreat pattern. Adaptability of this type of caving is 25%.

Based on the readiness to caving:

- Self caving[36]
- Induced caving.[1]

16.4.3.3 Methods of draw[13,24]

Gravity draw – this system requires a finely fragmented and easy flowing ore/product. It requires maximum amount of development openings and uses the minimum amount of mechanized equipment comparing to all other drawing schemes. It is unskilled, labor intensive and highly sensitive to fragmentation size. Percentage adaptability is around 30%. Some of the mines adopting this system with their production rates are:[24] San Manuel, Arizona, U.S.A (60000 t/d); El Teniente mine, Chile (60000 t/d); Rio Blanco mine, Chile (12000 t/d); Lutopan mine, Philippines (24000 t/d). This system has been illustrated in figure 16.29.[9] This figure shows the relative position of the various structures that are required to be developed prior to initiate the caving i.e. the stoping operations in a stope.

Slusher draw – this system can handle coarser fragments. It requires arrangement for electric power in the stoping areas. It requires more elaborate arrangement for the ventilation than is required in the gravity system but less than the LHD system. Percentage adaptability is up to 45%. Some of the mines adopting this system with their production rates are:[24] Climax mine, Colorado, U.S.A (31000 t/d); El Salvador mine, Chile (18000 t/d); Santo Thomas II mine, Philippines (24000 t/d); Ertsberg Eastmine Indonesia (10000 t/d). This system has been illustrated in figure 16.30. The ore discharged through the finger raises to the slusher drifts are ultimately loaded into the mine cars through the draw holes. Thus, this allows a continuous feed to the train, which can be hauled by the locomotives.

LHD/loader draw – this method is least sensitive to fragmentation size. Concept is new and can be used for small, irregular and large massive deposits. The development openings required are almost equal to the gravity system. Sufficient capital and mechanization is needed. Elaborate ventilation arrangements are necessary. Percentage adaptability is 25% but growing. Some of the mines adopting this system with their production rates are (after Pillar 1981): Henderson mine, Colorado, U.S.A (30000 t/d); El Teniente mine, Chile (40000 t/d). This system has been illustrated in figure 16.29.[9] The system allows ore fragments of large size unlike the other two system described above. Large bucket sized LHDs run by electric or diesel motive powers can be used to suit production and working conditions' requirements.

16.4.3.4   Stope preparation

First the block to be mined is connected to the main mine entries such as shafts, main levels, declines etc. in the usual manner by driving a network of mine roadways in the form of drives, crosscuts and raises to facilitate haulage, ventilation and other mine services.

Sublevel or sub-horizons are required to be developed between the undercut and the haulage level. This includes driving of finger raises, grizzly level and transfer raises in case of grizzly system of draw. In case of slusher system of draw, the development work between undercut and haulage level involves driving of draw or finger raises, scram/ slusher drifts and transfer raises/trenches. Similarly for LHD draw system a network of troughs and draw-points need to be developed.

Amongst these development activities, the most critical activity is undercutting. This needs careful removal of pillars or supports installed, so that caving can be initiated without danger of air blast, premature collapse etc.

The interval required between different development entries (sublevels, ore passes and undercut) is a function of the draw system adopted. A gravity system requires maximum development work whereas the mechanized loading by LHD etc. is the least.

Boundary weakening, other than undercutting, is rarely performed in block caving. Occasionally corner raises are driven on one side of undercut block, and slab (widened) to create a narrow slot.

Table 16.2(b) Parameters to be considered before implementing cave mining[49] (www. smenet.org).

CAVABILITY	PRIMARY FRAGMENTATION	DRAWPOINT/DRAWZONE SPACING
Rock mass strength (RMR/MRMR)	Rock mass strength (RMR/MRMR)	Particle size of ore and overlying rock
Rock mass structure-condition geometry	Geological structures	Overburden load and direction
In situ stress	Joint and fracture spacing, and geometry	Friction angles of caved particles
Induced stress	Joint condition ratings	Practical excavation size
Hydraulic radius of ore body	Stress or subsidence caving	Stability of host rock mass (MRMR)
Water	Induced stress	Induced stress
DRAW HEIGHTS	**LAYOUT**	**ROCK BURST POTENTIAL**
Capital	Particle size	Regional and induced stresses
Ore body geometry	Drawpoint spacing and size	Variations in rock mass strength, modulus
Excavation stability	Method of draw—gravity or LHD	Structures
Effect on ore minerals	Orientation of structures and joints	Mining sequence
Method of draw	Ventilation, ore handling, drainage	
SEQUENCE	**UNDERCUTTING SEQUENCE (pre/advance/post)**	**INDUCED CAVE STRESSES**
Cavability: poor to good or vice versa	Regional stresses	Regional stresses
Ore body geometry	Rock mass strength	Area of undercut
Induced stresses	Rock burst potential	Shape of undercut
Geological environment	Rate of advance	Rate of undercutting
Rock burst potential	Ore requirements	Rate of draw
Production requirements	Completeness of undercut	
Influence on adjacent operations	Shape (lead, lag)	
Water inflow	Height of undercut	
DRILLING AND BLASTING	**DEVELOPMENT**	**EXCAVATION STABILITY**
Rock mass strength	Layout	Rock mass strength (RMR/MRMR)
Rock mass stability (drillhole closure)	Sequence	Orientation of structures and joints
Required particle size	Production	Regional and induced stresses
Hole diameter, lengths, rigs	Drilling and blasting	Rock burst potential
Patterns and directions		Excavation size (orientation and shape)
Powder factor		Draw point
Swell relief		Mining sequence
SUPPORT	**PRACTICAL EXCAVATION SIZE**	**METHOD OF DRAW**
Excavation stability	Excavation stability	Fragmentation
Rock burst potential	Induced stress	Practical drawpoint spacing
Brow stability	Caving stresses	Practical size of excavation
Timing of support: initial, secondary, and production	Secondary blasting	Gravity or mechanical loading
	Equipment size	
RATE OF DRAW	**DRAWPOINT INTERACTION**	**DRAW COLUMN STRESSES**
Fragmentation	Drawzone spacing	Draw-column height
Method of draw	Critical distance across major apex	Particle size
Percentage hangups	Particle size	Homogeneity of ore particle size
Secondary breaking/blasting	Time frame of working drawpoints	Draw control
Seismic events		Draw-height interaction
Air blasts-drawpoint cover		Height-to-short axis base ratio
		Direction of draw
SECONDARY FRAGMENTATION	**SECONDARY BLASTING/BREAKING**	**DILUTION**
Rock, block shape	Secondary fragmentation	Ore body geometry
Draw height	Draw method	Mining geometry
Draw rate, time-dependent failure	Drawpoint size	Particle size distribution
Rock block workability and strength	Gravity grizzly aperture	Range of particles, unpay ore and waste
Range in particle size, fines cushioning	Size of equipment and grizzly spacing	Grade distribution of pay and unpay ore
Draw control program	Ore handling system, size restrictions	Mineral distribution in ore
		Drawpoint interaction
		Secondary breaking
		Draw control (techniques, predictions)
		Draw markers
TONNAGE DRAWN	**SUPPORT REPAIR**	**ORE/GRADE EXTRACTION**
Level interval	Tonnage drawn	Mineral distribution
Shut-off grade	Point and column loading	Method of draw
Drawpoint spacing	Brow wear	Rate of draw
Dilution percentage	Floor repair	Dilution percentage
Controls	Secondary blasting	Cut-off grade to lant
Redistribution		Ore losses
	SUBSIDENCE	
RMR/MRMR	Minimum and maximum spans	Depth of mining
Height of caved column	Major geological structures	Topography

16.4.3.5 Layouts

In figure 16.28 (top) a mine having a mass caving system has been illustrated. In block caving; to start with as shown in figure 16.28 (bottom) (a), the ore in-situ is under-cut. This initiates caving and production commences (fig. 16.28(b)). The effect of caving influencing the whole stope including capping can be seen in fig. 16.28(c); subsidence begins.

Parameters affecting caving operations[49]

The 25 parameters, cavability through to subsidence, as tabulated in table 16.2(b), should be considered before implementing cave mining operations. The parameters in

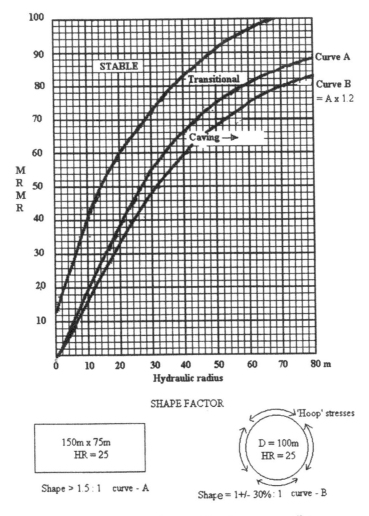

Figure 16.30   (e): Stability diagram based on world-wide experience.[49] (www.smenet.org)

capital letters are a function of the parameters that follow in the same box. Many parameters amongst them are uniquely defined by the ore body and the mining system, but the rest of the parameters need to be addressed, if any form of cave mining is contemplated.

Figure 16.30(e) based on worldwide experience illustrates caving and stable situations in terms of hydraulic radius (area divided by parameter) for a range of MRMR values. An additional curve has been added to account for the stability that occurs with equi-dimensional shapes. All rocks could cave but the manner of their caving and fragmentation needs to be predicted if cave mining is to be implemented successfully. The rate of caving can be slowed by control of the draw since the cave can be propagated only if there is space into which the rock can move. The rate of caving can be

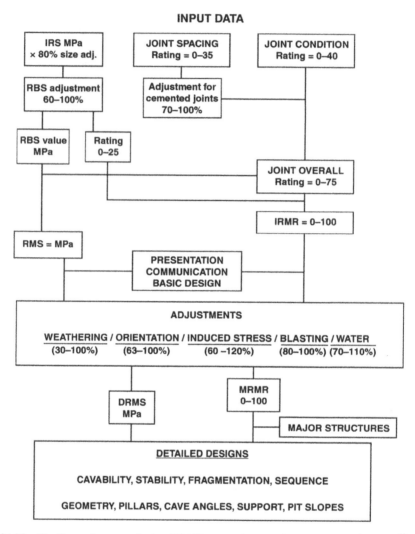

Figure 16.30   (f): Flow diagram of the MRMR procedure with recent modifications[49] (www. smenet.org).

increased by more rapid advance of undercut, but problems can arise if an air gap forms over a large area. In this situation, the intersection of major structures, heavy blasting, and influx of water can result in a damaging air blast. Rapid uncontrolled caving can result in an early influx of waste.

The Laubscher rock mass classification system provides both in situ rock mass rating (IRMR) and rock mass strength. Such a classification is necessary for design purposes. IRMR defines the geological environment, and the adjusted (modified) or mining rock mass rating (MRMR) considers the effects of mining operations on rock mass. Figure 16.30(f) is the flow sheet of the MRMR procedure with recent modifications. It takes into account cavability, subsidence angles, failure zones, fragmentation,

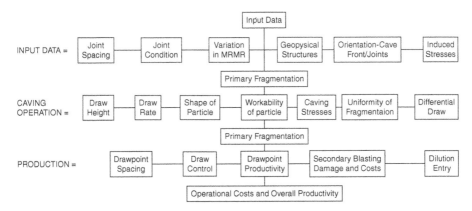

Figure 16.30   (g): Input data for calculation of fragmentation[49] (www.smenet.org).

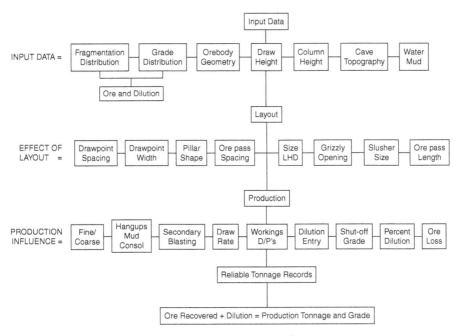

Figure 16.30   (h): Draw control requirements[49] (www.smenet.org).

cave front orientation, undercutting sequence, overall mining sequence and support design.

In conventional layouts, the rate of undercutting (RU) should be controlled so that rate of caving (RC) is faster than the rate of damage (RD) due to abutment stresses. Thus, RC > RU > RD. however, in the areas of high stress, the rate of caving must be controlled to maintain an acceptable level/amount of seismic activity in the cave back. Otherwise rock burst can occur in suitably stressed areas such pillars and rock contacts. As advance undercutting will be used in these situations, damage to the undercut and production level will not be a problem.

In caving operations, the degree of fragmentation has a bearing/impact on:

- draw-point spacing;
- dilution into draw column;
- draw control;
- draw productivity;
- secondary blasting and breaking costs;
- secondary blasting damage.

Figure 16.30(g) outlines the input data needed to calculate the primary fragmentation and the factors that determine secondary fragmentation as a function of caving operations, as shown line diagram 16.30(h).

Higher dilution has adverse impacts on cash-flow, as shown in table 16.2(c); as such it should be minimized. The draw control measures should be taken in order to reduce dilution, improve recovery and minimize damage to the pillars. Table 16.2(d) lists steps that should be taken in this regard.

Automated caving (using hydraulic fracturing) at Northparkes, Australia[4]: E26 mine belonging to Rio Tinto was Australia's first block cave mine. Construction of the first block known as lift 1 was commenced in 1993 and completed in 1997. It is known to be highly productive and low cost operation, for example, in 1999–2000; E26 produced 50,340t of copper-gold ore per underground employee, including contractors. E26 deposits has been divided into two sections (figs 16.31(a) and (b)),4 lift 1 extends to 480m below surface, and lift 2 consist of lower 350m of the deposit.

Table 16.2(c)   Effect of dilution on cash flow[49] (www.smenet.org).

Percent dilution	5%	10%	15%	20%	25%	30%	35%
Dry tons processed	360,000	360,000	360,000	360,000	360,000	360,000	360,000
Concentrator head grade	0.333	0.318	0.304	0.292	0.280	0.269	0.259
Concentrator recovery	95.5%	95.3%	95.0%	94.8%	94.6%	94.4%	94.2%
Gold sales—ounces	114,485	109,099	103,968	99,654	95,357	91,417	87,822
Revenue-$350/oz (000)	$40,069	$38,185	$36,389	$34,879	$33,375	$31,996	$30,741
Operating Costs—(000)	$30,125	$30,125	$30,125	$30,125	$30,125	$30,125	$30,125
Cash Flow—(000)	$9,944	$8,060	$6,264	$4,754	$3,250	$1,871	$616
Percent decrease in cash flow	0%	19.0%	37.0%	52.2%	67.3%	81.2%	93.3%

Table 16.2(d)   Draw control[49] (www.smenet.org).

Objective	Measure that should be taken
Reduce dilution	Calculation of tonnage
Improve ore recovery	Recording of tonnage produced
Avoid damage to pillars	Control draw

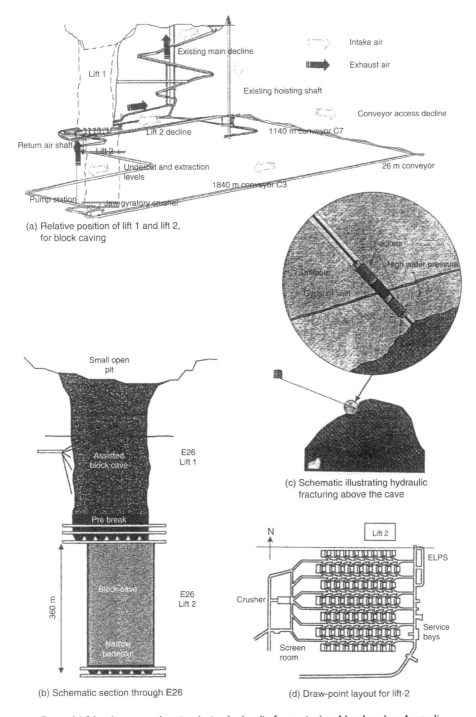

Figure 16.31 Automated caving (using hydraulic fracturing) at Northparkes, Australia.

The undercut of lift-1 dimensions measure 196 m long by 180 m wide. Continuous caving was never achieved so cave inducement was required to main caving and sustain production. Use of hydraulic fracturing to induce caving was tested in the existing boreholes and then it was applied. It yielded some 7 Mt of ore at significantly lower cost than conventional cave inducement techniques. An inflatable straddle packer system and diesel powered triple pump were used to induce hydraulic fracturing (fig. 16.31(c)). The straddle packers were connected to AQ drill rods and lowered down a selected borehole using a diamond drill. Once in position, the packers were inflated with water, usually around 5 MPa above the anticipated injection pressure. The triplex pump then pumped water under high pressure along an injection line and into the straddle section between the packers. Pressurization of rock between the straddle sections induces tensile stress along the walls of the hole and eventually fractures the rock, or opens existing fractures. Further injection forces water into these fractures causing them to extend into surrounding rock mass. Most hydraulic fracture treatments were characterized by increase seismicity both during and after injection. In several cases this increase in seismicity was followed by significant caving events.

Laubscher Mining Rock Mass Rating (MRMR) was chosen to identify cavability and it was found that with MRMR from 33 to 50 (lift 1) was suitable for caving. During planning Gemcom's PC-BC, a programme design and evaluation of block cave, was extensively used.

Draw points system as shown in figure 16.31(d) for use of Toro 450E LHD were used during lift 1 as well for lift 2 (with modification of making brows stronger by reinforcement).

16.4.3.5.1 Advantages[1,10,24,36]

- Productivity: fairly high, OMS in the range of 15–40 t/shift/man; maximum in the range of 40–50 t/shift/man.
- Production rate – high; it can be considered as one of the bulk mining methods.
- Recovery – with pillar extraction >90%, but with dilution sometimes it exceeds 100%.
- Commencement of regular production: On completion of undercutting and haulage layout production can be commenced. Drilling and blasting during stoping are completely eliminated.
- Cost – Comparatively lowest (20% relative cost). Cost is comparable with surface mining methods. Tonnage yield/m of development is the highest.
- Adaptability to mechanization: suitable for high degree of mechanization for mucking and transportation.
- Safety – good safety records.

16.4.3.5.2 Limitations[1,10,24,36]

- Dilution is high: it ranges 10–20% and needs control. This feature limits its application to high valued ores.
- Subsidence and caving occur over wide areas causing land degradation.
- Draw control is critical factor for the success of this method.
- Stope development is comparatively slow, tedious and costly.

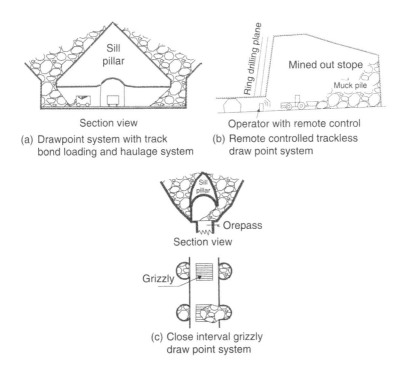

Figure 16.32 Summary of schemes that are applicable to draw muck from stopes. Selection of a particular system depends upon the stope layout and mucking unit deployed.

16.5 COMMON ASPECTS

In figure 16.33(a)[10], a comparison of technical aspects of the prevalent stoping methods has been made. Based on this it could be summarized that the block caving method could be considered as a bulk mining method and it is in operation at different parts of world to yield high outputs. Next in line are the methods such as: Sublevel Caving, Big Blasthloe Stoping (sublevel stoping), powered supported Longwall Mining and Mechanized Cut and Fill stoping. However, supported stoping methods are the costliest (fig. 16.34(a))[10]. Caving methods are prone to higher dilution and subsidence of the overlying strata that ultimately results into the surface ground degradation. Open stoping methods require more development, drilling and blasting which ultimately results in higher production costs. Ore fragmentation is the main task in these stopes. Use of timber as support is getting obsolete day by day due to its non-availability, costs and fire hazards. Today paste fill is often a default selection when planning new projects. It has been installed at old mines to to replace or supplement an existing backfill system. With the application of trackless units changes have been brought about in designing Room and Pillar system. It could be applied for a dips up to 60°, and even in wider orebodies and coal deposits; but the method suffers on account of poor recovery when pillars are not fully recovered. For mining thin and extensive coal seams longwall mining with self-advancing support and cutter-loader

UNDERGROUND MINING METHODS

Stoping method	ROOM & PILLAR	STOPE & PILLAR	SHRINKAGE STOPING	SUBLEVEL STOPING	CUT & FILL STOPING	STULL STOPING	SQUARE SET STOPING	LONGWALL MINING	SUBLEVEL CAVING	BLOCK CAVING
Relative cost % (w.r.t Sq. set stoping)	30	30	50	40	60	70	100	20	50	20
Productivity in tons./man-shift	30–80	30–50	5–10	15–30	10–20	n.a.	1–3	75–180	20–40	15–40
Recovery %	Av. 75	75	75–85	75	90–100	up to 90	up to 100	70–90	90–125	90–125
Dilution %	10–20	10–20	<10	up to 20	5–10	5–10	low	10–20	10–35	10–20
Powder factor kg/ton.	n.a.	n.a.	n.a.	0.15–0.30	0.25–0.60	n.a.	n.a.	n.a.	0.3–0.4	0.05–0.10
Development required	little	little	moderate	more, slow & expensive	little to moderate	low	low	low	highest	slow, extensive & costly
Capital required	large	moderate	low	moderate to high	moderate to high	low	low	moderate to high	moderate to high	moderate to high
Flexibility	flexible	flexible	fair	inflexible	flexible	flexible	flexible	non flexible	inflexible & rigid caving	high inflexible & rigid
Selectivity	selective	selective	fair	not selective	selective	selective	selective	not selective	rigid caving occurs	caving occurs
Ground control	needs monitoring	same as R & P	good	good	excellent	good	effective for bad strata	subsidence occurs	good safety & health	hang-ups are
Safety	poor ventilation	fair	rough foot hold under exposed back	little exposure to unsafe conditions	good safety records	Needs more support	very poor safety, fire hazards	good safety gob heating can cause fire	health	unsafe, delayed drawing can cause self-heating
Applicability & special features	Popular coal mining method	limited to metal mining	small scale mining method, ore tie up to 60%	very popular modern method	Method of all times & for varying conditions	Not popular	Costliest method in worst ground conditions	Popular coal mining method continuous mining possible	Early production but high development	bulk mining method effective draw control critical for success

Figure 16.33(a)   A comparison of stoping methods – based on operational parameters.

Figure 16.33(b) Selection of rock-drills and accessories for various stoping methods.

Mining method	Breast, large-scale (stope and pillar mining)		Overhand, small-scale (shrinkage stoping)		Cut-and-fill stoping		Overhand (sublevel stoping)		Overhand, large-scale (sublevel stoping)		Caving (sublevel caving)
Drilling and blasting technique	Drifting and slashing / Frontal benching	Vertical or downward benching	Overhand stoping	Frontal stoping, breasting	Roof drilling or overhand stoping	Frontal stoping-breasting	Ring drilling		Parallel drilling		Fan drilling
Applicable drilling equipment	Mechanized drifting jumbo	Mechanized airtrac	Hand-held stoper drill	Hand-held airleg drill	Mechanized light wagon drill	Mechanized drifting jumbo	Bar & arm rigging	Mechanized ring drill ring	Mechanized airtrac with downhole hammer and high pressure		Mechanized fan drill
Drilling data											
Hole diameter, in (mm)	1.5–2.0 (38–48)	2.5–3.0 (64–76)	1.1–1.3 (29–33)	1.1–1.3 (29–33)	(33–38)	(38–48)	(48–51)	(48–57)	4.0–4.5 (105–115)	6.0–6.5 (152–165)	1.9–2.0 (48–51)
Hole depth, ft (m)	10–18 (3.0–5.5)	as required	6.5–8.2 (2.0–2.5)	6.5–11.5 (2.0–3.5)	(3.0–4.0)	(3.0–4.0)	(10–20)	(10–25)	160–200 (50–60)	160–200 (50–60)	40–50 (12–15)
Drilling equipment performance											
With pneumatic rock drill ft/hr (m/hr)	200–250 (60–75)	50–80 (15–25)	25–40 (8–12)	30–50 (10–15)	(20–40)	(60–70)	(50–60)	(100–120)	160 (50)	160 (50)	650–800 (200–240)
With hydraulic rock drill ft/hr (m/hr)	300–360 (90–110)	(25–35)	na	na	-	(90–100)	-	(120–180)	na	na	800–1000 (240–300)
Drilling-blasting factor yd³ (m³) rock broken per drilled ft (m)	0.6–0.8 (1.5–2.0)	1.2–1.6 (3.0–4.0)	0.3–0.4 (0.7–0.9)	0.3–0.4 (0.7–0.9)	(0.9–1.2)	(1.0–1.4)	(1.5–2.5)	(1.5–2.5)	3–4 (8–10)	6–7 (14–18)	0.7–0.9 (1.8–2.3)

network is also at forefront in some of the countries. In table 16.8[38], the world's prominent mines having large outputs have been listed.

In figure 16.33(b) application of different types of drills together with drilling accessories as applicable to different stoping methods, have been illustrated. This is a useful guide[9] to select a drill for a particular job.

In tables 15.1, and 16.7, a summary of equipment used during different operations: mine development, stope preparation and stoping and production have been summarized. These tables also specify the services required.

Stope Reconciliation: In the recent years with application of heavy blasting it has become possible to design bulk mining methods as described in the preceding sections but recovery from such heavy blasts have been debatable. In order to know the deviation of ore recovered from the designed one, use of Optech (CMC) Survey could be made to obtain a 3D as build geometry of any stope. Figure 16.33(c) illustrates use of this technique that is being done at the Mount Isa Mines, Australia. The results are encouraging and helped in modifying designs to enhance recovery from the stopes.

Note: Apart from the guidelines given in the text above with each stoping method; for selection of equipment and services please refer section 15.10.

16.5.1   Stope design[33a,33b]

Stope is an ore block of proper size prepared out of a deposit to be mined, for the purpose of final exploitation of the orebody enclosed within it. Before stoping operations can be started, the stope needs to be developed or prepared. This task is often termed as stope preparation and the process of winning ore from a stope so prepared, by way of ore fragmentation and its handling, is known as stoping.

Stope dimensions differ from method to method and even for the same stoping method from mine to mine. While deciding stope dimensions the strength of orebody and its enclosing walls, rate of stoping, degree of mechanization, depth, size and shape of the orebody play important role. In absence of adequate geo-mechanical studies to know the ground stability pattern of a stope during its stoping operations, experience oftens play an important role to decide its size. And, therefore, usually the dimensions known to be safe are adopted.

Level interval usually decides the stope height, its width remains to be that of the orebody width, and the length is decided considering the ground stability parameters i.e. for how much span the orebody could be excavated, and also for how long duration it can be kept in stable (without its failure)?

Based on these considerations while designing the stope following parameters are laid out:

1. Model parameters
2. Design parameters.

16.5.1.1   *Model parameters*

This includes the parameters such as orebody profile in plan (at various horizons) and sections (along various latitudes and departures) which, in turn, depicts the orebody's thickness and dip. It means the stope is to be designed for the orebody presented in this manner.

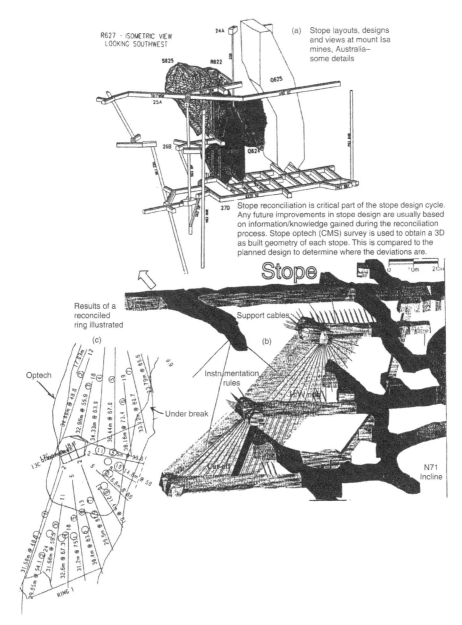

Figure 16.33(c)   Mount Isa Mines – Australia's largest mine to produce Cu, Zn, Ag. Use of Stope Optech (CMS) survey to obtain a 3D built geometry of each stope shown.

16.5.1.2   *Design parameters*

In order to mine out a stope, the following provisions need to be made:

a. Access to the stope for workers, machines, equipment and mine services (Service Raise or Manpass).

(a) Influence of supports on drivage costs. Influence
    of Mining System & Drives' size on costs also
    illustrated

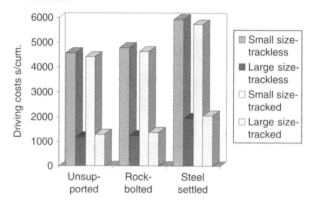

(b) Relative cost % of underground mining methods

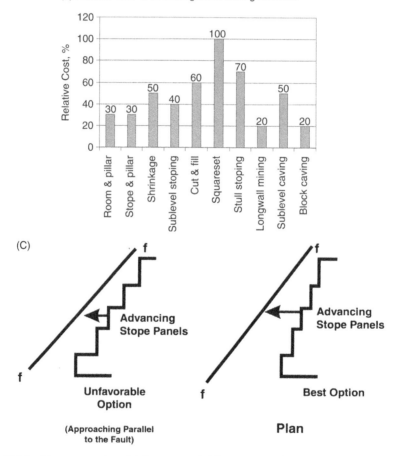

Figure 16.34   Cost comparison (a) Supports of different types. (b) Stoping methods. (c) Preferred
               approach to stoping through the geologic feature. Mining should take place at as large
               an acute angle as possible[43] (www.smenet.org).

b. Creation of free face for the stoping operations to start with (Slot).

c. Rock fragmentation by way of drilling and blasting the blast holes (Drill drives/ sublevels or crosscuts).

d. Handling fragmented ore muck (Extraction level layout).

e. Limiting and isolation of stope size or dimensions (Pillars of various kinds).

f. Approaching geological features.

a. *Access for workers, machines, back filling material, mine services such as water, compressed air, ventilation etc.*: This is achieved by putting a raise connecting two levels, commonly known as service raise. Services can be put at each of the extremities of a stope else it can be common between two or more number of stopes. It can be positioned within the orebody or away from it (may be in foot wall waste rock). The exact number and positioning is governed by the type of stoping method.

b. *Creation of free face for the stoping operations to start with*: This is achieved by putting a raise between two levels (some times terminating the raise just below the crown pillar) at the location where from stoping operations are to start with, when the ore winning pattern within the stope is along or across the longitudinal direction (i.e. along strike). In the former case the raise is widened up to the width of orebody and in the latter it is enlarged up to the stope length. Each of these excavations, so created, is known as 'slot'.

But when the ore winning pattern in the stope is from the lower horizon towards the upper horizon (in the form of horizontal slices) i.e. in the ascending order, the free face is created by driving an 'undercut level' and widening it up to the full width of the orebody, at the location where from the stoping operation is to start with.

c. *Rock fragmentation by way of drilling and blasting the blast holes*: In order to fragment the orebody within the stope, it needs to be drilled and blasted except in the block caving system. For carrying out these operations 'Drill Drives/Sub-levels' in case of longitudinal ore winning pattern or 'Drilling crosscuts' in case of transverse ore winning pattern need to be driven. Positioning of these drives and crosscuts with respect to orebody walls and horizons is mainly governed by the orebody thickness and capability of drills deployed for drilling purposes.

In the filled stopes this operation is carried out over the filled material by fragmenting the ore in slices using the different types of drills.

In block caving too some time to induce caving drilling and blasting is required to be carried out from the specially prepared cut raises and drives.

d. *Handling ore fragmented muck*: There are different practices (figs 16.29, 16.30 and 16.32)[41] that are being followed to achieve this task. The following are the prominent amongst them.

1. Trough – Draw point system (fig. 16.29, 16.30(c))
2. Funnel – Draw point system (fig. 16.9)
3. Grizzly – Finger raise system (fig. 16.29)
4. Ore pass system (fig. 16.30(b))
5. Scraper – millhole system (fig. 16.30(a)).

The dimension of these openings, covered under items (1) to (4), depends upon the type of equipment to be deployed during stope preparation and the stoping operations, whereas their location with respect to the orebody enclosed within the stope is governed by the orebody geometry (mainly thickness), stope dimensions and the capability of equipment used particularly for rock fragmentation purposes.

The shape of these openings is mainly governed by the working depth. At the great depth arched, circular or elliptical shaped openings need to be driven from the strata control point of view.

e. *Limiting and isolation of stope size or dimensions*: As mentioned in the preceding sections that the level interval decides the stope height but to support the upper level a pillar is required to be left. This pillar is known as crown pillar.

To limit the stope's size length wise, the pillar left in between the two consecutive stopes, is termed as rib pillar.

To hold the blasted muck, structures such as: funnels, finger raises, troughs etc. are required to be made at the tramming or extraction level. The rock mass involving these structures is often termed as sill pillar.

Taking into consideration the working depth, ground behavior and stoping rate, the size of these pillars is decided. In absence of proper rock mechanics studies usually based on the experience, the pillars' dimensions are decided.

f. *Approaching geological features*: The preferred approach to stoping through geologic features is shown in figure 16.34(c). Mining should take place at as large an acute angle as possible.

The various kinds of pillars (sec. 16.6.4) so left are usually recovered after the completing the stoping operations between two consecutive levels.

16.5.2   Application of computers in stope design and economic analysis

This can be demonstrated by describing a methodology together with a case study as described below.

16.5.3   Proposed methodology for selection of a stoping method for the base metal deposits with a case study

Methodology: The basic objectives in selecting a method to mine a particular deposit is to design an ore extraction system that will meet certain technical and economic criteria. This can be interpreted as aiming for maximum extraction with safe working conditions at low cost and maximum productivity. The methodology developed aims to examine some of these aspects. The proposed methodology was tested for copper deposits of a mining complex suitable to be won by underground stoping methods such as sublevel stoping, cut and fill stoping and their variants. In general, it provides a practical approach to both method selection and optimum stope design for base metal deposits. The methodology presented in this section undergoes the following features:

- *Estimating mineral inventory from the basic geological information.*
- *Formulating a basis for cutoff grade decisions to predict mineable ore reserves.*
- *Evaluating stope boundaries for different stoping methods.*

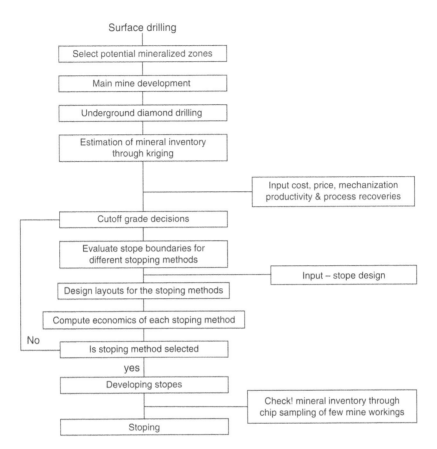

Surface drilling

Select potential mineralized zones

Main mine development

Underground diamond drilling

Estimation of mineral inventory
through kriging

Input cost, price, mechanization
productivity & process recoveries

Cutoff grade decisions

Evaluate stope boundaries for
different stopping methods

Input – stope design

Design layouts for the stoping methods

Compute economics of each stoping method

No

Is stoping method selected

yes

Developing stopes

Check! mineral inventory through
chip sampling of few mine workings

Stoping

Figure 16.35(a)   Flow diagram for the methodology to select stoping method for a base metal deposit.

- *Designing stopes and their economic evaluation for the different mining methods through the algorithms developed.*
- *Selection of stoping method.*

Use of computer i.e. development of algorithms is necessary to implement this methodology as illustrated in the following paragraphs. A flow diagram, showing the logical steps that should be followed for the proposed methodology is outlined in figure 16.35(a). In most cases, for the statistical and geostatistical analyses, available software were used to undertake above outlined steps; but for rest of the work algorithms were written by the author using Fortran77.

Estimating mineral inventory from the basic geological information
Based on the logic outlined above, any base metal deposit can be evaluated using kriging programs, which at present are available throughout the mineral industry. In order to test the algorithms developed, a small section of the deposit measuring 400 m along strike and 60 m vertically (between two main haulage levels of the mine) was chosen. Geological studies indicated mineralization in the strike direction to be

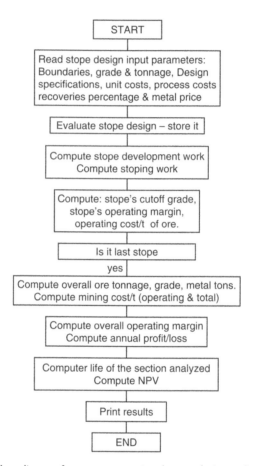

```
┌─────────────────┐
│      START      │
└─────────────────┘
         │
┌──────────────────────────────────────┐
│ Read stope design input parameters:   │
│ Boundaries, grade & tonnage, Design    │
│ specifications, unit costs, process costs │
│ recoveries percentage & metal price    │
└──────────────────────────────────────┘
         │
┌──────────────────────────────┐
│ Evaluate stope design – store it │
└──────────────────────────────┘
         │
┌──────────────────────────────────┐
│ Compute stope development work     │
│     Compute stoping work           │
└──────────────────────────────────┘
         │
┌──────────────────────────────────┐
│ Compute: stope's cutoff grade,     │
│  stope's operating margin,         │
│  operating cost/t  of ore.         │
└──────────────────────────────────┘
         │
┌──────────────────────────────┐
│      Is it last stope          │
└──────────────────────────────┘
        yes
┌──────────────────────────────────────────┐
│ Compute overall ore tonnage, grade, metal tons. │
│  Compute mining cost/t (operating & total)  │
└──────────────────────────────────────────┘
         │
┌──────────────────────────────────┐
│ Compute overall operating margin   │
│   Compute annual profit/loss       │
└──────────────────────────────────┘
         │
┌──────────────────────────────────┐
│ Computer life of the section analyzed │
│         Compute NPV                │
└──────────────────────────────────┘
         │
┌──────────────────────┐
│    Print results       │
└──────────────────────┘
         │
┌──────────────────────┐
│        END            │
└──────────────────────┘
```

Figure 16.35(b)   Flow diagram for computer assisted stope design and economic analysis.

continuous. The sampling data were obtained from the cores of diamond drill holes drilled underground in the pattern of 'fans' across the mineralization at 30 m interval along the strike. In this study a matrix of panels with 1 m × 1 m centers was created for each of the 15 sections. Grade value for each block/square was estimated by kriging. There were 4000 to 10,000 blocks indicating some mineralization in each of the 15 sections kriged. Figure 16.36 shows a display of kriged values of grade along a section and which forms its mineral inventory.

Formulating a basis for cutoff grade decisions to predict mineable ore reserves
It will not be practicable to mine out all the mineral blocks having a grade exceeding 0%, for example, as shown in figure 16.36. Hence determination of cutoff grade is necessary. In underground mining situation, the guidelines as described in section 12.2.2. and figures 12.4(a) (i to iv) could prove a useful guide. Keeping in view, these guidelines, care should be taken to calculate the cutoff grade at *certain level of productivity* and also at an appropriate *degree of mechanization*. In these calculations *variable costs* of the operations involved should be used.

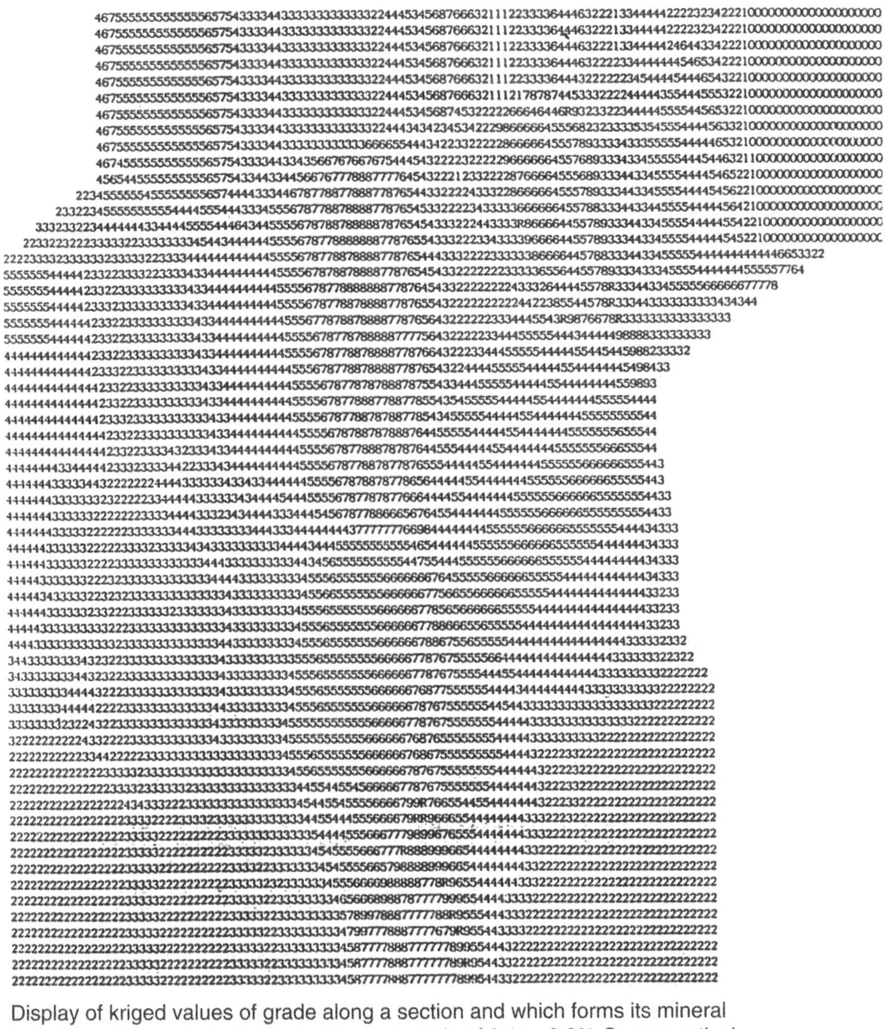

```
4675555555555555565754333344433333333333322444534568766632111223333644463222133444442222323422210000000000000000000
4675555555555555565754333344433333333333322444534568766632111223333644463222133444442222323422210000000000000000000
4675555555555555565754333344433333333333322444534568766632111223333644463222133444442464433422210000000000000000000
4675555555555555565754333344433333333333322444534568766632111223333644463222233444444546534222210000000000000000000
4675555555555555565754333344433333333333322444534568766632111223333644432222223454444544654322210000000000000000000
4675555555555555565754333344433333333333322444534568766632111217878744533322224444435544455532221000000000000000000
4675555555555555565754333344433333333333322444534568745322222666464646R9323322344444555554456532221000000000000000000
4675555555555555565754333344433333333333322444343423453422298666664559682322333535455544445632210000000000000000000
4675555555555555565754333344433333333333336666554443422332222286666664555789333343335555544446532100000000000000000000
46745555555555555565754333443434356676766767544543222232222296666664455768933334335555544454463211000000000000000000
45654455555555555565754334433445667677788877776454322212332222876666455689333443345555544454465222100000000000000000
22345555555454555555565744443334467877788778887787654433222243332286666645557893334433455555444545622210000000000000000000
2332223455555555555444455544433345556787788788887787654533222234333336666664455788333443344555554444564210000000000000000000
33323223244444443344445555446434455555678778888888778765453332224433338666664455789333443334555554444542210000000000000000000
2233223223333332332333333345454444444444555556787788888887787655433322234333396666644557893334433455555444454522210000000000000000000
22223333323333332233332233223334444444444455556787788888777876544433322233333866666445788333443345555444444444466653322
5555555444423322333322233332234433444444444455556787788888887787645433222222333336555644557893334333345555566666677764
555555544444423322333333333343444444444455556787788888887877876454332222224433326444455578R333443345555556666666677778
5555555444442332233333333333434444444444455556787788888778765432222222224422385554457R8R334433333333333434344
5555555444444233223333333334334444444444455567788788888778765643222222333444554R3R9876578R3333333333333333
55555554444442332233333333333434444444444455556787788887777564322222223344449888833333333
4444444444442332233333333333343344444444455555678778878888778766432223344455555444445544445988233332
444444444444423322333333333334334444444444455556787788788888778765432244455555444445544444445498433
444444444444423322333333333334334444444444455556787788887785433441445555444445544444559893
4444444444444233223333333333434344444444444555556787788877887785543455555444445544444445555554444
4444444444444233223333333343434444444444455556787788878788446555554444455444444445555555544
444444444444442332233333334323334434444444444555556787788787878764455555444445544444455555566665554
444444444423322333333334422333443444444444455555678788788787876555544444155444444455555555666666555543
4444444333334432222222244443333333433434444444455556787887877865644444554444444455555556666666555543
444444333332232222222222333444433332434444444334445456787788666656764554444444455555566666665555555554433
41444443333332222233332344443333333334444444444377777776698444444455555556666666555555544444433
4144444333333222223333333333444334444444444444345655555555555544755544455555556666666555554444444444433
414444333333222223333333333444443333333333434456555555555666666676455555556666666555554444444444433
44444343333322322323233333333334333333333334566555556666666775665556666666555554444444444433233
4144443333332322223333333223333433334533445565555555666666678565666666655555544444444444444433233
44444333333333322222333444333333334333333345555655555555666666778866665565555554444444444444433233
44443333333333222222222333334333233434444433324334445456787886665676455444444455555556666665555555554433
31443333333334323223333333333333333433333333345556565555555666666778767555566644444444444444333333322322
34333333334432322233333333333333343333333334556565555556666667787675555544544444444333333333222222
3333333344432222333333333333344333333333334556565555555666666768775555555443434444444333333333322222222
33333333424242333333333333333444333333333345565555555666666778767555555444433333333333333333222222222
3222222222243322223333333333344333333333455555555555666666768765555555544443333333333322222222222222
22222222223344222333333333333333333333334556555555555666666768765555555444443222233222222222222222222
22222222222222233333223333233333333333334545655555556666667876755555554444443222232222222222222222222
22222222222222233332332333233333333333334556555555555666666778767555555444443222233222222222222222222
2222222222222224343322233333333334444333333323434455455556666667787676555555544444443222232222222222222222222
222222222222222333332222333323333333333445564555555566666779R966655444444443332222322222222222222222
2222222222222223333322222222233333333333344555566677799967655544444443222222222222222222222
222222222222222333332222222222333332333334545555666777R888999665444444433222222222222222222222
222222222222222333332222222222333333323333345555566657988889996654444444333222222222222222222222
22222222222222223333222222222222333332323333334556666988887778R9655444443332222222222222222222222
2222222222222222333332222222222233333323333333346566689888787779995544444333222222222222222222222
222222222222222233333222222222222333333233333333578997888777778R955544443332222222222222222222222
222222222222222233333222222222223333323233333333347997788877776799R55443332222222222222222222222
2222222222222222333332222222222233333232333333333458777788877777789R954433222222222222222222222222
22222222222222223333322222222222233333232333333334587777R4R877777789R95443322222222222222222222222222
```

Display of kriged values of grade along a section and which forms its mineral inventory. Blocks marked 1 to 9 represent grade of 0.1 to 0.9% Cu respectively. Blocks marked R represent grade 1% Cu and above.

Figure 16.36   Mineral inventory of a deposit along a section.

Evaluating stope boundaries for different stoping methods using cutoff grade and incremental analysis criteria

An approach to identifying optimum reserves in the context of surface mining has been described by Halls et al. in 1969,[33b] and for underground mining situations by Tatiya and Allen.[33b] The principle adopted is to consider different mining excavation layouts. In underground situation this is interpreted as a number of 'envelopes' that could form possible stope boundaries. An envelope is defined as the mineralized zone bounded by the upper and lower development levels (i.e. level interval), the interface with the adjacent stope and the two side walls. Price of metal, revenues from the by-products, ore and metallurgical recoveries, total variable costs mining through to the process plants, royalty, excise duty and selling costs have been used to compute the operating margin of an envelope. The algorithms developed for the sublevel, DTH (Down-the-hole) and Flat back cut & fill stoping methods were used to carry out this task. The following procedure is adopted for defining the side walls for the different stoping methods:

1. The mineral inventory of the section to be studied is established by kriging, as shown in figure 16.36. or it could be established using other reserve estimation techniques.
2. The cutoff grade for the stoping method under consideration is then calculated using cost, price and recovery data (e.g. 0.4% for sublevel stoping in this case-study).
3. The next step establishes the slope of stope's side walls (for the sublevel and DTH methods). Initially these blocks, which are at and above cutoff grade are displayed. Refer figure 16.37(a). From this display the wall angles, for both the walls, are then selected manually so that angle chosen should include maximum number of blocks. The wall angles are increases incrementally, 3–5° and economics of each envelop calculated. That which yields the maximum operating margin thus establishes the wall angles of the stope.
4. Moving horizontally the stope walls by parallel increments of 1 m then derives the ultimate stope profile. Figure 16.37(b), shows how the maximum operating margin enables the optimum stope profile to be chosen for a sublevel stope.
5. Following this methodology, stope profiles for the sublevel and DTH stopings can be obtained, as shown in figure 16.38.
6. For cut & fill stoping better selectivity can be achieved. Figure 16.39, shows the display of mineral blocks which have grade values at or above cutoff grade (0.509% Cu in this study) of cut & fill method. In this figure, the walls for the stope with the inclusion of the subgrade material, have been also shown.

Thus, in figures 16.38 and 16.39, the stope boundaries for the three stoping methods from the same mineralized zone have been obtained. In this study the operating margin has been used to establish the optimum economic criterion.

Designing stopes and their economic evaluation for the different mining methods through the algorithms developed
Using the stope boundaries so defined (as outlined in the preceding paragraphs) or *even the hypothetical one*, procedure as outlined in the preceding section 16.5.5.1 could be used to design these stoping methods. Algorithms should be developed (as used in this case study) by using, 'model and design parameters', to design the layouts and carry out the economic analysis of the following stoping methods:

1. Sublevel stoping (longitudinal as well as transverse)
2. DTH (Down-the-hole) stoping
3. Flat back cut & fill stoping with posts and pillars.

The orebody profile (stope boundaries at various horizons including dips), its in-situ reserves and grade have been considered as the model parameters. The design parameters include the design specifications of a stope, the stope dimensions, equipment dimensions, dimension of the various stope workings and their orientations, production rate, process recoveries, mine costs, process costs and metal prices. The stope design output is displayed in figures 16.11 and 16.19.

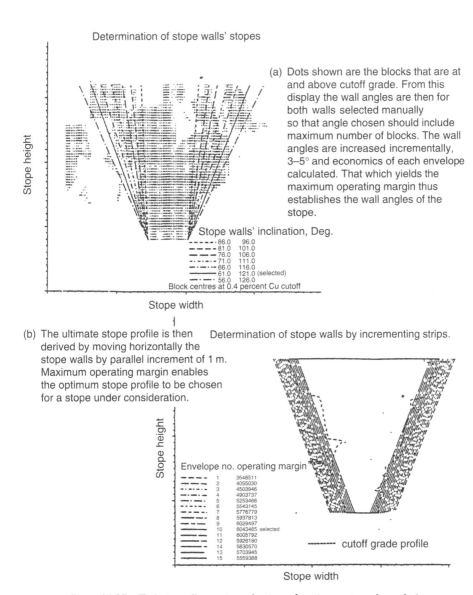

(a) Dots shown are the blocks that are at and above cutoff grade. From this display the wall angles are then for both walls selected manually so that angle chosen should include maximum number of blocks. The wall angles are increased incrementally, 3–5° and economics of each envelope calculated. That which yields the maximum operating margin thus establishes the wall angles of the stope.

(b) The ultimate stope profile is then derived by moving horizontally the stope walls by parallel increment of 1 m. Maximum operating margin enables the optimum stope profile to be chosen for a stope under consideration.

Figure 16.37   Technique illustrating selection of optimum stope boundaries.

Algorithm – stope design and economic analysis: In general, as shown in flow diagram figure 16.35(b), these algorithms have the following features:

- They are capable of designing and performing economic analysis for one or more stopes for a continuous orebody. The stope design and economic analysis of each individual stope are first obtained. The composite stope designs layout and overall economics of the stopes under study follow this.
- The stope design provides the layout of different stope workings in the form of plans, cross sections or longitudinal sections, as illustrated in figure 16.19.

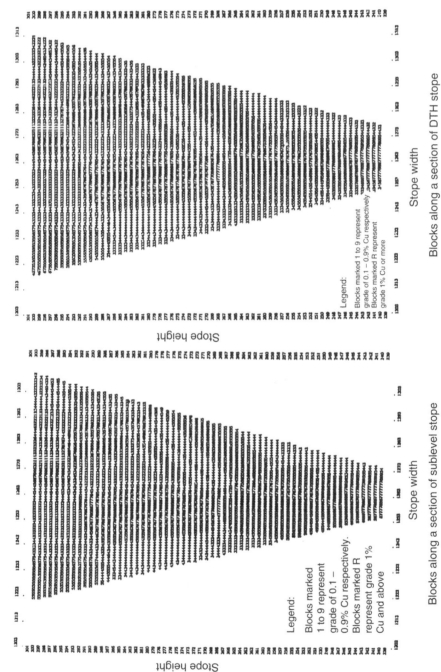

Legend:

Blocks marked
1 to 9 represent
grade of 0.1 –
0.9% Cu respectively.
Blocks marked R
represent grade 1%
Cu and above

Stope width

Blocks along a section of sublevel stope

Legend:

Blocks marked 1 to 9 represent
grade of 0.1 – 0.9% Cu respectively
Blocks marked R represent
grade 1% Cu or more

Stope width

Blocks along a section of DTH stope

Stope height

Stope height

Figure 16.38   Optimum stope boundaries along a sublevel & a DTH stope.

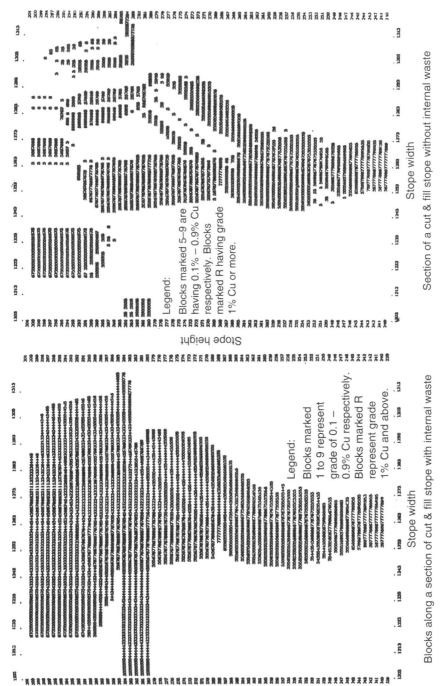

Figure 16.39 Optimum stope boundaries along a cut & fill stope with and without internal waste.

Table 16.3 A computer printout showing details of the development activities for a DTH stope.

Activities	Linear development work in m
Extraction level:	
Draw points	88.00
Slot crosscut	30.00
Extraction drive – 1	84.33
Extraction drive – 2	79.70
Trough drive – 1	40.50
Trough drive – 2	40.50
Foot wall drive	30.26
Drilling level:	
Drill crosscuts (all)	400.84
Slot crosscut/drive	30.00
Footwall drive	30.84
Decline (partly)	7.00
Crown level (optional)	75.16
Sub-total	937.13
Vertical development (optional)	61.84
Total stope development	998.97

- Based upon these stope design layouts, the computation of work involved to develop and mineout the stope, as shown in tables 16.3 and 16.4. The techno-economical details as shown in table 16.5 are also assessed.
- Using the annual rate of production (for which the mine has been designed) the life of the section is determined. Calculations for total mining cost per tonne of ore and operating margin follow. The difference in the overall revenues and costs for the year provides the annual profit. Assuming the annual profit to be uniform throughout the life of the section (deposit), the net present value of the profit/loss at specified rate of interest is ultimately assessed. This has been illustrated in table 16.5.

Output – Economic analysis:
These computer models provide the output of the economic analysis in the format as outlined in the tables 16.3 to 16.5.

Table 16.4 shows development work described in table 16.3. in m^3 (the volume) and the details of the other activities such as stope drilling, blasting, and mucking, transportation and hoisting. The unit costs for each of the activities shown are at 70% productivity with degree-3 mechanization. The total operating cost, cost per tonne of mining and cutoff grade, is also calculated.

Apart from the quantum of work, and its associated cost for each of the stopes analyzed, other technical details are estimated as shown in table 16.5.

A separate tablular output (not shown here) provides the details of the overall economics for the whole section analyzed. It shows the overall grades, mineable reserves, total and unit mining costs. The life of the section, the present worth (NPV) of the earnings (profit/loss) that will be derived by mining this section, have been

Table 16.4   A computer printout showing the activities and cost analysis for stope development and stoping for a DTH stope.

Operation	Quantum of work	cost/unit in Rs*	Total cost in Rs*
Development (horizontal)			
Extraction level, cu.m	5059.45	263.57	
Drilling level, cu.m	5655.64	263.57	
Crown sublevel, cu.m	676.48	263.57	
Decline (common), cu.m	122.50	263.57	
Total horizontal development, cu.m	11514.07		#
Total vertical development, cu.m	556.53	417.32	#
Longhole drilling, m	20931.61	102.14	#
DTH drilling (165 mm φ)	3952.01	235.71	#
Stoping (blasting, mucking & trans.)	243760.21	26.81	#
Hoisting tonnes	279972.00	7.50	#
Miscellaneous tonnes	279972.00	5.35	#
Total variable cost			#
Variable mining cost/t			#
Cutoff grade			0.352

*Currency units which could be as applicable to the country of application.
#Calculation for this column for the corresponding parameters proceeds but not shown here.

Table 16.5   A computer printout showing the techno-economical details for a DTH stope.

Parameters	Quantum
Percentage extraction	88
Stope reserves in tonnes	279972.00
Average grade in % Cu	1.10
Quantum of metal in tonnes	2556.23
Total revenues in Rs*	#
Total cost of mining in Rs*	#
Operating margin in Rs*	#

*Currency units which could be as applicable to the country of application.
#Calculation for this column for the corresponding parameters proceeds but not shown here.

determined, table 16.6, In a comparison of theses outputs for all the three methods considered for this analysis has been made.

Selection of stoping method: *Comparison – economics of the three stoping methods*: From this summary, *the inference drawn illustrates the utility of these computer models in taking decisions during the mine planning stage.* For the case study presented, DTH stoping allows a lower cutoff grade to be selected than the other two methods. These result in a higher ore tonnage, greater metal recovery and longer mine life. For the section analyzed losses by DTH stoping are the least. By changing the design

Table 16.6   Comparison of the techno-economics of the section analyzed for the three stoping methods.

Techno-economical parameters	Sublevel stoping (Alternative-1)	DTH stoping (Alternative-2)	Cut and fill stoping (Alternative-3)
Section's total reserves in million tons.	1.386	1.586	1.322
Average grade in % Cu	0.75	0.70	0.76
Cutoff grade in % Cu	0.43	0.35	0.51
Metal recovery in tons.	8736	9466	8616
Operating margin in millions Rs*./yr. (considering variable mining costs only)	278.86	289.52	258.90
Operating margin in millions Rs*./yr. (considering total mining costs only)	216.06	226.72	196.10
Operating margin in millions Rs*./yr. (considering mining costs mining through process plants)	−248.44	−217.75	−276.58
NPV of the section analyzed (in million Rs*.)	−209.51	−208.75	−222.83
Life of the section in years	0.924	1.057	0.881

*Currency units which could be as applicable to the country of application.

parameters different design spectrums can be obtained, and the stoping method, which meets the laid out objectives including the financial goals, can be selected in this manner. The salient features of this approach could be summarized:

- This methodology, in general, provides a basis to select the mineral inventory, estimated through geostatistical or any other technique, for base metal deposits suitable for mining.
- Using this methodology the internal waste or sub-grade material contained within a stope can be assessed. In addition, it provides a basis to compare the methods that could be technically feasible to mine out the same chunk of a deposit.
- The computer assisted stope design has automated many of the manual procedures and calculations involved in mine planning. This can allow evaluation of many layouts and mining strategies.
- The stope design computer models are flexible enough to run many times using a variety of economic and design parameters. This approach provides many options and alternatives to determine the best design for differing conditions.
- Mine production costs have been built up from the stope design and estimation of individual mine activities. This, in turn, forms the basis for mineral exploitation strategy.
- Finally, these programs could be written in any computer language but should be tested using the data sets obtained from a mining complex. Minor modifications, however, may be necessary to input and output routines when used on other systems and different mines.

Table 16.7 Details of unit operations during stope preparation and stoping. (C.air – compresses air; LPT – low profile truck; LPD – low profile dumper; DTH – down-the-hole drill; NG – nitroglycerine; ANFO – ammonium nitrate fuel oil; J/H – jack hammer).

Method	Drilling	Blasting	Mucking	Transportation	Services
Room & pillar	Rotary electric drills, cutter loaders in coalmines.	Permitted explosives with manual charging if coal.	In coal use of GAL and cutter loaders.	With continuous miners: mine-cars & conveyors, with GAL shuttle cars.	Ventilation, power, temp. support, water.
Stope & pillar	J/H drills, drill jumbos for flat deposits; J/H drills for inclined deposits.	NG based explosives and manual charging. ANFO if hole dia. larger than 40 mm.	Rocker shovel, LHDs etc. but for inclined deposits scraper haulage to discharge into chute.	In flat deposits use of LPT, LPD. For inclined deposits loading from chute to track or trackless units.	Ventilation, C.air, temp. support, water etc.
Shrinkage stoping	J/H and stoper drills, wagon drills if mechanized.	NG based explosives and manual charging. ANFO if hole dia. larger than 40 mm	Within stope gravity flow; at the draw points/funnels Rocker shovel, LHDs.	LPT or LPD if trackless loading; else rail haulage at the extraction level.	Ventilation, C.air, water etc.
Sublevel/blasthole stoping	J/H, stoper or jumbo drills during stope preparation. Drifter rigs and DTH drills during stoping	NG based explosives and manual charging. ANFO if hole dia. larger than 40 mm.	During stope preparation cavo, LHD; Within stope gravity flow; at the draw points Rocker shovel, LHDs.	LPT or LPD if trackless loading; else rail haulage at the extraction level.	Ventilation, C.air, water etc.
Stull stoping	J/H and stoper during stope development as well as stoping.	NG based explosives and manual charging.	Within stope gravity-flow, if not, use scraper haulage to draw muck into the chute.	Rail haulage at the discharge from chute.	Ventilation, C.air, support, water etc.
Cut & fill stoping	J/H, stoper or jumbo drills during stope preparation as well as during stoping.	NG based explosives and manual charging. ANFO if hole dia. larger than 40 mm.	During stope preparation as well as stoping LHDs, rocker shovel, cavo (scraper exceptionally).	LPT or LPD if trackless haulage at the main level, else rail haulage at the extraction level.	Ventilation, C. air, water, filling material, drainage etc.
Square-set stoping	J/H and stoper drills during stope development as well as stoping.	NG based explosives and manual charging. Use of very mild charge	Within stope gravity-flow, else use of scraper haulage to draw muck into chute.	Rail haulage from the chute loading.	Ventilation, C.air, power, support, water etc.
Sublevel caving	J/H, stoper or jumbo drills during stope preparation. Drifter rigs (fan drills) during stoping.	NG based explosives and manual charging. ANFO if hole dia. larger than 40 mm.	During stope preparation as well as stoping cavo, LHDs to discharge muck into ore pass.	To collect muck from chute of ore pass use of trackless or track haulage at the main level.	Ventilation, C.air, water etc.
Longwall mining	Rotary electric drills, cutter loaders in coal mines, else same as stope and pillar.	Continuous miners to cut coal else permitted explosives with manual charging.	In coal use of GAL, cutter loaders such as shearer, plough etc.	Conveyors, with GAL shuttle cars.	Ventilation, power, water, support etc.
Block caving	J/H, stoper or jumbo drills during stope preparation as well stoping.	NG based explosives and manual charging. ANFO if hole dia. larger than 40 mm.	During stope preparation cavo, LHD; Use of scraper haulage, LHDs during stoping.	Collecting muck from funnels, finger raises or draw-point system use of track or trackless units.	Ventilation, C.air, water, drainage etc.

Table 16.8 Mining methods at some of world's large capacity underground mines.

Company, location & mineral	Capacity TPD	Primary Mining Method	Mean Depth, ft	Years of operation & Remark	Ore flow system to surface	Ore transfer from stope
LKAB, Kiruna, Sweden. Fe.	52,000	78% SLC; 22% SLS	3000	30+. Recent expansion for 37,000 tpd.	Shafts	OP/rail
MIN Holdings, Mount Isa – Australia. Zn etc.;	31,000	70% SLC; 30% SLB	3600	70+. Electric trucks, Remote LHDs.	Shafts	
Western Mining, Olympic Dam, Australia. Cu, U.	20,000	Blasthole	2000	Under expansion. New shaft planned.	Shafts	OP/rail
JM Asbestos, Jeffrey (PQ), Canada, Asbestos.	20,000	Block Caving	2000	Under construction	Shafts	OP/rail
BHP (Magma), San Manuel USA (AZ), Cu.	68,000	Block Caving	2500	30. Peak production of 68,000 tpd in 1972	Shafts	OP/conveyor
BHP Lower (Magma) USA (AZ), Cu.	55,000	Block Caving	4000	5. Production (same mine as above)	Shafts	OP/conveyor
Confidential, USA, Cu.	60,000	Block Caving	5000	Planning. Pre-feasibility stage.	Shafts	
Cyprus Amex; Climax, (CO) USA. Mo.	36,000	Block Caving	2000	70. Closed 1986 (on standby)	Adits	OP/rail
Cyprus Amex; Henderson (CO) USA. Mo.	38,000	Block Caving	2000	10. New levels coming up, LHDs to ore passes to rail.	Tunnel	OP/rail
Noranda; Montanore – (MO) USA. Cu,Au.	20,000	Room & Pillar	2500	Development. Stopped (environmental objections)	Conveyor	LHD/truck
Molycorp; Questa,(NM) USA. Mo.	16,300	Block Caving	4000	12. Being prepared for re-opening	Conveyor	LHD
Codelco; El Teniente, Chile. Cu.	100,000	Block Caving	2000	100a. Everything used including remote-control LHDs.	Adits	OP/rail
Codelco – El Salvador Chile. Cu.	34,500	Block Caving	2000	30a. LHD to ore passes to trains.	Adits	OP/rail
Codelco,Andina/Rio Blanco,Chile. Cu.	15,000/45,000	Block Caving	2000	Expansion for ±45000. U/G Mill	Adits	OP/rail
Codelco Chukui Norte Chile. Cu;.	30,000	Block Caving	2500	Planning stage. Res.242 Mt. @ 0.7% Cu.		
Freeport, Ertsberg East – Indonesia.Cu. (Planning stage)	17,000	Block Caving		15a. Production varies (total production is 115000 tpd).	Ore passes	LHD/truck
Philex Minerals, Philippines. Cu.	28,000	Block Caving	3400	20a. Going concern	Adits	OP/rail
Atlas Carman, Philippines. Cu.	40,000	Block Caving		Planning stage	Adits	OP/rail
Lepanto Far SE, Philippines.Au.	17,000	Blasthole	5000	On hold. Feasibility stage.	Shafts	LHD/truck
RTZ Palabora, South Africa. Cu.	60,000	Block Caving	4000	Under construction for 80,000 tpd	Shafts	LHD/truck

16.6   MINE LIQUIDATION[8,22,26,29,30(a),42]

In underground mines the *worked out space* is the space created as a result of exploitation of a deposit. Every such space and its size have their effects on the surrounding rock mass. *Liquidation* is the systematic abandoning by way of achieving the desired changes in the state and the extent of the influencing zone corresponding to the size of the worked out spaces. It is undertaken not only to ensure ultimate safe and economic exploitation of the deposit but also to keep in mind the quantitative and qualitative aspects of conservation of a mineral wealth.

16.6.1   Liquidation of the stopes of different types

The size of worked out space and the period for which it can keep standing depends upon the strength characteristics (dynamic as well as static) of the orebody and its surrounding rock mass. Sometimes difficulties that arise while working the lower levels can also be attributed as a function of these characteristics. Liquidation operation can be grouped as per the stoping methods outline below:

- Liquidation of caving stopes (fig. 16.41(i)(c))
- Liquidation of supported/filled stopes (fig. 16.41(i)(b))
- Liquidation of open stopes (fig. 16.41(i)(a))
- Liquidation of standing stopes.

Liquidation of caving stopes: These types of the stopes exist where the ore bearing rocks and the superincumbent strata are prone to caving. The ore is extracted and the space is allowed to be replaced by the caved material. This substitution usually takes place simultaneous to the stoping operation. Thus, worked out space is liquidated by the caved material.

Liquidation of supported stopes: In the supported stopes the face advances with the walls and the back supported by timber, steel or hydraulic supports, roof bolts, waste fill, ore pillars or their combination, depending upon the local conditions.

The supports are either left out in the worked out space permanently or withdrawn at the time of liquidation. Thus, worked out space is not always rendered inaccessible immediately but after sometime depending upon the type of support used, in terms of its life. Magnitude to the damage to the protective structures depends mainly on the type of support used.

Liquidation of open stopes: The open stopes after their stoping constitute a skeleton of open spaces separated by crown pillars between the level and intervening pillars (in between two consecutive stopes) at the same level. These open spaces can be dealt in the following manner:

1. Leaving them standing without disturbing them.
2. Induce caving by blasting the intervening and/or crown pillars.
3. Isolating individual stopes or a group of stopes to localize the devastating effects, which may result due to sudden collapse of any one, or more than one of them.
4. Post filling the worked out space.

16.6.2 Planning liquidation[29]

When? In what order? And how? The liquidation of the worked space is to be carried, are some of the relevant questions that should be thought of when planning for the liquidation.

Relative time of liquidation: This means when liquidation should be carried out relative to the stoping operations. In some of the stoping methods, such as those based on the caving and supported systems, the liquidation form the part of a stoping cycle. But in case of filled stopes, opens stopes and shrinkage stopes this logic is not applicable. In such cases delayed liquidation is essential, in order to recover the blocked ore, and for protection against subsidence below the structures, to be protected.

Sequence of liquidation: Under this heading we consider two aspects:

1. The general direction of advancing the liquidation
2. Size of worked out space that should be taken for the purpose of liquidation – which in turn could be:

Unit liquidation: A stope or its fraction called unit liquidation. It is the size of the worked out space that must be liquidated before further excavation is made and the space is created. This could be even the space created by a square-set or slice in case of top slicing.

Block Liquidation: If two stopes are taken together and then liquidated, it is known as block liquidation. This is a usual practice to liquidate step by step the number of stopes situated between two levels. This practice can be made applicable in case of sublevel stoping, room & pillar stoping and shrinkage stoping.

Global Liquidation: This practice involves liquidation of more than two stopes at a time. In this practice the roof span increases beyond the self-supporting stage and the caving is resulted, if not, blasting the roof-rocks can induce it. The void is filled by the caved rock.

16.6.3 Liquidation techniques[29,30(a)]

Consideration of recovery, safety and economic of the operations are of prime importance while planning liquidation. The aim behind liquidation should be first of all clearly spelled out, as different aims would have different approaches. The techniques applied are:

- Caving (natural or induced)
- Filling the worked out space
- Isolation of the worked out spaces.

Caving (natural or induced): This is achieved by increasing the size of worked out space beyond the stability limits. In practice this can be achieved by robbing the intervening or crown pillars else by breaking overlying rocks. The caving operation should lead to release of stresses or their transfer to some other harmless areas.

Filling the worked out space: The filling is done as per the objective laid out, which could be protection of the surface structures, or just extraction of the locked ore with or without consideration of the subsidence of the overlying strata. If protection of

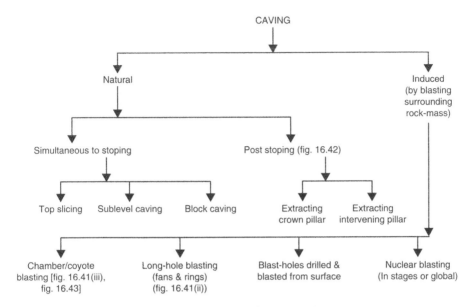

Figure 16.40   Classification of liquidation by caving.

surface structures is warranted in that case complete filling of the worked out space is essential, else it could be, partially filled without caring for the subsidence. To safeguard against air blast, sometimes filling or caving is essential. Use of mill tailings to fill the worked out space sometimes is purposely made where tailing disposal at the surface is a problem. In some of the countries such as Norway surface waste is sent underground for its disposal to minimize surface-land degradation.

Isolating the worked out spaces: After recovering the ore from the pillars, whatever is possible, the worked out area can be isolated if the deposit is small, shallow seated, strata not prone to rock bursts, low grade ore blocked in pillars and the pillars left are of sufficient strength to stand up to the desired period.

16.6.4   Pillar types & methods of their extraction[29,30(a)]

In mine formation, or leaving pillars of various kind; as described below, are almost mandatory. Following are the pillars of various types:

- Crown pillar
- Sill pillar
- Rib pillar/intervening pillar
- Remnant pillar
- Barrier pillar
- Boundary pillar.

Crown pillar: In order to support the workings of the immediately upper level of a stope, the horizontal pillar left is known as crown pillar. This pillar can be blasted in stages or at a time by designing a heavy blast. Sometimes crown pillars of the adjacent stopes are liquidated together. It can also be taken together with overlying stope's sill pillar.

Sill pillar: In order to provide a base to the blasted muck from the stope and to build the ore drawing system within a stope, the horizontal pillar left is known as sill pillar. In a stope it cannot be liquidated until its crown pillar has been liquidated. Heavy blasting in this pillar may damage the crown pillar of stope lying immediately below it. For this reason sometimes heavy blasting is planned to liquidate this pillar together with the crown pillar of the stope lying immediately below it.

Rib pillar/intervening pillar: To limit the size of a stope along its length of a continuous orebody, it needs to be separated by an ore pillar, known as rib or intervening pillar. This pillar can be blasted completely at a time by designing a heavy blast. Sometimes rib pillars of the adjacent open stopes are liquidated together. But the rib pillars of cut and fill stopes are usually mined in the similar manner as the stope has been mined i.e. by taking slice by slice of ore and filling the void by some kind of filling material.

Remnant pillars: Sometimes due to abnormal working conditions the complete stoping needs to suspend and the portion that has not been worked out is known as remnant. These conditions could be outbreak of fire, explosion, inundation, rock fall, rock bump or burst etc. Recovery from these pillars is usually attempted during the liquidation period.

Barrier pillar: These pillars are designed to isolate the panels from each other and applicable to stoping methods, such as, bord and pillar, room and pillar etc. Which are usually applicable to mine the flat dipping coal deposits.

Boundary pillar: If the deposit is very extensive it is divided into mines of suitable sizes. In order to isolate workings of one mine to another, the pillar left, is known as boundary pillar. These pillars limit the size of a mine.

16.6.4.1 *Pillar extraction methods*

It depends upon the position of a pillar within the stope and its surrounding rock-mass particularly those forming the immediate footwall and hanging-wall. It also depends where the working above has been liquidated or two or more levels are to be liquidated simultaneously.

Pillar extraction using supported or caving methods: Where surrounding worked out spaces have been filled with the waste or caved fill, it is not possible to use heavy blasting for pillar recovery. In such cases the intervening pillars could be recovered by any of the following methods. The choice is governed by the prevalent conditions.

- Top slicing
- Sublevel caving
- Cut and fill
- Square-set stoping.

Pillar extraction using heavy blasting: The drawback of the above mentioned technique when compared with heavy blasting is the low productivity and high cost of recovering them; hence these methods should be applied when it is absolutely necessary to extract the contents of the pillars with least loss of ore or contamination.

Moreover, when the worked out spaces surrounding the pillars are empty, the blasting technique can be applied. The heavy blasting method is suitable when caving of the overlying strata can be allowed. The pillars are blasted using longhole/ring drilling. Use of coyote (chamber) blasting (figs 16.41(iii) and fig. 16.43)), described below, is also made sometimes.

The consideration of damage to the surrounding due to heavy blasting is of prime importance. This can be assessed, and controlled by properly planning the blast. Peak particle velocity is the measure of such damages. This can be calculated using the following relation, which was developed by U.S.B.M for hard rock open pit mines.

$$\text{Peak Particle Velocity; } V = 1143\{(\sqrt{W})/D\}^{1.6} \tag{16.5}$$

Where: W – the maximum charge in kg/delay
D – the distance from center of blast to the recording site
V – peak particle velocity in mm/sec.

Coyote blasting: The process of blasting a huge rock mass against a free face by concentrating the explosive charge in a suitably developed chamber is known as coyote blasting. This technique is popular in surface mines to blast the over burden.

16.6.4.2 *Planning a heavy-blast for liquidation purpose*[26]

• Purpose of blast should be clearly spelled out and precautions required against the likely damages due to the blast should be made known to the crew involved for this purpose. Efforts should be made for the minimum damage. Likely impact on the neighboring workings should be thought of.
• Drill patterns' designs with suitable burden and spacing should be prepared. Due precautions should be taken to avoid any damage to the drilled holes. Holes' survey should be carried out prior to shifting the drills from the site to ensure proper quality of drilling.
• The planning with regard to charging and blasting the heavy-blast requires to assess parameters such as: The type of explosive to be used, charging time, effect of longer duration after charging on the quality of explosive, initiation method, type of delay, charge/delay etc. Sometimes it becomes necessary to carry out some trails/tests before undertaking the heavy-blast.
• Ground vibrations and post blasting fumes: Peak particle velocity should be calculated to limit the explosive/delay for the purpose of controlling the likely ground vibrations. Adequacy of the ventilation arrangements for sweeping the likely noxious gases after the blast should be checked.
• Induce caving: In some situations caving of the hanging wall is induced to allow the caved rocks to fill up the stoped out void, thereby, avoiding occurrence of an air blast due to sudden collapse of a stope.

16.6.5 Case studies

16.6.5.1 *Heavy blasting at a copper mine*[22]

On 26 Jan. 1979 a heavy blast using 22 tonnes of explosive to blast rock mass of about 105,000 tons. was undertaken. The explosive was charged in the rings designed to

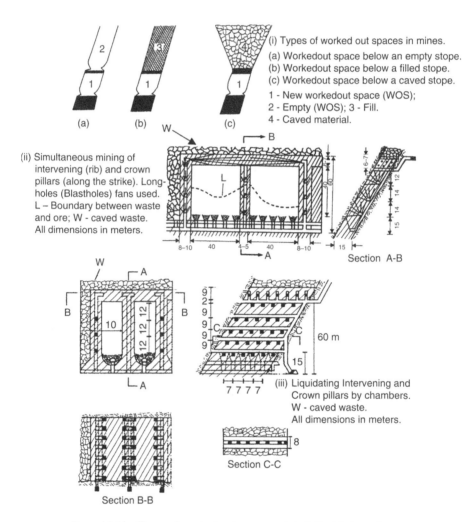

Figure 16.41 Types of worked outspaces and their liquidation schemes.

wreck the rib pillars between two worked out stopes, caving rings and coyote chambers.

The site of blast was the two uppermost stopes, adjacent to each other, covering a strike length of about 180 m. These were sublevel stopes having height of 60 m and width in the range of 15–20 m. Thus the volume of worked space after stoping them, amounts to be around 180,000 m³. With the passage of time they might collapse any moment causing a heavy air blast being open stopes. To avoid this disaster it was essential to block all the entries (accesses) to these stopes. The ore blocked in the rib pillar between these two stopes was also to be recovered. Taking into consideration of all these aspects this blast was planned. First the rib pillar between them was drilled using the blasthole drills in the form of rings and parallel holes. Then in the hanging wall side of these stopes the rings were drilled in planes almost parallel to the steeply

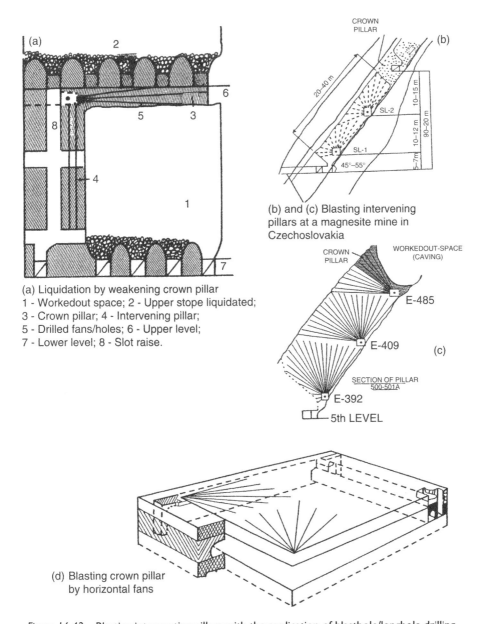

CROWN
PILLAR

(a)

2

6

5   3

8

4

1

7

(a) Liquidation by weakening crown pillar
1 - Workedout space; 2 - Upper stope liquidated;
3 - Crown pillar; 4 - Intervening pillar;
5 - Drilled fans/holes; 6 - Upper level;
7 - Lower level; 8 - Slot raise.

(b)

20–40 m

10–15 m

SL-2

10–12 m

90–20 m

SL-1

5–7 m

45°–55°

(b) and (c) Blasting intervening
pillars at a magnesite mine in
Czechoslovakia

CROWN
PILLAR

WORKEDOUT-SPACE
(CAVING)

E-485

E-409   (c)

SECTION OF PILLAR
500-501A

E-392

5th LEVEL

(d) Blasting crown pillar
by horizontal fans

Figure 16.42   Blasting interventing pillars with the application of blasthole/longhole drilling.

dipping orebody, which was present in these stopes. The various parameters used
during drilling and blasting of these rings and fans have been shown in table 16.9. To
avoid any risk of failure, caving rings, 4 coyote chambers (1.8 m × 1.8 m × 1.8 m)
at a distance of about 9 m from the last caving rings, were driven as shown in
figure 16.43. The rings were charged with explosive ANFO but the Coyote chambers
with NG based conventions explosive. The stemming and packing of the coyote

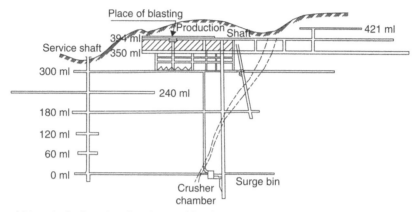

(a) Longitudinal section of workings at Khetri copper mine

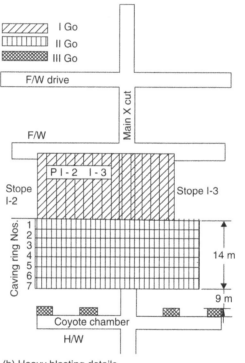

(b) Heavy blasting details

Figure 16.43 Details of heavy blasting at Khetri Cooper Complex using coyote and longhole/blast-hole blasting techniques. (a) Longitudinal section. (b) Details at 394 m level workings. Relative position of rings and fans drilled in ore; caving rings drilled in H/W waste; and coyote chambers driven have been shown. Blast sequence also marked.

chambers was carried out using sand bags. Packing was accompanied by alternate rows of gypsum and sand bags. Firing was planned in this sequence: First – pillar rings, next – the h/w caving rings, and lastly – the coyote chambers. This is to enable stopes' draw points and their bottom-most portions at the sill level to be filled with

Table 16.9   Details of heavy blast at KCC Mines.[22]

Particulars	Ore/ waste	Purpose	Long hole drilling	Explosive charged	Tons. of ore/ waste blocked
Pillar Rings	Ore	Blasting rib pillars	3770 m	4554 kg	16605
Caving Rings	Waste	To Induce caving	6106 m	8525 kg	44704
Coyote chamber	Waste	To induce caving and summing up	–	8975 kg	44000
Total			9876 m	22054 kg	105000

Table 16.10   Details of the design parameters for the heavy blast at KCC, Mines.

Items	Value
Diameter of blast/ring holes	57 mm
Ring design details:	
Pillar rings – burden, toe spacing	1.5 m; 1.5 m (max.)
Caving rings – burden, toe spacing	2.2 m; 3 m (max.)
Hole length range	3–25 m
Number of holes	656
Number of pillar rings	16
Caving rings	7 upward, 8 downward
Anodet Delay used and their nos.	0–29; 696
Drilled meters	9676 m
Charge length	5945 m
NG based explosive in 4 coyote chambers	8.975 ton (2.29 t/chamber)
Explosive consumed	22 ton
Electric detonators	60
Cordtex fuse consumed	600 m
Circuit resistance	320 ohms
Drill factor	6.2 tons/m
Rock blasted	105,000 ton

the *ore from the rib pillar*, and then it could be blanketed by the hanging wall waste rock. This waste rock cover was further blanketed by the waste-rock that was generated by blasting the portion between the last hanging wall ring and the coyote chambers.

The following observation was made by the author who happened to be the person in charge for this blast and was present during its execution.

There were some pipes of 6 m length and 15 cm dia. and some other material such as timber logs etc. lying near the mouth of 394 m level adit, which was the access to the blast site from the surface. These materials were not removed before blast with the expectations that nothing will happen to them. It was observed that immediately after the blast, the fumes and air, which could access through this adit was like a cloud moving out from it with very high speed to the extent that those pipes and materials were thrown to a distance of more than 200 m like bullets from a gun. In addition, all the electric-power and telephone lines, which were also at distance of more than 200–300 m from the blast site outside the mine at the surface, were also severely damaged.

Underground in the neighboring areas of these stopes there was heavy make of loose in the sides and backs of all drives crosscuts and entries. It was found that ore recovered was more than expected. This blast made these areas safe against the dangers of ground failure and air blast.

16.6.5.2   Remnant pillars' blast at lead-zinc mine[26]

A heavy blast consuming 39.7 tons. of explosives to blast about 130,000 tons of ore was successfully undertaken in 1988 at one of the units – Mochia mine of Hindustan Zinc Limited, India to mine out some of the remnant pillars which were left as crown, sill and rib while mining the upper levels at this mine with the application of conventional shrinkage and sublevel stoping. The area covered for this blast measured: 120 m (height) × 70 m (length) × 30 m (width). The available void to accommodate the blasted muck was somewhat 5 times than this area, which means it was much more than adequate. The blast covered three crowns and associated sills, one rib, and two partial rib pillars. The orebodies here at this mine had been steeply dipping with a maximum thickness of 45 m. The walls as well orebodies were competent. The pillar recovery was planned with an objective to allow stope wall closure in the worked out areas, which were standing uncaved for a very long period, thereby, redistribution of the stresses around the mine workings. And also to recover the ore blocked in the remnant pillars.

16.6.5.2.1   Blast planning

The peak particle velocity was computed using equation (16.5). Maximum charge/delay in this blast was 1805 kg. A comparison later on was made between the measured and predicted peak particle velocity, V, as shown in table 16.11.

Preparatory work: Before undertaking the blast the preparatory works that were undertaken include:

- Re-survey of the concerned areas to check the accuracy of the excavation.
- Drilling pilot holes to assess thickness of crown pillars.
- Setting of the instrument such as strain and stress gauges for the purpose of rock mechanics studies at the suitable locations such as: main and auxiliary shafts etc.
- Strengthening mine services such as ventilation, compressed air etc.
- Establishing extraction layout to handle the blasted muck likely to be generated after the blast.

Table 16.11   Comparison between recorded and predicted ground vibrations.

Recording site (RS)	Dist. bet. blast site & RS	Instrument used	Measured p. p. velocity 'V' in mm/sec.				Predicted* p.p.v. 'V' in mm/sec.
			Longit.	Trans.	Verti.	Result	
Adit – 3	305 m	Sprengnether	15.0	10.0	13.0	22.2**	48.8
Adit – 4	200 m	NOMIS	24.2	13.3	28.9	32.3***	95.0

*predicted from the relation used; **equal to the square root of the sum of the square of the longitudinal, transverse and vertical components. ***Instantaneous resultant.

Development work: These blasts require additional development work in the form of drill drives, chambers, access raises, drives in the hanging wall or foot wall for the safe access after the blast etc. The total development amounted in this blast was 790 m.

Blasthole drilling: The blast required drilling of the blastholes of 57 & 115 mm dia. Total holes drilled were 1500 involving 16620 m of 57 mm. dia. holes and 2160 m of 115 mm. dia. holes. The 57 mm. dia. holes were drilled in sub-horizontal to vertically up fans, whereas 115 mm dia. holes were drilled in the fans, which were sub-horizontal to vertically downward.

Figure 16.44(a) presents the longitudinal section of the mine, and the pillars that have been blasted have been shown in figure 16.44(b). The drill pattern for the sill, crown and vertical pillars have been shown in figures 16.44(c), 16.45(a) and 16.45(b).

Explosives including ANFO: Except in watery holes use of ANFO was made in almost all the holes. Pneumatic loading was carried out using pressure type ANFO loaders. In each hole in order to ensure continuity of the initiation and the charge, use of cordex detonating cord of 10 mg strength was made.

Use of ANODET anti-static detonators was made in all the holes. Each hole with ANFO was initiated at the toe using PRIMAX (cast Pentolite type) 20 gms and 250 gms in 57 mm and 115 mm. dia. holes respectively. For watery holes in place of ANFO, special gelatine explosive was used.

Few days before the charging was started a trail was undertaken by blasting the same number of detonators at the same locations after checking the detonators, blasting cables and exploder properly.

On the day of blasting each detonator, each reel of connecting wire, the main firing cable, the blasting ohmmeter and exploder were checked before engaging them for the blast. The connection work started four hours before the end of charging schedule and lasted about ten hours. The detonators were actuated by the use of capacitor-discharge type 200 shots exploder. The powder factor achieved was 0.3 kg/t.

16.6.5.2.2 Results of the blast

- All the 130,000 tonnes of blasted ore was fragmented and detached from the rock mass cleanly. Visual inspection after the blast indicated good fragmentation.
- Damage to the adjacent areas and the facilities were restricted to the minor spalling in the adjacent drives and crosscuts. No damage to main excavation drives such as ventilation roadways, shafts etc. were observed.
- Ground vibrations were measured by the two independent agencies, and results have been shown in table 16.11. The vibrations were well below the unsafe limit and also considerably lower than predicted.
- Post detonation fumes: The blast generated copious volume of nitrous fumes (orange brown colored). In addition to their high toxicity the oxides of nitrogen are low energy gas species. Therefore, their presence was undesirable from both cost and environmental point of view. The possible reason for this could be:
  (a) Decrease in the diesel content below its optimum value of 5.7%.
  (b) Water absorbing by ANFO from the blast hole.
  (c) Inadequate priming of ANFO.

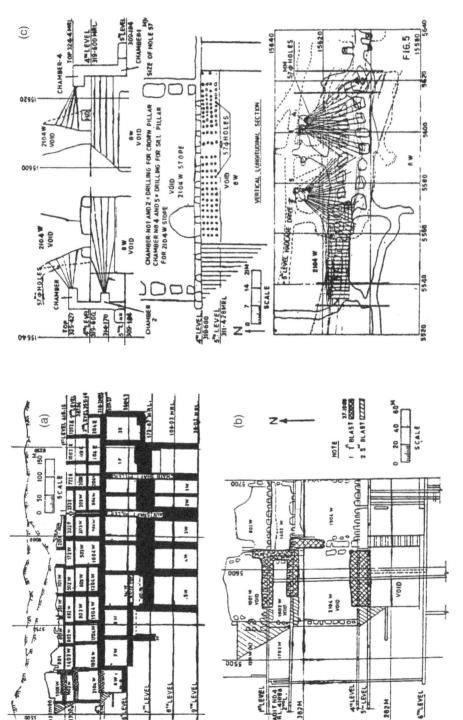

Figure 16.44 (a) Longitudinal section of Mochia (Pb–Zn Mine); (b) Blasting site/Pillars; (c) Drilling patterns for still and crown pillars, and unmined ore of a stope (210 4W, 8W stopes).

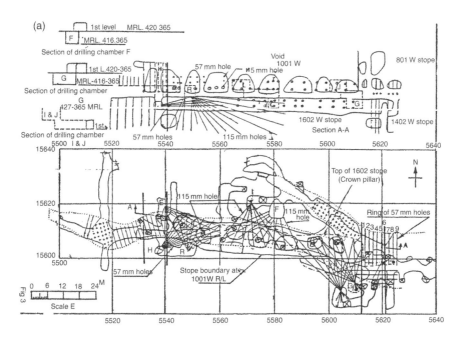

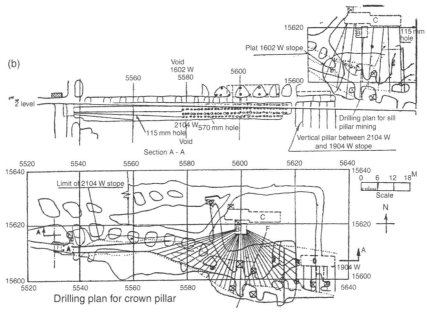

Figure 16.45 Details of workings together with drilling patterns shown. (a) Drilling patterns for sill, crown and vertical rib pillars (1001W & 1602W stopes). (b) Drilling patterns for sill, crown and vertical rib pillars (1602W & 2104W stopes).

The mine's ventilation system was operating for sixteen hours before anybody was allowed to enter into the mine.

16.7　PLANNING FOR MINE CLOSURE[8,42,46]

16.7.1　Introduction

As well as the various phases such as exploration, development and exploitation, mine closure is equally important. Regulations have been enforced in almost all countries to comply with this aspect. Mine owners are required to commit the execution of this phase during mine planning and while submitting the Environment Impact Statement (EIS) to the Government. This means it has got financial implications, and this is an extra cost, which was almost absent in the past, say prior to 1990.

The predicted adverse impacts must be reported in EIS and appropriate means for their mitigation must be proposed. A systematic and methodical approach is favored so that likely impacts are clearly identified and there is no doubt that all potential impacts have been considered. This involves preparation of a detailed mine closure programme that addresses all the issues associated with closure and rehabilitation.

The objective of the closure plan is to assure the controlling authorities that it will be successful and to release the mine owners and operators from their obligations so that the site can be disposed of in an appropriate manner.

16.7.2　Phases – Mine closure

Mine closure programmes comprise three phases:

- Closure Planning, immediately prior to cession of operations;
- Active Care, during which the mine is decommissioned and the whole site rehabilitated; and
- Passive Care, a period of monitoring to show the rehabilitation has been successful and that the site can be disposed of.

A bar chart shown in figure 16.46, illustrates these aspects. It is very unusual nowadays for a mining and mineral operations to get planning approval without the provision of some means of financial surety to indemnify the authorities against closure and rehabilitation costs. Different mechanisms are available, and the choice will depend on the preference and circumstance of the owner and the financial advice that he receives.

16.7.3　The Integrated Mine Closure Planning Guidelines (Toolkit)

The Integrated Mine Closure Planning Guidelines (Toolkits) were developed by ICMM to help site consideration of closure aspects in a holistic manner. The following text, with kind permission from ICCM, is presented herewith. This concept, and the supporting material in the report, apply equally well to both large and to small companies.

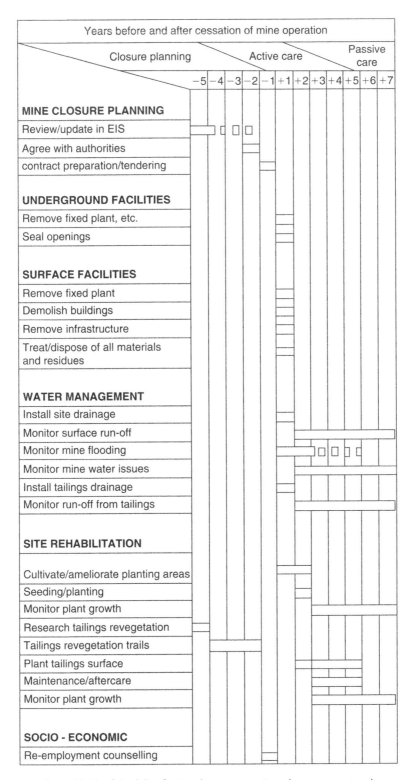

Figure 16.46 Schedule of mine closure operations (www.smenet.org).

16.7.3.1 Salient features (parameters to be considered) for closure planning

1. **Expected outcome:**Positive outcomes of effective closure planning should mean that:
    - engagement with affected and interested parties will be more consistent and transparent;
    - communities will participate in planning and implementing actions that underpin successful closure;
    - closure decisions will be better supported by stakeholders;
    - planning for closure will become easier to manage;
    - the accuracy of closure cost estimates will be improved;
    - the risk of regulatory non-compliance will be minimized;
    - potential problems will be identified in a timely manner;
    - there is more likely to be adequate funding for closure;
    - potential liabilities will be progressively reduced; and
    - opportunities for lasting benefits will be recognized and planned for adequately.
2. **Periodic checks and review:** Successful closure depends on setting, continually reviewing and validating and finally meeting closure goals that align with company and stakeholder requirements. There should be minimal residual risk to the company, and the community should realize benefits that will continue to exist without further input from the company.
3. **Number of phases within the lifecycle:** The life cycle of the operation has been characterized as having eight phases, as shown in figures 16.48 & 16.49: exploration, pre-feasibility, feasibility (which includes planning and design), construction, operation, decommissioning, closure and post closure (which may include relinquishment of tenure and liability).
4. **Basic steps to be considered:** There are three basic steps to developing an effective closure plan. If a mine makes closure planning part of its operational philosophy, these steps should blend into each other over time rather than being distinctive stages.
    I. Development of a **conceptual closure plan:** This plan is developed and used during exploration, pre-feasibility, feasibility/design and construction to guide the direction of activities. Its active life may be three to five years.
    II. Development and implementation of a **detailed closure plan:** This plan is used continuously during operations, and has an active life that could range from 5 to 30 years or more, during which time it is updated. A detailed plan will include milestones, detailed methodologies of achieving these, monitoring and validation processes.
    III. The final step is **decommissioning and post closure plan:** Its active life may be as little as a year or two, although depending on post closure responsibilities, it may extend many years past that time.
5. **Participants in effective closure planning**
   Effective closure planning involves bringing together the views, concerns, aspirations, efforts and knowledge of various internal and external stakeholders to achieve outcomes that are beneficial to the operating company and the community that hosts it. For a company, this involves:
    - Incorporating closure planning into the early stages of project development (nominally pre-feasibility and feasibility) and operations;

- Collating the goals and views of various stakeholders (project owner (internal stakeholder***), external stakeholders such as: local community, government, and non-governmental organizations (NGOs)) at the early stage of project development and operations to inform closure and post closure goals;
- Acting to meet the goals by working with the relevant stakeholders within and outside the project owner's organization;
- Using the concepts of risk and opportunity to both minimize liability and maximize benefits to all relevant parties; and
- Using multidisciplinary expertise and multi-stakeholder processes to ensure that mitigation of risk in one area does not increase risks in another.

***Internal stakeholders

The resources of an organization can have varying inputs and responsibilities, at different times, for the three stages of effective closure planning (conceptual closure planning, detailed closure planning and decommissioning/post closure planning). The need for integration between the various internal stakeholders at any one stage and between the stages is fundamental in ensuring that a mine operation is designed with closure in mind.

Table 16.12 shows an example of how important internal stakeholders can have a key input to the closure planning process throughout the operation's life cycle. Note that some organizations may not have all of these internal divisions or teams to provide the requisite support, but much of the knowledge and expertise is usually drawn upon during the various stages of the development and operation of the facility. Note too that these responsibilities may overlap to some extent, depending on the organizational structure of a company.

Table 16.12: Influence of project teams in delivering integrated closure planning. Definition of terms:

- A core activity is one in which the involved party is required to drive the process and may hold accountability for the success of that process.
- A support activity is one in which the involved party takes an active role, providing cross-functional expertise or management input.
- An advisory activity is one in which the involved party provides contextual information that is of value to the planning.
- A handover activity is a transitional role to ensure that continuity of the closure planning process is maintained.

6. This concept of continuous closure planning is not the same as concurrent rehabilitation. The former is a process that extends throughout the life of the mine. The latter is only one part of the closure planning process which is usually accounted to operations.
7. **Right time to start the closure planning:** Realizing closure goals requires a progressive reduction of risks and unknowns, as shown in figure 16.48. The earlier that risks and unknowns are reduced, the greater the potential for meeting specific objectives. This is one reason that planning for closure should begin at the earliest opportunity.
8. **Application of the Integrated Mine Closure Planning Guidelines (Toolkits):** These can be applied throughout the life of an operation, as shown in figure 16.49. These

Table 16.12 Key inputs by the internal stakeholders to the closure planning process throughout the operation's lifecycle (Permission ICMM).

| Stage of operation life cycle | Time interval | Typical organizational involvement (guide only) | | | | | | | Form of closure plan actively developed and used |
		Exploration team	Corporate team	Feasibility team	Construction closure team	Operational team	Decommissioning/ closure team	
Exploration	2–5 yrs	Core	Support	N/A	N/A	N/A	N/A	Conceptual closure plan
Pre-feasibility	1–5 yrs	Handover	Support	Core	N/A	N/A	N/A	Conceptual closure plan (developed further)
Feasibility	1–5 yrs	Handover	Core/Support	Core	Advisory	Advisory	N/A	Conceptual closure plan (developed further)
Construction	1–2 yrs	N/A	Advisory	Support	Core	Advisory	N/A	Conceptual closure plan (developed further)
Operation	5–30 yrs	N/A	Support	Advisory	Advisory (initial)	Core	N/A	Detailed closure plan
Decommissioning	0–1 yrs	N/A	Support	N/A	N/A	Core	Core	Decommissioning & post closure plan
Post-closure	5++ yrs	N/A	Support	N/A	N/A	N/A	Core/Support	Decommissioning & post closure plan

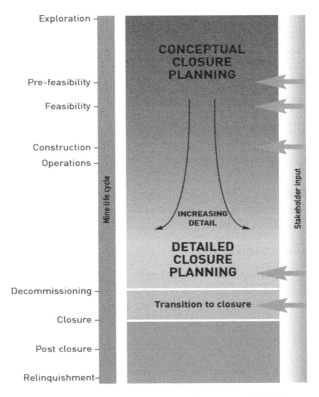

Figure 16.47   Closure planning (Permission: ICMM).

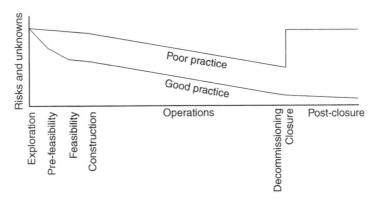

Figure 16.48   Reduction of risks and unknowns (Permission: ICMM).

13 Guidelines, as outlined below, provide the practitioner with practical work processes, examples and contexts within which to apply closure planning discipline:

- **Guideline (Toolkit) 1:** Stakeholder Engagement
- **Guideline (Toolkit) 2:** Community Development

- Guideline (Toolkit) 3: Company/Community Interactions to Support Integrated Closure Planning
- Guideline (Toolkit) 4: Risk/Opportunity Assessment and Management
- Guideline (Toolkit) 5: Knowledge Platform Mapping
- Guideline (Toolkit) 6: Typical Headings for Contextual Information in a Conceptual Closure Plan
- Guideline (Toolkit) 7: Goal Setting
- Guideline (Toolkit) 8: Brainstorming Support Table for Social Goal Setting
- Guideline (Toolkit) 9: Brainstorming Support Table for Environmental Goal Setting
- Guideline (Toolkit) 10: Cost Risk Assessment for Closure
- Guideline (Toolkit) 11: Change Management Worksheet
- Guideline (Toolkit) 12: The Domain Model
- Guideline (Toolkit) 13: Biodiversity Management

16.7.3.2   Guidelines/toolkit details

Guideline (Toolkit) 1: Stakeholder engagement
The application of stakeholder engagement to complex issues requires bringing experience and expertise to bear on the issues. However, the scoping and planning of stakeholder engagement can be guided by informed views on the various project development and operational processes.

Guideline (Toolkit) 2: Community development
The ICMM Community Development Guidelines (2006), which address one of the primary vehicles of lasting benefits, namely the development of host communities. (Individual Guidelines are highlighted here within Guideline 2 with the numbering system cross-referenced to the source document.)

Guideline (Toolkit) 2-1 (Stakeholder identification Guideline) allows the key stakeholders to be identified. This Guideline should be used at all stages of conceptual and detailed closure planning to initially identify stakeholders and then to ensure that the stakeholder list is kept current.

Guideline (Toolkit) 2-2 (Social baseline study Guideline) lists various aspects of social baseline studies that should be covered. The social baseline study contributes vital information relating to the sustainability of closure goals, particularly as they relate to the social and socio-economic conditions of the local area and the region. Note that for facilities with a long life (20 years or more) the baseline information collected may become outdated, and consideration should be given to updating specific social and socio-economic data sets to better inform the closure planner of changes and trends that may have occurred.

Guideline (Toolkit) 2-3 (Social impact and opportunities assessment Guideline) is particularly potent in developing closure planning goals and milestones. Positive and negative impacts are assessed, and this provides the closure planner with information that assists in addressing sustainable improvements in communities. Note the value of assessing positive social and socio-economic impacts and focusing the closure plan on ways to enhance or sustain them.

Guideline (Toolkit) 2-4 (Competencies assessment) is useful in planning for the delivery of community development programs and assists in identifying the competencies required within the community to continue the functions within these programs after the facility has closed and its support terminated. The competencies assessment can also be used in addressing skilled, semi-skilled and unskilled worker remobilization at the end of a facility's lifespan.

Guideline (Toolkit) 2-8 (Problem census) provides a rapid means of assessing the priorities within communities and is suited to early engagement at the exploration phase of mining, for example, as it allows the organization to focus on immediate issues. It is also a useful Guideline to use between the formal reviews of the detailed closure plan, as it provides an efficient way of confirming that the closure planning activities still meet community requirements or of indicating where deviations might begin to occur. The Problem Census is particularly effective when used prior to the commencement of capital or labor intensive programs that represent a significant cost to the company, as it may forestall wasted effort or wasted capital.

Guideline (Toolkit) 2-9 (Opportunity ranking) is useful in establishing priorities for community development programs, which allows better benefit/cost outcomes to be targeted early in the development of a mine. As with most social and community development agendas, priorities within communities can alter, and opportunities that previously represented high benefit/cost outcomes may lose their ranking.

Guideline (Toolkit) 2-10 (Stakeholder analysis) provides a useful set of questions to facilitate a strong foundation in the stakeholder engagement process. For ongoing processes such as closure planning, which need continuous engagement, revision and updates, the stakeholder analysis Guideline allows the closure planner to make informed decisions on which goals of the closure plan are likely to be stable, which are subject to external influences and the extent to which the external influences might dictate the work program and milestones set.

Guideline (Toolkit) 2-12 (Partnership assessment) allows the facility to more critically evaluate the prime vehicle for sustainable development in communities, the use of partnerships. The assessment Guideline is equally adaptable to community partners, industry partners, government and NGOs. Many community development programs falter because of inadequate partnering processes, requiring the company to take a more proactive stance and expend greater resources in achieving closure goals. The partnership assessment Guideline should be used to re-evaluate partnerships during review processes as well as at the outset of partnership development.

Guideline (Toolkit) 2-15 (Logical framework) should be used to develop the interim goals and milestones that underpin the overall goals and outcomes. Used in conjunction with **Guideline 2-14** (Overview of monitoring and evaluation Guidelines) and

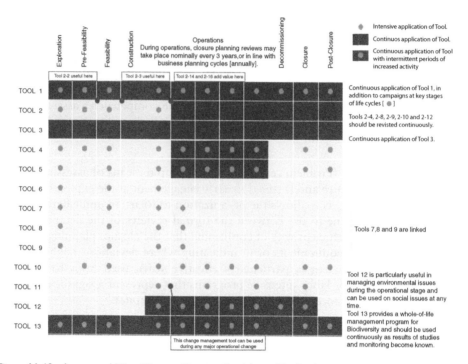

Figure 16.49 Integrated Mine Closure Planning Guidelines (Toolkits) as applicable throughout the life of an operation (Permission: ICMM).

Guideline 2-16 (Indicator development), it allows the systematic progression to post-closure outcomes and allows the user to define key performance indicators. The logical framework Guideline allows the execution strategy of closure planning to be managed. Because many of the closure planning activities are undertaken during operations, the logical framework Guideline suits a number of line management functions and can be used to complement ISO14001 programs for environmental management in operations when coupled with monitoring and evaluation Guidelines and indicator development.

Guideline (Toolkit) 3 Company/community interactions to support integrated closure planning

The relationship between a company and its neighbouring communities develops over time. In the mining industry, the company undergoes a series of transitions prior to its operational phase, and this requires the relationship to be built within a dynamic environment. Building trust and mutual respect, particularly during the usually rapid series of transitions between exploration and operation, is a challenging issue. It is also one of the most important activities to pursue. It is in this series of transitions that the tone of relationships for the first years of an operation's life is set.

Table 16.13 shows a suggested minimum platform for engagement and the actions that a company might take at the different phases of the operation's life to build trust and mutual respect in closure planning.

Table 16.13 Proposed minimum platform for engagement and the actions during different phases of the operation's life to build trust and mutual respect in closure planning (Permission: ICMM).

Phase	Characteristics	Project owner actions that support integrated closure planning	Exploration team	Corporate team	Pre/Feasibility team	Construction team	Operations team	Decommissioning team
Exploration	• Short time frame • Little knowledge of continuity • First impressions • Rising community expectations	Engage with communities	Y					
		Understand expectations	Y					
		Build rapport	Y					
		Provide direction and standards		Y				
		Provide background information		Y				
Pre-feasibility	• Limited time • Intensive studies • Scoping activity • Ability to discuss continuity • 'First strategies' • Rising community expectations	Handover	Y					
		Engage with communities and stakeholders			Y			
		Develop target closure outcome			Y			
		Develop objectives			Y			
		Research and data			Y			
		Build rapport			Y			
		Estimate closure costs			Y			
		Multidisciplinary decision-making			Y			
		Provide direction and standards		Y				
		Provide background information	Y	Y				
Feasibility	• Extensive studies • Extensive engagement • Discuss needs and priorities • Discuss mine options • Discuss development scenarios	Handover	Y					
		Engage with communities and stakeholders			Y			
		Fill knowledge gaps			Y			
		Refine target closure outcome			Y			
		Refine objectives			Y			
		Begin achieving objectives			Y			

Phase	Characteristics	Project owner actions that support integrated closure planning	Exploration team	Corporate team	Pro/Feasibility team	Construction team	Operations team	Decommissioning team
Feasibility	• Real information to hand • Opportunities to inform meaningfully • Rising community expectations	Build rapport			Y			
		Multidisciplinary decision-making			Y			
		Refine closure costs			Y			
		Input to viability (e.g., infrastructure location, constructability)				Y		
		Input to viability (resourcing, capacity, operability)					Y	
		Provide direction and standards		Y				
Construction	• Short and intensive time frame • Influx of construction workers • High stress period • Local labor use • Intensive cash flow into community • Exposure to uncontrolled changes in plans	Review process			Y			
		Activity controls			Y			
		Monitoring			Y			
		Adhere to plans				Y		
		Early advice on variations to plans that may affect closure				Y		
		Engage with communities and stakeholders			Y	Y	Y	
		Confirm construction maintains ability to close successfully				Y	Y	
		Take ownership of closure plan					Y	
Operation	• Stabilization • Opportunities for continuity • Opportunities to develop long-term programs • Local labor use • Long-term capacity building • Opportunities for sustainable partnerships • Opportunity to generate familiarity and stable relationships	Provide direction and standards	Y					
		Review and advise	Y					
		Review assumptions for closure			Y			
		Document as-built information				Y		
		Engage with community and stakeholders					Y	
		Develop detailed plan					Y	
		Update plan every two to three years					Y	
		Implement plan					Y	
		Measure against goals					Y	
		Progressively improve costing accuracy					Y	
		Multidisciplinary decision-making					Y	
		Readiness for sudden closure					Y	
Decommissioning	• Local partnerships important • Risk of decline in community income • 'Last impressions'	Engage with communities and stakeholders					Y	Y
		Update plan					Y	Y
		Implement plan					Y	Y
		Implement decommissioning						Y
		Measure against goals					Y	Y
		Provide direction and standards	Y					
		Review and advise	Y					
		Monitor	Y					
Post-closure	• Long-term monitoring horizon • Signoff on attainment of goals • Relinquishment period	Engage with communities and stakeholders						Y
		Measure against goals						Y
		Provide direction and standards	Y					
		Review and advise	Y					
		Monitor	Y					Y
		Report	Y					Y

Guideline (Toolkit) 4 Risk/opportunity assessment and management
Likelihood scales

Note [as shown in tables 16.14(a); (b) & (c)] that these are examples and suggestions; they can be redefined to suit the circumstances and risk sensitivity of any organization.

Guideline (Toolkit) 5 Knowledge platform mapping

Understanding how much is known and, equally important, how much is unknown is a crucial element in closure planning. It is important to acknowledge the validity or potential invalidity of assumptions made, because long-term closure planning should reduce the unknowns and the corresponding risk of invalid assumptions. In this simple model there are seven platforms of knowledge characterized by the information that is to hand to allow decisions to be made on whether goals are appropriate, whether they can be met, and whether the rate at which they are being met is adequate. These seven platforms are shown in table 16.15, with example numerical rankings assigned to these. By placing metrics on the levels of unknowns, the closure planner is able to track, between review periods, the robustness of information being used to refine closure plans. In addition, the closure planner is able to focus resources such as research activity on reducing unknowns in high-risk areas between review periods.

Guideline (Toolkit) 6 Typical headings for contextual information in a conceptual closure plan

The conceptual closure plan is primarily a vehicle for capturing and communicating the issues affecting closure. Contextual information is important in communicating these issues, as it lays out the constraints and opportunities under which the operation will develop and execute its closure plan. For practical purposes, although cross-references to key documents such as baseline studies and impact assessments should be used, a short synopsis of the key issues in the one closure planning document allows better access to these data by internal stakeholders (feasibility study managers, design teams, construction managers, etc.) as well as external ones. Select only those that apply:

1. Site history
2. Locality and geography
3. Land tenure
4. Land use
5. Settlement status
6. Transport networks
7. Geology
8. Hydrology
9. Hydrogeology
10. Population and demographics
11. Household composition, density and distribution
12. Languages
13. Culture and Heritage
14. Community groups
15. Community organizations
16. Indigenous people

Table 16.14 (a) Likelihood scales to assess to assess risk/opportunity. (b) Description of likelihood scales. (c) Description of consequence scale (Permission: ICMM).

Likelihood Scale	Consequence Scale				
	Consequential	Limited	Overt	Significant	Extreme
1 (improbable)	Low	Low	Medium	Medium	High
2 (unlikely)	Low	Low	Medium	Medium	High
3 (possible)	Low	Medium	High	High	High
4 (likely)	Medium	Medium	High	High	Peak/Very High
5 (almost certain)	Medium	High	High	Peak/Very High	Peak/Very High

Scale	Descriptor	Description
1	Improbable	It would require a substantial change in circumstances to create an environment for this to occur, and even then this is a rare occurrence in the mining and metals industry anywhere.
2	Unlikely	There are no specific circumstances to suggest this could happen, but it has happened before at least once in the mining and metals industry.
3	Possible	There is at least a 5 per cent chance it could happen, or it has happened occasionally in other areas before, or it has occurred (albeit infrequently) in the mining and metals industry in the recorded past or risk mitigation treatment cannot reduce the inherent likelihood further.
4	Likely	There is at least a 50 per cent chance it could happen, or it has happened several times in similar areas before, or this consequence is not uncommon in the mining and metals industry or any risk mitigation treatment cannot reduce the inherent likelihood further.
5	Almost certain	Has happened/will probably happen during mine life and there is no reason to suspect it will not happen again or it has occurred in this area before.

Scale	Negative Consequence	Positive Consequence
C Consequential	Related to, in consequence of. Not inconsequential, but no more severe than that.	Related to, in consequence of. Not inconsequential but no more substantial than that.
L Limited	Some consequence, generally reversible in the short term and/or with modest application of resources (similar to daily resources operating budget for mine, if financial comparisons are appropriate).	Some consequence, not sustainable without significant ongoing application of resources.
O Overt	Consequences may be reversible, usually requiring some time and/or significant application of resources. (Similar to monthly operating budget, if financial comparisons are appropriate).	Consequences may reversible but will generally be sustainable with only modest application of resources.
S Significant	Generally irreversible consequences, with impacts apparent for a prolonged period of time (similar time scale to mine life where time-scale comparisons are appropriate).	Generally sustainable consequences over a prolonged period of time, with little or no ongoing application of resources.
E Extreme	Irreversible consequences, impacts exceeding period similar to life of mine (where time-scale comparisons are appropriate).	Generally sustainable consequences exceeding period similar to life of mine.

Table 16.15 Details of knowledge platform ranking (Permission: ICMM).

Characteristic	Knowledge platform	Ranking
Common Knowledge	Decisions are based on information past history, namely similar sites, data on the experience of others and company's own experience.	20
General Data	Decisions are based on site-specific baseline information, including site specific social, environmental and economic data gathered from representative areas.	30
Focused Data	Decisions are based on good-quality site-specific baseline information gathered from every specific location/aspect around the site that could be affected by moderate consequences or worse.	40
General Analysis	Decisions are based on completed studies, theoretical process or dialogue in representative areas of concern.	60
Focused analysis	Decisions are based on completed studies, theoretical processes or dialogue in every specific location/aspect around the site that could be affected by moderate consequences or worse.	70
General proof	Decisions are based on completed physical and logistical experiments on processes or trialled certain models that provide real information supporting the likelihood of success on representative areas of concern.	80
Focused proof	Decisions are based on completed physical and logistical experiments on processes or trialled certain models that provide real information supporting the likelihood of success in every specific location/aspect around the site that could be affected by moderate consequences or worse.	100

17. Livelihood and income streams
18. Industry and yield
19. Agriculture and yield
20. Per capita income
21. Employment rates and patterns
22. Artisanal mining
23. Educational facilities
24. Literacy and numeracy levels
25. Vocational skills and capacity
26. Health facilities
27. Health statistics (including HIV, malaria and tuberculosis)
28. Maternal health
29. Infant mortality
30. Community infrastructure
31. Government planning schemes
32. Biodiversity
33. Existing social legacies
34. Existing environmental legacies

35. Existing economic legacies
36. Water quality in surface and aquifer resources
37. Air quality
38. Social values requiring protection
39. Environmental values requiring protection
40. Economic values requiring enhancement

Guideline (Toolkit) 7 Goal setting

The worksheet Guideline (tables 16.16 and 16.17) can be used to set goals in the majority of contextual areas (see Guideline 6) and new areas.

Where possible, the partial achievement of goals should be set as milestones, as the following example (shown in table 16.18) illustrates, so that there is a progressive means of assessing whether the facility is 'on track' to meet the defined closure goals.

Guideline (Toolkit) 8 Brainstorming support table for social goal setting (a new tool to support tool 7)

Tables 16.19(a) and 16.19(b) provide some suggested considerations for social goal setting. The list is not exhaustive, and elements should be added or deleted to suit the local conditions of the operation being considered. This format lends itself to facilitated workshops in multi-stakeholder forums and helps identify social risks and opportunities in closure planning.

Guideline (Toolkit) 9 Brainstorming support table for environmental goal setting (a new Guideline to support Guideline 7)

Table 16.20 provides some suggested considerations for environmental goal setting and some trigger words at the bottom of the table that can be used to formulate specific and quantifiable goals to assist in closure planning.

Guideline (Toolkit) 10 Cost risk assessment for closure

This Guideline should be read in conjunction with Financial Assurance for Mine Closure and Reclamation (ICMM, 2005). The costing of closure, particularly at the early stages of an operation's life, is aimed at adequate provisioning for closure. The contextual information in the financial assurance document should provide the practitioner with a sound understanding of why the cost of closure and the adequacy of provisioning are critical elements of closure planning.

There are four structural aspects to cost estimates for each element of the closure plan:

I. The quantity of activity (which may be subject to variance);
II. The cost rate per unit quantity of the closure activity (which may be subject to variance);
III. An allowance that is sensible to apply (for example, using the quantities and rates for topsoil application over a four-month period leading to closure, it is sensible to expect that some rain days may cause the application of topsoil to cease, the standby time incurring a cost that is not captured in quantities and rates); and
IV. A contingency, which is called into play if something unplanned occurs that causes a negative consequence to be triggered, such as the unavailability of local topsoil or statutory rejection of the standard of capping.

Table 16.16   Worksheet that could be used to set goals in the majority of contextual areas (see Guideline 6) and also for the new areas. (Permission: ICMM).

Aspect	What must be protected?	What can be enhanced?	Goals
Land tenure			
Land use			
Settlement status			
Transport networks			
Hydrology			
Hydrogeology			
Population and demographics			
Household composition, density and distribution			
Languages			
Culture and heritage			
Community groups			
Community organizations			
Indigenous people			
Livelihood and income streams			
Industry and yield			
Agriculture and yield			
Per capita income			
Employment rates and patterns			
Artisanal mining			
Educational facilities			
Literacy and numeracy levels			
Vocational skills and capacity			
Health facilities			
Health status (including HIV, malaria and tuberculosis)			
Maternal health			
Infant mortality			
Community infrastructure			
Government planning schemes			
Biodiversity			
Existing social legacies			
Existing environmental legacies			
Water resources			
Air			
Societal values			
Environmental values			
Economic values			

The product of quantities and rates is the base cost. An allowance is added to the base cost for conditions known from previous experience are likely to occur. A contingency is added to the base cost plus allowance to account for the possibility that something could go wrong.

Table 16.17   Comparison of vague and well-defined closure goals (Permission: ICMM).

Examples of vague closure goals	Examples of well-defined closure goals
Land suitable for grazing	Land able to sustain grazing for nine months of the year (non-winter) for up to 100 head of cattle
Two SMEs	Two SMEs with total employment of 100 full-time equivalent local staff
Permanent healthcare facilities in village	30-bed (1% of population) permanent health care facilities with outpatient facility and maternity unit
Water supply to village	Reticulated potable water to a minimum of six standpipes in the village, with sustainable supply at 45 liters per person per day for the population of the village
Improvement in primary education	Achievement of greater than 70% attendance to Grade 4, with equivalent demographic representation of boys and girls

Table 16.18   Setting milestones to assess progress of the defined goals (Permission: ICMM).

Closure activity	Milestone – 0%	Milestone – 25%	Milestone – 50%	Milestone goal–75%	Closure goal
Rehabilitation	Final land use plan formulated	100 hectares graded, topsoil applied and seeded	200 hectares graded, topsoil applied and seeded	300 hectares graded, topsoil applied and seeded	400 hectares graded, topsoil applied and seeded
Capacity building	Determine current and future skill base in village, including gender equality	Vocational training programs in place and full enrolment achieved	Transfer of vocational training admin. to government-owned corporation	Self-funded, self-administered vocational training institution running at full enrolment	Future skill base achieved
Tailings dam closure	Tailings impoundment location and dimensions scoped	Capping options determined and testing program commenced	Capping option chosen and material selection confirmed	Capping instituted on 50% of the (completed) cells in the tailings facility and under validation testing	Capping completed, monitoring in place

The worksheet (table 16.21) shows how this can be applied in the closure planning process to provide cost estimators and financial modellers with a context for conducting cost risk assessments for closure.

Table 16.22 shows how a specific closure cost element (tailings dam capping) can be broken down to provide a financial modeller with information to undertake a probabilistic cost analysis.

Table 16.19(a) Suggested considerations for social goal setting (Permission: ICMM).

	What social and socio-economic values or gains can be achieved
Poverty	In poverty reduction?
Hunger	In hunger reduction?
Education	In education?
Gender equality	In gender equality?
Child mortality	In child mortality?
Maternal health	In maternal health?
HIV/AIDS, malaria	The management of HIV/AIDS, malaria and other diseases?
Water supply	In water supply?
Health care	In health care management?
Employment	In the employment market?
Youth employment	In youth employment?
Employability	The employability of people in the community?

Table 16.19(b) Suggested considerations for social goal setting (Permission: ICMM).

Closure category	Typical open question
	What social and socio-economic values or gains can be achieved
Technology	In the application of technology?
Recreation	In recreation?
Infrastructure	Through the adaptation of infrastructure?
Indigenous	Losses are inherent to indigenous affairs?
Cultural	Losses are inherent to the cultural heritage of the community?
Enterprise	Through the generation of enterprise?

Table 16.20 Proposed considerations for environmental goal setting to formulate specific and quantifiable goals to assist in closure planning (Permission: ICMM).

Closure category	Typical open question
	What environmental values, gains or losses are inherent
Land resources	To land resources?
Water resources	To water resources?
Terrestrial flora	To terrestrial flora?
Terrestrial fauna	To terrestrial fauna?
Aquatic flora	To aquatic flora?
Aquatic fauna	To aquatic fauna?
Acid rock drainage	To acid rock drainage?
Air	To air?
Noise	To noise?
Waste	To waste?
Overburden dump	To the overburden dumps?
Tailings dam	To the tailings dams?

Table 16.21   The worksheet showing to guide cost estimators and financial modelers with a context for conducting cost risk assessments for closure (Permission: ICMM).

Activity	Quality Is the quantity of activity being:			Rate Is the unit of the cost being:		Allowance What variations, from experience, should be expected?	Contingency What could go wrong in this activity to increase costs?
	Measured/ surveyed?	Estimated through calculation?	Guessed?	Extracted from quotes?	Estimated from experience?		
Goal × Activity 1							
Goal × Activity 2							
Goal × Activity 3							
Goal × Activity 4							

Table 16.22   Illustration to understand a probabilistic cost analysis (Permission: ICMM).

Item	Quantity	−Variance in Quantity	+Variance in Quantity	Rate	−Variance in Rate	+Variance in Rate	Allowance	Contingency
Clay sealing material source silty clay (SC) to clay (CL) material	50,000 m³	−5%	+30%	$25/m³	−0%	+10%	+10%	+50%
Clay sealing material place at 95% compaction at optimum moisture content −1% +2% of OMC	50,000 m³	5%	+30%	$18/m³	−5%	+20%	+10%	+20%
Capillary material source gravely sand d50 = 3 mm	30,000 m³	−0%	+10%	$6/m³	−0%	+20%	+15%	+10%
Capillary material place, wheel loading & leveling only	30,000 m³	−0%	+10%	$7/m³	−0%	+10%	+10%	+10%
Topsoil material source from stockpile	25,000 m³	−10%	+25%	$5/m³	−5%	+5%	+5%	+10%

Notes
1. Wide variance here because the final tailings dam facility dimensions are unknown.
2. High contingency because the estimator has not yet identified where the material will be sourced from, and whether or not the appropriate material can be found on site.
3. As per Note 1.
4. Wide variance on labor costs because it has not been decided whether this will be an internal or external (contractor) cost – base rate assumes internal cost.
5. High contingency because it is uncertain (see Note 2) that material exists, and there is a possibility for processes such as gypsum stabilization of local or import material to meet specifications.
6. High variance because topsoil stockpile balance has not been carried out accurately.

Guideline (Toolkit) 11 Change management worksheet

At the juncture of the end of construction and the start of operations, many things may have changed between the pre-feasibility and operational stages. Although the impact assessment process may have been designed to scope out the majority of these changes, many actions deviate from their planned direction during the detailed design and construction periods. Design optimization, geotechnical constraints, social intervention, government statutes and other external factors can combine to ensure that one or more elements of the conceptual closure plan may have changed. Pl. refer table 16.23.

Guideline (Toolkit) 12 The domain model (Note: the figure is largely mining-activity-based).

A useful approach to dividing up the work to be carried out on closure is to segregate the facility into specific areas or domains. Each domain is treated as a separate detailed entity within an overall plan that deals with common issues like drainage and site monitoring. The following factors should be taken into account when developing a plan for each domain:

- The amount and area of disturbance
- Applicable legislation
- Hazardous areas and risk assessments
- A plan for deconstruction and decommissioning
- Contamination and mitigation
- End land use
- Required earthworks and capping
- Control of erosion
- A rehabilitation plan
- Monitoring
- Cost estimates
- Research

Waste materials should be viewed as a resource to be used to rehabilitate other areas. For instance, benign overburden can be used to cap potentially acid-producing material. Proactive thinking can save the operation future liabilities. Each domain should have its own plan (see example shown in table 16.24). Assumptions, inclusions and exclusions should be documented. Examples of domains at a mine are:

- Ore processing area
- Shaft hoisting and headgear (underground mines)

Table 16.23   Analysis for change in management to adopt corrective measures (Permission: ICMM).

Status change during construction	Extent of change	Could the ability to meet goals have changed [Y/N]	Should detailed closure plan be updated to capture [Y/N]	How?
Did in-migration occur?				
Were there changes in the biodiversity around the site?				
Were fauna corridors altered or transected?				
Were the courses of creeks or rivers changed?				
Did household incomes change?				
Could these changes have resulted in variability in income per household?				
Did average household sizes increase or decrease during construction?				
Were compensation payments made during construction?				
Was household relocation affected during construction?				
Could local health conditions, including the incidence of sexually transmitted diseases, have changed?				
Did local service industries benefit from, or increase as a result of, construction?				
Did local materials and supply industries benefit from, or increase as a result of, construction?				
Did employment patterns change?				
Did community organizations change, or did new community organizations form?				
Did the construction period result in new stakeholders entering the arena?				
Were there changes to the size, position or construction method applied to infrastructure during construction?				

Table 16.24   Illustration (an example showing how each domain should have its own plan) (Permission: ICMM).

xxx AREA	
Description	
Area of Disturbance	Void xx hectares; Waste Dumps xx hectares
Status	Active
Closure Date	xx void and waste rock dumps will be closed in 200x
Infrastructure to be Retained	Final void

OBLIGATIONS RELATING TO CLOSURE		
Subject	Obligation	Relationship to Closure
Regulatory Condition - Rehabilitation	(F4-1) Progressive rehabilitation must commence when areas become available within the operational land.	Incorporating details and requirements for leaving voids safe and stable after closure.

Final Land Use Objectives	Void
	(a) Safe with minimal risk to the public, native fauna and livestock
EXAMPLES	(b) Conceptual land use options include water bodies or partially filled water bodies. A decision will be based on the results of detailed geochemical and hydrological studies
	Waste Dumps
	(a) Provide an acceptable post-disturbance land use capability/suitability
	(b) Provide acceptable, stable post-disturbance landforms
	(c) Protect surface and groundwater quality on-site and leaving the mining lease
	(d) Rehabilitated using technically effective and cost-efficient methods and proven engineering practices to ensure that no long-term maintenance is required beyond the post-closure phase of 5 years
	Make the area safe with minimal risk to the public, native fauna and livestock

CLOSURE COMPLETION CRITERIA							
Description	xxx Void	Area (ha)	14	Photo No.	Photo 1	Timing	2008-12

Engineering and Rehabilitation Activities
(a)   Excavate and haul waste rock material to construct perimeter bund walls
(b)   Fencing of void perimeters
(c)   Purchase and erect warning signs
(d)   Final pit water balance and groundwater models
(e)   Geotechnical stability assessment for long-term pit wall stability

Statutory Sign-off			
No	Yes	Date	Document Reference No.
	✓	-	-

Description	xx Waste Dumps	Area (ha)	28	Photo No.	Photo 1	Timing	2008-12

Engineering and Rehabilitation Activities
(a)   Selective handling of acidic rock on outer dump face of xx Dump – dispose on top surface
(b)   Re-profile xx I/O stockpile to drain towards final void
(c)   Re-profile Central dump to drain towards final void
(d)   Xx I/O stockpile – stable, minor leaching, no activities
(e)   Excavate, load and haul inert oxide waste to the reshaped xx dumps
(f)   Moderate earthworks to place store and release cover system (inert oxide rock) over PAF, reshaped dump surface and rehabilitate
(g)   Re-profile and deep rip the balance of the waste dump surface
(h)   Minor erosion control works and seeding on the balance of the waste dump surface
(i)   Geotechnical assessment to demonstrate long-term dump stability

Table 16.24   Continued.

Statutory Sign-off			
No	Yes	Date	Document Reference No.
✓		-	-
Post-Closure Activities		Void (a) Continue surface and groundwater quality monitoring for 5 years, including void water qualities and monitoring void water levels Waste Dumps Monitoring and maintenance of: (a) Revegetation works; (b) Soil erosion and soil erosion control structures; (c) Weed control in and around the rehabilitation area; and (d) Surface water quality in leachate collection ponds as per current monitoring schedules.		
Specific Closure Assumptions EXAMPLES		Void (a) The void will not be backfilled (b) The void will be allowed to flood naturally (c) Bunding and fencing will occur at closure (d) There will be no impact on groundwater as a result of water accumulating in the final void (e) The closure strategy adopted for this closure plan will be accepted by all stakeholders Waste Dumps (a) Geochemical testwork of the dumps will confirm the applicability of proposed rehabilitation methods		
Closure Material Sources EXAMPLES		Void (a) 4500 m^3 of inert waste rock to construct 2 m high perimeter bund walls (b) 2500 m of fencing (c) 50 warning signs Waste Dump (a) selective handling machinery for acidic outer dump waste rock face (b) 80,000 m^3 of inert waste rock from xx dump (c) Seed and fertilizer to rehabilitate 28 ha		
Waste Disposal Sites		Not applicable		
Other Issues		None identified		
REHABILITATION COSTS ($) Engineering and Rehabilitation Cost Closure Administration Cost Post-closure Management Cost Total		$500,000 $20,000 $10,000 $530,000		
Costs Not Included		Consultant investigations on additional geochemical testwork		
Cost Saving Opportunities		(a) There may be an opportunity to generate additional cash flow by processing the I/O stockpile at the same time as reducing existing liabilities that would otherwise required re-profiling and rehabilitation. (b) Reduce existing liability by aiming to gain sign-off of the xx waste dump and final void as soon as regulators finalize the progressive rehabilitation policy.		
Further Investigation/Studies Required		Long-term water quality and groundwater impacts will be needed as well as investigations on geotechnical stability of the outer dump face and verification of cover depth to restrict infiltration before regulators accept the final landforms as they are		
Liabilities/Risks/Hazards		(a) The encapsulation method may not be effective in reducing acid leachate generation to acceptable levels (b) 'Hot spots' of PAF material may develop as acidic rock is exposed during any waste rock re-profiling		

- Workshops
- Tailings storage facility
- Process and raw water facilities
- Open voids and declines/shafts
- Roads (infrastructure and haul and exploration)
- Camps and other offices

For accuracy, the operation should use Geographical Information System (GIS) digital terrain models and aerial photos to illustrate the domain features and boundaries; 3D models of waste dumps, voids, tailings dams and other structures are also very useful.

Guideline (Toolkit) 13 Biodiversity management
The ICMM Good Practice Guidance for Mining and Biodiversity was published in 2006. In fact, Biodiversity impacts play a large part in the environmental scope of mining impacts. In many remote and rural areas, biodiversity management can become one of the key elements defining successful mine closure.

Figure 16.50 of the Good Practice Guidance provides a useful scoping Guideline (Toolkit) for capturing, for closure plan purposes, the potential impacts on biodiversity during the various stages of the facility life. Mapping of these potential impacts permits simpler access to the management, monitoring and evaluation processes that need to be included in the Conceptual Closure Plan. This figure, with site-specific impact mapping, could be included in conceptual closure plans to focus the issue.

Figure 16.51 of the Good Practice Guidance provides more detailed information on the intersection between various mining activities and potential biodiversity impacts. This figure should be a standard part of any detailed closure plan, as it distils the closure planning issues into a practical focal point. It is particularly useful when linked to Guideline 12, the Domain Model, as this figure is largely mining-activity-based.

16.7.3.3 *Glossary*

Care and maintenance – The period following temporary cessation of operations when infrastructure remains largely intact and the site continues to be managed.

Closure Planning – A process that extends over the mine life cycle and that typically culminates in tenement relinquishment. It includes decommissioning and rehabilitation. The term closure alone is sometimes used to indicate the point at which operations cease, infrastructure is removed and management of the site is largely limited to monitoring.

Decommissioning – The process that begins near or at the cessation of mineral production and ends with the removal of all unwanted infrastructure and services.

Deterministic estimates – Estimates of value (cost or benefit) of the outcome of an event occurring, expressed as a single mean or mode value and a range of single values (e.g., minimum, maximum).

Probabilistic estimates – Estimates of value (cost or benefit) that account for the likelihood of occurrence and the range of values of the outcomes. Values are expressed

POTENTIAL IMPACTS	Early stages of exploration	Exploration drilling	Access road construction	Land clearance (for construction, etc.)	Obtaining construction materials	Construction related infrastructure	Roads, rail & export infrastructure	Pipelines for slurries or concentrates	Energy/power & transmission lines	Water sources, wastewater treatment	Transport of hazardous materials
Impacts on terrestrial biodiversity											
Loss of ecosystems and habitats	●	●	●	●			●	●	●		●
Loss of rare and endangered species	●	●	●	●	●		●	●	●		●
Effects on sensitive or migratory species	●	●	●	●	●		●	●	●		●
Effects of induced development on biodiversity		●	●		●		●				●
Aquatic biodiversity & impacts of discharges											
Altered hydrologic regimes		●	●	●	●		●		●	●	●
Altered hydrogeological regimes	●		●								
Increased heavy metals, acidity or pollution	●		●	●	●		●		●	●	●
Increased turbidity (suspended solids)	●	●	●	●	●		●	●	●	●	●
Risk of groundwater contamination	●			●	●		●	●		●	●
Air quality related impacts on biodiversity											
Increased ambient particulates (TSP)	●	●	●	●	●		●		●		●
Increased ambient sulfur dioxide (SO$_2$)					●				●		●
Increased ambient oxides of nitrogen (NO$_x$)				●					●		●
Increased ambient heavy metals									●		
Social interfaces with biodiversity											
Loss of access to fisheries		●	●				●	●	●		
Loss of access to fruit trees, medicinal plants		●	●	●			●	●	●		
Loss of access to forage crops or grazing	●	●	●				●	●	●		
Restricted access to biodiversity resources		●	●				●	●	●		
Increased hunting pressures	●	●	●	●	●		●		●		●
Induced development impacts on biodiversity		●	●	●			●		●		●

Figure 16.50 A useful scoping Guideline for capturing, for closure plan purposes, the potential impacts on biodiversity during the various stages of the facility life (Permission: ICMM).

through a statistical analysis (e.g., Monte Carlo simulation) using a statistical distribution over the range of possible values accounting for the probability and timing of the event occurring.

Reclamation/Rehabilitation – Terms used interchangeably to mean the return of disturbed land to a stable and productive condition.

Relinquishment – Formal approval by the relevant regulating authority indicating that the completion criteria for the mine have been met to the satisfaction of the authority.

Stakeholder – A person, group or organization with the potential to affect or be affected by the process or outcome of mine closure.

Internal stakeholders: It includes various groups or teams within the mining company such as: exploration team, corporate team, feasibility team, construction closure team, operational team, decommissioning/closure team.

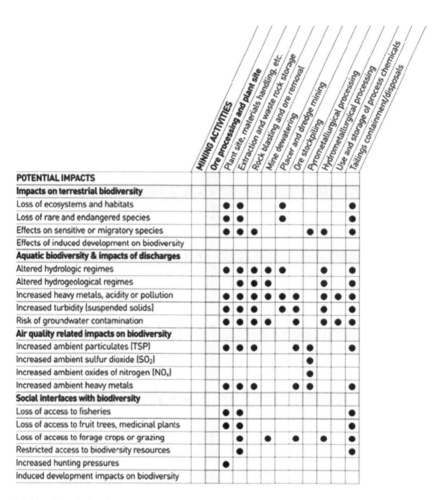

Figure 16.51 Detailed information on the intersection between various mining activities and potential biodiversity impacts (Permission: ICMM).

External stakeholders

Questions to ask when considering external stakeholders include:

- Who in the area may be directly affected by the construction and operation of the mine site?
- Who in the area may be indirectly affected by the construction and operation of the mine site?
- Who within or outside the area may be indirectly affected by upstream activities that support construction and operation?
- Who may influence the ability of the project to gain or retain its license to operate?
- Who are the interested stakeholders?

ICCM Mine Closure Toolkit, 2008. Source: Based on Department of Industry, Tourism and Resources, Government of Australia, Mine Closure and **Completion**, October 2006.

16.8   THE WAY FORWARD

- Present depths of stopes to mine out precious deposits are attaining ultra depths and mining them skilfully is a big challenge.
- For the rest of the ore and coal deposits; small to medium sized underground mines with the use of widely diversified tools, appliances and equipment; deeper, hotter, lower grade, highly stressed deposits, abnormally thick and thin coal seams, poor recoveries, adverse working conditions including rock-bumps and a few others are the constraints that need to be addressed with a sense of urgency. Bringing some innovative and effective solutions in line with global best practices to bring about a paradigm shift in this sector, is therefore, warranted.
- A thorough technology transfer from one country to another and adapting methods, techniques and equipment that incorporates economics, environment and modern management could meet this challenge.

QUESTIONS

A. Answer following questions:
1. "Stopes are the blocks of hidden wealth surround by environment that is risky to men, machines and equipment." Do you agree with this statement? How skilfully you could recover hidden wealth encompassing them?" "Success lies in the selection of proper stoping system and deploying matching methods, equipment and techniques." How you could achieve this?
2. A 2.2 m thick steeply dipping deposit of high grade multi-metal exists below a depth of 200 m or more. The ore is hard but the walls are weak. Propose a mining method. Illustrate it giving its layout. Write down the limitations of the method proposed.
3. A 2.5 m thick coal seam flatly dipping is to be mined at a depth of 250 m from the surface. The seam has a wide spread. The roof is weak. The surface needs protection. Suggest a mining method. Give a layout. Write down the stope preparation activities.
4. A 40 m thick Pb-Zn deposit exists below ground between 125–625 m datum. The walls of the deposit are regular and competent with an average dip of 70°. The ore is hard. Propose a method to yield a high rate of production with the application of modern equipment. Draw the layout of the method proposed and list the sets of equipment you shall need.
5. A coal seam 2.2 m thick dipping at 3° having weak and cavable roof but strong floor exists below a depth of 150 m from the surface. It has a large extension along the dip as well as the strike. Name the method you will apply. Give its layout. List the types of support which could be suitable to support the face.
6. A copper deposit has been identified in the form of steeply dipping lenses of varying thickness in the range of 4–12 m below a depth of 200 m from the surface. The h/w and f/w rocks are competent. Propose a suitable mining method, giving reasons. The method selected should give flexibility to store the ore within the stope. Give layout of the stope showing different views.
7. A galena (Pb ore) vein 3 m thick is dipping at 30°. The wall rocks are competent. Suggest the stoping method and draw its layout.

8. A part of a coal deposit is flatly dipping 2 m thick coal seam located at 100 m below the surface. It has a wide spread along and across the strike directions. The overlying strata are weak and cavable. Propose a suitable mining method. Give its layout and sets of equipment to carry out the unit operations, if a high rate of production is to be achieved. How much productivity you should expect so that the method selected work out to be cheapest?

9. A steeply dipping 10 m thick uranium ore shoot, which is fairly strong, has been investigated below a depth of 250 m from surface and its extension up to a depth of 1000 m has been proved. The walls are weak. Propose a suitable stoping method. Give its layout, if modern equipment to carry out various unit operations, are to be used to achieve a reasonably high rate of production.

10. A sulfide multi-metal lens (silver, lead and zinc) having a thickness of 40 m or more is dipping at 60°. This lens occurs between depths of 100 m–700 m or even more. The wall rocks are weak and cavable but the ore is competent and need fragmentation by drilling and blasting. Propose a stoping method and give its layout.

11. A sulfide multi-metal vein (silver, lead and zinc) 22 m thick dipping at 65° having weak hanging wall exists below a depth of 250 m. Its extension along depth continues up to 1000 m or more. Propose a stoping method. Draw the layout. Write down the percentage extraction, you expect and why?

12. A weak sulfide orebody with good percentage of silver, having varying thickness of 4–8 m is dipping at 60°. It exists between depths of 500–1000 m or even more. The wall rocks are weak. Which mining method will be most suitable? Give its layout. Write down the limitations of the method proposed.

13. Based on shape of the orebodies, classify them.

14. Based on the orebody thickness, the sublevel stoping can be classified as: Longitudinal sublevel stoping and Transverse sublevel stoping: List the conditions under which each of these variations of the sublevel stoping methods you would propose? Give a layout (at least two views) of any of these methods. Specify conditions, you have assumed, if any. List the equipment which you would deploy for these unit-operations: drilling, mucking and transportation.

15. Calculate the amount of stowing material/day if daily productions from a copper mine using cut-and-fill stoping is 1500 tonnes. The stowing material of 1.6 t/m^3 is to be used. Ore density is 3 t/m^3. Consider the value of K factor as 0.7.

16. Classify deposits as per the following criterion: (a) Thickness. (b) Dip. (c) Depth. (d) Strength.

17. Classify different underground mucking equipment. Also mention their field of application. Give performance curves for LHDs, Cavos and rocker shovels.

18. Compare important features of surface and underground mining. What is the limiting criteria for surface mining?

19. Compare the technical features of the various stoping methods you studied.

20. Consideration of recovery, safety and economy of the operations are of prime importance while planning liquidation. How you could achieve these objectives? Group liquidation operations as per the stoping methods.

21. Define 'spherical charge'. In a VCR stoping, the blasthole diameter selected is 165 mm. Calculate the length of spherical charge to be placed in the hole. If the density of explosive used is 1.6 g/cm^3, calculate weight of explosive to be charged.

22. Define the terms stope and stoping. Classify deposits based on their thickness.

23. Differentiate between longwall advancing and retreating and write down their merits and limitations. Write other variants of longwall mining and draw a layout for any one of them. Also work out the following for the layout you have chosen: stope preparations and a list of equipment to carry out various unit operations. How is the length of a long-wall coal face equipped with a cutter loader calculated?

24. Draw a layout of transverse sublevel caving stope. Mention suitable conditions for its application. List the stope preparation activities and three limitations of this method.

25. For room and pillar mining, sketch out the different types of rooms.

26. Give a general classification of orebodies based on their dip. How does dip of the deposit influence mining operations? Describe and illustrate wherever practicable.

27. Give a layout of a stope and pillar stope. Write down suitable conditions for its application.

28. Give the layout of a transverse sublevel caving stope. Write down suitable conditions for its application. What are the limitations of this method? Mention the range of dimensions used to space the sublevels (vertically as well as horizontally).

29. Give main and sub classification by a way of tabular chart of the underground stoping methods. Mention the governing parameters in this classification.

30. Group orebodies and country rocks (enclosing wall rocks) based on their stability.

31. How can you examine the facility a stoping method gives for rock fragmentation?

32. How you would take care of the direction of cleats during mining, to achieve best results? Illustrate it.

33. Illustrate the influence of dip in a sublevel caving system. Describe it briefly.

34. In a sublevel caving system if dip of deposit 50° in one case and 80° in another. Do you think it is going to matter and in what way?

35. In a VCR, stoping hole dia. is 165 mm. Calculate the length of spherical charge to be placed in a hole. If this charge weighs 34 kg, an explosive of what sp. gravity should be chosen?

36. In caving stopes name the techniques by which caving can be induced.

37. List considerations that should be looked at while planning a heavy-blast for liquidation purpose?

38. List the characteristics of deposits that should be considered in the selection of a stoping method. Give a classification of stoping methods based on deposit characteristics. What typical considerations are to be weighted in selecting a stoping method? (Hint: for example one of the considerations is that it should have maximum safety; likewise, you should write other considerations to be made)

39. List different ore drawing schemes from the stopes.

40. List different ore winning postures (i.e. direction of mining). Classify deposits based on their thickness. Also give a general classification of orebodies based on their dip.

41. List the drills and loaders of different types, giving utility of each one of them in underground mines.

42. List the desirable features a stoping method should be able to render, if selected.

43. List the parameters/factors which influence selection of a stoping method. Illustrate, wherever practicable, and describe each one of them in detail.

44. List out the stoping methods falls under the open stoping class. Also list out the variant of each one of these stoping methods.
45. List the variants of cut and fill stoping. Mention which variant of cut and fill stoping can allow maximum exposure of the stope of open space and which one the least.
46. List the problems of deep mining.
47. List the backfill systems. Write down materials used for filling purposes. Give their shrinkage coefficient. Which system is most suitable to minimize the risk of fire and rock burst?
48. List the desirable features a stoping method should be able to render, if selected. While mining sulfide ores what problems would you usually encounter?
49. List the loader used to charge explosives of different kinds. Write the hazards associated while charging ANFO pneumatically and how it can be minimized or dealt with. Give sketch of any one of the ANO loaders.
50. List the parameters /factors which influence selection of a stoping method.
51. List the prevalent systems of drawing ore from the block caving stopes. Draw sketch of any one of them.
52. List type of pillars known to you and mention the utility of each one of them. How they are finally recovered? What is the usual percentage recovery from them?
53. Mechanization means performing underground operations using machines. List three degrees (categories) of mechanization and write their influence on achieving productivity and cost of mining.
54. Mention the field of application of each of these techniques: bamboo blasting, plaster shooting, pop shooting, mechanical rock-breaker.
55. Mention the factors that govern the caving mechanism in sublevel caving system.
56. Mine closure programmes comprise three phases; list them. Describe them briefly.
57. Name the members of a square set. Where would you use: (i) A drainage tower. (ii) Bulk head.
58. Name types of underground mines known to you. Classify them as per their size or capacity and list types of equipment available for mucking in these mines and suggest their matching transportation equipment or system also.
59. Take an overview and prepare a list of equipment you know about for carrying out these unit operations: drilling, firing and charging explosives, mucking, transportation. Write locales (e.g. development, stoping, production) with respect to underground workings, and where they will be most suitable. Also write down matching equipment taking into consideration the degree of mechanization.
60. The present depth of a steeply dipping South African gold vein, 1.5 m thick is 2000 m and more. It has strong wall rocks. This vein has been already worked out from shallow depth to its present depth. Propose the mining method. Give a layout. Mention, will it be possible to achieve a high rate of production by the method you proposed?
61. Relative exposure of unfilled space within a cut-and-fill stope: good back/roof conditions allow conventional and mechanized methods; whereas bad back/roof conditions allows little unexposed space, and the costlier variants of cut and fill system are applicable. List the variants of cut-and-fill stoping in each of these scenarios.
62. Sketch out a section of sublevel caving operation showing relative position of sublevels (vertically as well as horizontally) with a range of dimensions usually used.

63. Sketch out any of the back filling systems known to you.
64. To determine the stability of strata it is important to know nature of caving; list the three modes of caving that are usually observed.
65. What are the main factors or points you will consider while selecting mucking and transportation equipment for their use in mines?
66. What is a 'Coyote Blasting'? Describe it. Where does it find its application?
67. What is the block caving system of mining (describe and illustrate)? Classify it. List the prevalent systems of drawing ore from the block caving stopes.
68. What is liquidation? What are its aims? Classify liquidation by caving.
69. Where (i.e. what type of deposit or mines) can you encounter the following problems: (i) Caking (ii) Oxidation (iii) Fire damp (methane).
70. Why in underground mining situations is selection of a mining method vital?
71. With regard to stability of ores and country rocks, no index is available which determines the permissible value of the exposure: time- and span-wise. However, they may be divided into five groups; list them.
72. Write limitations of (i) top slicing and (iii) stull stoping.

B. Propose a suitable stoping method for the following situations:
1. A 2.2 m thick steeply dipping deposit of high grade multi-metal exists below a depth of 200 m or more. The orebody is strong but the walls are weak.
2. A 2 m thick coal seam flatly dipping is to be mined at a depth of 250 m from the surface. The seam has a wide spread. Caving can be allowed.
3. A 3 m thick coal seam flatly dipping is to be mined at a depth of 250 m from the surface. The seam has a wide spread. The roof is weak. The surface needs protection.
4. A 43 m thick multi metal ore deposit is steeply dipping. It has strong wall rocks. The deposit exists between 50 m to 500 m below surface.
5. A 50 m thick Pb-Zn deposit exists below ground between 225–625 m datum. The walls of the deposit are weak and it has an average dip of 70°. The orebody is medium hard to strong.
6. A 70 m thick manganese deposit, flatly dipping, is to be mined from the depth of 80 m from the surface. The deposit has a wide spread. The roof is weak, fractured and cavable. The surface does not need protection.
7. A 7 m thick deposit of high grade multi-metal deposit exists below a depth of 200 m or more. The ore and walls are weak to very weak.
8. A coal seam 2.4 m thick dipping at 5° having a strong roof and floor has been investigated recently. The deposit exists below a depth of about 150 m from the surface and has large extension along dip as well as strike.
9. A coking quality coal having wide extension along depth and strike directions is flatly dipping seam of 1.5 m thickness. The roof is bad but floor is fairly strong. A high rate of production is required.
10. A copper deposit 15 m thick dipping at 65° having competent walls exists below a depth of 150 m. Its extension along depth continues up to 1000 m or more.
11. A copper deposit having wide extension along depth and strike directions is steeply dipping. Its thickness ranges from 4–10 m. Both ore and wall rocks are strong.
12. A copper vein 5 m thick is dipping at 30°. The wall rocks are competent.

13. A high grade multi metal, 35 m thick deposit dipping at 65°, having weak hanging wall exists below a depth of 250 m.
14. A Pb-Zn deposit having average thickness of 25 m and dip of 75°. The wall rocks are competent.
15. A porphyry copper deposit (weak and cavable), 55 m thick dipping at 85° having weak, fractured and cavable hanging wall exists below a depth of 150 m. Its extension along depth continues up to 1000 m or more.
16. A steeply dipping 16 m thick copper deposit, which is fairly strong, has been investigated recently below a depth of 150 m from surface and its extension up to a depth of 500 m has been proved. It has weak and cavable hanging wall.
17. A steeply dipping 40 m thick tin deposit having strong footwall but hanging wall is weak and cavable.
18. A sulphide multi metal vein (silver, lead and zinc) having varying thickness of 5–12 m is dipping at 60°. This vein exists between depth 500–1000 m or even more. The wall rocks are weak.
19. Present depth of a steeply dipping South African gold vein, 1.5 m thick is 2000 m and more. It has strong wall rocks. This vein has been already worked out from shallow depth to its present depth.

C. **Workout the following questions:**
1. For room and pillar mining sketch out the following:
   (a) Different types of rooms.
   (b) Different types of arrangements for driving rooms and entries.
   (c) Room and pillar advancing.
   (d) Room and pillar retreating.
   (e) Various schemes to extract pillars.
2. List out the variants of sublevel stoping. Draw their layouts explaining the main features associated with them. Define the term 'spherical charge'.
3. Work out the following in conjunction with sublevel stoping, in general:
   (a) Suitable conditions for application of this method.
   (b) List of equipment to carry out various unit operations.
   (c) Advantages and limitations.
4. List the variants of shrinkage stoping. Draw their layouts explaining the main features associated with them.
5. Work out the following in conjunction with shrinkage stoping, in general:
   (a) Suitable conditions for application of this method.
   (b) List of equipment to carry out various unit operations.
   (c) Advantages and limitations.
6. Work out, in conjunction with stull stoping:
   (a) Layouts showing different views. Also sketch a stull set and name its members.
   (b) Suitable conditions for application of this method.
   (c) Stope preparation/development.
   (d) List of equipment to carry out various unit operations.
   (e) Advantages and limitations.
7. List out the variants of cut and fill stoping. Draw their layouts (each one of them) explaining the specific situations/conditions under which application of each of

these methods will be most suitable. Mention which variant of cut and fill stoping can allow maximum exposure of the stope of open space and which one the least.

8. Work out the following in conjunction with cut-and-fill stoping, in general:
   (a) Suitable conditions for application of this method.
   (b) List of equipment to carry out various unit operations.
   (c) Advantages and limitations.

9. Work out, in conjunction with square set stoping:
   (a) Layouts showing different views. Also sketch square set and name its members.
   (b) Suitable conditions for application of this method.
   (c) Stope preparation/development.
   (d) List of equipment to carry out various unit operations.
   (e) Advantages and limitations.

10. Work out the following in conjunction with longwall mining:
    a) List out the variants of longwall mining. Draw their layouts (each one of them) explaining the specific situations/conditions under which application of each of these methods will be most suitable. Also work out the following in each case:
       I. Stope preparations and list of equipment to carry out various unit operations.
       II. Specify the advantages and limitations of longwall mining, in general.

11. (a) Sketch out a section of sublevel caving operation showing relative position of sublevels (vertically as well as horizontally) with a range of dimensions usually used.
    (b) Name the stoping method/methods with following features:
        a. Costliest
        b. Cheapest
        c. Highest dilution
        d. Least recovery
        e. Highest productivity
        f. Lowest productivity
        g. Early production but highest development
        h. Subsidence unavoidable
        i. Ore tie up during stoping
        j. Least exposure to unsafe conditions.

12. List out the variants of Room and Pillar mining. Draw the layout for any one of them. Workout the following in conjunction with room and pillar mining, in general:
    a. Suitable conditions for application of this method.
    b. List of equipment to carry out various unit operations.
    c. Advantages and limitations.

13. A sulfide ore free from clay depicts the following characteristics:
    Ore strength: moderate to strong
    Rock strength: fairly strong to strong.
    Deposit shape: tabular, regular dip and boundaries.
    Deposit dip: steep
    Average thickness: 20 m
    Ore grade: fairly uniform
    Depth: between 300–500 m.
    Ore value: comparatively low.

Propose a suitable method. Also draw its layout. Outline the stope preparation activities. Give advantages and limitations of the proposed method.

14. Name the costliest method and work out, in conjunction with it, the following:
    a. A layout.
    b. Suitable conditions for application of this method.
    c. Advantages and limitations.

15. List out the variants of sublevel caving. Draw their layouts (for each one of them showing different views) explaining the specific situations/conditions under which, the application of each of these methods will be most suitable. Also work out the following (for each of these variants):
    a. Stope preparation.
    b. List of equipment to carry out various unit operations.
    c. Advantages and limitations.

16. What is the block caving system of mining (describe and illustrate). Classify it, explaining salient features of each class. List , sketch and describe the prevalent systems of drawing ore from the block caving stopes. Draw a sketch showing the relative position of muck drawing from a series of draw points.

17. Work out the following in conjunction with the block caving system of mining, in general:
    a. Conditions suitable for its application.
    b. Stope preparation
    c. List of equipment to carry out various unit operations.
    d. Advantages and limitations.

18. Differentiate between following and mention where you will adopt them (in each case separately):
    a. Longwall advancing and longwall retreating.
    b. Transverse sublevel caving and longitudinal sublevel caving.
    c. VCR stoping and conventional large blasthole stoping.
    d. Top slicing and sublevel caving.
    e. In block caving – draw-point system and Finger raise system of drawing muck from stopes.

19. Give type of drills, mucking equipment and transportation units which are usually used for the following methods:
    a. Block caving
    b. Cut-and-fill stoping
    c. Room and pillar stoping
    d. Shrinkage stoping
    e. Sublevel caving
    f. Sublevel stoping/blasthole stoping and big blasthole stoping.

23. Where you will adopt the following ore winning patterns during the stoping operations, mention the stoping methods for which these systems are applicable:
    a. Longitudinal
    b. Transverse
    c. Upper level to lower level
    d. Lower level to upper level
    e. Breasting.

24. Differentiate between
    a. 'Slot' and 'Slot-Raise' and in which stoping methods they are driven/made?
    b. 'Toe-spacing' and 'Ring Burden' and in which stoping methods is 'Ring Drilling' done?
    c. Stopping and Stoping.
    d. Decline and Incline.
    e. Tunnel and Adit.
    f. Trough and Trough drive.
25. (a) Mention suitable conditions, in general, for the application of cut-and-fill stoping:
    • Ore strength
    • Rock strength
    • Deposit shape
    • Deposit dip
    • Deposit Size
    • Thickness
    • Ore grade
    • Depth (mention the max. depth up to which it could be applied)
    (b) List the activities/operations you would have to execute to prepare a 'cut-and-fill stope'.
    (c) List the mines in your own country, if any, where this method is being used and give details of any one of them including stope layout and production/day/stope.
26. Give following details in conjunction with 'Transverse Sublevel Caving' method:
    (a) Layout (at least 2 views: plan, longitudinal or transverse section)
    (b) Suitable conditions
    (c) Equipment deployment for different unit operations
    (d) Stope preparation activities.
27. (a) Given conditions as listed below; suggest a stoping method. Also list its variants known to you. Give lay out by including at least 2 views.
    • Ore strength: weak to moderate (its variants can mine out strong orebodies also).
    • Rock strength: moderate to strong.
    • Deposit shape: tabular.
    • Deposit dip: low (usually below 15°, but its variants can mine out orebodies up to 40° dip).
    • Size and thickness: large extent, thickness below 5 m, if more benching will be required.
    • Ore grade: moderate.
    • Depth: shallow to moderate (less than 500 m).
28. Describe the Vertical Crater Retreat (VCR) stoping method covering following aspects:
    a. Suitable conditions for its application
    b. Stope preparation work
    c. Equipment deployed for various unit operations during stope development as well as stoping.
    d. Advantages and limitations.

D. Attempt the following questions:

1. Calculate draw point (DP) length using the following data: Equipment length = 5 m, DP height = 3 m, Muck's angle of repose = 40°; Space for equipment movement = 2.5 m.

2. How does mine design change with the change in the dip of the deposit? Illustrate it.

3. Name parameters essential to design stopes and what you would include in each of these parameters.

4. Design a 'TROUGH' for a sublevel stope by considering an orebody dipping at 55° and having a thickness of 13 m. The height of this trough is to be kept at 10 m above (vertically) the back of the draw points. For this design, take the width of trough drive to be 4 m and its height as 3 m. After positioning the trough and the trough drive, with respect to the given orebody profile, design the FAN PATTERN for this trough, considering the toe spacing as per the hole length.

   If in the draw points connecting this trough drive, LHD of 6 m length, is to be used for mucking purposes, calculate the minimum length of the draw point considering height of draw point to be 3 m and the angle of draw of the muck to be 50°. Mark this draw point with respect to the trough drive, as designed above.

5. Calculate the length of a longwall coal face equipped with a cutter-loader using the following data:

   No. of stables = 2, Length of each stable = 4.5 m

   Duration of each shift = 8 hrs, Working shifts/day = 2

   Time spent in preparatory concluding works every shift = 2 hrs. 30 min

   Time lost for delay and auxiliary operations = 5 min/m

   Rate of advance of cutter loader = 0.6 m/min.

   After determining the face length in this manner does it need to be verified by some other calculations? If yes, then mention what is it?

6. Define 'Ring Burden' and 'Toe spacing'. Design a 'Ring Hole Pattern' to be drilled from the given sublevel of a sublevel stope. The dip of the deposit is 60° and its thickness is 8 m. The size of drill drive/sublevel is 3 m × 3 m. The sublevel is to be positioned in the f/w side of the deposit. The ground to be covered for the purpose of ring drilling is 8 m above it and 6 m below it, vertically. Mention the formula you shall use to compute the toe spacing.

7. For a blasthole stope (sublevel stope) a ring has been designed to cover an area of 425 sq.m of an orebody dipping at 65 deg. The ring burden for this ring is 1.8 m. Using tonnage factor of 3.2 tons/cu.m. Calculate the tonnage in this ring. If the drill factor of this ring is 3 tons/m, calculate the total amount of drilling. This ring is to be charged with ANFO, up to 2/3 of hole length. If the average diameter of the blast hole is 54 mm, and density of ANFO is 0.85 gms/c.c., calculate amount of ANFO that will be required and also the powder factor.

Answer following questions in conjunction with 'Mine Closure' operation:

1. Write down the salient features of the 'Integrated Mine Closure Planning' toolkit or Guidelines developed by ICMM. List these guidelines and briefly describe each one of them. Are these applicable throughout the life of an operation? Illustrate with help of a suitable diagram depicting the application of these toolkits at various phases during the life cycle of operations.

2. As per regulations in most countries: when should a document concerning this aspect be submitted to Government?
3. Compare vague and well-defined closure goals. Do you think well defined goals are a meaningful strategy?
4. Mine closure programmes comprise three phases; list them and prepare a bar chart scheduling mine closure operations.
5. The earlier that risks and unknowns are reduced, the greater the potential for meeting specific objectives. Do you agree with this statement? As such what is the right time to begin the closure planning?
6. Who are the participants in effective closure planning? Who are the internal stakeholders? What should be their contribution?
7. Why is mine closure important? Why it's planning important? Does it include parameters concerning decommissioning and rehabilitation?
8. Detail out 13 Guidelines (Toolkits) listed below to fit in your own mines closure planning operation:
Guideline 1: Stakeholder Engagement
Guideline 2: Community Development
Guideline 3: Company/Community Interactions to Support Integrated Closure Planning
Guideline 4: Risk/Opportunity Assessment and Management
Guideline 5: Knowledge Platform Mapping
Guideline 6: Typical Headings for Contextual Information in a Conceptual Closure Plan
Guideline 7: Goal Setting
Guideline 8: Brainstorming Support Table for Social Goal Setting
Guideline 9: Brainstorming Support Table for Environmental Goal Setting
Guideline 10: Cost Risk Assessment for Closure
Guideline 11: Change Management Worksheet
Guideline 12: The Domain Model
Guideline 13: Biodiversity Management

REFERENCES

1. Agoshkov, M., Borisov, S. and Boyarsky,V.: *Mining of Ores and Non-metallic Minerals.* Mir Publishers, Moscow, 1988, pp. 16–21; 74; 110; 134–156.
2(a). Anon: Coal Data: A Reference, US Dept. of Energy, *Govt. Print.* Off. Washington, DC, 1982a, pp. 69.
2(b). Anon: Contract documents, construction specifications, hazardous mine opening projects, phase, 21. *Abandoned Mine reclamation Bureau*, Helena, T, 1988C.
3. Boshkov, S.H. and Wright, F.D.: Underground mining system and equipment. In: Cummins & Given (eds.): *SME Mining Engineering Handbook.* AIME, New York, 1973, pp. 12: 1–13.
4. Chadwick, J.: Northparkes – a highly automated Block Caving leader. *Mining Magazine*, July, 2002, pp. 8–11.
5. Cokayne, E.W.: Sublevel caving. In: *Underground Mining Methods Handbook*, W.A. Hustrulid (edt.). SME-AIME, New York, 1982, pp. 874–875.
6. Deems, J.A.: Extending longwall faces beyond 600 feet. Coal Mng; vol. 21, no.8, 1984. pp. 58–61.

7. Donald, O.R and Ralph, C.S.: Filled and combined stopes. In: Cummins & Given (eds.): *SME Mining Engineering Handbook*. AIME, New York, 1973, pp. 12: 236–252.

8. Gray, T.A. and Gray, R.E.: Mine closure, sealing, and abandonment. In: Hartman (edt.): *SME Mining Engineering Handbook*. SMME, Colorado, 1992, pp. 666–670.

9. Hamrin, H.: *Guide to Underground Mining Methods and Application*. Atlas Copco, Sweden, 2002, pp. 20–27.

10. Hartman, H.L.: *Introductory Mining Engineering*. John Wiley & Sons, New York, 1987, pp. 338–460; 557–563.

11. Haycocks, C.: Stope and pillar mining. In: Hartman (edt.): *SME Mining Engineering Handbook*. SMME, Colorado, 1992, pp. 1705.

12. Henderson, K.J.: Shrinkage stoping at the Crean Hill Mine. In: *Underground Mining Methods Handbook*. W.A. Hustrulid (edt.), SME-AIME, New York, 1982, pp. 490–494.

13. Hohns, J.H.: Rubber tired equipment at Climax. In: D.R. Steward (edt.): *Design and operation of caving and sublevel stoping mines*. SME-AIME, New York, 1981, pp. 675–681.

14. Jackson, C.F. and Hedges, J.H.: Metal mining practices, *USBM, Bull 419*, Govt. Print. Off. Washington, DC, 1939, pp. 512.

15. Khetri copper complex. *Mining magazine*, 1984, vol. 151.

16. King, H.F.: A guide to the understanding of ore reserves. *Trans. Aus. Inst. Min. Metall*; 1982, pp. 11.

17. Krauland, N.: Development in Sweden of the rock mechanics of cut and fill mining. Innovation in mining back fill technology, Hassani et al. (eds), 1989, Balkema, Rotterdam, pp. 28–29.

18. Kvapil, R.: Sublevel caving. In: Hartman (edt.): *SME Mining Engineering Handbook*. SMME, Colorado, 1992, pp. 1797–1810.

19. Lang, L.C.: VCR an important new mining method. In: *Underground Mining Methods Handbook*. W.A. Hustrulid (edt.), SME-AIME, New York, 1982, pp. 456–63.

20. Lucas, J.R. and Haycocks, C.: Underground mining system and equipment. In: Cummins & Given (eds.): *SME Mining Engineering Handbook*. AIME, New York, 1973, pp. 12: 262.

21. Marklund, I.: Vein mining at LKAB Malnberget, Sweden. In: *Underground Mining Methods Handbook*. W.A. Hustrulid (edt.), SME-AIME, New York, 1982, pp. 443–446.

22. Mishra, H.C. and Ambastha, H.B.: Blasting techniques for induced caving by longhole and coyote chambers at Khetri mines. *Indian mining and Eng, Jou.*, May 1980, pp. 5–12.

23. Murray, J.W.: undercut and fill mining, Megma mine, Superiao. In: Cummins & Given (eds.): *SME Mining Engineering Handbook*. AIME, New York, 1973, pp. 12: 249–253.

24. Pillar, C.L.: A comparison of block caving methods. In: D.R. Steward (edt.): *Design and operation of caving and sublevel stoping mines*, SME-AIME, New york, 1981, pp. 87–97.

25. Rao and Sambasiva: Stope drive methods. *BGML Centenary souvenir*, 1980, pp. 158–171.

26. Rathore, S.S. and Joshi, A.: Successful firing of complex 130,000 tons. blast at Mochia mine, Zawar, Rajasthan. Jou. of Mines, Metal and Fuel, Aug. 1988, pp. 381–388.

27. Schroder, J.L.: Modern underground mining methods. In *Elements of practical coal mining*, S.M. Cassidy (edt.). SME-AIME, New York, 1981, pp. 346–476.

28. Shettigar.: Post and pillar stoping at Mosabani mines. *Int. Symp. on Mining with backfill*, Lulea, Sweden, 1983.

29. Singh, V.: Planning mine liquidation of worked out space in underground metal mines. *Jour. Mines, Metal and Fuels*, 1978.

30(a). Singh, V.: Pillar Blasting. *National sympo. on drilling and blasting*, Jodhpur University, India, 1981.

30(b). Stebbins, S.A. and Schumacher, O.L.: Cost estimating for underground mines. In: Hustrulid and Bullock (eds.): *SME Underground Mining Methods*. Colorado, 2001, pp. 67.

31. Stefanko, R. and Bise, C.J.: *Coal mining technology: Theory and practice*. SME-AIME, New York, 1983, pp. 410.
32. Stout, K.S.: *Mining methods and Equipment*. McGraw-Hill, New York, 1980, pp. 218.
33(a). Tatiya, R.R.: Computer Assisted Stope Design. *21st* APCOM, Lasvegas, USA, 1988, pp. 506–518.
33(b). Tatiya, R.R. and Allen, HEK.: An approach to model stoping boundaries using incremental analysis. *African Mining, 1987*, Harare, Zimbabwe.
34. Tamrock.: *Underground drilling and loading handbook*, Tuula Puhakka (edtr.), 1997, pp. 80–129.
35. Thomas, L.J.: *An introduction to mining*, rev. ed. Methuen of Australia, Sydney, 1978, pp. 471.
36. Trepanier, M.L. and Underwood, A.H.: Block caving at King Beaver Mine. In: *Design and operation of caving and sublevel stoping mines*, D.R. Steward (edt.). SME-AIME, New York, 1981, pp. 299–318.
37. Verobjev and Desmukh, R.T.: *Advance coal mining*. Mir publishers, Moscow, 1964, pp. 112–15, 291–95, 405–18.
38. Vergne, J.N.: *Hard Rock Miner's Handbook*. McIntosh Redpath Engineering, 2000, pp. 40–46.
39. Wallence, E.C.: Back filling methods. In: *SME Mining Engineering Handbook*, Hartman (edt.). SMME, Colorado, 1992, pp. 1756–1760.
40. Walter, A.P.: Excavation techniques. In: *SME Mining Engineering Handbook*, Hartman (edt.). SMME, Colorado, 1992, pp. 1749–1753.
41. Wise, J.J.: loading and haulage equipment for use in caving and sublevel stoping. In: *Design and operation of caving and sublevel stoping mines*, D.R. Steward (edt.). SME-AIME, New York, 1981, pp. 683–88.
42. *World mining equip*, Management report: Planning for closure., Dec. 1995, pp. 61–65.
43. Bullock, R.L.: General planning of the non-coal underground mine. In: Hustrulid and Bullock (eds.): *SME Underground Mining Methods*. Colorado, 2001, pp. 15–22.
44. Casey C. Grant: Spontaneous Combustion (Fires and Explosions in Mines). In: Chap. 74 – International Labor Organization (ILO) Encyclopaedia and CISILO database.
45. Holmberg, R., Hustrulid, H. and Cunningham, C.: Blast design for underground mining applications. In: Hustrulid and Bullock (eds.): *SME Underground Mining Methods*. Colorado, 2001, pp. 635–61.
46. ICCM Mine Closure Toolkit, 2008. Source: Based on Department of Industry, Tourism and Resources, Government of Australia, Mine Closure and Completion, October 2006.
47. James L. Weeks. Health Hazards of Mining and Quarrying. In: Chap. 74 – International Labor Organization (ILO) Encyclopaedia and CISILO database. Source: Chamberlain et al. 1970.
48. Landriault, D.: Backfill in underground mining. In: Hustrulid and Bullock (eds.): *SME Underground Mining Methods*. Colorado, 2001, pp. 601–10.
49. Laubscher, D.H.: Cave mining – the state of the art. In: Hustrulid and Bullock (eds.): *SME Underground Mining Methods*. Colorado, 2001, pp. 455–463.
50. Ovanic, J.: Mining operations at Pea ridge iron ore company – a case study. In: Hustrulid and Bullock (eds.): *SME Underground Mining Methods*. Colorado, 2001, pp. 229–234.
51. Perez, G. M.A. and Gonzalez, V.: Underhand Room and Pillar mining as applied at the Aurora mine, Charcus Unit, Grupo Mexixo. In: Hustrulid and Bullock (eds.): *SME Underground Mining Methods*. Colorado, 2001, pp. 121–23.
52. Ray, R.L.: Evolution of undercut and fill at SMJ's Jouac Mine, France. In: Hustrulid and Bullock (eds.): *SME Underground Mining Methods*. Colorado, 2001, pp. 355–58.
53. Vieira, F.M.C.C.; Diering, D.H. and Durrtheim, R.J.: Methods to mine the ultra-deep tabular gold-bearing reefs of the Witwatersrand basin, South Africa. In: Hustrulid and Bullock (eds.): *SME Underground Mining Methods*. Colorado, 2001, pp. 691–704.

Chapter 17

Surface excavations

Maximum productivity, safety and recovery with minimum costs at the desired rate of production should be aimed at to achieve the optimum results.

17.1 INTRODUCTION – SURFACE MINING METHODS

Civil as well as mining operations require surface excavations. Formation of slopes and benches over the hilly terrain that exists along the roadsides is an important civil work. This chapter deals with this topic as well as the surface excavations that are essential for surface mining to mine-out the mineral deposits that are outcropping to the surface, lies above surface datum, and extending to shallow depths.

When mining of a mineral deposit including stones of various kinds is undertaken by exposing them to the atmosphere (i.e. open air and sun), it is known as surface mining. Based on the location of the deposit w.r.t. the surface datum, the mines can be classified as open pit, opencast, quarrying or underground.

An open pit mine (fig. 17.2) is a mine to exploit the deposits which are outcropping to the surface, or those which are confined to a shallow depth, and the waste rock lying above (over burden) and at their sides (h/w and f/w) are removed and transported away from the place of their deposition.

Open cast is also a surface mine (figs 17.7 and 17.8) to mine out the flat deposits but the overburden is backfilled in the worked out area. When any deposit is extending beyond the break-even depth (i.e. the depth at which cost of mining is equal to price fetched), which could be attained by any of the surface mining methods; the underground mining could be applied.

The term quarrying of course is very loosely applied to any of the surface mining operations but it should be confined to a surface mining method to mine out the dimensional stones such as slate, marble, granite etc. (figs 17.22(a), (b), (c)).[19]

The deposits are sometimes located near the surface datum but covered by an aqueous body such as lake, tank, river, or even by seawater. Mining of such deposits is also a part of surface mining practices. These are known as aqueous extraction methods. In figure 17.1, a general classification of surface mining methods has been outlined.

17.2 OPEN PIT MINING

Elements of an open pit, and design parameters: In figure 17.1, the suitable conditions for the application of surface mining methods (mechanical as well as aqueous) have

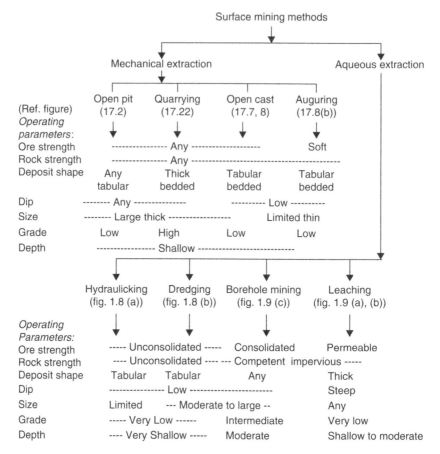

Figure 17.1 Classification of surface mining methods together with the desired parameters/conditions suitable for their applications. Rock is the overburden or hanging wall and footwall sides' rocks, or their combination.

been given. For open pit mining the key parameter is the dip of the deposit i.e. for the deposits having dip exceeding 20°, this system of mining is almost mandatory. In figures 17.2 different terms have been used to describe the structure of an open pit.[24] The main elements are described in the following sections.

17.2.1 Open pit elements

An excavation created to strip a deposit for the purpose of mining is called a pit and since this excavation is exposed to atmosphere; the resultant structure is known as open pit. Sometimes the deposit is outcropping to the surface and the rocks surrounding it cover it.

The rock masses on its hanging and footwall sides are termed as '*Hanging and Footwall Wastes*'. But if the same orebody is located at a certain depth, then the rock-mass covering top of the orebody is known as '*Over-Burden*'. Thus, to strip an

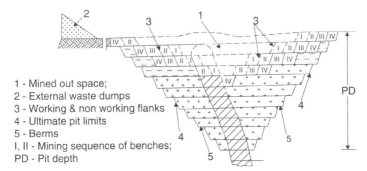

1 - Mined out space;
2 - External waste dumps
3 - Working & non working flanks
4 - Ultimate pit limits
5 - Berms
I, II - Mining sequence of benches;
PD - Pit depth

(a) For steep orebodies a suitable pit slope angle at foot wall side is also essential

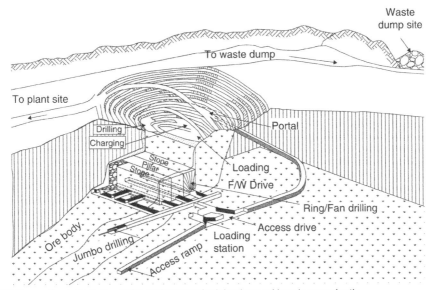

(b) Open pit mining followed by underground mining beyond break-even depth

Figure 17.2   (a): Open pit mining – Nomenclature. (b): Mining the outcropping or shallow seated deposits by open pit; followed by underground mining beyond the break-even depth. Illustration is typical example of copper mining at Sohar, Sultanate of Oman.

orebody suitable for open pit mining removal of hanging waste, footwall waste and the over burden is mandatory.

But the amount of waste rock enclosed in this envelope is a function of '*Overall Pit Slope Angle*', which can be defined as the angle formed while joining the *Toe* of the lowest bench (defined below) to the *Crest* of the top most bench of a pit with horizontal, when benches reach to their ultimate ends.

The amount of rocks need to strip the orebody increases as the depth is increased, and a situation arises when it becomes uneconomical to go beyond it. This is known as '*Break-Even Depth*'. This is also a function of pit slope angle; lower the over all pit slope angle lower would depth of pit and vise-versa.

The waste that need to be stripped cannot be taken at a stretch but it needs to be divided into convenient steps, which are safe and economical to be mined out, these steps so formed are called 'Benches'. *Bench Height* is a function of:

- *Ground competence* i.e. ground could be hard, compact, loose, friable, soft, consolidated, unconsolidated etc. In strata such as gravel, mourn, sand, alluvial soil, clay, running sand or any other similar strata, the bench height should not exceed 3 m.
- *Presence of water* – the ground or strata could be dry, wet, porous, non porous, above or below the water table etc.
- Presence of *geological disturbances* such as faults, folds, joints, cleavage or bedding planes etc.
- Height of the *boom or cutting height of the excavator* to be deployed for loading, mucking or excavation tasks.
- In general, the maximum allowable

$$Bench\ Height = Boom\ Height\ of\ Excavator + 3\,m \qquad (17.1)$$

Keeping the bench height more than this can prove unsafe.

- In case of *Dragline excavator*, it will depend upon its digging depth capabilities.

17.2.1.1   Bench angle or slope

It should be kept vertical but in practice it is difficult to maintain. Also it depends upon type of strata. Usually in practice it is kept to be 60°–80° to the horizontal for the working or active benches; and 45°–60° for non-working benches.

$$Minimum\ Bench\ Width = Working\ Berm\ Width + Non\text{-}Working\ Berm\ Width \qquad (17.2a)$$

Working berm width = 3 times the width of the truck/dumper to be operated on the bench.

$$(17.2b)$$

Non-working berm width = 3 m $\qquad (17.2c)$

Thus, Bench Width = 3 x Truck Width (Or width of largest equipment operating) + 3 m

$$(17.2d)$$

The Safety Berm is left when the bench reaches its '*ultimate end*'.

Safety Berm = 0.2 x Berm Interval (i.e. bench height)* $\qquad (17.3a)$

$$= (1/3)\ x\ Berm\ Interval\ (\ i.e.\ bench\ height)** \qquad (17.3b)$$

(* – Minimum Safety berm width as per Russian Safety Regulation)[24]
(** – Minimum Safety berm width as per Hustrulid, Kuchta, 1998)[14]

17.2.2   Overall pit slope angle

17.2.2.1   Computation of overall pit slope angle

Figure[14] 17.3 illustrates the geometry of an open pit. Geometrically overall pit slope angle can be computed as follows:[14]

$$OVERALL\ PITSLOPE\ (\pi) = \tan^{-1} \frac{N_B \times B_H}{\{(N_B - 1)B_W\} + \dfrac{N_B \times B_H}{\tan(B_A)} + (R_W)} \qquad (17.4)$$

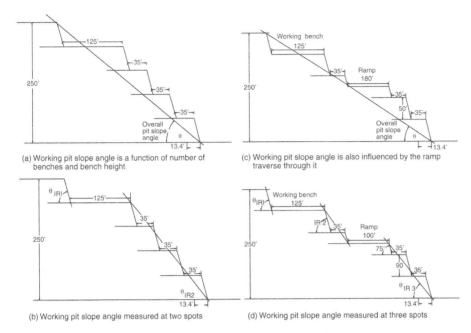

(a) Working pit slope angle is a function of number of benches and bench height

(c) Working pit slope angle is also influenced by the ramp traverse through it

(b) Working pit slope angle measured at two spots

(d) Working pit slope angle measured at three spots

Figure 17.3   Working pit slope angle's influencing parameters: bench height, ramp-width, number of benches. Final pit slope angle's influencing parameters: safety-berm's width and bench height.

Where: θ – overall pit slope angle; degrees
N_B – number of benches
B_H – bench height in meters;
B_W – bench width in meters;
B_A – bench angle in degrees;
R_W – ramp width in meters (if intersected).
Given data: (i) $N_B = 5$; $B_H = 16$ m; $B_W = 12$ m; $B_A = 75°$; $R_W = 0$ m; (answer: 49°)
(ii) $N_B = 8$; $B_H = 10$ m; $B_W = 12$ m; $B_A = 75°$; $R_W = 0$ m; (answer: 37.2°)
(iii) $N_B = 5$; $B_H = 16$ m; $B_W = 12$ m; $B_A = 75°$; $R_W = 30$ m; (answer: 38.8°)
(iv) $N_B = 5$; $B_H = 16$ m; $B_W = 3.2$ m; $B_A = 75°$; $R_W = 0$ m; (answer: 66.8°)
(v) $N_B = 5$; $B_H = 16$ m; $B_W = 0$ m; $B_A = 75°$; $R_W = 0$ m; (answer: 75.0°)

Overall pit profile (figs 17.5 and 17.10) is the function of rock massif surrounding the orebody, whereas the working pit-slope angle (fig. 17.3(a)) is the function of number of benches in operation at any time and also height of bench, which is governed by the factors given in the preceding sections. This is also influenced by the ramp if it traverses through it (fig. 17.3(c)) and (fig. 17.3(d)) and also condition of the orebody in terms of strength, presence of discontinuities and water. Any dry and strong orebody without discontinuities will allow higher bench height; thereby more working pit slope angle. This has implications on the productivity of the working pit, due to the fact that higher the bench height, the more is the scope of deploying bulky equipment fleets and that results in better productivity and lower overall costs. In figure 17.3(a) the pit slope at the beginning of the pit has been shown; when it reaches its final stage at the end of pit life it is termed as overall pit slope angle. Its value would be influenced by the width of safety berm and bench height.

As per Hudson[13] 12 leading parameters, as listed below, and their interaction forms a matrix of 12×12. This means there could be 132 ($12 \times 12 \times 12 - 12 = 132$) permutations and combinations i.e. scenarios that could influence the value of overall pit slope angle.

1. Overall environment – geology, climate, seismic risk, etc.
2. Intact rock quality – strong, weak, weathered (strong rocks permit high pit slopes)
3. Discontinuity geometry – set, orientations, apertures, roughness (increased number of unfavorable joint sets reduce the bench height)
4. Discontinuity properties – stiffness, cohesion, friction
5. Rock mass properties – deformability, strength, failure
6. In-situ rock mass stress – principal stresses' magnitudes/directions (Magnitude and direction of principal stresses are required to determine slope dimensions)
7. Hydraulique conditions – permeability, etc.
8. Slope orientation – dip direction, location, etc.
9. Slope dimensions – bench height/width and overall slope
10. Proximate engineering activities – blasting, etc.
11. Support and its maintenance – bolts, cables, grouting, etc.
12. Construction – excavation method, sequencing, etc.

In practice it may not be practicable to study in detail, the 12 parameters listed above. Rzhevsky[24] proposed guidelines, as given in table 17.1 to choose pit slope in the varying situations, which could be of significant importance and use.

In this table presence of water, if any, has not been considered and therefore the pit slope angle should be further reduced under wet conditions.

Reasons for pit slope failures:[6,11]

• Adopting a steep pit slope angle than appropriate.
• Presence of water and effective measures not taken to deal with it.
• Under-cutting of rock massif.
• Presence of geological disturbances.

Pattern of pit failures commonly known:[6,11]

• Slope failure (fig. 17.4(a))
• Base failure (fig. 17.4(b)).

These failures have been illustrated in figure 17.4 & they include:[12,14]

• Raveling (17.4(c))
• Rotational shear (17.4(d))
• Plane shear (17.4(e))
• Step path (17.4(f))
• Step wedge (17.4(g))
• Simple wedge (17.4(h)).

Table 17.1 A practical guide for selecting pit slope angle.

Rock types	Characteristic of rock massif	Final pit slope angle, degrees
Compact hard rocks, $\sigma_c > 8 \times 10^7 Pa$ * – Large values correspond to large dipping angle of weakness planes. σ_c – Compressive strength.	• Strong low fissured rocks without unfavorably oriented weakness planes	55
	• Strong low fissured rocks with steeply (above 60°) or gently (below 15°) dipping weakness planes	40–45
	• Strong long low and medium fissured rocks with weakness planes dipping at an angle 30–50° towards the open pit.	30–45*
	• Strong long low and medium fissured rocks with weakness planes dipping at angle 20–30° towards the open pit.	20–30*
Low strength compact & weathered hard rocks, $\sigma_c = 8 \times 10^6$ to $8 \times 10^8 Pa$	• Relatively stable rocks without unfavorably oriented weakness planes	40–45
	• Relatively stable rocks with weakness planes dipping at angle 30–55° towards the open pit	30–40*
	• Intensively weathered rocks in slopes	30–35*
	• All rocks of this group with weakness planes dipping at an angle of 20–30° towards the open pit.	20–30*
Soft and loose rocks, $\sigma_c < 8 \times 10^6 Pa$	• Plastic clays without glide planes, weak contacts between strata and other weakness planes	20–30
	• Plastic clays and other clayey rocks with weakness planes in the mid or bottom of slopes.	15–20

Remedial Measures:
1. Flatten pit slope angle
2. Strengthening the slope with use of R.C.C. piles (fig. 17.4(i)); Anchors, Retaining walls, Bulkheads.
3. Strengthening slope by (i) Bolting (fig. 17.4(j)); (ii) Flexible cables (fig. 17.4(k)).
4. Rock consolidation – cementation, injection of consolidating polymer solutions, tar bonding
5. Protective coatings for strong-fissured rocks liable to weathering; or leaching with the use of shotcreting, guniting, or bituminous grouting
6. Combination of above techniques.

17.2.3   Stripping ratio

In open pit mines in order to decide the depth of the pit, it is essential to carry out detailed calculations as how much waste rock will be required to remove to strip the orebody? The ratio between the amount of waste rock to be removed to mine out a

Common open pit failure patterns

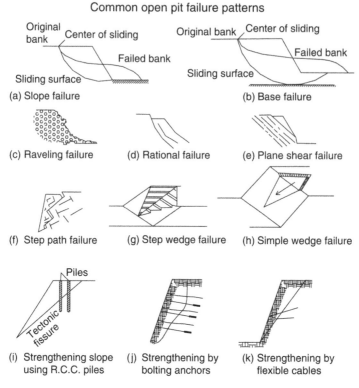

Figure 17.4   Open pit common failure patterns (a to h), and common slope strengthening techniques (i to k).

unit of ore is called stripping ratio. Since it is a ratio therefore it should be dimensionless. But in practice different connotations are being used;[11] e.g.

$$S.R = Total\ waste\ rock\ (tons.)/Total\ ore\ (tons.);\ within\ envelope\ considered;\quad (17.5a)$$

OR

$$S.R = Total\ waste\ (m^3)\ of/Total\ ore\ (tons.);\ within\ envelope\ considered;\quad (17.5b)$$

OR

$$S.R = Total\ waste\ (m^3)/Total\ ore\ (m^3);\ within\ envelope\ considered;\quad (17.5c)$$

To determine the maximum depth based on the profitability of the operation, it is essential to know about the overall costs and revenues that will be received by selling the ore and its bye-products, if any. In other words what will be cost of removing waste rock that is enclosing the orebody between surface datum and a particular depth, and also mining or exploiting the orebody itself that lies within this envelope.

$$SRmax = \frac{(Price\ or\ revenues\ received\ per\ tonne\ of\ ore - cost\ of\ mining\ per\ ton.\ of\ or}{(Waste\ removing\ Cost/t)}$$

$$(17.6)$$

Once the maximum allowable striping ratio is computed, then computation to find out an envelope fitting for the amount of rocks to be removed to strip the orebody up to the maximum allowable depth could be established.

17.2.4 Overall pit profile

The resultant envelope that will be created by removing the waste rocks that are surrounding the orebody is known as overall pit profile (fig. 17.6). It is also based on the geometry of the deposit. It should be projected during the planning stage. This could take a shape of a basket, a trapezoidal or a bathing tub, or any other configuration. This is the resultant profile that is likely to encompass the extent of excavation by the open pit mining. The peripheral infrastructure facilities such as rail, road, power-lines, buildings, offices, waste and ore dump yards, mills and plants etc. should be located outside this boundary.

17.2.4.1 Coning concept for open pit design

Assuming uniform grade of the ore that will be within the envelope that has been evolved based on the rock mechanics aspects; the profit function of an pit can be computed using the following relation:[4]

$$R_V \; x \; G \; x \; R_F \; = P_F + S_R \, C_W + C_O + C_M \qquad (17.7a)$$

Where: R_V = Revenue per unit of ore;
G = Grade of ore;
R_F = Recovery factor;
P_F = Profit per unit of ore;
S_R = Stripping ratio;
C_W = Waste mining cost/unit;
C_O = Ore Mining Cost/unit;
C_M = Processing Costs/unit.

If $P_F = 0$; Equation (17.7a), can be used to compute the minimum allowable grade, which is by definition is the cutoff grade, below which if mining is carried it will not be in profit.

This model can be further extended by considering the blocks of ore and waste rocks contained within the pit limits (figs. 17.5(a) to (c)) that can be allowed to arrive at the profit function of an open pit.

$$R_F \; x \; (O_1 G_1 + O_2 \, G_2 + ---- + On_{-1} \, Gn_{-1} + O_n \, G_n) \geq P_F + C_W (W_1 + W_2 + ----- + Wn_{-1} + W_n) + (C_O + C_M) \, (O_1 + O_2 + ------ + On_{-1} + O_n) \qquad (17.7b)$$

Where: $O_1 \, G_1 --- O_n \, G_n$ are the ore blocks having different grades; and $(W_1 --- W_n)$ are the waste rock blocks; Rest of the terms/symbols are the same as designated above.

An algorithm can be built for this model, which is dynamic in nature, as the cost and price are the parameters, which changes with time while the grade and size of blocks are static. Such dynamic models can be used to drive the profit function and cutoff grades scenarios of an open pit. Based on this logic different pit profiles can be obtained for different cutoff grades. The block values can be estimated using geostatistics. Author has carried out such an incremental analysis for deciding stope boundaries of different underground methods, as illustrated in figures 16.35 to 16.38.

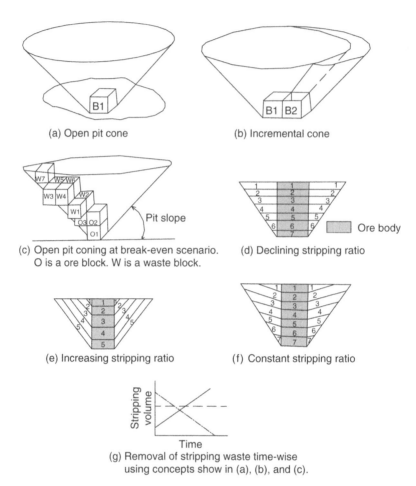

(a) Open pit cone

(b) Incremental cone

(c) Open pit coning at break-even scenario. O is a ore block. W is a waste block.

(d) Declining stripping ratio

Ore body

(e) Increasing stripping ratio

(f) Constant stripping ratio

(g) Removal of stripping waste time-wise using concepts show in (a), (b), and (c).

Figure 17.5 Top: Coning concept or models (a) to (c) to design open pit mines based on economical considerations. Bottom: Waste rock removal/stripping strategies/models (a) to (c).

17.2.5   Stripping sequence

It is interesting to look into the fact as to how the waste rock that will be required to be removed from the overall pit profile envelope should be sequenced? Following three schemes[7] as shown in could emerge:

- Constant Stripping Ratio (fig. 17.5(f))
- Declining Stripping Ratio (fig. 17.5(d))
- Increasing Stripping Ratio (fig. 17.5(e))

Each scheme has its own merits and demerits (fig. 17.5(g)). Declining SR method (fig. 17.5(d)) put up a heavy burden on production of waste in the beginning that may adversely affect the cash-flow in the earlier years. Increasing SR method (fig. 17.5(e)) produces large cash-flow in the initial period and covers the risk, and that is why, it is preferred in most of the cases. Sometimes it is impractical to operate large number of benches and faces, but keeping them open (active) gives an opportunity to blend the ore. The constant SR method (fig. 17.5(f)) is the compromise of the extreme conditions that are associated with the other two methods. In practice a method, which can generate high cash flow and require building up of resources (man and equipment) gradually, and reduction of resources gradually at the end of pit-life should be preferred.

17.3   HAUL ROADS

Width and number of lanes:[23] (fig. 17.6(g))

$$Wr_1 = T_W + 2y, \quad Wr_2 = 2(T_W + y) + x, \tag{17.8a}$$

$$\text{Road Width;} \quad Wr_2 \geq 4 \times T_W \tag{17.8b}$$

Where: Wr_1 = Road width for one lane traffic;
W_{r2} = Road width for two lane traffic;
T_W = truck width, m; $y = 0.5$;
$x = 0.5 + 0.005\ V$;
V is vehicle speed in km/hr.

Based on service life, the haul-road may be Permanent, Semi-permanent, or Temporary. Permanent routes are established mostly at the non-working flanks of the surface mines and semi-permanent on the portion of the working flanks of the mines which have been out of operation for certain period. Temporary routes are prepared at the working benches or flanks of the surface mines. The Spiral (fig. 17.6(a)); and Switchback (fig. 17.6(b)) are the two designs that are prevalent. Switchback is usually confined to Rail haulage and rarely used with automobile (Trucks) system. The spiral design follows the geometry of the pit and is run almost parallel to its longer axis, as shown in figure 17.6(f).[20] Sometimes a combination of two may be essential (fig. 17.6(d) and 17.6(e)).[24]

17.4   RAMP AND ITS GRADIENT[19,20]

Ramp gradient is governed by the statutory requirements of any country and usually it is in between 8–15% (5°–8.5°). For safety and drainage reasons long steep gradients should include a 50 m section of 2% (1° gradients at every 500–600 m of the severe gradient).[2] Provision of proper vertical and horizontal curves, with proper sight distance and line of sight, at the crossing with any other road or rail routes must be taken care. It is very important that proper drain is cut all along the haulage route to avoid wash outs, mudslides and saturation. This increases the road life, requires less maintenance and keeps the roads safer. Dry roads give longer tyre life and less undue stress on trucks.

Length of Ramp[24]

Theoretical length of Ramp; $R_{LT} = (E_S - E_E)/ \tan(G)$ (17.9a)

Actual length of Ramp; $R_{LA} = R_{LT} \times K_{EL}$ (17.9b)

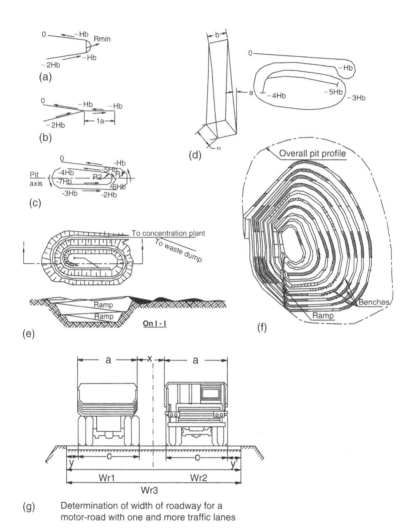

(g)   Determination of width of roadway for a
      motor-road with one and more traffic lanes

Figure 17.6   (a) to (e): Haul roads and ramp designs for surface mines. Ramp designs: Spiral, Shunt-
back and, Combination. (g): Determination of haul-road width; a = T_W = truck width;
Also a = c; c is out-to-out tyre width of truck.

Where: R_{LT} = Theoretical ramp length;
E_S = Elevation starting point;
E_E = Elevation at the bottom of pit (up to which ramp has to go);
G = Gradient of ramp in degrees; R_{LA} = Actual ramp length;
K_{EL} = Factor of elongation.

In practice the actual length of ramp is greater than theoretical, as at the curves on the
route gradient have to be reduced, and this makes the ramp length more than theoretically calculated.

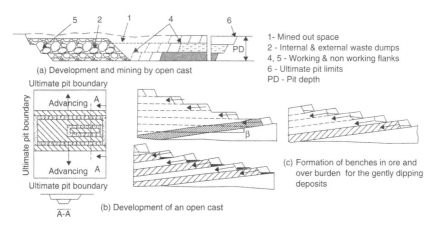

1- Mined out space
2 - Internal & external waste dumps
4, 5 - Working & non working flanks
6 - Ultimate pit limits
PD - Pit depth

(a) Development and mining by open cast

(c) Formation of benches in ore and
    over burden for the gently dipping
    deposits

(b) Development of an open cast

Figure 17.7   Development of opencast pits (a), and their deepening process (b). Formation and deepening of benches in orebody and overburden in gently dipping deposits (c).

17.5   OPEN CAST MINING/STRIP MINING[1,11,24]

17.5.1   Introduction

As the name indicates, this is a surface mining system in which a deposit is opened to the atmosphere and after removing the orebody, the over burden which was blanketing it, is cast back in the worked out area. This is also known as Strip Mining. In figures 17.7(a) and 17.8(a), nomenclature used has been illustrated. Thus, this system gives an added advantage of:

- Using the same land which was occupied by the deposit for dumping the waste rock and thereby a minimum land degradation (refer fig. 17.8(a)).[9a]
- The lead for waste dump from the working face is very little; thereby transportation cost is very much reduced. Waste rock recasting goes on simultaneously with ore mining that allows high production rate and almost continuous muck flow under suitable conditions.
- Application of high capacity and bulky equipment is practical for high outputs. This allows high productivity and low mining costs enabling mining of even low grade and deep-seated deposits with higher stripping ratios.
- Due to these inherent features today development of highest man-made equipment has become possible in the mining operations. In this system use of highly productive equipment such as bucket wheel excavators (section 6.15 fig. 6.10), draglines (sec. 6.13, figs 17.8(a), 6.9) and high capacity belt conveyors is therefore feasible.

17.5.2   Design aspects

This method is suitable for flat deposits, and deposits with gentle dip up to 15° or so (fig. 17.7). Thus, coal seams, layered deposits of clay and any other minerals can be most suitably mined by this method. This also allows for multi seams mining but the height

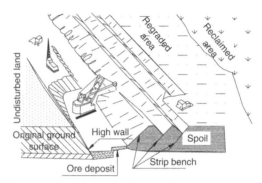

(a) An open cast mining. Ore mining, overburden
    stripping and casting and land reclamation operations shown.

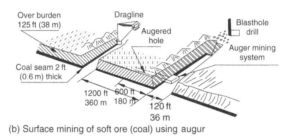

(b) Surface mining of soft ore (coal) using augur

Figure 17.8 Opencast ore/coal mining. Ore mining, overburden stripping and casting, and land reclamation operations shown.

of the bench is governed by the thickness of the coal seams. Height of the bench in thick overburden is governed by the equipment's digging depth capability, as described in section 17.2.1 (eq. 17.1); and the same logic is applicable to decide the bench height in the orebody or coal seams. In gently dipping deposits, the mining could proceed by dividing overburden and bedded deposit, into the benches, as shown in figure 17.7(c).[23,24]

The operation of open cast begins with the widening and deepening of the opening trench or box-cut and advancing in the manners shown in figure 17.7(b). The overburden in all cases is cast back. The systems could be:

- Direct casting back using draglines, shovel, or BWE i.e. internal dumping.
- Using conveyors, bridges and spreaders in case of stacking overburden minerals of different types at different locations but within the worked out areas.
- Combined system while working gently dipping deposits i.e. first dumping externally and then shifting back to the worked out space.

17.5.3   Operational details – surface mines

In surface mines the following logical steps should be followed to mine-out a deposit:

- Planning
- Site preparation
- Opening up the deposit

- Pit development
- Ore production
- Environment and land reclamation
- Liquidation phase and post mining operations.

17.5.3.1 Planning

Based on the feasibility studies, once a decision is taken to undertake mining by adapting a particular surface mining method, open pit, open cast, or quarrying, at the planning stage the following aspects should be taken care of:

- Development of the conceptual model for the mine and then going for the detailed engineering studies. It is ideal to prepare a 'Detailed Project Report (DPR)' after considering the various scenarios, alternatives, and options available to choose a particular system (sec. 12.1.2). During this phase liaison with different agencies such as government, bureau of mines and geology, exploration, market forces, financial institutes, equipment and raw material suppliers, contracting and construction companies, is established. Mineral rights – concession or lease, permission from government, environmental and safety authorities, and other local agencies are obtained.
- Sites that will be required, apart from the mines and plants such as office buildings and residential colonies, social and welfare amenities are chosen; and necessary rights are acquired. From the proposed mine site access and links to the available infrastructure facilities such as power, transport, communication, water etc. are established.
- Plans, sections, reports, drawings, contract documents to award different construction activities are prepared.
- Planning include details of the construction, development and final exploitation schedules. Phased manpower, equipment, material, energy and financial needs (budget), and likely cash flows are forecasted.
- During this phase details of the diversion plans of waterways such as river, drainage, or the catchments areas should be worked out. Sometimes evacuation of residential houses or diversion of some rail, road or power line becomes essential.

17.5.3.2 Site preparation

As shown in the flow chart, figure 17.9, a number of activities at the site selected for the mine, are undertaken. This includes removal of vegetation and cleaning the site from any obstruction. The topsoil is removed and stacked to the pre-determined site. Care should be taken that good soil is properly stacked so that it can be reused. This job can be contracted, if the magnitude of work is small. In some cases the initial overburden could be soft, semi-consolidated, or consolidated ground, which could be removed by dozing, ripping, or scrapping, as discussed in sec. 17.8, and figure 17.32.

17.5.3.3 Opening up the deposit

Based on the geometry of the deposit and its enclosing rocks which could be over burden capping, hanging wall and foot wall waste rocks (in case of inclined deposits) suitable for open pit mining, or it may be a cover of rock-mass over the flat deposits suitable for open cast mining; as discussed in section 17.5. This cover is known as overburden. In quarrying either of these two situations could exist.

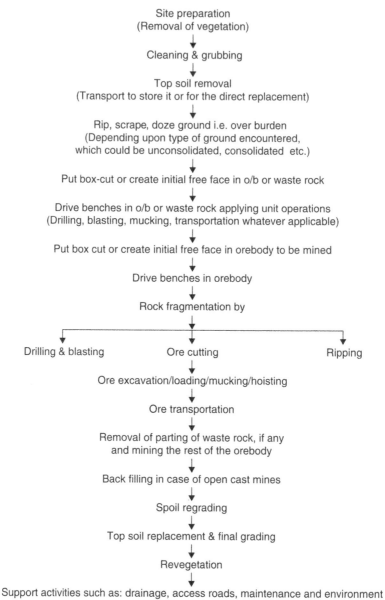

Site preparation
(Removal of vegetation)
↓
Cleaning & grubbing
↓
Top soil removal
(Transport to store it or for the direct replacement)
↓
Rip, scrape, doze ground i.e. over burden
(Depending upon type of ground encountered,
which could be unconsolidated, consolidated etc.)
↓
Put box-cut or create initial free face in o/b or waste rock
↓
Drive benches in o/b or waste rock applying unit operations
(Drilling, blasting, mucking, transportation whatever applicable)
↓
Put box cut or create initial free face in orebody to be mined
↓
Drive benches in orebody
↓
Rock fragmentation by
↓

Drilling & blasting   Ore cutting   Ripping
↓
Ore excavation/loading/mucking/hoisting
↓
Ore transportation
↓
Removal of parting of waste rock, if any
and mining the rest of the orebody
↓
Back filling in case of open cast mines
↓
Spoil regrading
↓
Top soil replacement & final grading
↓
Revegetation
↓
Support activities such as: drainage, access roads, maintenance and environment
management to be carried simultaneously

Figure 17.9   Operations need to be carried out while undertaking surface mining.

Any human being should never forget his last destination, which is known as a grave for Christians a kaber for Muslims; it is a small excavation or ditch dug in the ground to burry him or her. The same excavation is mandatory at the beginning of surface mining. This grave is first dug in the ground massif either manually using conventional tools, or using an excavator; and then it is extended or widened and deepened to reach up to the toe of the first bench. In case of rocks 'V', Wedge, or Pyramid cut of pattern of holes (sec. 9.3.1) are drilled and charged to create this initial excavation. This is known as initial 'box cut' or 'trench'. This trench can be extended in any direction along or across the longer axis of the open pit.[24] Following are the important features of these trenches:[24]

- They can be started from out side, or inside of the overall pit limits, as shown in figures 17.7(b), 17.22(b). The location of the box cut will be as per the sequence of mining a deposit. It could be at its either of the terminal points, or at the middle.
- A trench can serve one bench, or several benches, or all the benches up to the ultimate pit depth.

17.5.4 Development

The development work in surface mines is begun with the putting up the box cut, which gives way to development of ramps and benches in waste rock as well as in orebody. Pit geometry will be governed by the geometry of the orebody, in general, and to the dip of the deposit in particular. For inclined deposits it may not be essential to strip beyond the footwall contact of the orebody; whereas, for steep orebodies a suitable pit slope angle at foot wall side will also be essential figure 17.2. The construction of the ramp to access the deep levels of the pit is also a routine development activity. In figure 17.10,[24] open pit profiles with respect to orebody profiles have been shown. In figure 17.6(f), benches configuration together with the traversing of the ramp has been shown. Figure 17.10, depicts the different pattern of ore mining within the pit limits. It could be single sided, double sided, in longitudinal as well in the transverse directions. It could be centralized or decentralized. In figure 17.7(b) development of an open cast pit has been shown.

17.5.4.1 Waste rock dumps

The waste rock dump yards could also be located based on the geometry and suitability of the available land in terms of its techno-economical aspects. In figure 17.17,[24] various schemes of waste rock dumps have been illustrated.

17.5.5 Bench blasting design patterns[3,15,16,18,22,24]

Bench Geometry: In the preceding section details about the bench parameters such as height, slope and width have been described. In figure 17.11(a),[18] the terms used to describe bench blasting have been illustrated. In surface mining, bench blasting is one of the most important operations as it is based on a number of parameters and prominent amongst them are the type of rock (texture, structure and strength), hole diameter, terrain conditions, type of explosive and desired degree of fragmentation. In order to obtain proper fragmentation that can result in overall minimum cost, a careful designing of drilling and blasting pattern is essential. In bench blast design, the most important parameters are burden and spacing. Since the spacing is usually set either equal or at 1.25 times the burden (or even more under suitable conditions), it becomes all the more important to determine burden, which is the distance between the first row of holes running parallel to the free vertical surface of the rock, or it is the perpendicular (shortest)

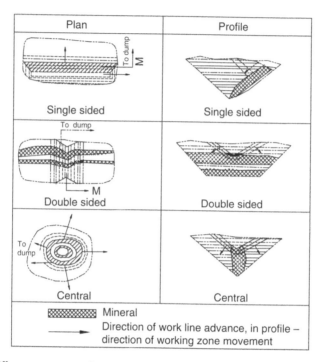

Figure 17.10 Different patterns of ore mining within the pit limits – single sided, double sided (in longitudinal as well as transverse directions); centralized and disconcentrated.

distance between two adjacent rows of holes. If the burden is too small, part of the explosive energy is used to obtain fine fragments and the rest will be lost in the form of noise, air-blast and throw. If the burden is too large, higher ground vibration and large fragments are generated. The optimum burden is the one that reduces overall mining cost, causes least over break, reduces vibrations, and produces proper fragmentation.

There are a number of empirical relationships that have been proposed to design bench blasting, but this section is confined to review those formulas in which burden can be calculated with respect to blasthole diameter. Prominent amongst them are the formulae which have the linear relation with blast-hole diameter. These formulas were commonly used in 1980s, but later on with the advent of high degree of mechanization and blasting techniques, it has been established that non-linear relations can give better results, particularly for the blasthole diameters in the range of 40–400 mm (Kou and Rustan, 1992). An overview on the subject is presented below.[16]

17.5.5.1 Linear formulas

Langefors et al. (1978) described the relation (17.10) to calculate the maximum burden for the blasthole diameters in the range of 0.03 to 0.089 m

$$B_m = 0.958d\sqrt{\frac{\rho_e s}{(S_b/B_b)c_0 f}}$$

(17.10)

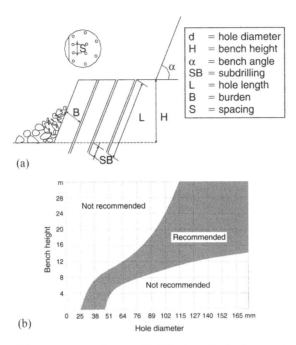

Figure 17.11   (a): Bench blasting: nomenclatures. (b): Selection of hole diameter based on bench height.

B_m = maximum burden for good breakage (m);
d = blasthole diameter (m);
ρ_e = explosive density (kg/m^3);
s = weight strength of explosive;
f = confinement of blasthole;
S_b = drilled spacing (m);
B_b = drilled burden (m);
c_0 = corrected blastability factor (kg/m^3);
c_0 = c + 0.5 for B ⩾ 1.4 to 1.5 m
but c_0 = c + 0.07/B for B_m < 1.4 m

c is the rock constant whose value varies from 0.2 to 0.4 depending upon type of rock; for brittle rocks 0.2, and for rest all other rocks it is 0.3 to 0.4.

Relation (17.10) is linear and when Rustan (1992) substituted the anticipated maximum and minimum values of the corresponding parameters, it can be written as

$$B_m = (14 \text{ to } 76) \, d \tag{17.11}$$

Ash (1963) suggested a similar formula (17.12a), in which value of constant k_b varies from 20–40 depending upon rock and explosives parameters.

$$B_m = k_b \, d. \tag{17.12a}$$

Ash (1968) also presented an empirical formula (17.12b), which he derived from Konya's formula

$$B_m = 38d \sqrt{\frac{\rho_e}{\rho_r}} \qquad\qquad (17.12b)$$

ρ_r = rock density (kg/m^3); other symbols have been explained earlier.

When Rustan put the maximum and minimum values in the range of explosives and rock density as considered by Lama and Vitukuri (1978), in the formula (17.12b) given by Ash, it could be expressed as formula (17.13a).

$$B_m = (15\ to\ 37)\ d. \qquad\qquad (17.13a)$$

Rustan (1992) derived Konya's formula and mentioned that practical burden (when ratio of burden and spacing equals one) it has linear relation with blasthole diameter, as presented in formula (17.13b).

$$B_{pl} = f\left(d \sqrt{\frac{\rho_e}{\rho_r}} \right) \qquad\qquad (17.13b)$$

B_{pl} practical burden (m), and other symbols remain the same as explained earlier.

17.5.5.2 *Power formulas derived by statistical analysis*

Rustan (1992) derived the following relation to calculate practical burden for open pit mines with blasthole diameters in the range of 0.089–0.381 m.

$$B_{pl} = 18.1\ d^{0.689} \qquad\qquad (17.14)$$

(With +52% Expected maximum; and with −37% minimum value; correlation coefficient R = 0.78).

17.5.5.3 *Formulas related to energy transfer in rock blasting,*
burden and blasthole diameter

In formulas (8) and (9), Kou and Rustan (1992) have tried to investigate energy transfer in rock blasting, and related burden with the blasthole diameter. The authors suggested formula (17.15) to calculate burden not exceeding 3 m, and formula (17.16) to calculate burden exceeding 3 m, but less than 10 m.

$$B = d\left(\frac{2K_0 Q_e \rho_e E}{\xi o^2 \ \tan(\theta/2)} \right)^{1/2} \qquad\qquad (17.15)$$

$K_0 = (\pi/4)\eta\ (R_D)^2$;
R_D = decoupling ratio;
η = energy transformation efficiency;
E = Young's modulus;
ρ_e = explosive density (kg/m^3);
ξ_o = a constant depending on the required fragmentation, confinement of blasthole as well as rock failure characteristics;
d = blasthole diameter (m).

θ = angle of breakage in degree;
Q_e = detonation heat (kJ/kg).

$$B = \left[\frac{K_0 Q_e}{\alpha g \tan(\theta / 2)} \right]^{1/3} \left(\frac{\rho_e}{\rho_r} \right) d^{2/3} \qquad (17.16)$$

α = a constant to take care of rock to be elevated against gravity in some circumstances.
g = acceleration of gravity (m/sec^2), and other symbols remain the same as explained earlier.

17.5.5.4   Tatiya and Adel's Formula to determine burden with respect to blasthole diameter[26]

Atlas Copco, Sweden, has plotted curves to estimate the burden as a function of blast-hole diameter for different rock blast abilities, based on its experience by keeping the spacing 1.25 times the burden, and bench height more than 2 times the burden, but not exceeding 20 m. But these curves don't specify the range of rock strength for which each one of them is applicable, and also the type of explosive that should be used.

Tatiya and Adel's relation (eq. 17.17) and curves (fig. 17.11(c)) could be used to compute burden in relation to rock strength and hole diameter.

$$B = a\, d^2 + b\, d + c \qquad (17.17)$$

Where: B = burden (m);
  d = hole diameter (m);
  a, b and c are constants and their values depend on rock strength.
  If the uniaxial compressive strength (σ_c) of the rock is:
σ_c < 55 MPa:
  a = −40, b = 35.9, c = 0.45
σ_c from 55 to 110 MPa:
  a = −30, b = 29.4, c = 0.35
σ_c > 110 MPa:
  a = −20, b = 24.1, c = 0.30
The proposed empirical formula has the following features:

- It is easy to use, as it is a function of hole diameter, and the uniaxial compressive strength of rock for the mines using ANFO as the main explosive charge.
- Through the field trials, prevalent blast designs of any mine may be checked.
- To use the model for any new deposit, using explosives other than ANFO, necessary change in the various constants used may be essential to achieve the desired results.
- The model has been tried for limestone deposits but may be calibrated to any other deposit.

17.5.5.5   Powder factor method

(i) *Sub-grade drilling,   J = 8d*, in meters   (17.18a)

(ii) *Stemming length, T = 25d*, in meters.   (17.18b)

(iii) Calculate the *length of charge in hole,   K = L + J − T*   (17.18c)

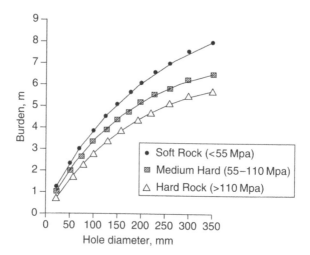

Figure 17.11   (c): Burden as a function of hole diameter and rock strength.

(iv) From the table or otherwise calculate the *explosive concentration i.e. charge* kg/m of hole length (L).

(v) Calculate *total charge Q, Q = K L,* in kg.   (17.18d)

(vi) Calculate *volume of rock broken/hole,* $V_H = Q/P,$ *in* m^3 (17.18e)

(vii) Calculate the *volume of rock/m of bench height* $V_1,$ $V_1 = V_H/L = Q/PL$
                                                                                      (17.18f)

(viii) Calculate the *burden* B, for the desired spacing to burden ratio $K_S,$

$$B = (V_1 / K_S)^{1/2}$$
                                                                                      (17.18g)

(ix) Calculate the spacing S, $S = K_S B,$ in meters.   (17.18h)

Where: d is hole diameter in meters.

   L is hole length; above bench's toe i.e. bench height, in meters.

   P is powder factor in kg/m^3

   K_S – burden to spacing ratio, which is usually 1:1.25, or even more in under suitable conditions.

17.5.6   Drilling and blasting operations[3,15]

Drills: selection of drills could be made as shown in figure 17.12(a) for construction and civil projects, and as per figure 17.12(b) for surface mines – opencast or open pits. Drilling accessories required are also shown in these illustrations. Application of hydraulic energy in the drilling unit results in energy saving as compared to pneumatic once; as shown in figures 4.6(a).

   Selection of proper hole diameter, bench height and matching explosive to perform a specific task and obtain the desired blasting results should be given due importance. Moreover, adverse results effects rock fragmentation, excavation's profile (contour); and generate undue noise, vibrations and fly rocks. In a stone quarry it may

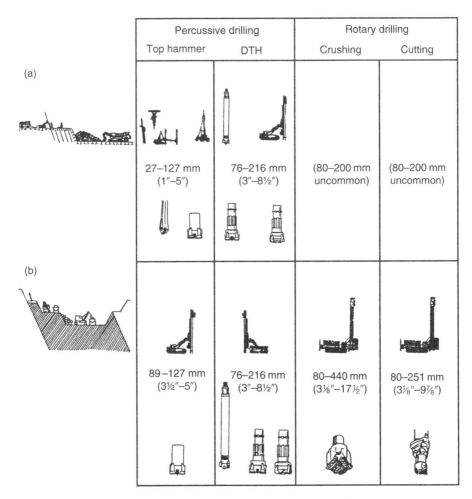

| | Percussive drilling | | Rotary drilling | |
	Top hammer	DTH	Crushing	Cutting
(a)	27–127 mm (1″–5″)	76–216 mm (3″–8½″)	(80–200 mm uncommon)	(80–200 mm uncommon)
(b)	89–127 mm (3½″–5″)	76–216 mm (3″–8½″)	80–440 mm (3⅛″–17½″)	80–251 mm (3⅛″–9⅞″)

Figure 17.12   Selection of rock drills and drilling accessories for; (a): construction projects. (b): surface mines (Courtesy: Atlas Copco).

spoil the valuable dimension stones themselves due to development of cracks and improper sized blocks. The guideline given in figure 17.11(b),[18] could be useful in deciding hole diameter in opencast and open pit mines.

In some specific cases controlled blasting is required to obtain proper bench geometry, reduce vibrations and over-break. In this book these techniques have been referred as smooth blasting, cushion blasting, pre-splitting, or buffer blasting. And the ones which do not require blasting but drilling very close or even skin-to-skin holes, is known as 'line Drilling'. The concept of contour-holes, or the contour-blast is the same as that of controlled blasting.

In routine blasting holes are not fully charged and the uncharged portion is filled with some stemming material (fig. 17.13(a)). This could be some clay-material, drill cutting or even sand. Stemming plugs are also available. Apart from stemming material

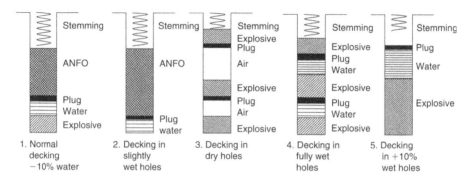

Figure 17.13(a)   Deck charging in dry and wet holes.

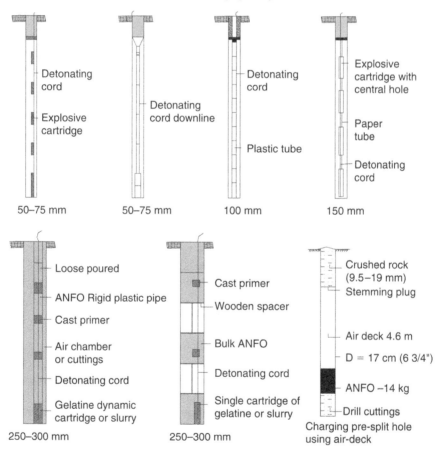

Figure 17.13(b)   Blast-hole charging schemes during normal and controlled blasting.

and explosive charge, providing a lighter charge to some length of holes becomes essential to achieve the desired powder factor. This is known as 'Deck-charging'. Deck charging could be used for the strata of varying rock-strength within the same bench-height. While undertaking controlled blasting use of this technique is almost mandatory

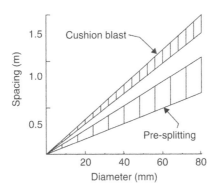

Figure 17.14 Hole spacing based on hole diameter during cushion blasting (CB) and pre-splitting (PS). In CB ratio of S/D is about 15 & for PS it is about 10.

(fig. 17.14). The space between the main charge and the stemming plug is filled with any combustible material. In practice use of wooden spacers, air bags (pneumatic), water bags or some chemicals, which are converted into foam within very short time (2–5 minutes), is frequently made to fill this space. Such plugs are available for hole diameter ranges of 75–380 mm (3"–15"). Figure 17.13(b)[15] illustrates charging schemes for holes 50 mm to 300 mm diameters. Use various types of explosives, spacers, detonating cords, and other accessories that are available and used in practice, have been shown.

If the space between stemming plug and explosive charge is kept empty, this is known as ADP (Air Deck Pre-splitting; fig. 17.13(b)) system and used in hole dia. between 127–300 mm (5"–12"). In figure 17.14 curves drawn could be used to determine hole spacing for cushion, or pre-spilling blasts.

Use of specially manufactured low-density explosives, placed in long and low diameter tubes, is used as the deck charge. In recent years use of high core-load (for example in Spain use of 40, 60, 100 g of pentrite per meter) detonating cords, for hole diameters in the range 76–89 mm (3"–3.5"), is made during contour blasting.[13]

17.5.7   Cast blasting[2]

In many cases the overburden is hard and compact and requires blasting prior to its casting. From the last two decades or more, with the use of cheaper explosive ANFO, which is also has a good heaving effect, advantage is taken to throw the muck into the worked out space of the working pit[15] Since heavy blasting in such cases is mandatory; and to effectively utilize the throw energy, a pre-split line is created prior to the main blast. This reduces the vibrations during the main blast and explosive energy can be effectively utilized. Thomas[2] proposed following relation to compute the % blast-over:

$$\% \ Blast \ Over = (57.5d/w) + 18 \tag{17.19}$$

Where: d is depth of pit and w is the exposed width;
    d/w varies from 0.4 to 0.9.

The following advantages of this technique could be advocated:

1. The access ramp/roads can be located on the high wall side of the pit, which in turn facilitates the movement of the equipment.
2. Equipment scheduling and placement is easy.
3. Surface reclamation is faster and easy.

Limitations:

1. When the dragline operates in the 'Chop Down' mode, its speed of operation becomes bit slowly.
2. Separate electrical cables etc. are required on both sides of pit i.e. on high wall side as well as spoil side.

17.5.8 Muck handling[5]

For excavating the ground suitable for digging with the use of earth movers such as dozers and scrapers are deployed (fig. 17.15(a) and (b)). But for the ground requiring blasting, the muck could be loaded using front-end-loader, hydraulic excavator, or dipper shovel (figs 17.15(c) to (g)). The selection of such units is a matter of production rate, matching hauling units and over all economics of the operation (as described in sec. 6.18 (fig. 6.12)). Based on the output requirement; the calculations for selecting a fleet of mucking and transportation units, as outlined in sec. 17.5.10 should be carried out.

17.5.9 Selection of excavator and transportation units

17.5.10 Calculations for selection of shovel/excavators[3,9]

To be practical, consideration of following three factors, to choose the right size of a shovel or any single-bucket excavator plays an important role.

- Time factor
- Operational factor
- Bucket fill factor.

17.5.10.1 Time factor

It is the percentage availability of the equipment in unit time, which could be an hour or a shift. If equipment is available 55 minutes/hour, and 7 hours in a shift of 8 hours, then it is considered as favorable situation. Under the average conditions availability of 50 minutes/hour is common but when it falls to 40 minutes/hour or below it, the condition is considered as unfavorable. The value of this factor is an overall reflection of management's efficiency.

Time factor T_f = Effective minutes available every hour/60

17.5.10.2 Operational factor (O_f)

This is a reflection of working conditions that includes layout, matching equipment meant for loading (mucking), transportation, crushing etc. This also takes into

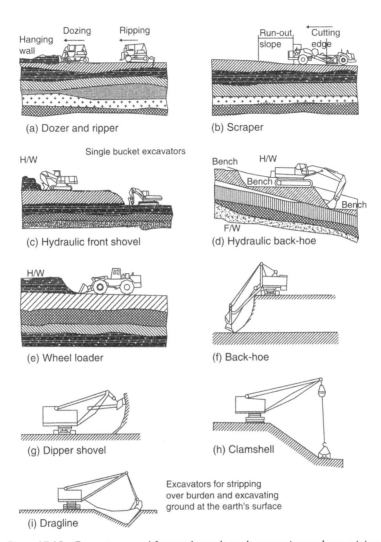

Figure 17.15   Excavators used for earth-work, rock extraction and ore mining.

account the services in terms of lighting, ventilation (comforts), heat, humidity or any other factor that effects the performance of the equipment including the operator's efficiency who operates it. To apply corrections, usual guideline is mentioned below:

Conditions	Correction
Favorable	80%
Average	70%
Unfavorable	60% and below

17.5.10.3   *Bucket fill factor (B_f)*

This is the percentage of rated bucket capacity (m^3) to the one which will actually be delivered in a working cycle. It is based on the degree of fragmentations, or size of

Table 17.2(a)   Factors to be considered in the calculations.

Material	Fill factor
Loose material	
Mixed moist aggregates	95–100%
Uniform aggregates up to 3 mm	95–100%
3 to 9 mm	90–95%
12 mm and above	85–90%
Blasted rocks	
Well blasted	80–95%
Average	75–90%
Poor	60–75%
Others	
Rock dirt mixture	100–120%
Moist loam	100–110%
Soil, boulders, roots	80–100%
Cemented material	85–95%
Factor to be considered	
Favorable	120%
Average	90%
Unfavorable	60%

Table 17.2(b)   Output/hour[11] from a dipper shovel, as given by the manufacturers to choose the shovel based on the desired output.

Bucket capacity (m³)	Rock output (m³/hr)	Earth bank output (m³/hr)
3.8	285–380	320–465
6.1	375–515	460–630
6.9	425–500	520–710
6.6	470–645	575–785
11.5	705–970	870–1185
19.1	1175–1585	1455–1910

Table 17.2(c)   Truck size details.[11]

Normal Truck Size: 22, 30, 35, 40, 55, 85, 100, 130 tons (20, 27, 32, 36, 50, 77, 90, 117 tonnes)

Giant size Trucks: 150, 175, 200, 250, 300, 350 tons (135, 158, 180, 225, 270, 315 tonnes)

material to be filled in, and also bucket penetration, breakout force, bucket profile and ground engaging tools such as bucket teeth or replaceable cutting edge or lip. Given below are the guidelines proposed by the Caterpillar.[3]

To select bucket size from the table (supplied by the manufacturer, such as table 17.2(b)), use the higher value (of the production range given in table) if expected fragmentation is good; the low value if it is going to be poor and use the average value (sum up both value and divide by 2) if the fragmentation is also going to be of average rank.

SELECTION OF SHOVEL/EXCAVATOR

Desired output/hr. = (Desired output in tons./shift)/Effective working hrs. per shift

$$OPD_h \;=\; OPD_s \,/\, W_h \tag{17.20a}$$

Where: OPD_h – the desired output/hr.

   OPD_s – the desires output/shift

   W_h – Effective working hours/shift

Actual output to be achieved/hr. = (Desired output per hr.)/Time factor $\times$ Operational factor $\times$ Bucket fill factor

$$OPA_h = OPD_h / (TF \; X \; OF \; X \, BF) \tag{17.20b}$$

Where: OPA_h – Actual output to achieved/hr.

   TF – Time factor

   OF – Operational factor

   BF – Bucket fill factor

   Actual output in m³/hr. = Output to be achieved per hr. in tons./Density of loose or blasted rock.

$$OPA_{cum} \;=\; OPA_h \,/\, D_{loose} \tag{17.20c}$$

Where: OPA_{cum} – Actual output to be achieved in m³/hr.

   D_{loose} – Density of the loose or blasted material in tons/m³

Based on this calculation select the matching shovel excavator bucket capacity from the performance table supplied by the manufacturer (e.g. table 17.2(b)

Check! output per shift/1200 < bucket capacity selected (17.20d)[9]

Actual material/pass, in tons. = Bucket capacity in m³ $\times$ Fill factor $\times$ Density of loose or (blasted) material

$$M_{PP} \;=\; B_{cap} \; X \; BF \, X \, D_{loose} \tag{17.20e}$$

Where: M_{PP} – Material/pass of the excavator in tons.

   B_{cap} – Bucket capacity in m³.

TRUCK/DUMPER SELECTION

TRUCK CAPACITY = 4 or 5 $\times$ Bucket capacity of excavator in tons.

$$T_{cap} = 4 \, or \, 5 \; X \; B_{capt} \tag{17.21a}$$

Where: T_{cap} – Truck capacity in tons.

   B_{capt} – Bucket capacity of the excavator/shovel in tons.

Select truck of matching capacity from table 17.2(c).

(*Continued*)

Determine the cycle time of truck selected. The cycle time consists of the following operations:

- Loading time
- Travel time of the loaded truck from loading site to the discharge site
- Discharge or unloading time
- Travel time of empty truck
- Spot or change over time from one truck to another.

Once the cycle time is known or calculated, calculate number of trips by a truck/shift, using the relation:

Number of trips/shift/truck = (Effective time available per shift in minutes)/ (cycle time in minutes)

$$NT_S = ETS_{min}/CYT_{min} \qquad (17.21b)$$

$$OPT_S = Nt_s \times T_{cap} \qquad (17.21c)$$

Where: OPT_S – Desired output per truck per shift in tons.
NT_S – Number of trips per shift per truck
T_{cap} – Truck capacity in tons.
ETS_{min} – Effective time available/shift in minutes
CYT_{min} – Cycle time in minutes

Number of trucks (NT) = (Desired Output/shift)/(Output/truck/shift
or $NT = OPD_s/OPTS \qquad (17.21d)$

Calculation of loading time of shovel into trucks:
$t_1 = (60 \times$ Truck capacity in m^3)/Average of the Rated output of shovel per hr. in m^3 * $\qquad (17.21e)$
* – From the table supplied by the manufacturer

Synchronization

Check! cycle time, $CYT_{min} \leq NT(t_1 + t_s)$ $\qquad (17.21f)$

Where: t_1 loading time of truck by the shovel or excavator, in minutes.
t_s – spot time in minutes.
NT – number of trucks.

17.5.11 Theoretical output from an excavator/hr

$$O_{th} = 3600 \times E/T, m^3/hour \qquad (17.22a)$$

O_{th} is the theoretical output/hour
E is bucket capacity in m^3 and T is cycle time in second.

In practice the factors such as bucket fill, time and bulking; need to be incorporated in the above mentioned formula and then it becomes:

$$O_{act} = 3600 \ E \ B_f \ T_f \ O_f \ / K \ T, \ m^3/hour \tag{17.22b}$$

Where: O_{act} – Actual output/hour;

B_f is bucket fill factor;

K is bulking factor;

T_f is time factor;

O_f – Operational factor.

Bulking factor (K) is the ratio of increase in volume of the material after getting fragmented to its volume in place or in-situ (original). It differs from material to material, depending upon their densities. Value of fill factor has been given in table 17.2(a) and also the time factor has been described in the preceding sections.

17.5.12   Output from a continuous flow unit

$$O_{act} = 3600 \ E \ B_f \ T_f \ O_f \ / K \ S_s, \quad m^3/hour \tag{17.23a}$$

$$S_s = V_c \ N_b \ / \ \pi D \tag{17.23b}$$

S_s is number of buckets discharged/sec;

V_c is cutting speed in m/sec;

N_b – Number of buckets in the wheel;

D – diameter of wheel in meters.

17.5.13   Transportation schemes[23,24]

Based on the hauling lead of the waste rock as well as the ore, terrain conditions and matching with mucking and excavating equipment various transportation systems could be applied, as shown in figures 17.16(a) to (f). It could be truck haulage, rail haulage or their combination. In deep pits a conveyor, rope haulage using skip hoist, or an ore pass system could be feasible 17.16(c) to (e). In hilly terrain cableways are most suitable for long distance hauling (fig. 17.16(f), Sec. 7.6, fig. 7.11). For placer deposit hydraulic transportation is mandatory (fig. 17.16(g)).

17.5.14   In-pit crushing and conveying[9]

Application of mobile crushers began in 1956 in German limestone mines.[8] A mobile crusher is the one which moves as the mining faces in an open pit advances, and it is directly fed by the excavators or trucks that are deployed at the working benches. This unit could be mounted on crawler track, walker or pneumatic tyres. Semi-mobile crushers are not that frequently moved as the mobile ones, and they are installed as near as possible to the common feeding point. Survey made[8] for such unit for the period 1956–1989, indicated that type of crusher used for this purpose could be jaw, hammer, gyratory or roller. The production range varies between 125–6000 t/h. The crushed material is fed to the belt conveyor unit that could ultimately discharge it to

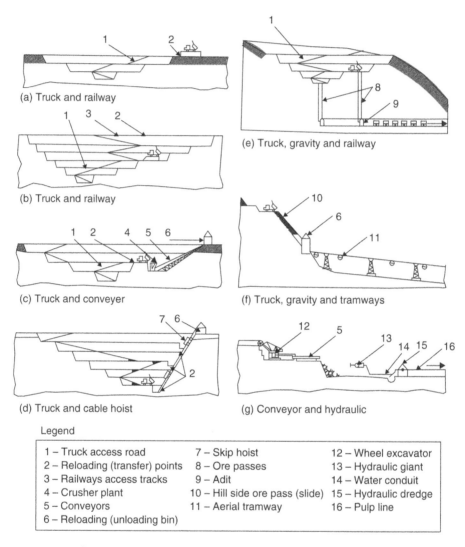

(a) Truck and railway

(b) Truck and railway

(c) Truck and conveyer

(d) Truck and cable hoist

(e) Truck, gravity and railway

(f) Truck, gravity and tramways

(g) Conveyor and hydraulic

Legend

1 – Truck access road	7 – Skip hoist	12 – Wheel excavator
2 – Reloading (transfer) points	8 – Ore passes	13 – Hydraulic giant
3 – Railways access tracks	9 – Adit	14 – Water conduit
4 – Crusher plant	10 – Hill side ore pass (slide)	15 – Hydraulic dredge
5 – Conveyors	11 – Aerial tramway	16 – Pulp line
6 – Reloading (unloading bin)		

Figure 17.16   Transportation schemes at surface mines and pits.

the feeding plant or factory. The advantage gained is the reduction in overall transport costs and increase in productivity of the system.

17.5.15   Dumping site[28]

In surface mines (open cast and open pit) the direction of advance could be parallel to the longer axis of the over all pit profile (fig. 17.17); single sided,[23,24] double sided; in the similar manner but transversely; also it could be in fanning shape, or in concentric circles.[23,24] In open cast mines the dumping would be within the worked out areas that are backfilled but in open pit mines it would be out side it as shown in

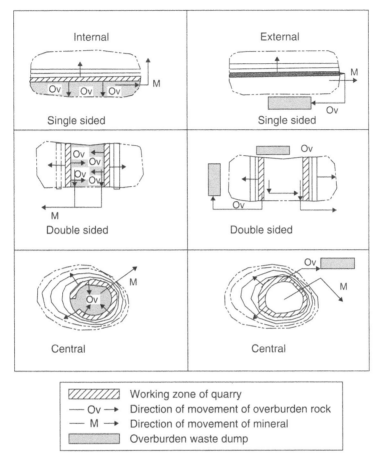

Figure 17.17 Various schemes of waste rock dumps. Dump locations are based on the geometry and suitability of the available land in terms of their techno-economical considerations. Direction of advance of these dumps is shown.

figure 17.17. Following are some of the guidelines that should be followed to achieve minimum land degradation and handling waste rocks systematically.

1. Separate dumping site for:
   (a) Waste dump
   (b) Subgrade mineral
   (c) Mineral of economic interest later on
   (d) Ore (temporary)
2. Site selection: care should be taken in the following aspects
   (a) Favorable surface topography
   (b) Nearest to the working pit
   (c) Devoid of any mineralization of economic interest

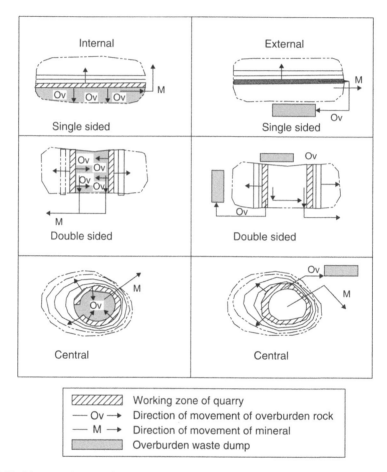

Figure 17.17   Various schemes of waste rock dumps. Dump locations are based on the geometry and suitability of the available land in terms of their techno-economical considerations. Direction of advance of these dumps shown.

(d) Devoid of any vegetation, plantation, forest area, agriculture land etc.
(e) Not an area of public utility and infrastructures
(f) Not the source of water or obstruction to the source of water. Also not any of the water bodies.
3. Procedure of dumping:
   (a) Keeping the height of dump to be 2.5 m to 3 m and angle of repose to be 35–40° (depending upon its angle of repose of the dumping rocks themselves)
   (b) Leveling the dumps so formed using bulldozer.
   (c) Dumping over the leveled heap, and again level it, till the dump yard/site permits.
   (d) Keeping record of dumping area-wise, date-wise and mineral-wise.
   (e) Putting fencing around the dump, and also keeping large sized boulders near the fencing to prevent the passage of silt during rainy season.

(f) Stabilization of dumps by growing vegetation. This will help to check the environment degradation in the area.

Types of dumping sites

1. Constructed on a flat terrain:
   (a) Heaped dump constructed in successive layers
   (b) Heaped dump constructed in a single layer.

The manner in which, the dump-yard could be developed is shown in figure 17.17.

2. Using a Valley:
   a) Valley fill by terracing
   b) Head-of-hollow fill – right up to full depth
   c) Cross valley fill – in an extensive valley area
3. Hill side dumps:
   a) Side hill fill – from a side of the hill starting at the predetermined elevation.
   b) Both sides of a ridge.

17.5.16 Integrated or matching equipment complex[24,25]

It is important that the type of equipment, methods and techniques match each other to obtain the desired results. The idea is the smooth flow of the ore and waste rocks from their places of generation to their final discharge destinations. The following are the salient points that could be considered:[24]

- The equipment chosen for the unit operations such as drilling, blasting, mucking, haulage and crushing are of the rated specifications and capacities, so that each of the units could be utilized optimally without keeping any equipment in the circuit under utilized. In this regard many software and modules have been developed. The idea is to optimize the resources for the best results. The flow diagram shown in figure 17.18 could be used to achieve this.[9]
- Matching climate and geological conditions.
- Matching layout with respect to bench geometry, pit slopes, ramps gradients and dumping sites.
- An equipment complex with a smaller number of operating machines and mechanisms operate more reliably, intensively, and results better efficiency and hence, the profitability.
- Wherever possible flow of material (ore and wastes) should be as continuous as possible. Use of information technology (IT) and communication system has begun. To site an example to automate the operations, Siemens, Germany[25] have installed Overall Process Network (OPN) at Laubag Lignite mines, Germany, where world's largest manmade equipment on earth – the Bucket Wheel Excavators, are in operation (fig. 6.10(b)). Figure 17.20 displays[25] line diagram of this system.
- The equipment used and methods applied should be safe and as per the statutory requirements. Safe and comfortable working conditions must be ensured.

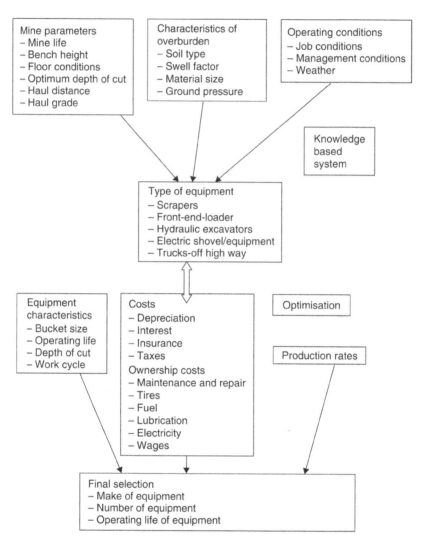

Figure 17.18   Flow chart for optimization of surface mining operations. Input required and output obtained are shown.

17.5.16.1   *Global Positioning System (GPS)*[20]

This system is an example of an integrated system to achieve automation and optimize resource utilization. In figure 17.19 'Total Mining System' and table 17.3;[20] salient features of the GPS have been described and illustrated.

17.5.17   Quarrying of dimension stones

Dimension stone – natural building stone that has been selected, trimmed or cut to specified shapes or/and sizes with or without one or more mechanically dressed surfaces. Common dimensions stones are tabulated in table 17.4.

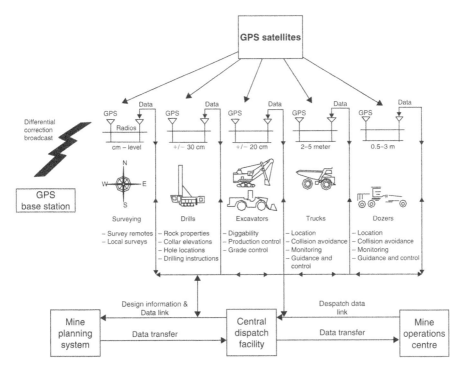

Figure 17.19   GPS monitoring system – a total monitoring system.

17.6   QUARRYING OF DIMENSION STONES[7,17,18,21,23,27]

Figure 17.21 classifies common quarrying methods. This also includes those which are prevalent from the past, and also those that have been developed in the recent past during a span of 10–20 years. Figure 17.22 illustrates the concept of quarrying. Its development scheme along the normal as well as hilly terrain has been shown in figure 17.23(a).

17.6.1   Drilling[7,21]

In Italy, USA, France, Belgium, Sweden and many countries use of different types of drills is made for this purpose. Drilling is used as an independent technology to undertake quarrying operations in all types of dimension stones. Pneumatically powered jack-hammers and rock drills of light, medium or heavy duty are used for drilling. Hydraulic drills also appeared in stone quarries from last two decades or more.

There are many types of auxiliary equipment or mountings that allow running of more than one rock drills simultaneously without any operator. One of such units is known as block-cutters (or drills) with one or more jackhammers. These mountings enable better productivity on account of:

• Reduction in drill maneuvering time as the positioning, shifting and other movements are all automatic.

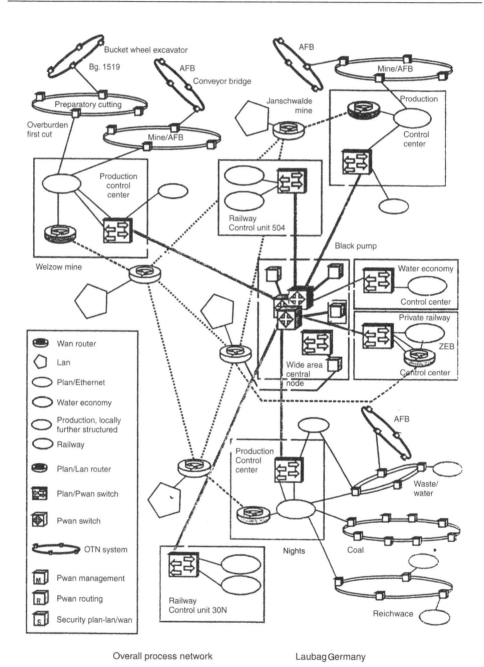

Overall process network   Laubag Germany

Figure 17.20   Lausitzer Braunkohle Open Cast Mining – Germany (world's largest man-made equipment), 'The Bucket Wheel Excavator' in operation. Open Transport Network (OTN) is used to automate the whole mining machinery. This ensures data communication between different processes, the telephone and radio communication, the interconnection of LANs and the video transmission Pwan – Private Wide Area Network (Courtesty: Siemens, Germany).

Table 17.3   GPS application in open pit mines.[20]

Equipment	Applications	Benefits	GPS Requirement
Ground surveying	– Replace and/or supplement current laser based two-man survey system – Ore volume calculations, road and bench profiles and limits exploration etc.	– One-man operation – Suited to all types of weathers and most pit configurations – Not restricted to daylight use	– High precision real-time surveying to +/−5 cm in 3D – Portable, rugged, lightweight, easy to operate system – Compatible with GPS system in mobile equipment – Data could be interfaced with the exiting mine planning softwares
Blast-hole drills	– Precise 3D positioning to designed blast-hole locations without surveying – Base platform for eventual development of autonomous capability	– Reduced blasting costs through improved fragmentation – Correct blasthole depths to target elevations – more even bench floors and least under/over breaks. – Reduced surveying needs	– High precision +/−30 cm in 3D in real time – Tilt, roll and heading incorporated with positions – 3D position displays to screen in operator's cab via graphic moving map display
Shovels (hydraulic or cable) & front-end loaders	– Maintain grade (elevations) within allowable limits – Correlate location of each dipper load with: (a) muck-pile diggability for improved blast design control; (b) material type for blending and stockpiling	– Improved pit floor profile – Reduced dilution – Improved equipment scheduling/dispatching and tracking of material movement – Improved ore-grade control – Grade-tonnage reconciliation	– 3D accuracy of 20 cm – Elevation, X, Y, are displayed to operator via graphic screen – Moving map displays shows ore/grade limits and bucket position w.r.t. shovel in plan view
Trucks	– Real time locations within open pit mine – Collision avoidance and autonomous operations	– Improved equipment scheduling/dispatching and tracking of material movement – All weather operations	– Real time accuracy more than 1 m – Positioning data not displayed to operator

- Regular drilling, especially when drilling deep holes; this also minimizes the deviations and provide better accuracy.
- Automatic checks on rock-drill depth; drill stops when the required depth is reached. Drills can be installed at any angle and orientation.

Table 17.4 Dimension stones details.[27]

Dimension stone	Definition	Origin/rock type
Granite	A fine to coarse grained, igneous rock formed by volcanic action consisting of quartz, feldspar, and mica with accessory minerals. Feldspar is usually in excess of quartz and accessory minerals. The granites used as dimension stones include gneiss, gneissic granite, granite gneiss, and rock species known as syenite, monozite and granodiorite.	It is a hard rock having four basic varieties:
Marble	A metamorphic (recrystallized) limestone composed mainly of crystalline grains of calcite or dolomite, or either having interlocking or mosaic texture. Marble containing $MgCO_3$ less than 5% – calcareous marble; $MgCO_3$ 5–40% – magnesium marble; $MgCO_3 > 40\%$ – dolomitic marble.	
Sandstone	A sedimentary rock consisting of mainly quartz cemented with silica, iron oxide or calcium carbonate. It is a durable rock with very high crushing and tensile strength. It has wide range of colours and textures. Its variety is based on the bonding and interstitial materials, this makes them as: siliceous, calcareous, ferruginous, feldspathic, conglomeratic sandstones.	
Slate	A very fined grained metamorphic rock derived from the sedimentary rock shale. It has excellent parallel cleavage entirely independent of original bedding, by virtue of this feature the rock may split into relatively thin slabs.	

Variety	Drilling (m/m³)	Relative cost (%)
Gray	<20	20–40
Red	20–50	15–40
Brown	20–50	15–40
Black	50–100	50–100

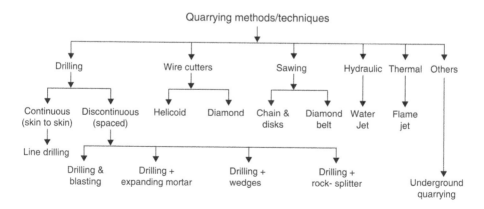

Figure 17.21 Classification of quarrying methods/techniques.

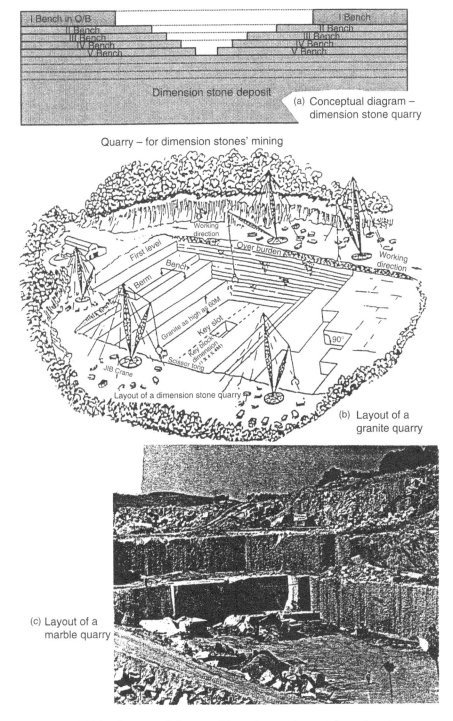

(a) Conceptual diagram –
dimension stone quarry

(b) Layout of a
granite quarry

(c) Layout of a
marble quarry

Figure 17.22   Conceptual diagrams: Quarrying to mine out dimension stones.

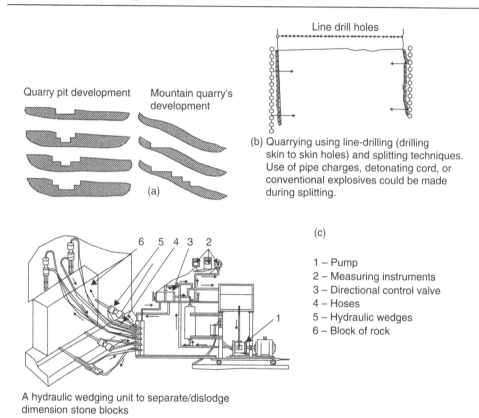

Quarry pit development   Mountain quarry's development

(a)

Line drill holes

(b) Quarrying using line-drilling (drilling skin to skin holes) and splitting techniques. Use of pipe charges, detonating cord, or conventional explosives could be made during splitting.

(c)

6 5 4 3 2

1

1 – Pump
2 – Measuring instruments
3 – Directional control valve
4 – Hoses
5 – Hydraulic wedges
6 – Block of rock

A hydraulic wedging unit to separate/dislodge dimension stone blocks

Figure 17.23   (a): Quarry development schemes. (b): Details of line drilling. (c): Hydraulic wedging.

- Optimizes drills' performance since they are maneuvered and guided automatically.
- This ultimately results into better work conditions for the quarry workers.

The latest development[21] in this area is the use of mobile hydraulic units run by a single operator and mounted on excavators (preferably with stabilizers). Telescopic arms can work at considerable distances (9 to 10 meters). Drilling operations can be controlled from the cabin, or by remote control, with radio commands for both the drill and the excavator (except its wheels). This level of mechanization and automation has considerably reduced personnel's exposure to dust, noise and vibrations, with overall improvements in working conditions and safety.

17.6.2   LIne drilling

When drilling is done skin to skin i.e. without any spacing (fig. 17.23(b)); the technique is known as line drilling. It was first time tried at Carrara, Italy in mid 1980s.[21] This is a refined method of block separation and mining but expensive. It is preferred where explosive and other methods don't work efficiently or, there are some technical restrictions.

17.6.3   Discontinuous or spaced drilling[21,23]

Drilling method of block separation from the massif consists of two inter-related processes (figs 17.23(b) and (c)): drilling of close rows of vertical, horizontal and possibly inclined holes in some cases, and then splitting off the stones with wedges. This is a very old method, which requires muscle power but is still very much in vogue at many quarries. This method is also known as '*Plugs and Feathers*'.

'*Drilling + rock-splitters*' are the mechanized version of 'Plug and Feather' method, in which 'Cylinders' are placed in holes, exerting enough pressure on the walls; causing rock to split along a preset plane.

Use of this technique is also made to reduce the size of granite monoliths into standard blocks. For this purpose, holes of 20–40 mm in dia., spaced at 5–10 cm are drilled 8–10 cm deep. Simple or composite (consists of two jaws and wedge proper) wedges are inserted into the holes. Individual standard blocks are separated from the monoliths by knocking upon the wedges with a sledgehammer. In marble the holes are drilled up to the full height and width of the block to be mined by spacing the holes at 10–20 cm apart 6–10 holes are drilled/m.

'*Drilling + expanding mortar*' is gaining popularity due to its inherent safety features; and also absence of noise, dust and vibrations. In this technique expanding mortar, or, cement or, even chemical demolition agents are used. These mixtures are added to water, and when they are active exert pressure up to 8000 ton/m^2; which is sufficient to break the traction resistance of any rock.[21] This technique is suitable where use of explosives is prohibitive and not satisfactory.

In '*Hydraulic wedging*' technique stresses are applied in a particular direction and the amount of drilling can be reduced. The unit to separate blocks has been shown in figure 17.23(c).[19] It saves in labor required, thereby, making the process productive and less costly.

Growth and development of line drilling techniques: Figure 17.24 illustrates how drilling operations, which used to be slow, tedious and noisy (fig. 17.24(A)) have been developed in the recent years that has come up as integrated units mounted on track, rigs and mobile jumbos. This has brought a revolution in quarrying operations. In figure 17.24 a complete range of drills that have been developed by Marini, Italy has been shown. Drilling operation could be now considered to be safe, productive, remote controlled, semi or fully automatic.

17.6.4   Drilling and blasting[17,21,27]

This is also one of the popular techniques in quarrying operations and particularly in granite quarries. This is also known as '*pre-splitting*', or '*dynamic-splitting*'. In this technique holes are drilled at a close spacing, and then mild blasting using plastic-pipe-cartridges; is carried out.

OY Forcit, Finland, has developed plastic pipe charge cartridges, which are charged in the holes dia. in the range of 27–32 mm. The charge density of the order of 60–150 g/m^3 is kept[27]. Their specially designed connectors automatically centralize the waterproof cartridges. This makes the charging operation quick and simple. The vertical and horizontal holes are fired simultaneously and that moves the stone-block about 150 mm off the face.

(a) Drilling using hand held Rock drill. It is slow, tedious, noisy and not very productive.

(b) Mounting single rock drill on Track. First step towards mechanization. Note improved working conditions.

(d) Mounting multi rock drill mounted on mobile jumbo. Drills could be pneumatic or hydraulic

(c) Mounting multi rock drill on Track. It improved quality of product, safety and productivity. It could be remote controlled.

(e) Mechanizing horizontal drilling

Figure 17.24(A)   Growth and development of drilling technology for quarrying (past to present) (Courtesy: Marini, Italy).

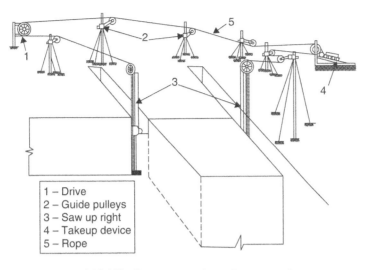

1 – Drive
2 – Guide pulleys
3 – Saw up right
4 – Takeup device
5 – Rope

Figure 17.24(B)   Rope saw – schematic presentation.

Use of 'K pipes' is also made to further split the blocks into smaller ones. For this holes are drilled with a spacing of 250 mm. The charge density in such blasts, which are tipped down to sand beds to minimize damage of the broken blocks, could be in the range of 30–80 g/m³. These blocks could be up to 30 m³ or more. They can be further spilt into smaller size by drilling holes precisely at closer spacing, and separating them by wedging actions, or with the detonating cord. Tamrock drills could be used for this purpose.

There are so many variables involved in these techniques but correct drilling and judicious use of explosive, are the keys to success. In absence of this approach, it may lead to generation of irregular blocks, a great amount of waste-material, and lower economic value for the saleable product. Sometimes it may create controversies between operators and buyers with regard to quality. The *advantages* of drilling and blasting method include:

• Simplicity and widespread; little preparatory work required.
• Mobility.
• Maximum utilization of natural fissures.
• Applying to any size, and strength of dimension stone.
• Low investment.

Disadvantages

• Manual labor intensive.
• Low productivity.
• High cost and less safety.
• More wastage. Low overall recovery when blasting methods used.
• Environment degradation.

Table 17.5   Explosives used in quarrying dimension stones.[18]

Explosive	Density (gms/cc)	Detonation velocity (m/sec)	Degree of packing (kg/m)	Relative strength
Dynamite d = 24 mm	1.5	6000	0.6	1
K-pipe explosives	0.95	1900	0.22	0.3
Detonating cord	1.25	6500	0.02	

17.6.4.1   Blast results at Vanga granite quarry in southern Sweden[29]

SveBefo has performed extensive tests (Olsson and Bergqvist 1993) where they have studied the crack length from blastholes. At Vanga granite quarry using the parameters as detailed below, hundreds of blast holes have been blasted; the observation and results for the use of explosives have been summarized below.

Burden and spacing used = 0.5 m × 0.5 m. Bench height = 5 m. Hole dia. tried = 38 mm; 51 mm and 64 mm. Four to five holes were shot by charging 4.5 m (the remaining 0.5 m was unstemmed); electronic detonators from Dyno Nobel were used. Types of explosive charged were as detailed below:

Swedish contour blasting explosive such as Gurit, Kimulux 42; Detonex 80 (80 g/m PETN Cord) and Emulet 20 (an emulsion styropore mix with 20% of volume strength relative to ANFO. The explosives tested had large variations in their velocity of detonation; for example: 2000 m/s Gurit; 4800 m/s Kimulux 42; 6500 m/s for Detonex 80.

Figure 17.24(C), details explosives that were used and comparison of the results have been tabulated for the single row as well as multi row blasting. The blocks were separated after blasting and cut horizontally using a circular cut-off saw. The crack patterns were highlighted using a conventional dye penetrant at one or several places along the hole axis.

Figure 17.24(D) details use of wooden spacers, Omega tubes and cords of different strengths which could be used during the contour/controlled blasting.

Some of the observations made are:

- Simultaneous initiation with electronic detonators gives much shorter cracks.
- Crack length increases with decreasing coupling**. A bulk explosive that is completely filled gives the longer cracks. ** – Separation between hole wall and surface of explosive cartridges (charge) is known as decoupling. It is the ratio of cartridge dia. to hole dia.
- Crack length increases with increased charge concentration.
- When delay times used were low as 1 ms, the crack length still looked more like the cracks from single hole blasts.
- Traditional smooth blasting procedure with long period delays give unnecessarily long cracks.

Thus, explosive energy which is the cheapest energy, if used in a planned way judiciously; it could be successfully used for dimensional stone mining thereby saving substantially compared with the use of cutting tools and equipment.

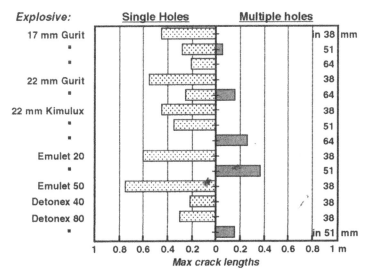

Figure 17.24(C)   Use of explosives and controlled blasting techniques to generate granite blocks at one of the quarries in Sweden. Note the crack development scenarios when adapoting different options (www.smenet.org).

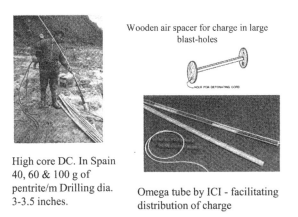

Figure 17.24(D)   Use of wooden spacers, Omega tubes and cords of different strengths which could be used during the contour/controlled blasting.

17.6.5   Wire cutter – helicoid and diamond[21,23]

These techniques are applied in marble quarries of Spain, Italy, USA, France, Portugal and many other countries. The sawing action is achieved by the abrasive action produced by the quartz sand that is continuously fed with water to the face. This is known as helicoids wire technique, or method (fig. 17.24(B)). Since the 1980s this method has been slowly and slowly replaced by a new technique, which used diamond wire (fig. 17.25(a)). The technique could be used for granites also. Table 17.6 compares important features of the two systems.

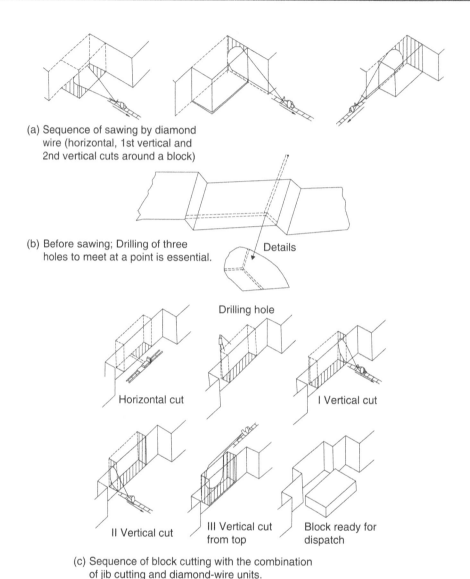

(a) Sequence of sawing by diamond
    wire (horizontal, 1st vertical and
    2nd vertical cuts around a block)

(b) Before sawing; Drilling of three   Details
    holes to meet at a point is essential.

Drilling hole

Horizontal cut   I Vertical cut

II Vertical cut   III Vertical cut   Block ready for
                       from top   dispatch

(c) Sequence of block cutting with the combination
    of jib cutting and diamond-wire units.

Figure 17.25   Quarrying techniques (a) and (b) – Diamond wire cutting unit. (c) – Jib (bar) cutting unit.

Normally the diamond wire is run in a closed loop around the rocky/stone mass (fig. 17.25(a)). Its speed is governed by the type of rock to be cut. The closed-loop arrangement is made possible by previously drilling two intersecting holes through which the wire can run. As it cuts the machine backs up, running on rails beneath it, and constant checks are made on the tightness of the wire. A diamond wire cutter can work at all angles and in various ways, depending on the type of cut to make.

Table 17.6 Comparison of Helicoid-wire and Diamond-wire technologies.[21]

Parameters	Helicoid wire	Diamond wire
Introduction Year	1895	After 1975
Wire dia.	4–6 mm	5 mm
Translation speed	4–15 m/sec	30–45 m/sec in Marble. 15–30 m/sec in granite
Productivity	0.5–1.6 m²/hr	3–12 m²/hr, and up to 18 m²/hr in Marble. 1–5 m²/hr, or up to 8–9 m²/hr in granite.
Wire length	Hundreds to Thousands of meters	Tens of meters.
Abrasive used	Selected siliceous sand	Sintered or electroplated beads.
Cut thickness	6–10 mm	9–12 mm
Motor Power	7–25 HP	25–70 HP (Electric); 80–100 HP (Diesel).
Merits and Limitations	Reduced cut thickness, very silent and safe. Longer setting & maintenance time; low productivity; Unsuitable for granite.	Efficient, productive, suitable for all stones, quicker setup and versatile to use. Preliminary drilling required.

In the more advanced versions[21] the wire machines have been replaced by electro-mechanical, automatically controlled units with power running between 25 and 75 HP (up to 100 HP when diesel motors are used). They have electronic devices (inverters), which can vary the wire's linear speed to suit the various cutting stages (and regulate it depending on the tool' s degree of wear). They also run checks in real time on wire tension, which maximizes the tool's yield/endurance ratio under all conditions, and have stopping systems should the cable break. The diamond wire has undergone many changes since its inception. Tables 17.6 and 17.7 detail its main features and locales of applications.

The *merits* of the system include:

• Simple design and operation.
• Possibility of generating blocks of required size and shape. No thermal or mechanical damage. Better volumetric fill-up on gangsaws.
• Better recovery and reduced discards: thinner cuts, more regularity on quarry fronts. For granites it is the only valid method in fractured deposits (the flame-jet is ineffective). No theoretical limitation to cut height.
• Environment friendly.
• Relatively low power consumption: low noise level, no dust or vibrations.
• Better productivity. Reduced labor as the same operator can undertake other jobs simultaneously.

Shortcomings include:

• Large preparatory work and skill labor required. Needs water.
• Difficulties when hard inclusions and more joints are encountered.

Table 17.7 Diamond – wire technology and its application in quarry operations. Procedures and average performance.[21]

Item	Marble quarries (including onyx, travertine, green marbles, soft sandstones and lime stones etc)	Granite quarries (including porphyry, serizzo, quartzite, etc.)
Type of wires and beads	Traditional cable, plastic coated, rubber coated with springs; sintered and electroplated beads	Only plastic or rubber coated cables, rarely with springs; almost solely sintered beads (although electroplated are sometimes used)
Configuration	Wires generally assembled with 28 to 34 beads/m	Wires generally assembled with 32 to 40 beads/m
Application	Primary and secondary cuts, block squaring	Primary and secondary cuts, not economical for block squaring
Cut length	$20\,m^2$ to over 350 to $400\,m^2$ (exceptionally 800 to $1000\,m^2$)	$20\,m^2$ to $150\,m^2$ rarely $200\,m^2$
Average productivity	3 to $12\,m^2$/hour up to $18\,m^2$/hour	1 to $5\,m^2$/hour up to 8 to $9\,m^2$/hour
Wire duration (yield)	15 to $40\,m^2$/m upto $120\,m^2$/m	An average of 2.5 to 7, upto 10 to $2\,m^2$/m
Linear wire speed	30 to 45 m/sec	15 to 30 m/sec

17.6.6 Cutter saw and rock channellers (impact cutting machines)

These machines, which are very common in the USA, Canada, France, Spain and many other countries, are capable of making vertical, horizontal and inclined kerfs (cuts). The working element of these machines is a set of reciprocating bits. When this channeller travels along a rail track, the bits knock upon the rock to break it forming a kerf up to 50–60 mm. wide and 6 m deep. The channeller's usual capacity is in the range of 0.8–1.2 m^2/h or 5–8 m^2/shift.

Classification: These stone cutting machines can be sub-divided into:[23]

1. *Disk cutters* (fig. 17.26(a)) – Disc cutters are used for cutting stones having ultimate compressive strength $\sigma_c = 10$–250 kgf/cm^2
2. *Jib (bar) cutters* (fig. 17.25(c))[7] – Machines with chain bars are suitable for stones with $\sigma_c = 10$–100 kgf/cm^2
3. *Cutters with annular milling cutter* – These machines find applications when dealing with stones of compressive strength $\sigma_c = 50$–1200 kgf/cm^2.

The preparation of sawing stone for extraction includes three operations:

- Formation of cross saw cuts
- Making horizontal saw cuts through the entire cut of the cutter
- Making the back saw cut and separation of stone from the massif.

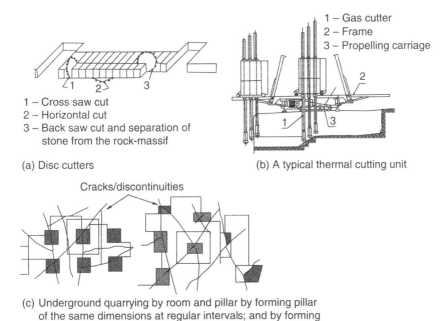

1 – Gas cutter
2 – Frame
3 – Propelling carriage

1 – Cross saw cut
2 – Horizontal cut
3 – Back saw cut and separation of
stone from the rock-massif

(a) Disc cutters

(b) A typical thermal cutting unit

Cracks/discontinuities

(c) Underground quarrying by room and pillar by forming pillar
of the same dimensions at regular intervals; and by forming
pillars of random size at random spacing

Figure 17.26 Quarrying techniques (a): Disc cutters. (b): Thermal cutting. (c): Underground quarrying using Room and Pillar system.

17.6.6.1 Merits[21]

- It gives better production rates and productivity due to reduced labor requirements
- Indispensable in underground quarrying through tunnel openings
- It works both dry as well as with water
- Production of better quality blocks for processing due to no thermal or mechanical damage; better gangsaw fill-up.

17.6.6.2 Disadvantages[21,23]

- An increased kerf width. Depth of cut is also limited
- Higher power consumption. May not prove always economical
- High blow on stones. Not suitable for granites and similar stones.

17.7 THE DIAMOND BELT SAW[21]

Use of diamond belt saws began in some of the American quarries in 1985. The concept and structure of the equipment is similar to the chain saw, as it is equipped with an arm carrying a belt rather than a chain. It can be used on marble, limestone and moderately hard stones but not on granites.

Its belt consists of a metal core made of steel cables about 3 mm in diameter assembled flat and covered with a very hard plastic. Attached to the cables are the abrasive sections, sintered, diamond-coated plaques as wide as the belt and about 15 mm thick. Diamond and bonding agent are chosen on the basis of the material to be cut. The cutting tools need no sharpening as the whole belt is replaced when its abrasive capabilities get reduced. The belt on the arm is lubricated and cooled solely by pressurized water. Eliminating use of grease or oil makes this equipment environment-friendly. The equipment is available in three versions:

(i) For vertical cuts only, (ii) For both vertical and horizontal cuts and (iii) For u/g quarrying (tunnel work).

Merits

- Usage of no grease or lubrication.
- Better productivity due to reduced labor requirements.
- Production of better quality blocks for processing due to no thermal or mechanical damage; better gangsaw fill-up.

Disadvantages

- Depth of cut (arm length) is also limited.
- May not prove always economical.
- Needs water. Not suitable for granites and similar stones.

17.7.1   Water jet technology[21,27]

In this technique with the application of a water jet (up to 350 MPa) the rock is cut. Such an installation cuts the rock by making the jet-carrying rod move back and forth along the cut lines, and penetrates them. On some models the jet's progress is completely automated so that the machine can work non-stop without supervision, and automatically stops if mishaps or emergencies arise. It can make cuts 2.5 to 3.6 m in depth and, if the arm is given an extension, even up to 8 meters.

The surfaces of the cuts are rather rough but very precise, and this feature is very important in improving the recovery from granite blocks. It is considered to be competitive with the flame-jet, and continuous drilling but still in its initial stage. This technique has great potential in future granite quarrying underground, in tandem with the diamond wire. It is not suitable for marble mining but could be used for other stones. Table 17.8, depicts performances recorded with different water-jet installations and different materials in some countries.

This technique could also be applied during processing of dimension stones and comprise of three major elements:

High-pressure water jet – capable of operating at 4000 kg/cm^2 pressure
Special table where the cutting of stone takes place
Computer with CAD/CAM Technology.

The system claims following advantages:

- Minimal adverse impact to environment, excellent work conditions for personnel (no dust, vibrations, fumes or noise)
- No preliminary preparations needed

Table 17.8 Performances recorded with different water-jet installations and different materials in some countries.[21] n.d. – no details available.

Material	Location	Pressure (Mpa)	Capacity l/min	Power (kW)	Cutting speed (m²/hour)
Granite	Elberton (USA)	280	11	52	1.17
Granite	Milbank (USA)	165	76	209	n.d.
Granite	Colorado (USA)	310	5	26	0.6
Granite	Quebec (Canada)	140	76	175	1.15
Granite	Lanhelin (France)	200	70	330	1.5
Sandstone	Rothbach (France)	80	60	160	6.5
Gneiss	Valdossol (Italy)	350	8	47	1.7
Granite	Sardinia (Italy)	200	18	60	2.4

- No limit to cut extension
- The only technology which together with the diamond wire, can work on granite underground
- Block shape and regularity: percentage recoup
- Does not damage the material
- Works independently and automatically.

However, the system has following limitations

- Heavy investment
- Limited cut depth (rod length)
- Not yet competitive; needs perfecting
- 'Works' well only in granite
- Requires a lot of water.

17.7.2 Thermal cutting[21,23]

Dimension stone such as granite can be cut into blocks by gas-flame type machines (fig. 17.26(b)). This technique is faster than conventional drilling and blasting. Quality of block is improved and less expenditure on manual labor and higher productivity of the stonecutter is obtained. In Russian mines it gives outputs of 1–2 m³/hr, which is 1.5–2 times faster than the conventional method of drilling and blasting. It is simple to operate as no preliminary work is required and those who are trained in this technique can efficiently work. However, this technique has these limitations:

- Usually incompatible with other quarry work as its needs a lot room around the work zone, thereby, restricting other activities. Cut height is generally not more than 6 m.
- It is suitable only for certain granites; as the performance strictly depends on the rock's chemical-mineralogical composition (cutting speed reduces as quartz content decreases). It damages the rock to a considerable depth.

- Low productivity since production depends on the deposit's fracturing pattern (energy is wasted in fractured zones and thereby the cutting rate get reduced). Overall higher energy costs are resulted.
- Adverse environmental impact due to generation of high noise level, gas and dusts.

17.7.3 Underground quarrying[7]

The concept of underground quarrying is a recent development and has potentialities in the years to come due to increasing depth of existing dimension deposits and the depth of quarries already attained. In situations of high overburden cover, it proves to be economical. Use of cutting saws and other technology could be made to mine-out stone deposits. Room and pillar with regular, and with irregular pillars (fig. 17.26(c)) are the suitable methods that could be applied.

Table 17.9 compares the salient features of the existing (conventional and traditional), alternatives and innovative techniques in stone mining locales. Table 17.10 describes the type of equipment that could be deployed to carry out various unit operation during mining by quarrying.

In figure 17.27 a comparison of surface mining methods,[11] taking into consideration, the important features for mechanical and aqueous methods, has been made.[9] This indicates that aqueous methods (solution mining) are the cheapest when applied under suitable conditions.

17.8 EARTH MOVERS[5]

Scraping, ripping and digging are the techniques (figs 17.15 & 17.27) commonly used for removing the soft and weak material such as clay, silt, sand, shale, weathered rock and topsoil. These operations are mandatory in civil works such as construction of roads, rail lines, dams, airports, buildings, and many others. Scrapers, rippers, dozers, graders, trenchers and excavators are the common earthmovers. Such equipment works best in ground that has a seismic velocity[3] lower than 1000 m/sec. Seismic velocity charts as recommended by Caterpillar Company or others, as shown in figure 17.28, provides a practical guide to select any of these earthmoving units.

A dozer (fig. 17.15(a)) is a crawler-mounted or wheeled tractor unit fitted with a blade. It is capable of excavating, moving and stock pilling the earth-rock (ground). Based on the duty this unit can perform, it is available as light, medium and heavy-duty units; and their power ranges from 20 to 300 h.p or more. Operating cycle of a bulldozer consists of cutting off a horizontal or inclined slice from the ground, formation of a dragging prism, moving the latter and dumping.

While selecting this unit a proper match between tractor (in terms of its horse-power (H.P) and weight) and the type of blade should be considered. Selection of the blade will depend upon the type of material to be moved. Most materials are doze-able. However, dozer performance will vary with the material characteristics such as: particle size and shape, presence of voids, and water content.

Scrapers (fig. 17.15(b)) are known as excavators but basically they are integrated load, haul and dumping units (LHDs). Their applications are many. In civil works prominent tasks are road, dam and dike constructions. Mining applications include

Table 17.9 Technologies used in extraction – a comparison. (*): Current status.[21]

Material	Traditional/conventional technologies					Alternative technologies		Innovative technologies	
	DD + E	FJ	DD + W	DW	BS	MD	DD + EM	WJ	DBS
Marbles (including onyx, travertine, green marbles, etc.)	X		XXX	XXXX	XXXX	X	XX		XX
Granites (including porphyry, basalt, quartzite, etc.)	XXXX	XXX	XXX	XXX		XX	XX	XX	
Sandstones, Slates	XX		XXX	XXX	XXX	X	XX	X	XX
Other stones ('split', soft limestone, peperino, etc.)	XXX		XXXX	XXX	XXX		XX		X

Traditional technologies:
DD + E – Discontinuous drilling + explosive; FJ – Flame jet; DD + W – Discontinuous drilling + wedges; DW – Diamond wire

Alternative technologies:
MD – Multi-drill/Line drilling; DD + EM – Discontinuous drilling + Expanding mortar

Innovative technologies:
WJ – Water Jet; DBS – Diamond Belt Saw

Use:
XXXX – Very Common; XXX – Frequent; XX – Scarce; X – Occasionnel

stripping overburdens, and at the mineral processing plants for moving minerals from stockpiles. Basically, scrapers units can be classified as:

• Wheel Tractor Scrapers
• Crawler Tractor Towed Scrapers.

Ripping is one of the methods of loosening the earth-rock. Ripping is a skilled task, and the efficiency of the operation depends upon the skill and experience of the operator and that is why this operation is still considered as an art and not a science. The ripped earth-rock needs to be removed from its original place to a predetermined destination, which could be a dump yard, a casting site, etc., any of these equipment (figs 17.15, 17.29, 17.30, 17.32) are deployed based on the type of job:

• Bulldozing
• Scraping

Table 17.10   Unit operations necessary for quarrying and stone processing.[21,23]

Symbol – operation	Equipment with its suitability and locales of applications	
Cutting and separation CS_{DS}	Stone cutting machines with chain jibs, disk saws, annular cutters, and diamond wire rope saws; percussive cutters Hand held and track mounted drills. Use of plastic pipe cartridges for mild blasting. Oxygen and gasoline-air thermal jet cutters; Hydraulic wedges and conventional wedges.	For cutting and separation of the dimension stones from the massif.
Extraction of blocks and lifting of blocks EB	Jib cranes – truck mounted, or wheel or crawler mounted; Derrick cranes (mobile or fixed); Forklifts; Mobile overhead cranes. Wheel loaders. Single bucket excavators, loaders, shovels.	Within the pit to extract and load the blocks of dimension stones. Extraction and loading of over burden and waste rocks other than dimension stones.
Transportation of blocks T_B	Dumpers, Trucks, Tractors with trailers, cage and skip hoists – higher depths. Excavators and bulldozers	Transportation of blocks and waste rocks. Waste rock pilling.
Block processing B_{PR}	Stone sawing machines; Machines fitted with disk, rope and belt saws; trimming machines, grinding-polishing machines; drilling (milling) machines;	Sawing of blocks.
Splitting of blocks B_{SP}	Stone splitting machines; thermal, hydraulic types. Tools: drills, wedges, jaws, bush hammers, chisels, mauls, groovers etc.	Processing blocks after mining
Auxiliary A_{OP}	Crushers of various types, screens, classifiers, mixers and other equipment Bulldozers scrapers, graders, rippers, jib cranes and other hoisting and earth moving equipment. Miscellaneous equipment	Processing waste rocks to obtain chips, raw material for cement or any other use. Road construction and maintenance. For various purposes, similar to those used in open pit, or open cast mines.

- Excavating by a bucket loader – Shovel, dragline, backhoe, bucket wheel excavator or any other loader
- Portable and movable belt loaders.

Tractor rippers are most commonly used with scrapers, and the ripping is undertaken parallel to the scraping path. Applications of excavators, loaders, transportation and earth moving units have been summarized in figure 17.32. The operation could be

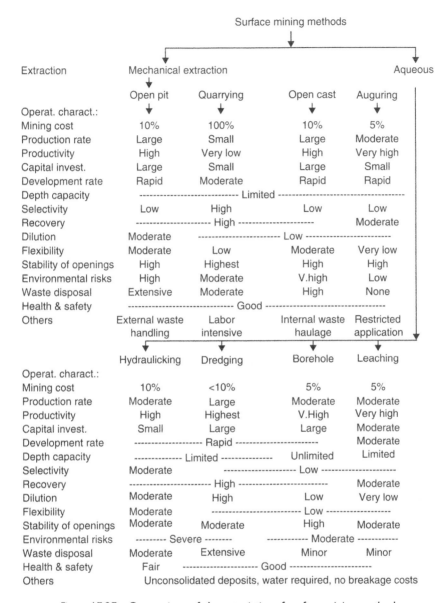

Surface mining methods

Extraction → Mechanical extraction / Aqueous

Operat. charact.:	Open pit	Quarrying	Open cast	Auguring
Mining cost	10%	100%	10%	5%
Production rate	Large	Small	Large	Moderate
Productivity	High	Very low	High	Very high
Capital invest.	Large	Small	Large	Small
Development rate	Rapid	Moderate	Rapid	Rapid
Depth capacity	----------------------- Limited -----------------------			
Selectivity	Low	High	Low	Low
Recovery	---------------- High ----------------			Moderate
Dilution	Moderate	----------------- Low -----------------		
Flexibility	Moderate	Low	Moderate	Very low
Stability of openings	High	Highest	High	High
Environmental risks	High	Moderate	V.high	Low
Waste disposal	Extensive	Moderate	High	None
Health & safety	----------------------- Good -----------------------			
Others	External waste handling	Labor intensive	Internal waste haulage	Restricted application

Operat. charact.:	Hydraulicking	Dredging	Borehole	Leaching
Mining cost	10%	<10%	5%	5%
Production rate	Moderate	Large	Moderate	Moderate
Productivity	High	Highest	V.High	Very high
Capital invest.	Small	Large	Large	Moderate
Development rate	----------------- Rapid -----------------			Moderate
Depth capacity	---------- Limited ----------		Unlimited	Limited
Selectivity	Moderate	----------------- Low -----------------		
Recovery	----------------- High -----------------			Moderate
Dilution	Moderate	High	Low	Very low
Flexibility	Moderate	----------------- Low -----------------		
Stability of openings	Moderate	Moderate	High	Moderate
Environmental risks	--------- Severe --------		----------- Moderate -----------	
Waste disposal	Moderate	Extensive	Minor	Minor
Health & safety	Fair	----------------- Good -----------------		
Others	Unconsolidated deposits, water required, no breakage costs			

Figure 17.27 Comparison of characteristics of surface mining methods.

continuous or cyclic.[2] Trenches have many applications in civil works such as laying pipes, cables, drainage and many others. They can be driven with and without aid of explosives. Wheel and ladder trenchers as shown in figure 17.31 can perform this operation. Applications of various earthmovers[30] to undertake different activities pertaining to construction and civil projects have been shown in table 17.11.

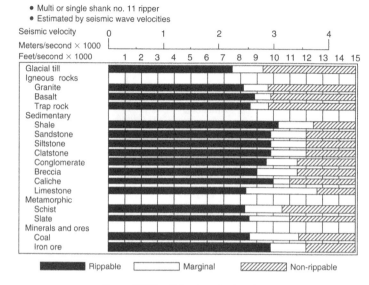

- Multi or single shank no. 11 ripper
- Estimated by seismic wave velocities

Seismic velocity	0		1		2		3		4
Meters/second × 1000									
Feet/second × 1000	1 2 3 4 5 6 7 8 9 10 11 12 13 14 15

Glacial till
Igneous rocks
  Granite
  Basalt
  Trap rock
Sedimentary
  Shale
  Sandstone
  Siltstone
  Clatstone
  Conglomerate
  Breccia
  Caliche
  Limestone
Metamorphic
  Schist
  Slate
Minerals and ores
  Coal
  Iron ore

■ Rippable   □ Marginal   ▨ Non-rippable

Figure 17.28 Seismic velocity chart.

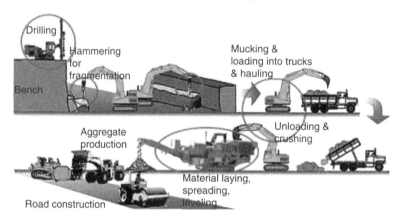

Drilling

Hammering for fragmentation

Mucking & loading into trucks & hauling

Bench

Aggregate production

Unloading & crushing

Material laying, spreading, leveling

Road construction

Figure 17.29 Operations involved and equipment deployed during road construction (Courtesy: Tamrock).

Sewerage projects

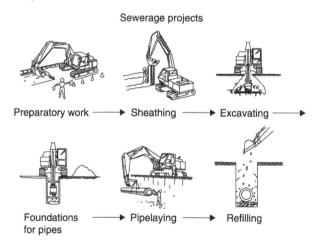

Preparatory work ⟶ Sheathing ⟶ Excavating ⟶

Foundations for pipes ⟶ Pipelaying ⟶ Refilling

Figure 17.30 Operations involved and equipment deployed while laying sewerage lines/Pipe lines (Courtesy: Fiat Hitachi).

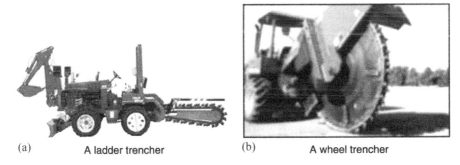

(a)   A ladder trencher

(b)   A wheel trencher

Figure/Photo 17.31   (a): A ladder trencher with other attachments. Interchangeable attachments increase the versatility and value of a unit. Attachments are available for trenching, vibratory plowing, pavement and rock sawing, and utility backhoe work. (b):A wheel trencher (Courtesy: Ditch Witch Co.).

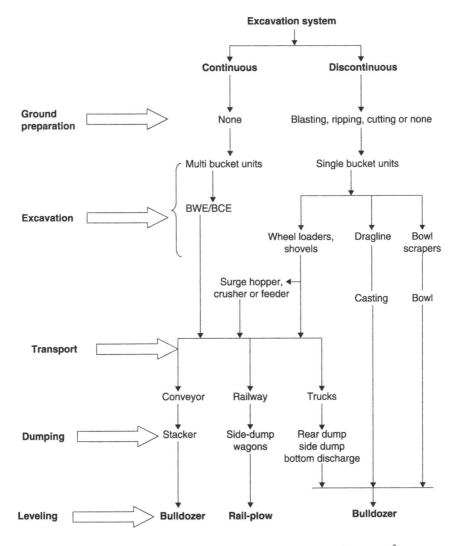

Figure 17.32   Material handling system classification for surface mines.[2]

Table 17.11 Application of various earthmovers[30] to undertake different activities pertaining to construction and civil projects. ■ – Used as main equipment. □ – Used to some extent. R.C. – Recycling equipment (Courtesy: Komatsu).

Activities	Hydraulic excavator (HE)	Mini HE	Bulldozer	Dozer shovel	Wheel loader (WL)	Mini WL	Dump truck	Crawler carrier	Motor grader crane	Trash comp.	Back hoe	Skid steer loader	R.C. Equip.
River mgmt.	■	□	□		□	□	□	□			□	■	□
Road construction	■	■	■	□	□	■	■	□	■		■		□
Harbor & airport const.	■	■	■	□	□	□	■	□	■		□	□	□
Building & demolition	■	■	□	□	□	□		□	□		□	□	■
Earth moving	■	■	■	□	□	■	■	■				□	■
Water main & sewer const.	□	■	□		□	■		□			■	■	□
Landscaping	■	■	□	□	□		□	□			□	□	□
Agricultural engineering	□	□	■	□		□		■	□		□	■	□
Livestock raising	■	□	■	□	□	■	□					□	□
Lumber and forestry	□	□	□	■	■	■			□			■	□
Cargo industry				□	■	□	■						■
Mining & quarrying	■		■		■	□	■	□	□		□		□
Waste management	■		■		□		□	□		■	□	■	■
Tunnel construction	■				■		■				□		□

17.9   THE WAY FORWARD[31]

Thanks to the increased attention from environmentalists around the world and the use of aeroplanes and aerial photography, mining enterprises are no longer free to "Dig and Run" when the extraction of the desired ore has been completed.

Laws and regulations have been promulgated in most of the developed countries and, through the activities of international organizations, are being urged where they do not yet exist. They establish an environmental management programme as an integral element in every mining project and stipulate such requirements as preliminary environmental impact assessments; progressive rehabilitation programmes, including restoration of land contours, reforestation, replanting of indigenous fauna, restocking of indigenous wild life and so on; as well as concurrent and long-term compliance auditing (UNEP 1991, UN 1992, Environmental Protection Agency (Australia) 1996, ICME 1996). It is essential that these be more than statements in the documentation required for the necessary government licenses. The basic principles must be accepted and practised by managers that the *mining activities should demonstrate to the stakeholders what will be left behind after mining is over; whether the area will be better than it was originally or worse!*

QUESTIONS

1. As a Blasting Engineer in an open pit Pb-Zn mine producing 5000 tpd having 10 m bench height; propose drills (numbers and type with hole dia. that would be drilled), explosive including detonators: type and quantity/month. Also give details of the man-power, drilling and blasting accessories to run the mine smoothly. Your answer should include blasting pattern and expected powder factor and drill factor from the mine. Write down the assumptions made, if any.
2. Based on the topography of the dumping site, name the different types of dumps that can be formed (i.e. patterns of dumping based on the topography).
3. Calculate the maximum allowable stripping ratio of an open pit producing zinc ore if the cost of removing waste rock is $1.10/t and the mining cost is $2.400. Zinc ore can be sold in the market at $5.5/t.
4. Classify quarrying methods/techniques.
5. Compare important aspects of the helicoid-wire and diamond-wire technologies.
6. Compare the important features of surface and underground mining. What are the limiting criteria for surface mining?
7. Cutter saw & rock channellers (impact cutting machines) – why are these machines very common in USA, Canada, France, Spain and many other countries? What are their main features and capabilities? What is the working element of these machines? How is a kerf formed and what is the capacity (m^2/shift) of this channeller?
8. Describe the diamond-wire technology and its application in quarry operations. Give the procedures and average performance achieved.
9. Describe these quarrying techniques (a) Disc cutters. (b) Thermal cutting. (c) Underground quarrying using the Room and Pillar system. Is there any future for underground quarrying? Where would this technique be most suitable?

10. Design an open pit (bench height, angle, width and safety berm) for limestone mining, given the following data: (i) Equipment height with boom = 12 m. (ii) Width of largest equipment = 3.5 m.
11. Draw 'deck charging' options that are usually used in mines.
12. Draw a flow diagram to suggest optimization of surface mining operations, mentioning inputs required and outputs obtained.
13. Draw a bench and illustrate its various components/elements and also give the usual nomenclature used to denote them.
14. Draw schemes to mine a deposit by the application of surface mining methods having the following mode of occurrences: (a) A quartz deposit outcropping to the surface and steeply dipping. (b) Flatly dipping clay deposit at depth of 20 m. (c) A lead-zinc-copper deposit outcropping at the slope of hill in the form of scattered lenses. (d) Steeply dipping thick ore deposit at a depth of 15 m from the surface.
15. The drilling method of block separation from the massif consists of two inter-related processes; describe them.
16. For pit slopes, Hudson and Harrison suggested 12 leading parameters; list them.
17. For surface mines list the logical steps that should be followed to mine-out a deposit.
18. For the haul road in the surface mines, the ramp gradient is governed by the statutory requirements of any country; what is its usual range?
19. Give a line diagram to classify the 'material handling system' for surface mines.
20. Give different schemes of stripping in open pit mines and mention situations where each of these schemes has advantages over others.
21. Give a line diagram to compare the characteristics of surface mining methods.
22. Illustrate quarry development schemes by drawing suitable sketches.
23. Give a schematic presentation of rope saws.
24. Give Tatiya and Adel's relation and curves. Are they capable to compute burden in relation to rock strength and hole diameter?
25. Given a line diagram to detail the steps that need to be carried out while under-taking surface mining operations.
26. GPS monitoring system – a total monitoring system in open pit mines. Tabulate its applications and show how it is beneficial comparing the conventional practices.
27. How does the development of opencast pits and their deepening process take place? Illustrate it.
28. How do you determine the ultimate pit slopes? What are the governing factors? List pit slope failure patterns and illustrate any four of them. How can pit slopes be repaired, give two figures to answer the question.
29. How do you select the hole diameter based on bench height; give the curve used to determine it.
30. Illustrate and detail these quarrying techniques: diamond wire cutting and jib (bar) cutting.
31. Illustrate operations involved and equipment deployed during road construction.
32. Illustrate the concept of 'terrace' used in conjunction with BWE. A flatly dipping lignite deposit has a thickness of 16 m and a cover of 49 m soft rock o/b. The o/b is worth mining with the use of BWE. Give a suitable layout illustrating the

process of winning this deposit. Mention the sets of equipment you will deploy for the different operations.

33. Illustrate the operations involved and equipment deployed while laying sewerage lines/pipe lines.

34. Illustrate the technique of 'deck charging'. Where would you apply this technique?

35. In a mechanised open pit mine the excavator to be deployed has height of 9 m. Find out the maximum allowable height of the bench. If the width of 35 ton-capacity truck to be used at this mine is 3.7 m; Calculate the minimum width of the bench.

36. In a surface mine ore reserves up to a depth of 50 m from surface are 3.85 million tons. It has got hanging wall waste, footwall waste and an overburden of 1.5 million tons, 50,000 tons, and 60,000 tons respectively. Calculate the stripping ratio.

37. In an open cast mine the depth of working, above the coal seam, has reached 35 m from the surface and the thickness of the coal seam being worked out is 1.75 m. Calculate its present ore to over-burden ratio.

38. In an open pit chromite mine a burden of 4 m and spacing of 5.5 m are being kept. If the height of bench is 9.5 m and 40 holes are to be blasted, calculate the amount of chromite this blast will yield if the density of rock is $2.5 t/m^3$. Calculate the powder factor if each hole is charged with 70 kg of explosive.

39. In an open pit limestone mine a burden of 3 m and spacing of 4.2 m is being kept. If the height of bench is 10 m and 30 holes are to be blasted, calculate the amount of limestone this blast will yield if the density of rock is $2.5 t/m^3$. Calculate the powder factor if each hole is charged with 50 kg of explosive.

40. In an open pit limestone mine during bench blasting, the spacing used is 1.25 times the burden and a burden of 5 m is kept. If bench height is 11 m and 50 holes are to be blasted, calculate the tonnage of limestone yield if its density is $2.9 t/m^3$. Calculate the powder factor if explosive charged in all holes is 3000 kg. Also calculate the limestone that will be yielded per hole.

41. It is important that the type of equipment, methods and techniques match each other to obtain the desired results. The idea is the smooth flow of the ore and waste rocks from their places of generation to their final discharge destinations. List the salient points that could be considered to achieve this goal.

42. List/draw various schemes of waste rock dumps. What considerations you would make to site them?

43. List and illustrate excavators used for earth-work, rock extraction and ore mining.

44. List auxiliary operations that need to be done in surface mines. What are the different motive powers (means energy in its pure or converted form) used to run equipment and provide services?

45. List auxiliary operations that need to be done in underground as well as in surface mines.

46. List surface mines known to you and suggest mucking equipment for each of these mines.

47. List the guidelines that should be followed to achieve minimum land degradation, and handling waste rocks systematically.

48. List the main elements important to an open pit design. Do you think some of these parameters are variable? If yes, list them.

49. List two types of scraper units.
50. Name the equipment or machines (known to you) to carry out unit operations while carrying out mining by surface or underground methods. List rock drilling methods.
51. Name the types of surface mines and write down the basic difference between them in not more than 3 sentences.
52. Sketch out the full block method and the partial block method of working the BWE (Bucket Wheel Excavator). Under what specific situations can each one of these techniques be applied?
53. Suggest a mining method for a flat clay deposit having a thickness of 6 m which is covered with an overburden of sandstone 15 m thick. Illustrate it by giving suitable sketch.
54. Suggest a mining method for a flat coal deposit having thickness of 4 m which is covered with an overburden of sandstone 24 m thick. Illustrate it by giving a suitable sketch. Calculate ore to overburden ratio.
55. Suggest a mining method for a steeply dipping fairly strong copper deposit outcropping to the surface. Its thickness is 20 m. Hanging and foot-wall are strong. Illustrate it by giving suitable sketch.
56. Suggest a mining method for a thick and almost vertically dipping granite deposit under the overburden of 10 m from the surface. Illustrate it by giving suitable sketch.
57. Tabulate common explosives used in quarrying dimension stones; incorporate important features.
58. Tabulate guidelines for selecting the pit slope angle with the consideration of: rock types and characteristics of the rock massif.
59. Tabulate to compare the salient features of the existing (conventional and traditional) alternatives and innovative techniques in stone mining locales.
60. There are different patterns of ore mining within the pit limits – single-sided, double-sided (in longitudinal as well as transverse directions); centralized and disconcerted – illustrate them giving suitable sketches.
61. What are dimension stones? How they can be mined? Name the common dimension stones describing important characteristics of each one of them.
62. What considerations should be taken into account while designing the haul road and ramps for surface mines? List various types of designs. How is the width of haul roads calculated as per number of lanes. Write formula used for this purpose.
63. What considerations will you give while selecting a site for the purpose of dumping the waste rock in open pit mines?
64. What do you understand by the technique of cast blasting or explosive casting? Mention the steps involved when this technique is applied for winning two seams. Illustrate your answer with suitable sketches. An o/c mine has attained a depth of 25 m up to the bottom of a coal seam. The width of exposed strip is 70 m. Calculate the percentage blast over that which is likely to be achieved by the application of explosive casting technique.
65. What are scraping, ripping and digging techniques commonly used for ? And for what type of formations?
66. What is a ladder trencher? What are its attachments/where do they find application?

67. What is a safety berm? Calculate it if the bench interval is 12 m.
68. What is an initial 'box cut' or 'trench'? What is its importance and how it is driven?
69. What is an overall pit profile? When it should be prepared? What purposes does it serve? What could be the possible shape/shapes of an open pit, ultimately?
70. What is the helicoid wire technique, or method? Where does it find its application? In which countries is this technique popular?
71. What is the in-pit crushing and conveying concept? What is a mobile crusher? When and where did application of mobile crushers begin?
72. What is line drilling and hydraulic wedging? Describe them.
73. What is the purpose of determining an ultimate open pit design? Describe the main elements important to an open pit design. What is the profit function concept in an open pit design? Give a generalized formula by which profit from an open pit can be determined at any stage during the life of the mine. What is the coning concept for designing an open pit?
74. What is the difference between open cast and open pit mines? Also between open pit and quarry? Give simple sketches to illustrate all these surface mines. What types of deposits are most suitable in each case?
75. What is the principal purpose for an early design of final pit limits of an open pit? Describe the physical and economical parameters that should be considered in designing an open pit mine. Once the physical and economical parameters of an open pit design have been defined, and an overall stripping ratio has been designated, write a generalized formula, which theoretically should be able to calculate the profit from open pit mining at any stage in the operation. List and illustrate the prevalent concepts for the removal of waste rock during open pit mining.
76. What is the use of diamond belt saws? Describe its utility for types of dimensional stones. Give its constructional features.
77. What is water jet technology? How is stone cutting achieved using this technique? Has automation been incorporated with technology? And how it is helpful? What is the usual depth size of cut it can make?
78. Where you would apply the concept of 'stripping sequence'. Illustrate different concepts and describe situations, in each case, for its suitability.
79. What is a working pit slope angle? How it is influenced? Compute it using following data and write factors that influence it. Given data:
    (i) $N_B = 5$; $B_H = 16$ m; $BW = 12$ m; $B_A = 75°$; $R_W = 0$ m
    (ii) $N_B = 8$; $B_H = 10$ m; $BW = 12$ m; $B_A = 75°$; $R_W = 0$ m:
    (iii) $N_B = 5$; $B_H = 16$ m; $BW = 12$ m; $B_A = 75°$; $R_W = 30$ m
    (iv) $N_B = 5$; $B_H = 16$ m; $BW = 3.2$ m; $B_A = 75°$; $R_W = 0$ m
    (v) $N_B = 5$; $B_H = 16$ m; $BW = 0$ m; $B_A = 75°$; $R_W = 0$ m
80. In an iron ore mine, rock density is 3.5 t/m^3, the drill hole diameter is 150 mm and the bench height is 12 m. Slurry explosives of density 1.15 g/cc are used. Calculate the B, S, J, T, Powder factor and drill factor using Ash's method.
81. In a limestone ore mine, the rock density is 2.5 t/m^3, the drill hole diameter is 150 mm and the bench height is 12 m. Slurry explosives of density 1.15 g/cc are used. Using the powder factor method of surface blast design, calculate subgrade

drilling, stemming length, length of hole, charge length/hole, amount of explosive to be charged/hole, if charge/m length amount is to be 21.5 kg/m. Also calculate the volume of rock that will be broken/hole, burden and spacing.

82. Use rock characteristic curves to design a surface mine blast, using the following data:
   – Hole diameter = 100 mm
   – Rock type: average type
   – Bench height = 12 m
   – Rock density = 2.5 t/m³
   Calculate spacing, burden, subgrade drilling, rock broken/hole, and drilling length/hole.

83. Repeat question (82) if *rock type* is: (i) Very easy to blast. (ii) Difficult to blast.

84. If density of explosive ANFO is 0.85 gms/c.c., calculate the explosive to be charged/m length of hole if the hole diameter is: (i) 55 mm, (ii) 100 mm, (iii) 150 mm, 200 mm, 250 mm and 300 mm.

85. Name the type of drills suitable for an opencast mine. Given the data set from an open cast mine, estimate the number of drills required for this mine.
   Ore to o/b ratio = 1:7
   O/b production/month = 700,000 cu.m.
   Blasthole design in o/b: Burden = 4 m, spacing = 6 m.
   Blasthole design in ore: Burden = 4 m, spacing 5 m.
   Drilling performance: 15 m/hr in o/b & 12 m/hr in ore.
   Workig schedule: 25 days/month, 2 shifts/day, 7 hrs/shift.

86. Given the following data from a cement company, work out the excavator-truck fleet for its open pit mine and synchronise the equipment you have selected.
   Deposit to be mined – Limestone
   Production required/shift = 5000 ton
   Effective working hours/shift = 7
   Time (load) factor = 50/60 minutes
   Operational factor = 0.79
   Bucket fill factor = 0.9
   Density of the blasted muck = 2.35 ton/cum.
   Cycle time = 15 min., spot time = 0.5 minute.
   The range of excavators (shovels) available in the market is as given below:

Bucket capacity in cum.	Rock output (in cum./hr)
3.8	285–380
6.1	375–515
6.9	425–500
7.6	470–645
11.5	705–970
19.1	1175–1585

   The trucks that are available in the market (listed below) have the following capacity (in tons): 20, 27, 32, 36, 50, 77, 90, 110.

REFERENCES

1. Aiken, G.: Surface mining – continuous methods. In: Cummins and Given (eds.): *SME Mining Engineering Handbook*, AIME, New York, 1973, pp. 17:50–51.
2. Atkinson, T.: Cast blasting of deep over-burden. In: Hartman (edt.): *SME Mining Engineering Handbook*. SMME, Colorado, 1992, pp. 1307–11.
3. Atlas Copco Manual. Atlas Copco, Sweden.
4. Barnes, P.: Computer Assisted Mineral Appraisal and feasibility, SME (AIME), New York, 1980, pp. 70.
5. Brealey, S.C.; Belley, J. and Rickus, J.E.: Mineral quality determination and control in stratified deposits. *IMM's International Symposium on Surface mining and Quarrying*, Bristol, England, 4–6 Oct. 1983, pp. 153–57.
6. Call, R.D. and Savely, J.P.: Open pit rock mechanics. In: B.A. Kennedy (edt.): *Surface Mining*. SMME, Colorado, 1990, pp. 872
7. Fantini, Italy.: *Leaflets and literature.*
8. Fourie, G.A. et al.: Open pit planning and design. In: Hartman (edt.): *SME Mining Engineering Handbook*. SMME, Colorado, 1992, pp. 1276–77.
9. Frizzell, E.M. and Martin, T.W.: In-pit crushing and conveying. In: Hartman (edt.): *SME Mining Engineering Handbook*. SMME, Colorado, 1992, pp.1343–47.
9a. Grim, E.C. and Hill, R.D.: Environmental protection in surface mining of coal, USEPA, EPA-670/2-74-093.
10. Haidar, A.D. and Naoum, S.G.: Open cast mine equipment selection using genetic algorithms. *Int. Jou. of Sur. Min. Rec. and Env.* 10, 1996, pp. 61–67.
11. Hartman, H.L.: *Introductory Mining Engineering*. John Wiley & Sons, New York, 1987, pp. 134–140, 167–205.
12. Hoek, E.: Estimating the stability of excavated slopes in open cast mines. Trans. IMM, 1970, 79(10): A109–132.
13. Hudson, J.A.: *Rock Engineering Systems*. Ellis Horwood, 1992, pp. 34–40.
14. Hustrulid, W. and Kuchta, M.: *Open Pit Mine Planning & Design*. A.A. Balkema, Netherlands, 1998, pp. 287–295.
15. Jimeno, C.L.; Jimeno, E.L. and Carcedo, F.J.A.: *Drilling and Blasting of Rocks*. A.A. Balkema, Netherlands, 1997, pp. 196–97, 252–66.
16. Kou, S.Q. and Rustan, P.A.: Burden related to blasthole diameter in rock blasting. *Int. J. Rock Mech. Min. Sci. & Geo-mechanics Abstr.* 1992, Vol. 29, No. 6, pp. 543–53.
17. Marini, Italy: *Leaflets and literature.*
18. Matti, H.: *Rock Excavation Handbook*. Sandvik – Tamrock, 1999, pp. 140–150, 309–22.
19. McCarter, M.K.: Design and operating considerations for waste embankments. In: B.A. Kennedy (edt.): *Surface Mining*. SMME, Colorado, 1990, pp. 890–95.
20. Peck, J. and Gray, J.H.: The total mining system (TMS): The basis of open pit simulation. CIM Bulleting, vol. 88, no. 993, 1995, pp. 38–44.
21. Primavori, P.: Technological development in machinery and installations for extracting and processing stone materials. *Seminars in Gulf area Dubai, Muscat*; 14–16 October 2002.
22. Rustan, Agne: Burden, spacing and borehole diameter at rock blasting. *Int. J. Surface Mining and Reclamation*, 1992, No. 6, pp. 141–49.
23. Rzhevsky, V.V.: *Opencast Mining Unit Operations*. Mir Publishers, Moscow, 1985, pp. 15–55; 131–135; 330–75; 431–34.
24. Rzhevsky, V.V.: *Opencast Mining – Technology and integrated Mechanization*. Mir Publishers, Moscow, 1987, pp. 12–92; 112–49; 300–07; 374–77.
25. Siemens, Germany: *Leaflets and liertaure.*

26. Tatiya, R. R. and Ajmi, A.: 'Evaluation of the Atlas Copco relation between burden and Blasthole diameter and rock strength at bench blasting – a case study', *Int. J. of Surface Mining and Reclamation.* Canada. Vol. 14; no. 2, 2000, pp. 151–60.
27. Vikram, K.: Pashan, 1990, pp. 60–80.
28. Zahl, E.G. et al.: Waste disposal and contaminant control. In: Hartman (edt.): *SME Mining Engineering Handbook.* SMME, Colorado, 1992, pp. 1170–75.
29. Holmberg, R.; Hustrulid, H. and Cunningham, C.: Blast design for underground mining applications. In: Hustrulid and Bullock (eds.): *SME Underground Mining Methods.* Colorado, 2001, pp. 635–61.
30. Komatsu Mining Systems, Inc; Japan – leaflets and literature.
31. Thomas A. Hethmon and Kyle B. Dotson: Surface mining methods. In: Chap. 74 – International Labor Organization (ILO) Encyclopaedia and CISILO database.

Chapter 18

Hazards, Occupational Health and Safety (OHS), environment and loss prevention

We all agree: "Prevention is always better than cure". Should we not make our utmost efforts in minimizing: emissions, incidents, wastage, injuries & illness; approaching their magnitude to almost zero? This is known as: 'zero-goal'. Respecting natural resources could achieve this.

18.1 INTRODUCTION

Imagine, if you want to produce gold, silver, marble, limestone, copper or any other mineral in the city or location where you are living; can you do so? The answer is no; because mining operations or civil excavations made for various purposes (as described in secs 1.7 and 1.8) are *Site/Location/Locale specific*. Hence, operations at mines and tunnelling sites are governed by the prevalent scenario in and around their *geographic locations*; and the rest of the parameters are inter-related as detailed below through a '9G – EVALUATION':

1. Geographic location – The location could be a hilly terrain, plain ground, desert, cropland, forests or any other terrain than these. It could be within urban land or in the countryside and even sometimes within water-bodies or in ground saturated with water. It could commence at, above or below the ground level or datum and extend in any direction: horizontal, inclined, vertically up or down.
2. Geology – how is the geological regime at that location in terms of rock/formation types; presence geological structure and discontinuities such as fault, folds, joints etc. and mode of occurrence of the ore or mineral of our interest or economic value?
3. Geometry of deposit – how does the ore-body occur: bedded/seam, vein, massive and what is its extent in all the 3 directions: along, across the dip direction and thickness?
4. Geo-mechanical properties – what is the strength of ore-body and its surrounding rock-mass, density, porosity, permeability etc.
5. Ground water and surface water regimes – what about presence of water and type of water that would be based on position of the water table and water-bodies at the surface?
6. Geo-environment – are mining/tunneling operations likely to disturb the prevalent flora, fauna, land, water and air regimes (bio-diversity)?
7. Grade (quality) & tonnage (quantity) – what is the commercial value of the deposit in terms of quantity (reserves) and quality (grade) and ultimately the detailed mineral inventory; which is an account of grade-wise tonnage (reserves) to access the commercial value of the mineral property (grade-tonnage curve).

8. Gross revenues – inputting the production costs and sales values would give the gross – revenues from a particular deposit enabling an investor to understand the rate of return on his investment – ROI and take decision accordingly.
9. God Gift – Thus, *it is a 'God-Gift' and it really matters how one handles it?*

18.2 POTENTIAL EXCAVATION HAZARDS[29,32]

Hazard means danger, risk; hazardous means dangerous, risky. Hazard is a condition with the potential of causing harm or damage to resources that include (fig. 18.1): man, property (movable & immovable such as buildings, plant, equipment, materials, machines etc.), air, water, land, flora and/or fauna. Sec. 1.7; details how mining industry differs from other industries with a potential to be hazardous due to following reasons:

- In an underground situation the working space is inherently tight, distorted, congested, isolated and inaccessible, of poor quality, and deteriorating. These adverse conditions endanger personnel, damage mobile equipment, and affect all activities.
- It is not only the confined space underground but also adverse working conditions such as darkness, heat, humidity; gassy and watery conditions that make the miner's job most difficult and risky.
- It is also the case while working at the surface mines/excavations under adverse climatic conditions.
- The miners are also liable to occupational diseases such as asbestosis, silicosis and few others. In addition, the risks of fire, explosion, inundation and ground failure are part and parcel of this industry.
- Figure 18.2 details potential hazards and includes the following (listed alphabetically with their further details in the line diagram).
  - ↓ Adverse ground conditions – collapses, deformations, swelling, squeezing, subsidence, bumps and bursts, slow penetration and progress, water inrush and seepage, fluctuation in hydrostatic pressure, etc.
  - ↓ Electricity-related hazards
  - ↓ Equipment & hardware hazards – mechanization and automation related
  - ↓ Exposure to the following, particularly when an allowable limit is exceeded: dust/fibers, radiations, fumes (blasting, welding & diesel), toxic gases (NO_X, SO_X, CO), extreme temperature, corrosive liquids (acids), toxic salts, aqueous effluents and others – not covered above.

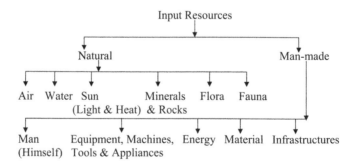

Figure 18.1 Resources – classification. Success lies in effectively utilizing these resources.

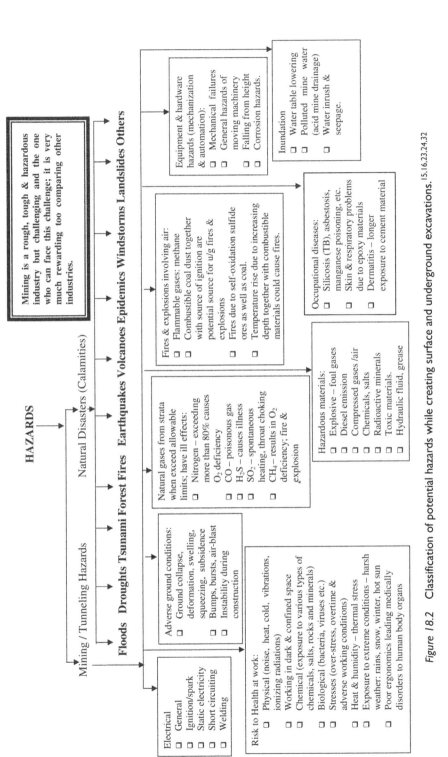

HAZARDS

Mining / Tunneling Hazards

Natural Disasters (Calamities)

Floods Droughts Tsunami Forest Fires Earthquakes Volcanoes Epidemics Windstorms Landslides Others

Mining is a rough, tough & hazardous industry but challenging and the one who can face this challenge; it is very much rewarding too comparing other industries.

Equipment & hardware hazards (mechanization & automation):
☐ Mechanical failures
☐ General hazards of moving machinery
☐ Falling from height
☐ Corrosion hazards.

Inundation
☐ Water table lowering
☐ Polluted mine water (acid mine drainage)
☐ Water inrush & seepage.

Fires & explosions involving air:
☐ Flammable gases: methane
☐ Combustible coal dust together with source of ignition are potential source for u/g fires & explosions
☐ Fires due to self-oxidation sulfide ores as well as coal.
☐ Temperature rise due to increasing depth together with combustible materials could cause fires.

Occupational diseases:
☐ Silicosis (TB), asbestosis, manganese poisoning, etc.
☐ Skin & respiratory problems due to epoxy materials
☐ Dermatitis – longer exposure to cement material

Natural gases from strata when exceed allowable limits; have ill effects:
☐ Nitrogen – exceeding more than 80% causes O_2 deficiency
☐ CO – poisonous gas
☐ H_2S – causes illness
☐ SO_2 – spontaneous heating, throat choking
☐ CH_4 – results in O_2 deficiency; fire & explosion

Hazardous materials:
☐ Explosive – foul gases
☐ Diesel emission
☐ Compressed gases /air
☐ Chemicals, salts
☐ Radioactive minerals
☐ Toxic materials.
☐ Hydraulic fluid, grease

Adverse ground conditions:
☐ Ground collapse, deformation, swelling, squeezing, subsidence
☐ Bumps, bursts, air-blast
☐ Instability during construction

Electrical
☐ General
☐ Ignition/spark
☐ Static electricity
☐ Short circuiting
☐ Welding

Risk to Health at work:
☐ Physical (noise, heat, cold, vibrations, ionizing radiations)
☐ Working in dark & confined space
☐ Chemical (exposure to various types of chemicals, salts, rocks and minerals)
☐ Biological (bacteria, viruses etc.)
☐ Stresses (over-stress, overtime & adverse working conditions)
☐ Heat & humidity – thermal stress
☐ Exposure to extreme conditions – harsh weather: rains, snow, winter, hot sun
☐ Poor ergonomics leading medically disorders to human body organs

Figure 18.2 Classification of potential hazards while creating surface and underground excavations.[15,16,23,24,32]

↓ Fires & explosions involving air
↓ Hazardous materials
↓ Inundation – water inflows/inrush – flooding, improper sealing, inadequate consolidation or water-table lowering measures. Adverse effects of raw water.
↓ Natural gases from strata when allowable limits are exceeded
↓ Risk to health at work due to:
  ❑ Physical (noise, heat, cold, vibrations, ionizing radiations)
  ❑ Working in dark & confined spaces
  ❑ Chemical (exposure to various types of chemicals, salts, rocks and minerals)
  ❑ Biological (bacteria, viruses etc.)
  ❑ Stresses (over-stress, overtime & adverse working conditions)
  ❑ Heat & humidity – thermal stresses
  ❑ Exposure to extreme conditions – harsh weather: rains, snow, winter, hot sun
  ❑ Poor ergonomics (sec. 18.4.3) leading to medical disorders to human body organs.

Thus, it is a fight against nature wherein the conditions change every moment and requires utmost safety and alertness, failing which may result in losses of various kinds to the input resources which could be as detailed in figure 18.1. The *most valuable resource amongst them is the human being, our workers*, who should be protected by all means to the extent that they live healthy when they retire.

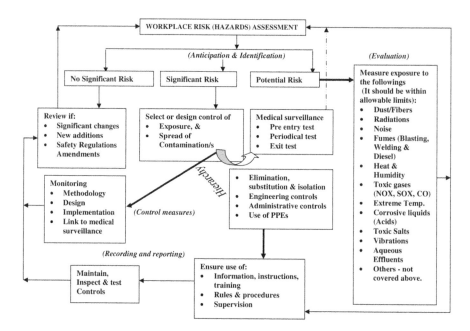

Figure 18.3 Important steps: anticipation & identification, evaluation, recording (including reporting, as appropriate) and control measures to assess the workplace hazards (risks) and remedial measures to combat/minimize them.[17,32,35,37]

18.2.1   Hazards (risks) analysis and management[3,17,32,35,37]

Steps for managing excavation hazards include:

- Anticipation
- Identification
- Evaluation: analyzing for quantity as well as quality (either by direct measurement or using appropriate methods of estimation)
- Recording and reporting, as appropriate
- Control measures to minimize them.

Figure 18.3 details above listed steps. In addition, the modern techniques such as: 'risk matrix' (tables 18.1, 18.2 and fig. 18.4) could be applied to identify magnitude of the potential hazards.[17,32,35,37]

Table 18.8 lists important phases of the working life of a worker/employee and the variables influencing his/her health during this tenure. The medical surveillance includes: pre entry test, periodical test and exit test. The outcome could be death to healthy exit when he or she retires, as shown in table 18.8. A thorough balance amongst occupational health, safety, environment (least pollution) and loss prevention strategy, as shown in figure 18.5, provides a 'proactive work-culture', which in turn could result in a 'healthy exit' and efficient operations. The following sections deal in detail with all these four elements:

- Safety and accidents
- Occupational health and surveillance
- Environment degradation and remedial measures
- Loss prevention.

Table 18.1   Probability Ranking.[27]

Category	Definition
A	Common or frequent occurrence (1 yr)
B	Is known to occur or it has happened (1:10 yr)
C	Could occur or heard of happening (1:100 yr)
D	Not likely to occur (1:1000 yr)
E	Practically impossible (<10,000 yr)

Table 18.2   Consequence Ranking.

	People	Equipment	Production loss	Environment (scenario)
1	Fatality	$10 m	2 weeks	Major
2	Disabled	$1–10 m	1 day–2 weeks	Serious
3	Major LTI (Weeks)	$100 k	1 shift	Moderate
4	LTI	$10 k	2 hr	Minor
5	Minor	$1 k	<1 hr	Insignificant

Probability

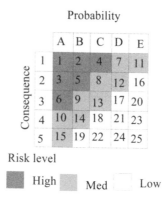

Risk level

High   Med   Low

Figure 18.4 Risk ranking (matrix) – combined consequence and probability consequence ranking. m – Million dollars; k – thousand dollars. These amounts could vary from one company to another.[27]

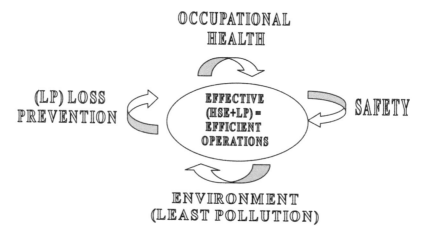

Figure 18.5 Due importance to the occupational health, safety, environment (least pollution) (HSE) and loss prevention (LP) results efficient operations.

18.3   SAFETY AND ACCIDENTS

18.3.1   Terminology

Safety: A condition that keeps mind, body or property free from injuries, damage or destruction.

Accident: Accident is an abnormal event or happening whether it causes injuries, damage or not. When it does not cause any harm, or it is a 'near-miss', it is known as an 'incident'.

Systemize: Unit Operations & Safe Practices

Avoid: Any risk including unsafe acts & unsafe conditions to avert accidents

Follow: Legal compliances, best practices, timely accomplishment & teamwork

Eradicate: Pollution & losses of all kinds

Train: Everybody from shop floor worker to the topmost executive

Yield: 'Goodwill & reputation' to your company by following the above guidelines.

Figure 18.6 The letters of word 'SAFETY' defines a well-balanced safety program/campaign/strategy that should be aimed at by every mine and tunneling project to promote its safety campaign.

Disaster: A major accident or natural event or natural calamity involving loss of lives (human and other creatures), property and resources (example given in fig. 18.2). It could be a natural or manmade disaster. Definition differs from country to country.

Loss – as given in Oxford dictionary: 'loss is diminution of ones possessions or advantage; detriment or disadvantage involved in being deprived of something, or resulting from change of condition'. All these parameters are tangible. 'Loss can also include failure to take advantage of something or failure to gain or obtain'.[23] Thus, loss is the result of interrupted practices or procedures that can lead to personal injury or property damage.[7]

The losses could be loss of life, injuries, damage to property, harm to the environment (pollution) and materials. They could be caused by hazards of various kinds as given in figure 18.2. It is not possible to eliminate all losses but counter measures are available to keep injuries less severe and losses smaller, as detailed in table 18.12.

18.3.2 Safety strategies

In figure 18.6 the characters of the word 'SAFETY' defines a well-balanced safety program/campaign that should be aimed at. The outcome of safety is visible and can evoke emotions, if results reverse.

What strategies should be followed that would prevent losses of all kinds and result in a safe environment at the workplace? Incident, accidents and losses are the result of hazards of various kinds, as illustrated in figure 18.2, Safety strategies, based on a literature survey, can be classified into four categories:

• Inherent
• Passive
• Active
• Procedural.

The inherent ones don't need any human intervention, just incorporate them and forget. For example, a lift will not start if its door is not closed (inter-locking), and this is how hazard of falling from a lift when it is in motion does not remain. Passive ones need some human intervention. Hazard remains but it is under control. The active ones need periodic testing and maintenance but a hazard could materialize if maintenance is not properly done. The procedural ones depend upon the training of people concerned and expecting they will do as trained. In most countries over-speeding of automobiles is the major cause of road accidents. The possible solutions are:

- Inherent – Design automobiles so they would not run beyond certain prescribed (within practical) limits.
- Passive – Install and maintain speed breakers/bumps (strategic locations, good design, warning signals and regular painting).
- Active – Synchronous road signals, police patrolling and radars; heavy fines to offenders.
- Procedural – Proper education to drivers.

Based on above cited example, one would agree that the inherent safety is the best but others are equally important and should be enforced as much as possible.

Relatively, a safety problem eliminated by use of inherent safety will cost $1 to fix it at the research stage; $10 at the process flow-sheet stage, $100 at the final design stage, $1000 at production stage and $10,000 at post incidental stage[16]. Thus, an early application of the inherent safety pays dividends.

A well-known strategy that is followed in industries is:

- Eliminate hazards
- Minimize them, if they cannot be eliminated
- Substitute or moderate them with less hazardous materials/operations
- Simplify – The principle of 'KISS': Keep It Safe and Simple – Design, Layout, Equipment, Process and Procedure; is the essence of this concept.
- Providing layers of protection, which include:
  - *Operational supervision, control systems, alarms, interlocks etc.*
  - *Physical protection devices (relief devices, dykes, pillars, barriers, etc.) and*
  - *Emergency response systems (plant/project emergency response, community emergency response), described in sec. 18.3.4.5.*

18.3.3   Safety elements

In the preceding paragraph the resources deployed to produce goods and services in any industry including mining and tunneling projects have been listed. *Most precious resource amongst them is the person. Utmost importance should be given to avoid their injury or illness (occupational disease) of any kind.*

The word 'health' in place of 'occupational health (OH)' has been used through-out the book to denote a healthy worker free from occupational diseases of any kind. The following three elements are responsible for the occupational health and safety:

1. People/mine workers
2. Systems developed to run the show
3. The working environment.

18.3.3.1   *People/mine workers*

The key element is the personnel, the people, the workers. Referring to the famous proverb: *It is the man behind the machine who really matters*; hence the quality of human resource should be checked at the entry level during recruitment itself and thereafter. People vary in age, gender, strength (health), skill [unskilled (raw), semi-skilled, skilled], capability, knowledge, experience, expertise, behavior (maturity, self-discipline, temperament), attitude, personality (leadership, initiative, capabilities), character (values, trustworthiness), needs and expectations.[14]

People have to get the work done from equipment (sets of machines) which are their faithful servants. Apart from the background, to do the operations in the right manner training aids are essential. Vocational training, refresher courses and application of behavioral science could help in inculcating safety habits amongst workers. Training needs are to be recognized using the *Why, Who, When, Where, What and How* concept to improve skill and knowledge.[13]

It has been published and established that 85% of accidents are caused by unsafe acts and 15% by unsafe conditions.[10] This percentage could be up to 90%; depending upon type of industry and prevalent safety culture within an industrial setup.

Universally, people involved in safety management have accepted these data. Heinrich (1931) suggested 3Es; Education, Engineering and Enforcement to establish proactive safety culture. In fact the solution lies in practicing truly the 5Es to establish proactive work culture including safety:

- Education (which includes training)
- Engineering
- Enforcement
- Engagement
- Eradication.

EDUCATION[12]

This refers to basic education that is required to carry out a function that is assigned to any individual, which is obtained from the universities and schools. On-the-job training then follows. As shown in figures 18.7, 18.11 and 18.22 people are responsible for accidents and losses of various kinds; 90% by unsafe acts and 10% by unsafe conditions. Research carried on this aspect show that *changes in human behavior with respect to safety* could improve the situation. DO IT practice, as described below, has given positive results in changing behavior of industrial workers.

Define/Describe/Demonstrate

The safe practices that need to be inculcated should be first explained, demonstrated and their outcome in terms of benefits should be made known. It has been found that in most mines and tunneling projects 'unsafe practices/conditions' that prevail include:

- Inadequate guarding, unguarded/not fenced
- Defective tools, appliances and equipment
- Improper design
- Improper layout of the working site/spot
- Defective or not using safety wear
- Inadequate ventilation – defective vision, air circulation and heat dispersion
- Inadequate illumination
- Improper roads, travel-ways, man-ways – where men and means transportation move

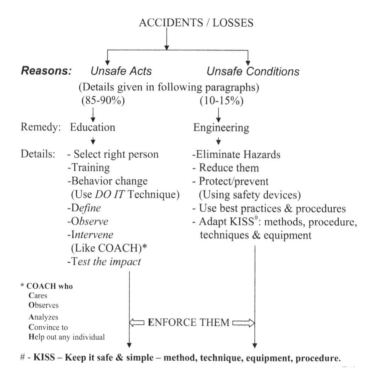

ACCIDENTS / LOSSES

Reasons: *Unsafe Acts*   *Unsafe Conditions*
(Details given in following paragraphs)
(85-90%)   (10-15%)

Remedy: Education   Engineering

Details: - Select right person   -Eliminate Hazards
          -Training   - Reduce them
          -Behavior change   - Protect/prevent
          (Use *DO IT* Technique)   (Using safety devices)
          -D*efine*   - Use best practices & procedures
          -O*bserve*   - Adapt KISS[#]: methods, procedure,
          -I*ntervene*   techniques & equipment
          (Like COACH)*
          -T*est the impact*

* COACH who
   Cares
   Observes
   Analyzes   ⇐ ENFORCE THEM ⇒
   Convince to
   Help out any individual

- KISS – Keep it safe & simple – method, technique, equipment, procedure.

Figure 18.7   Strategies to minimize accidents and losses.[14,18]

- Inadequate infrastructure
- Lack of proper maintenance of the equipment
- Use of defective or substandard material, consumables or spares
- Lack of proper communications
- Lack of standard practices and procedures
- Lack of means for health and hygiene – washrooms, toilets, drainage, sewage lines etc.
- Odd working hours
- Adverse climatic conditions
- Faulty system (as detailed in sec. 18.3.3.2).

Prevailing 'unsafe acts' include:

- Allowing unauthorized operations
- Allowing unauthorized operators/crews
- Lack of training and experience of the working crews
- Improper use of safety appliances
- Over confidence
- In adequate and vague (not clear) instructions
- Improper posture while working – handling, loading, disposal
- Lack of coordination and team work
- Lack of knowledge and non observance of rules, regulations, standard practices and standing orders

- Not using personal protective equipment (PPE) where warranted – masks, respirators, aprons, gloves etc.
- Using faulty, worn and rejected tools and appliances
- Tampering with materials, devices etc.
- Taking unsafe positions
- Lack of supervision or proper instructions and follow ups; it includes lack of:
  - Proper allocation
  - On-the-job training and instructions
  - Day-to-day observations and guidance for the correct procedure
  - Immediate corrections
  - Follow-up
  - Setting good example by demonstration
  - Disciplinary action – when required.

Observe: Like a COACH, who Cares for you, Observe your performance, Analyzes it and ask you to Comply with it to Help you out from the situation.

Intervene: After consistent watch and observations; Intervene him for the feedback, encouragement, suggesting modifications, additions, alterations etc.

Test: The change in safety behavior and if found OK. Proceed further for the next change/improvement.

As stated above apart from academic education and background, training is equally important

TRAINING[13]

Referring to figure 18.1, in any organization, the most important task is to appoint the best-fitted person for a job, and make use of this human resource, which is most valuable. He/she may be raw (fresher), semi-skilled, skilled or even highly skilled, or an executive of a company. With time, the methods, techniques, equipment, procedures, laws, regulations, policies, and job related parameters changes. To make oneself capable of performing the job in the right and efficient manner, thus, everyone needs some degree of training and education. It is an ongoing process, which starts from day one (while joining) till last day (while separating).

For a company, to extract the maximum from its employees, it becomes almost mandatory to train and educate them. Education means updating the knowledge that is required to run the organization in the best possible manner. As stated, training needs should be understood using the well-established concept of: *Why, Who, When, Where, What and How*.

WHY: Germain and Arnold (2001)[13] mentioned following benefits for effective safety and health education and training:

- Reduced accidents, personal harm, property damage, and related losses
- Increased awareness of the value of tools, appliances, materials, supplies and facilities.
- Decreased downtime and delays
- Improved morale and motivation
- Reduced mistakes and waste
- Optimum performance, productivity and profitability.

If you think education and training are expensive try ignorance and inefficiency (Germain & Arnold, 2001)[13]

WHO: As stated above everyone needs training and education. Subject Matter Experts (SMEs) could judge the magnitude (how much) and direction (what exactly). SMEs could be managers themselves or external consultants/experts who can give an independent evaluation of the training needs and make suggestions to the management accordingly.

WHEN: This starts with induction, orientation and on-the-job training in most of the cases for a newly joined employees. Induction introduces the company's structure, organizational setup, policies, mission, goals, objectives, prevailing facilities, procedures, introduction to service providers such as: computing, maintenance, finance and accounts, medical, housing, transport, recreational (clubs, community centers), welfare establishments, external agencies, such as banks, schools, hospitals, post and telegraph services, public transport etc.

Orientation – rotation amongst cross-functional department/sections within the mining (industrial) setup/organization, and also in many cases other units of the same organization located at other sites.

On-the-job training, as per the needs of the individual then follows. Apart from this there are some statutory requirements to provide training periodically. Refresher courses, attending workshops, symposiums, seminars and conferences add in updating knowledge, and may help in overcoming the inefficiencies, reducing mistakes and minimizing wastes and losses.

WHERE: Classroom and on-the-job training and education are equally important, as both have their specific merits and limitations. Classroom training with the help of modern aids such as audio, video, handouts and study material, drills and demonstrations are the effective ways of learning; particularly when experts and trained instructors are used for this purpose. Similarly on-the-job training by experienced supervisors and engineers proves to be very effective and useful.

WHAT: As mentioned by Germaine and Arnold (2001), it should cover Total Accident Control Training (TACT)[13]. The areas that should be covered include:

- Occupational Health, Safety and Environment (HSE) – Routine
- Occupational Health, Safety and Environment (HSE) – Specific
- Leadership and management specific to HSE.

Leadership and management specific to HSE should ensure control to:

- Prevent accidents and
- Minimizing losses.

HOW: The training and education imparted must be effective and it should be well received by the recipient so that best practices are followed at the work site. As mentioned, modern means/aids are useful in delivering the subject matter. The following steps are useful apart from the practice of '**DO IT**' described in sec. 18.3.3.1.

- Motivate the trainee by explaining its advantages to him as well as to the organization. Make him comfortable and at ease to receive whatever is delivered. The subject matter should be easy to understand and very simple and straightforward.
- Demonstrate its operation physically, wherever practical. Also give practical examples. Stress the key points.
- Test what has been learned including the key points explained.
- Check – while on the job, as whatever required is followed exactly, or does it require further explanation?

There are excellent guidelines available by almost all the Safety/HSE related government/regulatory bodies in your own country. For example: OSHA's 7 steps guideline model broadly covers:

1. Determine if training is needed: many performance problems can be solved using:[7]
   a. Hiring right man to the right job
   b. Improved design, layout and maintenance of the workplace
   c. Use of ergonomically improved/designed tools, appliances and facilities
   d. Job aids such as task procedures, flowcharts, decision tables, trouble-shooting guidelines, hotline etc.
   e. Effective communication and administration.
2. Identify training needs – as described in the preceding paragraphs.
3. Identify goals and objectives
4. Develop learning activities
5. Conduct the training
6. Evaluate program effectiveness
7. Follow up.

ENGINEERING

This aspect covers application of right method, technique, and equipment starting from the planning and design stage to execution stages. Sec. 19.5 and its 10 sub-sections, 19.5.1 to 19.5.10, and figures 19.13 and 19.21(b) details best practices, renovations and advancements which are required to be brought about in the engineering aspects of mining. While designing, it is essential to know the capabilities of human beings. One size does not fit all. Care should be taken as to how hazards can be eliminated or minimized in the design, layout, procedure, services and utilities. Proper engineering helps in evolving efficient systems to contain 'unsafe conditions' and 'unsafe acts' listed above. Sec. 18.3.3.1 details these aspects.

ENFORCEMENT

- Education and training as illustrated in figure 18.7 and table 18.3
- Engineering concepts as illustrated in figure 18.7 and section 18.3.3.1.
- Legal compliance – rules, regulations and statutory provisions – regulatory framework that prevails in the country where industry is located. Checklist as detailed in table 18.4, sec. 19.7 and figure 19.21(a)
- Best practices (refer sec. 19.5.1 to 19.5.10, 19.9.4, 19.10 and figs 19.12 to 19.15, 19.21(b))

Table 18.3   A Three-Day Supervisory Program – Total Accident Control Training (TACT).[13] *(After Germain, G.L. and Arnold, R. Source: Mine Health and Safety Management,* Michael Karmis (edt.), SME, AIME, Littleton, Colorado, 2001 (www.smenet.org)).

Three-Day Supervisory Program – Total Accident Control Training (TACT)

First Day

1. INTRODUCTION – getting acquainted; program explanation; small group discussion exercise to spotlight key areas of concern for participants.

2. PEOPLE, PROPERTY & PROFITS – business and social reasons for a more professional job of supervisory management than ever before; basic concepts of risk management and Total Accident Control (TAC).

3. PROFESSIONAL SUPERVISORY MANAGEMENT – marks of a "pro", "management/leadership work" vs. "operating work", guiding principles; benefits.

4. PROBLEM CAUSES, EFFECTS & CONTROLS – supervisory management as a key to control; difference between "symptoms" and basic causes; three stages in a managing control; twenty elements for successful accident control.

5. SUPERVISORY INVESTIGATION – a practical, professional approach to accident/incident investigation; how effective investigation saves time and reduces losses; how to measure the quality of investigation; tips on interviewing, analyzing, documenting and following-up.

Second Day

6. EFFECTIVE SUPERVISORY INSPECTIONS – kinds of inspections; why and how to inventory "critical parts," the importance of housekeeping; an effective hazard classification system; tips on using inspection reports to get more employee involvement and management action.

7. GROUP COMMUNICATION SKILLS – why supervisors should have group TAC meetings with workers; how to give a good talk (the 5P formula); when meetings should be held and how to make them most effective.

8. MEASUREMENT AS A MANAGEMENT TOOL – three key types of measurement (consequences – causes – controls); how to make best use of TAC measurements; values and benefits.

9. PERSONAL SUPERVISORY COMMUNICATION SKILLS – the supervisor's role in pre-job orientation, how to give key point tips for efficiency, safety and productivity; the "Motivate – Tell & Show – Test – Check" technique for proper job instruction; effective coaching guidelines.

10. THE SUPERVISOR'S ROLE IN DAMAGE CONTROL – bridging the gap between injury control and TAC; uncovering and reducing huge hidden costs; protecting people, preserving property and promoting profits.

Third Day

11. MAINTAINING EFFECTIVE DISCIPLINE – tips on obtaining proper use of personal protective equipment, compliance with rules and regulations and better handling of "problem cases", comparing punitive and positive discipline.

12. CRITICAL TASK PROCEDURES – how to use the three-column (steps – potential problem – controls) analysis worksheet; how to save time effort and money with the "improvement check"; how to write the procedure from the worksheet; eight ways to put procedures to work.

13. EMERGENCY PREPAREDNESS – doing a needs analysis; developing, implementing and monitoring a system, the use of emergency teams; key point tips on emergency response and emergency care.

(Continued)

Table 18.3   Continued.

14. GENERAL PROMOTION – variety is the spice of promotion; how to get the most benefit for the least expenditure, the place of "gimmicks"; double-barreled contests; employee involvement and supervisory leadership example.

15. MOTIVATING TAC PERFORMANCE – five basic guidelines for understanding motivation; six practical principles for managing motivation, basic aspects of a performance management and motivation system for supervisors.

The length of the program, the number of topics covered, what they are and the amount of time devoted to each can be changed in many ways ... to meet the needs of the sponsor and the participants.

Table 18.4   Checklist on: Legal, statutory & other compliances.[25]

Parameters aspects to be considered	Status
Do you have an **existing process** for identifying applicable legal and other requirements?	
If yes, does that process need to be revised? In what way?	
Who needs to be involved in this process within your organization? And what should be their responsibilities?	
What **sources of information** do you use to identify applicable legal and other requirements?	
Are these sources adequate and effective? How **often do you review** these sources for possible changes?	
How do you ensure that you have **access** to legal and other requirements? (List any methods used, such as on-site library, use of web sites, commercial services, etc.)	
How do you **communicate information** on legal and other requirements to people within the organization who need such information?	
Who is **responsible** for analyzing new or modified legal requirements to determine how you might be affected?	
How will you keep information on legal and other requirements **up-to-date**?	
Your next step on legal and other requirements is to ...	

- Empowering approach that says if I take responsibility of my safety, I can reduce workplace hazards. This will benefit my coworkers, my family, my life and me.[17]
- Incorporating HSE as a critical business activity at par with other critical activities such as production and productivity, as described in sec. 19.6.1.

Likewise such a checklist could be prepared for different aspects that need follow up, monitoring and keeping the status up to date.

ENGAGEMENT

After recruitment; the efficiency of management lies in extracting the maximum out of any individual. "An empty mind is the devil's workshop", as such first thing is to allocate the best-fitted job to any individual based on his/her background. Referring

Table 18.5 Letters of the word (the literal meaning) 'Manage' itself depicts (proposes) ways to manage any industrial set-up.

Managing any Industrial set-up by giving equal weightage to the critical Business activities: production, productivity, HSE and social responsibilities *by*

Always following laid-out standards, norms, rules & industrial benchmarks *and*

Nurturing best-practices, and reliable & trustworthy workforce (employees) *through*

Administrating them effectively and efficiently, *also*

Gearing up for the abnormalities in marketing & sales, failures and emergencies; security and secrecy; *and expect to*

Earn profit, goodwill, excellence and reputation by following above guidelines.

The Engagement Pyramid

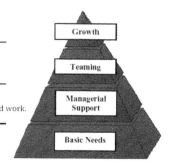

12) This last year, I have had opportunities at work to learn and grow.
11) In the last six months, someone at work has talked to me about my progress.

10) I have a best friend at work.
9) My associates or fellow employees are committed to doing quality work.
8) The mission or purpose of my company makes me feel my job is important.
7) At work, my opinions seem to count.

6) There is someone at work who encourages my development.
5) My supervisor, or someone at work, seems to care about me as a person.
4) In the last seven days, I have received recognition or praise for doing good work.
3) At work, I have the opportunity to do what I do best every day.

2) I have the materials and equipment I need to do my work right.
1) I know what is expected of me at work.

Figure 18.8 Ways to engage any individual fully to get the maximum for the benefit of both: the employer as well as the employee – A proactive work culture.[36]

to table 18.5; letters of word (the literal meaning) 'Manage' itself depicts ways to manage any industrial set-up. Figure 18.8 outlines a strategy to engage any individual fully to get the maximum for the benefit of both: the employer as well as the employee – a healthy situation for everybody.

ERADICATE[32]

Eliminating/minimizing unsafe acts and unsafe conditions could help in eradicating: injuries, incidents (sec. 18.3.4 and fig. 18.12), illness, emission and wastage (fig. 18.9). Incorporating best practices could be equally helpful. Following the loss prevention strategy as detailed in table 18.12, could be the right approach/solution to eradicate the mining and tunneling evils.

18.3.3.2 The systems[29,32]

A system incorporates following features

- It has set practices, procedures and norms to follow
- It follows rules, regulation and by-laws – statutory provisions
- It has certain inputs and yield output

Figure 18.9   **Minimizing**: **E**missions, **I**ncidents, **W**astage, **I**njuries & **I**llness; approaching their magnitude to almost **Z**ero? This is known as: **'Zero-Goal'**.[17,18]

- It interacts with other agencies
- It requires periodical checkups, maintenance and renovations.
- In this industry following systems need to be very efficient and free from fault, if any:
  - Material handling and disposal
  - Transportation
  - Movement horizontally or vertically
  - Drainage and pumping
  - Communication
  - Ventilation
  - Illumination
  - Traffic control/logistics
  - Making available the following devices, equipment, appliances, mechanisms or means (use as applicable to your own setup/mine):
    - Electrical hoists and cranes
    - Hook, chains, eyebolts, slings and cables
    - Pressure vessels, and reliefvalves
    - Digesters, cookers and the like
    - Temperature control devices
    - Fire detection and extinguishing equipment and apparatuses
    - Ladders, lifts and elevators
    - Proper crossovers.

The systems that need to be established[9] for a successful HSE program include:

- Training and placement (right man to the right job, as described in sec. 18.3.3.1)
- Auditing, checks, inspection by the statutory bodies and independent agencies
- Recognition and incentives to safe workers
- Policies and procedures that establishes best practices
- Effective communication
- Accountability and measurement of losses (as detailed in sec. 18.6.5).

Search your own body to find out how many systems need to function properly to make you fit, and what could happen if any one of them doesn't function properly. And this is also the case if systems related to safety, as listed above, fail. Incidents, *injuries, illness, inefficiency, insurance-charges, investment–returns (not as expected), irregularities are the yardsticks to assess safety-systems' effectiveness and failures.* Failures indicate that something is wrong with the system, and it is not simply the human error.

Equally important to understand that: '*We must compatible with the systems of nature; otherwise, the systems of nature will be unable to support us. We must learn ourselves as living within the systems of nature*'.[20]

18.3.3.3   *The working environment (conditions)*

Peoples' attitude, behavior and sense of empowerment towards a safety campaign play an important role. People are responsible for making systems and run effectively and also creating proper working conditions, as described in sec. 18.4.2. And the combination of these three aspects creates a proactive-working environment.

As described in sec. 18.4.4.1 "organizational culture" is what employees perceive a company's values to be. Everyone knows what is important. In a healthy organizational culture first step is to locate the sources of stresses and find solutions to reduce them. Workplace stress could be also result due to following circumstances:

• Organizational change
• Company mergers
• Uncertainty
• Employees 'kept in dark' for the change – the mushrooms?

In such cases change in behavior and attitude, as described in sec. 18.3.3.1; should be exercised.

In the USA mining is the oldest industry, but during 1997 alone more than 5000 people lost their lives in work related incidents and more than 3.8 million suffered disabled injuries and illness resulting in a direct cost to US industry nearly $128 billions with estimated direct and indirect costs of $600 billions.[9] Today it is considered amongst the safest industries as reflected by industries' fatality and injury rate. The measures that have been taken include:

• Improved methods and techniques
• Increased automation that limits interaction between man, machine and material

Figure/Photo 18.10   Workplace stresses – Should any one live with them (first 3 from left)? Or Like the two from right?[7] (After Cairney, M. 2007).

- Education and intensive training
- Empowering workers and adapting progressive labor practices
- Incorporating ergonomic-based designs and layouts (section 18.4.3)
- Treating HSE a critical business activity
- Greater recognition given to protect industry's greatest asset – its People.

18.3.4   Accidents

An accident is a three-step event:[13]

- **Initiation:** the event that starts the accident
- **Propagation:** the events that maintain or expand accidents
- **Termination:** the events that stop the accident or diminish it in size.

The initiation can be diminished through effective training, maintenance, process design and providing up-to-the-mark grounding, bounding, fire and explosion proof electrics, guide rails and guarding wherever required. Propagation could be diminished by reducing inventories of flammable material, providing effective mechanism for quick transfers in emergency (sec. 18.3.4.5, figs 18.13(a) & (b)) and by providing adequate space in the layout and using nonflammable material of construction. Quick termination of an accident could be achieved through the effective fire fighting, relief and sprinkler systems and also by installation of check and emergency shutoff valves. Inherently safer strategy, as described in sec. 18.3.2; can impact or influence the accident process at any of the three stages.

Accidents cannot be totally eliminated due to the fact that plants, logistics, operations and maintenance are designed, constructed, operated, and maintained by human beings and human beings are not perfect. All accidents can be traced back and one would find human failure, which could be poor judgment, being forgetfulness, ignorance, incapacitation, alcohol or drug dependence, fatigue etc., as shown in figure 18.11.

18.3.4.1   *Accidents/incident analysis & calculations*

Purpose of incident reporting and analysis is to:[14,21]

- Learn from mistakes
- Prevent reoccurrence
- Increase level of safety awareness
- Demonstrate commitment to continuous improvements.

Underlying causes include: substandard acts, substandard conditions and their underlying causes such as individual and work factors. 'Individual' refers to inadequate knowledge, skill, motivation, capability, strength, and attitude, etc. as described in sec 18.3.3.1. Accidents are like mosquitoes (fig. 18.12); it is best to drain the swamps in which they breed. Human failure (fig. 18.11 and fig. 18.22) is responsible for most of the incidents and the best way get rid of them is to attack the root causes which are causing them.

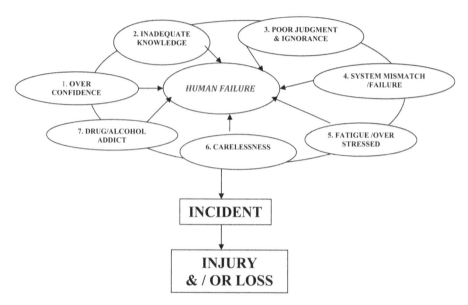

Figure 18.11 Human failure is responsible for most of the incidents which may result in injuries or/and losses.[2]

Figure 18.12 Accidents are like mosquitoes, it is best to drain the swamps in which they breed.[19]

Accident-related calculations[1,21]

Accidents are calculated as:

- Frequency rate: number of injuries per 200,000 man-hours.[#]
- Severity rate: number of lost days per 200,000 man-hours worked.[#]

– Definitions and calculations differ from one organization to other.

Frequency deals with likelihood of occurrence/hazard that will lead to an undesired event, incident, or accident. Severity deals with extend of damage or harm. Hazards with high frequency and high severity need greatest attention.

As per the prevalent practices; the frequency rate, for the following losses; is usually calculated:

- Lost time injury/illness
- First aid or minor injury/illness
- Reportable/recordable injury/illness (to government/statutory authorities)
- Property damage
- All accidents
- Near-miss.

Severity rate, as per the prevalent practices, for the following losses is usually calculated:

- Accident severity rate
- Property damage (monetary losses) severity rate.

Thus, frequency rate = Number of losses (as listed above) × 200,000 divide by total number of employee hours worked.

Employee hours = number of employee × hours worked per year.

Degree (Type) of injuries
Fatal – Causing death
Serious – Causing forceful absent from work
Minor Injury – Injury without loss of workdays.

18.3.4.2   *Common accident areas/heads*

Causes would differ from one project to another or one industry to another. Listed below are some common areas, or heads, to describe them (Also refer to figs 18.11, 18.22).

- Haulage – over speed, improper turns & gradient, inadequate safety fittings or their failures.
- Machines & equipment (hardware) – failures due improper maintenance & operation.
- Structure failures – inadequate structures, design defects etc.
- Slip or fall of workers – slippery roads, inadequate width of workings & roads.
- Material handling – toxic, nontoxic, hazardous such as explosives, tools and appliances.
- Fall or sliding of material – absence of fencing, barricade, guard etc.
- Improper supervision – incompetent, negligent, overstressed, miscommunication and coordination.
- Miscellaneous – not covered above.

In mining and tunneling operations many incidents happened not because of safety system engineering were lacking but safe procedures and preventive strategies were not followed.[8,13,16]

18.3.4.3 Accident costs

- Injury, loss of body's parts – disability; a greatest loss having practically no substitute. Any amount of compensation is just a token.
- Absence from duty, delays, loss of time.
- Loss of moral, loss of efficiency of crew/workers
- Loss of material, property, equipment
- Cost of treatment, cost of production loss, overtime payment.
- Cost of replacement, clean ups, repair, standby etc.
- Fines or penalties by the government/safety authorities.
- Cost of legal assistance.
- Cost of compensation.
- High premium by Insurance companies.

18.3.4.4 Remedial measures[21]

- Conceptual planning, detailed design and evaluation
- Compliance with design specifications
- Safe working conditions – lighting, ventilation, sanitation
- Safe equipment – fittings, design, maintenance
- Safety wear, detectors and warning mechanisms
- Precautions and measures against fires and explosions
- Training, education and refresher courses
- Emergency measures
- Welfare amenities and medical check ups
- Legislation – rules, regulations, code of practices
- Accident analysis and preventive measures
- Risk analysis.

Some of the items listed above have been already discussed in the preceding paragraphs and sections, the remaining are discussed below.

18.3.4.5 Measures/preparedness[8]

Given in table 18.6, are the checklists that should be prepared for the project in hand, to assess the likely hazards or events that could cause problems, and remedial measures that will have to be taken. An emergency management system is an integral part of industrial safety. Figure 18.13(a) outlines the events, occasions/situations which could be termed as an 'Emergency'. Standing orders (Emergency Development Plan, fig. 18.13(b))[8] should be prepared and workers should be trained to act as per these orders in the event of an emergency. It requires coordination, liaison and co-operation between different agencies. It requires enforcing scheduled training, refresher courses and demonstration and mock practices to the concerned – crews, rescue and recovery trained staff and others.

18.3.4.6 Hazards analysis methods[2,32]

There is number of methods/techniques, as detailed in table 18.7 illustrating which technique could be applied during a particular phase of an industrial setup. These techniques

Table 18.6 What if check list – A risk analysis.

WHAT IF CHECK-LIST   REVIEWER_____
OPERATION TYPE_____   LOCATION_____

What if? (1)	Cause	Consequences	Risk control measures	Monitoring measures	Likelihood	Severity	Risk	ERP recommondations
An event or situation that may arise. Refer table 18.7 & fig. 18.13(a) also	What could have potentially caused the event situation in column (1)	What could have potentially occurred as a result of the event/situation in column (1) (escalation)	What measures are in place to prevent, mitigate and recover an incident/ accident	What active/ reactive monitoring measures are in place? Are these measures are adequate & reliable.	The probability of frequency that the event/situation in column 1 will occur based on hazard matrix definitions.	The severity of the consequence based on the hazard matrix definitions	Taken from the hazard matrix Risk = consequence × likely hood	Any additional information or investigation required? Are there practicable changes which could reduce risk or eliminate hazards.

Table 18.7 Application of hazards analysis methods during various phases of an industrial setup.[2,32]

Different phases of an mines and tunneling projects	Hazards Analysis Methods											
	Safety Review	Check List	Relative Ranking	PHA – Preliminary Hazard Analysis	What-if Analysis	What-if Checklist	HAZOP – Hazards and Operability Analysis	FMEA – Failure Modes and Effects Analysis	FT – Fault Tree	ET – Event Tree	CCA – Cause Consequence Analysis	HRA – Human Reliability Analysis
Research & Development	o	o	●	●	●	o	o	o	o	o	o	o
Conceptual design	o	●	●	●	●	●	o	o	o	o	o	o
Pilot plant operation	o	●	o	●	●	●	●	●	●	●	●	●
Detailed Engineering	o	●	o	●	●	●	●	●	●	●	●	●
Construction/Start up	●	●	o	o	●	●	o	o	o	o	o	●
Routine Operation	●	●	o	o	●	·	●	●	●	●	●	●
Expansion or Modification	●	●	●	●	●	●	●	●	●	●	●	●
Incident investigation	o	o		o	●	o	●	●	●	●	●	●
Decommissioning (mine closure)	●	●	o	o	●	●	o	o	o	o	o	o

o - Rarely used or inappropriate. ● – Commonly used

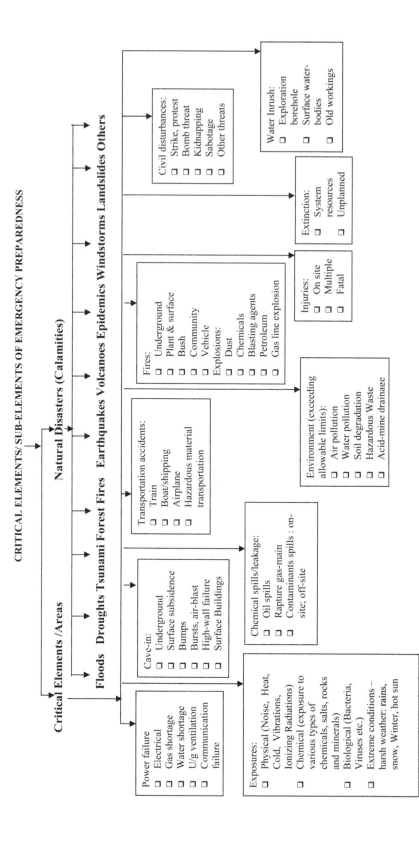

CRITICAL ELEMENTS/ SUB-ELEMENTS OF EMERGENCY PREPAREDNESS

Critical Elements /Areas

Natural Disasters (Calamities)

Floods Droughts Tsunami Forest Fires Earthquakes Volcanoes Epidemics Windstorms Landslides Others

Power failure
□ Electrical
□ Gas shortage
□ Water shortage
□ U/g ventilation
□ Communication failure

Exposures:
□ Physical (Noise, Heat, Cold, Vibrations, Ionizing Radiations)
□ Chemical (exposure to various types of chemicals, salts, rocks and minerals)
□ Biological (Bacteria, Viruses etc.)
□ Extreme conditions – harsh weather: rains, snow, Winter, hot sun

Cave-in:
□ Underground
□ Surface subsidence
□ Bumps
□ Bursts, air-blast
□ High-wall failure
□ Surface Buildings

Chemical spills/leakage:
□ Oil spills
□ Rapture gas-main
□ Contaminants spills : on-site; off-site

Transportation accidents:
□ Train
□ Boat/shipping
□ Airplane
□ Hazardous material transportation

Environment (exceeding allowable limits):
□ Air pollution
□ Water pollution
□ Soil degradation
□ Hazardous Waste
□ Acid-mine drainage

Fires:
□ Underground
□ Plant & surface
□ Bush
□ Community
□ Vehicle
Explosions:
□ Dust
□ Chemicals
□ Blasting agents
□ Petroleum
□ Gas line explosion

Injuries:
□ On site
□ Multiple
□ Fatal

Extinction:
□ System resources
□ Unplanned

Civil disturbances:
□ Strike, protest
□ Bomb threat
□ Kidnapping
□ Sabotage
□ Other threats

Water Inrush:
□ Exploration borehole
□ Surface water-bodies
□ Old workings

Figure 18.13(a)   Critical elements/sub-elements for Emergency preparedness (Concepts taken from Mine Accident Prevention Association, Ontario, Canada).

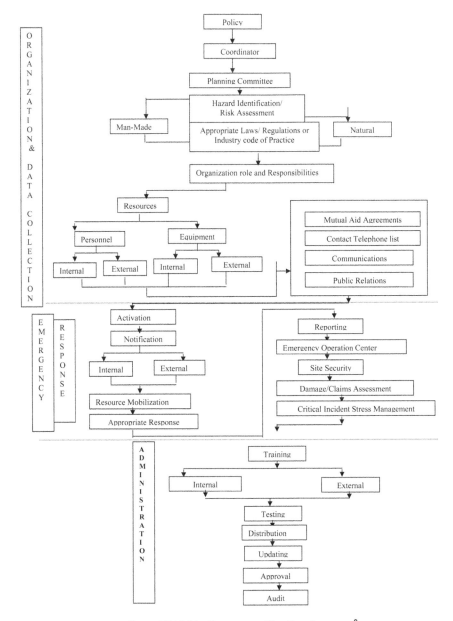

Figure 18.13(b)   Emergency Plan Development.[8]

could be divided into three groups: Broad Brush (for preliminary assessments of hazards), Design & Routine Operations (to access what could happen in the event of failures) and Special Situations (to analyze reasons for failures that have already occurred).

18.4　OCCUPATIONAL HEALTH AND SURVEILLANCE

Figures 18.2 and 18.3 detail hazards to which workers health is the first target and to the people living in the surrounding areas thereafter. The Bhopal disaster in 1984 in India could be cited as an example in which not only the employees of the company Union-Carbide but also about 2000 people living in the surrounding areas were affected badly.[20] This example illustrates that *Occupational Health and Safety (OHS)* is not confined to industrial workers but also it covers those living around including flora and fauna.

Elements: Occupational Health (OH)[1,3]
- Industrial hygiene including efficient effluent discharge
- Housekeeping & working conditions
- Ergonomics
- Occupational health surveillance.

18.4.1　Industrial hygiene[3]

Meaning of word hygiene means healthful/healthy. It addresses pollution and stresses in the workplace, which may cause sickness, impaired health and well being, or significant discomfort among workers, on local communities, flora and fauna.

18.4.1.1　*Aqueous effluents – permissible quality & efficient discharge*[24,33]

Aqueous effluent discharges have the potential to be hazardous to human health and/or it could harm the environment. They are resulted from the mineral processing activities such as washing, jigging, concentration, smelting and refining. The water-bearing formations or strata releases water, which could be acidic or contain heavy metals. Mining operations can influence quality and quantity of the fresh water in the following manner:

- In the absence of effective sealing, while working below water table, the fresh-water from the underground aquifers is taken away.
- At surface mines also, when mine workings cross the water table, it takes away the useful freshwater.
- The rainwater carries the silt from the waste rock dumps, if by constructing effective bunds around the waste rock-dumps does not prevent it.
- Water can become contaminated with toxic or radioactive materials from mine sites and abandoned hazardous waste sites.
- Acid Mine Drainage (AMD) or, Acid Rock Drainage (ARD) is usually confined to coal as well as metal mines from where outflow of acidic water takes place. This is due to presence of sulfur as pyrite (FeS_2), or sulfide ores of copper, zinc, nickel and few others including sulfide minerals. The mines could be active as well as abandoned. Copper mines usually have this problem. There are thousands of sites located different parts of the world including the mine discharges which are suffering from AMD.
- Tailing ponds have been also source for the surface as well as ground water contamination.

Thus, mining operations can influence water quality and quantity in the area of operation, if suitable measures such as: proper diversion of the waterways, provision of adequate catchments areas and efficient treatment before discharge are not taken.

18.4.1.2 House keeping[24,33,36]

Housekeeping: Activities necessary for cleanliness, orderliness, and neatness in all areas of work are known as housekeeping. Good housekeeping is the key to prevent accidents and, in maintaining good health standards in any industrial setup including mines and mining complexes, which is usually consisting of many buildings encompassing: offices, workshops, laboratories, plants, warehouses, vocational training centers, residential accommodations, community/recreation centers, health centers, clubs and all other buildings within its premises. It is considered to be first-line of defense against any illness or injury that could be caused while undertaking any operation.

18.4.1.3 The 5S concept[36]

'5S' is a Japanese concept involving the words that begin with 'S'. 5S is a philosophy and a way of organizing and managing the workspace and work flow with the intent to improve efficiency by eliminating waste, improving flow and reducing process unevenness. Its phases are as under:

- Phase 1 – **Seiri** Sorting: Going through all the tools, materials, etc., in the mine, plant and work area and keeping only essential items. Everything else is stored or discarded.
- Phase 2 – **Seiton** Straighten or Set in Order: Focuses on efficiency. When we translate this to "Straighten or Set in Order", it sounds like more sorting or sweeping, but the intent is to arrange the tools, equipment and parts in a manner that promotes work flow.
- Phase 3 – **Seisō** Sweeping or Shining or Cleanliness: Systematic Cleaning or the need to keep the workplace clean as well as neat. At the end of each shift, the work area is cleaned up and everything is restored to its place. This makes it easy to know what goes where and have confidence that everything is where it should be. The key point is that maintaining cleanliness should be part of the daily work – not an occasional activity initiated when things get too messy.
- Phase 4 – **Seiketsu** Standardizing: Standardized work practices or operating in a consistent and standardized fashion. Everyone knows exactly what his or her responsibilities are to keep above 3S's.
- Phase 5 – **Shitsuke** Sustaining the discipline: Refers to maintaining and reviewing standards. Once the previous 4S's have been established, they become the new way to operate. Maintain the focus on this new way of operating, and do not allow a gradual decline back to the old ways of operating. However, when an issue arises such as a suggested improvement, a new way of working, a new tool or a new output requirement, then a review of the first 4S's is appropriate.

The key targets of 5S are improved workplace morale, safety and efficiency. Advocates of 5S believe the benefits of this methodology come from deciding *what* should be kept, *where* it should be kept, *how* it should be stored and most importantly *how* the new order will be maintained. This decision making process usually comes from a dialog about standardization which builds a clear understanding, between employees, of how work should be done. It also instills ownership of the process in each employee.

18.4.2   Working conditions[28,30]

It includes perfection in the following aspects:

- General layout: Adequate working space and perfect layout includes provision for fencing, guards, dykes, barriers, emergency access, aisle ways, enough headroom and sufficient clearance. Buildings should have adequate ladders, stairways; escape ways, fire doors, safety glasses, fire-proof structure, overhead power lines, etc. wherever required.
- Effective ventilation (refer sec. 9.5 also) and air conditioning would mean:
  - ○ Contain and exhaust hazardous substances. Keep exhaust system under negative pressure
  - ○ Use properly designed hoods; use hoods for charging and discharging
- Wholesome water – Potable as well as non-potable
- Effective/Perfect drainage and sewerage
- Proper lighting and illumination
- Warning signals and alarms
- Safety and security
- Proper sanitation and housekeeping; for example:[5]
  - ○ Use wet method to minimize contamination with dust.
  - ○ Use water sprays for cleaning
  - ○ Clean areas frequently.
- Effective communication and information systems
- Adequate recreation facilities and amenities
- Use of IT and software/computing wherever applicable (ref. sec. 18.6.6 also)
- Automation wherever applicable (ref. sec. 19.5.8 also)
- Use and availability of PPEs; for example for effective hygiene control[5]
  - ○ Use safety glass and face shields
  - ○ Use aprons, arm shields and space suits
  - ○ Wear appropriate respirators, airline respirators are required when oxygen concentration is less than 19.5%
  - ○ Effective supervision and trained staff (ref. sec. 18.3.3.1 for training)
- Adequate measures to minimize pollution (ref. sec. 18.5 and tables 18.9 and 18.12)
- Noise abatement measures
- Well-maintained equipment, appliances, tools and tackles
- Effective services – transportation, communication, connectivity, power, water and sanitation
- Health and hygiene care as detailed in the preceding sections.
- Clarity of job profile, role and responsibility (fig. 18.8)
- Administrative controls – disciplinary actions, encouragement and rewards for exceptionally good performance; delegation of responsibilities. This can be described as Organizational Culture. A proactive culture brings Organizational Commitment (OC) that reduces workplace stresses, as described in sec. 18.4.4.1.

- Exposure to changes in technology and renovations
- Adequate medical facilities including effective first aid, rescue and recovery
- Routine checkups and medical surveillance
- Emergency preparedness and measures
- Security – personal as well as property.

In addition, all these aspects have been dealt in detail in various chapters.

18.4.3 Ergonomics[6,9,23,34]

18.4.3.1 *Introduction*

The term ergonomics has been derived from the Greek words *ergon* (meaning "work") and *nomos* (meaning "rules"). Thus, its literal meaning is 'the rules of work'. A good way to understand what ergonomics means is to think about the term "user-friendly." The two terms are actually synonymous; anything that is friendly is ergonomic, and anything that is unfriendly is un-ergonomic.

Ergonomics aims at making things more human compatible, which can bring improvement in productivity – a key to save money, and making the operation *'user friendly'*. In the workplace focus should be on making/improving the tools, appliances, equipment, work-methods/procedures/techniques, layouts and working environment *user-friendly.*

Humans have been doing "ergonomics" for a long time (that is, reducing the physical demands of jobs). Good ergonomic improvements include switching over from stone-age tools to modern tools and appliances (fig. 1.6).

18.4.3.2 *Impacts of poor ergonomics[6,9,23,34]*

Poor ergonomic conditions typically involve the bones, muscles, joints, tendons, and nerves. Symptoms include:

- Painful joints
- Pain, tingling or numbness in hands or feet
- Pain in wrists, shoulders, forearms, knees, etc.
- Back or neck pain
- Fingers or toes turning white
- Shooting or stabbing pains in arms or legs
- Swelling or inflammation
- Stiffness
- Weakness or clumsiness in hands
- Burning sensations
- Heaviness
- Sustained muscle exertion, which reduces blood flow to the muscles and causes muscle strains and sprains
- Contact stresses, which are injuries that occur due to repeated contact with a hard surface.

These symptoms could also be the result of other medical conditions, so checking with the doctor is important if there is any concern about any of these. Figure 18.14

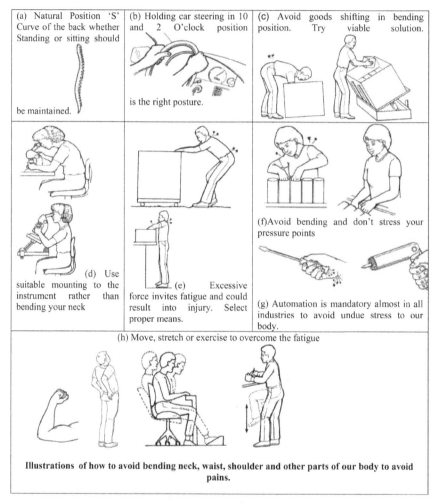

(a) Natural Position 'S' Curve of the back whether Standing or sitting should be maintained.	(b) Holding car steering in 10 and 2 O'clock position is the right posture.	(c) Avoid goods shifting in bending position. Try viable solution.
(d) Use suitable mounting to the instrument rather than bending your neck	(e) Excessive force invites fatigue and could result into injury. Select proper means.	(f)Avoid bending and don't stress your pressure points
		(g) Automation is mandatory almost in all industries to avoid undue stress to our body.

(h) Move, stretch or exercise to overcome the fatigue

Illustrations of how to avoid bending neck, waist, shoulder and other parts of our body to avoid pains.

Figure 18.14 How to avoid bending neck, waist, shoulder and other parts of our body to avoid pains.[6,9,23,34]

illustrates as how to avoid bending neck, waist, shoulder and other parts of our body to avoid pains.

Impacts of good ergonomics include:

- Improved labor relations
- Safeguarding skilled and experienced human resources
- Offsetting limitations on age of employees
- Reduced maintenance downtime.

18.4.4 Occupational health surveillance[7,13,22]

Preceding sections have dealt with the important elements that are essential for the good health of the industrial workers. Based on the experience of some of the

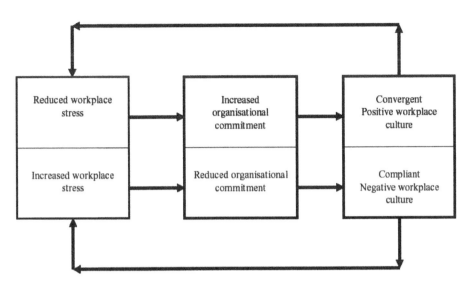

Figure 18.15   The integrated approach to bring changes in the Organizational Culture leading to an increase in Organizational Commitment (OC)[7] (After Cairney, M. 2007).

leading mining companies; paying attention to the following practices is equally important:

- Organizational culture and workplace stresses[7]
- Lost performance at work (presenteeism)[22]
- Occupational hygienic risk – exposure assessment and control measures.[18]

18.4.4.1   *Organizational culture and workplace stresses*

Organizational culture (OC) is what employees perceive a company's values to be. In a healthy Organizational Culture first step is to locate the sources of stresses and find solutions to reduce them. In fact we can understand stresses. Some of them could be removed whereas some may have little control to overcome them. In such cases change in behavior and attitude, as described in sec. 18.3.3.1; should be exercised. Mick Cairney (2007)[7] in the following paragraphs have proposed an integrated model, as shown in figure 18.15, which could be summarized as under:

- There is a link between OC and workplace stresses. An integrated approach, as illustrated in figure 18.15; could bring improvements.
- Stresses at workplace have been identified by research and can be targeted.
- Establishing a positive culture is desirable for positive business outcomes.
- Good leadership and management practices can deliver positive culture, which can achieve higher safety and business values.
- Releasing stresses through regular exercise, meditations and sound sleep is equally important.
- Solutions must be not only sustainable but also sustained.

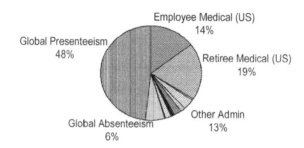

Figure 18.16 Parameters influencing productivity losses (After McDonald, R. 2006).[22]

18.4.4.2 *'Presenteeism' – lost performance at work*[22]

Presenteeism: The problem of workers being on the job but because of medical conditions, not fully functioning is known as 'presenteeism'. The health problems that result in presenteeism include such chronic or episodic ailments such as: depression, back pain, arthritis, heart disease, high blood pressure, and gastrointestinal disorders. Figure 18.16 shows impairment at work (presenteeism) is the largest single component of productivity losses in the workplace. The extent and cost of productivity losses within a workplace are much larger than previously thought. Solution lies in productivity-improvement projects that could be designed and implemented for success and have a positive ROI (Return On Investment).

18.4.4.3 *Periodic health surveillance: based on exposure-risk*[13]

Referring table 18.8, there is a definite relation between working life and health of any individual who is working in any industry. As per the demands of the job one has to keep fit and healthy to perform his/her job effectively right from the day one on joining till he/she leaves the job.

Assessing exposure to mine hazards and implementing control measures provides a sound footing to the preventive and proactive measures to the health of the mine workers or those concerned. It involves following steps:

- Workplace assessment is conducted and evaluated by competent professional hygienists using recognized standard methodology
- Initial observation, interview with operational personnel and professional judgment lead to basic characterization and qualitative assessment
- Through quantitative assessment, statistical analysis and interpretation of data, a baseline exposure risk profile is established. The baseline exposure risk profile directs:
  ○ Exposure control initiatives
  ○ Periodic review/re-evaluation
  ○ Health surveillance programs.

Table 18.8   Working life of an employee and the variables influencing his/her health during this tenure.[32,37]

Exposure to Hazards, Preventive & Proactive Measures During Working Life

Employee Joins	Exposure to Hazards	Periodic Health Surveillance	Employees' Exit
Medically check up for fitness to the Job Exposure to working culture, Systems Undergoing Status Changes: Promotion, Transfer and Change in Responsibilities	Dusts, Fibers Foul Gases, Smoke, Fumes, Mist Radiations, Ionization, Chemicals Thermal Stress – Heat & Humidity Noise, Vibrations, Poor Ergonomics Physical Injury; Likewise other hazards could be included	Basic Characterization Initiatives; Qualitative as well as Quantities Assessment of Exposures Monitoring Implement Controls & Periodic Review/Re-Evaluation Health surveillance programs such as: Health Promotion Management (HPM) to deal with issues such as: Impairment at work (Presenteeism)	**Either of the following happens:** • **Die** • **Injured** • **Sick** • **Medically Affected Exit** • **Employee Exit Sick** • **Employee Exit Healthy**
HR Related Issues	**Fatigue, Stresses & Organizational Culture**	**Preventive & Proactive Measures**	Outcome

18.4.4.4   *Notified diseases and preventive measures*[10,37]

In order to work systematically and to safeguard against the occupational diseases the following guidelines should be adhered to:

- Noise mapping should be made mandatory of various operations that generate noise along with personal noise dissymmetry of individual workmen exposed to noise level above 85 dbA.
- Vibration studies for any equipment that generate them should be done before its introduction as per the ISO standards.
- All equipment before their introduction should be checked for their suitability in terms of 'Ergonomical fitness', as per the ISO standards. These checks should assess:
  ○ Working procedure
  ○ Working aids/tools
  ○ Working posture.
- Drinking water supplied to the employees should be regularly tested, irrespective of its source, preferably after rainy seasons, the sample of water should be collected from the points of consumption
- Initial medical examination (prior to offering job), as described in table 18.8; is mandatory in any industrial setup irrespective it is permanent, temporary or contractual. It should be strictly adhered to.

- Special tests should be included in the Periodical Medical Examination (PME) for employees exposed to specific health hazard; for example:
  (a) For employees exposed to manganese, special emphasis should be given to behavioral and neurological disturbances such as speech defect, tremor, and impairment of equilibrium, adiadochokinesia H_2S and emotional changes.
  (b) For persons exposed to lead, PME should include blood lead analysis and delta aminolevulinic acid in urine, at least once in a year.
  (c) Employees engaged in food handling and preparation and handling of stemming material activities should undergo routine stool examination once in every six months and sputum for AFB and chest radiograph once in a year.
  (d) Employees engaged in driving Heavy Earth Moving Machines (HEMM) should undergo eye refraction test at least once in a year.
  (e) Employees exposed to ionizing radiation should undergo blood count at least once in a year.

The diseases, such as the listed below are usually recognized as occupational diseases in an excavation industry where from they could be caused:

a. All other types of pneumoconiosis (excluding coal workers pneumoconiosis), silicosis and asbestosis. this includes siderosis & berillyosis
b. Noise-induced hearing loss
c. Contact dermatitis caused by direct contact with chemicals
d. Pathological manifestations due to radium or radioactive substances.

There could be more in the above listing; and due recognition should be given to them. No doubt there are rules, regulations, laws and standard norms and practices that should be sufficient to guard against any occupational disease.

18.5   ENVIRONMENT DEGRADATION AND MITIGATION MEASURES

Why pollution? Environment means surroundings. As described in preceding sections tunneling, excavations and mining operations are bound to pollute air and water, and degrade land. To understand the basics of environment in relation to these industries, let us take example of gold mining. Gold is mined if its grade/concentration is 5 gms/ton. It means while mining 1 ton (1000 kg) of gold ore, only 5 gms (on an average) would be of useful and rest 999.995 kg of rock that would be generated is waste. In recovering this 5 gms gold it has gone through mining (breaking rock into small fragments from in-situ), concentration (crushing, grinding in to powder and separation from rest of the rock-mass using chemicals), smelting and lastly refining and casting into bars or any other shape. One can imagine how much energy it required, materials of different kinds it consumed, foul gases it produced and land it required to dispose of the wastes generated. This is the reason mining is blamed for pollution. But minerals are our basics needs we cannot do away with them. One should not forget that more than 75% of power is generated using fossil fuels (coal, oil and gases). Automobiles are run by fossil fuels. Fossil fuels contribute to the greenhouse effect. Exhaust from automobiles is responsible for bad ozone. Industrial pollutants are causing acid rains, for example in recent survey it was found that in China it is affecting its one third of the land.

Let us understand what has happened in the past. 'Imagine if a river becomes stagnant (doesn't flow), sky smoke shrouded (covered), dumping on the ground odoriferous and unsightly (ugly) waste. Populations can ignore all these things till they have an ill effect on their health and well-being. It went on for many years but ultimately man started experiencing a negative impact on health, aesthetic and cultural pleasures and economic opportunities'.

18.5.1 Balance system/equation

$$\text{Accumulation} = \text{Input} - \text{Output} \qquad\qquad (18.1(a))$$

$$\text{Accumulation Rate} = \text{Input Rate} - \text{Output Rate} +/- \text{Transformation Rate}$$
$$(18.1(b))$$

Imagine what approach we should follow? **ADD or Not ADD?** And up to what extent?

This has called for an awakening, which is known as: 'Sustainable Development'. The United Nations World Commission on Environment and Development defined the term 'sustainable development' in 1987 for Ecological Sustainable Development[28] as "Development that meets the needs of the present generation without compromising the ability of future generations to meet their own needs".

18.5.2 Environmental degradation[29,30,32]

Figure 18.17 details main sources to pollute air, water and degrade land environment.

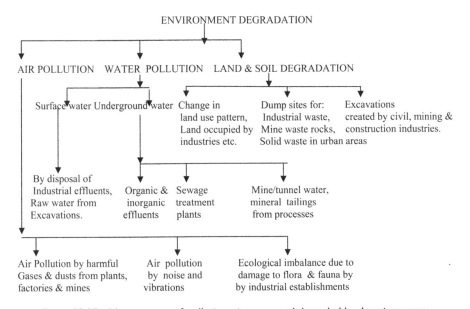

ENVIRONMENT DEGRADATION

AIR POLLUTION   WATER POLLUTION   LAND & SOIL DEGRADATION

Surface water   Underground water   Change in land use pattern, Land occupied by industries etc.   Dump sites for: Industrial waste, Mine waste rocks, Solid waste in urban areas   Excavations created by civil, mining & construction industries.

By disposal of Industrial effluents, Raw water from Excavations.   Organic & inorganic effluents   Sewage treatment plants   Mine/tunnel water, mineral tailings from processes

Air Pollution by harmful Gases & dusts from plants, factories & mines   Air pollution by noise and vibrations   Ecological imbalance due to damage to flora & fauna by by industrial establishments

Figure 18.17 Main sources of pollution: air, water and degraded land environment.

18.5.3   Environmental management[29,30,32]

This is a mechanism to mitigate environment degradation and it is a three steps process:

   I. Base line i.e. existing or pre-excavation scenario.
  II. Environmental Impact Assessment (EIA).
 III. Environmental Management Plan (EMP).

Table 18.9 briefly summarizes, steps that need to be taken during all the three phases, bulleted above.

18.5.4   Environmental system

In addition to above, minimizing pollution, companies should include the following guidelines in their system:

- Environmental policy
- Adequate resources
- Responsibilities and authorities
- Training
- System documentation
- Operational controls
- Document control
- System audits
- Management review.

18.6   LOSS PREVENTION

18.6.1   Classification – losses[28,31,36]

Direct Losses in various forms or types include followings:

1. Production related losses
2. Lack of effective utilization of time
3. Human efficiency (effectiveness) – man-management
4. Poor recovery (yield)
5. Imbalance amongst production, productivity and HSE
6. Equipment and plants performance
7. Use of substandard materials & auxiliary machines & spares
8. Inadequate internal as well as external infrastructures
9. Quality defects
10. Frequent interruptions, inconsistencies & shutdowns
11. Defective design & layouts; and lack of planning
12. Energy losses & excessive material consumption
13. Lack of supervision and accountability
14. Lack of tools, techniques (up to date technology) and renovations
15. Lack of attention to the abnormalities
16. Excessive waste generation.

Table 18.9 Summary of aspects to be looked into for preparing the management plan.[29,30,32]

I. Base-line information	II. Impact assessment (EIA)	III. Management plan (EMP)

1. Ground/Land Environment
1.1 Surface topography
    Plain land
    Hilly terrain
1.2 Soil
    Type (quality)
    Thickness
1.3 Land use pattern
Industrial
Residential
Grazing &forest
Occupied by
infrastructures
Barren land areas.
Others – not covered above
1.4 Aesthetic – Existing
    natural beauty.
	1. Likely Impacts
	
	1.1 Surface topography
	    Magnitude, location.
	
	1.2 Soil
	Quantity, Entent
	
	1.3 Land use pattern
	    Magnitude & location of
	    disturbance. Land occupied
	    by waste rock dumps,
	    effluents by the process
	    plants.
	
	1.4 Impacts to the existing
	    natural beauty
		• Diversion plans of existing land use pattern (if any) together with its mechanism, schedule and cost implications.
		• Excavation/mining sequence ensuring minimum land disturbance at a time
		• Waste rock dumping schemes with location and procedure ensuring minimum horizontalspread. Provision for bunds around dump-yards to prevent silt flow to the surrounding
		• Stacking different mineral wastes separately and keeping its record for future reference.
		• Land reclamation and beautification schemes. Useful soil reuse schemes.

2.Water environment
2.1 Existing water
    bodies and their
    proximity to the project
2.2 Position of water
    table Potable water
    availability
    (quality & quantity)
2.3 Sewerage water
    disposal
2.4 Industrial water
    disposal
2.5 Rain
    Average rain
    Rainy months
    Natural drainage
    system
	2. Impacts to water environment
	2.1 Diversion of water-bodies,
	    if any.
	
	2.2 Water pollution – quality,
	    quantity (resultant water
	table lowering).
	
	2.3 Sewerage water –
	    quantity.
	2.4 Effluents – types and
	    quantity
	2.5 Impacts of rain water to
	    the workings. Chances
	    of flooding, Water pumping
	    requirements. Impacts to
	    existing drainage system.
	    Chances of acid water,
	    acid rain, if any.
		• Water flow diversion if required, with costs, schedule and mechanisms.
		• Provision for effluent discharge
		• Provision for raw water treatment (including acid water, if any) before its discharge.
		• Assessment of pumping requirements including standby arrangements to avoid any flooding or inundation.
		• Water surveys, quality monitoring and plans to meeting potable water requirements.

3. Air
3.1 Existing climate
3.2 Air quality
    (existing industrial
    and general
    pollution)
	3. Air pollution assessment
	3.1 Assessing any rise in
	    temperature, humidity,
	    if any.
	3.2 Pollutants – types and
	    quantity.
		• Provision for effective ventilation – fan types, location. Meeting water-gauge and air quantity requirements.
		• Periodic air surveys.

(Continued)

Table 18.9 Continued.

I. Base-line information	II. Impact assessment (EIA)	III. Management plan (EMP)
3.3 Noise sources and sources. 3.4 Dust – airborne dust (quantity and quality)	3.3 Noise – magnitude and quantitative assessment. 3.4 Likely dust generation – sources, quality and quantity.	• Noise containment measures. • Dust sampling, analysis and suppression measures. • Measures to maintain temperature and humidity of the surroundings within allowable limits.
4. Seismic – existing (natural) Induced by industrial activities	4. Assessing vibrations due to pneumatic machines, and as a result of blasting operations.	• Measures to design blasts enabling peak particle velocity to be within allowable limits, so that damage to the surround structures is minimum. Scheduling blasts by considering minimum disturbance to the workers and local inhabitants.
5. Bio-Environment 5.1 Flora – vegetation type, tree Density, crops – types, 5.2 Fauna – birds, insects, animals (types and population)	5. Impacts to the existing flora and fauna. 5.1 Assessing magnitude of vegetation removal, deforestation. 5.2 Fauna likely to be effected by the project activities.	• Re-vegetation, plantation schemes: location, schedule, manner and costs. • Possibility of relocating existing fauna, if any.
6. Social Environment 6.1 Local inhabitants – population, trade and vocation 6.2 Places of worship & historical importance 6.3 Existing facilities – education, medical, housing, power, roads, recreation, playgrounds, transport, communication	6. Assessing likely impacts to the people and facilities within a radius of 5 km or so (as required by the prevalent regulations). 6.1 Relocation of existing inhabitants 6.2 Disturbance/demolition, shifting of places of public interests. 6.3 Diversion or damage to the existing facilities.	• Planning relocation of local inhabitants, facilities, if any, with schedules, costs and mechanisms. Assessment of compensation, if any. • Diversion plans of existing infrastructures, facilities with their details and costs involved. • Provisions to settle public disputes and grievances.
7. Existing legislation 7.1 Rules, regulations, laws regarding environment, health and safety. 7.2 Trade unions – type of laborers tests,	7.1 Implications of enforcing existing laws – impacts on production, productivity and costs. Necessity of formulating code of conduct, or procedures, in absence oflocal laws. 7.2 Assessing likely problems from local laborers and tradeunions in existence, if any.	• Provision for employing competent persons to comply with the regulations, and carry out activities with safety and least harm to the environment. • Provisions for periodical checkups and inspections. • Provision for safety wears, training, and basic health and welfare facilities.

The indirect losses include:

- Affects adversely the goodwill and reputation of the company
- Reduces talent retention
- Increase/rise in indirect costs
- High insurance premiums
- It increases frustration and lowers moral amongst working crews/workers.

18.6.2   Abnormalities[31,36]

Abnormalities and deviation from standards brings the quality down, adds costs and enhances losses of various kinds, as listed in the preceding paragraphs. Table 18.10

Table 18.10   Diagnosis of abnormalities and suggested remedial measures

Abnormality	Type	Remedial measures
Apparent Defects: Visible and invisible *Most important aspect is the effective cleaning regularly, and using our 5 senses: ear, nose, eyes, feel and touch, which would reveal the abnormalities that could be rectified.*	• Deposit & Accumulation • Damage & Deformation • Play & Gaps • Sway & Slackness • Abnormal Phenomenon • Adhesion & Restriction	• Know the equipment, its components & operational details fully. • Identify the abnormalities by thorough inspection using senses for abnormal look, noise, vibrations, and smell/odor, if any. • Search for Invisible/visible defects. • Note abnormality, if any, and establish its root cause. If need arise apply: Why-Why, or 5W-2H Analysis or any other technique. • Remove them applying appropriate measures; and where not feasible put a tag for the abnormality identified, and rectify it at the earliest.
Basic Non-fulfillment *Ignoring these defects could be the main causes of bottlenecks preventing rated output or planned production from the plants.*	• Lub./Coolant/Priming • Supply & Systems • Inadequate Gauges • Inadequate Tightening • Improper Measuring Instruments • Unmatched Equipment & Machines • Defective Design • Lack of Maintenance	Following guideline could be useful: • Ensure that all meters operate correctly & are clearly marked with specified values. • Investigate any leaks of product, steam, water, oil, compressed air etc. • Hunt for scaling blockages inside chutes etc. • Adhere to routine, scheduled, preventive and predictive maintenance.

(Continued)

Table 18.10 Continued.

Abnormality	Type	Remedial measures
		• Ensure Periodic overhaul, Checking, parts' replacement. Pay attention to slight defects for example while cleaning check for bolts – are they loose? Missing? Is length adequate and protrude 2–3 threads length from the nut? Are washers of suitable type used?
Non-approachable Places *They are the hindrances to run the plants and equipment for their smooth running, and causes delays.*	• For Cleaning • For Checking • For Tightening • For Operation • For Adjusting	• Note whether equipment is easy to clean, Lubricate, Inspect, Operate & adjust (Identify hindrances such as large obstructive covers, ill positioned lubricators etc.)
Outside Contaminants from: *They interferer smooth running of the plant and machinery, They could be also responsible for quality defects in some cases.*	• Raw Materials • Lubricants • Gases • Liquids • Scrap materials • Other Carriers	• Keep the vicinity of the equipment in order and tidy. • Remove unnecessary things and attachments, if any. Check while Cleaning Lubrication System • Storage: Is lubricant storage clean? • Is lubricant container always capped? • Lubricant Inlets: Are grease/lubrication points kept clean? • Are the inlets labeled with correct type/quantity? • Level Gauges: Are the gauge glasses/indicators kept clean? • Is the correct oil level clearly marked?
Materials Obsolete & Non-Urgent *There have been number of accidents when due attention has not been paid to guidelines, given in column 4.*	• Machinery & Equipment • Interface Connection & • Piping • Electrical Equipment • Jigs & Tools	These are auxiliary machines/equipment/tools/appliances which play vital role in the running of plants and equipment smoothly. Following guidelines are useful: • Their proper selection and matching with the system must be ensured. • Equally important is their maintenance and timely regular replacement. • Substandard quality of material when used for their construction often leads to failures and proves to be problematic and could cause losses of different types. • Used and worn-out items should not be used as replacement. • Outdated and mismatched items are often not reliable, and should not be used.

depicts abnormalities of various types and suggested measures to deal with them. It could be used as a guide for the industry including mines to which you belonging to.

18.6.3   5W-2H Analysis

Table 18.11   5W-2H details.[31,36]

5W-2H Analysis

Who? – It identifies individuals associated with the problem. Also clarify who is complaining. And which operators are having difficulty?

What? – Describe the problem adequately. Does the severity of the problem vary? Are operational definitions clear (e.g. defects)? Is the measurement mechanism is effective and accurate?

When? – Identifies the time when the problem started and its prevalence in earlier time periods. Do all production shifts experience the same frequency of the problem? What time of year does the problem occur?

Where? – If a defect occurs on a part, where is the defect located? A location check sheet may help. What is the geographic distribution of customer (or those who are concerned) complaints?

Why? – Any known explanation(s) contributing to the problem should be stated.

How? – In what mode or situation did the problem occurred? What procedures were used?

How Many? – What is the extent (frequency) of the problem?

Table 18.12   Losses of various kind, their reason and remedial measure to minimize them.

Losses types	Reasons & measures to minimize them
Production related issues	**Deviation from best practices or excellence to achieve planned (Rated Production).** To achieve targeted production, deviation from the practices, as outlined below could result losses, or reduced profits:
Increase in the cost of production is resulted due to reasons given in col. 2 Also refer sec. 18.6.5 and figures 18.19 and 18.20.	• Quick handling/removal of the finished goods is essential for safety and productivity, and therefore, methods allowing this feature should be preferred. • Concentrating the activities within a compact layout and deploying resources there It can yield better productivity i.e. output/man/shift due to effective supervision and better coordination and minimum movement. • Proper match of equipment, methods, techniques and layouts brings optimum results. • Production rates: Higher output rates can reduce expenses on services, over-heads and fixed costs. However, this is governed by the designed capacity and market demand. • Minimize cost of production by keeping minimum inventory. Selecting safe, eco-friendly, technological sound and economically viable equipment always pays. • Minimize the schedule required to achieve rated production i.e. pre-production or gestation period.

(Continued)

Table 18.12 Continued.

Losses types	Reasons & measures to minimize them
	• Maximize flexibility and adaptability at the manufacturing units. Less Production than targeted due to: • Shutdown or frequent stoppage • Frequent interruptions due to failure of equipment (keep standby ready, following maintenance schedules and practices strictly could minimize this problem) • Mismatch of unit operations. Excessive production: more than targeted • Profitable if market is favourable; else • Results in blocking capital. Usually creates storage and handling • Concentrating on marketing and sales' promotion strategy problems could solve this problem.
Lack of effective utilization of time***	Improper allocation, ambiguity in job profile and responsibility assigned, sequencing operations improperly, mismatch amongst equipment, layout and working conditions, and absence of clear instructions are some of the reasons for time-lost and delays. In addition, following guidelines are useful: • Punctuality always pays. • Every moment is precious*** and could be utilized to add value. • Time has direct relation with speed; accomplishing any task at a faster speed means adding to the productivity. • Waiting time should be minimized. • Undue delays and interruptions add costs and overall dissatisfaction amongst those involved. *** - **Time is precious as evident by the following facts:** *1) To realize the value of ONE YEAR – Ask a student who failed a Grade* *2) To realize the value of ONE MONTH – Ask a mother who gave birth to a pre-mature baby* *3) To realize the value of ONE WEEK – Ask the Editor of a weekly newspaper* *4) To realize the value of ONE HOUR – Ask the lovers who are waiting to meet* *5) To realize the value of ONE MINUTE – Ask a person who missed a train* *6) To realize the value of ONE SECOND – Ask a person who just avoided an accident* *7) To realize the value of ONE MILLI SECOND – Ask a person who won a silver medal in the Olympics*
Human Efficiency (Effectiveness) – Man-Management	It is a measure of effectiveness of a system, which includes personal efficiency; equipment or plant efficiency. • Maximize mechanization, wherever feasible

(Continued)

Table 18.12   Continued.

Losses types	Reasons & measures to minimize them
	• Maximize automation (deployment of remote controlled equipment) • Computer Aided Design (CAD) found to be effective, as such use wherever appropriate. • Adverse working conditions add to inefficiencies. Use of robotics and automation are the viable alternatives, if practicable in a given situation. Ref. sec 19.5.8.
Poor Recovery (Yield)	Recovery is a vital factor that measures overall yield out of the input resources. **Process Recovery = (Process losses)/(Input);** Whereas, **Process loss = Input Quantity − Output Quantity.** Process losses are the function of applied method, technique and equipment to accomplish an operation or task; particularly when applied in conjunction with extraction of natural resources such as minerals: Petroleum, natural-gas, and solid minerals. • Better is recovery higher would be the realization (i.e. the financial gains). • Optimize recovery by minimizing dilution and contaminants. Contaminants are due to over-break in mining. Contaminants are also brought by pollutants' ingress and wind storms.
Imbalance amongst production, productivity and HSE	Productivity is a function of effective utilization of available resources to accomplish a job. It is expressed as: **Productivity = Output/Man/Shift.** (18.2) Thus, it is a measure of amount of work per unit time by a man; and also by a machine or equipment. • Carry out scientific studies to reduce cycle time for various unit operations. • Maximize natural supports such as gravity, wind direction, surface terrain and topography to boost productivity. Equally important is the safety of man (directly involved and third-party), equipment and process. • Failing which, it directly affects production targets, productivity and ultimately the costs. • Inadequate safety results in accidents. Accident has indirect costs, it degrade moral of people involved and also brings bad reputation to the company. • Pollution has adverse impact on workers' health that reduces their efficiency.
Equipment and plants performance Refer sec. 18.6.6 and figure 18.21 also	As described above, equipment play most important role in prevention of losses and in cost reduction as well. Equipment availability and its effective utilization is the key for success. World Class Management (WCM) recommends Autonomous Maintenance System (AMS), which aims at: • Developing a team who could take care of plants and equipment. Team is trained to diagnoses abnormalities, troubles and deficiencies of the equipment and ways to remove them and bring the equipment its ideal condition. • And then running it to its full efficiency regularly without interruptions.

(Continued)

Table 18.12   Continued.

Losses types	Reasons & measures to minimize them
Man-Management	Table 18.5 describes meaning of word 'MANAGE', specifying as how efficiently this resource could be managed. The underlying reasons for it are: • When human resources are managed effectively overall gain to the company is obvious. • Mismanaged companies ultimately lands into losses, closures and industrial disputes and unrests.
Use of substandard Materials & auxiliary machines & spares	Input material and its quality play an important role in preventing losses. Substandard material has following demerits: • It can damage the process, equipment and plant. • It can endanger safety of not only those involved but to the third party. • Operational failures could be greater. • Thus, it could cause losses rather than savings. • It must be avoided. • Timely procurement of right-quality material, its proper storage, handling and issue system is equally essential to avoid losses.
Inadequate internal as well as external infrastructures	Any factory, mill, or mine could be compared to a busy city with a concern for water, light, power, communication, transportation, supplies, sewerage and construction. • As such any shortcoming or interruptions to these services results into losses.
Quality defects Refer table 18.9 for more details.	Quality of any product is as important as its quantity. It is a function of so many parameters and deviation from them result defects or a substandard product. In achieving quality product following pays important role: • Preventing (or minimizing) mishandling, dilution, foreign matters and moisture. • Effective controls on grain size (in case of solid products) and viscosity (in case of liquid products) should be adhered to. • Strict supervision, laid-out tests and checks by the internal as well as external agencies should be mandatory. • Process control mechanisms must work effectively. Malfunction of process control meters and gauges should not be tolerated. • Variability/non-uniformity in the process often leads to quality defects, and must be removed. • Defective products are either rejected or reprocessed which are non-value adding activities adding costs. They are non-productive and causes delays in delivery. It is usually becomes cause of customers' complaints and dissatisfaction. Product becomes non-reliable. • Producers' motive must be to improve the quality rather than leaving any scope of defects. Research and Development (R & D) should be part and parcel of quality improvement and control measures.

(Continued)

Table 18.12   Continued.

Losses types	Reasons & measures to minimize them
Frequent interruption, inconsistencies & shutdown **Refer sec. 18.6.6 and figure 18.21 also**	Unplanned and unscheduled layoffs, lockouts, shutdowns are the evils which must be prevented. • It brings non-value additions and unrest amongst those involved – the workers, management and stakeholders. • It could be due to mismanagement, lack of proper planning and foresight resulting huge losses. • It must be avoided. During planned or scheduled shutdown • Ensure delivery of supplies, effective communications and supplier's presence, to avoid delays. • Proper planning including use of Critical Path Method (CPM) is helpful to monitor the progress. It should include resources: men, tools, appliances and material handling mechanism, which should be readily available. • Use experienced crews during shutdowns to do right job within the time allocated.
Defective design & layouts; and lack of planning	It is always sensible to prepare different designs, alternatives and options during economical studies – feasibility and DPR stages, and select the best. • Spending a dollar during this phase could prevent losses of many dollars if design is defective and the equipment has been procured and plant/mine has been set. • Input from the experienced experts and professionals during conceptual, feasibility, engineering studies or detailed project report (DPR) stages pays. • Eco-friendly design should be incorporated wherever practical. Such designs pay in the long run, and are sustainable. • Inadequate internal as well as external infrastructures also add to poor layouts.
Energy losses & excessive material consumption	*Energy & material consumption:* • The methods, which involve higher consumption of materials and consumable, are often costlier; and should be avoided. • Another consumable item is energy/unit of output, which is required to carry out different unit operations and services such as ventilation, pumping, illumination etc • Minimize power distribution and transmission losses • Minimize wastage of water. It would add to savings beside judicious use of this precious resource. • Pollution means more energy has been already spent than required. Least pollution and clean environment are the keys of success. • It also includes selection of an appropriate energy source: electric, diesel, compressed air, hydraulic or non-conventional (if available), which could bring overall benefit to the company.

(Continued)

Table 18.12 Continued.

Losses types	Reasons & measures to minimize them
Lack of tools, techniques & renovations	• Lack of tools, techniques (up to date technology) and renovations is like a solder at the front without adequate and up-to-the-mark arms, ammunitions, weapons and know-how. And imagine if the enemy is equipped with all these fully. • **'Kaizen'** is applied to remove or minimize bottlenecks in the process, technique, equipment and machines by making use of existing resources. It should be encouraged. • **Renovation** is change in technology, method and equipment, which can bring improvement in the system and results into increase in profits as well as safe and clean environment. For the growth and to compete in the market renovations are essential. Periodical renovations are almost mandatory in any industry to improve quality as well as output rates.
Lack of attention to the abnormalities	Table 19.9 describes types of abnormalities that need to be given immediate attention and must be removed and rectified failing which losses of various kinds could occur.
Wastage	Abnormal wastage generation results losses. Details are given in sec. 18.6.4.

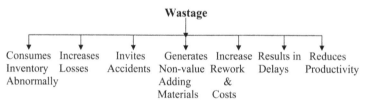

Figure 18.18 Impacts of wastage.

18.6.4 Wastage[28,31,36]

Waste is a natural phenomenon, which is part and parcel of any process or system. To understand let us consider human-body. We take in food and drinks which are converted into many useful products such as blood, bones, flesh and many more; that keeps us alive and out of the total-intake-quantity; the useless components are thrown out of body in the form of stool, urine, perspiration, etc. But imagine if we overeat and drink, can we keep our body balanced? Rather there would be side effects and illness.

In any industrial process also the same concept is applicable while inputting certain ingredients; a final product is obtained together with foul gases, heat, noise, and also solid waste in some cases. Efficiency lies in converting more and more input material into useful products and generating wastes as little as possible as it has got no value rather its handling, disposal and storage costs. When produced in excessive quantity, it degrades environment. The excessive production of waste (except in case of mining and petroleum industries where it is a natural phenomenon); indicates drawback in the technology, or systems. It is non-value adding activity that adds to the cost of production, and hence it is a loss. The line-diagram, figure 18.18, lists the impacts of wastage.

Structure of Cost-Loss Matrix:

Total Losses	Loss Category	Availability Loss			Performance Loss					Quality Loss				Management Loss			
		1	2	3	4	5	6	7	8	9	10	11	12	13	14	15	16
	Losses	Shutdown Losses	Production Adjustment Losses	Equipment Failure Losses	Normal Production Losses	Abnormal Production Losses				Quality defect Losses	Reprocessing Losses			High Inventory Losses	Substandard Material Losses	Substandard tool & Equipment Losses	Man Management Losses
	Stoppage Days																
	Qty Loss (MT)																
	Loss in (***)																
A] Total Variable Cost 1. Raw Material																	
2. Utilities																	
B] Total Fixed Expenses																	
1. Salary																	
2. Stores & Repair																	
3. Overhead																	
Total [A +B]																	

Figure 18.19   Assessment of abnormalities and the resulting financial losses on various accounts. *** – Numerical figures could be in the currency to which the mining/tunneling unit belongs to.

The abnormalities and losses of various types results into financial losses, which could be assessed as illustrated in figure 18.20, which is a graphical illustration for analyzing losses of four consecutive years.

18.6.5   Case-study illustrating computation of financial losses[18,31,36]

The abnormalities and losses of various types' results into financial losses, which could be assessed as illustrated in figure 18.19. Figure 18.20 is a graphical illustration for analyzing losses of four consecutive years in a manufacturing unit. It provides basis as how various types of losses could be taken into account to assess the financial implications of an industrial setup.

There is link amongst the inputs that includes Human Resources; Plant and machinery; Raw Materials – natural (such as minerals, fossil fuels, flora; and man-made materials which are innumerable); Energy including the natural resource air and water and infrastructures to accomplish production from any industrial set-up including the industry to which one may belongs to.

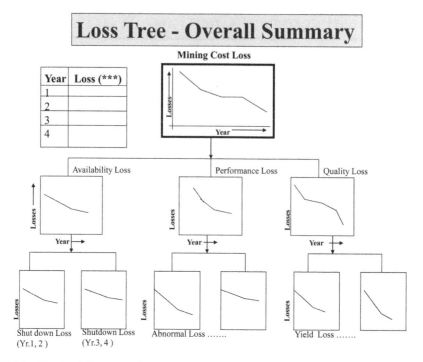

Figure 18.20   Graphical illustration for analyzing losses of four consecutive years. *** – Numerical figures could be in the currency to which the mining/tunneling unit belongs to.

18.6.6   Use of information technology (IT) in integrating processes and information[11]

Initiative to integrate processes and information has already begun which can reduce loses of various kinds as detailed preceding sections. More than 300 mining companies including those who are amongst the top 10 are using software popularly known as *'ERP – Entrepreneur Resources Planning'* which has potential to revolutionize processes and optimize resources. Such standardization and integration strategies should be considered by every mine owner. Figure 18.21, details ERP concept which could be helpful in achieving the following benefits:

- Operational costs reduction
- Faster and better decision making based on accurate on-time information
- Streamlined processes (inventory reduction, procurement cost reduction etc.)
- Improved plant operations with better asset availability and usage
- Total cost of ownership reduction
- Credit losses reduction
- Integrated visibility into financial, sales, manufacturing, and supply chain information
- Flexibility and visibility to better react to company growth and market changes.

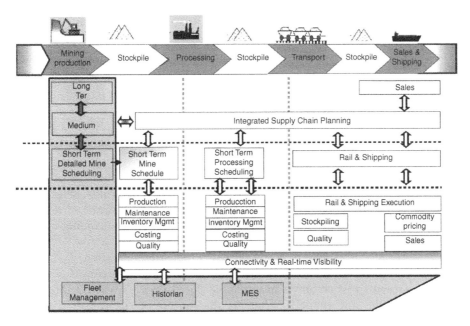

Figure 18.21   ERP concept[11] for faster and better decision making based on accurate on-time information – an efficient tool to optimize resources, as detailed in figure 18.1.

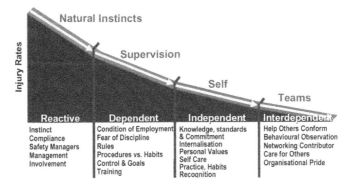

Figure 18.22   It is teamwork that could minimize injuries, which means losses (After Latus, M. 2006).[38]

18.7   THE WAY FORWARD[38]

The key-words shown in figure 18.22 have following meanings: Reactive – action in response (which is the natural human tendency); Dependent – one who depends upon others (it has element of fear that if not obeyed as instructed disciplinary action against the concerned could be there); Independent – not dependent (empowering with responsibilities works equally well. It provides an opportunity for self-development of the individual, and also to show his worth); Interdependent – mutually dependent; illustrating the fact that it is the *teamwork* which could bring desired results out of the various options available to us.

QUESTIONS

1. '5S' is a philosophy; list its phases, and outcome when implemented.
2. 'Working in the hazardous industries is a big challenge' – how can we meet this challenge?
3. 'Accidents are like mosquitoes, it is best to drain the swamps in which they breed'. And what are the swamps? Are they unsafe acts and unsafe conditions? How these could be minimized?
4. An accident is a three-step event – list those three phases/steps.
5. Are 'Benchmarking & Standardization' essential to run any mine or a tunneling project?
6. Changes in human behavior with respect to safety could improve the situation; in this conjunction the concept: DO IT could be helpful. Describe this concept in detail.
7. Classify losses of various types. How could 'minimizing losses of various kinds' help in achieving aims and objectives of an industrial setup including mines and tunneling projects?
8. Classify the resources used to accomplish any industrial operation.
9. Define accidents. List the main reasons. List the costs of accidents.
10. Define hazards. List common hazards associated with the excavation industry including mines and tunnels. What are the sources of electrostatic hazards in these industries? How they can be minimized?
11. Define terms: Severity Rate. Frequency Rate.
12. Describe as how substandard behaviour and workplace accidents are interrelated?
13. Develop an Emergency Plan for a mine or tunneling project and present it in the form of a line diagram.
14. Does underground air differs from the atmospheric air? If, yes, in what manner?
15. During design phase what techniques concerning safety of operations are usually applied. List them.
16. Ergonomics is the process of making things user-friendly – comment.
17. The ERP concept is meant for faster and better decision making based on accurate on-time information – an efficient tool to optimize resources. List the benefits that could be achieved through it.
18. Give a line diagram by including the critical elements/ sub-elements of emergency preparedness
19. Give a line diagram classifying the potential hazards while creating surface and underground excavations.
20. Heinrich (1931) suggested 3Es; Education, Engineering and Enforcement to establish proactive safety culture. In fact the solution lies, in practicing truly the 5Es to establish proactive work culture including safety; elaborate these 5Es.
21. How is application of ergonomic principles useful to us?
22. How is establishing the root causes of incidents leading to losses helpful in accident prevention?
23. How do explosions differ from fires? List hazards analysis methods and write their applications during various phases of an industrial setup.
24. How is good housekeeping the key to safety and good health?

25. How could safety education be imparted to mine workers and hydro power project workers?
26. How should senor management focus on HSE management? Do you agree successful HSE management reflects teamwork? Elaborate it.
27. Human failure is responsible for most of the incidents which may result in injuries or/and losses. Represent these failures by a flow/line diagram.
28. Illustrate by means of a line diagram the impacts of 'wastage'.
29. Illustrate how the letters of word (the literal meaning) 'Manage' itself depicts (proposes) ways to manage any mine or tunneling project?
30. In an 'Environmental Management Plan', give details of each of the following items covered under it: Base-line information, Environment Impact Assessment and Management Plan.
31. In the absence of proper education and training what could result? Do you think they are expensive?
32. In general, what are the main reasons or causes of accidents in the industry or stream you belong to: petroleum, civil, construction, mining, or any other (you belong to). What are the risks involved in accidents. Also list safety wear you will recommend for a worker working in your stream.
33. In the USA mining is the oldest industry, but during 1997 alone more than 5000 people lost their lives in work related incidents and more than 3.8 million suffered disabled injuries and illness resulting in a direct cost to US industry nearly $128 billion with estimated direct and indirect costs of $600 billion. Today it is considered amongst the safest industries as reflected by industries' fatality and injury rate. List the measures that have been taken.
34. In which circumstances does training become mandatory?
35. List the 'best practices' which should be implemented to run mines/tunneling projects.
36. List the 'Three Pillars of equal strength' for loss prevention.
37. List the degree (type) of injuries.
38. List fundamental principles of industrial hygiene. List steps for managing industrial hygiene.
39. List the hazards associate with underground mining and tunneling operations.
40. List the impacts of good ergonomics as well as poor ergonomics.
41. List the natural gases from underground strata and their impact when mixed with air.
42. List reasons of 'quality defects'.
43. List the reasons for pollution and prepare a line diagram to illustrate pollution caused by excavations / mining operations.
44. List the reasons of suffering from some occupational disease while working in civil, construction, mining and tunneling industries.
45. List the steps involved in periodic health surveillance based on exposure-risk.
46. List the steps that should be followed to run any operation or system effectively.
47. List the type of layers that are usually provided between hazards that may arise by any process; and the people, property and environment surrounding it.
48. List the type of abnormalities which are usually found in industries, in general. How is assessment of abnormalities and the resulting financial losses useful?

49. List the types of losses in mining /tunnelling project, and their reasons. Suggest measures to minimize them.
50. How you can achieve human efficiency? List ways to engage any individual fully to get the maximum for the benefit of both: the employer as well as the employee – A proactive work culture.
51. Mining operations or civil excavations made for various purposes are site/location/ locale specific. Is it true? How do these features differentiate them from other industries?
52. Operations at mines and tunnelling sites are governed by the prevalent scenario in and around their geographic locations; and the rest of the parameters are inter-related as detailed through a '9G – Evaluation'. What is this evaluation? Detail it.
53. Prepare 'What if' check list – for risk analysis.
54. Prepare an 'Environmental Management Plan' including a brief summary of steps that need to be taken during all the three phases; for petroleum, mining, or construction, or any other industry that you belong to.
55. Propose a 'Three-Day – Total Accident Control Training program' for shop-floor workers and supervisors considering industry you belong to.
56. Safety strategies can be classified into four categories: Inherent, Passive, Active and Procedural. Explain each one of them.
57. Summarize the integrated approach as proposed by Mick Cairney.
58. Tabulate the different phases of the working life of an employee and the variables influencing his/her health during this tenure.
59. Tabulate measures used to reduce or limit the consequences arising from hazards.
60. Tabulate various hazards analysis methods? Give Application of each one of them during various phases of an industrial setup.
61. We all agree: 'Prevention is always better than cure'. Should we not make our utmost efforts in minimizing: emissions, incidents, wastage, injuries & illness; approaching their magnitude to almost zero? How it could be achieved?
62. What are 'Notified Diseases'?
63. What are the sources of heat in underground mines and tunnels?
64. What are workplace stresses, their adverse impacts and way outs?
65. What features does a system incorporate – list them.
66. What is 'organizational culture' and what are its impacts?
67. What is 'Presenteeism' and what are its impacts?
68. What is significance of 'Diagnosis Technique – 5W-2H'?
69. What is the modern technique to assess risk?
70. What is the most valuable asset (resource) to run any industry efficiently?
71. What logic you would follow to determine the training needs of your subordinates?
72. What procedure should be followed in the event of outbreak of fire? Prepare 'Standing Orders' for the mine/organization you belong to.
73. What should be your process risk management strategies?
74. What types of dusts are responsible for fires and explosions?
75. What types of laws are necessary in mining? List them. Prepare a checklist on: legal, statutory & other compliances, as applicable to your mine/tunneling project.

REFERENCES

1. Arnold, R.M.: Measurement techniques in safety management. In: *Mine health and safety management*, Michael Karmis (edt.), Littleton, Colorado, SME, AIME, 2001, pp. 51–57. www.smenet.org

2. Amyotte, P.: Orientation course, Dalhousie University. Overview of Hazards methods, Supplementary notes, 1997.

3. Bailey, K.F.: Industrial hygiene in mining. In: *Mine health and safety management*, Michael Karmis (edt.), Littleton, Colorado, SME, AIME. 2001, pp. 263–73. www.smenet.org.

4. Baldermann, G.: Occupational Health & Safety – An Element of Effective Business Management *International conference focusing on safety and health on 14–16 November, 2006, Johannesburg, South Africa. London ICMM.*

5. Baruer, R.L.: Engineering for health and safety. In: *Mine health and safety management*, Michael Karmis (edt.). Littleton, Colorado, SME, AIME, 2001, pp. 83–86.

6. Bridger, R.S.: *Introduction to Ergonomics*. McGraw-Hill, Inc. 1995.

7. Cairney, M.: Maintain a safer mining industry and organizational commitment. Pres. In: *'Mining 2020', International Mining Conference, 5–6 Sept. 2007, Sydney, Australia. Organizer: AIMEX.*

8. Canadian Standards Association: Emergency Planning for industry, 1995.

9. Dan MacLeod's Ergonomics Website, http://www.danmacleod.com/ [Accessed 2009].

10. Directorate General of Mines' Safety, India; Circular on: Notified Diseases and preventive measures and interaction with authorities, 2008.

11. Gaba, V. & Chaturvedi, N.: Integrated ERP software solutions enable the Mining Industry to meet future business challenges. Pres. In: International Seminar: 'Mine Advan. Tech. Jodhpur, India, 14–16 February, 2009, pp. 229–36

12. Geller, E.S.; Carter, N.; DePasquale, J.; Pettinger, C. & Williams, J.: Application of behaviour science to improve mine safety. In: *Mine health and safety management*, Michael Karmis (edt.). Littleton, Colorado, SME, AIME. 2001, pp. 67–78.

13. Germain, G.L. & Arnold, R.: Management strategy and system for education and training. In: *Mine health and safety management*, Michael Karmis (edt.). Littleton, Colorado, SME, AIME. 2001, pp. 128–144. www.smenet.org

14. Gupta, J.P.: Safety course at Sultan Qaboos University, Oman. 2000.

15. Hethmon, T.A. and Doane, C.W.: Health and safety management. In: *Mine health and safety management*, Michael Karmis (edt.). Littleton, Colorado, SME, AIME. 2001, pp. 17–32. www.smenet.org

16. Inherently safer chemical processes. Daniel A. Crowl (edt.), New York, 1996; pp. 19–23.

17. Jager, K.D.: Wellness in the Workplace. Pres. In: *International conference focusing on safety and health on 14–16 November 2006, Johannesburg, South Africa.* London ICMM.

18. Jansen, J.: First Steps on the Journey to Zero Harm. *International conference focusing on safety and health on 14–16 November, 2006, Johannesburg, South Africa. London ICMM.* (Permission: Lonmin).

19. Kroemer, K.H.E.; Kroemer, H.B. & Kroemer, K.E.: *Ergonomics, how to design for easy and efficiency*, Prentice Hall, 1994.

20. Kupchella, C.E. & Hyland, M.C.: *Environmental Science*. Prentice-Hall International Inc. 1993, pp. 558.

21. Martin, D.K.: Incident reporting and analysis. . In: *Mine health and safety management*, Michael Karmis (edt.). Littleton, Colorado, SME, AIME. 2001, pp. 183–193. www.smenet.org

22. McDonald, R.: Health and Productivity Management. Pres. In: *International conference focusing on safety and health on 14–16 November 2006, Johannesburg, South Africa. London ICMM.*

23. Pegg, M.J.: Safety class-notes and course at Sultan Qaboos University, Oman, 2000–03.
24. Petroleum Development of Oman – HSE management, (1996–2004).
25. Philip J. Stapleton and Margaret, A.: Glover Environmental Management Systems: An Implementation Guide for Small and Medium-Sized Organizations. S. Petie Davis, NSF ISR; 789 N. Dixboro Road; Ann Arbor, MI 48158; 1-888-NSF-9000
26. Sevren, T.: Safe production. *International conference focusing on safety and health on 14–16 November, 2006, Johannesburg, South Africa. London ICMM.*
27. Standish, P.N. & Reardon, P.A.: Semi-Quantitative Risk Analysis for underground development projects, International Tunnelling Association, 2002.
28. Sustainable Development Framework – Indian Mining Initiative (Draft Report), Federation of Indian Mining Institute (FIMI), 2009
29. Tatiya, R.R.: *Civil Excavations and Tunneling – A Practical Guide.* Institution of Civil Engineers/Thomas Telford Ltd. London, U.K. 2005, pp. 273–305.
30. Tatiya, R.R.: Health, Safety and Environment (HSE) Management – Where do you stand? *Souvenir: 17th* Environment Week Celebration, Indian Bureau of mines, Rajasthan region, 2006, pp. 20–26.
31. Tatiya, R.R.: 'Loss Prevention' – A need of hour in the Indian Mining Sector. *National conference on latest trends in equipment, technology and management in mineral sector; Indian Mining and Engineering Journal, 11–12 May, 2009, Bhuwneswar, India.*
32. Tatiya, R.R.: *Elements of Industrial Hazards – Health, Safety, Environment and Loss Prevention* CRC Press, Taylor & Francis Group, London, U.K. 2011, pp. 179–385.
33. Tatiya, R.R.: *Course material for Petroleum and Natural Resources Engg; Chemical Engg. And Petroleum and Mineral Resources Engg.* Sultan Qaboos University, Oman.
34. *The Ergonomics Kit for General Industry*, Second Edition, Taylor & Francis, *2006.*
35. *Workplace safety.* John Ridley & John Channing (edts.), Butter Heinmann, 1999, pp. 30–80.
36. World Class Management (WCM) – Leaflets, displays, lectures, and talks on various topics – Essel Mining and Industries Limited, Aditya Birla group, India (2004–2008; through interaction, and participation)
37. Wrigley, D.: Silica Control in Southern Africa. Pres. In: *International conference focusing on safety and health on 14–16 November 2006, Johannesburg, South Africa. London ICMM.*
38. Latus, M.: Leadership in Safety – One Company's Approach. *International conference focusing on safety and health on 14–16 November, 2006, Johannesburg, South Africa. London ICMM.*

Chapter 19

Sustainable Development

The need of the hour is to develop 'Professionals' who could take care of society (stakeholder) by producing ore at the desired rate safely with maximum productivity and recovery together with minimum-cost to fulfill the needs of people. This approach would lead to "Sustainable Development (SD)" which is beneficial economically, ecologically and socially to the present as well as future generations. Should we not strive for it wholeheartedly?

19.1 SUSTAINABLE DEVELOPMENT (SD) IN MINING[15]

19.1.1 Sustainable Development

The meaning of the term 'Sustainable Development (SD)', which is in general acceptable development that is *beneficial economically, socially and ecologically* to the present as well as future generations (fig. 18.1(b)). It is easy to define SD, but extremely difficult to implement it; as its time-frame is unlimited, it involves natural as well as man-made resources (fig. 18.1) and the task is to be carried by direct as well as indirect stakeholders.

19.1.2 Global issues & backlog on sustainable development

It is the industrialization in the immediate past, say within the last two centuries that has given rise to global issues such as: acid rain, ozone depletion, bad ozone, global warming, air and water pollutants that are causing huge expenditure on health care and remedial measures. It is this negligence and carelessness of the past that is compelling implementation of sustainable development and hence, there is a backlog so far as SD is concerned in any industry, including mining, tunneling and construction industries.

19.1.3 Sustainable Development in mining[15,22]

Concerning SD in mining and construction industries, the task is made more difficult by virtue of inherent features of these industries being rough, tough and hazardous. The output from mines is not only minerals but also generation of gaseous emissions, liquid effluents, solid wastes, radiations, particulate matter, heat and noise. Equally associated with them are hazards such as fires, explosions, inundation, accidents, disasters and a few others. All these are detrimental to health not only to direct and

Figure 19.1 The four elements which are essential to achieve sustainable development include: Production at the desired (targeted) rate, productivity, HSE and social welfare and least losses of various kinds.

indirect stakeholders particularly when the allowable limits are exceeded but also to the biotic and abiotic components of nature. Growing health problems of the world's citizens and global issues (problems) such as acid rain, ozone depletion, photochemical smog, acid drainage and global warming are witnessing this fact.

Mining is also economically critical for millions of the world's poorest people with some 50 countries being significantly dependent on mining. "Politics, economics, and governance are all likely to have a bearing on the conduct of the mining project and its economic and social contribution." In order to enhance a positive contribution and minimize negative impact to society (stakeholders); the contribution of mining companies to the following areas becomes almost mandatory: poverty reduction; revenue generation; economic development: regional as well as local areas; mining & disputes resolution.

The mineral resources of today are the results of natural processes of billions of years and mineral wealth belongs to every one of us, no matter where it is located. And who owns it? But with the prevalent rate of consumption (as described in the following sections); they would be depleted within the next few centuries. Thus, its proper exploration, systematic development and exploitation and judicious utilization are our moral responsibilities.

Depleting mineral resources, Global completion and recession, increasing depths, poor recoveries, remoteness of mining sites, shortage of skilled man-power and few others (as described in sec. 19.3.6) are some of the challenges that the mining industry is already facing and warrant immediate solution. These are the challenges to those who want to remain with this industry. In fact SD in any industry including mining needs system, culture, education and discipline amongst those concerned – the producers and the consumers.

The solution lies in adapting a strategy which brings about a thorough balance amongst cost effective mining practices, its adverse impacts mitigation measures (ecological balance) & social welfare (fig. 19.1) which is a key to 'sustainable development'. This chapter attempts to cover these aspects so that the world community welcomes the 'mining community,' and sustainable development in mining industry becomes a reality.

19.2   STAKEHOLDERS AND SUSTAINABLE DEVELOPMENT[22]

As stated above, equally important is the welfare of society, the stakeholders who are directly or indirectly affected by the mining operations. For this aspect, the 10 principles as suggested by the International Council of Minerals and Metals (ICMM) could be a useful guide to satisfy both the employees as well as the stakeholders.

19.2.1   Principles/guidelines for SD by ICMM[13,22]

Principle 01: Implement and maintain ethical business practices and sound systems of corporate governance.

- Develop and implement company statements of ethical business principles, and practices that management is committed to enforcing.
- Implement policies and practices that seek to prevent bribery and corruption.
- Comply with or exceed the requirements of host-country laws and regulations.
- Work with governments, industry and other stakeholders to achieve appropriate and effective public policy, laws, regulations and procedures that facilitate the mining, minerals and metals sector's contribution to sustainable development within national sustainable development strategies.

Principle 02: Integrate sustainable development considerations within the corporate decision-making process.

- Integrate sustainable development principles into company policies and practices.
- Plan, design, operate and close operations in a manner that enhances sustainable development.
- Implement good practice and innovate to improve social, environmental and economic performance while enhancing shareholder value.
- Encourage customers, business partners and suppliers of goods and services to adopt principles and practices that are comparable to our own.
- Provide sustainable development training to ensure adequate competency at all levels among our own employees and those of contractors.
- Support public policies and practices that foster open and competitive markets.

Principle 03: Uphold fundamental human rights and respect cultures, customs and values in dealings with employees and others who are affected by our activities.

- Ensure fair remuneration and work conditions for all employees and do not use forced, compulsory or child labour.
- Provide for the constructive engagement of employees on matters of mutual concern.
- Implement policies and practices designed to eliminate harassment and unfair discrimination in all aspects of our activities.
- Ensure that all relevant staff, including security personnel, are provided with appropriate cultural and human rights training and guidance.
- Minimize involuntary resettlement, and compensate fairly for adverse effects on the community where they cannot be avoided.
- Respect the culture and heritage of local communities, including indigenous peoples.

Principle 04: Implement risk management strategies based on valid data and sound science.

- Consult with interested and affected parties in the identification, assessment and management of all significant social, health, safety, environmental and economic impacts associated with our activities.
- Ensure regular review and updating of risk management systems.
- Inform potentially affected parties of significant risks from mining, minerals and metals operations and of the measures that will be taken to manage the potential risks effectively.
- Develop, maintain and test effective emergency response procedures in collaboration with potentially affected parties.

Principle 05: Seek continual improvement of our health and safety performance

- Implement a management system focused on continual improvement of all aspects of operations that could have a significant impact on the health and safety of our own employees, those of contractors and the communities where we operate.
- Take all practical and reasonable measures to eliminate workplace fatalities, injuries and diseases among our own employees and those of contractors.
- Provide all employees with health and safety training, and require employees of contractors to have undergone such training.
- Implement regular health surveillance and risk-based monitoring of employees.
- Rehabilitate and reintegrate employees into operations following illness or injury, where feasible.

Principle 06: Seek continual improvement of our environmental performance

- Assess the positive and negative, the direct and indirect, and the cumulative environmental impacts of new projects – from exploration through closure.
- Implement an environmental management system focused on continual improvement to review, prevent, mitigate or ameliorate adverse environmental impacts.
- Rehabilitate land disturbed or occupied by operations in accordance with appropriate post-mining land uses.
- Provide for safe storage and disposal of residual wastes and process residues.
- Design and plan all operations so that adequate resources are available to meet the closure requirements of all operations.

Principle 07: Contribute to conservation of biodiversity and integrated approaches to land use planning

- Respect legally designated protected areas.
- Disseminate scientific data on and promote practices and experiences in biodiversity assessment and management.
- Support the development and implementation of scientifically sound, inclusive and transparent procedures for integrated approaches to land use planning, biodiversity, conservation and mining.

Principle 08: Facilitate and encourage responsible product design, use, re-use, recycling and disposal of our products

- Advance understanding of the properties of metals and minerals and their lifecycle effects on human health and the environment.
- Conduct or support research and innovation that promotes the use of products and technologies that are safe and efficient in their use of energy, natural resources and other materials.
- Develop and promote the concept of integrated materials management throughout the metals and minerals value chain.
- Provide regulators and other stakeholders with scientifically sound data and analysis regarding our products and operations as a basis for regulatory decisions.
- Support the development of scientifically sound policies, regulations, product standards and material choice decisions that encourage the safe use of mineral and metal products.

Principle 09: Contribute to the social, economic and institutional development of the communities in which we operate

- Engage at the earliest practical stage with likely affected parties to discuss and respond to issues and conflicts concerning the management of social impacts.
- Ensure that appropriate systems are in place for ongoing interaction with affected parties, making sure that minorities and other marginalized groups have equitable and culturally appropriate means of engagement.
- Contribute to community development from project development through closure in collaboration with host communities and their representatives.
- Encourage partnerships with governments and non-governmental organizations to ensure that programmes (such as community health, education, local business development) are well designed and effectively delivered.
- Enhance social and economic development by seeking opportunities to address poverty.

Principle 10: Implement effective and transparent engagement, communication and independently verified reporting arrangements with our stakeholders

- Report on our economic, social and environmental performance and contribution to sustainable development.
- Provide information that is timely, accurate and relevant.
- Engage with and respond to stakeholders through open consultation processes.

19.2.2   Status of SD in mining, based on stakeholders' views though a survey by GlobalScan[21,22]

The Executive Summary of the third stakeholder survey since 2004 that ICMM has conducted in association with GlobeScan has been bullet below. In fact the objective of this survey/project is to understand the broad issues of the international mining and metals industries and the findings are based on a sample of 847 respondents, drawn from ICMM's database, in 81 nations, across six pre-defined sectors (private sector, public sector, institutions, academia, non-governmental organizations, and media).

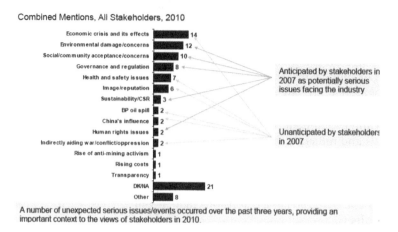

A number of unexpected serious issues/events occurred over the past three years, providing an important context to the views of stakeholders in 2010.

Figure 19.2   Most serious issues/events in past three years[21,22] (Permission: ICMM).

The online survey was carried out during 24 August to 18 September 2010. In 2010, the top five recommendations (by theme) that stakeholders have for the industry remain the same as those in 2007. However, the respective importance of these recommendations has changed; listed in descending order, they are:

- **More transparency/review** (Be transparent and accountable, releasing trustful data on ESG performance). [*NGO, South America*].
- **Communicate with/listen to communities/stakeholders** (Work more closely and transparently with stakeholders in a positive way to strike a balance between financial rewards and benefits to the economy and local people. This may seem to reduce returns only but, in the long run, should reduce risk thereby maintaining the same or even improving the return to risk ratio) [*Private Sector (Consultant), Europe)*].
- **Address social/community needs** (Respect and protect people and their rights and remember we have only one earth) [*Academia, Oceania*].
- **Impact on the environment** (Do not leave the agenda of climate change to the politicians. Let industry do what it can reasonably do and not under compulsion) [*Public sector (national government), Africa*].
- **Improve sustainable development efforts** (Walk the talk! There is little evidence that most mining and metals companies put environmental and social and governance (sustainability) issues on a reasonably equal footing as expediency and profit. Unless companies pay greater respect to host communities and deliver better balanced benefits sharing (economic, environmental and social), they will become increasingly unwelcome. On climate change the global community will eventually make decisions for business/industry; industry is incapable of taking voluntary action) [*Private sector (mining company), Oceania*].

Issues such as lingering economic uncertainty, high profile accidents/events, and the growing importance of China are important factors that may be causing stakeholders

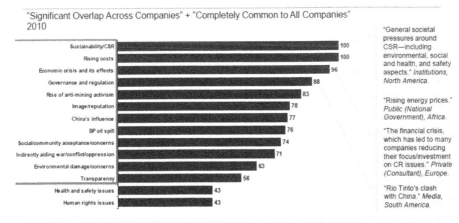

"Significant Overlap Across Companies" + "Completely Common to All Companies"
2010

Issue	Value
Sustainability/CSR	100
Rising costs	100
Economic crisis and its effects	96
Governance and regulation	88
Rise of anti-mining activism	83
Image/reputation	78
China's influence	77
BP oil spill	76
Social/community acceptance/concerns	74
Indirectly aiding war/conflict/oppression	71
Environmental damage/concerns	63
Transparency	56
Health and safety issues	43
Human rights issues	43

"General societal pressures around CSR—including environmental, social and health, and safety aspects." *Institutions, North America.*

"Rising energy prices." *Public (National Government), Africa.*

"The financial crisis, which has led to many companies reducing their focus/investment on CR issues." *Private (Consultant), Europe.*

"Rio Tinto's clash with China." *Media, South America.*

Of the 14 top-of-mind issues mentioned, 12 are perceived by majorities of stakeholders as common issues. The two issues perceived to be unique to affected companies within the industry are health and safety and human rights issues.

Figure 19.3 How common or unique were the most serious issues/events?[21,22] (Permission: ICMM).

to consider transparency and stakeholder engagement as the two top-of-mind recommendations this year. Environmental and social challenges clearly must still be addressed, but it seems that any efforts must be addressed with more transparency.

Delivering sustainable development will mean dealing with several issues at once. Ensuring access to electricity and supporting economic growth are essential to eradicating human poverty and supporting human development.[20]

In order to achieve these objectives participation and close liaison amongst governmental and non-governmental agencies, international and national agencies, mining companies, research and educational institutes are equally important.

19.3   SCENARIOS INFLUENCING MINING INDUSTRY (ALSO REF. SEC. 1.9)

19.3.1   Population growth and resulting impacts/implications

With the present growth rate of about 1.8%, the world's population is going to be doubled in the next 39 years. In the year 2000 world's population was about 6 billions, which is going to be doubled by the year 2039, as shown in figure 19.6.

Population problem is just not numbers but it has various impacts: social, political, economic and environmental. Providing the ever-increasing numbers with clean air, water, nourishing food, shelter, services and basic needs is becoming more and more difficult.

19.3.2   Use of minerals by world's citizens

Does the population (the world's citizens) need minerals? A funny question! Why minerals? Why mining? Sec. 1.4, figures 1.3, 1.4 and table 1.2; describe use of minerals with

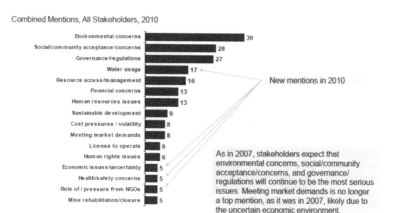

Figure 19.4   Most serious issues over the next three years[21,22] (Permission: ICMM).

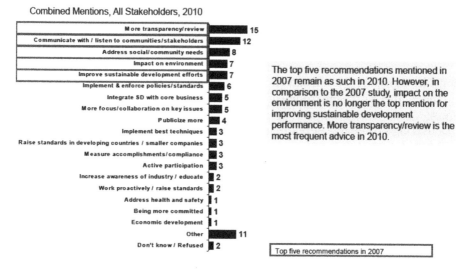

Figure 19.5   Advice for improving sustainable development performance based on surveys: 2007 & 2010[21,22] (Permission: ICMM).

the things we need most including energy. The bulb shown in the illustration (figure 19.7) is glowing by electricity. Now, try to analyze the various components of a bulb that includes filament, lead wires, fuse, stem press, tie wires, base, heat deflector and support wires. You would find that many minerals have been used to make them. A mixture of argon and nitrogen gas has been filled in this bulb and the current that lights it is generated from minerals, which are known as fossil fuels – coal, oil, or natural gas. It can also be produced from atomic minerals. In USA 57%, and in Asian countries such as India 68% of electricity is generated by coal. More than 90% of electricity all over the world is generated using these minerals. So what do you say?

• The world is dark without electricity/power or, is it without minerals?
• Factories would come to halt without electricity or, is it without minerals?

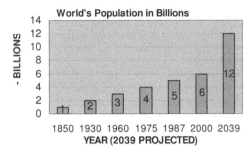

Figure 19.6   World Population statistics with future projections.[17,18]

- Services such as transportation, communication, information-technology and others would be jeopardized without electricity or, is it without minerals?

Energy crisis: The energy crisis has already begun in many countries. The consumption pattern/trend is likely to continue in the next 25 years. Access to energy is essential to addressing the problems that cause poverty. Without energy, people cannot access the opportunities provided by the modern world but 1.3 billion people lack access to electricity today (Mr Catelin, WCA, 2012)[20]. No need to mention that this deficiency could be met mainly by mining more and more coal and minerals.

19.3.3   Mineral consumption trends

How much minerals do we need every day? The consumption pattern is very erratic. In fact peoples' culture, customs and lifestyles govern mineral consumption patterns. In some of the developed countries such as USA (fig. 1.3), Western Europe and Japan people's habits and life-style require large consumption of goods, services and energy, and ultimately this results in huge mineral consumption per capita. In some under-developed or developing countries even though their population is considerable but their simple living and moderate life style requires much less natural resources including minerals; and it gives very little environment related problems.

Developed countries with a population share of 26% account for 79% of consumption of steel, and 86% of other metals. Currently the USA alone, with 5% of the world's population, uses 33% of world's mineral resources. Thus, the *consumption pattern is erratic*. Could we afford to continue with such a trend?

19.3.4   Status of quality, quantity, type of mineral and *Resources* depletion[17,18,20]

Equally important is to understand the qualitative and quantitative status of minerals (which is known as 'mineral inventory') in any country. The consumption pattern depicts an astonishing scenario. For example, India has 10% of the world's coal, about 92 billion tons, third only to USA and China in total reserves. At the current rate of consumption, India has enough coal for the next 217 years[1]. But is it realistic?

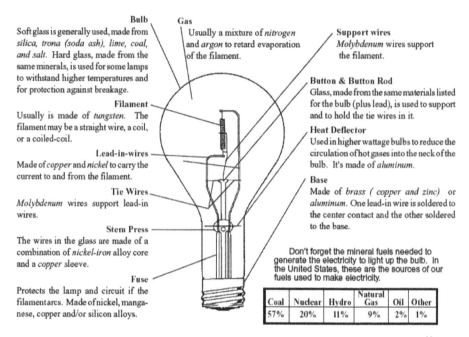

Bulb
Soft glass is generally used, made from *silica, trona (soda ash), lime, coal, and salt.* Hard glass, made from the same minerals, is used for some lamps to withstand higher temperatures and for protection against breakage.

Filament
Usually is made of *tungsten.* The filament may be a straight wire, a coil, or a coiled-coil.

Lead-in-wires
Made of *copper* and *nickel* to carry the current to and from the filament.

Tie Wires
Molybdenum wires support lead-in wires.

Stem Press
The wires in the glass are made of a combination of *nickel-iron* alloy core and a *copper* sleeve.

Fuse
Protects the lamp and circuit if the filament arcs. Made of nickel, manganese, copper and/or silicon alloys.

Gas
Usually a mixture of *nitrogen* and *argon* to retard evaporation of the filament.

Support wires
Molybdenum wires support the filament.

Button & Button Rod
Glass, made from the same materials listed for the bulb (plus lead), is used to support and to hold the tie wires in it.

Heat Deflector
Used in higher wattage bulbs to reduce the circulation of hot gases into the neck of the bulb. It's made of *aluminum.*

Base
Made of *brass (copper and zinc)* or *aluminum.* One lead-in wire is soldered to the center contact and the other soldered to the base.

Don't forget the mineral fuels needed to generate the electricity to light up the bulb. In the United States, these are the sources of our fuels used to make electricity.

Coal	Nuclear	Hydro	Natural Gas	Oil	Other
57%	20%	11%	9%	2%	1%

Figure 19.7   Is it electricity/power that is providing energy to us, or the minerals?[11]

Since the consumption pattern of minerals including coal is going to increase five times in the next five decades (as described in sec. 19.3.5) and, if recovery percentage continues to remains around 55%. In addition, the mines are getting deeper and left out reserves are thin, scattered and difficult to mine-out, which poses a further question mark on recovery percentage of the deposits! Analyzing Indian coal reserves as shown in figure 19.8; based on these arguments gives a warning signal. And what about the status of 1100 minerals that have been identified so far, in terms of their occurrences and mining; are they sufficient, deficient (requiring import) or surplus worth exporting based on their consumption pattern in your own country? *The solution lies in taking intensive measures in terms of huge and speedy exploration and also renovations leading to cost effective methods, techniques and equipment for the exploitation of ore reserve to improve upon this situation/scenario.*

Similarly, analyzing mineral resources of some minerals such as: coal, iron, manganese, bauxite and limestone based on the available data on minerals resources, geological and mineable reserves, as shown in figure 19.9; of a thickly populated Asian country; this depicts a very alarming situation with regard to availability of mineable reserves of these minerals and suggests measures such as: cost-effective techniques, systems and renovations to convert more and more mineral resources into reserves.

Taking into consideration the analyses as shown figures 19.8 and 19.9; it suggests that this situation may lead to crisis if appropriate measures are not taken. Analyses such as these are equally applicable to every country to understand status of this valuable resource/asset in their own country.

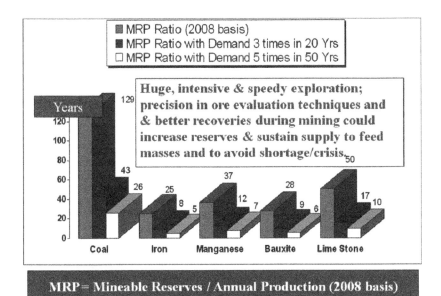

Figure 19.8   Disproportionate reserves & production patterns. Huge, speedy exploration and precision in ore evaluation techniques and better recoveries during mining could increase the reserve base of minerals to meet their mass consumption requirements on a sustainable basis. MRP = Mineable Reserves/Annual Production (2008 basis).[1]

19.3.5   Mineral consumption prediction

The foregoing discussion reveals that the *mineral consumption pattern is erratic and so is the status of the mineral inventory's quality, quantity and occurrences (e.g. out of 1100 minerals how many of them are available in your own country?).* Could we afford to continue with such a scenario? It would be appropriate, at this juncture, to quote the forecast/prediction by IMM – Australia.[17,18]

"Over the next 50 years the world will use 5 times the mineral resources that have been mined to the year 2000. To meet this predicted demand, the industry must grow as internationally competitive sector, underpinned by innovations and technology." In addition, the following facts emerge:

- Demand for minerals and fuels will remain cyclic but on a strong upward trajectory as China, India and others develop their economics.
- This demand will lead to the large-scale expansion of existing mines which would be having features such as: deeper, lower grade and the development of major new mines in politically risky locations.
- The mining industry will not be able to continue to operate on the scale demanded without technologies that do not yet exist.
- People do not want to live in remote, risky locations and this would result in application of high degree of automation, remote sensing & machine diagnostics mechanisms.
- Application of large-scaled deep mining technologies which do not yet exist.

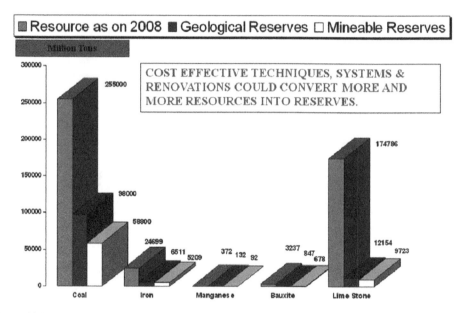

Figure 19.9 Adapting cost-effective techniques, systems and renovations could convert more and more mineral resources into reserves.[1]

19.3.6 Mining industry's inherent problems and challenges[6,18,19]

In spite of the impressive progress that has been made, as described in sec. 19.4.1, the mining industry, being the most hazardous as described in Chapter 18, has its inherent problems too as shown in figure 19.10 and also mining of past 150 years has resulted in the following scenario:

- Depleting assets and slow paced exploration
- Rise in remote & difficult areas
- Rising depths & difficult deposits to mine particularly those abnormally thin/thick, scattered, deep seated and geologically disturbed
- Poor recoveries
- Rising costs
- Disproportionate – consumption & production pattern (fig. 19.8)
- Small sized mines and lack of optimum sized mines
- Backlog on SD & Health, safety and environment (HSE) related issues
- And many others including (fig. 19.10):
  ○ Lack of adequate capital
  ○ Procedural delays
  ○ Lack of infrastructures
  ○ Shortage of skilled man-power
  ○ Lack of standardization, equipment manufacturing and maintenance facilities
  ○ Few others (other than listed above).

Figure 19.10 Constraints which mining sectors in most of the countries need to address.[6,18]

19.3.7 Global risk ranking and competitiveness in the mining sector

Lack of investment in the mining sector in most countries is a usual bottleneck. It has a long gestation period. It requires huge investment. Mining must be performed economically i.e. it must yield profit. In a scenario like this nobody wants to invest into it until a high rate of return (ROR) of the investment is ensured. There are MNCs who have expertise in mining and are willing to invest in mining ventures provided confidence in the legal and fiscal frame work is ensured by the country wishing their presence. They look into the parameters as detailed in table 19.1. And they also wish to invest in a country with least risk. Figure 19.11 depicts risk ranking in mining sector of few countries. This is equally applicable/true to encourage investors within the country as well.

19.4 IS MINING INDUSTRY EQUIPPED TO MEET THE CHALLENGES?[17,18]

19.4.1 Technological developments in mining

Looking at the global scenario, the progress that has been made in the process of mineral exploration and exploitation in the last five decades is far ahead of what has been achieved during the last five centuries. Since the 19th century there have been many important events, inventions and developments that have resulted in new techniques, methods and equipment. This includes giant sized equipment (as discussed in Chapters 5, 6, 7, 10 and 11) for rocks and ground fragmentation and their subsequent

Table 19.1   Global competitiveness in providing the suitable environment
to attract foreign investors and MNCs.[2]

Global Competitiveness Parameters	India	Canada	Australia	Chile
Economic	6	10	10	9
Political	6	9	10	8
Social issues	2	8	6	7
Permitting delays	3	8	5	7
Corruption	3	10	10	8
Currency Stability	5	8	8	7
Tax Regime	4	7	7	4
Total	**29**	**60**	**56**	**50**
Overall Ranking	***16***	***1***	***2***	***3***

Source: Behre Dolbears, Survey of Countries for Mining Investment 2009
*Countries surveyed has been a total points of 10 in each parameters

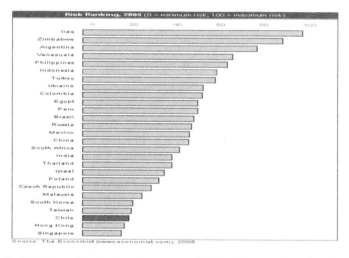

Figure 19.11   Global Risk ranking in the mining sector (2006); minimum risk = 0; and max. risk = 100.

disposal. Application of automation, modular system, and remote control has already begun. In addition it depicts the following scenario:

• Mines are producing minerals in bulk quantity both from surface as well as underground locales.
• Developments & renovations in underground mines are equally impressive. Use of giant sized equipment has begun. This includes large sized tunnel borers, giant sized drill jumbos, muck handling and transportation units as described in chapter 16.

- Design & operation of large sized mines having output 10 MTA or more with matching layouts to use giant sized equipment (draglines: 45–55 cum.); rope shovels (bucket cap. up to 42 cum.); 10–15 cum hydraulic shovels, have begun.
- Use of dumpers of 85 to 320 tons capacity; in-pit crushing and conveying; high wall mining for coal are gaining popularity.
- Remote sensing, automation in remote & adverse climatic conditions have begun.

19.4.2 Initiatives already taken globally to meet demand of minerals of mass consumption

To meet this challenge, mineral industry's initiatives in these directions have already begun, as evident by the following facts and detailed in sec. 1.9 also:

Globalization: Business has crossed the barriers/boundaries. Most of the countries are welcoming foreign investment in the mineral sector which was closed by many countries in the past. MNCs are looking for acquisition of mineral properties and also quieting mineral rights where conditions cease to remain favorable. Uneconomical mining ventures are no longer continuing. Foreign investors bring different systems, culture and managerial skills. They look for political and fiscal stability, prevalent regulations and mining culture at the country of operation. MNCs invest huge sums of money and expect its recovery at the earliest, and hence, the higher rate of returns.

Merging and out-sourcing: Use of IT (also ref. sec. 18.6.6 and fig. 18.21) and globalization in the mineral sector has compelled mineral producers to compete in the international market. Many mining and equipment manufacturing companies are merging. MNCs are looking for outsourcing and wish to take advantage of cheaper energy/power, human resources and know-how.

Buyer's market: Except for minerals of strategic importance such uranium and those used in the manufacturing of arms and ammunition minerals are available at a competitive price. There is no more monopoly in the mineral sector and thus a quality product at a cheaper rate is readily available.

This scenario, in turn, is asking for better recoveries and cost reduction at all levels – mining, processing and preparing salable product, and also taking care of sustainable development.

19.5 PROPOSED STRATEGY TO RUN MINES IN AN ECONOMICALLY VIABLE (BENEFICIAL) WAY[15]

The mining industry must sustain its long term 'welcome', which has begun within last decade. SD is ultimately judged by the quality of its performance technically, economically, environmentally, socially and ethically.

Every country has mineral resources supported by their basic infrastructures developed in the past based on their policies but most of them are facing constraints, as shown in figures 19.10, 19.11 and table 19.1; which have the potential to hamper/delay the expected growth in their mineral/mining sector. *The mining scenario is depicting an alarming situation with regard to:*

- Speed – exploration as well as exploitation – in mining
- Education, research & development
- Precision in ore evaluation, mineral exploitation (mining) and exploitation techniques
- HSE & SD together with social responsibilities
- Output rates, productivity and cost of mining
- Recovery percentage (losses of all kinds)
- Legal compliances
- Best practices and cost effective systems
- Skilled manpower and effective human resources management.

These constraints need to be addressed with a sense of urgency and by bringing some innovative and effective solutions in line with global practices to bring about a paradigm shift in this sector. The following sections briefly cover these aspects.

19.5.1 Exploration: Huge, intensive & speedy together with bringing precision in ore evaluation techniques[5]

Use of picks (a conventional way and means) to satellites (the modern technologies) has been made for searching minerals, as detailed in Chapter 3. It establishes a mineral resource base as shown in figure 12.3(f). The exploration companies/agencies establish mineral resources. Thus, many deposits have been already found and what is left would be harder to find. The success rate of mineral prospects that are ultimately converted into mining ventures is 8–10% that in turn is advocating for huge exploration. In a situation like this the task of exploration particularly in the under-developed and developing countries should be strengthened further. There is an urgent need for up-dating their database and adding new regimes. Exploration task must be speed up by engaging MNCs as well as domestic agencies that are expert in this task. Equally important is the precision in the evaluation process. Technologies such as:

- Geological vision (Radar) – ability to see through rock to locate an orebody in space, grade & geological structures (CRC-Australia) as shown in figure 19.12, could be used in this regard.

19.5.2 Establishing mineral inventory, cutoff grade and ore reserves

For any mining venture the first thing is to establish the mineral inventory in the form of Grade-Tonnage curve, as described in chapter 12. This means knowing the mineral assets fully. Converting mineral reserves into ore reserves involves 'Cut-off grade' which is dependent of *cost and price* of the mineral in question (fig. 12.3). Reduction in cost would mean increase in the ore reserves. As shown in figure 12.3(f) out of total minerals resources only those qualify as 'Ore Reserves' are worth mining and the rest are waiting to qualify as 'Ore'. Ultimately recovered are the 'Commercial Reserves (fig. 12.5(d))', which may be just 55% to 65% of the initially assessed geological reserves. Thus, we should be interested in mining of ore-reserves and not the whole

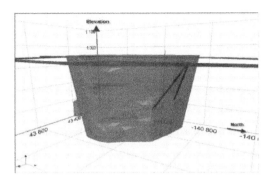

Figure 19.12   Geological vision (Radar) – ability to see through rock to locate orebody in space, grade & geological structures (Courtesy: CRC-Australia).[5]

mineral-deposit! And any deposit should have its Grade-Tonnage (total mineral inventory) details.

The key theme for SD in this conjunction is to establish minerals resources to the maximum through consistent exploration and evaluation. Prior to mining mineral inventory (quantitative and qualitative assessment) of any deposit must be established fully. How can one make a building on a shaky foundation?

19.5.3   Division of mineral property (i.e. orebody or coal deposits into level and panels)

This refers to size of prospect – the lease area (also known as 'concession'); and ultimately the mines. Small units often suffer on account of output rate, cost of production and productivity (fig. 12.4(b)). Should any country continue with small sized leases; particularly in the range of 1 to 5 hectares?

In figure 12.6 division of ore reserves into level system of mining and in figure 12.7 into panel system have been shown. As described in sec. 12.3, important mine design parameters such as level interval and size of panel, apart from other parameters as detailed in this section; are governed by the cost of mining. In the prevalent scenario and to compete in the global market, it becomes all the more important to review these design-parameters; particularly when the pricing is controlled by big companies. To cite an example of iron ore mining world over, it reveals that pricing is in the hand of big companies who have reserves exceeding 500 million tons (MT). Their output exceeds 25 million tons and they are close to port with low cost logistics.

In the prevalent scenario, therefore, would it be not logical for any country to revise, renovate, and update the mines' size and their production capacities? Mines producing 100,000 tons/day of ore/coal apart from handling enormous quantity of overburden are in operation in few countries but the rest have small to medium sized mines, which at present is a bottleneck. The success lies in the use of optimum sized equipment fleet ensuring its optimum utilization with the matching layout.

The key theme for SD in this conjunction is to optimize mine size and its further division into panels, levels or stopes to minimize wastage and to operate mines to their designed capacity. The idea is to use resources of today with respect to the needs of future generations judiciously.

19.5.4   Locale-specific challenges and proposed solutions/way-outs

19.5.4.1   Underground metalliferous mining challenges

Small to medium sized mines with the use of widely diversified tools, appliances and equipment which are lowering productivity; mining deeper, hotter, lower grade, highly stressed deposits and poor recoveries. To deal with these abnormalities solution lies in the application of:

- High speed shaft sinking & tunneling
- Bulk mining methods
- Reserves reconciliation surveys
- Mine supports & roof bolting
- Environmental monitoring
- Automated drilling & explosives.

19.5.4.2   Underground coal mining challenges[4,5]

Abnormally thick and thin seams, rising depths, poor recoveries, adverse working conditions including rock-bumps, small to medium sized mines with the use of widely diversified tools, appliances and equipment which are lowering productivity and few others. Proposed solutions include***:

- Developing technology for: pillar extraction; mining thick as well as thin seams; multiple seams' mining.
- Longwall mining for higher output in the range of 2–10 MTA.
- Automated haulage, traffic management.
- Needs-based changes in methods, techniques, equipment and mine size and output rates.

***– The technology being used in China in this conjunction could be cited as an example of the initiatives taken to address issues as listed above. China is a country with coal as a major energy source. Underground mining is the major method of Chinese coal mine exploitation, and more than 95% of coal is produced through underground mining. Long-wall mining is at the forefront. Its advantages as claimed by the Chinese producers include: Higher extraction (recovery) rate normally 70%~90%, high output (ranging 1 Mt to as high as 19 Mt annually), safety and high efficiency. It suits all kinds of geological conditions. Maximum support height is 7m (fig. 19.13(a)), average face length is 300 m (maximum 400 m). Run of mine coal production from a face is in the range of 6–13 Mt per year. Quite high initial investment is its disadvantage/limitation.[4] Equally impressive are the performance data of the longwall faces with the use of powered-supports, in the countries other than China, as shown in table 19.2.

Figure 19.13   (a) Use of power supports for mining thick seams. (b) Use of Powered supports for thin seam mining (Courtesy: Zhengzhou Coal Mining Machinery (Group) Co., Ltd. China).[4]

Equally impressive is the application of: low seam mechanized mining technology; coal cutting (mining) using: shearers and ploughs. It also includes automation in mining of thin seams with the application of supports having height as low as 0.5 m fig. 19.13(b)).

Room & Pillar exploitation could be made applicable to favor lower investment. In this case the face exploitation and gallery drivage can be done with the same sets of equipment. Disadvantage: lower extraction (recovery) rate, normally 50%~70%, bad ventilation condition.[4]

19.5.4.3   Open cast/open pit mines (coal & non coal) challenges

Small sized mines in most of the African and Asian countries having constraints such as:

- Lower grades, use of widely diversified tools, appliances and equipment which are lowering productivity (i.e. increase in costs).
- Low output, increasing depths & overburdens which are adding costs.
- Remote locations, adverse climatic conditions, inadequate infrastructures resulting high FOB Costs.

Table 19.2 Performance data on powered supported Longwall (PSLW). Mtpa – Million tons per annum.[23]

Country	No. of PSLW Faces	Best production (Mtpa)	Av. Production Mtpa
USA	49	7.2	4.3
China	600	8.9	0.8
Australia	24	5.7	2.8
Russia	120	4.1	0.5

19.5.5 Mining difficult deposits using non-conventional technologies

It is the right time to further develop and apply the non-conventional mining methods (figs 1.5, 1.9 and sec. 1.6) to mine out low grade, thin, geological disturbed, old mine-dumps and worked out stopes and mines. In addition, mining of low graded deposits having typical features (characteristics) require much more energy to remove over-burden and tailings (what is left after ore processed). Application of non-conventional methods could improve the situation; as they are less costly comparing conventional methods and yields better recoveries with least development. The technologies and methods available include:

- Gasification (of coal)
- Methane drainage
- Borehole mining
- Leaching, and solution mining
- Frasch Processing (for sulfur) and few others including some of the novel techniques that are in their initial phase and to name them they are:
  - ○ Automation and robotics
  - ○ Underground retorting
  - ○ Nuclear mining and
  - ○ Extraterrestrial mining.

19.5.6 Improved fragmentation – a better way to extract minerals (ore, waste rocks, overburden) to save energy[3,6]

Brining precision in the rock fragmentation techniques with and without use of explosives (drilling, blasting and cutting) results in reduction in energy spent, pollution that is caused, hazards that are associated. Use of Controlled (contour, smooth) blasting techniques, as detailed in sec. 9.3.3, should be an integral part while creating excavation of any kind. Some of the techniques that have been recently introduced include: Automated drilling & explosive placement, as being used at some of the Australian mines for improved fragmentation could be cited as an example to bring precision in the mine operations. Several mining companies actively developing fully-autonomous surface drills having advantages such as: accuracy, consistency, speed, improved reliability, less maintenance and also ability to measure rock properties in real-time and thereby improve blasting and fragmentation. It has challenges such as: drill auto-tramming and auto-leveling require high integrity navigation and guidance systems; auto-drilling requiring adaptive pull-down pressures and torque.

Autonomous digging means: the effectiveness of the fragmentation can be determined from digging forces so that the operator can continuously update blasting model. Autonomous digging also improves production rates; reduces machine stress – reducing maintenance; prevents overloading of vehicles and distributes loading.

All of the major Australian iron ore companies are planning to mine a significant fraction of their ore using rock cutting machines known as: 'Oscillating Disc Cutting' – a novel technology for cutting hard rock at high rates with low cutter forces. It is a dramatic shift away from drill-and-blast. Its advantages include: elimination of: drills, explosives, diggers, trucks (continuous haulage). It facilitates selective mining of orebodies. The challenges include: demonstration of ability to cut a significant fraction of orebodies productively and economically. This uses the concept of Prototype mining system now being built in South Africa for use in narrow vein platinum mines having potential for both rapid mine development and for surface mining.

The key theme for SD in this conjunction is to encourage intensive research, technology transfer, and needs – based changes in methods, techniques and equipment to achieve maximum recovery from mining through to the salable product.

19.5.7 Precision in operations – maximizing recovery[3,6,7]

This refers to better recoveries at all levels – mining, processing (for coal sizing and washing), smelting, refining, handling and disposal. It results in reduction in energy spent, pollution that is caused, hazards that are associated. Every mine should assess as how much overall recovery it is achieving? Tally it with the geological reserves that have been assessed, and you might be astonished! The solution lies in incorporating precision in ore extraction techniques. It results cost reduction, improved recovery and reduced dilution, and thereby quality of the product is enhanced. Automation, as already incorporated in some of the unit operations and most of the other industries could be a solution in this regard. The following paragraphs describe few important aspects of 'automation'.

Automation (ancient Greek: = *self dictated*): in the scope of industrialization, it is a step beyond mechanization. Whereas *mechanization* provided human operators with machinery to assist them with the *physical* requirements of work, *automation* greatly reduces the need for human *sensory* and *mental* requirements as well.

Currently, for manufacturing companies, the purpose of automation has shifted from increasing productivity and reducing costs, to broader issues, such as increasing quality and flexibility in the manufacturing process.

Hazardous operations, such as oil refining, the manufacturing of industrial chemicals and all forms of metal working were always early contenders for automation. This feature allows its application in mining which is even more hazardous than these industries.

A word of caution with regard to the safety issues of automation: the safety issue with automation is that while it is often viewed as a way to *minimize* human error in a system, *increasing the degree and levels of automation also increases the consequences of error.* For example, The Three Mile Island nuclear event was largely due to over-reliance on "automated safety" systems. Unfortunately, in the event, the designers had never anticipated the actual failure mode which occurred, so both the "automated safety" systems and their human overseers were inundated with vast amounts of largely irrelevant information. With automation we have machines designed by (fallible) people with high levels of expertise, which operate at speeds well beyond human ability to react, being operated by people with relatively more limited education (or other failings, as in the Bhopal disaster or Chernobyl disaster). Ultimately, with increasing levels of automation over ever larger domains of activities, when something goes wrong the consequences rapidly approach the catastrophic. This is true for all complex systems however, and one of the major goals of safety engineering for nuclear reactors, for example, is to make safety mechanisms as simple and as foolproof as possible.[7]

Automation in mining intends to improve quality as already being achieved in other industries. For example, automobile and truck pistons used to be installed into engines manually. This is rapidly being transitioned to automated machine installation, because the error rate for manual installment was around 1-1.5%, but has been reduced to 0.00001% with automation. In order to reduce wastage and losses which are key to reduce costs, automation in mining has begun and it is predicted that during next generation mine development, as per Peter Carter[3] and Michael Hood[6], there would be more and more automation covering most of the operations, as shown in the illustration in figure 19.14. However, McCarthy and Noort (2009)[12] suggested a critical path to full automation in underground mining, as shown in figure 19.15(d).

The self explanatory illustrations shown in figures 19.15(a) to (e) are describing various concepts that would be there in the days to come. It is predicted that there would be:

- More concentrated production centers
- More effective rock fragmentation techniques
- Continuous rock transport systems from the mine
- Geological vision – ability to see through rock to locate an orebody in space, ore grades, geological structures to minimize geological uncertainty (fig. 19.12).[6]
- Safety specific technologies[6]
- Precise vehicle navigation, narrow roads & ramps
- Small mine site team with multiple responsibilities involving practitioners, high skills and knowledge for quality outcomes.
- Greatly improved equipment reliability through much more highly automated systems
- Autonomous haulage – to remove cycle time variability, and also for the following:
  - Use of autonomous haul shuttle with no operator cabin, no spotting to improve bi-directional efficiency
  - Flexible engine and tire configurations
  - Precise placement of material
  - Reduced haul distances
  - Reduced reclamation costs.

19.5.8   The Critical Path to Full Automation[12]

Development is well underway on the technologies required for the transition to full automation of underground mines. However, examples gleaned from the automobile industry suggest that automation of underground mining is unlikely to offer significant

Automated drilling & explosive placement
(a) Automated drilling allows drill bit to be used as a sensor to monitor: rock strength, rock type and rock mass rating.
(b) It permits round design on-the-fly: burden, spacing, blast design on-the-fly: loading per hole & hole delays And also permits prediction of muck-pile shape and fragmentation.

(c) 'Oscillating Disc Cutting' – a novel technology for cutting hard rock at high rates with low cutter forces. In fact the conventional drills and cutting tools break hard rock using an indentation (compression) process (as described in sec. 4.5) whereas this new Oscillating Disc Cutting (ODC) approach uses robust disc cutters to undercut the rock; breaking it in tension. Rock is 10 times weaker in tension than compression. Cutter oscillates whilst undercutting rock. It helps to promote cracking through fatigue , which further reduces force. Technology demonstrated in series of exhaustive lab and field (quarry) tests.

Figure 19.14   (a) Several mining companies are actively developing fully-autonomous surface drills. (b) It permits round design on-the-fly. (c) 'Oscillating Disc Cutting' – a novel technology for cutting hard rock at high rates with low cutter forces.[6]

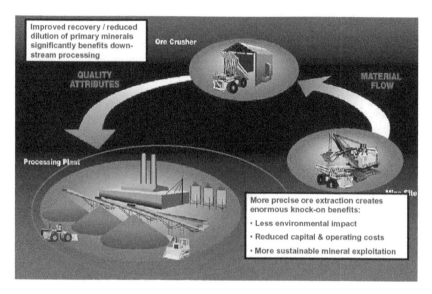

Figure 19.15(a)   Precision in ore exploitation techniques results in enormous knock-on benefits[3] (Courtesy: Modular Mining Systems).

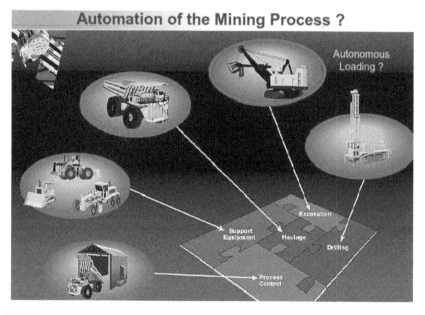

Figure 19.15(b)   Automation could be incorporated with various unit operations in mining[3] (Courtesy: Modular Mining Systems).

cost benefits or labor efficiencies. The necessary impetus must instead come from the point-of-view of improving safety.

The first critical milestone (fig. 19.15(d)) will be to remove people from exposure to the underground mine operating environment by enclosing them in specially

Autonomous Haulage – Mine Design Advances

• Precise vehicle navigation, narrow roads & ramps

• Reduced overburden stripping ratio & excavation footprint

• Precise placement of material

• Reduced haul distances

• Reduced reclamation costs

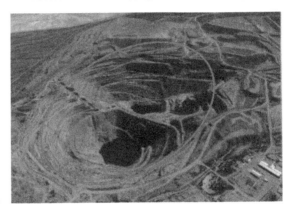

• Autonomous haul shuttle

• No operator cabin, no spotting

• Bi-directional efficiency

• Flexible engine and tire configurations

Precise Control of Autonomous Haul Fleet

• Strategically Safer Mining

• Extended Life of Components

• 100% Control of Material Destination

• Consistent Flow of Primary Material to Process

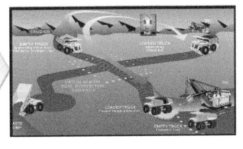

Figure 19.15(c)   Automation and future development to bring precision in operations at the forefront[3] (Courtesy: Modular Mining Systems).

designed cabs. This will probably achieve the majority of any productivity and safety benefits gained from full automation in any case. It then becomes a question of when (or if) progression to tele-remote operation and full automation is desirable. Substantial funding, strong leadership and belief in the vision of a fully automated mine system will be required from the industry to ensure its success.

There are still many jobs which are in no immediate danger of automation. *No device has been invented which can match the human eye for accuracy and precision in many tasks; nor the human ear. Even the admittedly handicapped human is able to identify and distinguish among far more scents than any automated device. Human pattern recognition, language recognition, and language production ability is well beyond anything currently envisioned by automation engineers. As such the best*

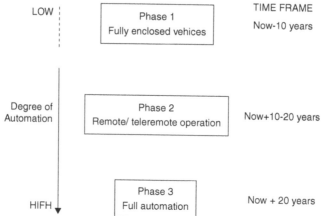

		TIME FRAME
LOW	Phase 1 Fully enclosed vehices	Now-10 years
Degree of Automation	Phase 2 Remote/ teleremote operation	Now+10-20 years
HIFH	Phase 3 Full automation	Now + 20 years

Figure 19.15(d) Schematic illustrating the phased approach to automation.[12]

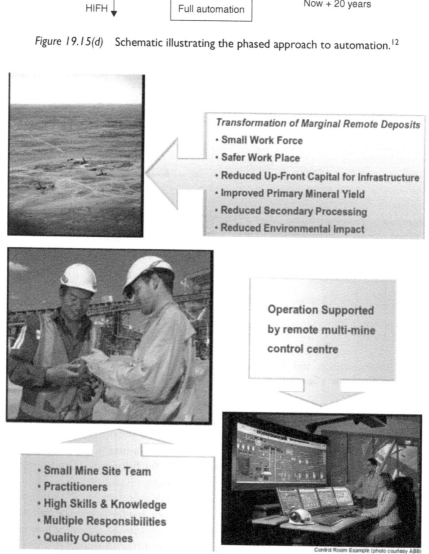

Transformation of Marginal Remote Deposits
- Small Work Force
- Safer Work Place
- Reduced Up-Front Capital for Infrastructure
- Improved Primary Mineral Yield
- Reduced Secondary Processing
- Reduced Environmental Impact

Operation Supported
by remote multi-mine
control centre

- Small Mine Site Team
- Practitioners
- High Skills & Knowledge
- Multiple Responsibilities
- Quality Outcomes

Control Room Example (photo courtesy ABB)

Figure 19.15(e) Proposed mine design models in the days to come advocating better mining practices for SD with cost effective measures[3] (Courtesy: Modular Mining Systems).

strategy is to use our 5 senses to deal with the abnormalities, as described in tables 18.9 and 18.11, and bring radical changes in the mining sector.

19.5.9   Effective utilization of resources through standardization & benchmarking[19]

The mining industry today uses widely diversified types of tools, appliances and equipment. This poses problems to both the manufacturers and the users. To illustrate this point, in metal mines rails of different gauges; 600 mm, 750 mm and 900 mm are used and in coal mines the rail gauges used are of 24″, 30″, 36″ and 42″. Similarly there is much variation in sets of equipment and services' specifications for the trackless system of mining. Imagine how difficult it is for the manufacturers of the locomotives, mine-cars, loaders and sets of trackless equipment to manufacture them and their spare parts.

The standardization of the machines and equipment used for carrying out the various unit operations such as drilling, blasting, mucking and transportation at national-level should be the utmost priority to effectively utilize the resources. Similar logic should be applied to train and carrying out these unit operations (formulating standard procedures). Keeping a record of the performance norms at every mine and national level helps to assess manpower requirement, production planning, equipment scheduling and budget forecasting. In the absence of a data bank of such norms, mine planning often goes erratic and unpractical.

Amongst various best practices and systems, standardization of equipment, operations, procedures and practices would mean following the laid out norms, guidelines and instructions. Standards are laid out taking into consideration the best performances and industrial benchmarks. They are based on experience, scientific studies, debates and consensus arrived amongst those involved and concerned. They are aimed at effective utilization of resources.

As shown in figure 18.21 and described in sec. 18.6.6 more than 300 mining companies are using software popularly known as *'ERP – Entrepreneur Resources Planning'* which has potential to revolutionize processes and optimize resources. Such standardization and integration strategies should be considered by every mine owner.

The key theme for SD in this conjunction is to bring standardization with regards to methods (procedures), equipment, techniques, tools, appliances and mine services that are appropriate and practical to implement to minimizing wastage.

19.5.10   Needs-based changes, research and development[3,6]

Changes for better are a continuous process. A thorough technology transfer from one country to another and adapting methods, techniques and equipment that incorporates economics, environment and modern management could accomplish this. In tomorrow's mines, as described in sec. 1.9; use of robotics to carryout repetitive tasks that too in hazardous and risky locales would be part of the process. Application of lasers for precise survey, measurements and monitoring would play an important role. Sensors installed at strategic locations would help to monitor mine atmosphere. Use of automation would be playing a leading role. There will be thorough technology transfer from one field of engineering to another such changes have already begun at some of the mines/mining companies, but at the rest they should be incorporated

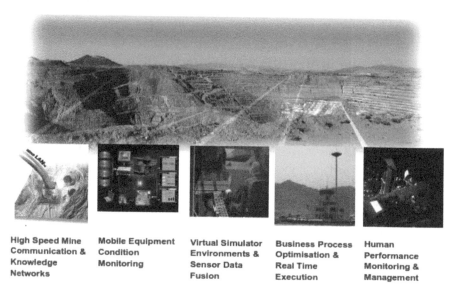

High Speed Mine	Mobile Equipment	Virtual Simulator	Business Process	Human
Communication &	Condition	Environments &	Optimisation &	Performance
Knowledge	Monitoring	Sensor Data	Real Time	Monitoring &
Networks		Fusion	Execution	Management

Figure 19.16   Focused research and development areas in the mineral sector in the days to come[3] (Courtesy: Modular Mining Systems).

at the earliest. The need of the hour is focused research in the areas as shown in figure 19.16.

The key theme for SD in this conjunction is to bring about need-based changes through R & D on conventional (on-going) Methods, Equipment and Techniques that reduces energy consumption.

19.6   MEASURES FOR SD THROUGH IMPROVEMENTS ENVIRONMENTALLY, SOCIALLY AND ETHICALLY

19.6.1   HSE – A critical business activity for sustainable development[15]

The meaning of the word 'sustain' is to keep in existence. Sustainable development is the one, which keeps in existence the ecological cycles that sustain renewable natural resources. To achieve this one of the aspects that needs attention is to recognize Occupational Health, Safety and Environment (HSE) together with welfare to the society (stakeholders) amongst the critical business activities that are set in achieving goals and objectives of a company. Thus, HSE together with welfare to the society (stakeholders) should be considered at par with production and productivity and it should be built in component while planning the life-cycle model of any mining venture (figs 19.17(a) & (b)). And any imbalance amongst these parameters could jeopardize business (fig. 19.17(b)). Case study described in sec. 19.6.2 Mining together with social welfare, as adapted at Rhineland Lignite Mining, Germany; since its inception in 1948 verifies the positive impacts of this concept/best practice.

Mining activities should demonstrate that they disturb the natural cycles least by way of reclamation and least damage to the environment. Environmentally and

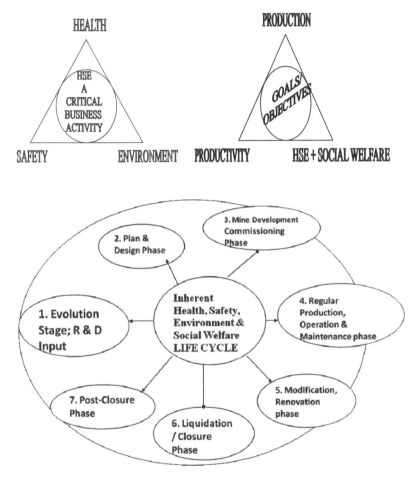

Figure 19.17   (a) (left) Interrelationship of occupational health, safety and environment. (b) (Right) Interrelationship of critical business activities. (c) (Lower) Life cycle approach by incorporating HSE and social welfare during all the stages of running a mining venture.[9,15,16]

socially – SD would be judged by the stakeholders, as what is left behind? Is the mine-area better than originally or worse!

19.6.2   Economic development regional as well as local – A case-study[8]

Rhineland Lignite Mining, is mining lignite deposits in Germany since 1948. It depicts following scenario:

- 3 Mines producing lignite coal @120MTA (Million tons. Annually)
- 22 Bucket Wheel excavators (BWEs) (each weighing 13,000 t) and capable of handling up to 240,000 m³/day of O/B say about 100 MTA

Figure 19.18   Economic Development regional as well as local at the Rhineland Lignite Mining, Germany since its inception in 1948 – a classical example of SD with production of lignite in bulk.[8]

- Stripping Ratio (SR) = 5 cum OB/Ton of coal
- Depths up to 350 m have been attained, and will reach 500 m in the future.
- 266 km of conveyors.
- 300 km of rail track for handling O/B & lignite using fleet of 19 stackers
- Since 1948, some 30,000 people have been moved, comprising around 70% of the entire population affected by the operations' progress. To date, over 170 km^2 have been re-cultivated. Mined out areas have been returned to farm land and for recreational use demonstrating the fact that the land-use after mining has been better than its original use – truly a case having mining with sustainable development.

Thus, after mining land have been reclaimed and having better use than original. This practice has been followed since beginning. Thus, it is a classic example of sustainable development which takes care of the local as well as regional development and the stakeholders, employees and employers – A goody-goody situation.

19.7   LEGAL COMPLIANCES AND MINING POLICY[18]

19.7.1   Mining laws – legislation

Mining laws/legislation, in general, are available in the form of Acts, Rules, Regulations, By-laws, Circulars and Standing orders or any other form of legislation. Every country has its own laws for the purpose of mineral exploration, development, exploitation, and conservation and to safeguard the environment. The recent amendments in the legislation/regulations in some of the countries, of course they are stringent, but give direction for the sustainable development of mineral properties.

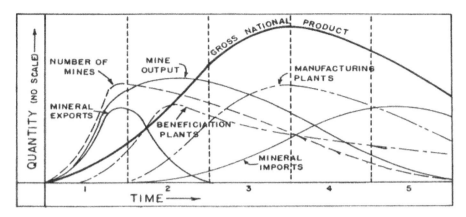

Figure 19.19   The 'Progressive growth of any country through mining' – the usual strategy followed by most of the developed countries. Find out where the country you belong to stands today?[10] (www.smenet.org).

Is it not the right time for any country to review the existing mining laws, particularly in those countries in which these were formulated long back (of course amendments are usually incorporated from time to time)? Should there be a combined legislation covering Occupational Health, Safety and Environment (HSE) together with welfare to the society (stakeholders)? Is it not the right time to recognize HSE together with welfare to the society, as critical business activity? Should strategy be formed, in the form of legislation, to give utmost importance for the maximum recovery of the mineral assets, which are nonrenewable?

It could be sensible to review mining legislation and bring changes that are necessary for SD by every nation. The policy change should ensure the three aspects of SD: Public participation in decision making, distribution of wealth and post-mining-closure environmental and social conditions.

19.7.2   Minerals & mining policy[5,10]

As described in the preceding sections that mining industry has *not come up to expectation so far as its sustainable development* is concerned. In addition, to whatever strategies that have been proposed in the preceding sections, the following paragraphs attempts to propose solutions and strategies that could ease the situation further.

Analyzing some of countries' *Import and Export status of Minerals,* for example, for a developing Asian country it amounted about $ 100,000 million in the year 2004–05, while the import amounts to be about $ 20,000 million. *Thus, export has been five times more than imports. Should this trend be continued?* Should this trend not be reversed, by producing value-aided good and services in the country itself – the strategy that has been followed by most of the developed countries, as shown in Figure 19.19? This illustration pertains to USA, depicting as how mining policy have changed in a phased manner from its initial phase of exporting minerals to the prevalent phase in which minerals are imported and value added goods and services are produced to increase GNP. Out of the wealth generated by following this strategy; a fair share of mining goes to 'have not's' in local communities and not disproportionate to the 'haves'.

Through mining; the emergent (Third-World) countries and developing countries can finance growth progressively by the export of raw mineral resources, then by processing these raw materials prior to export, and finally by achieving progressive industrial development to improve their GDP/GNP.

19.8   QUALITY OF HUMAN RESOURCES

Referring to the famous proverb: *"It is the man behind the machine that really matters"*. *Hence quality of human resources must be ensured. Imparting proper education at the Universities, practical training to the students in industry, organizing vocational training, refresher courses, symposium and seminars on a regular basis for the working crews, could achieve this."*

19.8.1   Academic (educational) status and standard of mining schools

Looking into the current *academic status and standard* of mining schools; is it not the right time to evaluate them for these aspects/parameters, Particularly the underdeveloped and developing countries and compare their status with those of developed countries, as bulleted below?

- Adequate qualified/competent staff and faculty
- Up-dated curriculum/syllabus for Undergraduate and Graduate studies
- Adequately equipped and well maintained laboratories
- Mechanism for thorough industrial training
- Research
  - Favorable setup to pursue higher studies including PhD
  - Library facilities and publication environment
  - Consultancy to industries.
  - Accreditation (recognition): by international agency such as ABET as how do they qualify themselves to meet the laid out academic standards?

The solution lies in close liaison amongst Government, industry/society and Universities on these issues, and a thorough feedback and communication, in terms of:

- Govt. policy and requirements
- Industries' requirements
- Universities' requirements.

The key theme for SD in this conjunction is to recognize and establish that the primary vehicles for SD training are the Mining Schools and their Professors.

Figure 19.20   Sustainable Development's three pillars of equal strength to run any mining venture.[14,16]

19.9   THE ULTIMATE AIM

Industries including mines require human resources, which are ample (fig. 19.6) but if the current trend of consumption pattern of the natural resources such as minerals and water, and man-made resources such as energy continues they would soon be depleted and their scarcity would be felt. Rather the present boom in the mineral market and oil prices are already influencing world's economy. Water scarcity has begun, and it is being felt world over.

There is a link amongst the inputs (fig. 18.1) that includes human resources; plant and machinery; raw materials – natural (such as minerals, fossil fuels, flora; and man-made materials which are innumerable); Energy, natural resources air and water; and infrastructures to accomplish production from any industrial set-up including mines and mills. Any mismatch amongst them could jeopardize production or the ultimate goal/objective of any company; or any hindrance in their functioning could lead to losses, inefficiencies, defects, delays, customers' complaints, stockholders' dissatisfaction, communities' complaints and dissatisfaction amongst those concerned – the workers and managers and executives.

Under its aims/objectives and mission, Company's Policy is formulated. There are three pillars of equal strength to deal with global recession and competition', as shown in figure 19.20 which should be given utmost importance. These are as outlined below.

19.9.1   Contented employees & stakeholders

It could be as a result of effective human resources' management. Figure 18.8, describes how to engage any individual fully and then get the maximum from she or he for the benefit of both: the employer as well as the employee. Section 19.2.1 describes principles/guidelines given by ICMM to satisfy the stakeholders.[22]

19.9.2 Efficient systems including best practices

Best practices, which are the result of effective systems that are developed to run mines efficiently. In this conjunction some of the best practices that should be considered are as under:

- Autonomous maintenance system (AMS)
- Quality management system (QMS)
- Effective training, competency and awareness (table 18.3)
- Effective communication
- Precision in operations (fig. 19.15(a) to (e))
- Emergency preparedness and response (fig. 18.13(a) & (b))
- Adapting WCM – world class management system (fig. 19.21)
- Incorporating HSE as a critical business activity at par with other critical activities such as production and productivity (fig. 19.17(a))
- There could be few others also.

19.9.3 Legal compliance including Environment Management Systems (EMS)

As described in sec. 19.7.1 and table 18.4; every industry including mines, institute, business or the organization work under a legal framework. A company can have its specific codes and practices too. All these are well thought steps and well documented, and which are in the interest of all: the company owner, stakeholders, employees and the public. Any noncompliance could cause damage to the environment and safety, which in-turn could result revenue loss and impact on goodwill and reputation of the company.

19.9.4 World Class Management (WCM)[19]

The Best Practices such as WCM, which has been adapted by the Fortune-500 companies, or the renowned industrial houses the world-over, could be a useful guide. This concept addresses eight focused areas to achieve Visual Management & Control – QCDIP – Control on Quality as well as Quantity, Cost, Delivery (JIT – Just-in-time), Innovations and Productivity. This is what is known as 'QCDIP' as shown in figure 19.21.

19.10 THE WAY FORWARD: PROPOSED MILESTONES/STRATEGY[3,6]

Eradication of: injuries, incidents, illness, emission and wastage (fig. 18.9); incorporating best practices as shown in figure 19.22(a) (upper); and adhering to the flow diagram (steps) shown in figure 19.22(b) (lower) to plan and implement SD strategy; could allow not only to compete in the global market but also it could help in image building and boosting goodwill to attract foreign investors, who are looking for the least risked country for their investments (table 19.1 and fig. 19.11).

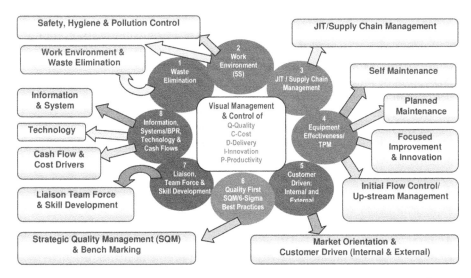

Figure 19.21 WCM concept addresses eight focused areas to achieve Visual Management Control on Quality as well as Quantity, Cost, Delivery, Innovations and Productivity (QCDIP).[19]

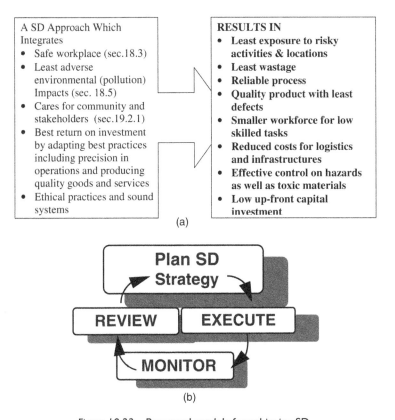

Figure 19.22 Proposed models for achieving SD.

QUESTIONS

1. 'No device has been invented which can match the human eye for accuracy and precision in many tasks'. Is this true? Do you agree that use of our five senses (name them) is the best strategy to deal with any abnormality?

2. 'SD would be judged by the stakeholders, as what is left behind? Is the mine-area better than originally or worse?' Could it be achieved by least disturbance to the natural cycles?

3. "Over the next 50 years the world will use 5 times the mineral resources that have been mined to the year 2000". To meet this predicted demand, how must the industry grow?

4. "Sustainable Development (SD) is development which is *beneficial economically, ecologically* and *socially* to the present as well as future generations". Is this statement true? How we should strive to achieve it?

5. A thorough balance amongst production, productivity and HSE and social welfare; could achieve SD? Do you agree? Should we consider HSE and social welfare – a critical business activity at par with production and productivity?

6. A word of caution with regard to the safety issues of automation: describe what exactly it means?

7. Assess the overall ore-recovery your mine is achieving? Tally it with the geological reserves. How you could improve recovery percentage?

8. Automated drilling & explosive placement – how does this combination work? Does it permit round design and blast planning and prediction of muck-pile shape and fragmentation?

9. Autonomous digging – what is this concept and how is it useful to the miners?

10. Briefly review the technological developments that have taken place within last five decades in mining.

11. Briefly review use of minerals with the things we need most including energy through a literature survey including making use of the related websites.

12. Bringing standardization with regards to methods (procedures), equipment, techniques, tools, appliances and mine services that are appropriate and practical to implement would minimize wastages. Is it true? How it could be achieved?

13. Changes for better are a continuous process. Is a thorough technology transfer from one country to another is an urgent need to bring radical changes in the mineral sector? How it could be achieved?

14. Could bringing need-based changes through R & D on conventional (on-going) methods, equipment and techniques that reduce energy consumption bring radical changes which are helpful in achieving the SD?

15. Currently do manufacturing companies use automation for increasing quality and flexibility in the manufacturing process? Does automation in mining also intend to achieve the same objective?

16. Describe geological vision (radar) – ability to see through rock to locate an ore-body in space, grade & geological structures. Could it be helpful in precise evaluation of a deposit? How?

17. Developed countries with a population share of 26% account for 79% of consumption of steel, and 86% of other metals. Thus, the consumption pattern is

erratic. Could we afford to continue with such a trend? Could you suggest viable solution to it?

18. Differentiate between automation and mechanization. Write down the importance of each one of them in mining.

19. Disproportionate – reserves, production pattern and consumption rates which are prevalent at most of the countries could result in to shortage of mineral. How it can it be avoided?

20. Economic development, regional as well as local, at the Rhineland Lignite Mining, Germany since its inception through SD is a classical example. Detail out as what measures they have taken to achieve SD?

21. Best practices are the result of effective systems that are developed to run mines efficiently. List the best practices known to you describing them their benefits, if implemented?

22. Give a schematic illustration for the phased approach to automation. Is it a rational approach to adapt automation in the mining sector, in general?

23. How could adapting cost effective techniques, systems and renovations convert more and more mineral resources into reserves?

24. How could bringing precision in the rock fragmentation techniques benefit us?

25. How could we establish minerals resources to the maximum? Why, prior to mining, must the mineral inventory (quantitative and qualitative assessment) of any deposit be established fully? How it could be accomplished?

26. Illustrate and describe the three pillars of Sustainable Development of equal strength to run any mining venture.

27. Initiatives have already been taken globally to meet demand for mineral mass consumption – list the initiatives taken.

28. Is it electricity/power that is providing energy to us, or the minerals?

29. Is it not the right time to evaluate academic standard of mining schools? List areas which they should focus upon to achieve academic excellence at par with the one achieved at some of the developed countries?

30. Is it the right time to further develop, and apply the non-conventional mining methods? List these methods and techniques and illustrate them with the aid of suitable sketches. Mention suitable conditions for their applications, operational details and scope of implementing in your own country. You could refer other chapters including Chapter1 for this purpose.

31. It is the industrialization in the immediate past, say within the last two centuries that has given rise to global issues; what are they?

32. List and display by means of suitable illustration, the constraints in the mining sector which most of the countries need to address with a sense of urgency.

33. List locale-specific challenges and proposed solutions/ways-out for: Underground metalliferous mining; underground coal mining; and surface mines for coal as well as non coal deposits.

34. List the four elements which are essential to achieve sustainable development.

35. Mineral resources of today are the results of natural processes of billions of years but with the prevalent rate of consumption they would be depleted within the next few centuries. Suggest what should be our strategy in this scenario to conserve this valuable resource?

36. 'The mining industry has not come to the expectations so far as its sustainable development (SD) is concerned'. Is this true? What is the backlog on SD? How this backlog is being felt?

37. The mining Industry must sustain its long term 'Welcome', which has begun within last decade. SD is ultimately judged by what?

38. The mining industry today uses widely diversified types of tools, appliances and equipment. How it poses problems to both the manufacturers and the users? What is the solution?

39. Oscillating Disc Cutting is a novel technology for cutting hard rock at high rates with low cutter forces. Describe this in detail mention its scope in the days to come.

40. Our ultimate aim is sustainable development, may it be whatever industry – how it can be achieved?

41. The population problem is just not numbers but has various impacts: list those impacts.

42. Precision in operations – does this refer to maximizing recovery? Cite few examples to justify your answer. Does it result in reduction of energy consumption, pollution and hazards?

43. The prevalent scenario, in turn, is asking for better recoveries and cost reduction at all levels – mining, processing and preparing salable product, and that too taking care of sustainable development – suggest strategy to achieve these objectives.

44. Propose an effective model to accomplish continual improvements.

45. Propose how speed in exploration as well as exploitation mining could be achieved.

46. Rank the leading countries that are at the forefront in achieving the global competitiveness. What are the parameters that are judged for the suitable environment to make investment by the foreign investors and MNCs?

47. Standards are laid down based on experience, scientific studies, debates and consensus arrived amongst those involved and concerned. Should we implement them and what are benefits we could expect?

48. What are the best practices such as WCM, which has been adapted by the Fortune-500 companies? Illustrate it by depicting the main activities which are focused upon when adapting this concept? What is the meaning of 'QCDIP'?

49. 'Through mining, the emergent (Third-World) countries and developing countries can finance growth progressively by the export of raw mineral resources, then by processing these raw materials prior to export, and finally by achieving progressive industrial development to improve their GDP/GNP'. Is this true? Illustrate giving a diagram as how developed nation such a USA has followed this strategy?

50. Underground mining is the major method of Chinese coal mine exploitation, and more than 95% of coal is produced through underground mining. Longwall mining is at the forefront. List its advantages and the extraction (recovery) rate and output range being achieved at the Chinese mines. How much is the maximum support height and average face length? Also give range of run of mine coal production from a longwall face.

51. Use of power supports for mining thick as well as thin seams is a remarkable achievement by the Chinese mining industry. How is it beneficial in improving the recovery of the coal deposits?

52. What are the fully-autonomous surface drills? What are the benefits due to this technique?

53. What should be eradicated from the mineral industry? What benefit could be expected if these evils could be minimized to an acceptable level?

54. Why should controlled (contour, smooth) blasting be an integral part while creating excavation of any kind? List the innovations that have been introduced recently in this area.

55. Why is lack of investment in the mining sector in most of the countries a usual bottleneck? What should MNCs look into before making investment? List the parameters that should be considered to attract the foreign as well as local investors.

56. Why is mining also economically critical for millions of the world's poorest people?

57. Why is optimizing the mine size and its further division into panels, levels or stopes essential? How it could be achieved? How could waste be minimized to preserve mineral resources judiciously for our future generations?

58. Why SD in mining and construction industries is more difficult than other industries? It is easy to define SD, but why is it extremely difficult to implement?

59. Why cannot people access the opportunities provided by the modern world without energy? Is this true? 1.3 billion people lack access to electricity today; Suggest a way out to resolve this issue.

60. The world's population would be doubled in how many years based on the prevalent growth rate?

61. Detail out the following concepts mentioning their utility in the days to come; you can make use of illustrations shown in figures 19.15(a) to (e).
    a. More concentrated production centers
    b. More effective rock fragmentation techniques
    c. Continuous rock transport systems from the mine
    d. Geological vision
    e. Safety specific technologies
    f. Precise vehicle navigation, narrow roads & ramps
    g. Small mine site team with multiple responsibilities
    h. Greatly improved equipment reliability through automated systems
    i. Autonomous haulage – to removes cycle time variability, and also achieve other benefits, list them.

62. The online survey on SD was carried during 24 August to 18 September 2010 by ICMM in collaboration with 'GlobalScan'. In fact this concept was initiated in 2004 by ICMM. Work out following in this regard:
    a. List the main points from this survey.
    b. Present/illustrate the advice for improving sustainable development performance based on surveys: 2007 & 2010.
    c. Present/illustrate the most serious issues/events in past three years.
    d. Describe the 10 principles/guidelines for SD proposed by ICMM.

REFERENCES

1. Ambesh, C.P.: 'Indian non-coal mining sector and its future directions.' *Proc. 1st Asian Mining Congress, MGMI*, 16–18 Jan. 2006, Kolkata, India.
2. Behre Dolberas. Survey of countries for Mining investment, 2009.

3. Carter, P.: The mine of the future. Pres. *Mining 2020, The AIMEX International Mining Conference, Sydney, Australia, Sept. 5–6, 2007.*

4. Chengyao, Jiao. (Zhengzhou Coal Mining Machinery (Group) Co., Ltd. China). 'Longwall mechanized mining technology and equipment in China.' *Proc. 3rd Asian Mining Congress, MGMI, Jan. 2010, Kolkata, India.*

5. CII Background Paper in: Global Mining Summit, 2008, Kolkata, India

6. Hood, Michael. Technologies for mining in the 21st century. Presentation in: *Mining 2020, The AIMEX International Mining Conference, Sydney, Australia, Sept. 5–6, 2007.*

7. http://en.wikipedia.org

8. http://www.miningtechnology.com/projects/Rhineland/Rhineland Lignite Mining, Germany.

9. Inherently safer chemical processes. Daniel A. Crowl (edt.), New York, 1996, pp. 19–23.

10. Lacy, W.C. and Lacy, J.C.: History of mining. In: Hartman (edt.): *SME Mining Engineering Handbook.* SMME, Colorado, 1992, pp. 20–23.

11. Mineral Information Institute (2001). www.mii.org; web site, Denver, Colorado.

12. McCarthy, P. & Noort, D. Automated Underground Mining. Pres. In: International Seminar: 'Mine Advan. Tech. Jodhpur, India, 14–16 February, 2009; pp. 201–06.

13. Sustainable Development Framework – Indian Mining Initiative (Draft Report), Federation of Indian Mining Institute (FIMI), 2009.

14. Tatiya, R.R.: 'Perfect bonding amongst Mining, its Adverse Impacts Mitigation Measures & Social Welfare is a key to Sustainable Development' Proc. National Seminar on: Corporate role in sustainable development in the Indian Mineral sector; MEAI Sept. 2011; Jodhpur, India.

15. Tatiya, R.R.: *Civil Excavations and Tunneling – A Practical Guide.* Institution of Civil Engineers/Thomas Telford Ltd. London, U.K. 2005.

16. Tatiya, R.R.: *Elements of Industrial Hazards – Health, Safety, Environment and Loss Prevention* CRC Press, Taylor & Francis Group, London, U.K. 2010.

17. Tatiya, R.R.: 'Why Mining and Minerals in the prevalent scenario?' *Proc. 1st Asian Mining Congress, MGMI, 16–18 Jan. 2006, Kolkata, India.*

18. Tatiya, R.R.: Key-Note-Address: 'Impacts of current scenario to Mining Industry – Proposed solutions'. *Proc. National Seminar on Mining and Processing Minerals; MEAI* 12–13 Aug. 2006, Jodhpur, India.

19. World Class Management (WCM) – Leaflets, displays, lectures, and talks on various topics – Essel Mining and Industries Limited, Aditya Birla group, India (2004–2008; through interaction, and participation)

20. World Coal Association (WCA). Report: Energy and sustainable development report, 2012, (Mr. Catelin). London. (http://worldcoal.org/).

21. www.globalscan.com

22. www.icmm.com. Stakeholders' Views of Mining and Sustainable Development– Highlights of report for survey respondents 22 November 2010, ICMM 35/38 Portman Square London W1H 6LR UK.

23. Singh, A.K.: 'Growth Perspective of coal sector in India.' *Proc. 3rd Asian Mining Congress, MGMI, Jan. 2010, Kolkata, India.*

Subject index

Printed and bound by CPI Group (UK) Ltd, Croydon, CR0 4YY

22/10/2024

01777611-0013